Stem Cells and Aging

Stem Cells and Aging

Edited by

Surajit Pathak, Ph.D.
Professor, Faculty of Allied Health Sciences,
Chettinad Hospital and Research Institute,
Chettinad Academy of Research and Education, Chennai, India

Antara Banerjee, Ph.D.
Associate Professor, Faculty of Allied Health Sciences,
Chettinad Hospital and Research Institute,
Chettinad Academy of Research and Education, Chennai, India

Academic Press is an imprint of Elsevier
125 London Wall, London EC2Y 5AS, United Kingdom
525 B Street, Suite 1650, San Diego, CA 92101, United States
50 Hampshire Street, 5th Floor, Cambridge, MA 02139, United States
The Boulevard, Langford Lane, Kidlington, Oxford OX5 1GB, United Kingdom

Copyright © 2021 Elsevier Inc. All rights reserved.

No part of this publication may be reproduced or transmitted in any form or by any means, electronic or mechanical, including photocopying, recording, or any information storage and retrieval system, without permission in writing from the publisher. Details on how to seek permission, further information about the Publisher's permissions policies and our arrangements with organizations such as the Copyright Clearance Center and the Copyright Licensing Agency, can be found at our website: www.elsevier.com/permissions.

This book and the individual contributions contained in it are protected under copyright by the Publisher (other than as may be noted herein).

Notices

Knowledge and best practice in this field are constantly changing. As new research and experience broaden our understanding, changes in research methods, professional practices, or medical treatment may become necessary.

Practitioners and researchers must always rely on their own experience and knowledge in evaluating and using any information, methods, compounds, or experiments described herein. In using such information or methods they should be mindful of their own safety and the safety of others, including parties for whom they have a professional responsibility.

To the fullest extent of the law, neither the Publisher nor the authors, contributors, or editors, assume any liability for any injury and/or damage to persons or property as a matter of products liability, negligence or otherwise, or from any use or operation of any methods, products, instructions, or ideas contained in the material herein.

Library of Congress Cataloging-in-Publication Data
A catalog record for this book is available from the Library of Congress

British Library Cataloguing-in-Publication Data
A catalogue record for this book is available from the British Library

ISBN 978-0-12-820071-1

For information on all Academic Press publications
visit our website at https://www.elsevier.com/books-and-journals

Publisher: Masucci, Stacy
Acquisitions Editor: Brown, Elizabeth
Editorial Project Manager: Allard, Samantha
Production Project Manager: Raviraj, Selvaraj
Cover Designer: Rogers, Mark

Typeset by SPi Global, India

Contents

Contributors   xi

1. Theories of stem cell aging

Anisur Rahman Khuda-Bukhsh, Sreemanti Das, and Asmita Samadder

1 Introductwion	1
2 Stem cells	1
3 Types of stem cells	1
3.1 Embryonic stem (ES) cells	1
3.2 Mesenchymal stem cells	2
3.3 Adult stem cells	2
4 Aging	2
4.1 Role of stem cells in aging	2
5 Theories of stem cell aging	3
5.1 Free radical theory of aging	3
5.2 Telomere shortening	4
5.3 DNA damage accumulation and mutations	5
5.4 Epigenetic alterations	5
References	6

2. Progress in human embryonic stem cell research and aging

Anjali P. Patni, Joel P. Joseph, D. Macrin, and Arikketh Devi

1 Introduction	9
2 Properties of hESCs	10
3 Signaling pathways involved in the regulation of hESCs	14
3.1 LIF signaling pathway	15
3.2 Wnt signaling pathway	15
3.3 FGF signaling and its interaction with several other pathways in hESCs	16
4 Metabolic and epigenetic regulation of hESCs	17
5 Sources and regulations for generating human embryonic stem cell lines for research	17
6 Aging and hESCs	18
6.1 Pathways modulating aging in hESCs	20
6.2 Factors contributing to the aging and senescence in hESCs	20
6.3 Beneficial molecules secreted by the hESCs	24
6.4 Current application of hESCs	25
7 Ethical constraints	32
8 Conclusion	33
References	33

3. Stem cell aging and wound healing

Vijayalakshmi Rajendran, Mayur Vilas Jain, and Sumit Sharma

1 Introduction	53
2 Stem cell aging and inflammation	53
3 Role in tissue degeneration	54
4 Role in fibrosis	55
5 Alternative stem cell rejuvenating strategies in aging	57
6 Summary	58
References	58

4. Stem cells and multiomics approaches in senescence: From benchside to bedside

Atil Bisgin

1 Introduction	61
2 "Omics" in senescence	61
3 Multiomics data integration and its application	64
4 Stem cell-based multiomics and data integration	65
5 Conclusion	66
References	66

5. Signaling pathways influencing stem cell self-renewal and differentiation

Mahak Tiwari, Sinjini Bhattacharyya, and Deepa Subramanyam

1 Introduction	69
2 LIF and JAK/STAT3 signaling pathway	69
3 TGF-β/Smad pathway	72

v

4 ERK1/2 signaling pathway	75		16 Consequence of aging HSCs	108
5 Wnt signaling pathway	76		17 Rejuvenating strategies for HSCs	109
6 Conclusions and future perspectives	78		18 Future perspective	110
References	79		Conflict of interest	110
			References	110

6. Immunity, stem cells, and aging

Ezhilarasan Devaraj, Muralidharan Anbalagan, R. Ileng Kumaran, and Natarajan Bhaskaran

1 Introduction	89
1.1 Epigenetics and aging stem cells	90
1.2 Sex differences in aging	90
2 Aging	90
3 Stem cells in aging	91
4 Immunity and immune cells	92
5 Immunity in aging	92
6 Protein damage, quality control mechanism, and aging	93
7 Molecules and mechanisms involved in stem cells and aging	94
8 Altered developmental pathways during aging	96
8.1 WNT signaling	96
8.2 Notch signaling	96
8.3 Cellular senescence	96
9 Drugs and methods to prevent aging	97
10 Conclusion	98
References	98

7. Aging of hematopoietic stem cells: Insight into mechanisms and consequences

Bhaswati Chatterjee and Suman S. Thakur

1 Introduction	103
2 Changes in hematopoietic stem cells (HSCs) due to aging	103
3 Differential expression of genes related to HSC aging	104
4 Mechanism of aging in HSCs	104
5 Cell-intrinsic mechanisms for HSC aging	105
6 DNA damage	105
7 Senescence and polarity	106
8 Impaired autophagy and mitochondrial activity	107
9 Epigenetics and aging	107
10 miRNA	107
11 Clonal hematopoiesis	107
12 Hypercholesterolemia	108
13 Signaling pathways and aging	108
14 Cell-extrinsic mechanism for HSCs aging	108
15 Single cell and HSCs	108

8. Ocular stem cells and aging

Neethi Chandra Thathapudi and Jaganmohan R. Jangamreddy

1 Introduction to stem cells	113
2 Stem cells of the cornea	114
2.1 Corneal stem cells	114
2.2 Limbal epithelial stem cells (LESC)	114
2.3 Corneal stromal stem cells (CSSC)	115
2.4 Effect of aging on the stem cells of the anterior cornea	115
2.5 Corneal endothelial stem cells	115
3 Stem cells of the conjunctiva	115
3.1 Conjunctival stem cells (CSC)	115
3.2 Aging in conjunctival stem cells	116
4 Stem cells of the lens	116
4.1 Lens epithelial stem/progenitor cells (LECs)	116
4.2 Cell cycle regulation in lens epithelial stem cells	117
4.3 Changes in the aging lens epithelial stem cells	117
5 Stem cells of the retina	118
5.1 Retinal stem cells	118
5.2 Pathways involved in maintenance of retinal stem cells	118
5.3 Pathway modifications in aged retinal stem cells	118
6 Stem cells of the trabecular meshwork	119
6.1 Trabecular meshwork stem cells	119
6.2 ER stress in aging TM stem cells	119
6.3 Regeneration in the aging TM	119
7 Conclusion	120
References	120

9. Skeletal muscle cell aging and stem cells

Shabana Thabassum Mohammed Rafi, Yuvaraj Sambandam, Sivanandane Sittadjody, Surajit Pathak, Ilangovan Ramachandran, and R. Ileng Kumaran,

1 Introduction	125
2 Skeletal muscle stem cells: Satellite cells	125
2.1 Origin of satellite cells	126
2.2 Transcription factors-mediated gene regulation in satellite cells	126
2.3 Quiescent state of satellite cells	128

3 Skeletal muscle stem cells in adult and aging ... 128
  3.1 Age-dependent changes in satellite cells ... 128
4 Aging and metabolism of satellite cells ... 129
  4.1 Mitochondria in satellite cells ... 129
  4.2 Autophagy in satellite cells ... 130
5 Aging and extrinsic factors of satellite cells ... 132
  5.1 Extracellular matrix of satellite cells ... 132
  5.2 Circulatory factors and satellite cells ... 132
  5.3 Growth factors and satellite cells ... 133
  5.4 Cytokines and satellite cells ... 133
6 Aging and intrinsic factors of satellite cells ... 133
  6.1 Telomeres in satellite cells ... 133
  6.2 Notch signaling pathway of satellite cells ... 134
  6.3 Wnt/ß-catenin signaling of satellite cells ... 135
7 Aging and epigenetic regulation of satellite cells ... 135
  7.1 Epigenetic events in quiescent state of satellite cells ... 135
  7.2 Epigenetic events in proliferative state of satellite cells ... 137
  7.3 Epigenetic events in differentiation state of satellite cells ... 137
8 Conclusions ... 139
Acknowledgements ... 139
Conflict of interest ... 139
References ... 139

10. Aging and stability of cardiomyocytes

Shouvik Chakravarty, Johnson Rajasingh, and Satish Ramalingam

1 Introduction ... 147
2 Cardiomyocytes and aging ... 148
  2.1 Stability of cardiomyocytes and potential regeneration ... 148
  2.2 Mechanism of aging in cardiomyocytes ... 148
3 Regeneration of cardiomyocytes—A mechanistic overview ... 149
  3.1 Cardiomyocyte survival and protection ... 149
  3.2 By means of inflammation reduction ... 150
  3.3 Cell-to-cell communication ... 150
  3.4 Through angiogenesis and vascularization ... 150
  3.5 Cardiomyogenesis ... 151
  3.6 Molecular mechanisms behind proliferation and cell cycle ... 151
  3.7 miRNA-mediated regeneration ... 152
  3.8 Cardiac-resident stem cell-mediated regeneration ... 152
4 Conclusion and future perspectives ... 153
References ... 153

11. Signaling pathways influencing stem cell self-renewal and differentiation—Special emphasis on cardiomyocytes

Selvaraj Jayaraman, Ponnulakshmi Rajagopal, Vijayalakshmi Periyasamy, Kanagaraj Palaniyandi, R. Ileng Kumaran, Sakamuri V. Reddy, Sundaravadivel Balasubramanian, and Yuvaraj Sambandam

1 Introduction ... 157
2 Stem cells ... 158
3 Stem cell culture and therapy ... 158
4 Signaling pathways and factors involved in the self-renewal and differentiation of cardiomyocytes ... 160
5 Wnt/ß-catenin signaling pathway ... 160
6 Differentiation of stem cells ... 161
7 Notch signaling pathway ... 161
8 Mechanotransduction pathways ... 164
9 Conclusion ... 164
References ... 164

12. Angiogenesis in aging hearts—Cardiac stem cell therapy

Vinu Ramachandran and Anandan Balakrishnan

1 Introduction ... 169
2 Cardiac aging ... 169
  2.1 Factors influencing cardiac aging ... 169
  2.2 Vascular changes (impaired angiogenesis) in cardiac aging ... 170
3 Therapy for aging heart ... 170
  3.1 Cardiac stem cells for therapy ... 170
  3.2 Targeting angiogenesis using cardiac stem cells for therapy ... 172
4 Conclusion ... 173
References ... 173

13. Gut stem cells: Interplay with immune system, microbiota, and aging

Francesco Marotta, Baskar Balakrishnan, Azam Yazdani, Antonio Ayala, Fang He, and Roberto Catanzaro

1 Fundamental background ... 177
2 Regulatory mechanisms ... 178
3 Biorhythm and gut stem cells ... 179
  3.1 Mitochondrial function in the gut and stem cell interplay ... 179
4 Aging gut, microbiota, immune system, and stem cells ... 180

5	Gut microbiome	180	9.3 DNA damage	208
	5.1 Microbiome on the immunity of stem cell aging	181	9.4 Mitochondrial dysfunction	208
			9.5 Proteostasis dysfunction	209
	5.2 Microbiome on epigenetics of stem cell aging	181	9.6 Epigenetics	210
			9.7 Changes in stem cell metabolic intake	211
6	Tentative interventional perspectives	183	9.8 Stem cell aging and gender	211

14. Cellular senescence and aging in bone

Manju Mohan, Sridhar Muthusami, Nagarajan Selvamurugan, Srinivasan Narasimhan, R. Ileng Kumaran, and Ilangovan Ramachandran

1	Introduction	187
2	Skeletal aging: An insight into bone cellular senescence	190
3	Key factors regulating senescence in bone cells	190
	3.1 Intrinsic factors	190
	3.2 Extrinsic factors	193
4	Role of senescence-associated secretory phenotype and senescent markers on skeletal aging	195
5	Key pathways regulating the skeletal aging	196
	5.1 Wnt signaling pathway	196
	5.2 p53/p21 signaling pathway	197
6	Cellular senescence as a therapeutic target to improve bone frailty	198
7	Conclusions	199
	Acknowledgments	199
	Conflict of interest	199
	References	199

15. Aging-induced stem cell dysfunction: Molecular mechanisms and potential therapeutic avenues

Yander Grajeda, Nataly Arias, Albert Barrios, Shehla Pervin, and Rajan Singh,,

1	Introduction	203
2	Hematopoietic cells	203
3	Intestinal stem cells	204
4	Germline stem cells	204
5	Skeletal muscle stem cells	204
6	Neural stem cells	205
7	Mesenchymal stem cells	205
8	Models used to study stem cell dysfunction	206
9	Factors responsible for stem cell dysfunction	206
	9.1 Stem cell exhaustion	206
	9.2 Microenvironment	206

10	Common therapeutic approaches for the treatment of aging-associated stem cell dysfunction	211
	10.1 Parabiosis	211
	10.2 Retrotransposons	213
	10.3 Cellular reprogramming of induced pluripotent stem cells (iPSCs)	213
	10.4 Telomere lengthening	214
	10.5 Caloric restriction	214
11	Conclusions	215
	Acknowledgments	215
	References	215

16. Therapeutic approaches for the treatment of aging-induced stem cell dysfunction

Debora Bizzaro, Francesco Paolo Russo, and Patrizia Burra

1	Introduction	223
2	The past of rejuvenation research	223
3	Therapeutic approaches targeting extracellular aging factors	224
	3.1 Rejuvenation of stem cell niche	224
4	Intrinsic aging factors and therapeutic approaches	225
	4.1 Improving protein homeostasis and regulation of autophagy	225
	4.2 Regulation of oxidative metabolism and improving mitochondrial function	226
5	Nuclear aging as potential therapeutic targets	226
	5.1 Targeting DNA damage repair to restore stem cell function	226
	5.2 Epigenetic reprogramming	227
6	Conclusions	228
	References	228

17. Role of biological markers in stem cell aging and its implications in therapeutic processes

Sivanandane Sittadjody, Aamina Ali, Thilakavathy Thangasamy, M. Akila, R. Ileng Kumaran, and Emmanuel C. Opara

1	Introduction: Stem cells and their therapeutic values	231
2	Hallmarks of stem cell aging	232

3 Biological markers (biomarkers) of stem
   cell aging   232
   3.1 Exhaustion of stem cell pools   233
   3.2 Oxidative stress   234
   3.3 Genomic instability   234
   3.4 Telomere attrition   236
   3.5 Epigenetic modifications   236
   3.6 Altered microRNA profile   237
   3.7 RNA splicing and defective
        ribosomal machineries   237
   3.8 Loss of proteostasis   237
   3.9 Change in cell polarity   238
   3.10 Mitochondrial dysfunction   238
   3.11 Deregulated nutrient sensing and
        cellular metabolism   238
   3.12 Niche deterioration   239
   3.13 Circulating factors   240
   3.14 Altered intercellular
        communications and accumulation
        of inflammasomes   240
   3.15 Cell cycle arrest and cellular
        senescence   241
4 Potential therapeutic interventions   241
5 Conclusion and perspectives   242
References   242

18. Alternative stromal cell-based therapies for aging and regeneration

Dikshita Deka, Alakesh Das, Meenu Bhatiya, Surajit Pathak, and Antara Banerjee

1 Introduction   251
2 Aging in unicellular organisms   251
3 Aging in multicellular organisms   251
4 Age-related genes and their role   252
5 Age-related diseases   255
   5.1 Alzheimer's disease (AD)   255
   5.2 Parkinson's disease   256
   5.3 Cancer   256
   5.4 Colon cancer   256
   5.5 Skin cancer   258
6 Factors affecting stromal cell quality
   due to aging   259
7 Intrinsic limitations of aged stem cells   260
8 Role of stromal cells in aging   261
9 The pros and cons of stromal cell therapy   262
10 Pitfalls in stem cell therapy for aging
   and regeneration   262
11 Stem cell-derived conditioned medium
   (CM) as therapy   263
12 Cell-free therapy with extracellular
   vesicles   265
13 Tissue microenvironmental cues and
   influences in cell-free therapies   265
14 Next generation of cell-based therapies
   for aging and regeneration   266
15 Conclusion and perspectives   266
Acknowledgments   266
Funding   266
References   267

19. Stem cell-based therapeutic strategy in delaying prion disease

Sanjay Kisan Metkar, Koyeli Girigoswami, and Agnishwar Girigoswami

1 Introduction   271
2 The innovation of the prion protein   271
3 Therapeutic approaches for managing
   prion disease   273
4 Adult neural stem cells in therapeutic
   interventions of prion disease   274
5 Fetal neural stem cell-based therapeutic
   approaches in prion diseases   274
6 Embryonic stem cells (ESCs) and induced
   pluripotent stem cells   276
7 Mesenchymal stem cells   276
8 Microglial cells   277
9 Discussion   278
Acknowledgment   278
References   278

20. Traditional medicine for aging-related disorders: Implications for drug discovery

Antara Banerjee, M.S. Pavane, L. Husaina Banu, A. Sai Rishika Gopikar, K. Roshini Elizabeth, and Surajit Pathak

1 Introduction   281
   1.1 Siddha medicine   281
   1.2 Unani medicine   283
   1.3 Ayurveda   283
2 Aging   284
3 Aging-related disorders   286
   3.1 Colon and rectal cancer   286
   3.2 Neurodegenerative diseases   287
   3.3 Facial wrinkling   287
   3.4 Graying of hair   288
4 Clinical applications of traditional medicine   288
5 Applications of traditional medicine in
   aesthetics   288
   5.1 *Ocimum tenuiflorum*   288
   5.2 *Tinospora cordifolia*   289
   5.3 *Centella asiatica*   289
   5.4 *Panax ginseng*   289
   5.5 Wolfberry   289
   5.6 *Atriculum lappa*   289

5.7 Chamomile	290	7 Conclusion	293
5.8 Soy	290	8 Future perspectives	294
5.9 Plant extracts in treating diseases	290	9 Approaches toward drug discovery	294
6 Insights from traditional medicine in translational stem cell research	291	Acknowledgment	295
		Funding	295
6.1 Mesenchymal stem cells in aging and role of traditional medicines	291	**Consent for publication**	295
		Conflict of interest	295
6.2 Potential role of herbal remedies in stem cell therapy: Proliferation and differentiation	292	**References**	295
		Index	299

Contributors

Numbers in parenthesis indicate the pages on which the authors' contributions begin.

M. Akila (231), College of Nursing, JIPMER, Puducherry, India

Aamina Ali (231), Wake Forest Institute for Regenerative Medicine, Wake Forest University School of Medicine, Winston-Salem, NC, United States

Muralidharan Anbalagan (89), Department of Structural and Cellular Biology, Tulane University School of Medicine, New Orleans, LA, United States

Nataly Arias (203), California State University Dominguez Hills, Los Angeles, CA, United States

Antonio Ayala (177), Department of Molecular Biochemistry and Biology, University of Seville, Seville, Spain

Anandan Balakrishnan (169), Department of Genetics, Dr. ALM PG Institute of Basic Medical Sciences, University of Madras, Taramani Campus, Chennai, Tamil Nadu, India

Baskar Balakrishnan (177), Department of Immunology, Mayo Clinic, Rochester, MN, United States

Sundaravadivel Balasubramanian (157), Department of Radiation Oncology, Hollings Cancer Center, Medical University of South Carolina, Charleston, SC, United States

Antara Banerjee (251, 281), Department of Medical Biotechnology, Faculty of Allied Health Sciences, Chettinad Academy of Research and Education (CARE), Chettinad Hospital and Research Institute (CHRI), Chennai, India

L. Husaina Banu (281), Department of Medical Biotechnology, Faculty of Allied Health Sciences, Chettinad Academy of Research and Education (CARE), Chettinad Hospital and Research Institute (CHRI), Chennai, India

Albert Barrios (203), California State University Dominguez Hills, Los Angeles, CA, United States

Natarajan Bhaskaran (89), Department of Biomedical Sciences, Sri Ramachandra Institute of Higher Education and Research, SRMC, Chennai, Tamil Nadu, India

Meenu Bhatiya (251), Department of Medical Biotechnology, Faculty of Allied Health Sciences, Chettinad Academy of Research and Education (CARE), Chettinad Hospital and Research Institute (CHRI), Chennai, India

Sinjini Bhattacharyya (69), National Centre for Cell Science; Savitribai Phule Pune University, Pune, India

Atil Bisgin (61), Cukurova University, Faculty of Medicine, Medical Genetics Department of Balcali Hospital and Clinics; Cukurova University AGENTEM (Adana Genetic Diseases Diagnosis and Treatment Center), Adana, Turkey

Debora Bizzaro (223), Department of Surgery, Oncology and Gastroenterology, Gastroenterology/Multivisceral Transplant Section, University Hospital Padova, Padova, Italy

Patrizia Burra (223), Department of Surgery, Oncology and Gastroenterology, Gastroenterology/Multivisceral Transplant Section, University Hospital Padova, Padova, Italy

Roberto Catanzaro (177), Department of Clinical and Experimental Medicine, Section of Gastroenterology, University of Catania, Catania, Italy

Shouvik Chakravarty (147), Department of Genetic Engineering, School of Bio-Engineering, SRM Institute of Science and Technology, Kanchipuram, Tamil Nadu, India

Bhaswati Chatterjee (103), National Institute of Pharmaceutical Education and Research, Hyderabad, India

Alakesh Das (251), Department of Medical Biotechnology, Faculty of Allied Health Sciences, Chettinad Academy of Research and Education (CARE), Chettinad Hospital and Research Institute (CHRI), Chennai, India

Sreemanti Das (1), West Bengal State Health & Family Welfare, Krishnanagar, India

Dikshita Deka (251), Department of Medical Biotechnology, Faculty of Allied Health Sciences, Chettinad Academy of Research and Education (CARE), Chettinad Hospital and Research Institute (CHRI), Chennai, India

Ezhilarasan Devaraj (89), Department of Pharmacology, Biomedical Research Unit and Laboratory Animal Center, Saveetha Dental College and Hospital, Saveetha Institute of Medical and Technical Sciences, Chennai, Tamil Nadu, India

Arikketh Devi (9), Stem Cell Biology Lab, Department of Genetic Engineering, School of Bioengineering, Faculty of Engineering and Technology, SRM Institute of Science and Technology, Chennai, Tamil Nadu, India

K. Roshini Elizabeth (281), Department of Medical Biotechnology, Faculty of Allied Health Sciences, Chettinad Academy of Research and Education (CARE), Chettinad Hospital and Research Institute (CHRI), Chennai, India

Agnishwar Girigoswami (271), Medical Bionanotechnology, Faculty of Allied Health Sciences, Chettinad Hospital and Research Institute (CHRI), Chettinad Academy of Research and Education (CARE), Kelambakkam, Tamil Nadu, India

Koyeli Girigoswami (271), Medical Bionanotechnology, Faculty of Allied Health Sciences, Chettinad Hospital and Research Institute (CHRI), Chettinad Academy of Research and Education (CARE), Kelambakkam, Tamil Nadu, India

A. Sai Rishika Gopikar (281), Department of Medical Biotechnology, Faculty of Allied Health Sciences, Chettinad Academy of Research and Education (CARE), Chettinad Hospital and Research Institute (CHRI), Chennai, India

Yander Grajeda (203), California State University Dominguez Hills, Los Angeles, CA, United States

Fang He (177), Department of Nutrition, Food Safety and Toxicology, West China School of Public Health, Sichuan University, Chengdu, People's Republic of China

R. Ileng Kumaran (89, 125, 157, 187, 231), Biology Department, Farmingdale State College, Farmingdale, NY, United States

Mayur Vilas Jain (53), Department of Molecular Medicine and Gene Therapy, Lund Stem Cell Center, Lund University, Lund, Sweden

Jaganmohan R. Jangamreddy (113), UR Advanced Therapeutics Private Limited, Aspire-BioNEST, University of Hyderabad, Hyderabad, India

Selvaraj Jayaraman (157), Department of Biochemistry, Saveetha Dental College & Hospitals, Saveetha Institute of Medical and Technical Sciences (SIMATS), Chennai, Tamil Nadu, India

Joel P. Joseph (9), Stem Cell Biology Lab, Department of Genetic Engineering, School of Bioengineering, Faculty of Engineering and Technology, SRM Institute of Science and Technology, Chennai, Tamil Nadu, India

Anisur Rahman Khuda-Bukhsh (1), Formerly at Cytogenetics and Molecular Biology Lab., Department of Zoology, University of Kalyani, Kalyani, India

D. Macrin (9), Stem Cell Biology Lab, Department of Genetic Engineering, School of Bioengineering, Faculty of Engineering and Technology, SRM Institute of Science and Technology, Chennai, Tamil Nadu; Department of Bioinformatics, Saveetha School of Engineering, Saveetha Institute of Medical and Technical Services, Saveetha Nagar, Chennai, India

Francesco Marotta (177), ReGenera R&D International for Aging Intervention and Vitality & Longevity Medical Science Commission, Femtec, Milano, Italy

Sanjay Kisan Metkar (271), Medical Bionanotechnology, Faculty of Allied Health Sciences, Chettinad Hospital and Research Institute (CHRI), Chettinad Academy of Research and Education (CARE), Kelambakkam, Tamil Nadu, India

Manju Mohan (187), Department of Endocrinology, Dr. ALM PG Institute of Basic Medical Sciences, University of Madras, Taramani Campus, Chennai, Tamil Nadu, India

Sridhar Muthusami (187), Department of Biochemistry, Karpagam Academy of Higher Education; Karpagam Cancer Research Centre, Karpagam Academy of Higher Education, Coimbatore, Tamil Nadu, India

Srinivasan Narasimhan (187), Department of Allied Health Sciences, Chettinad Hospital & Research Institute, Chettinad Academy of Research and Education, Kelambakkam, Chennai, Tamil Nadu, India

Emmanuel C. Opara (231), Wake Forest Institute for Regenerative Medicine, Wake Forest University School of Medicine, Winston-Salem, NC, United States

Kanagaraj Palaniyandi (157), Department of Biotechnology, School of Bioengineering, Cancer Science Laboratory, SRM Institute of Science and Technology, Chennai, Tamil Nadu, India

Surajit Pathak (125, 251, 281), Chettinad Hospital and Research Institute, Chettinad Academy of Research and Education, Kelambakkam, Tamil Nadu; Department of Medical Biotechnology, Faculty of Allied Health Sciences, Chettinad Academy of Research and Education (CARE), Chettinad Hospital and Research Institute (CHRI), Chennai, India

Anjali P. Patni (9), Stem Cell Biology Lab, Department of Genetic Engineering, School of Bioengineering, Faculty of Engineering and Technology, SRM Institute of Science and Technology, Chennai, Tamil Nadu, India

M.S. Pavane (281), Department of Medical Biotechnology, Faculty of Allied Health Sciences, Chettinad Academy of Research and Education (CARE), Chettinad Hospital and Research Institute (CHRI), Chennai, India

Vijayalakshmi Periyasamy (157), Department of Biotechnology & Bioinformatics, Holy Cross College (Autonomous), Trichy, Tamil Nadu, India

Shehla Pervin (203), Department of Biology, California State University; Division of Endocrinology and Metabolism, Charles R. Drew University of Medicine and Science; Department of Obstetrics and Gynecology, UCLA School of Medicine; Johnson Comprehensive Cancer Center, UCLA School of Medicine, Los Angeles, CA, United States

Shabana Thabassum Mohammed Rafi (125), Department of Endocrinology, Dr. ALM PG Institute of Basic Medical Sciences, University of Madras, Taramani Campus, Chennai, Tamil Nadu, India

Ponnulakshmi Rajagopal (157), Central Research Laboratory, Meenakshi Academy of Higher Education and Research, Chennai, Tamil Nadu, India

Johnson Rajasingh (147), Bioscience Research, Medicine-Cardiology, University of Tennessee Health Science Center, Memphis, TN, United States

Vijayalakshmi Rajendran (53), Division of Ophthalmology, Department of Clinical Sciences, Lund University, Lund, Sweden

Ilangovan Ramachandran (125, 187), Department of Endocrinology, Dr. ALM PG Institute of Basic Medical Sciences, University of Madras, Taramani Campus, Chennai, Tamil Nadu, India

Vinu Ramachandran (169), Department of Genetics, Dr. ALM PG Institute of Basic Medical Sciences, University of Madras, Taramani Campus, Chennai, Tamil Nadu, India

Satish Ramalingam (147), Department of Genetic Engineering, School of Bio-Engineering, SRM Institute of Science and Technology, Kanchipuram, Tamil Nadu, India

Sakamuri V. Reddy (157), Darby Children's Research Institute, Department of Pediatrics, Medical University of South Carolina, Charleston, SC, United States

Francesco Paolo Russo (223), Department of Surgery, Oncology and Gastroenterology, Gastroenterology/Multivisceral Transplant Section, University Hospital Padova, Padova, Italy

Asmita Samadder (1), Cytogenetics and Molecular Biology Lab., Department of Zoology, University of Kalyani, Kalyani, India

Yuvaraj Sambandam (125, 157), Department of Surgery, Comprehensive Transplant Center, Northwestern University, Feinberg School of Medicine, Chicago, IL, United States

Nagarajan Selvamurugan (187), Department of Biotechnology, College of Engineering and Technology, SRM Institute of Science and Technology, Kattankulathur, Tamil Nadu, India

Sumit Sharma (53), Division of Molecular Medicine and Virology, Department of Biomedical and Clinical Sciences, Linköping University, Linköping, Sweden

Rajan Singh (203), Division of Endocrinology and Metabolism, Charles R. Drew University of Medicine and Science; Department of Obstetrics and Gynecology, UCLA School of Medicine; Johnson Comprehensive Cancer Center, UCLA School of Medicine, Los Angeles, CA; Research Program in Men's Health: Aging and Metsabolism, Brigham and Women's Hospital, Harvard Medical School, Boston, MA, United States

Sivanandane Sittadjody (125, 231), Wake Forest Institute for Regenerative Medicine, Wake Forest University School of Medicine, Winston-Salem, NC, United States

Deepa Subramanyam (69), National Centre for Cell Science, Pune, India

Suman S. Thakur (103), Centre for Cellular and Molecular Biology, Hyderabad, India

Thilakavathy Thangasamy (231), Department of Human Biology, Forsyth Tech Community College, Winston-Salem, NC, United States

Neethi Chandra Thathapudi (113), L V Prasad Eye Institute, Hyderabad, India

Mahak Tiwari (69), National Centre for Cell Science; Savitribai Phule Pune University, Pune, India

Azam Yazdani (177), Department of Anesthesiology, Perioperative and Pain Medicine, Brigham and Women's Hospital, Harvard Medical School, Boston, MA, United States

Chapter 1

Theories of stem cell aging

Anisur Rahman Khuda-Bukhsh[a], Sreemanti Das[b], and Asmita Samadder[c]

[a]*Formerly at Cytogenetics and Molecular Biology Lab., Department of Zoology, University of Kalyani, Kalyani, India,* [b]*West Bengal State Health & Family Welfare, Krishnanagar, India,* [c]*Cytogenetics and Molecular Biology Lab., Department of Zoology, University of Kalyani, Kalyani, India*

1 Introduction

Aging is a continuous and unavoidable process that occurs naturally in all living organisms. However, aging might be delayed to some extent by deploying modern tools and techniques. Numerous approaches to uphold the antiaging process have been developed so far, which have drawn the interests of researchers particularly keen on taking up approaches in the area of global beauty and fashion world. Stem cell therapy has proven itself as promising potential alternatives to regular antiaging agents/tools used today. Adipose-derived stem cells (ADSC), mesenchymal stem cells (MSC), and bone marrow-derived mesenchymal stem cells (BMMSC) have exhibited their ability to rejuvenate the skin, which was aging out, thereby slowing down/reducing the rate of overall aging process. The potential of stem cells to proliferate and differentiate helps in the regeneration and repair of different tissues and organs. In turn, this leads to regulate the secretion of various growth factors and cytokines to maintain the tissue microenvironment related to regeneration. As the process of aging continues, there is a decline in tissue regeneration capacity. Therefore, the ability of regeneration and repair of tissues even in elderly persons can be achieved/improved by applying advanced stem cell technology that is used nowadays to delay the aging process and/or treat age-related diseases.

2 Stem cells

Stem cells are group of undifferentiated cells that undergo self-renewal and differentiation. All multicellular organisms undergo a gradual loss of tissue homeostasis and organ function with time and age. These populations of all adult stem cells maintain major human tissue systems like blood, intestinal epithelium. Consequently, with growing age, these tissues shrink and their homeostasis decreases, resulting in the possibility of an age-dependent decline in the capability of these stem cells to replace the damaged cells. Stem cells can divide asymmetrically to generate two daughter cells. These daughter cells remain completely identical to the parent stem cells that further undergo different fates. These stem cells maintain the undifferentiated state as well as initiate self-renewal of the stem cell population. The second set of daughter cells commits toward differentiating and producing progenitor cells and terminally differentiated cells [1].

3 Types of stem cells

Adult stem cells are found throughout the body in every tissue and organ because the self-renewing cell pools are destined to replenish the dying cells and revive the damaged tissues throughout life [2]. The stem cell functioning also diminishes with the growing age. This fact justifies the decline of regenerative power of cells/tissues with age; the injuries in older individuals heal more slowly than in childhood. For example, when a bone is fractured, it takes much longer to heal in an elderly person than in any young individual [3]. There is a substantial amount of evidence showing that deterioration of adult stem cells in the adult phase can become an important player in the initiation of several diseases in aging [4]. Some of the examples of aging-associated effects on various types of stem cells are mentioned in the following sections.

3.1 Embryonic stem (ES) cells

These are pluripotent cells and can be differentiated into three germ layers [5]. ES cells have recently been adopted as a resource for treatment of various diseases. But these pluripotent stem cells pose a certain risk of having uncontrolled growth after transplantation and the threat of emergence of tumorigenicity remains [6]. Therefore, careful attention is needed to observe the behavior of cells after transplantation of pluripotent cells. Although new cells replenish or self-renew and differentiate in their unique lineage, many physiological changes affect the functional ability of these stem cells and progenitor cells. These cells that differentiate into several cell types are called adult stem cells and have multipotency.

3.2 Mesenchymal stem cells

The cells isolated from mouse bone marrow upon culture exhibit the plastic adherent property and form spindle-shaped colonies. These are referred to as colony-forming unit fibroblasts [7]. Due to their ability to differentiate into specialized cells developing from mesoderm, they were named as mesenchymal stem cells (MSCs). MSCs, also known as multipotent cells, exist in adult tissues of different sources, ranging from murine to humans. They are self-renewable, multipotent, easily accessible, and culturally expandable in vitro with exceptional genomic stability and few ethical issues, marking its importance in cell therapy, regenerative medicine, and tissue repairing [8].

Since the first description of hMSCs derived from bone marrow [9, 10], they have been isolated from almost all tissues including perivascular area [11]. However, there is neither a single definition nor a quantitative assay to help in the identification of MSCs in mixed population of cells. However, the International Society for Cellular Therapy has proposed minimum criteria to define MSCs. These cells should (a) exhibit plastic adherence, (b) possess specific set of cell surface markers, i.e., cluster of differentiation (CD)73, D90, CD105, and lack expression of CD14, CD34, CD45, and human leukocyte antigen-DR (HLA-DR), and (c) have the ability to differentiate in vitro into adipocyte, chondrocytes, and osteoblast [12]. These characteristics are valid for all MSCs, although few differences exist in MSCs isolated from various tissue origins.

3.3 Adult stem cells

Adult stem cells can be found in various adult tissues and organs such as brain, pancreas, bone marrow, skin, and liver, which range from multipotent to unipotent cells and lead to a restricted unidirectional diversity of progeny based on the tissue in which they reside [13].

4 Aging

Aging is progressive decline in physiological functioning inside the body. The enigmatic nature of this problem is highlighted by the fact that immortality is able to be maintained at the cellular level; there is nothing inevitable about aging. There are various views on aging. Actually, this term may be defined as an increasing process in the probability of death with passing time. On the contrary, athletes can define aging on a compressed time scale that is mostly shorter than their reproductive years. Evolutionary biologists have not yet delineated the events that occur afterward or postreproduction, since postreproductive mass of a population hardly contributes to the gene pool of any species.

Aging has especially powerful implications for humans. Advances in public health and medicine have greatly increased the average life span of people living in societies with access to those benefits—so much so that the reproductive years of most people's lives are but a small fraction of their total life span. In these societies, the average age of the population is rapidly increasing, leading to new requirements in healthcare planning and a revamping of economic considerations to accommodate a populace older than the age at which they traditionally have retired from the workplace. With the extension of life span, there is an increasing interest in slowing or reversing the deleterious effects, some real and some perceived, of aging. The yearning for eternal youth plays an impressive role in the economies of developed countries; the robust growth of cosmetics industry, the demand for cosmetic surgery, and growing interest in being cryopreserved after death are some of the vivid signs of craze for possessing eternal youth. Increasingly, people are trying to redefine the limit to longevity, and at the same time, trying to take measures to maintain quality of life while pursuing its extension.

Several causes of aging have been delineated over the past few years. The major causes of aging have been determined as epigenetic alteration, telomere shortening, dysfunction of the mitochondria, and stem cell exhaustion. Among these, stem cells have generated attention as one of the highest potential sources for cell-based therapeutic strategies owing to their self-renewal and proliferative properties [14, 15]. Among the different lineages of stem cells, mesenchymal stem cells (MSCs) are known to contribute to maintain tissue regeneration and homeostasis. As a result, the gradual quantitative decrease in stem cell function during the process of aging has been marked as one of the causes that advocate the process of aging [16].

4.1 Role of stem cells in aging

In view of the fact that stem cells show extensive ability of self-renewal, their possible role on the aging process has been scrutinized. We all know that germ-line stem cells continue their propagation to differentiate from one generation to the other. So, an original miniscule population of stem cells can produce mature progeny in excess of the original donor's life span [17], leading to the emergence of several other questions relating to the specific role of stem cells in aging. Moreover, some fatal diseases in humans (e.g., Parkinson's disease, Alzheimer's disease), have etiologies that apparently do not involve the role of stem cells. Further, it has been observed that the diseases associated with loss in stem cells like aplastic anemia or bone marrow failure are relatively rare and not especially age-dependent [18]. Experimental comparison of the engraftment properties of young and old marrow in a dog (used as a large animal model) failed to show any decrease in stem cell function with age [19]. Nevertheless, regardless of the degree to which stem cells may be involved in particular disease processes, current stem cell concepts must be incorporated into any comprehensive consideration of aging.

There are also several pieces of evidence that favor stem cell aging. There is increasing evidence that the aging process can have adverse effects on stem cells. As stem cells age, their renewal ability deteriorates and their ability to differentiate into the various cell types is altered. Accordingly, it is suggested that aging-induced deterioration of stem cell functions may play a key role in the pathophysiology of the various aging-associated disorders. Previous kinds of literature also reveal significant age-related changes in stem cell populations. While it is true that stem cells are capable of serial passage through a succession of mouse recipients, they are not immortal [20, 21]. Despite the apparently complete and long-term recovery of progenitor cell numbers and the return to normalcy of mature blood cell counts, stem cells recover to only a small fraction of the total number found in an unmanipulated animal [22]. But to reiterate, even an incomplete restoration of stem cell numbers is nonetheless impressive and attests to stem cells' extensive powers of self-renewal. There are few factors related to failure of stem cells to fully regenerate their population. This may be due to extrinsic factor that is associated with the transplant procedure, particularly the reassociation with bone marrow stromal components and their integuments, including cytokines and extracellular matrices that they produce. It is for these reasons that to obtain successful engraftment, successively larger numbers of marrow cells must be transplanted with each passage. Recently, a report suggested that the process of apoptosis is involved in age-related changes in hematopoietic stem cells and gradual aging has stress on the replicative process in the body, thus causing an increase in apoptosis. The p53-mediated apoptotic pathway that leads to apoptosis removes the compromised stem cells from the pool. These stem cells are thus prevented from initiating proliferation with potentially dysfunctional or tumorigenic progeny. Simultaneously, the antiapoptotic gene bcl-2 overexpression occurs that markedly increases the number of hematopoietic stem cells in these mice under steady-state conditions and enhances their engraftment potential. These results argue for the importance of apoptosis in the physiological regulation of hematopoietic stem cell population size and suggest that this mechanism may be a target of aging.

Stem cells are also involved in tissue repair, which is known to deteriorate with age. It has been shown that bone marrow-derived endothelial progenitor cells (EPCs) play a role in the maintenance of endothelial function by contributing to re-endothelialization and neovascularization [23, 24]. Similar event occurs in muscles. The muscle stem cells are usually quiescent and mobilized when required for repair, for example, when an injury occurs. The numbers of satellite cells do not appear to change much with age [25]. However, their repair, i.e., differentiation ability is dramatically reduced. Thus, the satellite cells are altered in at least one of the major features of stem cells.

In summary, adult stem cells, with an exception of epidermal stem cells, undergo age-related changes. The age-related decline is mainly functional, but in some cases, a decline in stem cell numbers can be observed as mentioned earlier. Therefore, the molecular mechanism underlying the observed stem cell deficits in the aged organism remains an issue to be clarified in future studies.

5 Theories of stem cell aging

The theories of stem cell aging can be conveniently discussed under the following four subsections.

5.1 Free radical theory of aging

Mitochondria are known to be ubiquitous intracellular organelle in mammals and are the major storehouse of cellular adenosine triphosphate (ATP) that plays a vital role in a variety of cellular processes. When mitochondria produce most of the cellular energy, there are three major causes of mitochondrial dysfunction, namely aging-related ROS generation, disruption in Ca^{2+} homeostasis, and increased cellular apoptosis. These three processes directly affect age-related diseases [26]. Several kinds of literature suggest that there lies a direct relationship between mitochondrial dysfunction and stem cell aging [27, 28]. Accordingly, in several cell systems, mitochondrial dysfunction has been shown to lead to respiratory chain dysfunction, which may be the result of the accumulation of mutations in mitochondrial DNA. Several environmental factors like exposure to sunlight, chemicals and heavy metals, or other contaminants, or having unhealthy lifestyles, like smoking, alcohol consumption, and malnutrition, influence the generation of reactive oxygen species (ROS) [9], which affect various intrinsic factors like genetics, metabolism, and the passage of time, resulting in the repair functions to become sluggish or defective [29]. If the generation of ROS overwhelms the endogenous antioxidant defense shield inside the body, it subsequently leads to oxidative stress or damage [30]. Thus, oxidative stress is a process caused by ROS. These unstable molecules damage or "oxidize" cells in the body, similar to an oxidized sliced apple turning brown when exposed to air. Most of the ROS that are generated, consist of superoxide anions (O^{2-}), hydrogen peroxide (H_2O_2), and hydroxyl radicals (OH•), generated in cells by the mitochondrial respiratory chain. Among antioxidants, superoxide dismutase (SOD), catalase, peroxiredoxin, thioredoxin, and glutathione systems are well-known antioxidative enzymes. Nuclear factor erythroid 2-related factor 2 (NRF2) is known as a master regulator of these genes. Emerging evidence suggests that oxidative stress plays a key role in stem cell-aging induction and the progression of various diseases. Cellular redox status significantly influences stem cell homeostasis. The existence of oxidative stress frequently demands an adaptive response from the endogenous antioxidant stress machinery, which in turn significantly modulates the level of oxidative stress. Under a mild stress condition, cells mainly regulate apoptosis-related gene expression, antioxidant enzyme activity, and defensive transduction pathways to fulfill antioxidative needs. In contrast, sustained and intense oxidative stress dampens stem cell proliferation and promotes premature aging. The fact that ROS may lead to stem cell dysfunction with age draws precedence from the free radical theory of aging, described by Oh et al. [31]. This theory proposes that accumulated cellular damage and declining mitochondrial integrity in aged cells lead to elevated ROS production, which in turn drives a vicious cycle that further damages cellular macromolecules and disrupts mitochondrial oxidative phosphorylation, leading to eventual cellular decomposition. Till now, the role of oxidative damage in the aging process remains inconclusive in part because of the absence of a clear correlation between the efficacy of antioxidant defense mechanisms and extended cell function or longevity. There have been several pieces of evidence based on the fact that ROS generation may promote stem cell aging. Several studies have shown that the frequency of hematopoietic stem cells (HSCs) with low ROS levels declines with age in mice. In addition, in HSCs and neural stem cells (NSCs) of mice, excessive cellular ROS concentrations lead to abnormal proliferation, malignancy, and compromised stem cell self-renewal [31]. Conditional ablation in the mouse hematopoietic system of the transcription factors FoxO1, FoxO3, and FoxO4, downstream effectors of the insulin and insulin-like growth factor 1 (IGF-1) signaling pathways, induces a marked increase in ROS accumulation in HSCs. A decline in long-term regeneration capacity of HSCs in mice with conditional deletion of phosphatase and tensin homolog (PTEN) and double deficiency of protein kinase B (AKT) 1 and 2 also indicates that signaling through the PTEN-AKT-mammalian target of rapamycin (mTOR pathway, upstream of FoxO), senses and controls ROS and regulates HSC self-renewal and survival [3]. It has been suggested that the Wnt/b-catenin pathway is implicated in the process of stem cell aging brought on by microenvironmental changes [32].

5.2 Telomere shortening

Telomere is the end of mammalian chromosomal DNA that has G-rich sequences. The importance of these sequences lies in protecting the end of the chromosome from damage and degradation. Therefore, telomeres play a crucial role in maintaining chromosome stability. Telomerase, a DNA polymerase enzyme, is a ribonucleoprotein complex. The telomere length regulation is mainly done by telomerase. Also the G-rich sequences are linked with a protein complex each of which has a unique role in regulating the length of telomere.

Telomere theory is one of the most important theories of stem cell aging and senescence that is based on the mechanisms of shortening of telomeres. This acts a critical biomarker in terms of cellular aging [33]. Based on the telomere theory, it was indicated that aging is irreversible, programmed cell cycle arrest happens in response to the telomerase activity, and the total number of cell divisions cannot exceed a particular limit termed the Hayflick limit. According to a previous study, telomerase activity and telomere length maintenance are related with immortality of embryonic stem cells and germ-line cells [34]. Overexpression of telomerase reverse transcriptase (TERT), the enzymatic subunit of telomerase, in later stages of life in mice increases cancer incidence, thus suggesting that telomere length contributes to late-life survival [35]. This activity helps to prevent the replication-dependent cellular senescence and loss of telomere length in highly proliferative cells including the majority of tumor cells and germ-line cells [36]. The stem cells having multilineage differentiation potential and self-renewal capacity have rapid cell expansion, thus they should have a mechanism that maintains telomere length during cell replication [37]. On the contrary, some authors reported that in hematopoietic and nonhematopoietic stem cell (HSC)-derived mammary epithelial, kidney, skin tissues as well as mesenchymal stem cells, low levels of telomerase activity have been detected. Because of the fact that stem and progenitor cells play a significant role in homeostasis and tissue repair, senescence of these cells is considered as an important factor in the aging process. Despite considerable evidence that telomeres play a role in aging, it is not clear how impactful their shortening is for species that begin life with long telomeres and have shorter life spans, such as mice. For example, laboratory mice lacking telomerase RNA component (TERC) have no miserable phonotypical effect for five generations. So, more clear understanding of the mechanisms involved in aging of these cells would need further in-depth research.

5.3 DNA damage accumulation and mutations

Hematopoietic stem cells maintain the tissue homeostasis throughout lifetime. It is therefore essential for stem cells to maintain their genomic integrity. One of the major factors driving stem cell aging is DNA damage. The DNA damage theory based on stem cell aging states that there are age-related changes in DNA repair system and the cell cycle regulation mechanism. These result from increased DNA mutations that contribute to decrease in stem cell functioning with age. Earlier, studies related to this theory showed that mice lacking a distinct set of proteins related to DNA damage repair showed the decreased function of stem cells. This subsequently causes an overall depletion of the stem cell pool [38–40]. There are diverse types of DNA damage like oxidative DNA damage, hydrolytic DNA damage, and ultraviolet and other radiation damages. The endogenous sources for DNA damage mainly include replication and recombination errors, spontaneous hydrolysis, and other reactive metabolites formed as a by-product of cellular metabolism like ROS. These form many basic sites and cause base deamination, 8-oxoguanine lesions, oxidation of bases, and single/double-stranded DNA breaks. Earlier kinds of literature suggest that mutations in the proteins involved in DNA repair system cause symptoms of premature aging [41]. DNA damages are also caused due to exogenous factors, like ultraviolet radiations from the sun, harmful chemicals, X-rays and gamma rays, and chemotherapeutic drugs causing base modifications, which subsequently form cross-links between the strands, dimer formation, and single- and double-strand breaks in DNA. Some studies showed that an XPD deficiency did not result in the depletion of hematopoietic stem cells with age, but the proliferative potential of hematopoietic stem cell was severely affected [42]. Some studies also showed that the impairment of the NHEJ DNA repair pathway caused a progressive loss in hematopoietic stem cell pool with aging. In human hematopoietic stem cells and progenitors, DNA damage occurs during aging, and this accumulation of DNA is independent of telomere length [43]. Accumulating studies also demonstrated that a component of serum is altered during aging. Conboy and coworkers [44] showed that the age-related decline of muscle stem cell activity can be modulated by systemic factors that change with age. Naito et al. [45] identified complement C1q as a canonical activator of Wnt signaling that is upregulated in aged mice serum. They also showed that C1q binds Frizzle (a receptor for Wnt) and activate canonical Wnt pathway. Zhang et al. [46] also demonstrated that Wnt/β-catenin signaling induced aging in mesenchymal stem cells that caused DNA damage. Therefore, acceleration of the canonical Wnt signaling pathway might affect both stem cell compartments and mesenchymal progenitors in aged tissues.

5.4 Epigenetic alterations

Hematopoietic stem cells (HSCs) are a small population of cells capable of producing mature myeloid and lymphoid cells through self-renewal and differentiation [47]. Aging being a degenerative change, several epigenetic alterations like DNA methylation, histone modifications, and noncoding RNA expressions contribute to aging. Several experimental studies using human lymphocytes [48] and blood cell tissues [49, 50] confirmed that DNA methylation diminishes with age whereas certain genes like estrogen receptor [51], insulin-like growth factor-II [52]; p14ARF [53] are found to be hypermethylated. The aging process also lowers the level of SIRT1, a mammalian HDAC deacetylating H4 and H3 at lysine 16 and 9 positions, respectively [54], having a role in DNA repair and transcriptional repression. Studies also reported that reduction of SIRT6 in mice correlates with aging accompanied by significantly reduced subcutaneous fat, retardation of growth, and even death at an early age [55]. Transcriptional regulation by histone methyltransferase causing histone modifications gets altered in aging as revealed from the experimental data performed in senescence-accelerated prone mouse 8 (SAMP8) model using brain tissues. A marked reduction in methylation of H4K20me and H3K36me3 sites [56] was observed, which signifies the aforementioned hypothesis of epigenetic regulation. Alternatively, H4K20me3 was abundant in kidney and liver tissues of aging rats [57]. Further, aged HSCs show increased expression of the H3K27me3 methylation markedly [58]. Thus, methylated/demethylated histone turns on/off genes by activating or repressing transcription that contributes to aging disorders. Additionally, few more studies also revealed that aging upregulates miRNA expressions [59]. Therefore, epigenetic errors are known to be an important factor affecting the aging process by regulating the factors for age-related defects or malfunctioning of genes.

With this brief introductory idea about the different theories of stem cell aging, various other important aspects such as cellular aging and senescence, signaling pathways influencing stem cells' self-renewal and differentiation, progress in the field of human embryonic stem cell research and aging, hematopoietic stem cells and aging, stem cells and neurodegenerative disorders, aging and neural stem cells, aging and stability or potential regeneration of cardiomyocytes, muscle cells aging and stem cells, and therapeutic approaches for the treatment of aging-induced stem cell dysfunction will be dealt with in greater detail in the subsequent chapters, to give readers a comprehensive picture of the various hot topics related to the stem cells in relation to the aging process.

References

[1] F.Q. Alenzi, B.Q. Alenazi, S.Y. Ahmad, M.L. Salem, A.A. Al-Jabri, R.K. Wyse, The haemopoietic stem cell: between apoptosis and self renewal, Yale J. Biol. Med. 82 (2009) 7–18.

[2] L.B. Boyette, R.S. Tuan, Adult stem cells and diseases of aging, J. Clin. Med. 3 (2014) 88–134.

[3] A.D. Ho, W. Wagner, U. Mahlknecht, Stem cells and ageing: the potential of stem cells to overcome age-related deteriorations of the body in regenerative medicine, EMBO Rep. 6 (2005). Spec No: S35-S38.

[4] E. Mansilla, V. Díaz Aquino, D. Zambón, G.H. Marin, K. Mártire, G. Roque, T. Ichim, N.H. Riordan, A. Patel, F. Sturla, G. Larsen, R. Spretz, L. Núñez, C. Soratti, R. Ibar, M. van Leeuwen, J.M. Tau, H. Drago, A. Maceira, Could metabolic syndrome, lipodystrophy, and aging be mesenchymal stem cell exhaustion syndromes? Stem Cells Int. (2011) 1–10. Article ID 943216 https://doi.org/10.4061/2011/943216.

[5] J.T. Do, H.R. Scholer, Regulatory circuits underlying pluripotency and reprogramming, Trends Pharmacol. Sci. 30 (2009) 296–302.

[6] R.C. Addis, J.W.M. Bulte, J.D. Gearhart, Special cells, special considerations: the challenges of bringing embryonic stem cells from the laboratory to the clinic, Clin. Pharmacol. Ther. 83 (2008) 386–389.

[7] A.J. Friedenstein, J.F. Gorskaja, N.N. Kulagina, Fibroblast precursors in normal and irradiated mouse hematopoietic organs, Exp. Hematol. 4 (1976) 267–274.

[8] E.M. Horwitz, K. Le Blanc, M. Dominici, I. Mueller, I. Slaper-Cortenbach, F.C. Marini, R.J. Deans, D.S. Krause, A. Keating, Clarification of the nomenclature for MSC: the international society for cellular therapy position statement, Cytotherapy 7 (2005) 393–395.

[9] M.F. Pittenger, A.M. Mackay, S.C. Beck, R.K. Jaiswal, R. Douglas, J.D. Mosca, M.A. Moorman, D.W. Simonetti, S. Craig, D.R. Marshak, Multilineage potential of adult human mesenchymal stem cells, Science 284 (1999) 143–147.

[10] A. Pansky, B. Roitzheim, E. Tobiasch, Differentiation potential of adult human mesenchymal stem cells, Clin. Lab. 53 (2007) 81–84.

[11] M. Crisan, S. Yap, L. Casteilla, C.W. Chen, M. Corselli, T.S. Park, G. Andriolo, B. Sun, B. Zheng, L. Zhang, C. Norotte, P.N. Teng, J. Traas, R. Schugar, B.M. Deasy, S. Badylak, H.J. Buhring, J.P. Giacobino, L. Lazzari, J. Huard, B. Péault, A perivascular origin for mesenchymal stem cells in multiple human organs, Cell Stem Cell 3 (2008) 301–313.

[12] M. Dominici, K. Le Blanc, I. Mueller, I. Slaper-Cortenbach, F. Marini, D. Krause, R. Deans, A. Keating, D. Prockop, E. Horwitz, Minimal criteria for defining multipotent mesenchymal stromal cells, the international society for cellular. Therapy position statement, Cytotherapy 8 (2006) 315–317.

[13] T.A. Rando, Stem cells, ageing and the quest for immortality, Nature 441 (2006) 1080–1086.

[14] E. Fathi, R. Farahzadi, Isolation, culturing, characterization and aging of adipose tissue-derived mesenchymal stem cells: a brief overview, Braz. Arch. Biol. Technol. 59 (2016) e16150383.

[15] S. Gholizadeh-Ghalehaziz, R. Farahzadi, E. Fathi, M. Pashaiasl, A mini overview of isolation, characterization and application of amniotic fluid stem cells, Int. J. Stem Cells 8 (2015) 115–120.

[16] R. Ren, A. Ocampo, G.H. Liu, J.C.I. Belmonte, Regulation of stem cell aging by metabolism and epigenetics, Cell Metab. 26 (2017) 460–474.

[17] D.E. Harrison, Normal function of transplanted mouse erythrocyte precursors for 21 months beyond donor life spans, Nat. New Biol. 237 (1972) 220–222.

[18] S.B. Marley, J.L. Lewis, R.J. Davidson, I.A. Roberts, I. Dokal, J.M. Goldman, M.Y. Gordon, Evidence for a continuous decline in haemopoietic cell function from birth: application to evaluating bone marrow failure in children, Br. J. Haematol. 106 (1999) 162–166.

[19] J.M. Zaucha, C. Yu, G. Mathioudakis, K. Seidel, G. Georges, G. Sale, M.T. Little, B. Torok-Storb, R. Storb, Hematopoietic responses to stress conditions in young dogs compared with elderly dogs, Blood 98 (2001) 322–327.

[20] M. Bartucci, R. Dattilo, D. Martinetti, M. Todaro, G. Zapparelli, A.D. Virgilio, M. Biffoni, R.D. Maria, A. Zeuner, Prevention of chemotherapy-induced anemia and thrombocytopenia by constant administration of stem cell factor, Clin. Cancer Res. 17 (2011) 6185–6191.

[21] A.P. Beltrami, D. Cesselli, C.A. Beltrami, At the stem of youth and health, Pharmacol. Ther. 129 (2011) 3–20.

[22] V. Paradis, N. Youssef, D. Dargere, N. Bâ, F. Bonvoust, J. Deschatrette, P. Bedossa, Replicative senescence in normal liver, chronic hepatitis C, and hepatocellular carcinomas, Hum. Pathol. 32 (2001) 327–332.

[23] T. Asahara, T. Murohara, A. Sullivan, M. Silver, R. van der Zee, T. Li, B. Witzenbichler, G. Schatteman, J.M. Isner, Isolation of putative progenitor endothelial cells for angiogenesis, Science 275 (1997) 964–967.

[24] M. Reyes, A. Dudek, B. Jahagirdar, L. Koodie, P.H. Marker, C.M. Verfaillie, Origin of endothelial progenitors in human postnatal bone marrow, J. Clin. Invest. 109 (2002) 337–346.

[25] A.S. Brack, T.A. Rando, Intrinsic changes and extrinsic influences of myogenic stem cell function during aging, Stem Cell Rev. 3 (2007) 226–237.

[26] G.C. Kujoth, A. Hiona, T.D. Pugh, S. Someya, K. Panzer, S.E. Wohlgemuth, T. Hofer, A.Y. Seo, R. Sullivan, W.A. Jobling, J.D. Morrow, H.V. Remmen, J.M. Sedivy, T. Yamasoba, M. Tanokura, R. Weindruch, C. Leeuwenburgh, T.A. Prolla, Mitochondrial DNA mutations, oxidative stress, and apoptosis in mammalian aging, Science 309 (2005) 481–484.

[27] A. Bratic, N.G. Larsson, The role of mitochondria in aging, J. Clin. Invest. 123 (2013) 951–957.

[28] T.G. Fellous, S. Islam, P.J. Tadrous, G. Elia, H.M. Kocher, S. Bhattacharya, L. Mears, D.M. Turnbull, R.W. Taylor, L.C. Greaves, P.F. Chinnery, G. Taylor, S.A. McDonald, N.A. Wright, M.R. Alison, Locating the stem cell niche and tracing hepatocyte lineages in human liver, Hepatology 49 (2009) 1655–1663.

[29] E.D. Lephart, Skin aging and oxidative stress: Equol's anti-aging effects via biochemical and molecular mechanisms, Ageing Res. Rev. 31 (2016) 36–54.

[30] K.H. Al-Gubory, Environmental pollutants and lifestyle factors induce oxidative stress and poor prenatal development, Reprod. BioMed. Online 29 (2014) 17–31.

[31] J. Oh, Y.D. Lee, A.J. Wagers, Stem cell aging: mechanisms, regulators and therapeutic opportunities, Nat. Med. 20 (2014) 870–880.
[32] S. Fujimaki, T. Wakabayashi, T. Takemasa, M. Asashima, T. Kuwabara, The regulation of stem cell aging by Wnt signaling, Histol. Histopathol. 30 (2015) 1411–1430.
[33] A. Bernadotte, V.M. Mikhelson, I.M. Spivak, Markers of cellular senescence, telomere shortening as a marker of cellular senescence, Aging (Albany NY) 8 (2016) 3–11.
[34] B. Brazvan, A. Ebrahimi-Kalan, K. Velaei, A. Mehdipour, Z.A. Serej, A. Ebrahimi, M. Ghorbani, O. Cheraghi, H.N. Charoudeh, Telomerase activity and telomere on stem progeny senescence, Biomed. Pharmacother. 102 (2018) 9–17.
[35] B.B. de Jesus, E. Vera, K. Schneeberger, A.M. Tejera, E. Ayuso, F. Bosch, M.A. Blasco, Telomerase gene therapy in adult and old mice delays aging and increases longevity without increasing cancer, EMBO Mol. Med. 4 (2012) 691–704.
[36] M.A. Blasco, Telomeres and human disease: ageing, cancer and beyond, Nat. Rev. Genet. 6 (2005) 611–622.
[37] E. Hiyama, K. Hiyama, Telomere and telomerase in stem cells, Br. J. Cancer 96 (2007) 1020–1024.
[38] J.M. Prasher, A.S. Lalai, C. Heijmans-Antonissen, R.E. Ploemacher, J.H.J. Hoeijmakers, I.P. Touw, L.J. Niedernhofer, Reduced hematopoietic reserves in DNA interstrand crosslink repair-deficient Ercc/mice, EMBO J. 24 (2005) 861–871.
[39] A. Nijnik, L. Woodbine, C. Marchetti, S. Dawson, T. Lambe, C. Liu, N.P. Rodrigues, T.L. Crockford, E. Cabuy, A. Vindigni, T. Enver, J.I. Bell, P. Slijepcevic, C.C. Goodnow, P.A. Jeggo, R.J. Cornal, DNA repair is limiting for haematopoietic stem cells during ageing, Nature 447 (2007) 686–690.
[40] K. Parmar, J. Kim, S.M. Sykes, A. Shimamura, P. Stuckert, K. Zhu, A. Hamilton, M.K. Deloach, J.L. Kutok, K. Akashi, D.G. Gilliland, A. D'andrea, Hematopoietic stem cell defects in mice with deficiency of Fancd2 or Usp1, Stem Cells 28 (2010) 1186–1195.
[41] S. Maynard, E.F. Fang, M. Scheibye-Knudsen, D.L. Croteau, V.A. Bohr, DNA damage, DNA repair, aging, and neurodegeneration, Cold Spring Harb. Perspect. Med. 5 (2015) a025130.
[42] F. Rossi, R. Moschetti, R. Caizzi, N. Corradini, P. Dimitri, Cytogenetic and molecular characterization of heterochromatin gene models in *Drosophila melanogaster*, Genetics 175 (2007) 595–607.
[43] C.E. Rube, A. Fricke, T.A. Widmann, T. Furst, H. Madry, M. Pfreundschuh, C. Rube, Accumulation of DNA damage in hematopoetic stem and progenitor cells during human aging, PLoS One 6 (2011) e17487.
[44] I.M. Conboy, M.J. Conboy, A.J. Wagers, E.R. Girma, I.L. Weissman, T.A. Rando, Rejuvenation of aged progenitor cells by exposure to a young systemic environment, Nature 433 (2005) 760–764.
[45] A.T. Naito, T. Sumida, S. Nomura, M.L. Liu, T. Higo, A. Nakagawa, K. Okada, T. Sakai, A. Hashimoto, Y. Hara, I. Shimizu, W. Zhu, H. Toko, A. Katada, H. Akazawa, T. Oka, J.K. Lee, T. Minamino, T. Nagai, K. Walsh, A. Kikuchi, M. Matsumoto, M. Botto, I. Shiojima, I. Komuro, Complement C1q activates canonical Wnt signalling and promotes aging related phenotypes, Cell 149 (2012) 1298–1313.
[46] D.Y. Zhang, H.J. Wang, Y.Z. Tan, Wnt/B catenin signalling induces the aging of mesenchymal stem cells through the DNA damage response and the p 53/p21 pathway, PLoS One 6 (2011) e21397.
[47] D. Bryder, D.J. Rossi, I.L. Weissman, Hematopoietic stem cells: the paradigmatic tissue-specific stem cell, Am. J. Pathol. 169 (2006) 338–346.
[48] R.D. Drinkwater, T.J. Blake, A.A. Morley, D.R. Turner, Human lymphocytes aged in vivo have reduced levels of methylation in transcriptionally active and inactive DNA, Mutat. Res. 219 (1989) 29–37.
[49] H.T. Bjornsson, M.I. Sigurdsson, M.D. Fallin, R.A. Irizarry, T. Aspelund, H. Cui, W.W. Yu, M.A. Rongione, T.J. Ekstrom, T.B. Harris, L.J. Launer, G. Eiriksdottir, M.F. Leppert, C. Sapienza, V. Gudnason, A.P. Feinberg, Intra-individual change over time in DNA methylation with familial clustering, JAMA 299 (2008) 2877–2883.
[50] C. Fuke, M. Shimabukuro, A. Petronis, J. Sugimoto, T. Oda, K. Miura, T. Miyazaki, C. Ogura, Y. Okazaki, Y. Jinno, Age related changes in 5-methylcytosine content in human peripheral leukocytes and placentas: an HPLC-based study, Ann. Hum. Genet. 68 (2004) 196–204.
[51] J.P. Issa, Y.L. Ottaviano, P. Celano, S.R. Hamilton, N.E. Davidson, S.B. Baylin, Methylation of the oestrogen receptor CpG island links ageing and neoplasia in human colon, Nat. Genet. 7 (1994) 536–540.
[52] J.P. Issa, P.M. Vertino, C.D. Boehm, I.F. Newsham, S.B. Baylin, Switch from monoallelic to biallelic human IGF2 promoter methylation during aging and carcinogenesis, Proc. Natl. Acad. Sci. U. S. A. 93 (1996) 11757–11762.
[53] L. Shen, Y. Kondo, S.R. Hamilton, A. Rashid, J.P. Issa, P14 methylation in human colon cancer is associated with microsatellite instability and wild-type p53, Gastroenterology 124 (2003) 626–633.
[54] K. Pruitt, R.L. Zinn, J.E. Ohm, K.M. McGarvey, S.H. Kang, D.N. Watkins, J.G. Herman, S.B. Baylin, Inhibition of SIRT1 reactivates silenced cancer genes without loss of promoter DNA hypermethylation, PLoS Genet. 2 (2006) e40.
[55] R. Mostoslavsky, K.F. Chua, D.B. Lombard, W.W. Pang, M.R. Fischer, L. Gellon, P. Liu, G. Mostoslavsky, S. Franco, M.M. Murphy, K.D. Mills, P. Patel, J.T. Hsu, A.L. Hong, E. Ford, H.L. Cheng, C. Kennedy, N. Nunez, R. Bronson, D. Frendewey, W. Auerbach, D. Valenzuela, M. Karow, M.O. Hottiger, S. Hursting, J.C. Barrett, L. Guarente, R. Mulligan, B. Demple, G.D. Yancopoulos, F.W. Alt, Genomic instability and aging-like phenotype in the absence of mammalian SIRT, Cell 124 (2006) 315–329.
[56] C.M. Wang, S.N. Tsai, T.W. Yew, Y.W. Kwan, S.M. Ngai, Identification of histone methylation multiplicities patterns in the brain of senescence accelerated prone mouse 8, Biogerontology 11 (2010) 87–102.
[57] B. Sarg, E. Koutzamani, W. Helliger, I. Rundquist, H.H. Lindner, Postsynthetic trimethylation of histone H4 at lysine 20 in mammalian tissues is associated with aging, J. Biol. Chem. 277 (2002) 39195–39201.
[58] D. Sun, M. Luo, M. Jeong, B. Rodriguez, Z. Xia, R. Hannah, H. Wang, T. Le, K.F. Faull, R. Chen, H. Gu, C. Bock, A. Meissner, B. Göttgens, G.J. Darlington, W. Li, M.A. Goodell, Epigenomic profiling of young and aged HSCs reveals concerted changes during aging that reinforce self-renewal, Cell Stem Cell 14 (2014) 673–688.
[59] O.C. Maes, J. An, H. Sarojini, E. Wang, Murine micro RNAs implicated in liver functions and aging process, Mech. Ageing Dev. 129 (2008) 534–541.

Chapter 2

Progress in human embryonic stem cell research and aging

Anjali P. Patni[a], Joel P. Joseph[a], D. Macrin[a,b], and Arikketh Devi[a]
[a]Stem Cell Biology Lab, Department of Genetic Engineering, School of Bioengineering, Faculty of Engineering and Technology, SRM Institute of Science and Technology, Chennai, Tamil Nadu, India, [b]Department of Bioinformatics, Saveetha School of Engineering, Saveetha Institute of Medical and Technical Services, Saveetha Nagar, Chennai, India

1 Introduction

Stem cell research has created an interest to scientists worldwide for its promise in cell-based therapeutics and regenerative medicine. Innumerable studies on embryonic and adult stem cells have determined the role of stem cells in tissue renewal, regeneration, and therapy. In 1981, Evans and Gail Martin performed individual studies on mouse blastocyst, and the first multipotent cells were collected from the mouse embryos [1]. These cells were later named embryonic stem cells (ESCs) by Gail Martin [2]. The blastocyst is a stage in the embryogenesis of mammals that consists of an inner cell mass (ICM) [3], which later forms the embryo and the outer trophoblast that later gives rise to the extramembryonic membranes and the placenta [4,5]. The trophoblast layer encompasses the inner cell mass and the blastocoel cavity. Stem cells from the embryonic origin, extracted from the inner cell mass (ICM) of the blastocyst, are pluripotent. These stem cells possess the properties of being unspecialized, self-renewing and the capability of differentiating into any cells of the three germ layers such as the ectoderm, mesoderm, and endoderm, except for the placenta and the extraembryonic membranes [6–9]. While embryonic stem cells are pluripotent, the adult stem cells are multipotent and can differentiate into fewer cell types. Furthermore, ESCs can divide indefinitely in vitro under appropriate conditions.

Nevertheless, the isolation of nonhuman primate cells, especially murine ESCs (mESCs), took more than 17 years to pave the way for the derivation of embryonic stem cells from primate blastocysts [10]. In 1998, the Thomson group became the first to extract human ESCs from blastocyst [11]. Conventionally, hESCs were derived, by microsurgical detachment of ICM from the blastocyst stage preimplantation embryo [12]. Human ESC cultures consist of morphologically heterogeneous populations of cells from which a subset of fibroblast-like cells spontaneously differentiated from hESCs usually surrounds the colonies [13]. hESCs can self-renew for a long time, divide, and proliferate into different cell types, thus deriving new cell lines under different conditions [14]. Also, hESCs research from different laboratories have identified two main states of hESCs: naïve and primed, which correspond to the in vivo cell stages of preimplantation and postimplantation blastocysts. While the cells in these states are similar in their expression of pluripotency markers like Oct4, Sox2, and Nanog, they exhibit differences in their morphology, dependence on growth factors, developmental potential, metabolism, and transcriptomes. Buecker et al. demonstrated the importance of Oct4 in maintaining the naïve state and showed how, redirecting Oct4 to new enhancer sites plays a role in the transition of naïve to a primed state of pluripotency. The work also analyzed various genomic and biochemical parameters to identify key candidates for primed-state pluripotency [12].

Moreover, studies conducted with hESCs during the last few decades suggest that these cells could contribute in a significant way to regenerative therapy in humans [15]. The essential steps before exploring the therapeutic value of hESCs is: to optimize the culture conditions; to sustain hESCs in the undifferentiated state for an extended period in vitro, and this requires detailed knowledge of molecular pathways influencing self-renewal, pluripotency, apoptosis, and differentiation. Additionally, under suboptimal conditions, hESCs were shown to adjust slowly to the long-term culture with improvements in survival and proliferation capability [16]. In stem cell research, lineage-restricted progenitor cell-line isolation is the most notable feature of hESCs that can differentiate into postmitotic cells [17, 18]. Moreover, because of the potential to differentiate indefinitely, hESCs have a nearly never-ending supply for different cell populations [19].

Although with an advancement in the knowledge of ESCs culture and proliferation, there are various factors [20] that influence the maintenance of uniformity in hESCs culture propagation and differentiation [21, 22], the functionality of hESCs is governed by several intrinsic factors which includes cell-cycle activity [23], senescence [24], quiescence [25], different cytokine signals [26], and various signaling pathway expression [27], followed by the extrinsic processes that involve signals acquired by their surrounding niche microenvironment of hESCs [28]. However, the intrinsic factors of ESCs is regulated mainly by the process of aging [29, 30].

From the layman who thinks of aging as a deterioration of his/her looks, to the philosophers who relate it to death and the liberation of soul, to scientists who are eager to understand its complexity by unraveling the molecular mechanisms underlying it, aging has been a topic of debate from time immemorial. Aging is a degenerative process that causes a progressive deterioration leading to dysfunction of the different organ systems of the body and finally to the death of the individual.

Aging poses as a major risk factor for several diseases such as cancer, cardiovascular disorders, diabetes, dementia, idiopathic pulmonary fibrosis, arthritis, osteoporosis, neuropathy, stroke, obesity, glaucoma, and neurodegenerative disorders. Although several studies have attempted to identify antiaging genes and the importance of stem cells during aging, these studies are mostly restricted to adult stem cells such as the hematopoietic stem cells (HSCs) and the mesenchymal stem cells (MSCs). Work on hESCs and aging is too limited. Knowledge of molecular pathways essential for the longevity of hESCs is of the foremost factor in designing treatment strategies aimed at optimizing organ health and functioning throughout the aging process. This chapter highlights some of the advances in this area, assesses the regulatory mechanisms governing hESCs, and discusses the impact of aging on both stem cells and the microenvironment. We also give an overview of some of the potential implications of these age-associated alterations and how they can be possibly treated with hESCs therapy.

2 Properties of hESCs

Most of the hESCs potential is an assessment of the revolutionary mouse model analysis [31]. Human ESCs mature in flat-compact colonies with well-defined morphology and display increased nucleus to cytoplasm ratio with prominent nucleoli, exhibiting a population doubling rate of 30–48 h [32, 33]. In the presence of bovine serum or animal-derived serum, hESCs were initially harvested and cultured on murine embryonic fibroblast feeder cells [11]. These feeder cells constitute a viable, adherent growth-resistant and a bioactive group of cells that act as a substratum to the medium that supports the growth of other cell populations with low or clonal densities [34]. The application of human feeder cells was later reduced [35] as the use of xeno (nonhuman cells)-derived feeder cells and components for hESC lines resulted in xenohazards and immunologic rejection ultimately making them unfit for clinical transplantation [36]. However, to counteract the possible impact of xeno-derived feeder cells, a conditioned medium was used as an alternative to culture the hESC cell lines. Matrigel, isolated from mouse sarcoma tumor Engelbreth-Holm-Swarm (EHS) cells, is a dynamic, undefined blend of the basement membrane, cellular proteins, and growth factors, and it emerged as the major source for hESCs culture conditions [37, 38]. Several protocols previously documented the use of Matrigel coupled with human foreskin fibroblast feeder cells containing hESCs [39]. Xeno-free techniques further established and encouraged extremely robust hESCs culture [40, 41]. Also, in vitro hESCs culturing is a challenging process and should be characterized for genetic stability and genomic integrity by assessing its pluripotency.

In humans, relevant evidences have been documented concerning the biochemical processes influencing the lineage decisions within the blastocyst [42]. Orchestrated activities of extrinsic signaling pathways such as FGF, BMP4, TGF-β/Activin, and IGF, as well as an intrinsic network of core factors such as Oct4, Sox2, and Nanog, stabilize the pluripotent state in hESCs. The expression of hESC-associated pluripotency markers also involves cell-surface markers such as SSEA-1, SSEA-3, SSEA-4, Tra-1-60, Tra-1-81, and many other well-established factors that play a key role in prolonging the self-renewal process and maintenance of the undifferentiated state of hESCs population (Table 1). The pluripotent state of self-renewing murine EpiSCs and hESCs named as "primed" under suitable cultural circumstances, suggests that these cells were developmentally more restricted or less potent than the more primitive and less restricted "naïve" hESCs. The derived naïve and primed pluripotent hESCs can be further classified as either transgenic-dependent or transgenic-independent generation, depending on the growth conditions applied [104]. The conventional ICM-derived hESCs retain a wide context of primed pluripotent characteristics, viz.: (1) low expression of naive pluripotency markers; (2) accumulation of trimethylated H3K27 across developmental genes; (3) TFE3 (transcription factor E3) nuclear localization absence; (4) loss of pluripotency upon inhibition of MEK/ERK pathway; and (5) the early X-inactivation failure in ESCs lines [105].

TABLE 1 Key transcriptional regulatory circuits of hESCs involved in self-renewal and lineage-specific differentiation.

Sr no.	Transcription factor	Transcription factor family	Biological mechanism	References
Major core factors				
1.	Oct4/POU5F1 (Octamer-binding transcription factor 4/ POU-domain, class 5, transcription factor 1)	A 38 kDa protein belonging to POU (PIT/OCT/UNC) transcription factor family	Oct4 POU domain may participate in cell fate decisions during the early stages of embryonic development by generating heterodimers through binding to the promoters or enhancing the regulatory factors of their target genes	[43, 44]
2.	Sox2 (SRY (sex-determining region Y)-box 2)	A 34 kDa protein belonging to HMG-box (Sox) transcription factor family	Sox2 establishes a binary complex by engaging with Oct4, which then employs other nuclear factors triggering pluripotent gene expression, thereby restricting genes associated with differentiation	[45]
3.	Nanog	A 34 kDa protein belonging to Nanog homeobox transcription factor family	Nanog acts in parallel with cytokine stimulation of STAT3 to facilitate self-renewal and sustain pluripotency in undifferentiated cells and plays a key role in ESC's fate specification	[46, 47]
Self-renewal promoters				
1.	c-Myc (Cellular Myc)	A 48 kDa protein belonging to MYC proto-oncogenes bHLH (basic helix-loop-helix) transcription factor family	Enhanced c-Myc level is essential for hypoxia-mediated increase in reprogramming efficiency vital for hESC self-renewal and enriched pluripotent-state maintenance	[48]
2.	TBX3 (T-box 3)	A 79 kDa protein belonging to T-box transcription factor family	TBX3 facilitates proliferation of hESCs by repressing cell-cycle regulator expression and maintains the hESCs primed state	[49]
3.	KLF (Krüppel-like factors)	This group of proteins belong to zinc-finger Krüppel-like transcription factor family	Overexpressed KLF4 tends to develop cancer. KLF4 helps in the telomerase activity maintenance and display low rates of KLF protein expression, which increases gradually after hESC differentiation. It is expressed in both primed- and naïve-state of hESCs	[50, 51]
4.	GDF3 (Growth differentiation factor 3)	A 26 kDa protein belonging to the transforming growth factor-beta (TGF-β) superfamily	GDF3 functions in hESCs to maintain pluripotency signatures by regulating BMP signaling pathway	[52]
5.	SSEA3/4 (Stage-specific embryonic antigen3/4)	A 43 kDa belonging to Src family kinases and small G-protein family	These are carbohydrate-associated cell-surface antigen glycolipid molecules expressed in the hESC pluripotent cells of the inner cell mass that is lost upon differentiation	[53, 54]
6.	TRA1–60/80 (T-cell receptor alpha locus/ podocalyxin)	A 200 kDa glycoprotein belonging to sialomucin family	These are the keratan sulfate molecules that recognize carbohydrate epitopes on the glycoprotein moiety of undifferentiated hESCs, embryonic carcinoma cells (EC), and embryonic germ cells (EG)	[55, 56]
7.	Rex1/ZFP42 (Reduced expression 1/zinc-finger protein 42 homolog)	A 132 kDa protein belonging to Kruppel C2H2-type zinc-finger protein family	Rex1 is crucial in self-renewal and pluripotency as it facilitates proliferation and cell survival while restricting differentiation, apoptosis, and cell-cycle arrest in order to sustain hESC's cellular functions	[57]
8.	DICER1 (dsRNA endoribonuclease)	A 218 kDa protein belonging to helicase family, Dicer subfamily	DICER1 plays a prosurvival function in hESCs by breaking pre-miRNA hairpins into mature miRNAs, its removal contributes to enhanced death-receptor-mediated apoptosis and self-renewal failure	[58]

Continued

TABLE 1 Key transcriptional regulatory circuits of hESCs involved in self-renewal and lineage-specific differentiation—cont'd

Sr no.	Transcription factor	Transcription factor family	Biological mechanism	References
10.	TFCP2L1 (Transcription factor CP2-like protein 1)	This group of protein belongs to Grh/CP2 (grainyhead/CCAAT box-binding protein 2) transcription factor family	Tfcp2l1 encourages the maintenance of pluripotency in both primed and naïve state of hESC and delays differentiation while Tfcp2l1 suppression instantly disrupts self-renewal and associates the signaling pathway Wnt/β-catenin to induce lineage-specific differentiation	[59]
11.	LEFTY2 (Left-right determination factor 2)	A 41 kDa glycoprotein belonging to TGF-β superfamily	Lefty functions through proteolytic cleavage and glycosylation process in embryonic patterning and sustains self-renewal and is implicated in the differentiation of hESCs	[60]
12.	hTERT (Human telomerase reverse transcriptase)	A 127 kDa protein belonging to the reverse transcriptase-telomerase subfamily	Telomerase central element, hTERT, is strongly expressed throughout the hESC's culture regulated by Wnt/β-catenin signaling-pathway	[61]
13.	PRDM14 (PR-domain zinc-finger protein 14)	A 64 kDa protein belonging to the class V-like SAM-binding methyltransferase superfamily	PRDM14 is active in sustaining self-renewal by recruiting PRC2 to suppress the modification of H3K27me3 that inhibits differentiation-related marker genes in primed-state hESCs	[62, 63]
14.	DPPA5 (Developmental pluripotency associated 5)	A 13 kDa protein belonging to KHDC1 (KH-domain containing 1) family	DPPA5 promotes stemness and reprogramming by regulating NANOG turnover via a post-transcriptional mechanism in both primed and naïve-state hESCs	[64, 65]
Naïve-state hESC promoters				
1.	ESRRB (Estrogen-related receptor-beta)	A 48 kDa protein belonging to the nuclear hormone receptor 3 subfamily	ESRRB plays an important role in the reprogramming of human primedstate cells to a naive state existence in hESCs	[66]
2.	TFAP2C (Transcription factor AP-2-gamma)	A 49 kDa protein belonging to the AP-2 (activator protein-2) family	TFAP2C operates to maintain hESC's pluripotency and restrains neuroectodermal differentiation during the shift from primed to naïve, by enabling the opening of proximal enhancers to pluripotency factors	[67]
3.	DPPA3	A 17 kDa protein belonging to KHDC1 (KH-domain containing 1) family	DPPA3 plays a role in the early embryogenesis and also in the maintenance of naive hESCs pluripotency	[68, 69]
4.	XIST (X-inactive specific transcript)	A noncoding RNA transcript on the X chromosome of the placental mammals	Indirect markers such as a decline in the number of cells with H3K27me3 foci and even the presence of XIST clouds or XIST promoter methylation are being assessed for generating naïve hESC reactivation	[70]
Lineage-specific differentiation markers				
Ectodermal lineage				
1.	NES (Nestin)	A 177 kDa protein belonging to the intermediate filament family	Nestin is associated with important stem cell functions including self-renewal, proliferation, and differentiation, particularly in the production of neural lineage	[71, 72]
2.	β-III tubulin/Tuj-1 (Tubulin-beta-3 chain)	A 50 kDa protein belonging to the tubulin family	Human brain develops healthy neurons from stem cell reservoirs, and in several research studies, beta-tubulin III has been used as a predictor of positive neuronal identification	[73]
3.	Pax6 (Paired box protein Pax-6)	A 46 kDa protein belonging to the paired homeobox family	Pax6 performs the neural inductive role by repression of hESCs pluripotent genes that promote the activation of neuroectoderm (NE) genes	[74, 75]

TABLE 1 Key transcriptional regulatory circuits of hESCs involved in self-renewal and lineage-specific differentiation—cont'd

Sr no.	Transcription factor	Transcription factor family	Biological mechanism	References
Retinal differentiation markers				
1.	Otx2 (Orthodenticle homeobox 2)	A 31 kDa protein belonging to the paired homeobox family, bicoid subfamily	The OTX2 gene serves a vital function in eye and brain development, such as the nerves that bring sensory input from the eyes to the brain (optic nerves) and mediates retinal glial cell differentiation in hESCs via Notch signaling pathway	[76]
2.	CRX (cone-rod homeobox)	A 32 kDa protein belonging to the paired homeobox family	The CRX gene supplies instructions for the development of photoreceptor precursor proteins located in the eyes, in the light-sensitive tissue at the rear of the eye called the retina	[76, 77]
3.	LHX2 (LIM/homeobox protein 2)	A 44 kDa protein belonging to LIM-domain large protein family	An eye field marker LHX2 plays a role in the long-term survival and differentiation of retinal neurons derived from hESCs by regulating neocortical region patterning through downstream epigenetic regulators that cause molecular properties to be retained in cortical plate neurons	[78, 79]
Neural differentiation markers				
1.	Musashi1/MSI1 (RNA-binding protein Musashi homolog 1)	A 39 kDa protein belonging to the Musashi family	Musashi1 is a strongly expressed RNA-binding protein during the cell cycle which enables versatility in the regulation of target mRNA translation in neural progenitor cells like neural stem cells	[80, 81]
3.	SOX1 (SRY-box transcription factor 1)	A 39 kDa of protein belonging to SOX (SRY-related HMG-box) transcription factors family	Sox1 serves as a mechanism in neuronal growth through maintaining neural cells undifferentiation by alleviating proneural protein production and preventing neuronal differentiation	[82]
5.	CXCR4 (C-X-C chemokine receptor type 4)	A protein belonging to the G-protein-coupled receptor 1 family	CXCR4 plays a key role in the neural induction of hESCs through its activation by the agonist stromal cell-derived factor 1 (SDF-1) and may mediate hippocampal-mature neuronal survival	[83]
Mesodermal lineage				
1.	Brachyury/T (T-box transcription factor T)	A 47 kDa protein belonging to T-box protein family	Brachyury protein binds to a palindromic site (called T site) and triggers the modulation of gene transcription necessary for mesoderm formation and differentiation when bound to such a site	[84, 85]
2.	MIXL1 (Homeobox protein MIXL1)	A 24 kDa belonging to the paired homeobox family	Mixl1 homologues are required for intermediate ventral mesoderm patterning and differentiation of hESCs induced by BMP4	[86, 87]
Cardiomyocyte differentiation markers				
1.	Nkx2.5 (NK2 transcription factor-related, locus 5)	A 35 kDa protein belonging to NK-2 homeobox family	Nkx2.5 is an early cardiac marker. Nkx2.5 in coordination with GATA4 is involved in the commitment and/or differentiation of the myocardial lineage in hESCs through transcriptional activation of atrial natriuretic factor (ANF)	[88]
2.	GATA-4 (GATA binding protein 4 transcription factor)	A 44 kDa protein belonging to zinc-finger transcription factor family	GATA-4 is an early cardiac marker. It interacts with Nkx2.5 in cardiomyocyte stretch response and also helps in the initiation of cardiac-specific gene expression regulated by bone morphogenetic protein by linking it to the BMP consensus sequences within cardiac-specific domains	[88, 89]

Continued

TABLE 1 Key transcriptional regulatory circuits of hESCs involved in self-renewal and lineage-specific differentiation—cont'd

Sr no.	Transcription factor	Transcription factor family	Biological mechanism	References
3.	TNNT2 (Cardiac muscle troponin T type2)	A 36 kDa protein belonging to troponin T family	TNNT2 is a late cardiac marker. The variable N-terminal segment of TNNT2 imparts a high calcium response of actomyosin ATPase activity and myofilament growth to modulate cardiac and skeletal muscle contractility throughout development	[90, 91]
4.	MYH6 (Myosin heavy chain 6)	A 224 kDa protein belonging to myosin family	MHY6 is a late cardiac marker. The MYH6 gene encodes a protein known as the cardiac alpha (α)-myosin heavy chain that is responsible for myocardial contraction kinetics	[90, 92]
Endodermal lineage				
1.	SOX17 (Transcription factor SOX-17)	A 44 kDa of protein belonging to SRY-related HMG-box transcription factor family	Sox17 activates the hESC differentiation system for definite endoderm development and endodermal-derived organogenesis	[93, 94]
2.	GATA6 (GATA-binding factor 6)	A 60 kDa of protein belonging to zinc-finger transcription factor family	GATA6 is a key regulator for the development of definitive human endoderm from pluripotent stem cells by direct regulation of endoderm gene expression	[95]
3.	Wnt3 (Proto-oncogene Wnt3)	A 39 kDa of protein belongin to the Wnt family	Recent study introduced Wnt3 as a biomarker capable of assessing the definitive endoderm differentiation ability of hESCs	[96]
Insulin-producing cells (IPCs) differentiation markers				
1.	FoxA2/HNF-3B (Forkhead box protein A2/hepatocyte nuclear factor 3-beta)	A 48 kDa protein belongings to FoxA subfamily	In mammals, the ventral layer of the endoderm develops under FOXA2, a transcription factor reported to play a crucial role in the differentiation of pancreatic cells derived from hESCs	[97, 98]
2.	PDX1 (Pancreas/duodenum homeobox protein 1)	A 30 kDa protein belongin to Antp (antennapedia) homeobox family	The Pdx1 gene is involved in the induction of hESCs into IPCs via the direct delivery of human Pdx1 protein and acts as a pancreatic transcription factor triggering a group of genes related to the growth, synthesis, and release of pancreatic β-cell lineage in response to glucose challenges	[97, 99]
3.	NKX6.1 (Homeobox protein Nkx-6.1)	A 37 kDa protein belonging to NK homeobox (Drosophila), family	The transcription factor NKX6.1 binding to different A/T-rich DNA sequences in the promoter regions of an insulin gene is necessary for the production of highly functional mature β-cells in the Langerhans islets during the derivation of pancreatic progenitor cells from hESCs	[100, 101]
4.	SOX9 (SRY-box transcription factor 9)	A 56 kDa protein belonging to SRY-related HMG-box family	Sox9 promotes endoderm and pancreatic lineage cell differentiation by stimulating their survival in early hESCs fate determination by interplaying between FGF, Activin A, and BMP signaling and regulating the Notch-effector HES1 (hairy and enhancer of split-1)	[102, 103]

3 Signaling pathways involved in the regulation of hESCs

Embryonic stem cell pluripotency exhibits different cell states with a distinct cellular, metabolic, and epigenetic state being regulated by the surrounding microenvironment [106]. The inner cell mass (ICM) of the blastocyst results in the primitive endoderm formation, which gives rise to the visceral and parietal yolk sacs and epiblast, shortly before implantation [107]. The epiblast becomes the precursor of all tissues of the body, including the germ line, as the implantation phase progresses [108]. The preimplantation epiblast represents a naive state of pluripotency, before any lineage specification, and the postimplantation epiblast represents a primed pluripotent state that likely derives hESCs [33, 109]. hESC features and tissue culture criteria are quite distinct from mESCs [110]. The first lineage decisions between trophectoderm, primitive endoderm, and epiblast were identified by studying transcriptional profiles of embryonic developmental stages from Day 3 (E3) to Day 7 (E7) of human preimplantation embryos [111].

The molecular properties of ESCs rely on the transcription factor regulatory network such as Oct4 (octamer-binding transcription factor 4), Nanog (homeobox protein), Sox2 (SRY (sex-determining region Y)-box 2), and the growth factors such as FGF (fibroblast growth factor) and BMP (bone morphogenetic protein) [112]. Numerous signaling pathways linked with pluripotency have been determined, namely LIF (leukemia inhibitory factor)/STAT3 (signal transducer and activator of transcription 3) pathway, Wnt (Wingless-related integration site)/β-catenin pathway, FGF/MAPK (mitogen-activated protein kinase)/ERK (the extracellular signal-regulated kinase) pathway, TGF-β (transforming growth factor β)/SMAD (Drosophila MAD (mothers against decapentaplegic)) pathway, and PKC (protein kinase C) pathway [113, 114]. Human ESCs were recognized to adapt actively to the amplification of signaling mechanisms known to facilitate postimplantation gastrulation by pathways of FGF, TGF-β, BMP, and Wnt more prominently [115].

3.1 LIF signaling pathway

Leukemia inhibitory factor (LIF), a cytokine of the IL6 (interleukin 6) superfamily, is utilized in cell culture to maintain murine ESCs in an undifferentiated state. LIF binds to glycoprotein 130 (gp130) receptor and triggers JAK (Janus kinase) activation in the cell interior and phosphorylates STAT3. LIF activates at least three separate signaling pathways in mouse ESC: the JAK/STAT3 pathway, PI3K (phosphoinositide 3-kinase)/AKT (protein kinase β) pathway, and the SHP2 (SH2-domain-containing tyrosine phosphatase 2)/MAPK (mitogen-activated protein kinase) pathway. But LIF/STAT3 does not help in the self-renewal of human ESCs. This idea is backed by the fact that during normal blastocyst growth, hESCs do not exhibit diapause and fail to react to LIF/STAT3. Interestingly, STAT3 hyperactivation transforms EpiSCs (epiblast stem cells) that possess multiple characteristics of hESCs to naive pluripotency that triggers elevated LIF/STAT3 activation [70, 116]. Human blastocyst ICM expresses Tfcp2l1 (transcription factor CP2-like 1), which subsides during the hESC derivation and then elevates gradually by the incorporation of the following factors in culture: Klf2-Klf4 (Kruppel-like factor) or KLF4-Oct4 all through the naive hESC generation [117, 118]. Tfcp2l1 may serve a vital part, downstream of LIF/STAT3, in the maintenance of naive pluripotency [119].

3.2 Wnt signaling pathway

In the embryonic state, Wnt signaling influences key development processes [120]. It regulates the maintenance and self-renewal of stem cells in adult mammalian tissues [121]. Wnt ligands are autocrine and paracrine cell signaling molecules that belong to a conserved secreted glycoprotein family [122]. The absence of Wnt ligand prompts GSK3 (glycogen synthase kinase 3) to form a degradation complex, comprising of CK1 (casein kinase1), APC (adenomatous polyposis coli), and Axin (axis inhibition protein) that phosphorylates β-catenin, initiating ubiquitination and proteasome-mediated β-catenin degradation [123]. Conversely, with the activation of the Wnt ligands, the signal is sent to Fz (frizzled) and LRP5/6 (low-density lipoprotein receptor protein receptors 5/6) to disrupt the degradation complex formation, contributing to the build-up of β-catenin in the cytoplasm. This excess β-catenin enters the nucleus, communicates with the transcription factors TCF/LEF (T-cell factor/lymphoid enhancer factor), and attaches to the consensus motif that initiates the transcript activation for specific genes identified as Axin2 [124], Cdx1 (caudal-type homeobox 1) [125], and T (brachyury) [126].

STAT3, the key LIF effector, and the Oct4 activator in ESCs are regulated via the β-catenin mechanism. These findings relate to a potential functional continuity or regulation of feedback between the core factors and the adhesion molecules in mESCs. Wnt signaling facilitates self-renewal in ESCs through dose-dependent inhibition of differentiation into murine EpiSCs [127]. The signaling of Wnt/β-catenin causes different effects on naive cells as compared to primed cells [128]. Wnt proteins stimulate the self-renewal of mESCs and promotes the naïve state. However, the precise role of the canonical Wnt/β-catenin system in hESCs remains unknown. RNA profiling demonstrated the presence of 19 Wnt genes along with 10 Fz receptors comprising both LRP5 and LRP6 coreceptors in H7 hESCs [129]. Studies in hESCs utilizing small molecules directed at various Wnt-β-catenin signal pathway components have given fresh perspectives into the role of Wnt/β-catenin signaling [130]. The Wnt/β-catenin pathway activation utilizing either Wnt3A or GSK3 inhibitor BIO (2′Z,3′E)-6-bromoindirubin-3′-oxime) [131] only promotes growth in short-term culture but not continued pluripotent hESC expansion [132]. Signals disrupted by porcupine enzyme inhibition or by stabilizing the Axin1/2 to block β-catenin nuclear translocation promotes hESC's self-renewal and proliferation by establishing the importance of this enzyme in Wnt ligand secretion. A recent study determined that Wnt triggers autocrine or paracrine signals for the promotion of self-renewal in naive hESCs, where Wnt/β-catenin signals were found actively involved in naive-state pluripotency but absent in primed hESCs [133]. Current study on the self-renewal and early differentiation of hESCs revealed that the transcription factor Bach1 (BTB and CNC homology 1) directly targets Usp7 (deubiquitinase ubiquitin-specific processing protease 7), thereby disrupting mesodermal gene expression by procuring PRC2 (polycomb repressive complex 2) to the gene promoters by triggering the signaling pathways Wnt/β-catenin and Nodal/SMAD2/SMAD3, respectively [134].

3.3 FGF signaling and its interaction with several other pathways in hESCs

Growth factors like FGF and ERK help in maintaining the pluripotency of hESCs. The interaction between FGF and its tyrosine kinase-mediated FGF receptors (FGFR1, FGFR2, FGFR3, and FGFR4) supervises many development processes: proliferation, survival, migration, and differentiation [135]. The ERK corresponds to the MAPK family, where the ERK/MAPK pathway contributes to the phosphorylation of Raf (rapidly accelerated fibrosarcoma) kinase, MEK, and ERK family members via a three-stage kinase cascade [136]. This activated ERK moves to the nucleus, where it phosphorylates the downstream partners and plays a role in cell growth [137], proliferation, and differentiation. Transcriptional analysis of hESC-derived cell lines reported FGF2 signaling activation via upregulation of several crucial components of the MAPK/ERK cascade for the maintenance of viability in hESCs [138]. FGF/ERK signaling stabilizes the primed cell state by facilitating the shift from a naive into a primed state and prevents primed cells from returning to naive state [135, 139]. Goke et al. explored the function and the mechanism of the ERK-mediated response across the genome to classify the ERK2 targets and highlighted the extracellular signaling stimulating self-renewal and intrinsic suppression of differentiation to retain a pluripotent status of hESCs [140].

Several studies have shown that cell survival pathway PI3K signaling performs a vital function in the proliferation and suppression of apoptosis in mESCs [141]. Insulin and insulin-like growth factors (IGF1 and IGF2) binding to insulin-receptor family members such as IGF1R and IGF2R [142] activate PI3K and subsequently AKT that upregulates the downstream substrates such as FOXO1 (forkhead box protein O1) [143], and mTOR (mammalian target of rapamycin) [144] signals enabling self-renewal and proliferation in hESCs. Similarly, elimination of FOXO3a (forkhead class O3a) expression through the activation of the PI3K/AKT pathway promotes pluripotency in hESCs [145, 146]. The study also revealed the cross talk among PI3K/AKT signaling, MEK (also known as MAPK-extracellular signal-regulated protein kinase)/ERK, and FGF pathway in two hESC lines, (H1, and H9) suggesting that both ERK1/2 and PI3K/AKT cascades are a downstream target of the FGF pathway to regulate hESC self-renewal and undifferentiated state in a cooperative manner [147]. However, the inhibitors of PI3K discourage self-renewal and promote loss of pluripotency markers followed by lineage-specific differentiation in hESCs [148, 149]. Also, exogenous FGF-2 stimulation assists in the tyrosine phosphorylation of downstream proteins such as PI3K, MAPK, and other members of the Src (proto-oncogene tyrosine-protein kinase) family [150].

Moreover, a previous study demonstrated the combination of two pathways in hESC and induced pluripotent stem cells to be an effective means of differentiating hESC into stable CD34(+) progenitor cells by the inhibition of MEK/ERK signaling and activating BMP4 pathway [151]. BMPs are known to have fate in the extraembryonic lineage differentiation of hESCs [86]. The TGF-β superfamily consists of two main types of ligand signals: a) the BMP and GDF ligand signals which activate SMAD1/5/8 via the ALK1, ALK2, ALK3, and ALK6 type I receptors and b) the TGF-β/activin/nodal ligand signals activating SMAD2/3 via ALK4, ALK5, and ALK7 [152]. A previous study examined the role of TGF-β superfamily and determined an interaction between the TGF-β and Wnt signaling essential to preserve the undifferentiated state in hESCs [153]. Furthermore, FGF2 enhances the effect of the BMP4-mediated hESC differentiation into mesendoderm, marked by the uniform T (brachyury) expression by the maintenance of NANOG levels via MEK/ERK pathway [154]. hESCs derived from the primed pluripotent population substantially relies on Activin/Nodal and FGF signaling [155]. FGF2 also facilitates the NANOG gene-guided pluripotency in hESCs through interaction with TGF-β/SMAD2 signaling pathway [156, 157]. The specific role of three signaling pathways: the TGF-β/SMAD2, BMP/SMAD1, and the FGF/ERK, is to promote the self-renewal of hESCs while removing these pathways led to the effective development of the neuroectoderm through the expression of neuroectodermal fate determinant PAX6 [158].

A group of small molecules, some of which act as pathway inhibitors or activators, used in ESC's culture tweak crucial signaling mechanisms that promote self-renewal and inhibit differentiation [159]. Van der Jeught et al. assessed the synergistic effect of small-molecule inhibitors (PD0325901 and CHIR99021) on the FGF/MEK/ERK and GSK3β pathways in the hESC culture medium, which showed enhanced pluripotency through high OCT3/4 and NANOG-positive cells in the human ICM [160]. A study demonstrated a cocktail of small molecules and growth factors (2i/LIF/basic/FGF/ascorbic acid/forskolin) in culture medium-activating FGF signaling pathway via PI3K/AKT/mTOR promoting the conversion of primed hESCs toward naive pluripotency state enabling an unbiased lineage-specific differentiation in hESCs [161]. In contrast, the inhibition of any of such pathways involving BMP, Wnt, MEK/ERK, PI3K/AKT, and TGF-β may impedes the self-renewal capacity and encourages the lineage-specific differentiation in hESCs (Fig. 1).

4 Metabolic and epigenetic regulation of hESCs

For progressive development, changes that include cell fate decisions, pluripotency, and lineage-specific differentiation involving complex epigenetic processes are an important hallmark for hESC's maintenance. Interaction of metabolic and epigenetic processes determine the association of the physiological state, epigenetics, and ESC's survival [162]. Oxygen is another vital factor inside the stem cell niche [163]. A recent finding unveiled the role of oxygen in the modulation of intracellular metabolic flux that determines carbon fate and influences the methyltransferase and demethylase activity in hESCs [164]. ESCs primarily assist biosynthesis by aerobic glycolysis [165]. Murine ESCs employ oxidative phosphorylation (OXPHOS) to meet their energy demands, while hESCs with very weak mitochondrial respiration capability mostly rely on glycolysis [166]. The HIFs (hypoxic inducible factors) maintained under physiological oxygen conditions mainly consist of (1) GLUT1 (glucose transporter 1) that facilitates the transport of glucose to the cell and (2) PDK (pyruvate dehydrogenase kinase) that prevents the processing of pyruvate to acetyl-CoA by PDH (pyruvate dehydrogenase) in the mitochondrion [167].

Metabolism in hESCs is described by a glycolysis-dependent process that converts the majority of the glucose to lactate that produces ATP rapidly [168]. The NADPH (nicotinamide adenine dinucleotide phosphate) thus derived gets metabolized by the oxidative pentose phosphate pathway (PPP) that helps to generate metabolites enabling nucleotide and lipid biosynthesis accompanied by a steady antioxidant depletion to maintain the balance between pluripotency and initial differentiation [169]. Also, the presence of protein UCP2 (uncoupling protein 2) in the mitochondrial inner membrane partly regulates pyruvate flux by shunting glucose-derived carbon back from mitochondrial oxidation into the PPP [170]. Zhang et al. reported that retinoic acid-induced differentiation of the hESCs affects the production of UCP2 protein [171], accompanied by decreased glycolysis and elevated OXPHOS. However, hESCs have restricted ability to access pyruvate-derived citrate to produce ATP by OXPHOS [172]. This restrictive pyruvate oxidation may act plausibly by stabilizing ROS (reactive oxygen species) formation, leading to excessive use of glutamine as a source of TCA cycle (tricarboxylic acid cycle) intermediates, promoting NAD+ (nicotinamide adenine dinucleotide) reuse through glycolysis for quick self-renewal and proliferation of ESCs [173]. Glutamine is the most utilized source of nutrients in hESCs culture apart from glucose [172]. hESCs also utilize ATP, antioxidants such as glutathione (GSH), and NADPH through a robust glutaminolysis process. Glutamine-derived GSH inhibits OCT4 oxidation, and degradation, that helps OCT4 to bind DNA [174]. The use of glutamine in mESCs culture has demonstrated that it helps in the maintenance of α-ketoglutarate pools [175]. Enrichment with α-ketoglutarate enhances the expression of neuroectoderm or endodermal lineage during primed hESCs differentiation indicating that elevated levels of α-ketoglutarate may ameliorate primed-state differentiation [176]. Besides, these findings suggest that a supply of nutrients can significantly influence metabolic pathways and cellular conditions.

The intermediate metabolites obtained by carbon metabolism (glycolysis and TCA cycle) mainly function as epigenetic enzyme modifiers [177]. The epigenetic makeup and maintenance depend on the epigenetic promoter activities controlling chromatin arrangement, DNA methylation, and histone modification. Also, PRC2 (polycomb repressive complex 2), a multiprotein epigenetic complex, may function as a molecular signature for discriminating naive over primed state in hESCs by regulating the quantity of PRC2 substrates influencing methylation and hESC's lineage commitment decision [178, 179]. Likewise, hESCs culture requires high methionine levels [180]. The enzyme methionine adenosyltransferase triggers the production of SAM (S-adenosylmethionine) from methionine [181]. S-adenosylmethionine is a central regulator for preserving and controlling the undifferentiated status of hESCs [182]. The depletion of methionine induces a drastic decrease in SAM levels with decreased trimethylation of H3K4me3 (lysine-4 on the histone H3 protein) markings impacting the NANOG expression that ultimately triggers differentiation in hESCs [182]. During the naive-to-primed-state transition in hESCs, as the epigenetic environment shifts owing to increased H3K27me3 repressive marker, the nicotinamide N-methyltransferase (NTMT) enzyme, regulates the SAM level [183, 184]. Besides lactate, acetyl-CoA is also catalyzed by glycolysis in hESCs [185]. Glucose-derived acetyl-CoA, produced during the TCA cycle, serves as a cofactor for histone acetyltransferases (HAT), fine-tunes histone acetylation, and maintains the self-renewal of hESCs [185]. This observation illustrates the evolving cell's metabolic demand with pluripotency, and differentiation signifies the need to relate chromatin metabolism to pluripotency. It would be interesting to understand to what extent the presence/absence of metabolite influences the differentiation in stem cell population.

5 Sources and regulations for generating human embryonic stem cell lines for research

Human ESCs arise in a transition from early embryos as the cell cycle rapidly enhances the cell-type diversifications [186]. The collaborative advancement of isolation and treatment approaches for hESCs provides the likelihood of cell transplant strategies, through somatic cell-nuclear transfer to enucleated egg cells, for the extraction of pluripotent cells from the blastocyst stage [187]. Also, there is a growing need for new human embryonic strains for therapeutic use due to the lack of epigenetic flexibility in the hESCs culturing process [188]. However, human embryos have been a subject of political and moral debate [189] and the controversy on hESC's embryo-destructive derivation also deals with the researcher's ability to study without ethical restrictions [190].

Across certain European and Asian countries, it is illegal to use federal funding for the generation of new hESCs as that will contribute to the loss of human embryos or extract hESCs from surplus embryos [191]. Since the emphasis of hESCs is scientific, legislative actions are evolving rapidly [192]. The executive order on reducing obstacles in ethical scientific work that involves hESCs was issued by the US President, Barack H. Obama, on March 9, 2009, where the current regulatory structure of the institutional review board (IRB) outlined the core ethical principles necessary for proper monitoring of the hESCs derivation [193]. The International Society for Stem Cell Research (ISSCR) facilitates collaborations and sharing of stem cell knowledge and insights and encourages awareness in all areas of stem cell and health science [194].

Consequently, several derived measures were introduced to ensure that hESC's origins do not require an embryo. Specific approaches were performed to create genetically unmodified hESCs that do not compromise with the human embryo's developmental ability. A single-blastomere biopsy is a technique developed by Lanza and colleagues in which a single blastomere isolated from the embryo can be utilized for the derivation of ESCs [195]. Single blastomere and a parthenogenetic embryo ICM were also extracted and developed by other scientists. [196]. Since the cells are homozygous for major human lymphocyte antigen (HLA) alleles, it may help avoid the immunologic rejection that may occur in hESCs transplantation therapies, and parthenogenetic embryos, produced by artificial fertilization of donor oocytes are highly desirable sources, and these viable embryos are not developed nor destroyed [197]. Therefore, blastomere-derived hESCs bypass the ethical disagreements, since the elimination of a single blastomere may not affect the capacity of the remaining blastomeres to form a healthy embryo. Innumerous hESC lines formed from different embryonic origins and are now used across the world in basic clinical science [198].

hESC-derived lines are now characterized by (1) the expression of stem cell pluripotency markers, (2) HLA typing, (3) G-banding karyotyping, (4) in vitro/in vivo differentiation with EBs (embryoid bodies) formation, (5) teratoma assay, and (6) comparative genomic hybridization arrays [199]. These approaches help in the successful preservation of hESC lines and qualify for therapeutic use in stem cell banks. Currently, there are clinical trials employing hESC lines generated under Good Manufacturing Practice guidelines [199, 200].

6 Aging and hESCs

During normal conditions, in response to injury, several tissues may expand and regenerate with the aid of tissue-resident cell populations. Tissue-resident cells migrate to the site of injury driven by a complex interplay of molecular factors involving chemokines, cellular interactions, and signaling molecules. However, as multicellular organisms age, the functionality of the tissue and organ declines. For some organ systems, even at an early age, endogenous repair processes become sluggish since the resident cell pool lacks or has a minimal functionality [201]. Aging correlates with a decline in the ability to regenerate cells, which causes joints, blood vessels, and other physiological areas to behave differently than when young. The cycle of aging is influenced by the changes taking place in the cellular morphology, growth, and differentiation process. Most aging transitions are inherent, given that stem cells derived from elderly patients develop differently in culture in comparison with their younger equivalents. Moreover, stem cells are arranged in a distinct niche, which retains them in an undifferentiated and self-renewing condition. The physiological and molecular dynamics of stem cell survival are fundamental to homeostasis regulation and may potentially lead to a severe condition in later life unless altered. The aging cycle has a dual effect on stem cell growth and development. Multiple factors could lead to the age-related malfunction of the stem cells and are the root cause of several age-associated diseases. Numerous factors drive aging in stem cells and involves both intrinsic and extrinsic factors, *viz.* a reduction in stem cell count and proliferative ability [202], telomere shortening [203], mitochondrial dysfunction [204], epigenetic modifications [205], oxidative damage to DNA [206], and mutation accumulation [207] usually accompanied by remodeling of the extracellular matrix toward more fibrotic pattern associated with chronic inflammation [208], intensifying these processes. Stem cells play an essential part in the reintegration of degenerated ailments, being one of the key modules of biological therapy. Adult stem cells have so far been extensively exploited, for laboratory tests and clinical applications in tissue engineering. Adult stem cells, nevertheless, exhibit certain limitations such as a minimal capacity for growth and differentiation, a steady deterioration of functional features during in vitro culture, teratoma formation following in vivo implantation and subsequent age-related variations in cellular health [209–212]. The aging cycle degrades the function of stem cells and also affects the capacity for regeneration, the ability to self-renew, and interaction with extrinsic signals. Human ESCs,

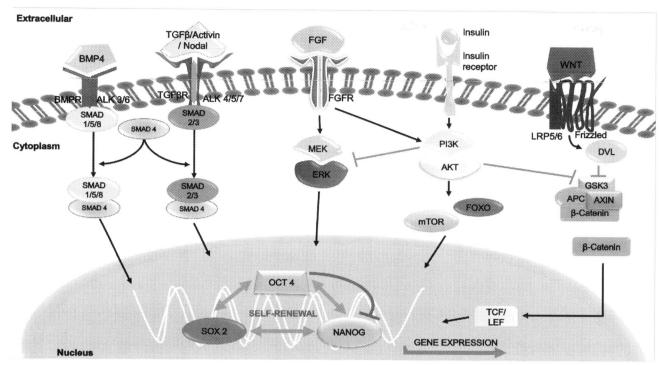

FIG. 1 Signaling pathways involved in the self-renewal and differentiation of human embryonic stem cells (hESCs)

on the contrary, permit unrestricted repertoire of replacement/reparative cells. hESCs trigger lineage-specific progenitor cells to have restricted differentiation ability within the germ layer which mitigates the possibility of teratoma formation during in vivo implantation. A limited number of research studies exist about regeneration and homeostasis transitions with age. Understanding the role of the aging process in the hESCs function is essential, not only in identifying the pathophysiology of aging-related disorders but also in creating new stem cell-based therapies for treating aging-related diseases (Fig. 2).

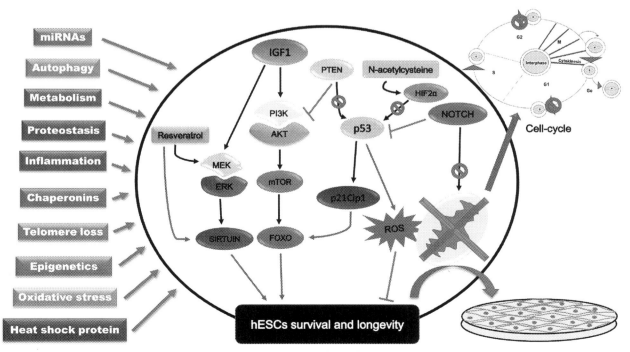

FIG. 2 Factors involved in the regulation of pathways promoting survival and longevity in human embryonic stem cells (hESCs)

6.1 Pathways modulating aging in hESCs

The aging process gets influenced by numerous signaling pathways. Several studies documented the pathways operating in the cytosol that may regulate aging and expand longevity in hESCs via insulin-like growth factor 1 (IGF-1) signaling pathway, p53 pathway, mTOR pathway PI3K/Akt pathway, Notch pathway, and MEK/ERK signaling pathway. The insulin-like growth factor 1 (IGF-1) signaling pathway is the most established aging phenomenon that influences the existence of an evolutionary process [213]. For instance, impaired insulin receptor activity and its downstream component promote longevity in humans and insulin, IGFs are essential for the maintenance of hESCs [39, 40]. To establish self-renewal and longevity in hESCs, both IGF and insulin binding to IGF/insulin receptor triggers the endogenous signaling cascade of PI3K/Akt enabling the phosphorylation of FoxO (forkhead box O, a longevity-promoting gene) protein by the Akt protein kinase within the conserved regions of FoxO protein [214, 215]. Moreover, the activation of PI3K/Akt is crucial to stimulate self-renewal and maintenance of pluripotency by neutralizing the caspase cleavage and senescence, which further promotes activation of mTOR, FoxO1, ribosomal S6 kinase (S6K), and GSK3β in hESCs [142, 214]. Few studies have attributed mTOR signaling components to facilitate survival and self-renewal in hESCs through apoptosis inhibition [144, 216]. The role of PTEN (phosphatase and tensin homolog of chromosome 10) implicates in promoting cellular aging by negatively impacting the PI3K signaling cascade to suppress AKT function [217, 218]. PTEN loss promotes senescence due to elevated p53 translation encouraging self-renewal [219], survival, and proliferation [220] and triggers the Akt/mTOR signal route by elevating p-S6 levels without impacting GSK3 function in hESCs [221]. However, in hESC-derived neural stem cells, PTEN encourages dopamine-mediated neuronal differentiation through the regulation of ERK-related suppression of S6K signaling [222]. Besides, a recent study demonstrated correlation between the PI3K/Akt and MAPK/ERK signaling pathways with low survival [223]. Under stress conditions, long-lived species exhibited altered ERK phosphorylation patterns, further indicating a correlation between ERK behavior and longevity [224]. Several signaling pathways involving IGF1/PI3K/Akt/mTOR and ERK together participate in early human embryonic development, and hESCs can expand in vitro even in the absence of FGF2, when PI3K activators and Activin A are available, suggesting a signaling pathway cross talk which may encourage hESCs survival [225].

Sirtuin (SIRT), a nicotinamide dinucleotide (NAD+)-dependent deacetylase, is another longevity-promoting factor that prevents cellular aging by retarding telomere erosion, maintaining the genome integrity and facilitating repair of DNA throughout by involving with several signaling molecules such as FoxO and p53 including certain signaling (insulin/IGF1 and adenosine monophosphate-triggered protein kinase signals) pathways [226–229]. The pathway of p53 consists of a tumor suppressor protein p53, considered to be a genome protector that gets triggered under a multitude of stress conditions contributing to rapid p53 aggregation in stressed cells, which speeds up ROS generation, thereby impacting cellular longevity [230]. A direct link, between SIRT1 (a class III histone deacetylase) and the proteasome pathway promotes cellular longevity through p53 inactivation due to enhanced acetylation of p53 at lysine residues, which effectively deactivates DNA damage-induced programmed cell death in hESCs [44, 231]. A recent study established ERK signaling in context of cell survival as a potent-resveratrol-inducible element for sirtuin activation [232]. Similarly, SIRT1 activated by dietary antioxidants, namely resveratrol, exerts enhanced cell survival through modulating MEK/ERK signaling pathway [233]. Another conserved pathway, namely Notch signaling pathway, integrates with several inflammatory and hypoxia-mediating pathways that contribute to age-related deterioration [234]. Studies revealed that the expression of the longevity regulator SIRT1 in endothelial cells enhances Notch signaling to suppress aging damage caused by vasodilator responses [235]. However, Notch signaling under normal oxygen supply is not required for hESCs expansion but promotes the lineage commitment differentiation as opposed to hESCs cultured under a hypoxic condition that allows long-term self-renewal via activation of Notch signaling with delay in differentiation [236–238]. Thus, accelerated exploration of essential regulatory pathways may help to determine the impact of genetic and environmental approaches to improve the survival of hESCs which might benefit therapeutic applications.

6.2 Factors contributing to the aging and senescence in hESCs

In tissue and organ development, and regeneration, the role of ESCs is likely related to the proliferation, differentiation, and secretion potential that metabolizes various growth factors and cytokines for tissue microenvironment regulation.

6.2.1 Telomere loss

Telomeres are small guanine-rich repeats of 5 to 20 kb at the end of the chromosome that shields DNA from terminal fragmentation and chromosome fusion by providing genomic stabilization to the chromosome structure [239]. It consists of telomerase holoenzyme, which is a cluster of ribonucleoproteins comprising of reverse transcriptase (TERT), a protein component, and an RNA component (TERC) that acts as a framework for telomere growth [240]. TERT forms an indispensable element for the assembly of an active telomerase complex. Interaction of telomerase and telomere synthesis is not only associated with exponential growth ability of embryonic stem cells and germline cells [241] but is also associated with tumor cells [242]. The steady decline of telomeric length along with a depletion in TERT expression relates to one of the processes causing aging, and progressively shorter telomeres contribute to the loss of cell viability and senescence [243]. The telomere spans are closely related to the development of ESCs, since the short telomeres lower the ESC's proliferative rate and modify telomeric epigenetic traits [244]. hESCs exhibit elevated telomerase activity along with TERT expression in an undifferentiated state [245]. However, the rate of telomerase activity decreases as hESCs differentiate to specific cell lineages [203]. The posttranscriptional TERT regulatory framework, under distinct oxygen microenvironment, may play a significant role by expressing extratelomeric isoforms of telomerase that facilitates hESCs survival, self-renewal, and differentiation functionality [246]. Certain factors such as poly-(ADP-ribose) polymerase 1 (PARP1), Krüppel-like transcription factor 4 (KLF4), FGF-2 also regulate the TERT expression that is associated with differentiation and aging in hESCs [247, 248]. The high-mobility group box proteins [249] such as HMGB1 (stimulatory factor) [250] and HMGB2 (an inhibitory factor) [251] known to be implicated in a range of cellular processes like DNA damage repair, replication, proliferation, differentiation, cell migration, and inflammation also participate in the regulation of telomerase activity in hESCs [252]. A recent study on the long-term impact of cell culture on the growth and differentiation of young and aged hESCs population indicated that continuous culturing negatively impacts the mitochondrial function by increasing mitochondrial membrane capacity, mitochondrial morphology, and ROS content of the hESCs population [204].

6.2.2 Oxidative stress

Several main routes of the cellular antioxidation systems exist where the cells scavenge the ROS by activating endogenous antioxidant enzymes, e.g., SOD (superoxide dismutase), GPX2 (glutathione peroxidase) and other nonenzymatic entities (e.g., vitamin C). Also, early aging factors appear to be triggered through stressful conditions that inhibit cell-cycle progression affecting the cell microenvironment. Remarkably, unresolved oxidative stress and subsequent ROS, produced by cellular metabolic processes or by extrinsic factors, contribute majorly in stem cell aging and other diseases [253]. A large proportion of cytoplasmic ROS emerges from mitochondrial superoxide radicals, but it least impacts the genomic identity in hESCs [254]. Early response to ROS minimizes the copy number of mitochondrial DNA [255] in hESCs and activates the antioxidant defense mechanism [256] that downregulates the expression of GSR (glutathione reductase), GSTA3 (glutathione S-transferase), GPX2, MGST1 (microsomal glutathione S-transferase 1), MAPK26 (mitogen-activated kinase 26), SOD2, and HSPA1B (heat-shock protein 1B) accompanied by an elevated level of HSPB1 (heat-shock protein 1) [257–259]. Recently, the role of forkhead (FoxM1 and FoxH1) transcription factors as an active gene in hESCs have been identified in a recent study [260]. Moreover, FoxM1 (forkhead box M1) regulates hESCs proliferation and oxidative stress defense, while FoxH1 (forkhead box protein H1) participates with Oct4 in NODAL signaling, developmental pattern modulation, and pluripotency regulation in hESCs [261, 262].

6.2.3 Autophagy

The transcription factor Nrf2 (nuclear factor erythroid 2-related factor 2), a regulator of antioxidant machinery in stem cells [263], maintains self-renewal of hESCs and other cells and coordinates cellular responses [264]. Autophagy is a process of clearing up damaged cells that serve a significant task in regulating the Nrf2 function and cellular aging [265]. A synchronized cellular regulatory mechanism of a cilium mediated autophagy Nrf2 axis, and transition in cell-cycle proliferation drives hESCs toward early neuroectodermal lineage [266]. Many findings have suggested the clinical application of antioxidants in targeting autophagy to overcome cellular senescence [267, 268]. Treatment with N-acetylcysteine, an antioxidant, improved hESC's properties and sustained cellular homeostasis by regulating the p53 activity inhibition by HIF2α, since depletion of p53 activity switch hESCs from a quiescent state to normal state [269]. N-acetylcysteine may also block cell apoptosis and rescue the function of the Nrf2 signaling pathway through the scavenging of ROS generation in response to toxic stress [270]. A study on ABCG2 (ATP-binding cassette (ABC) transporter) suggested its protective effect in hESCs from physical stress, drugs, and exposure to UV radiation, and uneven expression of ABCG2 displayed relatively lower autophagic activity in cells [271, 272].

6.2.4 Proteostasis of hESCs in aging

Most proteins mostly tend to fold into well-defined overall three-dimensional structures to sustain a stable proteome state over the life span in mammalian cells to fulfill their biological functions [273]. In recent years, it has become apparent that cellular aging also characterized by a progressive loss of proteostasis accompanied by aggregation of misfolded proteins [274].

The nexus of proteostasis comprises three main processes: (1) protein synthesis; (2) protein structure maintenance; (3) regulated intracellular protein degradation via the ubiquitin-proteasome system (UPS) or autophagy by lysosomes. Proteostasis is crucial to maintain the undifferentiated state of hESCs [275]. The transcription factor of FoxO (forkhead box transcription factors of the class O) family promotes longevity and stem cell function by maintaining protein homeostasis, activating genes that code for chaperones that enable intracellular clearance of misfolded proteins [276]. FoxO1 is critical in maintaining the hESCs pluripotency through the regulation of ROS, by binding to the regulatory regions of OCT4 and SOX2 markers [277]. Similarly, the antioxidant resveratrol facilitates osteogenic differentiation through the upregulation FoxO3A/SIRT1 (forkhead box O3A/sirtuin1) but restricts adipogenesis in mesenchymal progenitor cells derived from hESCs [278]. In particular, FoxO4 enhances proteasomal activity in hESCs, leading to the degradation of the damaged proteins, and decreases their ability for neural differentiation [279, 280].

6.2.5 Role of heat-shock proteins and chaperones in aging

Heat-shock proteins (HSPs) are evolutionarily conserved proteins that shield cells and tissues from stress in the growing and aging organism. HSPs are classified into three major groups based on molecular weight. The high molecular weight HSP, comprising of the HSP60, HSP70, HSP90, and HSP110 family makes up Group I. Group II consists of those that are triggered by glucose deficiency conditions and classified as the "minor HSPs," like glucose-regulated proteins (GRPs). A low molecular weight HSP contains at least 10 members (HSPB1-B10) of 12 to 30 kDa proteins and constitutes the Group III HSPs. These ubiquitous proteins manage stress by triggering apoptosis and facilitates normal metabolism. The absence or failure of HSPs is fatal throughout the short-run or long-run instances of stress. The inflammatory and cellular senescence alterations arise if the rate of HSPs rises above the usual range. The role of HSPs as surprisingly flexible mediators of health and disease is becoming evident. HSPs are well known as molecular chaperones that reprogram the conformational state of cellular proteins and translocate proteins as the protein gets older. Stem cell features of self-renewal, differentiation, longevity, and aging are closely monitored by various intrinsic and external cues, which suggests that HSPs have a crucial role to play in its regulation. HSP60 identified as Group I mitochondrial chaperonin known as HSPD1, functions in cellular stress. HSP60 can serve accordingly as an inducer or inhibitor of apoptosis. The loss of HSP60 triggers mitochondrial abnormalities and is also known to be engaged in the cellular aging process. The latest study also demonstrated the role of HSP60 in mESCs deficiency which restricts proliferation and self-renewal by a reduced expression, thereby initiating mESCs differentiation [281]. LIF with gp130 receptors stimulates the JAK/STAT pathway via phosphorylation of STAT3. Moreover, the study showed that STAT3 comprises a huge chaperone and cochaperone protein group that preserves self-renewal property in ESCs [282]. However, the role of HSP60 and STAT3 in modulating the hESCs stemness and aging needs more clarity. The family of HSP70 proteins express abundantly under stress conditions in cells. Owing to its antiapoptotic nature, HSP70 triggers premature senescence phenotype in hESC-derived cells [259]. The HSP70 assists in the early differentiation of hESCs by triggering JNK, and PI3K/AKT pathways, modulated by epigenetic histone deacetylase [283]. HSP70 plays a protective role by regulating chaperone-mediated autophagy in multiple neurodegenerative diseases linked with aging; also, its availability decreases with age in neuronal tissues suggesting insights in stem cell aging studies [284]. MAP kinases hyperactivate during stress conditions in hESCs, by enabling the H1 (heat-shock factor 1) factor, whereas the HSP90 releases H1 that attaches to the OCT4 promoter region contributing to the repression of MAP kinases [285]. HSP90 affects not only STAT3 behavior but also OCT4 and NANOG levels, which are vital for the maintenance of pluripotency in both human and mouse ESCs [286]. HSP90 associates with OCT4 and NANOG in the same cellular complex and protects them from proteasomal degradation. The misfolded protein aggregation inside ESCs can influence the organism's aging cycle as the transfer of these proteins to the progenitor cells could impact growth during uneven division [287]. Similarly, hESCs facilitate the optimal assembly of the TRiC (T-complex protein ring complex) chaperonin complex by increasing the expression of one of its eight CCT8 (Chaperonin Containing TCP-1) subunits, where it helps to sustain a balanced proteomic state throughout its entire existence [288].

6.2.6 Metabolic and epigenetic alterations in hESCs and aging

Metabolic and epigenetic changes are best-described by an age-related decline in stem cell regenerative capacity [289–291]. Early evidence indicates that basic-cell metabolism could modify epigenetic conditions and might impact the population aging process [292]. hESCs possess a well-defined characteristic in the contexts of cellular metabolism, epigenetic changes, gene expression, and development [290]. Posttranslational modifications of histone tails are important epigenetic regulators, where the young hESCs population demonstrates minimal methylation errors modulated by methionine and SAM metabolic signatures in earlier passages [180, 293]. Sirtuin, a histone deacetylase protein, displays a contrasting feature in regulating the lineage-specific differentiation of hESCs. SIRT1 inhibition not only promotes neuronal differentiation in hESCs but also enables chondrogenic differentiation via expression of extracellular matrix genes and chondrogenic transcription factors [294, 295].

Proliferating cells show a higher metabolic rate than nonproliferating cells of the same nature, which promotes biosynthesis. Likewise, intracellular deposition of damaged macromolecules may be crucial in aging, via promoting cells with impaired function [205]. For instance, the accumulation of asymmetrically distributed degraded proteins in cultured cells such as HEK293 (human embryonic kidney cell line) or hESCs suggested more unhealthy damaged protein-receiving cells affecting proliferation [296]. hESCs also exhibit irreversible aerobic glycolysis [297]. The differentiation of hESCs delays the rate of division, reduces the rate of aerobic glycolysis, and transitions to oxidative phosphorylation as a result of which the respiratory capacity of mitochondria gets adequately utilized [298, 299]. In general, the two states of hESCs: the naive state and the primed state, have been stabilized in vitro [300]. Human naïve-state ESCs have significantly higher glycolysis and active mitochondria as compared to primed-state hESCs. In naive hESCs, ATP is produced primarily through aerobic glycolysis instead of mitochondrial oxidative phosphorylation. Current research based on transcriptomic analysis has begun to explore the variations among naïve-state and primed-state ESCs through determining the chromosomal, epigenetic, and metabolic changes. Studies show that hESCs have low mitochondrial respiratory ability while having a comparatively more evolved and extended mitochondrial content [301]. The mitochondrial transition between naïve- and primed-state hESC lines is observed in the same defined media, albeit hESCs can alter their metabolism of fatty acids depending on culture conditions. Notably, an RNA-binding protein Lin28 (protein lin-28 homolog A) drives transition between naïve- and primed-state pluripotency through regulating the mitochondrial function, organization, and one-carbon metabolism in hESCs [302]. Several studies reported the epigenetic alteration to the effect of the X chromosome inactivation (XCI) state, and XIST (X-inactive specific transcript) expression on the differentiation pattern of hESCs [303]. A recent study reported no XCI initiation and XIST expression in primed-state hESCs, but the enhanced X-linked gene expression in naïve-state hESCs contributed to extensive cell death under differentiated conditions [304, 305]. Earlier studies have also shown that low oxygen tension influences glucose metabolism, lactate production rate, amino acid turnover, and oxygen uptake, suggesting a role in the self-renewal potential of hESCs [306]. Likewise, a high-glucose environment may mediate an adaptive response to increased oxidative stress via joint FoxO3/β-catenin complex to promote p21Cip1 (cyclin-dependent kinase interacting protein 1) expression triggering ROS removal that ultimately affects proliferation of hESCs [307]. The evolving role of HIFs in the management of hESCs proliferation also highlights the significance of the primed-state metabolism [308, 309]. A recent study on the implication of a proinflammatory gene, known as the downstream regulatory element antagonist modulator (DREAM) in the hESC's aging process, suggested that it controls the pluripotency and differentiation of hESCs by employing CREB (cyclic AMP-responsive element-binding protein) phosphorylation at the serine 133 residue that initiates cAMP response element activity [310].

6.2.7 Role of miRNAs in hESC aging

MicroRNAs (miRNAs) are short single-stranded noncoding RNAs that regulate cell fate decisions through translational or epigenetic modulation of gene expression involved in cell-cycle progression [311], apoptosis [312], numerous signaling pathways [313], aging process, and diseases [314]. The involvement of miRNAs in the process of stem cell regulation, function, maintenance, and differentiation could lead to the future therapeutic application of stem cells in regenerative medicine [315, 316]. In embryonic stages, the expression of evolutionary miRNA is barely noticeable; but, it gradually becomes evident during the development of the organism [317]. miRNAs play an integral role through their complexity in a diverse cellular senescence network of biological processes [318]. A recent study recognized the function of RNA exosome nuclease complex to regulate the hESCs differentiation, suggesting the significance of RNA decay pathway for pluripotency maintenance [319]. The miRNA's function is to sustain the undifferentiated and differentiated states [320, 321]. miR-302 retains hESC's stemness through the suppression of the G1 phase and stimulating the S phase of the cell cycle through cyclin D1 repression [322]. Similarly, the DNA damage repair mechanism prompts miR-302 to negatively regulate the p21 activity in hESCs cell-cycle regulation [323]. In the same way, Oct4 and miR-302 function together through regulating the essential gene NR2F2 (nuclear receptor subfamily 2, group F, member 2) activity during neural differentiation in hESCs [324]. Also, the enhanced expression of miR-145 negatively regulates the pluripotency markers such as Oct4, Sox2, and Klf4 of hESCs upon differentiation [325]. Likewise, Kim et al. demonstrated that miR-6086 and miR-6087 suppresses differentiation by impairing the expression of CDH5 (Cadherin-5) and endoglin in endothelial cells derived from hESCs [326]. The role of heterochronic let-7 miRNA in the development of stem cells fosters cell fate mechanisms during aging has also been reported lately [327]. Moreover, mature let-7 miRNA fine-tunes a miRNA repressor protein-Lin28 interaction with Oct-4 and Sox2 in a complex pluripotency network during hESC's differentiation [328–330]. A recent study revealed that miR-363-3p retains pluripotency in both naïve- and primed-state hESCs by blocking Notch-induced differentiation through targeting the Notch1 and Psen1 (Presenilin1) receptors [331]. Besides, a hESC-derived mature sequence of mmu-miR-291a-3p has been shown to alleviate senescent cell phenotypes in human dermal fibroblasts, through a TGF-β receptor 2/p21 pathway, thereby improving wound repair mechanism in aged mice suggesting its role against aging [332].

6.3 Beneficial molecules secreted by the hESCs

After embryogenesis, the ESCs undergo multiple stages of self-renewal and differentiation and replication. The asynchronous action of multiple exocrine and paracrine factors that regulate these stages of ESCs yet remains elusive. Further down the development cascade, ESCs produce multiple factors such as growth factors, growth hormones, and enzymes that drive the differentiation and replication of developed cells. In the past decade, researchers have tried to understand and explore the possibility of applying these beneficial paracrine factors secreted by embryonic stem cells to treat human diseases.

6.3.1 Role of growth factors in the proliferation and maintenance of stem cells

Growth factors other than insulin-like growth factors (IGFs) also play a role in cellular proliferation and differentiation. Basic fibroblast growth factor (bFGF) has been used to grow embryonic stem cells in a feeder-free culture in vitro [37]. bFGF enhances the adipocytic differentiation of mesenchymal stem cells derived from ESCs [333]. Synergistically, bFGF is known to promote the proliferation and maintenance of mesenchymal stem cells derived from embryonic stem cells by influencing the oxidative stress in these cells [334]. FGFs through multiple molecular mechanisms are known to promote proliferation, self-renewal, and maintenance of three major classes of stem cells: ESCs, HSCs, and neuronal stem cells (NSCs) [335].

Insulin-like growth factors (IGFs)

Insulin-like growth factors, especially IGF-II, play a key role in multiple processes of embryonic stem cells such as cellular metabolism and mitogenesis [336, 337]. In vitro and in vivo experiments using functional assays have correlated the secretion of IGFs with basic cellular functions such as proliferation, differentiation, and apoptosis of the cultured embryonic cells. Insulin and insulin-like growth factors (IGFs) are associated with multiple levels of cellular as well as systemic aging. Both of these factors are involved in diverse functions such as cell cycle, metabolism, and epigenetics of stem cells and the normal cells. In mammals, insulin and IGFs use distinct receptors to act. However, they extensively overlap in the downstream signals and pathways to the extent that it is difficult to differentiate between them [338]. Contextually, genomic analysis detected IGF peptides and transcripts in the early stages of embryogenesis to the final maturation stages [339]. IGF1-based therapies were tested to be beneficial for multiple degenerative diseases such as ALS (amyotrophic lateral sclerosis), diseases of RPE (retinal pigment epithelium), and AMD (age-related macular degeneration) [340–342].

Fetal insulin

Fetal insulin plays a synchronous role in fetal development with IGFs. However, it can directly affect the proliferation of somatic cells like adipose progenitor cells [343].

Fetal growth factors

Growth factors involve and command basic aspects of ESCs, like their proliferation, differentiation, and maturation of the embryonic progenitors [344]. There are three major growth factors secreted by embryonic stem cells: EGFs, TGFs, and FGFs. In vitro analysis has revealed that the EGF (epidermal growth factor) interacts with multiple pathways such as the STAT pathway, ERK pathway, Akt pathway, and MAPK pathway. In long-term cultured adipose stem cells, EGF maintains proliferation and multipotency by preventing senescence and terminal differentiation [345]. EGF systemically upregulates cyclin D1 and downregulates p16 in hair follicle-derived MSCs, by increasing their proliferation through ERK/Akt signaling pathway [346]. EGF elevates the paracrine activity of multipotent stromal cells by enhancing the secretion of beneficial factors like VEGF (vascular endothelial growth factor) and HGF (hepatocyte growth factor) by promoting wound healing in multiple organs [347].

TGF represents a superfamily of proteins that play a role in embryonic and somatic cell differentiation by driving tissue morphogenesis and organogenesis [348]. In the stages of embryogenesis, TGF functions primarily through the SMAD signaling pathway [344]. The members of the TGF superfamily utilize the serine/threonine kinase receptors, by initiating the biosynthesis of a heterotetrameric complex of type 1 and 2 receptors [349]. TGF-β is a pleiotropic cytokine factor that is involved in multiple counts of embryonic functions such as regeneration, homeostasis, and proliferation. In ESCs, TGF members maintain heterogeneity through a dynamic equilibrium regulated by other signaling pathways [350]. In general, TGF-β acts as a potent growth inhibitor associated with degenerative diseases like Alzheimer's disease [351, 352]. Other members of the TGF-β superfamily, viz. Activin and Nodal, are well-reported morphogens and are involved in the differentiation of ESCs into cell types of all three germ layers. [353].

FGF contributes to the proliferative aspects such as self-renewal, pluripotency, and senescence of the stem cells. The family of FGFs are initiated by 22 known FGF receptors (FGFRs). The mediation of FGF functions is arbitrated by numerous pathways such as MAPK, PI3K, and STAT. These pathways in turn closely interact with Wnt (Wingless family), RA (retinoic acid), and TGF-β pathway [354].

6.3.2 Embryonic cholinesterase (ChE)

The significant role of cholinesterase in development is morphogenesis. ChE is biosynthesized in different developmental stages by the expression of muscarinic cell-surface receptors that bind to acetylcholine. ChE enzymes, particularly AChE (acetyl cholinesterase), provide the initiation of neuronal development in human embryonic stem cells. In experiments done by Paraoanu et al., it is demonstrated that ESCs expressed all the cholinergic components including Ach (acetylcholine). The expression of AChE increased proportionally as the neuronal differentiation progressed [355, 356]. In the adult, the human brain AChE signaling system maintains behavioral traits such as mood and anxiety. Diseases such as seizures and status epilepticus (SE) can cause inhibition of cholinesterase in the brain by altering the behavioral trait of an individual. The recovery of functioning AChE and other cholinesterase is considered the first step toward restoring brain function [357]. The role of AChE also found to be synchronous with the levels of BDNF (brain-derived neurotrophic factor) which is a key molecule in neuronal regeneration [358].

6.3.3 Role of glucocorticoids in the development and preservation of adult stem cell pool

Fetal glucocorticoids primarily play a role in the later stages of development like organogenesis. Apart from transient cellular effects such as proliferation and differentiation, fetal glucocorticoids induce long-term fetal programming. They commence programming of the hypothalamic-pituitary-adrenal (HPA) axis necessary for the development of complex organs such as lung and liver development. The interruption to the HPA axis can cause long-term effect's impairment in brain development and susceptibility to chronic diseases [359–361]. In adults, glucocorticoids are involved in the homeostatic engagement of rest and stress. The functions of glucocorticoids are maintained through specific transcription factors and receptors.

The role of glucocorticoids is established to be a major effector of adult neuronal stem cells. In the embryonic stem cells, glucocorticoids alter the degradation of Cyclin D1 through ubiquitin, thus decreasing their proliferation. [362]. In adult neuronal stem cells, the ultradian secretion of glucocorticoids controlled by the circadian rhythm maintains them in their quiescent state and prevents activation [363]. Ironically, in adult MSCs, glucocorticoids enable the differentiation of chondrocytes [364]. Glucocorticoids are primarily linked with brain and neuronal aging in adult mammals, which again is dependent on the metabolism mediated by 11β-hydroxysteroid dehydrogenases (11β-HSD) [365].

6.4 Current application of hESCs

6.4.1 hESCs in tissue regeneration

The ability of ESCs to differentiate into tissues of any germ layer has made it a potential source of tissue regeneration. Several groups have studied the ability of ESCs to differentiate into cell lineages originating from the three germ layers—neurons, chondrocytes [366], adipocytes [367], cardiomyocytes [368, 369], hepatocytes [370, 371], hematopoietic progenitors [372], and beta-cells [373, 374]. In addition to the ability to differentiate, homing—the ability of the cells to interact with stromal cells and repopulate the niches that they are transplanted in—is an essential criterion for successful tissue regeneration [375]. For this to occur, the cytokines used in the differentiation process plays a crucial role. The coculture of ESCs with mature cells or tissues drive differentiation toward required lineages [375, 376]. For instance, Vranken et al. showed that the coculture of ESCs with murine embryonic pulmonary mesenchyme enhances the differentiation of ESCs toward pneumocytes [376].

Accordingly, ESCs have been explored as a source to engineer tissues that can recapitulate one or more functions of the corresponding tissues in vivo. The use of ESCs, or progenitors derived from them, has been explored to enable the tissues to regenerate better for tissue engineering applications. For instance, Marolt et al. cultured mesenchymal progenitors that were derived from hESCs on 3D scaffolds, and it facilitated the formation of large and compact bone constructs in bioreactors [377]. The bone implants, thus engineered, maintained the bone matrix and caused its maturation for 8 weeks in immunodeficient mice. In other words, the tissue-engineered bone constructs underwent the process of maturation, vascularization, and remodeling. Further, the implant did not form teratomas—characteristically in an undifferentiated hESC's implantation. ESCs also hold an application in neural tissue engineering [377]. Some groups have derived dopaminergic neurons—that get degenerated in Parkinson's disease (PD)—and shown that their transplantation into Parkinson's disease rat model improved the disease condition [378]. Cho et al. also developed an efficient protocol to produce dopaminergic neurons from hESCs on a large scale drawing the science closer to compliance with the regulatory standards [379]. Song et al. differentiated ESCs to neural progenitors and transplanted them into a rat model of Huntington's disease (HD) and found an improvement in animals that were treated [380]. Similar approaches applied to spinal cord injury models have shown promising results. Rossi et al. demonstrated that motor neurons derived from hESCs could promote the recovery of function in a rat model of spinal cord injury [381]. This approach could also be promising in treating motor neuron disease, which is characterized by the loss of motor neurons. Human ESCs, therefore, constitute a potential source for tissue regeneration in disorders related to aging and neurodegeneration.

6.4.2 hESCs in age-related diseases

Aging-evoked impairment plays a crucial role in the pathophysiology of various disorders [382]. Considering that the potential to combat and cure aging-associated diseases is progressively becoming a primary focus in healthcare sectors [383]. Tissue repair therapies may depend on the in vivo transplantation of hESC-derived tissue progenitor cells. Successful functional grafting of hESC-derived tissue progenitors onto the tissue of the hosts relies on the effectiveness of these therapies [384]. In the past few decades, hESCs have become a popular source for regenerative medicine [385]. Several studies suggested a positive outcome in promoting differentiation of hESCs into pancreatic β-cells, osteogenic cells, cardiomyocytes, adipocyte cells, and neuronal cells [386]. Thus, the properties of pluripotency and self-renewal of hESCs hold applications in the field of developmental biology, genomics gene therapy, tissue engineering, and transgenic processes as well as in the therapeutic approach of immune-genetic diseases, tumors and disorders, juvenile diabetes, neurodegenerative disease, spinal cord injury and eye-related disorders [387].

Macular degeneration

In several degenerative diseases, including those involving the brain and eye, ESCs have been acknowledged for their potential benefits [388]. Neuroectoderm-derived RPE (retinal pigment epithelial cells) supports the function and survival of overlying photoreceptor cells through mechanisms, including recycling of visual pigment and phagocytosis of outer segments of rhodopsin [389, 390]. Age-related degeneration of macular retinal areas (AMD) is a disease with multifactorial consequences, distinguished by degeneration of RPE, by progressive degradation of central vision [391]. It is clinically graded as early (medium-sized drusen or retinal pigmentary changes) to the late (neovascular and atrophic) stages [392]. There are mainly two types, dry (atrophic) and wet (neovascular or exudative). Gene defects mediated by a cellular stressor, viz. lipofuscin-component-di-retinoid-pyridinium-ethanolamine (A2E), potentiate the deposition of a putative toxic metabolite, within the underlying RPE, triggering cell degradation, potential cell death, and progressive central atrophy [393, 394]. The most severe cause of macular degeneration in children and young adults is Stargardt's disease (STGD1) [395]. The syndrome arises from variants of the ABCA4 (ATP-binding cassette, subfamily A, member 4) gene that contributes to a sequence of disease symptoms and impaired vision [396].

No retinal regeneration therapies are available till date. However, regenerating RPE cells with hESCs provide a benefit, thereby increasing the survival of overlying photoreceptor cells and maintaining a vision for a time-restricted period [397]. RPE cells are derived from hESCs in the presence of nicotinamide and Activin A under a serum-free environment [398]. Pigmented cells extracted from hESCs display the morphological and functional features of RPE cells following transplantation in an animal macular degeneration model [399]. The atrophic age-based macular degeneration, with key features shared by STGD1, including progressive atrophy, may gain from subretinal management of hESC-derived RPE cells [400]. A recent clinical trial showed that hESCs are well tolerated for up to 37 months after transplantation in individuals with atrophic age-related macular degeneration and Stargardt's macular dystrophy suggesting a potentially safe new source of cells for the early treatment in amenable retinal disorders caused by tissue loss or dysfunction [401]. A similar study on transplantation of hESC-derived RPE cells demonstrated the development of subretinal hyperpigmentation in patients affected with advanced Stargardt's disease (STGD1) of macular degeneration [402, 403]. The clinical trial commencement would further lead the research of hESCs to therapy for degenerative diseases, with results awaited with great hope in the areas of stem cell biology and geriatric therapy.

Neurodegenerative diseases

Aging impacts numerous patterns in brain development driven by neuronal and metabolic modifications, misfolded protein accumulation, elevated oxidative stress, autophagy, impaired mitochondrial function, ROS-mediated DNA damage, RNA-mediated toxicity, and reduced self-repair capacity [404]. Psychological aging abnormalities emerge through various systemic and functional modifications at specific neuronal circuits leading to cognitive impairment [405], diminished sensorimotor responses, and dramatically enhances neurodegenerative disease vulnerability [406]. This includes Parkinson's disease (PD) [407], amyotrophic lateral sclerosis (ALS) [408], frontotemporal dementia (FTD) [409], Huntington's disease (HD) [410] and Alzheimer's disease (AD) [411]. Numerous molecular pathways, including Nrf2, sirtuin, IGF, mTOR, ROS signaling, and TGF-β, characterize the etiology of the aging processes implicated in neurodegenerative disorders [412, 413]. The probability of contracting these diseases and prevalence is difficult to determine since a defined genetic factor is absent in the vast majority of cases [414].

Stem cell therapy comprising MSCs and ESCs has developed in recent decades as a potential clinical alternative to disorders of the central nervous system (CNS) associated with aging [415]. The neuronal stem cells derived from hESCs demonstrated successful implantation into the spinal cord injury site and activation of the spinal motor neuron occurs through enhanced PGC-1α (peroxisome proliferator-activated receptor-gamma coactivator 1α) respiratory subunit expression and enhanced mitochondrial biogenesis [416–418]. The hESCs are useful in the treatment of numerous functional degenerative pathologies including Parkinson's disease (PD), which is characterized by selective depletion of dopaminergic neurons influencing nigrostriatal degradation in the midbrain [419]. Thus, in PD therapy, one could direct the differentiation of hESCs into dopaminergic phenotype to substitute the degenerated cells [420]. Our perception of the ALS and FTD disorders has reformed as a consequence of C9orf72 (chromosome 9 open reading frame 72) gene expression causing repetitive stretches that display defective localization of the Ran guanine nucleotide exchange factor for nuclear transport [421, 422]. Likewise, FUS (fused in sarcoma) mutations in ALS and FTD patients displayed disrupted nucleocytoplasmic localization of the low complexity protein FUS and were unable to establish neuromuscular junctions [423, 424]. These thus, suggest that both the factors can diverge at the nuclear pore level to encourage protein delocalization, neuronal dysfunction, and ultimately neurodegeneration. hESC-derived motor neuron progenitors transplanted into the ALS animal model with motor neuron dysfunction demonstrated the capacity to revive dying motor neurons by the acquisition of motor neuron-specific neurotrophic reinforcement as a possible effective therapeutic technique for ALS [425]. There are clinical trials conducted (Table 2) that explore the usage of astrocytes extracted from hESCs (called AstroRx) administered into the spinal cord in the early-stage illness of ALS [426, 427]. Hardly any therapy is currently available to cure this disease. Similarly, current research has outlined the application of cerebral organoid models extracted from ESCs that may address the clinical manifestations of neurodegenerative disorders [428, 429]. Preclinical trials involving hESCs may help combat neurodegenerative diseases but the weak correlation between animal experiments and research for humans is noteworthy here.

Degenerative cardiovascular diseases

Cardiovascular repair mechanisms become progressively impaired with age and pose a significant risk factor for cardiovascular disease (CVD) [430]. The involvement of conditions such as cardiac fibrosis and amyloidosis contributes to CVD pathology [431]. Though the prognosis for people with CVD has strengthened, the fatality rate over the past few years has remained unaltered [430]. Therefore, there is a need for alternative strategies to treat CVDs [432]. The suggestion that the heart will regenerate has increased the prospect of cell therapies providing an option to new therapy in recent years [433]. Cardiomyocytes (CM) and vascular cell generation, observed in both adult heart and peripheral tissue, can replace the impaired myocardium and vascular tissue, like endothelial progenitor cells [434]. However, in the event of age-associated alterations that may lead to the dysregulation of endogenous cardiovascular reparation systems, amidst the findings, cardiovascular stem cell/precursors seem to be inadequate to safeguard against the cardiovascular condition in old people [430]. A key player in CVD includes PI3K/Akt, glucocorticoid, Wnt signaling, mTOR, FoxO, inflammatory, and autophagy-mediated pathway [435–440]. Recent advances in research have made it possible to derive cardiomyocytes through the application of glucose-depleted, lactate-rich media or through sorting for high mitochondrial membrane potential, successfully from adequately differentiated hESCs [441, 442]. The expression of cardiomyocyte markers reported to generate a pure cardiomyocyte population from hESCs can further aid in distinguishing the cardiomyocytes from other cell population [443]. hESCs have been widely utilized to derive CMs in vitro in order to study the multitude of genes/epigenetic factors responsible for the implication of CVD. However, the injections of hESCs extracted CMs into failed myocardium of nonhuman primates could promote heart regeneration but the recurrent occurrence of incidence of ectopic arrhythmia by abnormal phenotype of injected hESCs-CMs cells affected the outcomes [444]. However, there has also been some advancement in this area, where the Japanese government sanctioned the trial of cardiac sheet transplantation of CMs obtained from human pluripotent cells for the treatment of patients with cardiac diseases [445].

Endometrial degeneration

A shift in the hormone levels in the female reproductive system is primarily associated with the age-related processes [446]. Late gestational age coincides with decreased fertility and the negative effect of birth [447]. Aging poses a detrimental effect on the female reproductive system [448]. Aging pregnant women with existing morbidity are more prone to an elevated risk of contracting new diseases concerning psychological demands [449]. PI3K/Akt, PTEN, and mTOR signaling pathways are primarily involved in oocyte development [450–452]. Destruction of oocytes initiates alteration in any of these pathways associated with aging. However, the defective aged maternal mitochondria may raise a potential risk

TABLE 2 Summary of published and ongoing clinical trials utilizing hESC-derived cells.

Sr no.	Clinical application	Condition	Study title	Intervention	Phase	NCT number	Status
1	Age-related cardiovascular disease	Ischemic heart disease	Transplantation of Human Embryonic Stem Cells derived Progenitors in Severe Heart Failure	Human embryonic stem cells derived progenitors	I	NCT02057900	Completed
2	Age-related eye disease	Macular degenerative disease	A Safety Surveillance Study in Subjects with Macular Degenerative Disease Treated with Human Embryonic Stem Cells derived Retinal Pigment Epithelial Cell Therapy	Human embryonic stem cells derived retinal pigment epithelial cells	I/II	NCT03167203	Enrolling by invitation
		Dry age-related macular degeneration	Treatment of Dry Age-Related Macular Degeneration Disease with Retinal Pigment Epithelium Derived from Human Embryonic Stem Cells	Retinal pigment epithelium transplantation	I/II	NCT03046407	Recruiting
		Dry macular degeneration, geographic Atrophy	Study of Subretinal Implantation of Human Embryonic Stem Cells Derived RPE Cells in Advanced Dry AMD	CPCB-RPE1	I/II	NCT02590692	Active, not recruiting
		Dry age-related macular degeneration	Subretinal Transplantation of Retinal Pigment Epitheliums in Treatment of Age-related Macular Degeneration Diseases	Retinal pigment epithelium transplantation	I/II	NCT02755428	Recruiting
		Dry age-related macular degeneration	A Phase I/IIa, Open-Label, Single-Center, Prospective Study to Determine the Safety and Tolerability of Sub-retinal Transplantation of Human Embryonic Stem Cells Derived Retinal Pigmented Epithelial (MA09-hRPE) Cells in Patients with Advanced Dry Age-related Macular Degeneration (AMD)	MA09-hRPE	I/II	NCT01674829	Active, not recruiting
		Dry age-related macular degeneration	Safety and Tolerability of Sub-retinal Transplantation of hESC Derived RPE (MA09-hRPE) Cells in Patients with Advanced Dry Age-Related Macular Degeneration	MA09-hRPE	I/II	NCT01344993	Completed
		Age-related macular degeneration	Safety and Efficacy Study of OpRegen for Treatment of Advanced Dry-Form Age-Related Macular Degeneration	OpRegen	I/II	NCT02286089	Recruiting
		Age-related macular degeneration	Long Term Follow Up of Sub-retinal Transplantation of hESCs derived RPE Cells in Patients With AMD	MA09-hRPE		NCT02463344	Completed
		Age-related macular degeneration	A Study of Implantation of Retinal Pigment Epithelium In Subjects with Acute Wet Age Related Macular Degeneration	PF-0520 6388	I	NCT01691261	Active, not recruiting
		Age-related macular degeneration	Retinal Pigment Epithelium Safety Study for Patients in B4711001			NCT03102138	Active, not recruiting

Condition	Title	Intervention	Phase	NCT Number	Status
Dry age-related macular degeneration	The Safety and Tolerability of Sub-retinal Transplantation of SCNT-hES-RPE Cells in Patients with Advanced Dry AMD	SCNT-hES-RPE cells	I	NCT03305029	Unknown status
Stargardt's Macular Dystrophy (SMD)	A Follow up Study to Determine the Safety and Tolerability of Sub-retinal Transplantation of Human Embryonic Stem Cells Derived Retinal Pigmented Epithelial (hESC-RPE) Cells in Patients with Stargardt's Macular Dystrophy (SMD)	hESC-RPE		NCT02941991	Completed
Stargardt's Macular Dystrophy (SMD)	Safety and Tolerability of Sub-retinal Transplantation of Human Embryonic Stem Cell Derived Retinal Pigmented Epithelial (hESC-RPE) Cells in Patients with Stargardt's Macular Dystrophy (SMD)	MA09-hRPE	I/II	NCT01469832	Completed
Stargardt's Macular Dystrophy (SMD)	Sub-retinal Transplantation of hESCs Derived RPE(MA09-hRPE) cells in Patients with Stargardt's Macular Dystrophy	MA09-hRPE	I/II	NCT01345006	Completed
Stargardt's Macular Dystrophy (SMD)	Safety and Tolerability of MA09-hRPE Cells in Patients with Stargardt's Macular Dystrophy (SMD)	MA09-hRPE	I	NCT01625559	Unknown status
Stargardt's Macular Dystrophy (SMD)	Long Term Follow Up of Sub-retinal Transplantation of hESCs Derived RPE Cells in Stargardt Macular Dystrophy Patients	MA09-hRPE		NCT02445612	Completed
Macular degeneration, Stargardt's Macular Dystrophy (SMD)	Clinical Study of Subretinal Transplantation of Human Embryo Stem Cells Derived Retinal Pigment Epitheliums in Treatment of Macular Degeneration Diseases	Subretinal transplantation	I/II	NCT02749734	Unknown status
Macular degeneration, Stargardt's disease	Stem Cell Therapy for Outer Retinal Degenerations	Procedure: injection of hESCs-RPE in suspension Procedure: injection hESCs-RPE seeded in a substrate	I/II	NCT02903576	Unknown status
Macular degeneration, Stargardt's Macular Dystrophy (SMD)	Clinical Study of Subretinal Transplantation of Human Embryo Stem Cells Derived Retinal Pigment Epitheliums in Treatment of Macular Degeneration Diseases	Procedure: subretinal transplantation	I/II	NCT02749734	Unknown status
Myopic macular degeneration	Research with Retinal Cells Derived from Stem Cells for Myopic Macular Degeneration	MA09-hRPE cellular therapy	I/II	NCT02122159	Withdrawn
Retinitis pigmentosa	Safety and Efficacy of Subretinal Transplantation of Clinical Human Embryonic Stem Cells Derived Retinal Pigment Epitheliums in Treatment of Retinitis Pigmentosa	Retinal pigment epitheliums transplantation	I	NCT03944239	Recruiting
Retinitis pigmentosa	Interventional Study of Implantation of hESCs derived RPE in Patients with RP due to Monogenic Mutation	Human embryonic stem cells derived retinal pigment epithelium (RPE)	I/II	NCT03963154	Recruiting

Continued

TABLE 2 Summary of published and ongoing clinical trials utilizing hESC-derived cells—cont'd

Sr no.	Clinical application	Condition	Study title	Intervention	Phase	NCT number	Status
3	Neurodegenerative disorder	ALS (amyotrophic lateral sclerosis)	A Study to Evaluate Transplantation of Astrocytes Derived from Human Embryonic Stem Cells, in Patients with Amyotrophic Lateral Sclerosis (ALS)	AstroRx		NCT03482050	Recruiting
		Parkinson's disease	Safety and Efficacy Study of Human ESCs derived Neural Precursor Cells in the Treatment of Parkinson's Disease	NPC transplantation, Drug: Levodopa	I/II	NCT03119636	Recruiting
		Neurodegenerative disorders	Development of iPSCs From Donated Somatic Cells of Patients with Neurological Diseases			NCT00874783	Recruiting
4	Degenerative meniscus	Meniscus injury	Safety Observation on hESCs Derived MSC Like Cell for the Meniscus Injury	hESCs derived MSC-like cells	I	NCT03839238	Active, not recruiting
5	Uterus degeneration	Intrauterine adhesion	Clinical Safety Study of Human Embryonic Stem Cells Derived Mesenchymal Cells in the Treatment of Moderate and Severe Intrauterine Adhesions	Inject a solution of stem cell preparation, inject stem cells	I	NCT04232592	Not yet recruiting
6	Chronic degenerative disease	Type 1 diabetes	Stem Cell Educator Therapy in Type 1 Diabetes	Device: stem cell educator	II	NCT01350219	Recruiting
7	Cell-line derivation and differentiation	TBX3, cell differentiation	The Role of TBX3 in Human ES Cells Differentiation			NCT00581152	Unknown status
		Infertility	The Derivation of Human Embryonic Stem Cell Lines from PGD Embryos			NCT00353210	Recruiting
		Infertility	Derivation of New Human Embryonic Stem Cell Lines for Clinical Use			NCT00353197	Recruiting
		Normal healthy embryos	Derivation of New Human Embryonic Stem Cell Lines: Identification of Instructive Factors for Germ Cells Development			NCT01165918	Unknown status

of chromosomal anomalies, and several metabolic disorders imply the association of late maternal age with an elevated risk of trisomy 21 [453]. Aging-ovaries is a gradual depletion of the primordial follicle supply, as the ovary functions halt marginally earlier than half the lifetime of a female. A rather prevalent endometrial condition, namely Asherman syndrome (AS) [454], known as intrauterine adhesions (IUAs), or intrauterine synechiae, is identified as a potential candidate for regenerative stem cell-driven intervention in the area of obstetrics and gynecology. Asherman syndrome (AS) is a gynecological aberration wherein intrauterine adhesions crowd the cavity and wall of the uterus that mainly arises due to the surgical uterine interventions that typically require the placement of surgical tools into the uterus [449]. AS is characterized by a variety of signs and symptoms, namely menstrual irregularity or discomfort and infertility. Hysteroscopy is the most common procedure, requiring the surgical extraction of the fibrous strands [455]. In AS therapy, numerous animal studies and clinical trials concentrate on utilizing biomaterials to change stem cells. Following Asherman syndrome therapy, age factor correlates to the infertility problems. Females below the age of 35 exhibit higher chances of pregnancy than females over the age of 35 following extensive intrauterine adhesion treatment [456]. The possible emergence of stem cells to be able to rebuild the primordial follicle, and the oocyte supply, provides the exciting prospect of new approaches in vivo and in vitro to overcome the unending drop of female reproductive efficiency. Although adult stem cell therapy is advancing, research suggested that it not only leads to the proliferation and regeneration of normal or damaged endometrium but may even promote the production of ectopic endometriosis, which poses a potential threat to treat the condition [457]. The first human oogonia were generated using activator of activin A and Wnt pathway in IPSCs culture [458]. This research ultimately leads to the implementation of regenerative hESC-based treatment and the possibility of achieving e a stable and efficient approach for regulating the development of age-related ovarian failure and menopause when clinically favorable.

Diabetes

Diabetes mellitus (DM) is an endocrine illness with an abnormally high serum glucose level either by absolute (type 1) or relative (type 2) insulin deficiency culminating in hyperglycemia. A progressive degenerative type 1 diabetes displays severe glycemic dysregulation, followed by the autoimmune depletion of β-cell activity and a long-term hyperglycemic complication [459]. Multiple-organ system failures occur equally in diabetes as it does during normal-physiological aging, but it also develops with diabetes at an earlier age triggered by different biological mechanisms. Aging, [460] a risk factor for diabetes, which addresses basic aging pathways, such as cellular senescence [461], might have a significant impact on growing health problems and increasing the burden of this disease. The pathogenesis along with the progression of diabetes primarily involves a reduction of progenitor pools, abnormal mRNA processing, perturbed proteostasis, failure of DNA damage repair, cellular senescence, reduced differentiation capability, immune cell invasion, activation of proinflammatory cytokines in the lack of a particular pathogen, and fibrosis correlating with the fundamental aging processes [462, 463]. Today, islet cell transplantation is a promising procedure in a selected group of type I diabetic patients with the Edmonton protocol, showcasing that transplanted cells could reduce complications with diabetes [464]. However, the inadequate availability with islets remains a crucial problem for islet transplantation effectiveness. Thus, the supply of pancreatic β-cell function restoration has led to the examination of the way of how insulin-producing cells (IPCs) are produced from stem cells for the clinical assessment of diabetes. Retinoic acid supplementation reported encouraging the growth and maturation of β-cells via differentiation of ESCs into endocrine cells in vitro under hypoxic conditions [465]. hESCs have specific immune-privileged features, and therefore, their transplantation has less possibility of immune-mediated rejection [466]. Preclinical studies have demonstrated the role of alginate-encapsulated hESCs derived mature β-cells, which, when transplanted to a diabetic mouse model induced successful glycemic modulation [467]. A recent study showed that insulin-producing cells, derived from hESCs using a novel differentiation protocol in vivo and in vitro under hypoxic environment, sustained hyperglycemia and decreased the levels of inflammatory cytokine IL-1β, [465].

6.4.3 Organoids from hESCs

In a multicellular organism, differentiated cells organize into tissues that further organize based on their structure and function into organs and functions together as organ systems. To recapitulate organs in a dish, one could generate organoids that mimic the biochemical and physical cues of tissue development (Fig. 3) [468]. The process of organoid generation from ESCs involves the differentiation of ESCs into specific cell lineages and providing them the proper scaffold and biochemical cues to self-organize and form tissue-specific organoids [469–474].

Several genetic, molecular, and cellular factors are critical for the normal function of cells and tissues. In addition to the soluble factors and the transcription factors that regulate the molecular signaling, spatial interactions of a cell with its neighboring cells in the niche and the extracellular matrix are critical to its normal function. These interactions are coordinated by a set of membrane proteins called adhesion molecules [475]. For these reasons, in vitro 3D cultures make a substantial contribution. Tissue engineering scaffolds made from natural or synthetic materials help in morphogenesis in vitro and also

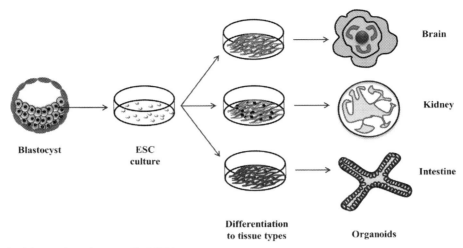

FIG. 3 Organoids derived from embryonic stem cells (ESCs)

maintain the structure and function of the construct. These aspects are crucial as the construct would get integrated with the host tissues after implantation. These tenets, along with the molecular signals, play a role in generating organoids in vitro.

Organoid cultures are particularly helpful in modeling human development and disease conditions. Generating organized tissues of different organ systems through organoid cultures paves the way to a better understanding of organogenesis, functions of genes implicated in different diseases, and interaction of transplanted tissues within the host [476]. Numerous applications are explored by various research groups across the globe. For instance, Takasato et al. generated kidney organoids that recapitulated the kidney tissue architecture, where segmented nephrons connected to collecting ducts, surrounded by renal interstitial cells and an endothelial network demonstrated to be functionally mature [471]. Lancaster et al. developed cerebral organoids—a 3D organoid culture system of different regions of the brain—from hESCs [470]. The cerebral organoids included a cerebral cortex with progenitor populations that could organize to form mature cortical neurons and could model human cortical development [469, 470]. The group also modeled microencephaly using patient-derived induced pluripotent stem cells (iPSCs) [469]. Matsui et al. generated organoids from hESCs cultivated for 6 months and observed mature cerebral organoids that contained neural stem cells, astrocytes, oligodendrocytes, and functional neurons [477]. These research developments emphasize the potential of organoid cultures in modeling aging, neurodegeneration, and neurodevelopment. These models would be useful in screening of drug candidates and also for studies to understand the molecular pathways in neurological diseases.

7 Ethical constraints

Although hESCs have immense potential in medicine, because of their pluripotent nature, there have been a lot of ethical constraints associated with their use. Traditionally, the sources of hESCs are cells retrieved from the preimplantation embryos cultured for in vitro fertilization-embryo transfer (IVF-ET). hESCs are also isolated from single blastomeres developed from the 4-cell, 8-cell, and 16-cell stages of an embryo [478].

One of the ethical concerns that arise in the utilization of hESCs is related to getting informed consent from all the stakeholders involved—the gamete donor, the embryo donor, and the recipient of the stem cells [479]. Ethical constraints in the exploitation of hESCs are circumvented by getting informed consent so that the uncertainty in the process gets minimized to the greatest extent possible. Another way to get around the ethical dilemma of using hESCs was proposed by Landry and Zucker, who suggested the use of dead embryos to obtain pluripotent stem cells. This paper defined the nonviability of an embryo describing its inability to develop into live embryo and distinguished it from organismic death [480]. As Laverge et al. reported, many embryos show cleavage arrest and are unfit for uterine implantation [481]. Accordingly, these embryos get shelved for IVF-ET purposes, but the resulting cells are a mosaic of cells where some are free of any chromosomal aberrations, and it is from these normal cells that hESCs can be derived [480, 481]. However, the assurance that the embryo has no harm when one cell—a blastomere—is used to generate hESCs is another crucial aspect in performing hESCs research [478, 482]. By applying these measures in hESCs research, ethical concerns around it must get addressed. Further, case-by-case investigation and approval of individual research proposal by ethical committees constituted at various levels—institutional, national, and international—can act as bodies that provide guidelines for ethically acceptable processes for research involving hESCs [478] (Table 3).

TABLE 3 Regulatory bodies around the globe that provide ethical guidelines for research involving hESCs.

Country	Regulatory framework/guidelines
USA	National Institutes of Health (NIH); Food and Drug Administration (FDA)
UK	National Research Ethics Service (NRES); Medicines and Healthcare Products Regulatory Agency (MHRA)
Europe	European Medicines Agency (EMA)
Australia	Australian Health Ethics Committee (AHEC)
New Zealand	Ministry of Health
India	Indian Council of Medical Research (ICMR); Department of Biotechnology (DBT)
China	Beijing Ministry of Health Medical Ethics Committee; Southern Chinese Human Genome Research Centre Ethical, Legal, and Social Issues Committee (ELSI)
Japan	Council on Science and Technology (CST) Bioethics Committee
Singapore	Bioethics Advisory Committee (BAC)
South Africa	National Health Act; Medicines Control Act

Thus, ESCs are employed as a cell resource for regenerative therapeutics that might prove valuable in regenerative therapy for CNS conditions. Nevertheless, ethical issues have contributed to studies of alternative sources of stem cells. Moreover, ESCs are being taken over by iPSCs and somatic cell therapies to address the ethical concerns associated with the use of embryo-derived materials and provide a source of autologous cells for transplant. Besides, for stem cell-based therapy, the identical differentiation into several types of somatic cells could be a disadvantage. Several studies have shown that ESCs grafted into animals show poor differentiation and inadequate migration abilities as well as teratomas or tumor formation, in some instances. Allogeneic transplantation of ESCs could also result in the rejection by the host tissue of the implanted cells. These issues have deterred translational medicine from implementing ESCs.

8 Conclusion

Understanding how hESC's distinctive property can be modified to prevent it from aging has become evident in recent years and is deemed as an expanding topic in the field of biological aging. In conclusion, hESCs hold great potential in the field of regenerative therapies for studying age-associated disorders like neurodegeneration, macular degeneration, neurodegeneration, and degenerative cardiovascular diseases. The intrinsic properties of self-renewal and pluripotency of hESCs, and the beneficial molecules secreted by them, provide the basis for designing cell-based therapies for these degenerative diseases. In this direction, several research groups deciphered the metabolic and epigenetic changes in the undifferentiated and differentiated states of the hESCs, designed scaffolds to facilitate differentiation, and also developed strategies to make organoids in a dish. Efforts to determine the function of hESCs in the aging process remains a big challenge, despite its application in the field of age-related diseases. Regenerative therapies using stem cells for antiaging would pave the way to improve health conditions for the aged and sick. Research on hESCs would probably reinvigorate the controlled expression of genes that could promote longevity by slowing down aging and end up making huge strides in age-related disease therapy along with addressing the ethical concerns around it. Unwinding the molecular mechanisms involved in hESCs would thus aid to develop novel therapies for antiaging which can prove beneficial to reverse the aging cycle and help many to stay younger and healthier.

References

[1] M.J. Evans, M.H. Kaufman, Establishment in culture of pluripotential cells from mouse embryos, Nature 292 (1981) 154–156, https://doi.org/10.1038/292154a0.

[2] G.R. Martin, Isolation of a pluripotent cell line from early mouse embryos cultured in medium conditioned by teratocarcinoma stem cells, Proc. Natl. Acad. Sci. U. S. A. 78 (1981) 7634–7638. https://www.ncbi.nlm.nih.gov/pmc/articles/PMC349323/. (Accessed 5 June 2020).

[3] K.S. Richter, D.C. Harris, S.T. Daneshmand, B.S. Shapiro, Quantitative grading of a human blastocyst: optimal inner cell mass size and shape, Fertil. Steril. 76 (2001) 1157–1167, https://doi.org/10.1016/s0015-0282(01)02870-9.

[4] R.L. Gardner, T.J. Davies, Trophectoderm growth and bilateral symmetry of the blastocyst in the mouse, Hum. Reprod. 17 (2002) 1839–1845, https://doi.org/10.1093/humrep/17.7.1839.

[5] T.M. Mayhew, B.L. Barker, Villous trophoblast: morphometric perspectives on growth, differentiation, turnover and deposition of fibrin-type fibrinoid during gestation, Placenta 22 (2001) 628–638, https://doi.org/10.1053/plac.2001.0700.

[6] F. Tang, C. Barbacioru, S. Bao, C. Lee, E. Nordman, X. Wang, K. Lao, M.A. Surani, Tracing the derivation of embryonic stem cells from the inner cell mass by single-cell RNA-Seq analysis, Cell Stem Cell 6 (2010) 468–478, https://doi.org/10.1016/j.stem.2010.03.015.

[7] J.F. Nicolas, P. Dubois, H. Jakob, J. Gaillard, F. Jacob, Mouse teratocarcinoma: differentiation in cultures of a multipotential primitive cell line, author's transl, Ann. Microbiol. 126 (1975) 3–22.

[8] P.P. Tam, R.S. Beddington, Establishment and organization of germ layers in the gastrulating mouse embryo, CIBA Found. Symp. 165 (1992) 27–41. Discussion 42-49 https://doi.org/10.1002/9780470514221.ch3.

[9] H. Niwa, J. Miyazaki, A.G. Smith, Quantitative expression of Oct-3/4 defines differentiation, dedifferentiation or self-renewal of ES cells, Nat. Genet. 24 (2000) 372–376, https://doi.org/10.1038/74199.

[10] J.A. Thomson, J. Kalishman, T.G. Golos, M. Durning, C.P. Harris, R.A. Becker, J.P. Hearn, Isolation of a primate embryonic stem cell line, Proc. Natl. Acad. Sci. U. S. A. 92 (1995) 7844–7848. https://www.ncbi.nlm.nih.gov/pmc/articles/PMC41242/. (Accessed 7 June 2020).

[11] J.A. Thomson, J. Itskovitz-Eldor, S.S. Shapiro, M.A. Waknitz, J.J. Swiergiel, V.S. Marshall, J.M. Jones, Embryonic stem cell lines derived from human blastocysts, Science 282 (1998) 1145–1147, https://doi.org/10.1126/science.282.5391.1145.

[12] C. Buecker, R. Srinivasan, Z. Wu, E. Calo, D. Acampora, T. Faial, A. Simeone, M. Tan, T. Swigut, J. Wysocka, Reorganization of enhancer patterns in transition from naive to primed pluripotency, Cell Stem Cell 14 (2014) 838–853, https://doi.org/10.1016/j.stem.2014.04.003.

[13] C. Xu, J. Jiang, V. Sottile, J. McWhir, J. Lebkowski, M.K. Carpenter, Immortalized fibroblast-like cells derived from human embryonic stem cells support undifferentiated cell growth, Stem Cells 22 (2004) 972–980, https://doi.org/10.1634/stemcells.22-6-972.

[14] R. Eiges, Genetic manipulation of human embryonic stem cells, Methods Mol. Biol. 1307 (2016) 149–172, https://doi.org/10.1007/7651_2014_155.

[15] D. Cyranoski, How human embryonic stem cells sparked a revolution, Nature 555 (2018) 428–430, https://doi.org/10.1038/d41586-018-03268-4.

[16] T. Enver, S. Soneji, C. Joshi, J. Brown, F. Iborra, T. Orntoft, T. Thykjaer, E. Maltby, K. Smith, R. Abu Dawud, M. Jones, M. Matin, P. Gokhale, J. Draper, P.W. Andrews, Cellular differentiation hierarchies in normal and culture-adapted human embryonic stem cells, Hum. Mol. Genet. 14 (2005) 3129–3140, https://doi.org/10.1093/hmg/ddi345.

[17] T. Vazin, W.J. Freed, Human embryonic stem cells: derivation, culture, and differentiation: a review, Restor. Neurol. Neurosci. 28 (2010) 589–603, https://doi.org/10.3233/RNN-2010-0543.

[18] M. Greenwood-Goodwin, J. Yang, M. Hassanipour, D. Larocca, A novel lineage restricted, pericyte-like cell line isolated from human embryonic stem cells, Sci. Rep. 6 (2016) 24403, https://doi.org/10.1038/srep24403.

[19] P.J. Gokhale, J.K. Au-Young, S. Dadi, D.N. Keys, N.J. Harrison, M. Jones, S. Soneji, T. Enver, J.K. Sherlock, P.W. Andrews, Culture adaptation alters transcriptional hierarchies among single human embryonic stem cells reflecting altered patterns of differentiation, PLoS One 10 (2015), https://doi.org/10.1371/journal.pone.0123467, e0123467.

[20] H. Darr, N. Benvenisty, Human embryonic stem cells: the battle between self-renewal and differentiation, Regen. Med. 1 (2006) 317–325, https://doi.org/10.2217/17460751.1.3.317.

[21] B.V. Johnson, N. Shindo, P.D. Rathjen, J. Rathjen, R.A. Keough, Understanding pluripotency—how embryonic stem cells keep their options open, Mol. Hum. Reprod. 14 (2008) 513–520. https://academic.oup.com/molehr/article/14/9/513/1168808. (Accessed 6 June 2020).

[22] R.A. Young, Control of embryonic stem cell state, Cell 144 (2011) 940–954, https://doi.org/10.1016/j.cell.2011.01.032.

[23] T. Barta, D. Dolezalova, Z. Holubcova, A. Hampl, Cell cycle regulation in human embryonic stem cells: links to adaptation to cell culture, Exp. Biol. Med. (Maywood) 238 (2013) 271–275, https://doi.org/10.1177/1535370213480711.

[24] X. Zeng, M.S. Rao, Human embryonic stem cells: long term stability, absence of senescence and a potential cell source for neural replacement, Neuroscience 145 (2007) 1348–1358, https://doi.org/10.1016/j.neuroscience.2006.09.017.

[25] S.-H. Song, K. Kim, J.J. Park, K.H. Min, W. Suh, Reactive oxygen species regulate the quiescence of CD34-positive cells derived from human embryonic stem cells, Cardiovasc. Res. 103 (2014) 147–155, https://doi.org/10.1093/cvr/cvu106.

[26] K. Chadwick, L. Wang, L. Li, P. Menendez, B. Murdoch, A. Rouleau, M. Bhatia, Cytokines and BMP-4 promote hematopoietic differentiation of human embryonic stem cells, Blood 102 (2003) 906–915, https://doi.org/10.1182/blood-2003-03-0832.

[27] H. Niwa, The principles that govern transcription factor network functions in stem cells, Development 145 (2018), https://doi.org/10.1242/dev.157420, dev157420.

[28] M.H. Stewart, S.C. Bendall, M. Bhatia, Deconstructing human embryonic stem cell cultures: niche regulation of self-renewal and pluripotency, J. Mol. Med. 86 (2008) 875–886, https://doi.org/10.1007/s00109-008-0356-9.

[29] W.-L. Tam, Y.-S. Ang, B. Lim, The molecular basis of ageing in stem cells, Mech. Ageing Dev. 128 (2007) 137–148, https://doi.org/10.1016/j.mad.2006.11.020.

[30] R. Jaenisch, R. Young, Stem cells, the molecular circuitry of pluripotency and nuclear reprogramming, Cell 132 (2008) 567–582, https://doi.org/10.1016/j.cell.2008.01.015.

[31] L. Turnpenny, S. Brickwood, C.M. Spalluto, K. Piper, I.T. Cameron, D.I. Wilson, N.A. Hanley, Derivation of human embryonic germ cells: an alternative source of pluripotent stem cells, Stem Cells 21 (2003) 598–609, https://doi.org/10.1634/stemcells.21-5-598.

[32] A.-M. Courtot, A. Magniez, N. Oudrhiri, O. Féraud, J. Bacci, E. Gobbo, S. Proust, A.G. Turhan, A. Bennaceur-Griscelli, Morphological analysis of human induced pluripotent stem cells during induced differentiation and reverse programming, Biores. Open Access 3 (2014) 206–216, https://doi.org/10.1089/biores.2014.0028.

[33] Y.B. Park, Y.Y. Kim, S.K. Oh, S.G. Chung, S.-Y. Ku, S.H. Kim, Y.M. Choi, S.Y. Moon, Alterations of proliferative and differentiation potentials of human embryonic stem cells during long-term culture, Exp. Mol. Med. 40 (2008) 98–108, https://doi.org/10.3858/emm.2008.40.1.98.

[34] S. Llames, E. García-Pérez, Á. Meana, F. Larcher, M. del Río, Feeder layer cell actions and applications, Tissue Eng. B Rev. 21 (2015) 345–353, https://doi.org/10.1089/ten.teb.2014.0547.

[35] O. Hovatta, M. Mikkola, K. Gertow, A.-M. Strömberg, J. Inzunza, J. Hreinsson, B. Rozell, E. Blennow, M. Andäng, L. Ahrlund-Richter, A culture system using human foreskin fibroblasts as feeder cells allows production of human embryonic stem cells, Hum. Reprod. 18 (2003) 1404–1409, https://doi.org/10.1093/humrep/deg290.

[36] M.J. Martin, A. Muotri, F. Gage, A. Varki, Human embryonic stem cells express an immunogenic nonhuman sialic acid, Nat. Med. 11 (2005) 228–232, https://doi.org/10.1038/nm1181.

[37] C. Xu, M.S. Inokuma, J. Denham, K. Golds, P. Kundu, J.D. Gold, M.K. Carpenter, Feeder-free growth of undifferentiated human embryonic stem cells, Nat. Biotechnol. 19 (2001) 971–974, https://doi.org/10.1038/nbt1001-971.

[38] H.K. Kleinman, Preparation of basement membrane components from EHS tumors, Curr. Protoc. Cell Biol. (2001), https://doi.org/10.1002/0471143030.cb1002s00. Chapter 10, Unit 10.2.

[39] S.K.W. Oh, A.K. Chen, Y. Mok, X. Chen, U.-M. Lim, A. Chin, A.B.H. Choo, S. Reuveny, Long-term microcarrier suspension cultures of human embryonic stem cells, Stem Cell Res. 2 (2009) 219–230, https://doi.org/10.1016/j.scr.2009.02.005.

[40] D. Ilic, E. Stephenson, V. Wood, L. Jacquet, D. Stevenson, A. Petrova, N. Kadeva, S. Codognotto, H. Patel, M. Semple, G. Cornwell, C. Ogilvie, P. Braude, Derivation and feeder-free propagation of human embryonic stem cells under xeno-free conditions, Cytotherapy 14 (2012) 122–128, https://doi.org/10.3109/14653249.2011.623692.

[41] S. Yao, S. Chen, J. Clark, E. Hao, G.M. Beattie, A. Hayek, S. Ding, Long-term self-renewal and directed differentiation of human embryonic stem cells in chemically defined conditions, Proc. Natl. Acad. Sci. U. S. A. 103 (2006) 6907–6912, https://doi.org/10.1073/pnas.0602280103.

[42] G. Keller, Embryonic stem cell differentiation: emergence of a new era in biology and medicine, Genes Dev. 19 (2005) 1129–1155, https://doi.org/10.1101/gad.1303605.

[43] G.J. Pan, Z.Y. Chang, H.R. Schöler, D. Pei, Stem cell pluripotency and transcription factor Oct4, Cell Res. 12 (2002) 321–329, https://doi.org/10.1038/sj.cr.7290134.

[44] Z.-N. Zhang, S.-K. Chung, Z. Xu, Y. Xu, Oct4 maintains the pluripotency of human embryonic stem cells by inactivating p53 through Sirt1-mediated deacetylation, Stem Cells 32 (2014) 157–165, https://doi.org/10.1002/stem.1532.

[45] S. Zhang, W. Cui, Sox2, a key factor in the regulation of pluripotency and neural differentiation, World J. Stem Cells 6 (2014) 305–311, https://doi.org/10.4252/wjsc.v6.i3.305.

[46] I. Chambers, D. Colby, M. Robertson, J. Nichols, S. Lee, S. Tweedie, A. Smith, Functional expression cloning of Nanog, a pluripotency sustaining factor in embryonic stem cells, Cell 113 (2003) 643–655, https://doi.org/10.1016/s0092-8674(03)00392-1.

[47] M.H. Allouba, A.M. ElGuindy, N. Krishnamoorthy, M.H. Yacoub, Y.E. Aguib, NaNog: a pluripotency homeobox (master) molecule, Glob. Cardiol. Sci. Pract. 2015 (2015), https://doi.org/10.5339/gcsp.2015.36.

[48] E. Närvä, J.-P. Pursiheimo, A. Laiho, P. Rahkonen, M.R. Emani, M. Viitala, K. Laurila, R. Sahla, R. Lund, H. Lähdesmäki, P. Jaakkola, R. Lahesmaa, Continuous hypoxic culturing of human embryonic stem cells enhances SSEA-3 and MYC levels, PLoS One 8 (2013), https://doi.org/10.1371/journal.pone.0078847.

[49] T. Esmailpour, T. Huang, TBX3 promotes human embryonic stem cell proliferation and neuroepithelial differentiation in a differentiation stage-dependent manner, Stem Cells 30 (2012) 2152–2163, https://doi.org/10.1002/stem.1187.

[50] A.B. Bialkowska, V.W. Yang, S.K. Mallipattu, Krüppel-like factors in mammalian stem cells and development, Development 144 (2017) 737–754, https://doi.org/10.1242/dev.145441.

[51] C.-W. Wong, P.-S. Hou, S.-F. Tseng, C.-L. Chien, K.-J. Wu, H.-F. Chen, H.-N. Ho, S. Kyo, S.-C. Teng, Krüppel-like transcription factor 4 contributes to maintenance of telomerase activity in stem cells, Stem Cells 28 (2010) 1510–1517, https://doi.org/10.1002/stem.477.

[52] A.J. Levine, A.H. Brivanlou, GDF3, a BMP inhibitor, regulates cell fate in stem cells and early embryos, Development 133 (2006) 209–216, https://doi.org/10.1242/dev.02192.

[53] J.S. Draper, C. Pigott, J.A. Thomson, P.W. Andrews, Surface antigens of human embryonic stem cells: changes upon differentiation in culture, J. Anat. 200 (2002) 249–258, https://doi.org/10.1046/j.1469-7580.2002.00030.x.

[54] A.J. Wright, P.W. Andrews, Surface marker antigens in the characterization of human embryonic stem cells, Stem Cell Res. 3 (2009) 3–11, https://doi.org/10.1016/j.scr.2009.04.001.

[55] W.M. Schopperle, W.C. DeWolf, The TRA-1-60 and TRA-1-81 human pluripotent stem cell markers are expressed on podocalyxin in embryonal carcinoma, Stem Cells 25 (2007) 723–730, https://doi.org/10.1634/stemcells.2005-0597.

[56] S. Natunen, T. Satomaa, V. Pitkänen, H. Salo, M. Mikkola, J. Natunen, T. Otonkoski, L. Valmu, The binding specificity of the marker antibodies Tra-1-60 and Tra-1-81 reveals a novel pluripotency-associated type 1 lactosamine epitope, Glycobiology 21 (2011) 1125–1130, https://doi.org/10.1093/glycob/cwq209.

[57] M.-Y. Son, H. Choi, Y.-M. Han, Y.S. Cho, Unveiling the critical role of REX1 in the regulation of human stem cell pluripotency, Stem Cells 31 (2013) 2374–2387, https://doi.org/10.1002/stem.1509.

[58] V. Teijeiro, D. Yang, S. Majumdar, F. González, R.W. Rickert, C. Xu, R. Koche, N. Verma, E.C. Lai, D. Huangfu, DICER1 is essential for self-renewal of human embryonic stem cells, Stem Cell Rep. 11 (2018) 616–625, https://doi.org/10.1016/j.stemcr.2018.07.013.

[59] H. Sun, Y. You, M. Guo, X. Wang, Y. Zhang, S. Ye, Tfcp2l1 safeguards the maintenance of human embryonic stem cell self-renewal, J. Cell. Physiol. 233 (2018) 6944–6951, https://doi.org/10.1002/jcp.26483.

[60] Z. Khalkhali-Ellis, V. Galat, Y. Galat, A. Gilgur, E.A. Seftor, M.J.C. Hendrix, Lefty glycoproteins in human embryonic stem cells: extracellular delivery route and posttranslational modification in differentiation, Stem Cells Dev. 25 (2016) 1681–1690, https://doi.org/10.1089/scd.2016.0081.

[61] S. Zeng, L. Liu, Y. Sun, P. Xie, L. Hu, D. Yuan, D. Chen, Q. Ouyang, G. Lin, G. Lu, Telomerase-mediated telomere elongation from human blastocysts to embryonic stem cells, J. Cell Sci. 127 (2014) 752–762, https://doi.org/10.1242/jcs.131433.

[62] N. Tsuneyoshi, T. Sumi, H. Onda, H. Nojima, N. Nakatsuji, H. Suemori, PRDM14 suppresses expression of differentiation marker genes in human embryonic stem cells, Biochem. Biophys. Res. Commun. 367 (2008) 899–905, https://doi.org/10.1016/j.bbrc.2007.12.189.

[63] Y.-S. Chan, J. Göke, X. Lu, N. Venkatesan, B. Feng, I.-H. Su, H.-H. Ng, A PRC2-dependent repressive role of PRDM14 in human embryonic stem cells and induced pluripotent stem cell reprogramming, Stem Cells 31 (2013) 682–692, https://doi.org/10.1002/stem.1307.

[64] S.-K. Kim, M.R. Suh, H.S. Yoon, J.B. Lee, S.K. Oh, S.Y. Moon, S.-H. Moon, J.Y. Lee, J.H. Hwang, W.J. Cho, K.-S. Kim, Identification of developmental pluripotency associated 5 expression in human pluripotent stem cells, Stem Cells 23 (2005) 458–462, https://doi.org/10.1634/stemcells.2004-0245.

[65] X. Qian, J.K. Kim, W. Tong, L.G. Villa-Diaz, P.H. Krebsbach, DPPA5 supports pluripotency and reprogramming by regulating NANOG turnover, Stem Cells 34 (2016) 588–600, https://doi.org/10.1002/stem.2252.

[66] F. Kisa, S. Shiozawa, K. Oda, S. Yoshimatsu, M. Nakamura, I. Koya, K. Kawai, S. Suzuki, H. Okano, Naive-like ESRRB+ iPSCs with the capacity for rapid neural differentiation, Stem Cell Rep. 9 (2017) 1825–1838, https://doi.org/10.1016/j.stemcr.2017.10.008.

[67] W.A. Pastor, W. Liu, D. Chen, J. Ho, R. Kim, T.J. Hunt, A. Lukianchikov, X. Liu, J.M. Polo, S.E. Jacobsen, A.T. Clark, TFAP2C regulates transcription in human naive pluripotency by opening enhancers, Nat. Cell Biol. 20 (2018) 553–564, https://doi.org/10.1038/s41556-018-0089-0.

[68] T. Messmer, F. von Meyenn, A. Savino, F. Santos, H. Mohammed, A.T.L. Lun, J.C. Marioni, W. Reik, Transcriptional heterogeneity in naive and primed human pluripotent stem cells at single-cell resolution, Cell Rep. 26 (2019) 815–824.e4, https://doi.org/10.1016/j.celrep.2018.12.099.

[69] S. Zhao, J. Xu, S. Liu, K. Cui, Z. Li, N. Liu, Dppa3 in pluripotency maintenance of ES cells and early embryogenesis, J. Cell. Biochem. 120 (2019) 4794–4799, https://doi.org/10.1002/jcb.28063.

[70] C.B. Ware, A.M. Nelson, B. Mecham, J. Hesson, W. Zhou, E.C. Jonlin, A.J. Jimenez-Caliani, X. Deng, C. Cavanaugh, S. Cook, P.J. Tesar, J. Okada, L. Margaretha, H. Sperber, M. Choi, C.A. Blau, P.M. Treuting, R.D. Hawkins, V. Cirulli, H. Ruohola-Baker, Derivation of naïve human embryonic stem cells, Proc. Natl. Acad. Sci. U. S. A. 111 (2014) 4484–4489, https://doi.org/10.1073/pnas.1319738111.

[71] S. Shin, M. Mitalipova, S. Noggle, D. Tibbitts, A. Venable, R. Rao, S.L. Stice, Long-term proliferation of human embryonic stem cell-derived neuroepithelial cells using defined adherent culture conditions, Stem Cells 24 (2006) 125–138, https://doi.org/10.1634/stemcells.2004-0150.

[72] R. Pal, S. Totey, M.K. Mamidi, V.S. Bhat, S. Totey, Propensity of human embryonic stem cell lines during early stage of lineage specification controls their terminal differentiation into mature cell types, Exp. Biol. Med. (Maywood) 234 (2009) 1230–1243, https://doi.org/10.3181/0901-RM-38.

[73] J.A. Harrill, T.M. Freudenrich, D.W. Machacek, S.L. Stice, W.R. Mundy, Quantitative assessment of neurite outgrowth in human embryonic stem cell-derived hN2 cells using automated high-content image analysis, Neurotoxicology 31 (2010) 277–290, https://doi.org/10.1016/j.neuro.2010.02.003.

[74] X. Zhang, C.T. Huang, J. Chen, M.T. Pankratz, J. Xi, J. Li, Y. Yang, T.M. LaVaute, X.-J. Li, M. Ayala, G.I. Bondarenko, Z.-W. Du, Y. Jin, T.G. Golos, S.-C. Zhang, Pax6 is a human neuroectoderm cell fate determinant, Cell Stem Cell 7 (2010) 90–100, https://doi.org/10.1016/j.stem.2010.04.017.

[75] R. Taléns-Visconti, I. Sanchez-Vera, J. Kostic, M.A. Perez-Arago, S. Erceg, M. Stojkovic, C. Guerri, Neural differentiation from human embryonic stem cells as a tool to study early brain development and the neuroteratogenic effects of ethanol, Stem Cells Dev. 20 (2011) 327–339, https://doi.org/10.1089/scd.2010.0037.

[76] S.H. Chung, W. Shen, K.C. Davidson, A. Pébay, R.C.B. Wong, B. Yau, M. Gillies, Differentiation of retinal glial cells from human embryonic stem cells by promoting the notch signaling pathway, Front. Cell. Neurosci. 13 (2019), https://doi.org/10.3389/fncel.2019.00527.

[77] K.Y. Chau, S. Chen, D.J. Zack, S.J. Ono, Functional domains of the cone-rod homeobox (CRX) transcription factor, J. Biol. Chem. 275 (2000) 37264–37270, https://doi.org/10.1074/jbc.M002763200.

[78] A. Roy, J. de Melo, D. Chaturvedi, T. Thein, A. Cabrera-Socorro, C. Houart, G. Meyer, S. Blackshaw, S. Tole, LHX2 is necessary for the maintenance of optic identity and for the progression of optic morphogenesis, J. Neurosci. 33 (2013) 6877–6884, https://doi.org/10.1523/JNEUROSCI.4216-12.2013.

[79] R.K. Singh, R.K. Mallela, P.K. Cornuet, A.N. Reifler, A.P. Chervenak, M.D. West, K.Y. Wong, I.O. Nasonkin, Characterization of three-dimensional retinal tissue derived from human embryonic stem cells in adherent monolayer cultures, Stem Cells Dev. 24 (2015) 2778–2795, https://doi.org/10.1089/scd.2015.0144.

[80] C. Liu, Y. Zhong, A. Apostolou, S. Fang, Neural differentiation of human embryonic stem cells as an in vitro tool for the study of the expression patterns of the neuronal cytoskeleton during neurogenesis, Biochem. Biophys. Res. Commun. 439 (2013) 154–159, https://doi.org/10.1016/j.bbrc.2013.07.130.

[81] A.M. MacNicol, L.L. Hardy, H.J. Spencer, M.C. MacNicol, Neural stem and progenitor cell fate transition requires regulation of Musashi1 function, BMC Dev. Biol. 15 (2015), https://doi.org/10.1186/s12861-015-0064-y.

[82] H. Li, Z. Hu, H. Jiang, J. Pu, I. Selli, J. Qiu, B. Zhang, J. Feng, TET1 deficiency impairs morphogen-free differentiation of human embryonic stem cells to neuroectoderm, Sci. Rep. 10 (2020), https://doi.org/10.1038/s41598-020-67143-x.

[83] L. Zhang, Q. Hua, K. Tang, C. Shi, X. Xie, R. Zhang, CXCR4 activation promotes differentiation of human embryonic stem cells to neural stem cells, Neuroscience 337 (2016) 88–97, https://doi.org/10.1016/j.neuroscience.2016.09.001.

[84] B.L. Martin, D. Kimelman, Brachyury establishes the embryonic mesodermal progenitor niche, Genes Dev. 24 (2010) 2778–2783, https://doi.org/10.1101/gad.1962910.

[85] T. Faial, A.S. Bernardo, S. Mendjan, E. Diamanti, D. Ortmann, G.E. Gentsch, V.L. Mascetti, M.W.B. Trotter, J.C. Smith, R.A. Pedersen, Brachyury and SMAD signalling collaboratively orchestrate distinct mesoderm and endoderm gene regulatory networks in differentiating human embryonic stem cells, Development 142 (2015) 2121–2135, https://doi.org/10.1242/dev.117838.

[86] P. Zhang, J. Li, Z. Tan, C. Wang, T. Liu, L. Chen, J. Yong, W. Jiang, X. Sun, L. Du, M. Ding, H. Deng, Short-term BMP-4 treatment initiates mesoderm induction in human embryonic stem cells, Blood 111 (2008) 1933–1941, https://doi.org/10.1182/blood-2007-02-074120.

[87] C. Song, F. Xu, Z. Ren, Y. Zhang, Y. Meng, Y. Yang, S. Lingadahalli, E. Cheung, G. Li, W. Liu, J. Wan, Y. Zhao, G. Chen, Elevated exogenous pyruvate potentiates mesodermal differentiation through metabolic modulation and AMPK/mTOR pathway in human embryonic stem cells, Stem Cell Rep. 13 (2019) 338–351, https://doi.org/10.1016/j.stemcr.2019.06.003.

[88] M.-X. Zhu, J.-Y. Zhao, G.-A. Chen, L. Guan, Early embryonic sensitivity to cyclophosphamide in cardiac differentiation from human embryonic stem cells, Cell Biol. Int. 35 (2011) 927–938, https://doi.org/10.1042/CBI20110031.

[89] J.M. Gallagher, H. Komati, E. Roy, M. Nemer, B.V. Latinkić, Dissociation of cardiogenic and postnatal myocardial activities of GATA4, Mol. Cell. Biol. 32 (2012) 2214–2223, https://doi.org/10.1128/MCB.00218-12.

[90] S. Mazzotta, A.T. Lynch, S. Hoppler, Cardiomyocyte differentiation from human embryonic stem cells, Methods Mol. Biol. 1816 (2018) 67–78, https://doi.org/10.1007/978-1-4939-8597-5_5.

[91] B. Wei, J.-P. Jin, TNNT1, TNNT2, and TNNT3: isoform genes, regulation, and structure-function relationships, Gene 582 (2016) 1–13, https://doi.org/10.1016/j.gene.2016.01.006.

[92] M.G. Posch, S. Waldmüller, M. Müller, T. Scheffold, D. Fournier, M.A. Andrade-Navarro, B. De Geeter, S. Guillaumont, C. Dauphin, D. Youssef, K.R. Schmitt, A. Perrot, F. Berger, R. Hetzer, P. Bouvagnet, C. Özcelik, Cardiac alpha-myosin (MYH6) is the predominant sarcomeric disease gene for familial atrial septal defects, PLoS One 6 (2011), https://doi.org/10.1371/journal.pone.0028872.

[93] Y.-J. Liang, B.-C. Yang, J.-M. Chen, Y.-H. Lin, C.-L. Huang, Y.-Y. Cheng, C.-Y. Hsu, K.-H. Khoo, C.-N. Shen, J. Yu, Changes in glycosphingolipid composition during differentiation of human embryonic stem cells to ectodermal or endodermal lineages, Stem Cells 29 (2011) 1995–2004, https://doi.org/10.1002/stem.750.

[94] P. Wang, R.T. Rodriguez, J. Wang, A. Ghodasara, S.K. Kim, Targeting SOX17 in human embryonic stem cells creates unique strategies for isolating and analyzing developing endoderm, Cell Stem Cell 8 (2011) 335–346, https://doi.org/10.1016/j.stem.2011.01.017.

[95] J.B. Fisher, K. Pulakanti, S. Rao, S.A. Duncan, GATA6 is essential for endoderm formation from human pluripotent stem cells, Biol. Open 6 (2017) 1084–1095, https://doi.org/10.1242/bio.026120.

[96] W. Jiang, D. Zhang, N. Bursac, Y. Zhang, WNT3 is a biomarker capable of predicting the definitive endoderm differentiation potential of hESCs, Stem Cell Rep. 1 (2013) 46–52, https://doi.org/10.1016/j.stemcr.2013.03.003.

[97] N. Lavon, O. Yanuka, N. Benvenisty, The effect of overexpression of Pdx1 and Foxa2 on the differentiation of human embryonic stem cells into pancreatic cells, Stem Cells 24 (2006) 1923–1930, https://doi.org/10.1634/stemcells.2005-0397.

[98] Y. Duan, X. Ma, X. Ma, W. Zou, C. Wang, I.S. Bahbahan, T.P. Ahuja, V. Tolstikov, M.A. Zern, Differentiation and characterization of metabolically functioning hepatocytes from human embryonic stem cells, Stem Cells 28 (2010) 674–686, https://doi.org/10.1002/stem.315.

[99] Q.L. Liang, Z. Mo, X.F. Li, X.X. Wang, R.M. Li, Pdx1 protein induces human embryonic stem cells into the pancreatic endocrine lineage, Cell Biol. Int. 37 (2013) 2–10, https://doi.org/10.1002/cbin.10001.

[100] A. Rezania, J.E. Bruin, J. Xu, K. Narayan, J.K. Fox, J.J. O'Neil, T.J. Kieffer, Enrichment of human embryonic stem cell-derived NKX6.1-expressing pancreatic progenitor cells accelerates the maturation of insulin-secreting cells in vivo, Stem Cells 31 (2013) 2432–2442, https://doi.org/10.1002/stem.1489.

[101] R. Tran, C. Moraes, C.A. Hoesli, Controlled clustering enhances PDX1 and NKX6.1 expression in pancreatic endoderm cells derived from pluripotent stem cells, Sci. Rep. 10 (2020) 1190, https://doi.org/10.1038/s41598-020-57787-0.

[102] P.A. Seymour, K.K. Freude, M.N. Tran, E.E. Mayes, J. Jensen, R. Kist, G. Scherer, M. Sander, SOX9 is required for maintenance of the pancreatic progenitor cell pool, Proc. Natl. Acad. Sci. U. S. A. 104 (2007) 1865–1870, https://doi.org/10.1073/pnas.0609217104.

[103] X. Xu, V.L. Browning, J.S. Odorico, Activin, BMP and FGF pathways cooperate to promote endoderm and pancreatic lineage cell differentiation from human embryonic stem cells, Mech. Dev. 128 (2011) 412–427, https://doi.org/10.1016/j.mod.2011.08.001.

[104] A. De Los Angeles, Y.-H. Loh, P.J. Tesar, G.Q. Daley, Accessing naïve human pluripotency, Curr. Opin. Genet. Dev. 22 (2012) 272–282, https://doi.org/10.1016/j.gde.2012.03.001.

[105] J. Betschinger, J. Nichols, S. Dietmann, P.D. Corrin, P.J. Paddison, A. Smith, Exit from pluripotency is gated by intracellular redistribution of the bHLH transcription factor Tfe3, Cell 153 (2013) 335–347, https://doi.org/10.1016/j.cell.2013.03.012.

[106] A.J. Harvey, J. Rathjen, D.K. Gardner, Metaboloepigenetic regulation of pluripotent stem cells, Stem Cells Int. 2016 (2015), https://doi.org/10.1155/2016/1816525, e1816525.

[107] J. Artus, A. Piliszek, A.-K. Hadjantonakis, The primitive endoderm lineage of the mouse blastocyst: sequential transcription factor activation and regulation of differentiation by Sox17, Dev. Biol. 350 (2011) 393–404, https://doi.org/10.1016/j.ydbio.2010.12.007.

[108] M.Z. Ratajczak, D.-M. Shin, R. Liu, W. Marlicz, M. Tarnowski, J. Ratajczak, M. Kucia, Epiblast/germ line hypothesis of cancer development revisited: lesson from the presence of Oct-4+ cells in adult tissues, Stem Cell Rev. 6 (2010) 307–316, https://doi.org/10.1007/s12015-010-9143-4.

[109] J. Nichols, A. Smith, Naive and primed pluripotent states, Cell Stem Cell 4 (2009) 487–492, https://doi.org/10.1016/j.stem.2009.05.015.

[110] I. Ginis, Y. Luo, T. Miura, S. Thies, R. Brandenberger, S. Gerecht-Nir, M. Amit, A. Hoke, M.K. Carpenter, J. Itskovitz-Eldor, M.S. Rao, Differences between human and mouse embryonic stem cells, Dev. Biol. 269 (2004) 360–380, https://doi.org/10.1016/j.ydbio.2003.12.034.

[111] S. Petropoulos, D. Edsgärd, B. Reinius, Q. Deng, S.P. Panula, S. Codeluppi, A. Plaza Reyes, S. Linnarsson, R. Sandberg, F. Lanner, Single-cell RNA-Seq reveals lineage and X chromosome dynamics in human preimplantation embryos, Cell 165 (2016) 1012–1026, https://doi.org/10.1016/j.cell.2016.03.023.

[112] S. Ohtsuka, S. Dalton, Molecular and biological properties of pluripotent embryonic stem cells, Gene Ther. 15 (2008) 74–81, https://doi.org/10.1038/sj.gt.3303065.

[113] M. Kinehara, S. Kawamura, D. Tateyama, M. Suga, H. Matsumura, S. Mimura, N. Hirayama, M. Hirata, K. Uchio-Yamada, A. Kohara, K. Yanagihara, M.K. Furue, Protein kinase C regulates human pluripotent stem cell self-renewal, PLoS One 8 (2013), https://doi.org/10.1371/journal.pone.0054122, e54122.

[114] D. Dutta, Signaling pathways dictating pluripotency in embryonic stem cells, Int. J. Dev. Biol. 57 (2013) 667–675, https://doi.org/10.1387/ijdb.130064dd.

[115] S.P. Medvedev, A.I. Shevchenko, S.M. Zakian, Molecular basis of mammalian embryonic stem cell pluripotency and self-renewal, Acta Nat. 2 (2010) 30–46. https://www.ncbi.nlm.nih.gov/pmc/articles/PMC3347565/. (Accessed 8 June 2020).

[116] A.L. van Oosten, Y. Costa, A. Smith, J.C.R. Silva, JAK/STAT3 signalling is sufficient and dominant over antagonistic cues for the establishment of naive pluripotency, Nat. Commun. 3 (2012) 817, https://doi.org/10.1038/ncomms1822.

[117] X. Wang, X. Wang, S. Zhang, H. Sun, S. Li, H. Ding, Y. You, X. Zhang, S.-D. Ye, The transcription factor TFCP2L1 induces expression of distinct target genes and promotes self-renewal of mouse and human embryonic stem cells, J. Biol. Chem. 294 (2019) 6007–6016, https://doi.org/10.1074/jbc.RA118.006341.

[118] D. Qiu, S. Ye, B. Ruiz, X. Zhou, D. Liu, Q. Zhang, Q.-L. Ying, Klf2 and Tfcp2l1, two Wnt/β-catenin targets, act synergistically to induce and maintain naive pluripotency, Stem Cell Rep. 5 (2015) 314–322, https://doi.org/10.1016/j.stemcr.2015.07.014.

[119] G. Martello, P. Bertone, A. Smith, Identification of the missing pluripotency mediator downstream of leukaemia inhibitory factor, EMBO J. 32 (2013) 2561–2574, https://doi.org/10.1038/emboj.2013.177.

[120] B.T. MacDonald, K. Tamai, X. He, Wnt/β-catenin signaling: components, mechanisms, and diseases, Dev. Cell 17 (2009) 9–26, https://doi.org/10.1016/j.devcel.2009.06.016.

[121] R. Nusse, Wnt signaling and stem cell control, Cell Res. 18 (2008) 523–527, https://doi.org/10.1038/cr.2008.47.

[122] J.K. Sethi, A. Vidal-Puig, Wnt signalling and the control of cellular metabolism, Biochem. J. 427 (2010) 1–17, https://doi.org/10.1042/BJ20091866.

[123] J. Behrens, B.-A. Jerchow, M. Würtele, J. Grimm, C. Asbrand, R. Wirtz, M. Kühl, D. Wedlich, W. Birchmeier, Functional interaction of an axin homolog, conductin, with β-catenin, APC, and GSK3β, Science 280 (1998) 596–599, https://doi.org/10.1126/science.280.5363.596.

[124] E. Jho, T. Zhang, C. Domon, C.-K. Joo, J.-N. Freund, F. Costantini, Wnt/beta-catenin/Tcf signaling induces the transcription of Axin2, a negative regulator of the signaling pathway, Mol. Cell. Biol. 22 (2002) 1172–1183, https://doi.org/10.1128/mcb.22.4.1172-1183.2002.

[125] P. Prinos, S. Joseph, K. Oh, B.I. Meyer, P. Gruss, D. Lohnes, Multiple pathways governing Cdx1 expression during murine development, Dev. Biol. 239 (2001) 257–269, https://doi.org/10.1006/dbio.2001.0446.

[126] S.J. Arnold, J. Stappert, A. Bauer, A. Kispert, B.G. Herrmann, R. Kemler, Brachyury is a target gene of the Wnt/β-catenin signaling pathway, Mech. Dev. 91 (2000) 249–258, https://doi.org/10.1016/S0925-4773(99)00309-3.

[127] D. ten Berge, D. Kurek, T. Blauwkamp, W. Koole, A. Maas, E. Eroglu, R.K. Siu, R. Nusse, Embryonic stem cells require Wnt proteins to prevent differentiation to epiblast stem cells, Nat. Cell Biol. 13 (2011) 1070–1075, https://doi.org/10.1038/ncb2314.

[128] A.M. Singh, M. Bechard, K. Smith, S. Dalton, Reconciling the different roles of Gsk3β in "naïve" and "primed" pluripotent stem cells, Cell Cycle 11 (2012) 2991–2996, https://doi.org/10.4161/cc.21110.

[129] U.C. Okoye, C.C. Malbon, H. Wang, Wnt and frizzled RNA expression in human mesenchymal and embryonic (H7) stem cells, J. Mol. Signal. 3 (2008) 16, https://doi.org/10.1186/1750-2187-3-16.

[130] X. Lian, X. Bao, A. Al-Ahmad, J. Liu, Y. Wu, W. Dong, K.K. Dunn, E.V. Shusta, S.P. Palecek, Efficient differentiation of human pluripotent stem cells to endothelial progenitors via small-molecule activation of WNT signaling, Stem Cell Rep. 3 (2014) 804–816, https://doi.org/10.1016/j.stemcr.2014.09.005.

[131] L. Meijer, A.-L. Skaltsounis, P. Magiatis, P. Polychronopoulos, M. Knockaert, M. Leost, X.P. Ryan, C.A. Vonica, A. Brivanlou, R. Dajani, C. Crovace, C. Tarricone, A. Musacchio, S.M. Roe, L. Pearl, P. Greengard, GSK-3-selective inhibitors derived from Tyrian purple indirubins, Chem. Biol. 10 (2003) 1255–1266, https://doi.org/10.1016/j.chembiol.2003.11.010.

[132] N. Sato, A.H. Brivanlou, Manipulation of self-renewal in human embryonic stem cells through a novel pharmacological GSK-3 inhibitor, Methods Mol. Biol. 331 (2006) 115–128, https://doi.org/10.1385/1-59745-046-4:115.

[133] Z. Xu, A.M. Robitaille, J.D. Berndt, K.C. Davidson, K.A. Fischer, J. Mathieu, J.C. Potter, H. Ruohola-Baker, R.T. Moon, Wnt/β-catenin signaling promotes self-renewal and inhibits the primed state transition in naïve human embryonic stem cells, Proc. Natl. Acad. Sci. U. S. A. 113 (2016) E6382–E6390, https://doi.org/10.1073/pnas.1613849113.

[134] X. Wei, J. Guo, Q. Li, Q. Jia, Q. Jing, Y. Li, B. Zhou, J. Chen, S. Gao, X. Zhang, M. Jia, C. Niu, W. Yang, X. Zhi, X. Wang, D. Yu, L. Bai, L. Wang, J. Na, Y. Zou, J. Zhang, S. Zhang, D. Meng, Bach1 regulates self-renewal and impedes mesendodermal differentiation of human embryonic stem cells, Sci. Adv. 5 (2019), https://doi.org/10.1126/sciadv.aau7887, eaau7887.

[135] E. Arman, R. Haffner-Krausz, Y. Chen, J.K. Heath, P. Lonai, Targeted disruption of fibroblast growth factor (FGF) receptor 2 suggests a role for FGF signaling in pregastrulation mammalian development, Proc. Natl. Acad. Sci. U. S. A. 95 (1998) 5082–5087. https://www.ncbi.nlm.nih.gov/pmc/articles/PMC20217/. (Accessed 18 June 2020).

[136] R. Treisman, Regulation of transcription by MAP kinase cascades, Curr. Opin. Cell Biol. 8 (1996) 205–215, https://doi.org/10.1016/s0955-0674(96)80067-6.

[137] J.-C. Chambard, R. Lefloch, J. Pouysségur, P. Lenormand, ERK implication in cell cycle regulation, Biochim. Biophys. Acta 1773 (2007) 1299–1310, https://doi.org/10.1016/j.bbamcr.2006.11.010.

[138] L. Armstrong, O. Hughes, S. Yung, L. Hyslop, R. Stewart, I. Wappler, H. Peters, T. Walter, P. Stojkovic, J. Evans, M. Stojkovic, M. Lako, The role of PI3K/AKT, MAPK/ERK and NFkβ signalling in the maintenance of human embryonic stem cell pluripotency and viability highlighted by transcriptional profiling and functional analysis, Hum. Mol. Genet. 15 (2006) 1894–1913, https://doi.org/10.1093/hmg/ddl112.

[139] F. Lanner, J. Rossant, The role of FGF/Erk signaling in pluripotent cells, Development 137 (2010) 3351–3360, https://doi.org/10.1242/dev.050146.

[140] J. Göke, Y.-S. Chan, J. Yan, M. Vingron, H.-H. Ng, Genome-wide kinase-chromatin interactions reveal the regulatory network of ERK signaling in human embryonic stem cells, Mol. Cell 50 (2013) 844–855, https://doi.org/10.1016/j.molcel.2013.04.030.

[141] K. Takahashi, M. Murakami, S. Yamanaka, Role of the phosphoinositide 3-kinase pathway in mouse embryonic stem (ES) cells, Biochem. Soc. Trans. 33 (2005) 1522–1525, https://doi.org/10.1042/BST20051522.

[142] C. Godoy-Parejo, C. Deng, W. Liu, G. Chen, Insulin stimulates PI3K/AKT and cell adhesion to promote the survival of individualized human embryonic stem cells, Stem Cells 37 (2019) 1030–1041, https://doi.org/10.1002/stem.3026.

[143] F. Yu, R. Wei, J. Yang, J. Liu, K. Yang, H. Wang, Y. Mu, T. Hong, FoxO1 inhibition promotes differentiation of human embryonic stem cells into insulin producing cells, Exp. Cell Res. 362 (2018) 227–234, https://doi.org/10.1016/j.yexcr.2017.11.022.

[144] J. Zhou, P. Su, L. Wang, J. Chen, M. Zimmermann, O. Genbacev, O. Afonja, M.C. Horne, T. Tanaka, E. Duan, S.J. Fisher, J. Liao, J. Chen, F. Wang, mTOR supports long-term self-renewal and suppresses mesoderm and endoderm activities of human embryonic stem cells, Proc. Natl. Acad. Sci. U. S. A. 106 (2009) 7840–7845, https://doi.org/10.1073/pnas.0901854106.

[145] Y. Wang, C. Tian, J.C. Zheng, FoxO3a contributes to the reprogramming process and the differentiation of induced pluripotent stem cells, Stem Cells Dev. 22 (2013) 2954–2963, https://doi.org/10.1089/scd.2013.0044.

[146] S. Mathew, S. Sundararaj, H. Mamiya, I. Banerjee, Regulatory interactions maintaining self-renewal of human embryonic stem cells as revealed through a systems analysis of PI3K/AKT pathway, Bioinformatics 30 (2014) 2334–2342, https://doi.org/10.1093/bioinformatics/btu209.

[147] J. Li, G. Wang, C. Wang, Y. Zhao, H. Zhang, Z. Tan, Z. Song, M. Ding, H. Deng, MEK/ERK signaling contributes to the maintenance of human embryonic stem cell self-renewal, Differentiation 75 (2007) 299–307, https://doi.org/10.1111/j.1432-0436.2006.00143.x.

[148] A.K.K. Teo, S.J. Arnold, M.W.B. Trotter, S. Brown, L.T. Ang, Z. Chng, E.J. Robertson, N.R. Dunn, L. Vallier, Pluripotency factors regulate definitive endoderm specification through eomesodermin, Genes Dev. 25 (2011) 238–250, https://doi.org/10.1101/gad.607311.

[149] T. Nii, T. Marumoto, H. Kohara, S. Yamaguchi, H. Kawano, E. Sasaki, Y. Kametani, K. Tani, Improved hematopoietic differentiation of primate embryonic stem cells by inhibition of the PI3K-AKT pathway under defined conditions, Exp. Hematol. 43 (2015) 901–911.e4, https://doi.org/10.1016/j.exphem.2015.06.001.

[150] A.D. Zoumaro-Djayoon, V. Ding, L.-Y. Foong, A. Choo, A.J.R. Heck, J. Muñoz, Investigating the role of FGF-2 in stem cell maintenance by global phosphoproteomics profiling, Proteomics 11 (2011) 3962–3971, https://doi.org/10.1002/pmic.201100048.

[151] S.-W. Park, Y. Jun Koh, J. Jeon, Y.-H. Cho, M.-J. Jang, Y. Kang, M.-J. Kim, C. Choi, Y. Sook Cho, H.-M. Chung, G.Y. Koh, Y.-M. Han, Efficient differentiation of human pluripotent stem cells into functional CD34+ progenitor cells by combined modulation of the MEK/ERK and BMP4 signaling pathways, Blood 116 (2010) 5762–5772, https://doi.org/10.1182/blood-2010-04-280719.

[152] Y. Shi, J. Massagué, Mechanisms of TGF-β signaling from cell membrane to the nucleus, Cell 113 (2003) 685–700, https://doi.org/10.1016/S0092-8674(03)00432-X.

[153] D. James, A.J. Levine, D. Besser, A. Hemmati-Brivanlou, TGFbeta/activin/nodal signaling is necessary for the maintenance of pluripotency in human embryonic stem cells, Development 132 (2005) 1273–1282, https://doi.org/10.1242/dev.01706.

[154] P. Yu, G. Pan, J. Yu, J.A. Thomson, FGF2 sustains NANOG and switches the outcome of BMP4-induced human embryonic stem cell differentiation, Cell Stem Cell 8 (2011) 326–334, https://doi.org/10.1016/j.stem.2011.01.001.

[155] L. Vallier, M. Alexander, R.A. Pedersen, Activin/nodal and FGF pathways cooperate to maintain pluripotency of human embryonic stem cells, J. Cell Sci. 118 (2005) 4495–4509, https://doi.org/10.1242/jcs.02553.

[156] R.-H. Xu, T.L. Sampsell-Barron, F. Gu, S. Root, R.M. Peck, G. Pan, J. Yu, J. Antosiewicz-Bourget, S. Tian, R. Stewart, J.A. Thomson, NANOG is a direct target of TGFbeta/activin-mediated SMAD signaling in human ESCs, Cell Stem Cell 3 (2008) 196–206, https://doi.org/10.1016/j.stem.2008.07.001.

[157] B. Greber, H. Lehrach, J. Adjaye, Control of early fate decisions in human ES cells by distinct states of TGFbeta pathway activity, Stem Cells Dev. 17 (2008) 1065–1077, https://doi.org/10.1089/scd.2008.0035.

[158] B. Greber, P. Coulon, M. Zhang, S. Moritz, S. Frank, A.J. Müller-Molina, M.J. Araúzo-Bravo, D.W. Han, H.-C. Pape, H.R. Schöler, FGF signalling inhibits neural induction in human embryonic stem cells, EMBO J. 30 (2011) 4874–4884, https://doi.org/10.1038/emboj.2011.407.

[159] H. Kiyonari, M. Kaneko, S. Abe, S. Aizawa, Three inhibitors of FGF receptor, ERK, and GSK3 establishes germline-competent embryonic stem cells of C57BL/6N mouse strain with high efficiency and stability, Genesis 48 (2010) 317–327, https://doi.org/10.1002/dvg.20614.

[160] M. Van der Jeught, T. O'Leary, S. Ghimire, S. Lierman, G. Duggal, K. Versieren, D. Deforce, S.C. de Sousa Lopes, B. Heindryckx, P. De Sutter, The combination of inhibitors of FGF/MEK/Erk and GSK3β signaling increases the number of OCT3/4- and NANOG-positive cells in the human inner cell mass, but does not improve stem cell derivation, Stem Cells Dev. 22 (2013) 296–306, https://doi.org/10.1089/scd.2012.0256.

[161] G. Duggal, S. Warrier, S. Ghimire, D. Broekaert, M. Van der Jeught, S. Lierman, T. Deroo, L. Peelman, A. Van Soom, R. Cornelissen, B. Menten, P. Mestdagh, J. Vandesompele, M. Roost, R.C. Slieker, B.T. Heijmans, D. Deforce, P. De Sutter, S.C. De Sousa Lopes, B. Heindryckx, Alternative routes to induce Naïve pluripotency in human embryonic stem cells, Stem Cells 33 (2015) 2686–2698, https://doi.org/10.1002/stem.2071.

[162] A. Harvey, G. Caretti, V. Moresi, A. Renzini, S. Adamo, Interplay between metabolites and the epigenome in regulating embryonic and adult stem cell potency and maintenance, Stem Cell Rep. 13 (2019) 573–589, https://doi.org/10.1016/j.stemcr.2019.09.003.

[163] M.C. Simon, B. Keith, The role of oxygen availability in embryonic development and stem cell function, Nat. Rev. Mol. Cell Biol. 9 (2008) 285–296, https://doi.org/10.1038/nrm2354.

[164] J.G. Lees, T.S. Cliff, A. Gammilonghi, J.G. Ryall, S. Dalton, D.K. Gardner, A.J. Harvey, Oxygen regulates human pluripotent stem cell metabolic flux, Stem Cells Int. 2019 (2019), https://doi.org/10.1155/2019/8195614, e8195614.

[165] C. Simón, A. Pellicer, R.R. Pera, Stem Cells in Reproductive Medicine: Basic Science and Therapeutic Potential, Cambridge University Press, 2013.

[166] W. Zhou, M. Choi, D. Margineantu, L. Margaretha, J. Hesson, C. Cavanaugh, C.A. Blau, M.S. Horwitz, D. Hockenbery, C. Ware, H. Ruohola-Baker, HIF1α induced switch from bivalent to exclusively glycolytic metabolism during ESC-to-EpiSC/hESC transition, EMBO J. 31 (2012) 2103–2116, https://doi.org/10.1038/emboj.2012.71.

[167] C.E. Forristal, D.R. Christensen, F.E. Chinnery, R. Petruzzelli, K.L. Parry, T. Sanchez-Elsner, F.D. Houghton, Environmental oxygen tension regulates the energy metabolism and self-renewal of human embryonic stem cells, PLoS One 8 (2013), https://doi.org/10.1371/journal.pone.0062507, e62507.

[168] T. Pfeiffer, S. Schuster, S. Bonhoeffer, Cooperation and competition in the evolution of ATP-producing pathways, Science 292 (2001) 504–507, https://doi.org/10.1126/science.1058079.

[169] C.D.L. Folmes, P.P. Dzeja, T.J. Nelson, A. Terzic, Metabolic plasticity in stem cell homeostasis and differentiation, Cell Stem Cell 11 (2012) 596–606, https://doi.org/10.1016/j.stem.2012.10.002.

[170] J.D. Ochocki, M.C. Simon, Nutrient-sensing pathways and metabolic regulation in stem cells, J. Cell Biol. 203 (2013) 23–33, https://doi.org/10.1083/jcb.201303110.

[171] J. Zhang, I. Khvorostov, J.S. Hong, Y. Oktay, L. Vergnes, E. Nuebel, P.N. Wahjudi, K. Setoguchi, G. Wang, A. Do, H.-J. Jung, J.M. McCaffery, I.J. Kurland, K. Reue, W.-N.P. Lee, C.M. Koehler, M.A. Teitell, UCP2 regulates energy metabolism and differentiation potential of human pluripotent stem cells, EMBO J. 30 (2011) 4860–4873, https://doi.org/10.1038/emboj.2011.401.

[172] S. Tohyama, J. Fujita, T. Hishiki, T. Matsuura, F. Hattori, R. Ohno, H. Kanazawa, T. Seki, K. Nakajima, Y. Kishino, M. Okada, A. Hirano, T. Kuroda, S. Yasuda, Y. Sato, S. Yuasa, M. Sano, M. Suematsu, K. Fukuda, Glutamine oxidation is indispensable for survival of human pluripotent stem cells, Cell Metab. 23 (2016) 663–674, https://doi.org/10.1016/j.cmet.2016.03.001.

[173] T.G. Fernandes, M.M. Diogo, A. Fernandes-Platzgummer, C.L. da Silva, J.M.S. Cabral, Different stages of pluripotency determine distinct patterns of proliferation, metabolism, and lineage commitment of embryonic stem cells under hypoxia, Stem Cell Res. 5 (2010) 76–89, https://doi.org/10.1016/j.scr.2010.04.003.

[174] G. Marsboom, G.-F. Zhang, N. Pohl-Avila, Y. Zhang, Y. Yuan, H. Kang, B. Hao, H. Brunengraber, A.B. Malik, J. Rehman, Glutamine metabolism regulates the pluripotency transcription factor OCT4, Cell Rep. 16 (2016) 323–332, https://doi.org/10.1016/j.celrep.2016.05.089.

[175] B.W. Carey, L.W.S. Finley, J.R. Cross, C.D. Allis, C.B. Thompson, Intracellular α-ketoglutarate maintains the pluripotency of embryonic stem cells, Nature 518 (2015) 413–416, https://doi.org/10.1038/nature13981.

[176] T. TeSlaa, A.C. Chaikovsky, I. Lipchina, S.L. Escobar, K. Hochedlinger, J. Huang, T.G. Graeber, D. Braas, M.A. Teitell, α-Ketoglutarate accelerates the initial differentiation of primed human pluripotent stem cells, Cell Metab. 24 (2016) 485–493, https://doi.org/10.1016/j.cmet.2016.07.002.

[177] A. Nieborak, R. Schneider, Metabolic intermediates—cellular messengers talking to chromatin modifiers, Mol. Metab. 14 (2018) 39–52, https://doi.org/10.1016/j.molmet.2018.01.007.

[178] J.D. Moody, S. Levy, J. Mathieu, Y. Xing, W. Kim, C. Dong, W. Tempel, A.M. Robitaille, L.T. Dang, A. Ferreccio, D. Detraux, S. Sidhu, L. Zhu, L. Carter, C. Xu, C. Valensisi, Y. Wang, R.D. Hawkins, J. Min, R.T. Moon, S.H. Orkin, D. Baker, H. Ruohola-Baker, First critical repressive H3K27me3 marks in embryonic stem cells identified using designed protein inhibitor, Proc. Natl. Acad. Sci. U. S. A. 114 (2017) 10125–10130, https://doi.org/10.1073/pnas.1706907114.

[179] Y. Shan, Z. Liang, Q. Xing, T. Zhang, B. Wang, S. Tian, W. Huang, Y. Zhang, J. Yao, Y. Zhu, K. Huang, Y. Liu, X. Wang, Q. Chen, J. Zhang, B. Shang, S. Li, X. Shi, B. Liao, C. Zhang, K. Lai, X. Zhong, X. Shu, J. Wang, H. Yao, J. Chen, D. Pei, G. Pan, PRC2 specifies ectoderm lineages and maintains pluripotency in primed but not naïve ESCs, Nat. Commun. 8 (2017), https://doi.org/10.1038/s41467-017-00668-4.

[180] N. Shiraki, S. Kume, Methionine metabolism regulates maintenance and differentiation of human ES/iPS cells, Nippon Rinsho 73 (2015) 765–772.

[181] G.D. Markham, M.A. Pajares, Structure-function relationships in methionine adenosyltransferases, Cell. Mol. Life Sci. 66 (2009) 636–648, https://doi.org/10.1007/s00018-008-8516-1.

[182] N. Shiraki, Y. Shiraki, T. Tsuyama, F. Obata, M. Miura, G. Nagae, H. Aburatani, K. Kume, F. Endo, S. Kume, Methionine metabolism regulates maintenance and differentiation of human pluripotent stem cells, Cell Metab. 19 (2014) 780–794, https://doi.org/10.1016/j.cmet.2014.03.017.

[183] H. Sperber, J. Mathieu, Y. Wang, A. Ferreccio, J. Hesson, Z. Xu, K.A. Fischer, A. Devi, D. Detraux, H. Gu, S.L. Battle, M. Showalter, C. Valensisi, J.H. Bielas, N.G. Ericson, L. Margaretha, A.M. Robitaille, D. Margineantu, O. Fiehn, D. Hockenbery, C.A. Blau, D. Raftery, A.A. Margolin, R.D. Hawkins, R.T. Moon, C.B. Ware, H. Ruohola-Baker, The metabolome regulates the epigenetic landscape during naive-to-primed human embryonic stem cell transition, Nat. Cell Biol. 17 (2015) 1523–1535, https://doi.org/10.1038/ncb3264.

[184] H. Sperber, J. Mathieu, Y. Wang, A. Ferreccio, J. Hesson, Z. Xu, K.A. Fischer, A. Devi, D. Detraux, H. Gu, S.L. Battle, M. Showalter, C. Valensisi, J.H. Bielas, N.G. Ericson, L. Margaretha, A.M. Robitaille, D. Margineantu, O. Fiehn, D. Hockenbery, C.A. Blau, D. Raftery, A. Margolin, R.D. Hawkins, R.T. Moon, C.B. Ware, H. Ruohola-Baker, The metabolome regulates the epigenetic landscape during naïve to primed human embryonic stem cell transition, Nat. Cell Biol. 17 (2015) 1523–1535, https://doi.org/10.1038/ncb3264.

[185] A. Moussaieff, M. Rouleau, D. Kitsberg, M. Cohen, G. Levy, D. Barasch, A. Nemirovski, S. Shen-Orr, I. Laevsky, M. Amit, D. Bomze, B. Elena-Herrmann, T. Scherf, M. Nissim-Rafinia, S. Kempa, J. Itskovitz-Eldor, E. Meshorer, D. Aberdam, Y. Nahmias, Glycolysis-mediated changes in acetyl-CoA and histone acetylation control the early differentiation of embryonic stem cells, Cell Metab. 21 (2015) 392–402, https://doi.org/10.1016/j.cmet.2015.02.002.

[186] S. Pauklin, L. Vallier, The cell-cycle state of stem cells determines cell fate propensity, Cell 155 (2013) 135–147, https://doi.org/10.1016/j.cell.2013.08.031.

[187] M. Tachibana, P. Amato, M. Sparman, N.M. Gutierrez, R. Tippner-Hedges, H. Ma, E. Kang, A. Fulati, H.-S. Lee, H. Sritanaudomchai, K. Masterson, J. Larson, D. Eaton, K. Sadler-Fredd, D. Battaglia, D. Lee, D. Wu, J. Jensen, P. Patton, S. Gokhale, R.L. Stouffer, D. Wolf, S. Mitalipov, Human embryonic stem cells derived by somatic cell nuclear transfer, Cell 153 (2013) 1228–1238, https://doi.org/10.1016/j.cell.2013.05.006.

[188] J. Lewandowski, M. Kurpisz, Techniques of human embryonic stem cell and induced pluripotent stem cell derivation, Arch. Immunol. Ther. Exp. 64 (2016) 349–370, https://doi.org/10.1007/s00005-016-0385-y.

[189] R.P. George, P. Lee, Embryonic human persons. Talking Point on morality and human embryo research, EMBO Rep. 10 (2009) 301–306, https://doi.org/10.1038/embor.2009.42.

[190] B. Lo, L. Parham, Ethical issues in stem cell research, Endocr. Rev. 30 (2009) 204–213, https://doi.org/10.1210/er.2008-0031.
[191] E. Russo, Follow the money—the politics of embryonic stem cell research, PLoS Biol. 3 (2005), https://doi.org/10.1371/journal.pbio.0030234.
[192] N.D. Acosta, S.H. Golub, The new federalism: state policies regarding embryonic stem cell research, J. Law Med. Ethics 44 (2016) 419–436, https://doi.org/10.1177/1073110516667939.
[193] M.L. Condic, M. Rao, Alternative sources of pluripotent stem cells: ethical and scientific issues revisited, Stem Cells Dev. 19 (2010) 1121–1129, https://doi.org/10.1089/scd.2009.0482.
[194] M. Lévesque, J.R. Kim, R. Isasi, B.M. Knoppers, A. Plomer, Y. Joly, Stem cell research funding policies and dynamic innovation: a survey of open access and commercialization requirements, Stem Cell Rev. Rep. 10 (2014) 455–471, https://doi.org/10.1007/s12015-014-9504-5.
[195] M. Geens, I. Mateizel, K. Sermon, M. De Rycke, C. Spits, G. Cauffman, P. Devroey, H. Tournaye, I. Liebaers, H. Van de Velde, Human embryonic stem cell lines derived from single blastomeres of two 4-cell stage embryos, Hum. Reprod. 24 (2009) 2709–2717, https://doi.org/10.1093/humrep/dep262.
[196] Q. Mai, Y. Yu, T. Li, L. Wang, M. Chen, S. Huang, C. Zhou, Q. Zhou, Derivation of human embryonic stem cell lines from parthenogenetic blastocysts, Cell Res. 17 (2007) 1008–1019, https://doi.org/10.1038/cr.2007.102.
[197] J.E. Lee, M.S. Kang, M.H. Park, S.H. Shim, T.K. Yoon, H.M. Chung, D.R. Lee, Evaluation of 28 human embryonic stem cell lines for use as unrelated donors in stem cell therapy: implications of HLA and ABO genotypes, Cell Transplant. 19 (2010) 1383–1395, https://doi.org/10.3727/096368910X513991.
[198] C. Eguizabal, B. Aran, S.M.C. de Sousa Lopes, M. Geens, B. Heindryckx, S. Panula, M. Popovic, R. Vassena, A. Veiga, Two decades of embryonic stem cells: a historical overview, Hum. Reprod. Open 2019 (2019), https://doi.org/10.1093/hropen/hoy024.
[199] J. Ye, N. Bates, D. Soteriou, L. Grady, C. Edmond, A. Ross, A. Kerby, P.A. Lewis, T. Adeniyi, R. Wright, K.V. Poulton, M. Lowe, S.J. Kimber, D.R. Brison, High quality clinical grade human embryonic stem cell lines derived from fresh discarded embryos, Stem Cell Res Ther 8 (2017) 128, https://doi.org/10.1186/s13287-017-0561-y.
[200] D. Ilic, C. Ogilvie, Concise review: human embryonic stem cells-what have we done? What are we doing? Where are we going? Stem Cells 35 (2017) 17–25, https://doi.org/10.1002/stem.2450.
[201] J. Neves, P. Sousa-Victor, H. Jasper, Rejuvenating strategies for stem cell-based therapies in aging, Cell Stem Cell 20 (2017) 161–175, https://doi.org/10.1016/j.stem.2017.01.008.
[202] K. Yoshida, Y. Hirabayashi, F. Watanabe, T. Sado, T. Inoue, Caloric restriction prevents radiation-induced myeloid leukemia in C3H/HeMs mice and inversely increases incidence of tumor-free death: implications in changes in number of hemopoietic progenitor cells, Exp. Hematol. 34 (2006) 274–283, https://doi.org/10.1016/j.exphem.2005.11.016.
[203] L. Armstrong, G. Saretzki, H. Peters, I. Wappler, J. Evans, N. Hole, T. von Zglinicki, M. Lako, Overexpression of telomerase confers growth advantage, stress resistance, and enhanced differentiation of ESCs toward the hematopoietic lineage, Stem Cells 23 (2005) 516–529, https://doi.org/10.1634/stemcells.2004-0269.
[204] X. Xie, A. Hiona, A.S. Lee, F. Cao, M. Huang, Z. Li, A. Cherry, X. Pei, J.C. Wu, Effects of long-term culture on human embryonic stem cell aging, Stem Cells Dev. 20 (2011) 127–138, https://doi.org/10.1089/scd.2009.0475.
[205] A.R. Mendelsohn, J.W. Larrick, Aging stem cells lose the capability to distribute damaged proteins asymmetrically, Rejuvenation Res. 18 (2015) 581–584, https://doi.org/10.1089/rej.2015.1800.
[206] D.B. Lombard, K.F. Chua, R. Mostoslavsky, S. Franco, M. Gostissa, F.W. Alt, DNA repair, genome stability, and aging, Cell 120 (2005) 497–512, https://doi.org/10.1016/j.cell.2005.01.028.
[207] F. Blokzijl, J. de Ligt, M. Jager, V. Sasselli, S. Roerink, N. Sasaki, M. Huch, S. Boymans, E. Kuijk, P. Prins, I.J. Nijman, I. Martincorena, M. Mokry, C.L. Wiegerinck, S. Middendorp, T. Sato, G. Schwank, E.E.S. Nieuwenhuis, M.M.A. Verstegen, L.J.W. van der Laan, J. de Jonge, J.N.M. IJzermans, R.G. Vries, M. van de Wetering, M.R. Stratton, H. Clevers, E. Cuppen, R. van Boxtel, Tissue-specific mutation accumulation in human adult stem cells during life, Nature 538 (2016) 260–264, https://doi.org/10.1038/nature19768.
[208] M.G. Kapetanaki, A.L. Mora, M. Rojas, Influence of age on wound healing and fibrosis, J. Pathol. 229 (2013) 310–322, https://doi.org/10.1002/path.4122.
[209] A. Hermann, C. List, H.-J. Habisch, V. Vukicevic, M. Ehrhart-Bornstein, R. Brenner, P. Bernstein, S. Fickert, A. Storch, Age-dependent neuroectodermal differentiation capacity of human mesenchymal stromal cells: limitations for autologous cell replacement strategies, Cytotherapy 12 (2010) 17–30, https://doi.org/10.3109/14653240903313941.
[210] J.-K. Roh, K.-H. Jung, K. Chu, Adult stem cell transplantation in stroke: its limitations and prospects, Curr. Stem Cell Res. Ther. 3 (2008) 185–196, https://doi.org/10.2174/157488808785740352.
[211] E.D. Mariano, M.J. Teixeira, S.K.N. Marie, G. Lepski, Adult stem cells in neural repair: current options, limitations and perspectives, World J. Stem Cells 7 (2015) 477–482, https://doi.org/10.4252/wjsc.v7.i2.477.
[212] A. Musiał-Wysocka, M. Kot, M. Majka, The pros and cons of mesenchymal stem cell-based therapies, Cell Transplant. 28 (2019) 801–812, https://doi.org/10.1177/0963689719837897.
[213] L. Fontana, L. Partridge, V.D. Longo, Extending healthy life span—from yeast to humans, Science 328 (2010) 321–326, https://doi.org/10.1126/science.1172539.
[214] K. Siddle, Signalling by insulin and IGF receptors: supporting acts and new players, J. Mol. Endocrinol. 47 (2011) R1–10, https://doi.org/10.1530/JME-11-0022.
[215] F. Zanella, W. Link, A. Carnero, Understanding FOXO, new views on old transcription factors, Curr. Cancer Drug Targets 10 (2010) 135–146, https://doi.org/10.2174/156800910791054158.
[216] A. Yilmaz, M. Peretz, A. Aharony, I. Sagi, N. Benvenisty, Defining essential genes for human pluripotent stem cells by CRISPR-Cas9 screening in haploid cells, Nat. Cell Biol. 20 (2018) 610–619, https://doi.org/10.1038/s41556-018-0088-1.

[217] H. Sun, R. Lesche, D.M. Li, J. Liliental, H. Zhang, J. Gao, N. Gavrilova, B. Mueller, X. Liu, H. Wu, PTEN modulates cell cycle progression and cell survival by regulating phosphatidylinositol 3,4,5,-trisphosphate and Akt/protein kinase B signaling pathway, Proc. Natl. Acad. Sci. U. S. A. 96 (1999). https://www.ncbi.nlm.nih.gov/pubmed/10339565. (Accessed 21 April 2020).

[218] A. Ortega-Molina, M. Serrano, PTEN in cancer, metabolism, and aging, Trends Endocrinol. Metab. 24 (2013) 184–189, https://doi.org/10.1016/j.tem.2012.11.002.

[219] I.S. Tait, Y. Li, J. Lu, PTEN, longevity and age-related diseases, Biomedicine 1 (2013) 17–48, https://doi.org/10.3390/biomedicines1010017.

[220] W. Wang, G. Lu, X. Su, C. Tang, H. Li, Z. Xiong, C.-K. Leung, M.-S. Wong, H. Liu, J.-L. Ma, H.-H. Cheung, H.-F. Kung, Z.-J. Chen, W.-Y. Chan, Pten-mediated Gsk3β modulates the naïve pluripotency maintenance in embryonic stem cells, Cell Death Dis. 11 (2020) 1–15, https://doi.org/10.1038/s41419-020-2271-0.

[221] J.A. Alva, G.E. Lee, E.E. Escobar, A.D. Pyle, Phosphatase and tensin homolog regulates the pluripotent state and lineage fate choice in human embryonic stem cells, Stem Cells 29 (2011). https://www.ncbi.nlm.nih.gov/pubmed/21948699. (Accessed 21 April 2020).

[222] J.E. Lee, M.S. Lim, J.H. Park, C.H. Park, H.C. Koh, PTEN promotes dopaminergic neuronal differentiation through regulation of ERK-dependent inhibition of S6K Signaling in human neural stem cells, Stem Cells Transl. Med. 5 (2016) 1319–1329, https://doi.org/10.5966/sctm.2015-0200.

[223] T. Jin, D. Li, T. Yang, F. Liu, J. Kong, Y. Zhou, PTPN1 promotes the progression of glioma by activating the MAPK/ERK and PI3K/AKT pathways and is associated with poor patient survival, Oncol. Rep. 42 (2019) 717–725, https://doi.org/10.3892/or.2019.7180.

[224] N. Elbourkadi, S.N. Austad, R.A. Miller, Fibroblasts from long-lived species of mammals and birds show delayed, but prolonged, phosphorylation of ERK, Aging Cell 13 (2014) 283–291, https://doi.org/10.1111/acel.12172.

[225] S.E. Wamaitha, K.J. Grybel, G. Alanis-Lobato, C. Gerri, S. Ogushi, A. McCarthy, S.K. Mahadevaiah, L. Healy, R.A. Lea, M. Molina-Arcas, L.G. Devito, K. Elder, P. Snell, L. Christie, J. Downward, J.M.A. Turner, K.K. Niakan, IGF1-mediated human embryonic stem cell self-renewal recapitulates the embryonic niche, Nat. Commun. 11 (2020), https://doi.org/10.1038/s41467-020-14629-x.

[226] S. Yamashita, K. Ogawa, T. Ikei, M. Udono, T. Fujiki, Y. Katakura, SIRT1 prevents replicative senescence of normal human umbilical cord fibroblast through potentiating the transcription of human telomerase reverse transcriptase gene, Biochem. Biophys. Res. Commun. 417 (2012) 630–634, https://doi.org/10.1016/j.bbrc.2011.12.021.

[227] T. Anwar, S. Khosla, G. Ramakrishna, Increased expression of SIRT2 is a novel marker of cellular senescence and is dependent on wild type p53 status, Cell Cycle 15 (2016) 1883–1897, https://doi.org/10.1080/15384101.2016.1189041.

[228] C. Cantó, L.Q. Jiang, A.S. Deshmukh, C. Mataki, A. Coste, M. Lagouge, J.R. Zierath, J. Auwerx, Interdependence of AMPK and SIRT1 for metabolic adaptation to fasting and exercise in skeletal muscle, Cell Metab. 11 (2010) 213–219, https://doi.org/10.1016/j.cmet.2010.02.006.

[229] F. Liang, S. Kume, D. Koya, SIRT1 and insulin resistance, Nat. Rev. Endocrinol. 5 (2009) 367–373, https://doi.org/10.1038/nrendo.2009.101.

[230] V. Gambino, G. De Michele, O. Venezia, P. Migliaccio, V. Dall'Olio, L. Bernard, S.P. Minardi, M.A. Della Fazia, D. Bartoli, G. Servillo, M. Alcalay, L. Luzi, M. Giorgio, H. Scrable, P.G. Pelicci, E. Migliaccio, Oxidative stress activates a specific p53 transcriptional response that regulates cellular senescence and aging, Aging Cell 12 (2013) 435–445, https://doi.org/10.1111/acel.12060.

[231] J. Jang, Y.J. Huh, H.-J. Cho, B. Lee, J. Park, D.-Y. Hwang, D.-W. Kim, SIRT1 enhances the survival of human embryonic stem cells by promoting DNA repair, Stem Cell Rep. 9 (2017) 629–641, https://doi.org/10.1016/j.stemcr.2017.06.001.

[232] D.S. Yoon, D.S. Cha, Y. Choi, J.W. Lee, M. Lee, MPK-1/ERK is required for the full activity of resveratrol in extended lifespan and reproduction, Aging Cell 18 (2019), https://doi.org/10.1111/acel.12867.

[233] Z. Safaeinejad, M. Nabiuni, M. Peymani, K. Ghaedi, M.H. Nasr-Esfahani, H. Baharvand, Resveratrol promotes human embryonic stem cells self-renewal by targeting SIRT1-ERK signaling pathway, Eur. J. Cell Biol. 96 (2017) 665–672, https://doi.org/10.1016/j.ejcb.2017.08.002.

[234] C.R. Balistreri, R. Madonna, G. Melino, C. Caruso, The emerging role of notch pathway in ageing: focus on the related mechanisms in age-related diseases, Ageing Res. Rev. 29 (2016) 50–65, https://doi.org/10.1016/j.arr.2016.06.004.

[235] Y. Guo, C. Xu, A.W.C. Man, B. Bai, C. Luo, Y. Huang, A. Xu, P.M. Vanhoutte, Y. Wang, Endothelial SIRT1 prevents age-induced impairment of vasodilator responses by enhancing the expression and activity of soluble guanylyl cyclase in smooth muscle cells, Cardiovasc. Res. 115 (2019) 678–690, https://doi.org/10.1093/cvr/cvy212.

[236] S.A. Noggle, D. Weiler, B.G. Condie, Notch signaling is inactive but inducible in human embryonic stem cells, Stem Cells 24 (2006) 1646–1653, https://doi.org/10.1634/stemcells.2005-0314.

[237] X. Yu, J. Zou, Z. Ye, H. Hammond, G. Chen, A. Tokunaga, P. Mali, Y.-M. Li, C. Civin, N. Gaiano, L. Cheng, Notch signaling activation in human embryonic stem cells is required for embryonic, but not trophoblastic, lineage commitment, Cell Stem Cell 2 (2008) 461–471, https://doi.org/10.1016/j.stem.2008.03.001.

[238] S.M. Prasad, M. Czepiel, C. Cetinkaya, K. Smigielska, S.C. Weli, H. Lysdahl, A. Gabrielsen, K. Petersen, N. Ehlers, T. Fink, S.L. Minger, V. Zachar, Continuous hypoxic culturing maintains activation of notch and allows long-term propagation of human embryonic stem cells without spontaneous differentiation, Cell Prolif. 42 (2008) 63–74, https://doi.org/10.1111/j.1365-2184.2008.00571.x.

[239] W.E. Wright, J.W. Shay, Telomere biology in aging and cancer, J. Am. Geriatr. Soc. 53 (2005) S292–S294, https://doi.org/10.1111/j.1532-5415.2005.53492.x.

[240] A.S. Venteicher, E.B. Abreu, Z. Meng, K.E. McCann, R.M. Terns, T.D. Veenstra, M.P. Terns, S.E. Artandi, A human telomerase holoenzyme protein required for Cajal body localization and telomere synthesis, Science 323 (2009) 644–648, https://doi.org/10.1126/science.1165357.

[241] B. Choudhary, A.A. Karande, S.C. Raghavan, Telomere and telomerase in stem cells: relevance in ageing and disease, Front. Biosci. (Schol. Ed.) 4 (2012) 16–30, https://doi.org/10.2741/248.

[242] A. Shervington, C. Lu, R. Patel, L. Shervington, Telomerase downregulation in cancer brain stem cell, Mol. Cell. Biochem. 331 (2009) 153–159, https://doi.org/10.1007/s11010-009-0153-y.

[243] R.C. Allsopp, E. Chang, M. Kashefi-Aazam, E.I. Rogaev, M.A. Piatyszek, J.W. Shay, C.B. Harley, Telomere shortening is associated with cell division in vitro and in vivo, Exp. Cell Res. 220 (1995) 194–200, https://doi.org/10.1006/excr.1995.1306.

[244] J. Huang, F. Wang, M. Okuka, N. Liu, G. Ji, X. Ye, B. Zuo, M. Li, P. Liang, W.W. Ge, J.C. Tsibris, D.L. Keefe, L. Liu, Association of telomere length with authentic pluripotency of ES/iPS cells, Cell Res. 21 (2011) 779–792, https://doi.org/10.1038/cr.2011.16.

[245] C. Yang, S. Przyborski, M.J. Cooke, X. Zhang, R. Stewart, G. Anyfantis, S.P. Atkinson, G. Saretzki, L. Armstrong, M. Lako, A key role for telomerase reverse transcriptase unit in modulating human embryonic stem cell proliferation, cell cycle dynamics, and in vitro differentiation, Stem Cells 26 (2008) 850–863, https://doi.org/10.1634/stemcells.2007-0677.

[246] L. Radan, C.S. Hughes, J.H. Teichroeb, F.M. Vieira Zamora, M. Jewer, L.-M. Postovit, D.H. Betts, Microenvironmental regulation of telomerase isoforms in human embryonic stem cells, Stem Cells Dev. 23 (2014) 2046–2066, https://doi.org/10.1089/scd.2013.0373.

[247] Y. Zou, H.J. Tong, M. Li, K.S. Tan, T. Cao, Telomere length is regulated by FGF-2 in human embryonic stem cells and affects the life span of its differentiated progenies, Biogerontology 18 (2017) 69–84, https://doi.org/10.1007/s10522-016-9662-8.

[248] M.-H. Hsieh, Y.-T. Chen, Y.-T. Chen, Y.-H. Lee, J. Lu, C.-L. Chien, H.-F. Chen, H.-N. Ho, C.-J. Yu, Z.-Q. Wang, S.-C. Teng, PARP1 controls KLF4-mediated telomerase expression in stem cells and cancer cells, Nucleic Acids Res. 45 (2017) 10492–10503, https://doi.org/10.1093/nar/gkx683.

[249] M. Stros, T. Ozaki, A. Bacikova, H. Kageyama, A. Nakagawara, HMGB1 and HMGB2 cell-specifically down-regulate the p53- and p73-dependent sequence-specific transactivation from the human Bax gene promoter, J. Biol. Chem. 277 (2002) 7157–7164, https://doi.org/10.1074/jbc.M110233200.

[250] D. Tang, R. Kang, H.J. Zeh, M.T. Lotze, High-mobility group box 1, oxidative stress, and disease, Antioxid. Redox Signal. 14 (2011) 1315–1335, https://doi.org/10.1089/ars.2010.3356.

[251] A. Zirkel, M. Nikolic, K. Sofiadis, J.-P. Mallm, C.A. Brackley, H. Gothe, O. Drechsel, C. Becker, J. Altmüller, N. Josipovic, T. Georgomanolis, L. Brant, J. Franzen, M. Koker, E.G. Gusmao, I.G. Costa, R.T. Ullrich, W. Wagner, V. Roukos, P. Nürnberg, D. Marenduzzo, K. Rippe, A. Papantonis, HMGB2 loss upon senescence entry disrupts genomic organization and induces CTCF clustering across cell types, Mol. Cell 70 (2018) 730–744. e6, https://doi.org/10.1016/j.molcel.2018.03.030.

[252] M. Kučírek, A.J. Bagherpoor, J. Jaroš, A. Hampl, M. Štros, HMGB2 is a negative regulator of telomerase activity in human embryonic stem and progenitor cells, FASEB J. 33 (2019) 14307–14324, https://doi.org/10.1096/fj.201901465RRR.

[253] S. Pervaiz, R. Taneja, S. Ghaffari, Oxidative stress regulation of stem and progenitor cells, Antioxid. Redox Signal. 11 (2009) 2777–2789, https://doi.org/10.1089/ars.2009.2804.

[254] L. Armstrong, K. Tilgner, G. Saretzki, S.P. Atkinson, M. Stojkovic, R. Moreno, S. Przyborski, M. Lako, Human induced pluripotent stem cell lines show stress defense mechanisms and mitochondrial regulation similar to those of human embryonic stem cells, Stem Cells 28 (2010) 661–673, https://doi.org/10.1002/stem.307.

[255] J.C.S. John, A. Amaral, E. Bowles, J.F. Oliveira, R. Lloyd, M. Freitas, H.L. Gray, C.S. Navara, G. Oliveira, G.P. Schatten, E. Spikings, J. Ramalho-Santos, The analysis of mitochondria and mitochondrial DNA in human embryonic stem cells, Methods Mol. Biol. 331 (2006) 347–374, https://doi.org/10.1385/1-59745-046-4:347.

[256] G. Saretzki, T. Walter, S. Atkinson, J.F. Passos, B. Bareth, W.N. Keith, R. Stewart, S. Hoare, M. Stojkovic, L. Armstrong, T. von Zglinicki, M. Lako, Downregulation of multiple stress defense mechanisms during differentiation of human embryonic stem cells, Stem Cells 26 (2008) 455–464, https://doi.org/10.1634/stemcells.2007-0628.

[257] B. Dannenmann, S. Lehle, D.G. Hildebrand, A. Kübler, P. Grondona, V. Schmid, K. Holzer, M. Fröschl, F. Essmann, O. Rothfuss, K. Schulze-Osthoff, High glutathione and glutathione Peroxidase-2 levels mediate cell-type-specific DNA damage protection in human induced pluripotent stem cells, Stem Cell Rep. 4 (2015) 886–898, https://doi.org/10.1016/j.stemcr.2015.04.004.

[258] C.-K. Wang, S.-C. Yang, S.-C. Hsu, F.-P. Chang, Y.-T. Lin, S.-F. Chen, C.-L. Cheng, M. Hsiao, F.L. Lu, J. Lu, CHAC2 is essential for self-renewal and glutathione maintenance in human embryonic stem cells, Free Radic. Biol. Med. 113 (2017) 439–451, https://doi.org/10.1016/j.freeradbiomed.2017.10.345.

[259] L.L. Alekseenko, V.I. Zemelko, V.V. Zenin, N.A. Pugovkina, I.V. Kozhukharova, Z.V. Kovaleva, T.M. Grinchuk, I.I. Fridlyanskaya, N.N. Nikolsky, Heat shock induces apoptosis in human embryonic stem cells but a premature senescence phenotype in their differentiated progeny, Cell Cycle 11 (2012) 3260–3269, https://doi.org/10.4161/cc.21595.

[260] B. Mair, J. Tomic, S.N. Masud, P. Tonge, A. Weiss, M. Usaj, A.H.Y. Tong, J.J. Kwan, K.R. Brown, E. Titus, M. Atkins, K.S.K. Chan, L. Munsie, A. Habsid, H. Han, M. Kennedy, B. Cohen, G. Keller, J. Moffat, Essential gene profiles for human pluripotent stem cells identify uncharacterized genes and substrate dependencies, Cell Rep. 27 (2019) 599–615.e12, https://doi.org/10.1016/j.celrep.2019.02.041.

[261] C.T.D. Kwok, M.H. Leung, J. Qin, Y. Qin, J. Wang, Y.L. Lee, K.-M. Yao, The Forkhead box transcription factor FOXM1 is required for the maintenance of cell proliferation and protection against oxidative stress in human embryonic stem cells, Stem Cell Res. 16 (2016) 651–661, https://doi.org/10.1016/j.scr.2016.03.007.

[262] W.T. Chiu, R.C. Le, I.L. Blitz, M.B. Fish, Y. Li, J. Biesinger, X. Xie, K.W.Y. Cho, Genome-wide view of TGFβ/Foxh1 regulation of the early mesendoderm program, Development 141 (2014) 4537–4547, https://doi.org/10.1242/dev.107227.

[263] J.J. Tsai, J.A. Dudakov, K. Takahashi, J.-H. Shieh, E. Velardi, A.M. Holland, N.V. Singer, M.L. West, O.M. Smith, L.F. Young, Y. Shono, A. Ghosh, A.M. Hanash, H.T. Tran, M.A.S. Moore, M.R.M. van den Brink, Nrf2 regulates haematopoietic stem cell function, Nat. Cell Biol. 15 (2013) 309–316, https://doi.org/10.1038/ncb2699.

[264] J. Jang, Y. Wang, H.-S. Kim, M.A. Lalli, K.S. Kosik, Nrf2, a regulator of the proteasome, controls self-renewal and pluripotency in human embryonic stem cells, Stem Cells 32 (2014) 2616–2625, https://doi.org/10.1002/stem.1764.

[265] Y. Ichimura, S. Waguri, Y.-S. Sou, S. Kageyama, J. Hasegawa, R. Ishimura, T. Saito, Y. Yang, T. Kouno, T. Fukutomi, T. Hoshii, A. Hirao, K. Takagi, T. Mizushima, H. Motohashi, M.-S. Lee, T. Yoshimori, K. Tanaka, M. Yamamoto, M. Komatsu, Phosphorylation of p62 activates the Keap1-Nrf2 pathway during selective autophagy, Mol. Cell 51 (2013) 618–631, https://doi.org/10.1016/j.molcel.2013.08.003.

[266] J. Jang, Y. Wang, M.A. Lalli, E. Guzman, S.E. Godshalk, H. Zhou, K.S. Kosik, Primary cilium-autophagy-Nrf2 (PAN) Axis activation commits human embryonic stem cells to a neuroectoderm fate, Cell 165 (2016) 410–420, https://doi.org/10.1016/j.cell.2016.02.014.

[267] J. Hou, Z.-P. Han, Y.-Y. Jing, X. Yang, S.-S. Zhang, K. Sun, C. Hao, Y. Meng, F.-H. Yu, X.-Q. Liu, Y.-F. Shi, M.-C. Wu, L. Zhang, L.-X. Wei, Autophagy prevents irradiation injury and maintains stemness through decreasing ROS generation in mesenchymal stem cells, Cell Death Dis. 4 (2013), https://doi.org/10.1038/cddis.2013.338, e844.

[268] T.-C. Chang, M.-F. Hsu, K.K. Wu, High glucose induces bone marrow-derived mesenchymal stem cell senescence by upregulating autophagy, PLoS One 10 (2015), https://doi.org/10.1371/journal.pone.0126537.

[269] B. Das, R. Bayat-Mokhtari, M. Tsui, S. Lotfi, R. Tsuchida, D.W. Felsher, H. Yeger, HIF-2α suppresses p53 to enhance the Stemness and regenerative potential of human embryonic stem cells, Stem Cells 30 (2012) 1685–1695, https://doi.org/10.1002/stem.1142.

[270] L. Jin, J. Ni, Y. Tao, X. Weng, Y. Zhu, J. Yan, B. Hu, N-acetylcysteine attenuates PM2.5-induced apoptosis by ROS-mediated Nrf2 pathway in human embryonic stem cells, Sci. Total Environ. 666 (2019) 713–720, https://doi.org/10.1016/j.scitotenv.2019.02.307.

[271] Z. Erdei, B. Sarkadi, A. Brózik, K. Szebényi, G. Várady, V. Makó, A. Péntek, T.I. Orbán, Á. Apáti, Dynamic ABCG2 expression in human embryonic stem cells provides the basis for stress response, Eur. Biophys. J. 42 (2013) 169–179, https://doi.org/10.1007/s00249-012-0838-0.

[272] R. Ding, S. Jin, K. Pabon, K.W. Scotto, A role for ABCG2 beyond drug transport: regulation of autophagy, Autophagy 12 (2016) 737–751, https://doi.org/10.1080/15548627.2016.1155009.

[273] A.K. Dunker, I. Silman, V.N. Uversky, J.L. Sussman, Function and structure of inherently disordered proteins, Curr. Opin. Struct. Biol. 18 (2008) 756–764, https://doi.org/10.1016/j.sbi.2008.10.002.

[274] C.L. Klaips, G.G. Jayaraj, F.U. Hartl, Pathways of cellular proteostasis in aging and disease, J. Cell Biol. 217 (2018) 51–63, https://doi.org/10.1083/jcb.201709072.

[275] C.F.L. de Fernandes, R.P. Iglesia, M.I. Melo-Escobar, M.B. Prado, M.H. Lopes, Chaperones and beyond as key players in pluripotency maintenance, Front. Cell Dev. Biol. 7 (2019), https://doi.org/10.3389/fcell.2019.00150.

[276] A.E. Webb, A. Brunet, FOXO transcription factors: key regulators of cellular quality control, Trends Biochem. Sci. 39 (2014) 159–169, https://doi.org/10.1016/j.tibs.2014.02.003.

[277] X. Zhang, S. Yalcin, D.-F. Lee, T.-Y.J. Yeh, S.-M. Lee, J. Su, S.K. Mungamuri, P. Rimmelé, M. Kennedy, R. Sellers, M. Landthaler, T. Tuschl, N.-W. Chi, I. Lemischka, G. Keller, S. Ghaffari, FOXO1 is an essential regulator of pluripotency in human embryonic stem cells, Nat. Cell Biol. 13 (2011) 1092–1099, https://doi.org/10.1038/ncb2293.

[278] P.-C. Tseng, S.-M. Hou, R.-J. Chen, H.-W. Peng, C.-F. Hsieh, M.-L. Kuo, M.-L. Yen, Resveratrol promotes osteogenesis of human mesenchymal stem cells by upregulating RUNX2 gene expression via the SIRT1/FOXO3A axis, J. Bone Miner. Res. 26 (2011) 2552–2563, https://doi.org/10.1002/jbmr.460.

[279] D. Vilchez, L. Boyer, I. Morantte, M. Lutz, C. Merkwirth, D. Joyce, B. Spencer, L. Page, E. Masliah, W.T. Berggren, F.H. Gage, A. Dillin, Increased proteasome activity in human embryonic stem cells is regulated by PSMD11, Nature 489 (2012) 304–308, https://doi.org/10.1038/nature11468.

[280] D. Vilchez, L. Boyer, M. Lutz, C. Merkwirth, I. Morantte, C. Tse, B. Spencer, L. Page, E. Masliah, W.T. Berggren, F.H. Gage, A. Dillin, FOXO4 is necessary for neural differentiation of human embryonic stem cells, Aging Cell 12 (2013) 518–522, https://doi.org/10.1111/acel.12067.

[281] N.-H. Seo, E.-H. Lee, J.-H. Seo, H.-R. Song, M.-K. Han, HSP60 is required for stemness and proper differentiation of mouse embryonic stem cells, Exp. Mol. Med. 50 (2018), https://doi.org/10.1038/emm.2017.299, e459.

[282] E. Prinsloo, M.M. Setati, V.M. Longshaw, G.L. Blatch, Chaperoning stem cells: a role for heat shock proteins in the modulation of stem cell self-renewal and differentiation? BioEssays 31 (2009) 370–377, https://doi.org/10.1002/bies.200800158.

[283] J.-A. Park, Y.-E. Kim, H.-J. Seok, W.-Y. Park, H.-J. Kwon, Y. Lee, Differentiation and upregulation of heat shock protein 70 induced by a subset of histone deacetylase inhibitors in mouse and human embryonic stem cells, BMB Rep. 44 (2011) 176–181, https://doi.org/10.5483/BMBRep.2011.44.3.176.

[284] R. Kiffin, C. Christian, E. Knecht, A.M. Cuervo, Activation of chaperone-mediated autophagy during oxidative stress, Mol. Biol. Cell 15 (2004) 4829–4840, https://doi.org/10.1091/mbc.e04-06-0477.

[285] K. Byun, T.-K. Kim, J. Oh, E. Bayarsaikhan, D. Kim, M.Y. Lee, C.-G. Pack, D. Hwang, B. Lee, Heat shock instructs hESCs to exit from the self-renewal program through negative regulation of OCT4 by SAPK/JNK and HSF1 pathway, Stem Cell Res. 11 (2013) 1323–1334, https://doi.org/10.1016/j.scr.2013.08.014.

[286] E. Bradley, E. Bieberich, N.F. Mivechi, D. Tangpisuthipongsa, G. Wang, Regulation of embryonic stem cell pluripotency by heat shock protein 90, Stem Cells 30 (2012) 1624–1633, https://doi.org/10.1002/stem.1143.

[287] D. Vilchez, M.S. Simic, A. Dillin, Proteostasis and aging of stem cells, Trends Cell Biol. 24 (2014) 161–170, https://doi.org/10.1016/j.tcb.2013.09.002.

[288] A. Noormohammadi, A. Khodakarami, R. Gutierrez-Garcia, H.J. Lee, S. Koyuncu, T. König, C. Schindler, I. Saez, A. Fatima, C. Dieterich, D. Vilchez, Somatic increase of CCT8 mimics proteostasis of human pluripotent stem cells and extends C. elegans lifespan, Nat. Commun. 7 (2016) 13649, https://doi.org/10.1038/ncomms13649.

[289] M. Tatar, J.M. Sedivy, Mitochondria: masters of epigenetics, Cell 165 (2016) 1052–1054, https://doi.org/10.1016/j.cell.2016.05.021.

[290] C.A. Gifford, M.J. Ziller, H. Gu, C. Trapnell, J. Donaghey, A. Tsankov, A.K. Shalek, D.R. Kelley, A.A. Shishkin, R. Issner, X. Zhang, M. Coyne, J.L. Fostel, L. Holmes, J. Meldrim, M. Guttman, C. Epstein, H. Park, O. Kohlbacher, J. Rinn, A. Gnirke, E.S. Lander, B.E. Bernstein, A. Meissner, Transcriptional and epigenetic dynamics during specification of human embryonic stem cells, Cell 153 (2013) 1149–1163, https://doi.org/10.1016/j.cell.2013.04.037.

[291] S. Pal, J.K. Tyler, Epigenetics and aging, Sci. Adv. 2 (2016), https://doi.org/10.1126/sciadv.1600584.
[292] C. Lu, C.B. Thompson, Metabolic regulation of epigenetics, Cell Metab. 16 (2012) 9–17, https://doi.org/10.1016/j.cmet.2012.06.001.
[293] M.G. Guenther, G.M. Frampton, F. Soldner, D. Hockemeyer, M. Mitalipova, R. Jaenisch, R.A. Young, Chromatin structure and gene expression programs of human embryonic and induced pluripotent stem cells, Cell Stem Cell 7 (2010) 249–257, https://doi.org/10.1016/j.stem.2010.06.015.
[294] Y. Zhang, J. Wang, G. Chen, D. Fan, M. Deng, Inhibition of Sirt1 promotes neural progenitors toward motoneuron differentiation from human embryonic stem cells, Biochem. Biophys. Res. Commun. 404 (2011) 610–614, https://doi.org/10.1016/j.bbrc.2010.12.014.
[295] C.A. Smith, P. Humphreys, N. Bates, M. Naven, S. Cain, M. Dvir-Ginzberg, S.J. Kimber, SIRT1 activity orchestrates ECM expression during hESC-chondrogenic differentiation through SOX5 and ARID5B, BioRxiv (2020), https://doi.org/10.1101/2020.05.12.087957.
[296] L.C. Fuentealba, E. Eivers, D. Geissert, V. Taelman, E.M. De Robertis, Asymmetric mitosis: unequal segregation of proteins destined for degradation, Proc. Natl. Acad. Sci. U. S. A. 105 (2008) 7732–7737, https://doi.org/10.1073/pnas.0803027105.
[297] N. Shyh-Chang, Y. Zheng, J.W. Locasale, L.C. Cantley, Human pluripotent stem cells decouple respiration from energy production, EMBO J. 30 (2011) 4851–4852, https://doi.org/10.1038/emboj.2011.436.
[298] C.D. Folmes, H. Ma, S. Mitalipov, A. Terzic, Mitochondria in pluripotent stem cells: stemness regulators and disease targets, Curr. Opin. Genet. Dev. 38 (2016) 1–7, https://doi.org/10.1016/j.gde.2016.02.001.
[299] S. Varum, A.S. Rodrigues, M.B. Moura, O. Momcilovic, C.A. Easley, J. Ramalho-Santos, B. Van Houten, G. Schatten, Energy metabolism in human pluripotent stem cells and their differentiated counterparts, PLoS One 6 (2011), https://doi.org/10.1371/journal.pone.0020914, e20914.
[300] J. Mathieu, H. Ruohola-Baker, Metabolic remodeling during the loss and acquisition of pluripotency, Development 144 (2017) 541–551, https://doi.org/10.1242/dev.128389.
[301] J.G. Lees, D.K. Gardner, A.J. Harvey, Pluripotent stem cell metabolism and mitochondria: beyond ATP, Stem Cells Int. 2017 (2017), https://doi.org/10.1155/2017/2874283, e2874283.
[302] J. Zhang, S. Ratanasirintrawoot, S. Chandrasekaran, Z. Wu, S.B. Ficarro, C. Yu, C.A. Ross, D. Cacchiarelli, Q. Xia, M. Seligson, G. Shinoda, W. Xie, P. Cahan, L. Wang, S.-C. Ng, S. Tintara, C. Trapnell, T. Onder, Y.-H. Loh, T. Mikkelsen, P. Sliz, M.A. Teitell, J.M. Asara, J.A. Marto, H. Li, J.J. Collins, G.Q. Daley, LIN28 regulates stem cell metabolism and conversion to primed pluripotency, Cell Stem Cell 19 (2016) 66–80, https://doi.org/10.1016/j.stem.2016.05.009.
[303] C. Vallot, C. Patrat, A.J. Collier, C. Huret, M. Casanova, T.M. Liyakat Ali, M. Tosolini, N. Frydman, E. Heard, P.J. Rugg-Gunn, C. Rougeulle, XACT noncoding RNA competes with XIST in the control of X chromosome activity during human early development, Cell Stem Cell 20 (2017) 102–111, https://doi.org/10.1016/j.stem.2016.10.014.
[304] S. Patel, G. Bonora, A. Sahakyan, R. Kim, C. Chronis, J. Langerman, S. Fitz-Gibbon, L. Rubbi, R.J.P. Skelton, R. Ardehali, M. Pellegrini, W.E. Lowry, A.T. Clark, K. Plath, Human embryonic stem cells Do not change their X inactivation status during differentiation, Cell Rep. 18 (2017) 54–67, https://doi.org/10.1016/j.celrep.2016.11.054.
[305] T.W. Theunissen, M. Friedli, Y. He, E. Planet, R.C. O'Neil, S. Markoulaki, J. Pontis, H. Wang, A. Iouranova, M. Imbeault, J. Duc, M.A. Cohen, K.J. Wert, R. Castanon, Z. Zhang, Y. Huang, J.R. Nery, J. Drotar, T. Lungjangwa, D. Trono, J.R. Ecker, R. Jaenisch, Molecular criteria for defining the naive human pluripotent state, Cell Stem Cell 19 (2016) 502–515, https://doi.org/10.1016/j.stem.2016.06.011.
[306] D.R. Christensen, P.C. Calder, F.D. Houghton, Effect of oxygen tension on the amino acid utilisation of human embryonic stem cells, Cell. Physiol. Biochem. 33 (2014) 237–246, https://doi.org/10.1159/000356665.
[307] D.L. McClelland Descalzo, T.S. Satoorian, L.M. Walker, N.R.L. Sparks, P.Y. Pulyanina, N.I. Zur Nieden, Glucose-induced oxidative stress reduces proliferation in embryonic stem cells via FOXO3A/β-catenin-dependent transcription of p21(cip1), Stem Cell Rep. 7 (2016) 55–68, https://doi.org/10.1016/j.stemcr.2016.06.006.
[308] J. Brocato, Y. Chervona, M. Costa, Molecular responses to hypoxia-inducible factor 1α and beyond, Mol. Pharmacol. 85 (2014) 651–657, https://doi.org/10.1124/mol.113.089623.
[309] B. Mahato, P. Home, G. Rajendran, A. Paul, B. Saha, A. Ganguly, S. Ray, N. Roy, R.H. Swerdlow, S. Paul, Regulation of mitochondrial function and cellular energy metabolism by protein kinase C-λ/ι: a novel mode of balancing pluripotency, Stem Cells 32 (2014) 2880–2892, https://doi.org/10.1002/stem.1817.
[310] A. Fontán-Lozano, V. Capilla-Gonzalez, Y. Aguilera, N. Mellado, A.M. Carrión, B. Soria, A. Hmadcha, Impact of transient down-regulation of DREAM in human embryonic stem cell pluripotency: the role of DREAM in the maintenance of hESCs, Stem Cell Res. 16 (2016) 568–578, https://doi.org/10.1016/j.scr.2016.03.001.
[311] M.J. Bueno, M. Malumbres, MicroRNAs and the cell cycle, Biochim. Biophys. Acta 1812 (2011) 592–601, https://doi.org/10.1016/j.bbadis.2011.02.002.
[312] M. Li, Y. Yang, Y. Kuang, X. Gan, W. Zeng, Y. Liu, H. Guan, miR-365 induces hepatocellular carcinoma cell apoptosis through targeting Bcl-2, Exp. Ther. Med. 13 (2017) 2279–2285, https://doi.org/10.3892/etm.2017.4244.
[313] S. Peng, D. Gao, C. Gao, P. Wei, M. Niu, C. Shuai, MicroRNAs regulate signaling pathways in osteogenic differentiation of mesenchymal stem cells (review), Mol. Med. Rep. 14 (2016) 623–629, https://doi.org/10.3892/mmr.2016.5335.
[314] M. Malumbres, miRNAs and cancer: an epigenetics view, Mol. Asp. Med. 34 (2013) 863–874, https://doi.org/10.1016/j.mam.2012.06.005.
[315] U. Lakshmipathy, J. Davila, R.P. Hart, miRNA in pluripotent stem cells, Regen. Med. 5 (2010) 545–555, https://doi.org/10.2217/rme.10.34.
[316] A.K. Murashov, RNAi and MicroRNA-mediated gene regulation in stem cells, Methods Mol. Biol. 1622 (2017) 15–25, https://doi.org/10.1007/978-1-4939-7108-4_2.
[317] B.R.M. Schulman, A. Esquela-Kerscher, F.J. Slack, Reciprocal expression of lin-41 and the microRNAs let-7 and mir-125 during mouse embryogenesis, Dev. Dyn. 234 (2005) 1046–1054, https://doi.org/10.1002/dvdy.20599.

[318] D. Xu, H. Tahara, The role of exosomes and microRNAs in senescence and aging, Adv. Drug Deliv. Rev. 65 (2013) 368–375, https://doi.org/10.1016/j.addr.2012.07.010.

[319] C. Belair, S. Sim, K.-Y. Kim, Y. Tanaka, I.-H. Park, S.L. Wolin, The RNA exosome nuclease complex regulates human embryonic stem cell differentiation, J. Cell Biol. 218 (2019) 2564–2582, https://doi.org/10.1083/jcb.201811148.

[320] H.B. Houbaviy, M.F. Murray, P.A. Sharp, Embryonic stem cell-specific microRNAs, Dev. Cell 5 (2003) 351–358, https://doi.org/10.1016/s1534-5807(03)00227-2.

[321] M.-R. Suh, Y. Lee, J.Y. Kim, S.-K. Kim, S.-H. Moon, J.Y. Lee, K.-Y. Cha, H.M. Chung, H.S. Yoon, S.Y. Moon, V.N. Kim, K.-S. Kim, Human embryonic stem cells express a unique set of microRNAs, Dev. Biol. 270 (2004) 488–498, https://doi.org/10.1016/j.ydbio.2004.02.019.

[322] D.A.G. Card, P.B. Hebbar, L. Li, K.W. Trotter, Y. Komatsu, Y. Mishina, T.K. Archer, Oct4/Sox2-regulated miR-302 targets cyclin D1 in human embryonic stem cells, Mol. Cell. Biol. 28 (2008) 6426–6438, https://doi.org/10.1128/MCB.00359-08.

[323] D. Dolezalova, M. Mraz, T. Barta, K. Plevova, V. Vinarsky, Z. Holubcova, J. Jaros, P. Dvorak, S. Pospisilova, A. Hampl, MicroRNAs regulate p21(Waf1/Cip1) protein expression and the DNA damage response in human embryonic stem cells, Stem Cells 30 (2012) 1362–1372, https://doi.org/10.1002/stem.1108.

[324] A. Rosa, A.H. Brivanlou, A regulatory circuitry comprised of miR-302 and the transcription factors OCT4 and NR2F2 regulates human embryonic stem cell differentiation, EMBO J. 30 (2011) 237–248, https://doi.org/10.1038/emboj.2010.319.

[325] N. Xu, T. Papagiannakopoulos, G. Pan, J.A. Thomson, K.S. Kosik, MicroRNA-145 regulates OCT4, SOX2, and KLF4 and represses pluripotency in human embryonic stem cells, Cell 137 (2009) 647–658, https://doi.org/10.1016/j.cell.2009.02.038.

[326] J.K. Yoo, J. Kim, S.-J. Choi, H.M. Noh, Y.D. Kwon, H. Yoo, H.S. Yi, H.M. Chung, J.K. Kim, Discovery and characterization of novel microRNAs during endothelial differentiation of human embryonic stem cells, Stem Cells Dev. 21 (2012) 2049–2057, https://doi.org/10.1089/scd.2011.0500.

[327] H. Toledano, The role of the heterochronic microRNA let-7 in the progression of aging, Exp. Gerontol. 48 (2013) 667–670, https://doi.org/10.1016/j.exger.2012.08.006.

[328] C. Qiu, Y. Ma, J. Wang, S. Peng, Y. Huang, Lin28-mediated post-transcriptional regulation of Oct4 expression in human embryonic stem cells, Nucleic Acids Res. 38 (2010) 1240–1248, https://doi.org/10.1093/nar/gkp1071.

[329] F. Cimadamore, A. Amador-Arjona, C. Chen, C.-T. Huang, A.V. Terskikh, SOX2-LIN28/let-7 pathway regulates proliferation and neurogenesis in neural precursors, Proc. Natl. Acad. Sci. U. S. A. 110 (2013) E3017–E3026, https://doi.org/10.1073/pnas.1220176110.

[330] N. Rahkonen, A. Stubb, M. Malonzo, S. Edelman, M.R. Emani, E. Närvä, H. Lähdesmäki, H. Ruohola-Baker, R. Lahesmaa, R. Lund, Mature Let-7 miRNAs fine tune expression of LIN28B in pluripotent human embryonic stem cells, Stem Cell Res. 17 (2016) 498–503, https://doi.org/10.1016/j.scr.2016.09.025.

[331] I.M. de Souza Lima, J.L.S. dos Schiavinato, S.B.P. Leite, D. Sastre, H.L.O. de Bezerra, B. Sangiorgi, A.C. Corveloni, C.H. Thomé, V.M. Faça, D.T. Covas, M.A. Zago, M. Giacca, M. Mano, R.A. Panepucci, High-content screen in human pluripotent cells identifies miRNA-regulated pathways controlling pluripotency and differentiation, Stem Cell Res Ther 10 (2019), https://doi.org/10.1186/s13287-019-1318-6.

[332] Y.-U. Bae, Y. Son, C.-H. Kim, K.S. Kim, S.H. Hyun, H.G. Woo, B.A. Jee, J.-H. Choi, H.-K. Sung, H.-C. Choi, S.Y. Park, J.-H. Bae, K.-O. Doh, J.-R. Kim, Embryonic stem cell-derived mmu-miR-291a-3p inhibits cellular senescence in human dermal fibroblasts through the TGF-β receptor 2 pathway, J. Gerontol. A Biol. Sci. Med. Sci. 74 (2019) 1359–1367, https://doi.org/10.1093/gerona/gly208.

[333] X. Song, Y. Li, X. Chen, G. Yin, Q. Huang, Y. Chen, G. Xu, L. Wang, bFGF promotes adipocyte differentiation in human mesenchymal stem cells derived from embryonic stem cells, Genet. Mol. Biol. 37 (2014) 127–134, https://doi.org/10.1590/S1415-47572014000100019.

[334] D. Nawrocka, K. Kornicka, J. Szydlarska, K. Marycz, Basic fibroblast growth factor inhibits apoptosis and promotes proliferation of adipose-derived mesenchymal stromal cells isolated from patients with type 2 diabetes by reducing cellular oxidative stress, Oxidative Med. Cell. Longev. 2017 (2017) 1–22, https://doi.org/10.1155/2017/3027109.

[335] J.S.G. Yeoh, G. de Haan, Fibroblast growth factors as regulators of stem cell self-renewal and aging, Mech. Ageing Dev. 128 (2007) 17–24, https://doi.org/10.1016/j.mad.2006.11.005.

[336] J. Nakae, Y. Kido, D. Accili, Distinct and overlapping functions of insulin and IGF-I receptors, Endocr. Rev. 22 (2001) 818–835, https://doi.org/10.1210/edrv.22.6.0452.

[337] A.A. Butler, D. LeRoith, Minireview: tissue-specific versus generalized gene targeting of the igf1 and igf1r genes and their roles in insulin-like growth factor physiology, Endocrinology 142 (2001) 1685–1688, https://doi.org/10.1210/endo.142.5.8148.

[338] A.A. Akintola, D. van Heemst, Insulin, aging, and the brain: mechanisms and implications, Front. Endocrinol. 6 (2015) 1–13, https://doi.org/10.3389/fendo.2015.00013.

[339] G.D. Agrogiannis, S. Sifakis, E.S. Patsouris, A.E. Konstantinidou, Insulin-like growth factors in embryonic and fetal growth and skeletal development (review), Mol. Med. Rep. 10 (2014) 579–584, https://doi.org/10.3892/mmr.2014.2258.

[340] C. Edward, Y. Huaitao, Biologically active human IGF-1 fusion gene vectors for in vitro and in vivo gene transfer, Mol. Ther. 5 (2002) S79, https://doi.org/10.1016/S1525-0016(16)43067-4.

[341] C. Raoul, P. Aebischer, ALS, IGF-1 and gene therapy: 'it's never too late to mend, Gene Ther. 11 (2004) 429–430, https://doi.org/10.1038/sj.gt.3302204.

[342] F. Talebpour Amiri, F. Fadaei Fathabadi, M. Mahmoudi Rad, A. Piryae, A. Ghasemi, A. Khalilian, F. Yeganeh, N. Mosaffa, The effects of insulin-like growth factor-1 gene therapy and cell transplantation on rat acute wound model, Iran Red Crescent Med J 16 (2014) 1–7, https://doi.org/10.5812/ircmj.16323.

[343] P.A.J. Adam, K. Teramo, N. Raiha, D. Gitlin, R. Schwartz, Human fetal insulin metabolism early in gestation: response to acute elevation of the fetal glucose concentration and placental transfer of human insulin-I-131, Diabetes 18 (1969) 409–416, https://doi.org/10.2337/diab.18.6.409.

[344] H. Tiedemann, M. Asashima, H. Grunz, W. Knochel, Pluripotent cells (stem cells) and their determination and differentiation in early vertebrate embryogenesis+, Develop. Growth Differ. 43 (2001) 469–502, https://doi.org/10.1046/j.1440-169X.2001.00599.x.

[345] G. Ai, X. Shao, M. Meng, L. Song, J. Qiu, Y. Wu, J. Zhou, J. Cheng, X. Tong, Epidermal growth factor promotes proliferation and maintains multipotency of continuous cultured adipose stem cells via activating STAT signal pathway in vitro, Medicine 96 (2017), https://doi.org/10.1097/MD.0000000000007607, e7607.

[346] T. Bai, F. Liu, F. Zou, G. Zhao, Y. Jiang, L. Liu, J. Shi, D. Hao, Q. Zhang, T. Zheng, Y. Zhang, M. Liu, S. Li, L. Qi, J.Y. Liu, Epidermal growth factor induces proliferation of hair follicle-derived mesenchymal stem cells through epidermal growth factor receptor-mediated activation of ERK and AKT signaling pathways associated with upregulation of cyclin D1 and downregulation of p1, Stem Cells Dev. 26 (2017) 113–122, https://doi.org/10.1089/scd.2016.0234.

[347] K. Tamama, H. Kawasaki, A. Wells, Epidermal growth factor (EGF) treatment on multipotential stromal cells (MSCs). Possible enhancement of therapeutic potential of MSC, J. Biomed. Biotechnol. 2010 (2010), https://doi.org/10.1155/2010/795385.

[348] A.C. Mullen, J.L. Wrana, TGF-β family signaling in embryonic and somatic stem-cell renewal and differentiation, Cold Spring Harb. Perspect. Biol. 9 (2017) a022186, https://doi.org/10.1101/cshperspect.a022186.

[349] A.P. Hinck, Structural studies of the TGF-βs and their receptors—insights into evolution of the TGF-β superfamily, FEBS Lett. 586 (2012) 1860–1870, https://doi.org/10.1016/j.febslet.2012.05.028.

[350] K.E. Galvin-Burgess, E.D. Travis, K.E. Pierson, J.L. Vivian, TGF-β-superfamily signaling regulates embryonic stem cell heterogeneity: self-renewal as a dynamic and regulated equilibrium, Stem Cells 31 (2013) 48–58, https://doi.org/10.1002/stem.1252.

[351] K. Tominaga, H.I. Suzuki, TGF-β signaling in cellular senescence and aging-related pathology, Int. J. Mol. Sci. 20 (2019) 5002, https://doi.org/10.3390/ijms20205002.

[352] N. Oshimori, E. Fuchs, The harmonies played by TGF-β in stem cell biology, Cell Stem Cell 11 (2012) 751–764, https://doi.org/10.1016/j.stem.2012.11.001.

[353] T. Fei, Y.-G. Chen, Regulation of embryonic stem cell self-renewal and differentiation by TGF-β family signaling, Sci. China Life Sci. 53 (2010) 497–503, https://doi.org/10.1007/s11427-010-0096-2.

[354] M. Mossahebi-Mohammadi, M. Quan, J.-S. Zhang, X. Li, FGF signaling pathway: a key regulator of stem cell pluripotency, Front. Cell Dev. Biol. 8 (2020) 1–10, https://doi.org/10.3389/fcell.2020.00079.

[355] L. Paraoanu, J. Boutter, D. Landgraf, M. Barth, P. Layer, Cholinesterases and cholinergic system in embryonic stem cell regulation: data on gene expression and functions, J. Stem Cells Regen. Med. 2 (2007) 139.

[356] L.E. Paraoanu, G. Steinert, A. Koehler, I. Wessler, P.G. Layer, Expression and possible functions of the cholinergic system in a murine embryonic stem cell line, Life Sci. 80 (2007) 2375–2379, https://doi.org/10.1016/j.lfs.2007.03.008.

[357] E.M. Prager, V. Aroniadou-Anderjaska, C.P. Almeida-Suhett, T.H. Figueiredo, J.P. Apland, F. Rossetti, C.H. Olsen, M.F.M. Braga, The recovery of acetylcholinesterase activity and the progression of neuropathological and pathophysiological alterations in the rat basolateral amygdala after soman-induced status epilepticus: relation to anxiety-like behavior, Neuropharmacology 81 (2014) 64–74, https://doi.org/10.1016/j.neuropharm.2014.01.035.

[358] M. Miranda, J.F. Morici, M.B. Zanoni, P. Bekinschtein, Brain-derived neurotrophic factor: a key molecule for memory in the healthy and the pathological brain, Front. Cell. Neurosci. 13 (2019) 1–25, https://doi.org/10.3389/fncel.2019.00363.

[359] V.G. Moisiadis, S.G. Matthews, Glucocorticoids and fetal programming part 1: outcomes, Nat. Rev. Endocrinol. 10 (2014) 391–402, https://doi.org/10.1038/nrendo.2014.73.

[360] E.A. Rog-Zielinska, R.V. Richardson, M.A. Denvir, K.E. Chapman, Glucocorticoids and foetal heart maturation; implications for prematurity and foetal programming, J. Mol. Endocrinol. 52 (2014) R125–R135, https://doi.org/10.1530/JME-13-0204.

[361] E.J. Agnew, J.R. Ivy, S.J. Stock, K.E. Chapman, Glucocorticoids, antenatal corticosteroid therapy and fetal heart maturation, J. Mol. Endocrinol. 61 (2018) R61–R73, https://doi.org/10.1530/JME-18-0077.

[362] M. Sundberg, Glucocorticoid hormones decrease proliferation of embryonic neural stem cells through ubiquitin-mediated degradation of cyclin D1, J. Neurosci. 26 (2006) 5402–5410, https://doi.org/10.1523/JNEUROSCI.4906-05.2006.

[363] M. Schouten, P. Bielefeld, L. Garcia-Corzo, E.M.J. Passchier, S. Gradari, T. Jungenitz, M. Pons-Espinal, E. Gebara, S. Martín-Suárez, P.J. Lucassen, H.E. De Vries, J.L. Trejo, S.W. Schwarzacher, D. De Pietri Tonelli, N. Toni, H. Mira, J.M. Encinas, C.P. Fitzsimons, Circadian glucocorticoid oscillations preserve a population of adult hippocampal neural stem cells in the aging brain, Mol. Psychiatry 25 (2020) 1382–1405, https://doi.org/10.1038/s41380-019-0440-2.

[364] A. Derfoul, G.L. Perkins, D.J. Hall, R.S. Tuan, Glucocorticoids promote chondrogenic differentiation of adult human mesenchymal stem cells by enhancing expression of cartilage extracellular matrix genes, Stem Cells 24 (2006) 1487–1495, https://doi.org/10.1634/stemcells.2005-0415.

[365] J.L.W. Yau, J.R. Seckl, Local amplification of glucocorticoids in the aging brain and impaired spatial memory, Front. Aging Neurosci. 4 (2012) 1–15, https://doi.org/10.3389/fnagi.2012.00024.

[366] R.A. Oldershaw, M.A. Baxter, E.T. Lowe, N. Bates, L.M. Grady, F. Soncin, D.R. Brison, T.E. Hardingham, S.J. Kimber, Directed differentiation of human embryonic stem cells toward chondrocytes, Nat. Biotechnol. 28 (2010) 1187–1194, https://doi.org/10.1038/nbt.1683.

[367] C. Xiong, C. Xie, L.I. Zhang, J. Zhang, K. Xu, Derivation of adipocytes from human embryonic stem cells, Stem Cells Dev. 675 (2005) 671–675.

[368] C.L. Mummery, J. Zhang, E.S. Ng, D.A. Elliot, A.G. Elefanty, T.J. Kamp, Differentiation of human ES and iPS cells to cardiomyocytes: a methods overview, Circ. Res. 111 (2013) 344–358, https://doi.org/10.1161/CIRCRESAHA.110.227512.Differentiation.

[369] I. Batalov, A.W. Feinberg, Differentiation of cardiomyocytes from human pluripotent stem cells using monolayer culture, Biomark. Insights 10 (2015) 71–76, https://doi.org/10.4137/BMI.S20050.Received.

[370] S. Mitani, K. Takayama, Y. Nagamoto, K. Imagawa, F. Sakurai, M. Tachibana, R. Sumazaki, H. Mizuguchi, Human ESC/iPSC-derived hepatocyte-like cells achieve zone-specific hepatic properties by modulation of WNT signaling, Mol. Ther. 25 (2017) 1420–1433, https://doi.org/10.1016/j.ymthe.2017.04.006.

[371] C. Du, Y. Feng, D. Qiu, Y. Xu, M. Pang, N. Cai, A.P. Xiang, Highly efficient and expedited hepatic differentiation from human pluripotent stem cells by pure small-molecule cocktails, Stem Cell Res Ther 9 (2018) 1–15.

[372] L. Wang, P. Menendez, C. Cerdan, M. Bhatia, Hematopoietic development from human embryonic stem cell lines, Exp. Hematol. 33 (2005) 987–996, https://doi.org/10.1016/j.exphem.2005.06.002.

[373] J.H. Shim, S.E. Kim, D.H. Woo, S.K. Kim, Directed differentiation of human embryonic stem cells towards a pancreatic cell fate, Diabetologia 50 (2007) 1228–1238, https://doi.org/10.1007/s00125-007-0634-z.

[374] K. Osafune, L. Davidow, S. Chen, M. Borowiak, J.L. Fox, K. Lam, L.F. Peng, S.L. Schreiber, L.L. Rubin, D. Melton, A small molecule that directs differentiation of human ESCs into the pancreatic lineage, Nat. Chem. Biol. 5 (2009) 258–265, https://doi.org/10.1038/nchembio.154.

[375] A. Vats, N.S. Tolley, A.E. Bishop, J.M. Polak, Embryonic stem cells and tissue engineering: delivering stem cells to the clinic, J. R. Soc. Med. 98 (2005) 346–350.

[376] B.E. Van Vranken, H.M. Romanska, J.M. Polak, H.J. Rippon, J.M. Shannon, A.E. Bishop, Coculture of embryonic stem cells with pulmonary mesenchyme: a microenvironment that promotes differentiation of pulmonary epithelium, Tissue Eng. 11 (2005) 1177–1187.

[377] D. Marolt, I. Marcos, S. Bhumiratana, A. Koren, P. Petridis, G. Zhang, Engineering bone tissue from human embryonic stem cells, Proc. Natl. Acad. Sci. 109 (2012) 1–5, https://doi.org/10.1073/pnas.1201830109.

[378] T. Ben-hur, M. Idelson, H. Khaner, M. Pera, E. Reinhartz, A. Itzik, B.E. Reubinoff, Transplantation of human embryonic stem cell-derived neural progenitors improves behavioral deficit in parkinsonian rats, Stem Cells 22 (2004) 1246–1255, https://doi.org/10.1634/stemcells.2004-0094.

[379] M. Cho, D. Hwang, D. Kim, Efficient derivation of functional dopaminergic neurons from human embryonic stem cells on a large scale, Nat. Protoc. 3 (2008) 1888–1894, https://doi.org/10.1038/nprot.2008.188.

[380] J. Song, S. Lee, W. Kang, J. Park, K. Chu, S. Lee, T. Hwang, H. Chung, M. Kim, Human embryonic stem cell-derived neural precursor transplants attenuate apomorphine-induced rotational behavior in rats with unilateral quinolinic acid lesions, Neurosci. Lett. 423 (2007) 58–61, https://doi.org/10.1016/j.neulet.2007.05.066.

[381] S.L. Rossi, G. Nistor, T. Wyatt, H.Z. Yin, A.J. Poole, J.H. Weiss, M.J. Gardener, S. Dijkstra, D.F. Fischer, H.S. Keirstead, Histological and functional benefit following transplantation of motor neuron progenitors to the injured rat spinal cord, PLoS One 5 (2010) 1–15, https://doi.org/10.1371/journal.pone.0011852.

[382] D.E. Vaillancourt, K.M. Newell, Changing complexity in human behavior and physiology through aging and disease, Neurobiol. Aging 23 (2002) 1–11, https://doi.org/10.1016/s0197-4580(01)00247-0.

[383] M. Ullah, Z. Sun, Stem cells and anti-aging genes: double-edged sword—do the same job of life extension, Stem Cell Res Ther 9 (2018), https://doi.org/10.1186/s13287-017-0746-4.

[384] D.E. Cohen, D. Melton, Turning straw into gold: directing cell fate for regenerative medicine, Nat. Rev. Genet. 12 (2011) 243–252, https://doi.org/10.1038/nrg2938.

[385] W. Zakrzewski, M. Dobrzyński, M. Szymonowicz, Z. Rybak, Stem cells: past, present, and future, Stem Cell Res Ther 10 (2019), https://doi.org/10.1186/s13287-019-1165-5.

[386] M. Amit, J. Itskovitz-Eldor (Eds.), Atlas of Human Pluripotent Stem Cells: Derivation and Culturing, Humana Press, 2012, https://doi.org/10.1007/978-1-61779-548-0.

[387] B. Larijani, E.N. Esfahani, P. Amini, B. Nikbin, K. Alimoghaddam, S. Amiri, R. Malekzadeh, N.M. Yazdi, M. Ghodsi, Y. Dowlati, M.A. Sahraian, A. Ghavamzadeh, Stem cell therapy in treatment of different diseases, Acta Med. Iran. 50 (2012) 79–96.

[388] A. Ul Hassan, G. Hassan, Z. Rasool, Role of stem cells in treatment of neurological disorder, Int. J. Health Sci. (Qassim) 3 (2009) 227–233. https://www.ncbi.nlm.nih.gov/pmc/articles/PMC3068820/. (Accessed 18 July 2020).

[389] O. Strauss, The retinal pigment epithelium in visual function, Physiol. Rev. 85 (2005) 845–881, https://doi.org/10.1152/physrev.00021.2004.

[390] L. Ruggiero, S.C. Finnemann, Rhythmicity of the retinal pigment epithelium, in: G. Tosini, P.M. Iuvone, D.G. McMahon, S.P. Collin (Eds.), The Retina and Circadian Rhythms, Springer, New York, NY, 2014, pp. 95–112, https://doi.org/10.1007/978-1-4614-9613-7_6.

[391] W.L. Wong, X. Su, X. Li, C.M.G. Cheung, R. Klein, C.-Y. Cheng, T.Y. Wong, Global prevalence of age-related macular degeneration and disease burden projection for 2020 and 2040: a systematic review and meta-analysis, Lancet Glob. Health 2 (2014) e106–e116, https://doi.org/10.1016/S2214-109X(13)70145-1.

[392] F.L. Ferris, C.P. Wilkinson, A. Bird, U. Chakravarthy, E. Chew, K. Csaky, S.R. Sadda, Beckman initiative for macular research classification committee, clinical classification of age-related macular degeneration, Ophthalmology 120 (2013) 844–851, https://doi.org/10.1016/j.ophtha.2012.10.036.

[393] A. Iriyama, R. Fujiki, Y. Inoue, H. Takahashi, Y. Tamaki, S. Takezawa, K. Takeyama, W.-D. Jang, S. Kato, Y. Yanagi, A2E, a pigment of the lipofuscin of retinal pigment epithelial cells, is an endogenous ligand for retinoic acid receptor, J. Biol. Chem. 283 (2008) 11947–11953, https://doi.org/10.1074/jbc.M708989200.

[394] L. Perusek, B. Sahu, T. Parmar, H. Maeno, E. Arai, Y.-Z. Le, C.S. Subauste, Y. Chen, K. Palczewski, A. Maeda, Di-retinoid-pyridinium-ethanolamine (A2E) accumulation and the maintenance of the visual cycle are independent of Atg7-mediated autophagy in the retinal pigmented epithelium, J. Biol. Chem. 290 (2015) 29035–29044, https://doi.org/10.1074/jbc.M115.682310.

[395] G.A. Fishman, Historical evolution in the understanding of Stargardt macular dystrophy, Ophthalmic Genet. 31 (2010) 183–189, https://doi.org/10.3109/13816810.2010.499887.

[396] R.S. Molday, M. Zhong, F. Quazi, The role of the photoreceptor ABC transporter ABCA4 in lipid transport and Stargardt macular degeneration, Biochim. Biophys. Acta 1791 (2009) 573–583, https://doi.org/10.1016/j.bbalip.2009.02.004.

[397] Z.-B. Jin, M. Takahashi, Generation of retinal cells from pluripotent stem cells, Prog. Brain Res. 201 (2012) 171–181, https://doi.org/10.1016/B978-0-444-59544-7.00008-1.

[398] M. Idelson, R. Alper, A. Obolensky, E. Ben-Shushan, I. Hemo, N. Yachimovich-Cohen, H. Khaner, Y. Smith, O. Wiser, M. Gropp, M.A. Cohen, S. Even-Ram, Y. Berman-Zaken, L. Matzrafi, G. Rechavi, E. Banin, B. Reubinoff, Directed differentiation of human embryonic stem cells into functional retinal pigment epithelium cells, Cell Stem Cell 5 (2009) 396–408, https://doi.org/10.1016/j.stem.2009.07.002.

[399] S. Wang, B. Lu, S. Girman, T. Holmes, N. Bischoff, R.D. Lund, Morphological and functional rescue in RCS rats after RPE cell line transplantation at a later stage of degeneration, Invest. Ophthalmol. Vis. Sci. 49 (2008) 416–421, https://doi.org/10.1167/iovs.07-0992.

[400] B. Diniz, P. Thomas, B. Thomas, R. Ribeiro, Y. Hu, R. Brant, A. Ahuja, D. Zhu, L. Liu, M. Koss, M. Maia, G. Chader, D.R. Hinton, M.S. Humayun, Subretinal implantation of retinal pigment epithelial cells derived from human embryonic stem cells: improved survival when implanted as a monolayer, Invest. Ophthalmol. Vis. Sci. 54 (2013) 5087–5096, https://doi.org/10.1167/iovs.12-11239.

[401] S.D. Schwartz, C.D. Regillo, B.L. Lam, D. Eliott, P.J. Rosenfeld, N.Z. Gregori, J.-P. Hubschman, J.L. Davis, G. Heilwell, M. Spirn, J. Maguire, R. Gay, J. Bateman, R.M. Ostrick, D. Morris, M. Vincent, E. Anglade, L.V. Del Priore, R. Lanza, Human embryonic stem cell-derived retinal pigment epithelium in patients with age-related macular degeneration and Stargardt's macular dystrophy: follow-up of two open-label phase 1/2 studies, Lancet 385 (2015) 509–516, https://doi.org/10.1016/S0140-6736(14)61376-3.

[402] M.S. Mehat, V. Sundaram, C. Ripamonti, A.G. Robson, A.J. Smith, S. Borooah, M. Robinson, A.N. Rosenthal, W. Innes, R.G. Weleber, R.W.J. Lee, M. Crossland, G.S. Rubin, B. Dhillon, D.H.W. Steel, E. Anglade, R.P. Lanza, R.R. Ali, M. Michaelides, J.W.B. Bainbridge, Transplantation of human embryonic stem cell-derived retinal pigment epithelial cells in macular degeneration, Ophthalmology 125 (2018) 1765–1775, https://doi.org/10.1016/j.ophtha.2018.04.037.

[403] R.F. Spaide, L. Yannuzzi, K.B. Freund, R. Mullins, E. Stone, Eyes with subretinal drusenoid deposits and no drusen: progression of macular findings, Retina (Philadelphia, PA) 39 (2019) 12–26, https://doi.org/10.1097/IAE.0000000000002362.

[404] G.V. Mendonca, P. Pezarat-Correia, J.R. Vaz, L. Silva, K.S. Heffernan, Impact of aging on endurance and neuromuscular physical performance: the role of vascular senescence, Sports Med. 47 (2017) 583–598, https://doi.org/10.1007/s40279-016-0596-8.

[405] R. Sala-Llonch, D. Bartrés-Faz, C. Junqué, Reorganization of brain networks in aging: a review of functional connectivity studies, Front. Psychol. 6 (2015), https://doi.org/10.3389/fpsyg.2015.00663.

[406] M.P. Mattson, T. Magnus, Aging and neuronal vulnerability, Nat. Rev. Neurosci. 7 (2006) 278–294, https://doi.org/10.1038/nrn1886.

[407] L.V. Kalia, A.E. Lang, Parkinson's disease, Lancet 386 (2015) 896–912, https://doi.org/10.1016/S0140-6736(14)61393-3.

[408] A. Chiò, G. Logroscino, B.J. Traynor, J. Collins, J.C. Simeone, L.A. Goldstein, L.A. White, Global epidemiology of amyotrophic lateral sclerosis: a systematic review of the published literature, Neuroepidemiology 41 (2013) 118–130, https://doi.org/10.1159/000351153.

[409] R. Rademakers, M. Neumann, I.R. Mackenzie, Advances in understanding the molecular basis of frontotemporal dementia, Nat. Rev. Neurol. 8 (2012) 423–434, https://doi.org/10.1038/nrneurol.2012.117.

[410] D.M. Hatters, Protein misfolding inside cells: the case of huntingtin and Huntington's disease, IUBMB Life 60 (2008) 724–728, https://doi.org/10.1002/iub.111.

[411] P. Scheltens, K. Blennow, M.M.B. Breteler, B. de Strooper, G.B. Frisoni, S. Salloway, W.M. Van der Flier, Alzheimer's disease, Lancet 388 (2016) 505–517, https://doi.org/10.1016/S0140-6736(15)01124-1.

[412] K.A. Jellinger, Basic mechanisms of neurodegeneration: a critical update, J. Cell. Mol. Med. 14 (2010) 457–487, https://doi.org/10.1111/j.1582-4934.2010.01010.x.

[413] R. Hussain, H. Zubair, S. Pursell, M. Shahab, Neurodegenerative diseases: regenerative mechanisms and novel therapeutic approaches, Brain Sci. 8 (2018), https://doi.org/10.3390/brainsci8090177.

[414] E. Sturm, N. Stefanova, Multiple system atrophy: genetic or epigenetic? Exp. Neurobiol. 23 (2014) 277–291, https://doi.org/10.5607/en.2014.23.4.277.

[415] R. Sakthiswary, A.A. Raymond, Stem cell therapy in neurodegenerative diseases, Neural Regen. Res. 7 (2012) 1822–1831, https://doi.org/10.3969/j.issn.1673-5374.2012.23.009.

[416] L.C. O'Brien, P.M. Keeney, J.P. Bennett, Differentiation of human neural stem cells into motor neurons stimulates mitochondrial biogenesis and decreases glycolytic flux, Stem Cells Dev. 24 (2015) 1984–1994, https://doi.org/10.1089/scd.2015.0076.

[417] G. Nistor, M.M. Siegenthaler, S.N. Poirier, S. Rossi, A.J. Poole, M.E. Charlton, J.D. McNeish, C.N. Airriess, H.S. Keirstead, Derivation of high purity neuronal progenitors from human embryonic stem cells, PLoS One 6 (2011), https://doi.org/10.1371/journal.pone.0020692, e20692.

[418] W. Han, Y. Li, J. Cheng, J. Zhang, D. Chen, M. Fang, G. Xiang, Y. Wu, H. Zhang, K. Xu, H. Wang, L. Xie, J. Xiao, Sitagliptin improves functional recovery via GLP-1R-induced anti-apoptosis and facilitation of axonal regeneration after spinal cord injury, J. Cell. Mol. Med. (2020), https://doi.org/10.1111/jcmm.15501.

[419] Z. Lv, H. Jiang, H. Xu, N. Song, J. Xie, Increased iron levels correlate with the selective nigral dopaminergic neuron degeneration in Parkinson's disease, J. Neural Transm. (Vienna) 118 (2011) 361–369, https://doi.org/10.1007/s00702-010-0434-3.

[420] C. Henchcliffe, H. Sarva, Restoring function to dopaminergic neurons: progress in the development of cell-based therapies for Parkinson's disease, CNS Drugs 34 (2020) 559–577, https://doi.org/10.1007/s40263-020-00727-3.

[421] H.J. Kim, J.P. Taylor, Lost in transportation: nucleocytoplasmic transport defects in ALS and other neurodegenerative diseases, Neuron 96 (2017) 285–297, https://doi.org/10.1016/j.neuron.2017.07.029.

[422] R. Balendra, A.M. Isaacs, C9orf72-mediated ALS and FTD: multiple pathways to disease, Nat. Rev. Neurol. 14 (2018) 544–558, https://doi.org/10.1038/s41582-018-0047-2.

[423] M.-L. Liu, T. Zang, C.-L. Zhang, Direct lineage reprogramming reveals disease-specific phenotypes of motor neurons from human ALS patients, Cell Rep. 14 (2016) 115–128, https://doi.org/10.1016/j.celrep.2015.12.018.

[424] H.A. Bowden, D. Dormann, Altered mRNP granule dynamics in FTLD pathogenesis, J. Neurochem. 138 (Suppl. 1) (2016) 112–133, https://doi.org/10.1111/jnc.13601.

[425] T.J. Wyatt, S.L. Rossi, M.M. Siegenthaler, J. Frame, R. Robles, G. Nistor, H.S. Keirstead, Human motor neuron progenitor transplantation leads to endogenous neuronal sparing in 3 models of motor neuron loss, Stem Cells Int. 2011 (2011) 207230, https://doi.org/10.4061/2011/207230.

[426] M. Izrael, S.G. Slutsky, T. Admoni, L. Cohen, A. Granit, A. Hasson, J. Itskovitz-Eldor, L. Krush Paker, G. Kuperstein, N. Lavon, S. Yehezkel Ionescu, L.J. Solmesky, R. Zaguri, A. Zhuravlev, E. Volman, J. Chebath, M. Revel, Safety and efficacy of human embryonic stem cell-derived astrocytes following intrathecal transplantation in SOD1G93A and NSG animal models, Stem Cell Res Ther 9 (2018) 152, https://doi.org/10.1186/s13287-018-0890-5.

[427] L. Barbeito, Astrocyte-based cell therapy: new hope for amyotrophic lateral sclerosis patients? Stem Cell Res Ther 9 (2018), https://doi.org/10.1186/s13287-018-1006-y.

[428] T.H. Kwak, J.H. Kang, S. Hali, J. Kim, K.-P. Kim, C. Park, J.-H. Lee, H.K. Ryu, J.E. Na, J. Jo, H.S. Je, H.-H. Ng, J. Kwon, N.-H. Kim, K.H. Hong, W. Sun, C.H. Chung, I.J. Rhyu, D.W. Han, Generation of homogeneous midbrain organoids with in vivo-like cellular composition facilitates neurotoxin-based Parkinson's disease modeling, Stem Cells 38 (2020) 727–740, https://doi.org/10.1002/stem.3163.

[429] Y. Chang, J. Kim, H. Park, H. Choi, J. Kim, Modelling neurodegenerative diseases with 3D brain organoids, Biol. Rev. Camb. Philos. Soc. (2020), https://doi.org/10.1111/brv.12626.

[430] J.B. Strait, E.G. Lakatta, Aging-associated cardiovascular changes and their relationship to heart failure, Heart Fail. Clin. 8 (2012) 143–164, https://doi.org/10.1016/j.hfc.2011.08.011.

[431] S. Tapio, Pathology and biology of radiation-induced cardiac disease, J. Radiat. Res. 57 (2016) 439–448, https://doi.org/10.1093/jrr/rrw064.

[432] J.D. Schwalm, M. McKee, M.D. Huffman, S. Yusuf, Resource effective strategies to prevent and treat cardiovascular disease, Circulation 133 (2016) 742–755, https://doi.org/10.1161/CIRCULATIONAHA.115.008721.

[433] V.L.T. Ballard, J.M. Edelberg, Stem cells and the regeneration of the aging cardiovascular system, Circ. Res. 100 (2007) 1116–1127, https://doi.org/10.1161/01.RES.0000261964.19115.e3.

[434] Y.Y. Leong, W.H. Ng, G.M. Ellison-Hughes, J.J. Tan, Cardiac stem cells for myocardial regeneration: they are not alone, Front. Cardiovasc. Med. 4 (2017), https://doi.org/10.3389/fcvm.2017.00047.

[435] B. Liu, L. Wang, W. Jiang, Y. Xiong, L. Pang, Y. Zhong, C. Zhang, W. Ou, C. Tian, X. Chen, S.-M. Liu, Myocyte enhancer factor 2A delays vascular endothelial cell senescence by activating the PI3K/p-Akt/SIRT1 pathway, Aging (Albany NY) 11 (2019) 3768–3784, https://doi.org/10.18632/aging.102015.

[436] B. Martin, B. Gabris, A.F. Barakat, B.L. Henry, M. Giannini, R.P. Reddy, X. Wang, G. Romero, G. Salama, Relaxin reverses maladaptive remodeling of the aged heart through Wnt-signaling, Sci. Rep. 9 (2019) 18545, https://doi.org/10.1038/s41598-019-53867-y.

[437] K. Chang, P. Kang, Y. Liu, K. Huang, T. Miao, A.P. Sagona, I.P. Nezis, R. Bodmer, K. Ocorr, H. Bai, TGFB-INHB/activin signaling regulates age-dependent autophagy and cardiac health through inhibition of MTORC2, Autophagy (2019) 1–16, https://doi.org/10.1080/15548627.2019.1704117.

[438] W.E. Fibbe, Y. Shi, FOXO3, a molecular search for the fountain of youth, Cell Stem Cell 24 (2019) 351–352, https://doi.org/10.1016/j.stem.2019.02.008.

[439] D. Cruz-Topete, R.H. Oakley, J.A. Cidlowski, Glucocorticoid signaling and the aging heart, Front. Endocrinol. (Lausanne) 11 (2020) 347, https://doi.org/10.3389/fendo.2020.00347.

[440] J. Li, D. Zhang, B.J.J.M. Brundel, M. Wiersma, Imbalance of ER and mitochondria interactions: prelude to cardiac ageing and disease? Cells 8 (2019), https://doi.org/10.3390/cells8121617.

[441] F. Hattori, H. Chen, H. Yamashita, S. Tohyama, Y.-S. Satoh, S. Yuasa, W. Li, H. Yamakawa, T. Tanaka, T. Onitsuka, K. Shimoji, Y. Ohno, T. Egashira, R. Kaneda, M. Murata, K. Hidaka, T. Morisaki, E. Sasaki, T. Suzuki, M. Sano, S. Makino, S. Oikawa, K. Fukuda, Nongenetic method for purifying stem cell-derived cardiomyocytes, Nat. Methods 7 (2010) 61–66, https://doi.org/10.1038/nmeth.1403.

[442] S. Tohyama, F. Hattori, M. Sano, T. Hishiki, Y. Nagahata, T. Matsuura, H. Hashimoto, T. Suzuki, H. Yamashita, Y. Satoh, T. Egashira, T. Seki, N. Muraoka, H. Yamakawa, Y. Ohgino, T. Tanaka, M. Yoichi, S. Yuasa, M. Murata, M. Suematsu, K. Fukuda, Distinct metabolic flow enables large-scale purification of mouse and human pluripotent stem cell-derived cardiomyocytes, Cell Stem Cell 12 (2013) 127–137, https://doi.org/10.1016/j.stem.2012.09.013.

[443] D.A. Elliott, S.R. Braam, K. Koutsis, E.S. Ng, R. Jenny, E.L. Lagerqvist, C. Biben, T. Hatzistavrou, C.E. Hirst, Q.C. Yu, R.J.P. Skelton, D. Ward-van Oostwaard, S.M. Lim, O. Khammy, X. Li, S.M. Hawes, R.P. Davis, A.L. Goulburn, R. Passier, O.W.J. Prall, J.M. Haynes, C.W. Pouton, D.M. Kaye, C.L. Mummery, A.G. Elefanty, E.G. Stanley, NKX2-5(eGFP/w) hESCs for isolation of human cardiac progenitors and cardiomyocytes, Nat. Methods 8 (2011) 1037–1040, https://doi.org/10.1038/nmeth.1740.

[444] J.J.H. Chong, X. Yang, C.W. Don, E. Minami, Y.-W. Liu, J.J. Weyers, W.M. Mahoney, B. Van Biber, S.M. Cook, N.J. Palpant, J.A. Gantz, J.A. Fugate, V. Muskheli, G.M. Gough, K.W. Vogel, C.A. Astley, C.E. Hotchkiss, A. Baldessari, L. Pabon, H. Reinecke, E.A. Gill, V. Nelson, H.-P. Kiem, M.A. Laflamme, C.E. Murry, Human embryonic-stem-cell-derived cardiomyocytes regenerate non-human primate hearts, Nature 510 (2014) 273–277, https://doi.org/10.1038/nature13233.

[445] D. Cyranoski, "Reprogrammed" stem cells approved to mend human hearts for the first time, Nature 557 (2018) 619–620, https://doi.org/10.1038/d41586-018-05278-8.

[446] J. Cui, Y. Shen, R. Li, Estrogen synthesis and signaling pathways during ageing: from periphery to brain, Trends Mol. Med. 19 (2013) 197–209, https://doi.org/10.1016/j.molmed.2012.12.007.

[447] R.W. Loftin, M. Habli, C.C. Snyder, C.M. Cormier, D.F. Lewis, E.A. DeFranco, Late preterm birth, Rev. Obstet. Gynecol. 3 (2010) 10–19. https://www.ncbi.nlm.nih.gov/pmc/articles/PMC2876317/. (Accessed 20 July 2020).

[448] K. Shirasuna, H. Iwata, Effect of aging on the female reproductive function, Contracept. Reprod. Med. 2 (2017), https://doi.org/10.1186/s40834-017-0050-9.

[449] S.M. Nelson, E.E. Telfer, R.A. Anderson, The ageing ovary and uterus: new biological insights, Hum. Reprod. Update 19 (2013) 67–83, https://doi.org/10.1093/humupd/dms043.

[450] S.-Y. Kim, T. Kurita, New insights into the role of phosphoinositide 3-kinase activity in the physiology of immature oocytes: lessons from recent mouse model studies, Eur. Med. J. Reprod. Health 3 (2018) 119–125. https://www.ncbi.nlm.nih.gov/pmc/articles/PMC6147255/. (Accessed 20 July 2020).

[451] Z. Guo, Q. Yu, Role of mTOR signaling in female reproduction, Front. Endocrinol. (Lausanne) 10 (2019), https://doi.org/10.3389/fendo.2019.00692.

[452] M. Maidarti, R.A. Anderson, E.E. Telfer, Crosstalk between PTEN/PI3K/Akt signalling and DNA damage in the oocyte: implications for primordial follicle activation, oocyte quality and ageing, Cells 9 (2020), https://doi.org/10.3390/cells9010200.

[453] P. May-Panloup, L. Boucret, J.-M. Chao de la Barca, V. Desquiret-Dumas, V. Ferré-L'Hotellier, C. Morinière, P. Descamps, V. Procaccio, P. Reynier, Ovarian ageing: the role of mitochondria in oocytes and follicles, Hum. Reprod. Update 22 (2016) 725–743, https://doi.org/10.1093/humupd/dmw028.

[454] F. Liu, S. Hu, S. Wang, K. Cheng, Cell and biomaterial-based approaches to uterus regeneration, Regen. Biomater. 6 (2019) 141–148, https://doi.org/10.1093/rb/rbz021.

[455] M. Wallwiener, H. Brölmann, P.R. Koninckx, P. Lundorff, A.M. Lower, A. Wattiez, M. Mara, R.L. De Wilde, Adhesions after abdominal, pelvic and intra-uterine surgery and their prevention, Gynecol. Surg. 9 (2012) 465–466, https://doi.org/10.1007/s10397-012-0762-4.

[456] H. Fernandez, F. Al-Najjar, A. Chauveaud-Lambling, R. Frydman, A. Gervaise, Fertility after treatment of Asherman's syndrome stage 3 and 4, J. Minim. Invasive Gynecol. 13 (2006) 398–402, https://doi.org/10.1016/j.jmig.2006.04.013.

[457] J. Verdi, A. Tan, A. Shoae-Hassani, A.M. Seifalian, Endometrial stem cells in regenerative medicine, J. Biol. Eng. 8 (2014) 20, https://doi.org/10.1186/1754-1611-8-20.

[458] C. Yamashiro, K. Sasaki, S. Yokobayashi, Y. Kojima, M. Saitou, Generation of human oogonia from induced pluripotent stem cells in culture, Nat. Protoc. 15 (2020) 1560–1583, https://doi.org/10.1038/s41596-020-0297-5.

[459] D.M. Nathan, S. Russell, The future of care for type 1 diabetes, CMAJ 185 (2013) 285–286, https://doi.org/10.1503/cmaj.130011.

[460] N. Møller, L. Gormsen, J. Fuglsang, J. Gjedsted, Effects of ageing on insulin secretion and action, Horm. Res. 60 (2003) 102–104, https://doi.org/10.1159/000071233.

[461] A.K. Palmer, B. Gustafson, J.L. Kirkland, U. Smith, Cellular senescence: at the nexus between ageing afnd diabetes, Diabetologia 62 (2019). https://link.springer.com/article/10.1007/s00125-019-4934-x. (Accessed 12 April 2020).

[462] M.K. Harishankar, S. Logeshwaran, S. Sujeevan, K.N. Aruljothi, M.A. Dannie, A. Devi, Genotoxicity evaluation of metformin and glimepiride by micronucleus assay in exfoliated urothelial cells of type 2 diabetes mellitus patients, Food Chem. Toxicol. 83 (2015) 146–150, https://doi.org/10.1016/j.fct.2015.06.013.

[463] A. Sapra, P. Bhandari, Diabetes mellitus, in: StatPearls, StatPearls Publishing, Treasure Island (FL), 2020. http://www.ncbi.nlm.nih.gov/books/NBK551501/. (Accessed 20 July 2020).

[464] M. McCall, A.M. James Shapiro, Update on islet transplantation, Cold Spring Harb. Perspect. Med. 2 (2012), https://doi.org/10.1101/cshperspect.a007823.

[465] P. Rattananinsruang, C. Dechsukhum, W. Leeanansaksiri, Establishment of insulin-producing cells from human embryonic stem cells underhypoxic condition for cell based therapy, Front. Cell Dev. Biol. 6 (2018), https://doi.org/10.3389/fcell.2018.00049.

[466] L. Li, M.L. Baroja, A. Majumdar, K. Chadwick, A. Rouleau, L. Gallacher, I. Ferber, J. Lebkowski, T. Martin, J. Madrenas, M. Bhatia, Human embryonic stem cells possess immune-privileged properties, Stem Cells 22 (2004) 448–456, https://doi.org/10.1634/stemcells.22-4-448.

[467] A.J. Vegas, O. Veiseh, M. Gürtler, J.R. Millman, F.W. Pagliuca, A.R. Bader, J.C. Doloff, J. Li, M. Chen, K. Olejnik, H.H. Tam, S. Jhunjhunwala, E. Langan, S. Aresta-Dasilva, S. Gandham, J.J. McGarrigle, M.A. Bochenek, J. Hollister-Lock, J. Oberholzer, D.L. Greiner, G.C. Weir, D.A. Melton, R. Langer, D.G. Anderson, Long-term glycemic control using polymer-encapsulated human stem cell-derived beta cells in immune-competent mice, Nat. Med. 22 (2016) 306–311, https://doi.org/10.1038/nm.4030.

[468] X. Yin, B.E. Mead, H. Safaee, R. Langer, J.M. Karp, O. Levy, Engineering stem cell organoids, Cell Stem Cell 18 (2016) 25–38, https://doi.org/10.1016/j.stem.2015.12.005.

[469] M.A. Lancaster, M. Renner, C. Martin, D. Wenzel, L.S. Bicknell, M.E. Hurles, T. Homfray, J.M. Penninger, A.P. Jackson, J.A. Knoblich, Cerebral organoids model human brain development and microcephaly, Nature 501 (2013) 373–379, https://doi.org/10.1038/nature12517.

[470] M.A. Lancaster, J.A. Knoblich, Generation of cerebral organoids from human pluripotent stem cells, Nat. Protoc. 9 (2014) 2329–2340, https://doi.org/10.1038/nprot.2014.158.

[471] M. Takasato, P.X. Er, H.S. Chiu, M.H. Little, Generation of kidney organoids from human pluripotent stem cells, Nat. Protoc. 11 (2016) 1681–1692, https://doi.org/10.1038/nprot.2016.098.

[472] T. Takebe, K. Sekine, M. Enomura, H. Koike, M. Kimura, T. Ogaeri, R.R. Zhang, Y. Ueno, Y.W. Zheng, N. Koike, S. Aoyama, Y. Adachi, H. Taniguchi, Vascularized and functional human liver from an iPSC-derived organ bud transplant, Nature 499 (2013) 481–484, https://doi.org/10.1038/nature12271.

[473] M. Eiraku, N. Takata, H. Ishibashi, M. Kawada, E. Sakakura, S. Okuda, K. Sekiguchi, T. Adachi, Y. Sasai, Self-organizing optic-cup morphogenesis in three-dimensional culture, Nature 472 (2011) 51–58, https://doi.org/10.1038/nature09941.

[474] J.R. Spence, C.N. Mayhew, S.A. Rankin, M.F. Kuhar, J.E. Vallance, K. Tolle, E.E. Hoskins, V.V. Kalinichenko, S.I. Wells, A.M. Zorn, N.F. Shroyer, J.M. Wells, Directed differentiation of human pluripotent stem cells into intestinal tissue in vitro, Nature 470 (2011) 105–110, https://doi.org/10.1038/nature09691.

[475] C.M. Metallo, S.M. Azarin, L. Ji, J.J. De Pablo, S.P. Palecek, Engineering tissue from human embryonic stem cells, J. Cell. Mol. Med. 12 (2008) 709–729, https://doi.org/10.1111/j.1582-4934.2008.00228.x.

[476] H.A. Mccauley, J.M. Wells, Pluripotent stem cell-derived organoids: using principles of developmental biology to grow human tissues in a dish, Development 144 (2017) 958–962, https://doi.org/10.1242/dev.140731.

[477] T.K. Matsui, M. Matsubayashi, Y.M. Sakaguchi, R.K. Hayashi, C. Zheng, K. Sugie, M. Hasegawa, T. Nakagawa, E. Mori, Six-month cultured cerebral organoids from human ES cells contain matured neural cells, Neurosci. Lett. 670 (2018) 75–82, https://doi.org/10.1016/j.neulet.2018.01.040.

[478] D. Ghosh, N. Mehta, A. Patil, J. Sengupta, Ethical issues in biomedical use of human embryonic stem cells (hESCs), J. Reprod. Health Med. (2016), https://doi.org/10.1016/j.jrhm.2016.09.002.

[479] J. Sugarman, Ethical issues in stem cell research and treatment, Cell Res. (2008), https://doi.org/10.1038/cr.2008.266.

[480] D.W. Landry, H.A. Zucker, Embryonic death and the creation of human embryonic stem cells, J. Clin. Invest. 114 (2004) 9–11, https://doi.org/10.1172/JCI200423065.1184.

[481] H. Laverge, J. Van der Elst, P. De Sutter, M.R. Verschraegen-Spae, A. De Paepe, M. Dhont, Fluorescent in-situ hybridization on human embryos showing cleavage arrest after freezing and thawing, Hum. Reprod. 13 (1998) 425–429.

[482] K. Kirkegaard, J.J. Hindkjaer, H.J. Ingerslev, Human embryonic development after blastomere removal: a time-lapse analysis, Hum. Reprod. 27 (2012) 97–105, https://doi.org/10.1093/humrep/der382.

Chapter 3

Stem cell aging and wound healing

Vijayalakshmi Rajendran[a], Mayur Vilas Jain[b], and Sumit Sharma[c]

[a]*Division of Ophthalmology, Department of Clinical Sciences, Lund University, Lund, Sweden,* [b]*Department of Molecular Medicine and Gene Therapy, Lund Stem Cell Center, Lund University, Lund, Sweden,* [c]*Division of Molecular Medicine and Virology, Department of Biomedical and Clinical Sciences, Linköping University, Linköping, Sweden*

1 Introduction

Aging is a natural physiological process that occurs in all living organisms. Although the structural degradation of all living tissues due to aging is physically evident, the accompanying functional deterioration causes the major impairment of crucial functions of a living body [1, 2]. Generally, the prevalence and susceptibility of most diseases, disorders, and degenerative conditions are known to be strongly associated with aging. One such major dysfunction due to aging is deficits in wound healing [3]. Wound healing is a complex physiological response to tissue injury involving influx of different cell types, growth factors, and cytokines. In a normal state, the healing process mediated by the early-mentioned factors goes through four important phases—hemostasis, inflammation, proliferation (replication and synthesis), and remodeling to restore tissue integrity [4]. However, this capacity of the tissue to repair and regenerate itself in response to an injury decreases substantially with aging. Patients with other underlying problems like diabetes, obesity, and most importantly aging are found to be strongly correlated with poor wound healing due to decreased strength, elasticity, and neovascularization of the injured tissue [5].

Under steady-state conditions, the pool of stem cells either are maintained and remain quiescent in the specific niches or are active in high-turnover tissues and proliferate in order to regenerate the injured tissue by differentiating into a population of committed progenitors [6]. This endogenous reparative system that contributes to the timely engagement in tissue regenerative program is less effective as the intrinsic stem cells undergo senescence [7]. Structural and functional changes in an organ postsenescence trigger unregulated fibrosis leading to an undesirable scar formation that in turn escalates the risk of morbidity and mortality [8].

Aging instigates multiple cascades of detrimental events including significant declination in stem cell production, maintenance of homeostasis, multilineage differentiation, extracellular matrix (ECM) synthesis and remodeling, intercellular communication, and immunomodulation in our body [7, 8]. Stem cell aging is one of the central contributors to age-related tissue degeneration. The triggered switch from quiescence to senescence due to aging initiates damage of stem cell niches and declination of stem cells. However, the pace and effect of stem cell aging differ in different tissues [9].

This chapter comprises the current understanding of stem cell aging in association with compromised wound healing. The critical aspects of aging with regard to different stem cell types are reviewed in the context of inflammation, tissue degeneration, and fibrosis. In addition, we have briefly discussed the extrinsic strategies by presenting optimal cues that regulate stem cell physiology in vivo to evade aging-related deficits in tissue repair and regeneration.

2 Stem cell aging and inflammation

During a wound healing process, the defined accurate levels of inflammatory cytokines and growth factors mediate cell-cell/cell-matrix interactions and execute a scarless healing [10]. However, the transient inflammation pattern expressed, which is critical for normal wound healing phases, can repeatedly get exaggerated due to age-mediated fueling of damage-associated molecular patterns (DAMPs) upon injury [11]. The occurrence of cellular senescence induces a chronic inflammatory microenvironment by elevating the levels of reactive oxygen species (ROS), inflammatory cells infiltration, cytokines, growth factors, metalloproteinases which compromises ECM networks and result in poor healing [10]. The persistence of systemic proinflammatory factors like interleukin-6 (IL-6), tumor necrosis factor alpha (TNF-α), interferon gamma (IFN-γ), and C-reactive protein (CRP) during aging has shown to be highly linked to several chronic degenerative diseases like atherosclerosis, rheumatoid arthritis, and neurodegeneration in the elderly population [12, 13]. In an age-based patient study, the "age-associated inflammation—inflammaging" has been noticed and a direct correlation has been demonstrated between decline in stem cell number and time taken for tissue regeneration [7, 14]. The altered inflammatory stimuli induce immune senescence that accelerates aging and aging-related regenerative failure [14].

To understand the stem cell behavior in wound healing, it is important to comprehend the cross talk between the age-associated proinflammatory environment and the tissue-specific stem cell niche [15]. The inflammatory signals disrupt the homeostasis and their balance in their clonal expansion in the major stem cell niche, bone marrow (BM). The differentiation and proliferative capacity of the mesenchymal and the hematopoietic stem cells get skewed through inflammaging mechanisms, which are identifiable from the noncorrelation between the functionality and phenotype of the BM origin stem cells. It has been previously shown that aged mesenchymal stem cells (MSC) favor adipogenesis differentiation than osteogenesis and chondrogenesis [16]. This age-dependent shift in the commitment of adipogenesis among the BM-residing MSC has been shown to be coupled with poor bone formation in elderly patients [16, 17]. Additionally, autophagy, a fundamental mechanism involving the degradation of detrimental cellular components, has been shown to be linked with osteogenesis and is declined with aging [18]. Likewise, the delay or impairment of bone regeneration has shown to be associated with NF-κB (nuclear factor kappa-light-chain-enhancer of activated B cells)-mediated inflammaging [7]. Moreover, the reduction in the skeletal stem cell frequency supports impaired bone healing [7]. The indigenous immunomodulatory function of the MSC has also been compromised in aging as the secretome of the senescent MSC includes IL-6, IL-8, IFN-γ, monocyte chemoattractant protein (MCP)-1, and matrix metalloproteinases (MMP) instead of the typical antiinflammatory cytokines and exosomes [19].

Similarly, the inflammaging has also been shown to influence the differentiation behavior and mobilization of hematopoietic stem cells (HSCs) in BM. Although aging influences both innate and adaptive immunity, the aged HSCs get skewed toward subsets of myeloid lineage rather than lymphoid progenitors (B and T cell precursors) [16, 20]. Treatment of aged mice with lipopolysaccharide (LPS) demonstrated an impaired B-lymphopoiesis [21]. The age-related inflammatory milieu postulates surge in the expression of transcription factors such as Klf5 (Kruppel-like factor 5), Ikzf1 (Ikaros family zinc finger protein 1), and Stat3 (signal transducer and activator of transcription 3) responsible for the myeloid tilt [22]. Notably, the activation of toll-like receptors and the inflammatory signaling pathways like NFκB and RANTES (regulated on activation, normal T cell expressed and secreted)—mTOR (mammalian target of rapamycin) due to aging has been seen along with the biased increase in myelopoiesis [22]. Thus, the presence of senescence-associated secretory phenotype (SASP) in the local microenvironment stimulates functional decline in the stem cell niche and dysregulation in their differentiation fate that in turn impedes regeneration of the wound tissue [13]. Furthermore, the age-related stem cell phenotypes that seen in myeloid precursors and MSC and the systemic release of their associated factors have deleterious effect in wound healing.

3 Role in tissue degeneration

As stem cell aging causes a progressive decline in the regenerative properties of different tissues, it is pivotal to understand the physical and chemical cues arising from the local and the systemic milieu contributing to tissue homeostasis and repair [1]. However, the cues controlling stem cell activity and the number of stem cells reserved in each tissue diminish with age [23]. The age-related aberrant cues in the system drive the damaged tissue to skew toward pathological differentiation into fibrotic tissue [24]. In the case of age-related muscular dysfunctions, severely defective functional cues for muscle stem cells (MuSC) were found to have poor engraftment due to alterations in signal transduction and myogenic differentiation factors [13]. Under in vivo experimental settings, a comparative study for MuSC from young and aged mice demonstrated that old MuSC displayed much pronounced deficits in self-renewal and myofiber formation [25]. In the pathogenesis of age-associated muscle disorder—sarcopenia, the old MuSC exhibited senescence phenotype along with accelerated self-repair dysfunction [13, 26].

The mounting of metabolic and epigenetic changes in aging causes accumulation of toxic metabolites, damaged macromolecules, and error-prone disturbance to the regulatory cycle of the synthesis and turnover of the different tissue-resident quiescent stem cells [27]. DNA damage in the quiescent stem cells causes short telomeres resulting in initiating senescence [27]. The SASP-related generation of retrotransposons (mobile DNA fragments) and the disruptions in the chromatin organization induce aberrant gene expressions and genomic arrangements among quiescent cells. Although, in general, the epigenetic alterations activate the stem cells from quiescence, in aging the complexity of the molecular circuitry involved in the activation of quiescent stem cells becomes vulnerable for DNA damage and the corresponding transcriptional/translational dysfunctions and mitochondrial aberrations [27]. In addition, the drastic shifts in metabolic signaling are strongly shown to be associated with limiting metabolic activity and subdued rates of oxygen consumption. The age-associated functional decline of stem cells is also found to be highly responsive to mitochondrial activity as they can influence the DNA methylation rates in a cell [9, 28, 29]. Both aged HSC and MSC exhibit altered DNA methylation patterns, telomere attrition, and accumulated ROS due to the inhibition of autophagy leading to structural/functional defects in stem cell differentiation [30].

Furthermore, endothelial progenitors are also shown to be dysregulated due to the extrinsic cues linked to aging. Their impaired proangiogenic functions are shown to be mediated by elevated oxidative stress, ROS accumulation, and downregulation of nitric oxide (NO) that affects vasculogenesis and wound healing activity [16]. Senescent MSC were found to have reduced expression of vascular endothelial growth factor (VEGF), stromal cell-derived factor 1, and protein kinase B, and thus, their angiogenic potential has been shown to be vastly reduced [23].

Collectively, the declination in the functional stem cell population and the associated paracrine factors contribute to tissue damage and delayed tissue repair. The mechanism for functional deterioration of stem cells in aging can be either due to the irreversible arrest in stem cell division causing frailty/senescence or due to stem cell exhaustion depleting the tissue-specific stem cell reserve [29]. The age-associated increase in intrinsic (DNA damage, mitochondrial dysfunction) and extrinsic alterations (inflammatory factors, stem cell niche defects, angiogenic soluble factors) that cause biochemical and biomechanical changes in a tissue orchestrate the progressive dysfunction and impairment causing tissue degeneration.

4 Role in fibrosis

Aging is known to be an important predisposing factor for fibrotic diseases [3]. The characteristic feature of fibrosis is scarring and hardening of tissues due to excessive ECM accumulation [31]. In general, wound healing involves both noncellular and cellular components. The noncellular ECM, particularly collagen or other fibrous proteins like fibronectin, elastin, laminins, proteoglycans, thrombospondins, tenascin, and vitronectin, provides a temporary scaffold, while the recruitment of other cell types to the wound site such as monocytes/macrophages, neutrophils, fibroblasts, and endothelial cells mediate interaction with newly generated ECM to restore tissue homeostasis and integrity [4, 10]. The ECM is a dynamic component of a cell that orchestrates various intercellular and intracellular activities [8]. The alterations in the components of ECM and their highly organized biochemical interactions with cells are critical in wound healing process [32].

The primary enzymatic proteins that regulate the remodeling of ECM are matrix metalloproteinases (MMP) [8, 10]. MMP are endopeptidases that cleave peptide bonds in the ECM proteins and lead to successive activation of different biochemical pathways involving cell-cell interactions, cell-matrix interactions, and release of growth factors [10]. Under steady-state conditions, the unique proportion of MMP and other fibrous proteins present in the ECM determine the rate of tissue degeneration and remodeling (Fig. 1) [10]. The accuracy in the maintenance of ECM synthesis/degradation balance determines the fate of a cell to be either physiological or pathological [9]. This delicate balance dictates the resident adult stem cells in the tissue to play role in proper tissue repair and functioning via paracrine secretion of cytokines and growth factors [4].

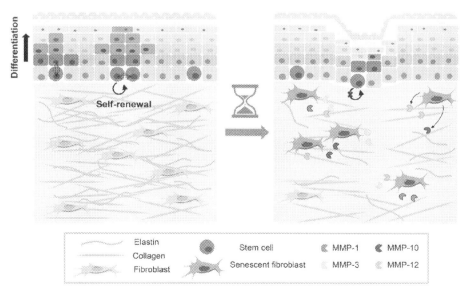

FIG. 1 Tissue remodeling in aging depicts how the induction of senescence alters the expression of matrix metalloproteinases (MMP) and homeostasis of stem cells that is critical for a functional tissue. The expression and accumulation of the senescent fibroblasts and MMP noticed in aging have shown to compromise proper wound healing. *(Reproduced with permission from S. Freitas-Rodriguez, A.R. Folgueras, C. Lopez-Otin, The role of matrix metalloproteinases in aging: tissue remodeling and beyond, Biochim. Biophys. Acta Mol. Cell Res. 1864 (11 Pt A) (2017) 2015–2025.)*

In age-related tissue remodeling, the disorganization of the ECM proteins, collagen, and elastin mainly triggers structural alterations that bring in critical modifications in their composition and loss of elasticity [10]. The aged ECM also modulates the behavior of the tissue-resident stem cells by which the stiffness of the ECM gets upregulated through a cascade of events mediated by MMP [10]. The recruitment of stem cells from the nearby tissue reservoir and migration of fibroblasts that assist in tissue remodeling would not be optimal in aged tissues, and thus, the structural integrity of the newly formed temporary matrix gets compromised leading to scar tissue formation [33]. Most of the age-related diseases and dysfunctions involve connective tissue, cartilage, bone, blood vessels, and skin. For example, wound revascularization that is critical for a functional tissue is a serious problem in aging [34]. If the stretchability of these tissues is limited, it will result in fragmented fibrous proteins in the ECM. Such age-associated ECM modifications, endothelial dysfunction, improper recruitment of endogenous progenitors/stem cells have been noticed to compromise resilience and contribute to increased stiffness in vasculature [8, 35].

Multiple fibrosis-related dysfunctions are prevalent in different tissues as age-related phenotypes and seen to be dramatically increasing with age (Table 1). In the case of idiopathic pulmonary fibrosis (IPF), an irreversible functional loss due to increased levels of SASP seen in alveolar epithelium leads to intra-alveolar fibrosis along with progression and accumulation of scar tissue [39, 40]. BM-MSC from aged IPF patients were shown to have significant differences in mitochondrial dysfunction and DNA damage resulting in defective cellular function compared to age-matched healthy controls [39]. The IPF BM-MSC were also shown to induce paracrine senescence in normal aged fibroblasts confirming the possibility of easy disease progression due to aging [39]. The age-dependent fibrogenic conversion of muscle stem cells was found to have facilitated impaired skeletal muscle regeneration and correlated to increased Wnt signaling in the myogenic progenitors in vivo [41]. A significant regression of tissue regeneration has been noticed in aging-related renal injuries, where the BM-derived stem cells from young mice were shown to have progress in regeneration due to increased expression of antifibrotic factors like transforming growth factor-beta and plasminogen activator inhibitor-1 [42]. The occurrence of age-related cellular senescence was shown to exhaust the regenerative potential in cardiac tissue, thereby resulting in pathological fibrosis [31].

Overall, many regenerative deficits are accompanied by aging in different tissues. The exhaustion of stem cell reserve and declination of stem cell function in aging are due to the diminution of the optimal cues required for tissue regeneration of an injured tissue, thus resulting in a fibrotic response.

TABLE 1 Functional interferences due to stem cell aging in wound healing models.

Stem cell type	Self-renewal	Stem cell fate	Differentiation potential/fibrosis	Tissue repair	Model	Reference
Airway basal stem/progenitor cells	Decreased	Airway epithelium	Reduced epithelial differentiation	Reduced re-epithelialization of lung airways	Human COPD	[36]
Hematopoietic stem cells	Maintained	Myeloid progenitors	Decreased lymphoid cell and increase in myeloid cells numbers	Reduced functional diversity and impaired lymphoid cell recruitment	Aged murine model	[20, 37]
BM-derived adipocyte progenitors	Increased	Adipocytes	Increased accumulation of adipocytes at the expense of osteoblast regeneration	Impaired bone regeneration	Fracture model in aged mouse	[38]
Muscle stem cells	Decreased	Myogenic fibers	Increased cellular senescence and fibrosis after injury	Activity of MuSC is limited to repair of muscle tissue and impaired osteogenesis	Human elderly subjects	[7]
Bone marrow mesenchymal stem cells	Very limited	Increased senescence and telomere shortening	Increased idiopathic pulmonary fibrosis	Functionally defective	Lung injury model in mice	[39]

5 Alternative stem cell rejuvenating strategies in aging

The rapidly aging global population has shown to have a decreased responsiveness to regenerate and restore various damaged tissue functions [43]. As aging has no cure, it is crucial to know how to modulate the detrimental effects of aging and maintain healthy aging. The reversal of stem cell aging is one of the currently emerging promising strategies to develop alternative treatment protocols for age-related chronic diseases and dysfunctions opposed to the conventional surgical treatments and lifelong immunosuppression [44]. In the context of wound healing, the total healing of an injured tissue and maintenance of tissue homeostasis can be addressed by either inducing functional cues to improve action of tissue-specific endogenous stem cells or replacing the nonfunctional native stem cells by exogenously derived ones [30].

The exogenous interventions can be targeted toward intrinsic tissue-specific stem cell properties, prevention of cellular senescence and modulation of microenvironment in aged tissues to optimize reparative functions. Many regenerative strategies have currently been experimented to reduce the fibrotic pathogenesis causing end-stage tissue dysfunction especially in vital organs such as lungs, kidneys, and liver [45]. Transcriptomic profiling of the stem cells in aged tissues has been done to understand the intrinsic changes in the molecular signatures [46]. Many recent studies have shed light on age-related differences in the regulation of genetic properties that dictate the stemness, self-renewal, and multipotency of the adult stem cells [47, 48]. Interestingly, although these studies suggest age-related decline in stem cell responsiveness, their potential can be brought back by applying targeted rejuvenating strategies.

Some of the therapeutic interventions that contribute to revert age-induced stem cell dysfunction are discussed in the following:

(i) **Cellular reprogramming**

Embryonic stem cells (ESC) are pluripotent in nature, derived from blastocysts that can be potentially differentiated into any specialized cell type. In a study involving patients with large burns, reconstruction of a homogeneous and fully functional human epidermis in vitro from ESC was indeed remarkable [49]. Although the advances in this field seemed to be promising initially, the potential tumorigenicity of ESC and the ethical concerns related to obtaining ESC from embryos delayed further development [50]. These controversies can be circumvented by using induced pluripotent stem cells (iPSC). A great range of somatic adult cells can be genetically reprogrammed into iPSC by incorporating four transcriptional factors, such as Oct 3/4, Sox2, *c-Myc*, and Klf4, that are responsible for maintaining cells in the pluripotent state [51]. Especially for elderly patients who are often associated with chronic wounds and diseases like macular degeneration, iPSC reprogrammed from dermal fibroblasts has shown to increase the ECM production and rate of wound healing [44].

The other population of stem cells that can be therapeutically approached for age-related dysfunction are adult stem cells from bone marrow, which are of mesenchymal and hematopoietic origin [52]. Bone marrow is a very limited source of stem cells and an age-associated reduction in stem cell number and migratory ability have always been noticed. Also, several in vitro studies have shown previously that the proliferation and differentiation potential is also restricted to certain passages [53]. The age-induced declination in the multipotency of these BM-derived adult stem cells can be replaced by using alternative sources of MSC such as umbilical cord, cord blood, adipose tissue, skin, orbital fat, and corneal limbus [53, 54]. A number of preclinical and clinical studies using MSC from multiple sources have shown promising results of differentiation into specific cell types that are required for reducing pathogenesis in the targeted fibrotic tissue [40]. Additionally, lentiviral transduced BM-MSC with pluripotent genes like Nanog and Oct-4 have shown to suffice the differentiation and homeostasis in aged tissues [54].

(ii) **Microenvironment**

The systemic microenvironment of the stem cells plays a vital role in determining their behavior [52, 55]. The continuous exposure to biological toxins, hormones, chemicals, physical stress, and ROS may lead to premature senescence [43]. With age, stem cells acquire apoptotic resistance and senescence-associated secretory phenotype and recruit leukocytes, cytokines, and other trophic factors leading to proinflammatory milieu accelerating cell cycle arrest [44]. In a mice model of idiopathic pulmonary fibrosis, it has been clearly shown that the senescent BM-MSC have the capacity to induce senescence in normal fibroblasts through paracrine mechanism and found to be less effective in reducing lung fibrosis [39]. Alterations in the cellular microenvironment can potentially delay senescence and telomere shortening [56]. Modulation of the molecular architecture of a microenvironment in aged animals can be reverted by incorporating systemic factors from young animals [55]. The CRISPR/Cas9-based gene editing has been reported to modify the telomere length of the stem cells to increase their proliferation rate [57].

(iii) Biomaterials

The combined approach of biomaterials and stem cells has shown to be very successful in different wound healing models [58, 59]. To ameliorate the effect of aging in stem cells related to wound healing, bioscaffold-based delivery of reprogrammed stem cells on chronic wounds is an increasingly promising approach. Bioscaffold can be composed of collagen, hyaluronic acid, fibronectin, laminin, or synthetic matrices that are loaded with stem cells or stem cell-derived factors to re-establish a functional niche for wound healing [44]. Alterations in the structure and composition of the bioscaffolds such as bioactive hydrogels and functionalized peptides can improve stem cell delivery and transplantation efficacy. The encapsulation of MuSC in 3D bioscaffold has shown to enhance MuSC proliferation, engraftment, and survival in aged mouse with dystrophic skeletal muscle [60].

Although regenerative stem cell therapy seems promising and several preclinical and clinical studies have been performed, many issues still need to be addressed. Among them are effective dose and route of administration, efficient type of stem cells—adult or iPSC, syngeneic or allogenic source, stem cells should be targeting early inflammatory phase or late fibrotic phase in a wound healing model [61–64].

6 Summary

In this chapter, we have reviewed the effects of stem cell aging in different tissues on wound healing properties. Aging imposes many detrimental effects on stem cells related to its activation, maintenance, and differentiation. Furthermore, inflammaging, persistence of cellular senescence, aberrations in DNA repair mechanism, and mitochondrial dysfunctions favor exaggerated tissue degeneration in a damaged tissue. Thus, a compromised wound healing process is determined as one of the obvious consequences of stem cell aging. However, understanding the mechanisms behind the functional deterioration of stem cells in aging and wound healing has paved way for several novel strategies to reverse, enhance, or repair the impaired function of wound tissue.

References

[1] J. Oh, Y.D. Lee, A.J. Wagers, Stem cell aging: mechanisms, regulators and therapeutic opportunities, Nat. Med. 20 (8) (2014) 870–880.

[2] A.D. Ho, W. Wagner, U. Mahlknecht, Stem cells and ageing. The potential of stem cells to overcome age-related deteriorations of the body in regenerative medicine, EMBO Rep. 6 (Suppl. 1) (2005) S35–S38.

[3] T. Niccoli, L. Partridge, Ageing as a risk factor for disease, Curr. Biol. 22 (17) (2012) R741–R752.

[4] P. Olczyk, L. Mencner, K. Komosinska-Vassev, The role of the extracellular matrix components in cutaneous wound healing, Biomed. Res. Int. 2014 (2014) 747584.

[5] S. Kanji, H. Das, Advances of stem cell therapeutics in cutaneous wound healing and regeneration, Mediat. Inflamm. 2017 (2017) 5217967.

[6] P. Munoz-Canoves, J. Neves, P. Sousa-Victor, Understanding muscle regenerative decline with aging: new approaches to bring back youthfulness to aged stem cells, FEBS J. 287 (3) (2020) 406–416.

[7] A.M. Josephson, V. Bradaschia-Correa, S. Lee, K. Leclerc, K.S. Patel, E. Muinos Lopez, H.P. Litwa, S.S. Neibart, M. Kadiyala, M.Z. Wong, M.M. Mizrahi, N.L. Yim, A.J. Ramme, K.A. Egol, P. Leucht, Age-related inflammation triggers skeletal stem/progenitor cell dysfunction, Proc. Natl. Acad. Sci. U. S. A. 116 (14) (2019) 6995–7004.

[8] S. Freitas-Rodriguez, A.R. Folgueras, C. Lopez-Otin, The role of matrix metalloproteinases in aging: tissue remodeling and beyond, Biochim. Biophys. Acta Mol. Cell Res. 1864 (11 Pt A) (2017) 2015–2025.

[9] M.B. Schultz, D.A. Sinclair, When stem cells grow old: phenotypes and mechanisms of stem cell aging, Development 143 (1) (2016) 3–14.

[10] J.M. Phillip, I. Aifuwa, J. Walston, D. Wirtz, The mechanobiology of aging, Annu. Rev. Biomed. Eng. 17 (2015) 113–141.

[11] N.X. Landen, D. Li, M. Stahle, Transition from inflammation to proliferation: a critical step during wound healing, Cell. Mol. Life Sci. 73 (20) (2016) 3861–3885.

[12] M. Bonafe, G. Storci, C. Franceschi, Inflamm-aging of the stem cell niche: breast cancer as a paradigmatic example: breakdown of the multi-shell cytokine network fuels cancer in aged people, BioEssays 34 (1) (2012) 40–49.

[13] H.M. Blau, B.D. Cosgrove, A.T. Ho, The central role of muscle stem cells in regenerative failure with aging, Nat. Med. 21 (8) (2015) 854–862.

[14] B.C. Lee, K.R. Yu, Impact of mesenchymal stem cell senescence on inflammaging, BMB Rep. 53 (2) (2020) 65–73.

[15] C. Kizil, N. Kyritsis, M. Brand, Effects of inflammation on stem cells: together they strive? EMBO Rep. 16 (4) (2015) 416–426.

[16] L.V. Kovtonyuk, K. Fritsch, X. Feng, M.G. Manz, H. Takizawa, Inflamm-aging of hematopoiesis, hematopoietic stem cells, and the bone marrow microenvironment, Front. Immunol. 7 (2016) 502.

[17] I. Bellantuono, A. Aldahmash, M. Kassem, Aging of marrow stromal (skeletal) stem cells and their contribution to age-related bone loss, Biochim. Biophys. Acta 1792 (4) (2009) 364–370.

[18] A. Infante, C.I. Rodriguez, Osteogenesis and aging: lessons from mesenchymal stem cells, Stem Cell Res. Ther. 9 (1) (2018) 244.

[19] I.H. Schulman, W. Balkan, J.M. Hare, Mesenchymal stem cell therapy for aging frailty, Front. Nutr. 5 (2018) 108.

[20] R.H. Cho, H.B. Sieburg, C.E. Muller-Sieburg, A new mechanism for the aging of hematopoietic stem cells: aging changes the clonal composition of the stem cell compartment but not individual stem cells, Blood 111 (12) (2008) 5553–5561.

[21] R.L. Riley, Impaired B lymphopoiesis in old age: a role for inflammatory B cells? Immunol. Res. 57 (1–3) (2013) 361–369.
[22] M. Mann, A. Mehta, C.G. de Boer, M.S. Kowalczyk, K. Lee, P. Haldeman, N. Rogel, A.R. Knecht, D. Farouq, A. Regev, D. Baltimore, Heterogeneous responses of hematopoietic stem cells to inflammatory stimuli are altered with age, Cell Rep. 25 (11) (2018) 2992–3005.e5.
[23] E. Gibon, L. Lu, S.B. Goodman, Aging, inflammation, stem cells, and bone healing, Stem Cell Res. Ther. 7 (2016) 44.
[24] S. Ancel, O. Mashinchian, J.N. Feige, Adipogenic progenitors keep muscle stem cells young, Aging (Albany NY) 11 (18) (2019) 7331–7333.
[25] B.D. Cosgrove, P.M. Gilbert, E. Porpiglia, F. Mourkioti, S.P. Lee, S.Y. Corbel, M.E. Llewellyn, S.L. Delp, H.M. Blau, Rejuvenation of the muscle stem cell population restores strength to injured aged muscles, Nat. Med. 20 (3) (2014) 255–264.
[26] I.M. Conboy, T.A. Rando, Aging, stem cells and tissue regeneration: lessons from muscle, Cell Cycle 4 (3) (2005) 407–410.
[27] S. Tumpel, K.L. Rudolph, Quiescence: good and bad of stem cell aging, Trends Cell Biol. 29 (8) (2019) 672–685.
[28] D.C. Zank, M. Bueno, A.L. Mora, M. Rojas, Idiopathic pulmonary fibrosis: aging, mitochondrial dysfunction, and cellular bioenergetics, Front. Med. (Lausanne) 5 (2018) 10.
[29] R. Ren, A. Ocampo, G.H. Liu, J.C. Izpisua Belmonte, Regulation of stem cell aging by metabolism and epigenetics, Cell Metab. 26 (3) (2017) 460–474.
[30] J. Neves, P. Sousa-Victor, H. Jasper, Rejuvenating strategies for stem cell-based therapies in aging, Cell Stem Cell 20 (2) (2017) 161–175.
[31] L.A. Murtha, M. Morten, M.J. Schuliga, N.S. Mabotuwana, S.A. Hardy, D.W. Waters, J.K. Burgess, D.T. Ngo, A.L. Sverdlov, D.A. Knight, A.J. Boyle, The role of pathological aging in cardiac and pulmonary fibrosis, Aging Dis. 10 (2) (2019) 419–428.
[32] S. Albeiroti, A. Soroosh, C.A. de la Motte, Hyaluronan's role in fibrosis: a pathogenic factor or a passive player? Biomed. Res. Int. 2015 (2015) 790203.
[33] S. Pacelli, S. Basu, J. Whitlow, A. Chakravarti, F. Acosta, A. Varshney, S. Modaresi, C. Berkland, A. Paul, Strategies to develop endogenous stem cell-recruiting bioactive materials for tissue repair and regeneration, Adv. Drug Deliv. Rev. 120 (2017) 50–70.
[34] J. Liu, D. Saul, K.O. Boker, J. Ernst, W. Lehman, A.F. Schilling, Current methods for skeletal muscle tissue repair and regeneration, Biomed. Res. Int. 2018 (2018) 1984879.
[35] L. Duca, S. Blaise, B. Romier, M. Laffargue, S. Gayral, H. El Btaouri, C. Kawecki, A. Guillot, L. Martiny, L. Debelle, P. Maurice, Matrix ageing and vascular impacts: focus on elastin fragmentation, Cardiovasc. Res. 110 (3) (2016) 298–308.
[36] D.M. Ryan, T.L. Vincent, J. Salit, M.S. Walters, F. Agosto-Perez, R. Shaykhiev, Y. Strulovici-Barel, R.J. Downey, L.J. Buro-Auriemma, M.R. Staudt, N.R. Hackett, J.G. Mezey, R.G. Crystal, Smoking dysregulates the human airway basal cell transcriptome at COPD risk locus 19q13.2, PLoS One 9 (2) (2014) e88051.
[37] K. Sudo, H. Ema, Y. Morita, H. Nakauchi, Age-associated characteristics of murine hematopoietic stem cells, J. Exp. Med. 192 (9) (2000) 1273–1280.
[38] T.H. Ambrosi, A. Scialdone, A. Graja, S. Gohlke, A.M. Jank, C. Bocian, L. Woelk, H. Fan, D.W. Logan, A. Schurmann, L.R. Saraiva, T.J. Schulz, Adipocyte accumulation in the bone marrow during obesity and aging impairs stem cell-based hematopoietic and bone regeneration, Cell Stem Cell 20 (6) (2017) 771–784.e6.
[39] N. Cardenes, D. Alvarez, J. Sellares, Y. Peng, C. Corey, S. Wecht, S.M. Nouraie, S. Shanker, J. Sembrat, M. Bueno, S. Shiva, A.L. Mora, M. Rojas, Senescence of bone marrow-derived mesenchymal stem cells from patients with idiopathic pulmonary fibrosis, Stem Cell Res. Ther. 9 (1) (2018) 257.
[40] A. Serrano-Mollar, Cell therapy in idiopathic pulmonary fibrosis, Med. Sci. (Basel) 6 (3) (2018) 64.
[41] A.S. Brack, M.J. Conboy, S. Roy, M. Lee, C.J. Kuo, C. Keller, T.A. Rando, Increased Wnt signaling during aging alters muscle stem cell fate and increases fibrosis, Science 317 (5839) (2007) 807–810.
[42] H.-C. Yang, A.B. Fogo, Fibrosis and renal aging, Kidney Int. Suppl. 4 (1) (2014) 75–78.
[43] A.K. Shetty, M. Kodali, R. Upadhya, L.N. Madhu, Emerging anti-aging strategies - scientific basis and efficacy, Aging Dis. 9 (6) (2018) 1165–1184.
[44] D. Duscher, J. Barrera, V.W. Wong, Z.N. Maan, A.J. Whittam, M. Januszyk, G.C. Gurtner, Stem cells in wound healing: the future of regenerative medicine? A mini-review, Gerontology 62 (2) (2016) 216–225.
[45] R. Lim, S.D. Ricardo, W. Sievert, Cell-based therapies for tissue fibrosis, Front. Pharmacol. 8 (2017) 633.
[46] B.E. Keyes, E. Fuchs, Stem cells: aging and transcriptional fingerprints, J. Cell Biol. 217 (1) (2018) 79–92.
[47] P.M. Helbling, E. Pineiro-Yanez, R. Gerosa, S. Boettcher, F. Al-Shahrour, M.G. Manz, C. Nombela-Arrieta, Global transcriptomic profiling of the bone marrow stromal microenvironment during postnatal development, aging, and inflammation, Cell Rep. 29 (10) (2019) 3313–3330.e4.
[48] S. Shaik, E.C. Martin, D.J. Hayes, J.M. Gimble, R.V. Devireddy, Transcriptomic profiling of adipose derived stem cells undergoing osteogenesis by RNA-Seq, Sci. Rep. 9 (1) (2019) 11800.
[49] H. Guenou, X. Nissan, F. Larcher, J. Feteira, G. Lemaitre, M. Saidani, M. Del Rio, C.C. Barrault, F.X. Bernard, M. Peschanski, C. Baldeschi, G. Waksman, Human embryonic stem-cell derivatives for full reconstruction of the pluristratified epidermis: a preclinical study, Lancet 374 (9703) (2009) 1745–1753.
[50] A. Nourian Dehkordi, F. Mirahmadi Babaheydari, M. Chehelgerdi, S. Raeisi Dehkordi, Skin tissue engineering: wound healing based on stem-cell-based therapeutic strategies, Stem Cell Res. Ther. 10 (1) (2019) 111.
[51] K. Takahashi, S. Yamanaka, Induction of pluripotent stem cells from mouse embryonic and adult fibroblast cultures by defined factors, Cell 126 (4) (2006) 663–676.
[52] J.E. Lim, Y. Son, Endogenous stem cells in homeostasis and aging, Tissue Eng. Regen. Med. 14 (6) (2017) 679–698.
[53] M.S. Hu, M.R. Borrelli, H.P. Lorenz, M.T. Longaker, D.C. Wan, Mesenchymal stromal cells and cutaneous wound healing: a comprehensive review of the background, role, and therapeutic potential, Stem Cells Int. 2018 (2018) 6901983.
[54] E.H. Rogers, J.A. Hunt, V. Pekovic-Vaughan, Adult stem cell maintenance and tissue regeneration around the clock: do impaired stem cell clocks drive age-associated tissue degeneration? Biogerontology 19 (6) (2018) 497–517.

[55] A.S. Ahmed, M.H. Sheng, S. Wasnik, D.J. Baylink, K.W. Lau, Effect of aging on stem cells, World J. Exp. Med. 7 (1) (2017) 1–10.

[56] S.K. Pazhanisamy, Stem cells, DNA damage, ageing and cancer, Hematol. Oncol. Stem Cell Ther. 2 (3) (2009) 375–384.

[57] A.C. Brane, T.O. Tollefsbol, Targeting telomeres and telomerase: studies in aging and disease utilizing CRISPR/Cas9 technology, Cell 8 (2) (2019) 186.

[58] K.C. Rustad, V.W. Wong, M. Sorkin, J.P. Glotzbach, M.R. Major, J. Rajadas, M.T. Longaker, G.C. Gurtner, Enhancement of mesenchymal stem cell angiogenic capacity and stemness by a biomimetic hydrogel scaffold, Biomaterials 33 (1) (2012) 80–90.

[59] V. Rajendran, M. Netukova, M. Griffith, J.V. Forrester, L. Kuffova, Mesenchymal stem cell therapy for retro-corneal membrane - a clinical challenge in full-thickness transplantation of biosynthetic corneal equivalents, Acta Biomater. 64 (2017) 346–356.

[60] R.N. Judson, F.M.V. Rossi, Towards stem cell therapies for skeletal muscle repair, npj Regen. Med. 5 (1) (2020) 10.

[61] L.A. Ortiz, F. Gambelli, C. McBride, D. Gaupp, M. Baddoo, N. Kaminski, D.G. Phinney, Mesenchymal stem cell engraftment in lung is enhanced in response to bleomycin exposure and ameliorates its fibrotic effects, Proc. Natl. Acad. Sci. U. S. A. 100 (14) (2003) 8407–8411.

[62] M.K. Glassberg, J. Minkiewicz, R.L. Toonkel, E.S. Simonet, G.A. Rubio, D. DiFede, S. Shafazand, A. Khan, M.V. Pujol, V.F. LaRussa, L.H. Lancaster, G.D. Rosen, J. Fishman, Y.N. Mageto, A. Mendizabal, J.M. Hare, Allogeneic human mesenchymal stem cells in patients with idiopathic pulmonary fibrosis via intravenous delivery (AETHER): a phase I safety clinical trial, Chest 151 (5) (2017) 971–981.

[63] N. Kosaric, H. Kiwanuka, G.C. Gurtner, Stem cell therapies for wound healing, Expert. Opin. Biol. Ther. 19 (6) (2019) 575–585.

[64] Y.Z. Huang, M. Gou, L.C. Da, W.Q. Zhang, H.Q. Xie, Mesenchymal stem cells for chronic wound healing: current status of preclinical and clinical studies, Tissue Eng. B Rev. (2020).

Chapter 4

Stem cells and multiomics approaches in senescence: From benchside to bedside

Atil Bisgin

Cukurova University, Faculty of Medicine, Medical Genetics Department of Balcali Hospital and Clinics, Adana, Turkey, Cukurova University AGENTEM (Adana Genetic Diseases Diagnosis and Treatment Center), Adana, Turkey

1 Introduction

Aging in humans refers to a multidimensional process of physical, psychological, and social changes. Biologically, aging is considered to begin approximately at the age of 60 due to United Nation's (UN) report of 1980. According to the World Health Organization (WHO) report in 1963, adulthood comes in three steps: 45–59 is called middle adulthood, 60–74 is adulthood, and older than 75 is the so-called late adulthood. Early scientific knowledge on the aging processes comes from longitudinal studies of the youth to adulthoods comparison. But besides these comparison studies, multiomics data including the genomics, proteomics and even phenomics together with environmental and lifestyle/quality become the most important and sometimes controversial factors that affect the changes in human by aging [1–3].

Senescence-related research studies showed significant process in the last two decades in which aging is defined as the process of irreversible proliferative deterioration by time that related to fertility and survival [4, 5]. The human lifespan has been prolonged in the last few decades. Though the aging process, progressive loss of physical capabilities and motor functions start from 60 years of age. All this involution is the sign of impaired tissue and organ function through losing physiological integrity [6, 7].

The most recent technological progress established methodological approaches to identify and study the mechanisms responsible for the aging process. These involve heterogeneous structural and functional impairment, and the decline of repair system maintenance associated with cellular and molecular mechanisms of aging [8–10].

Among all the areas mentioned above, this chapter is focused on senescence (cellular and/or biological aging) and the multiomics approach to the complex aging process, directing attention to hallmarks and biomarker discoveries.

2 "Omics" in senescence

The recent disruptive concept of "genomic medicine" in life science studies takes into account both the population characteristics as precision medicine and individual variability as personalized medicine. It comprises the customization of healthcare for an individual on the basis of providing effective and personalized treatment options by measurements and clinical findings obtained at the individual level. Hence, it is "omics" and omics-related technologies that have recently made precision medicine utilizable [2, 8].

Genes as well as their expression products (i.e., transcripts and proteins) and metabolites are the main biomarkers that underlie omics. The increase in the throughput of omics technologies together with efficiency and reliability make it possible to analyze all perspectives of biological systems in a hypothesis-free model [11].

Senescence is an appealing field for the clinical utilization of omics technologies that allows understanding the contextual kinetics of cell senescent processes while the pathophysiology divides it into two main categories—acute (programmed and transient) and chronic (not programmed and persistent). However, gene-environment interactions should be considered as well as population-individual interactions for a better descriptive approach and possible therapeutic interventions.

The domain "omic" reflects the representing type of data obtained from biological systems. Omics technologies enable overviews of molecules in cells, tissues, and organs as well as determining the genetic sequence of humans by whole genome sequencing. During senescence, the deterioration of repair mechanisms becomes progressive while there is also a decline in beneficial functions. Thus, the multiomics strategy helps to understand the decline of tissue regenerative capacity, the abandonment of organ maintenance, and the lack of tissue repair by adult stem cells. Moreover, senescence and the senescent cells have a relation to the inflammatory processes interfering with tissue health and repair via the senescence-associated secretory phenotype (SASP) [12, 13]. The understanding of the complexity of the human genome was revealed by the data from the Human Genome Project, but its effect on senescence is still under research.

Genomics approaches have helped to identify the genes involved in senescence widely from organ failures to tumor suppression. Within the most recent studies, new omics technologies have been developed to measure other biomolecules such as *epigenomics* for epigenetic biomarkers, *proteomics* for proteins, and *metabolomics* for low-molecular-weight metabolites, especially in tissue regeneration, wound healing, and their relation to stem cells [14–19].

Next-generation sequencing technologies make the targeted multigene panel studies possible, in addition to the whole exome sequencing and whole genome sequencing in routine clinical practice. Although Sanger sequencing has been used effectively and intensively in DNA sequencing and played a main role in the Human Genome Project, next-generation sequencing technologies are becoming widely available for many studies in health sciences [20]. These techniques have the ability to conserve the longer DNA sequence information in a short period of time with high accuracy. As an example in cancer, the variants detected by multigene sequencing panels provide insights into disease diagnosis as well as treatment, including targeted treatment strategies and prevention [21]. Eventually, genomics-based technologies within the context of single cell studies performed from stem cells should be integrated and introduced into transcriptomics and proteomics. Cell cycle regulators, repair mechanisms, and chromatin remodeling mechanism-related genes or even cancer stem cells can be used as the main biomarkers in senescence. The algorithm to identify and quantify the factors that may be associated with senescence and any disease-related biomarkers is shown in Fig. 1. However, it still remains difficult to identify senescent cell types or cell subpopulations within a complex tissue or organ structure due to the individual's health status.

The epigenome comprises chemical modifications of DNA, histones/nonhistone proteins, and nuclear RNA in senescent cells. The variations in the epigenome change gene expression without altering the genetic alteration in the DNA sequence that may be transient or inherited. The main and well-known epigenetic machinery includes DNA methylation, histone modification, microRNA (miRNA), and chromatin condensation. The most recent well-known studies monitoring senescent cells over the long term revealed that the progressive proteolysis of histones 3 and 4 without DNA loss has an important effect when compared to nonsenescent cells [22]. Moreover, chromatin condensation mechanisms and

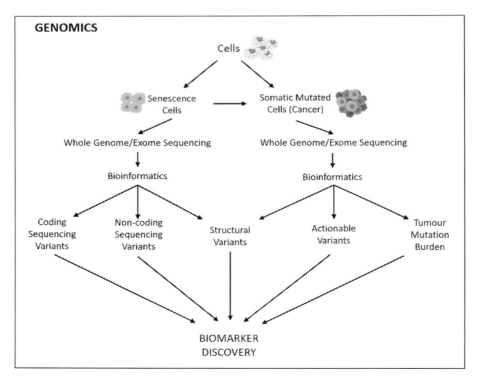

FIG. 1 Genomics.

structural changes of chromatin in senescent cells have been reported [22–24]. Thus, next-generation sequencing strategies could be used in *epigenomics* to have a clear perspective of the epigenetic machinery in senescence and the role of stem cells.

The other possible use of next-generation sequencing is transcriptome analysis, which is the sequencing of a complete set of RNA transcripts in cells or tissues called *transcriptomics*. The characteristic of senescence is the SASP, whereby proinflammatory cytokines, chemokines, matrix metalloproteinases, growth factors, and angiogenic factors are produced. However, the most recent studies are limited to the specific experimental models of cellular senescence in several cell types [25–28]. The most valuable data that can be obtained from transcriptomic studies are the identifying subsets of transcripts, coding and noncoding groups displaying shared expression patterns across a range of senescent cell models. The algorithm for transcriptomics to identify and quantify the factors that may be associated with senescence and any disease-related biomarkers is shown in Fig. 2. The modulation of age-associated processes by the senescent cells has been well identified in literature. Though, immune system and lysosomal functions are activated in senescent cells as the key factors [25, 29, 30]. Through these studies, the value of computational and statistical modeling methods to clarify the hidden structures beyond high-dimensional transcriptomic data together with the genomic data is highlighted [31].

The proteome consists of all the proteins expressed by a biological system. The protein structure and amount may differ in each single cell or organism due to the posttranslational modifications, which explains the generation of different proteins. These modifications are the most important pathways through the life span from infancy to adulthood whether having an illness or being healthy, thus, that broadens the complexity and functionalities of senescence [32].

Proteins may exhibit different conformations, localizations in a cell, and protein-protein interactions depending on the phase of the cell cycle. The proteome can mainly be analyzed in cells/tissues or organisms using mass spectrometry or

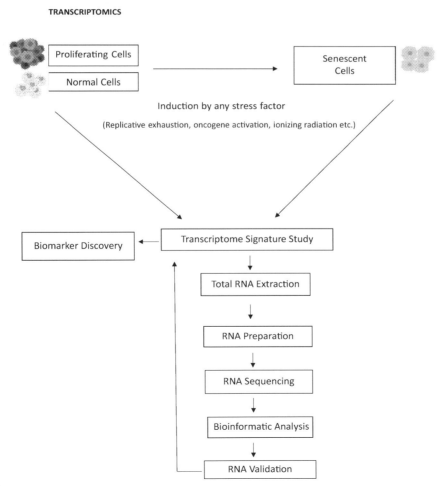

FIG. 2 Transcriptomics.

64 Stem cells and aging

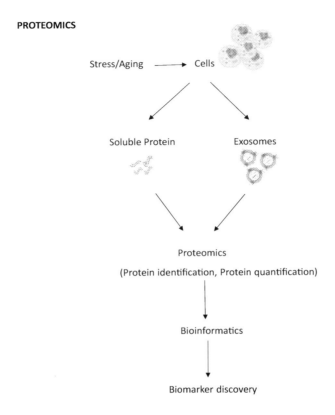

FIG. 3 Proteomics.

protein microarrays. Recent *proteomics* studies showed the relation of a few dozen secreted proteins called secretomes, which had been mostly underestimated because the novel techniques have been used in the last decade in senescence [33]. The novel SASP Atlas presents a comprehensive proteomic database of soluble proteins and exosomal SASP factors originating from different cell types, including the stem cells and senescence inducers [34]. The algorithm to identify and quantify the proteomic factors that may be associated with senescence and any disease-related biomarkers is shown in Fig. 3. Such a comprehensive profiling analysis is a valuable source for describing the senescent phenotypes to develop senescence biomarkers and moreover, to identify the individuals to treat and/or the efficiency of senescence-targeted therapeutics. However, translating the proteomic data in clinical use is still under investigation throughout the transcriptomic data of senescent cells. Thus, the number and the nature of proteins will change as the researchers interrogate different models and senescence inducers together with a multiomics approach.

The other important molecules in a human body are the metabolites, which have a relatively low molecular weight compared to proteins. These molecules can be characterized by *metabolomics* or metabolic profiling. Metabolites fluctuate due to both genetic and environmental factors. However, mass spectrometry has already been adopted in daily medical routines. Also, new advances in technologies such as nuclear magnetic resonance and ion mobility spectrometry provide diagnostic information by screening several diseases, although this doesn't yet apply to senescence and related processes [35].

Finally, the interactions between genes and the environment influence the phenotypes, which is explored by *phenomics* as a part of good clinical practice through measurable outcomes. After all, the translation of these omics technologies in a clinical context on the basis of a specific algorithm for better diagnosis and treatment from infancy to adulthood should be considered by using all the omics technologies in an integrative manner.

3 Multiomics data integration and its application

A comprehensive analysis of multiomics data is still only limited to human health and several diseases. The main problem of understanding senescence by the multiomics approach is the high requirements of interpretation at any level of omics:

TABLE 1 Multiomics databases listed in relation to disease.

Database name	Web address	Disease
The Cancer Genome Atlas (TCGA)	https://www.cancer.gov/about-nci/organization/ccg/research/structural-genomics/tcga	Cancer
International Cancer Genomics Consortium (ICGC)	https://icgc.org/	Cancer
Clinical Proteomic Tumor Analysis Consortium (CPTAC)	https://proteomics.cancer.gov/programs/cptac	Cancer
Therapeutically Applicable Research to Generate Effective Treatments (TARGET)	https://ocg.cancer.gov/programs/target	Pediatric cancers
Omics Discovery Index	https://www.omicsdi.org/	Consolidated datasets in a uniform framework
Molecular Taxonomy of Breast Cancer International Consortium (METABRIC)	https://europepmc.org/	Breast cancer
Cancer Cell Line Encyclopedia (CCLE)	https://portals.broadinstitute.org/ccle	Cancer cell line

genome, epigenome, transcriptome, proteome, and metabolome. There is still limited available multiomics databases along with clinical information, as listed in Table 1.

Various studies more focused on cancer combined omics datasets for better understanding and a clearer view of the complex biology. The most widely used database to make new discoveries about the progression, manifestation, and treatment is the Pan-Cancer Atlas (The Cancer Genome Atlas, TCGA) and the International Cancer Genomics Consortium (ICGC) [36]. However, the omics spectrum is beginning to extend much deeper such as lipidomes, phosphoproteomes, and glycol-proteomes. However, many more studies in senescence are needed to derive actionable and meaningful insights to interpret and integrate the multiomics data in clinical practice.

4 Stem cell-based multiomics and data integration

Over the past decades, much evidence has been accumulated from the omics-based studies in stem cells with great advances in knowledge. However, the senescence mechanisms and their relation to stem cells are not entirely clear yet. Thus, it is crucial to explore novel strategies from single stem cell studies to tissue- and disease-based studies via a multiomics approach to understand the mechanism of stem cell differentiation and proliferation.

Most recent studies provide an overview of the mesenchymal stem cells (MSC) that contributed:

i. The molecular mechanisms of senescence, including epigenetic changes, autophagy, mitochondrial dysfunction, and telomere shortening [37–39].
ii. Gene modifications and pretreatment strategies [40–42].
iii. Rejuvenation of senescent MSCs (e.g., autologous MSC-based therapy) [43–45].

On the other edge of senescence, there is also the senescence of MSCs that occurs after multiple divisions by entering replicating senescence state results in growth arrest. But more than that, based on the most recent published data, MSC senescence has been described in four types: replicative senescence, stress-induced senescence, oncogene-induced senescence, and developmental senescence [46].

The integration of multiomics data is not still as easy as moving toward the bedside. The primary key is the right choice of methodology to address senescence with the biological and molecular questions about stem cells for integrative analysis. There are some visualization portals that perform multiomics data analysis and benchmarking, but they are still not comprehensive enough (Table 2).

Additionally, the other most important dimension of stem cell studies that could add value to multiomics data interpretation is the clinical information. However, currently there are not enough clinical studies or a robust method to integrate omics data with the clinical metadata as the nonomics data.

TABLE 2 List of portals with source repository and supported omics data.

Visualization portal	Web address	Supported omics data
cBioPortal	https://www.cbioportal.org	Genomics, epigenomics, transcriptomics, proteomics, and clinical data
LinkedOmics	http://www.linkedomics.org	Genomics, epigenomics, transcriptomics, proteomics, clinical data, and phosphoproteome and glycoproteome data
Firebrowse	http://firebrowse.org/	Genomics, epigenomics, transcriptomics, proteomics, and clinical data
UCSC Xena	https://xena.ucsc.edu/	Genomics (also somatic), epigenomics, transcriptomics (also tissue specific), proteomics, and phenomics
3Omics	https://3omics.cmdm.tw	Transcriptomics, proteomics, and metabolomics
OASIS	https://www.oasisgenetics.com/	Genomics and transcriptomics
Paintomics 3	http://www.paintomics.org/	Transcriptomics, metabolomics, and epigenomics
NetGestalt	http://www.netgestalt.org/	Genomics and transcriptomics

5 Conclusion

To sum up, several limitations restrain the applications of stem cell studies but not the multiomics approach from bench to bedside. The identification of key pathways regulating senescence from the genomic level to the phenomic and environmental levels is the most important issue. Therefore, more clinical studies will be provided by the suppression or activation of these pathways. Thus, more studies have to be weighed against the side effects and the therapeutic efficiency by interventions.

References

[1] J. Zierer, C. Menni, G. Kastenmüller, et al., Integration of 'omics' data in aging research: from biomarkers to systems biology, Aging Cell 14 (2015) 933–944.
[2] A. Cellerino, A. Ori, What have we learned on aging from omics studies? in: Seminars in Cell & Developmental Biology, Elsevier, 2017, pp. 177–189.
[3] A.D. Yalcin, A. Bisgin, R.M. Gorczynski, Immunosenescence, Erciyes Med. J. 33 (2011) 229–234.
[4] P. Sousa-Victor, S. Gutarra, L. García-Prat, et al., Geriatric muscle stem cells switch reversible quiescence into senescence, Nature 506 (2014) 316–321.
[5] J. Duan, J. Duan, Z. Zhang, et al., Irreversible cellular senescence induced by prolonged exposure to H2O2 involves DNA-damage-and-repair genes and telomere shortening, Int. J. Biochem. Cell Biol. 37 (2005) 1407–1420.
[6] P.D. Adams, Healing and hurting: molecular mechanisms, functions, and pathologies of cellular senescence, Mol. Cell 36 (2009) 2–14.
[7] M.V. Blagosklonny, Aging-suppressants: cellular senescence (hyperactivation) and its pharmacologic deceleration, Cell Cycle 8 (2009) 1883–1887.
[8] R.M. Naylor, D.J. Baker, J.M. Van Deursen, Senescent cells: a novel therapeutic target for aging and age-related diseases, Clin. Pharmacol. Ther. 93 (2013) 105–116.
[9] J.M. Sedivy, G. Banumathy, P.D. Adams, Aging by epigenetics—a consequence of chromatin damage? Exp. Cell Res. 314 (2008) 1909–1917.
[10] K. Itahana, G. Dimri, J. Campisi, Regulation of cellular senescence by p53, Eur. J. Biochem. 268 (2001) 2784–2791.
[11] T. Nagano, M. Nakano, A. Nakashima, et al., Identification of cellular senescence-specific genes by comparative transcriptomics, Sci. Rep. 6 (2016) 1–13.
[12] A.R.J. Young, M. Narita, SASP reflects senescence, EMBO Rep. 10 (2009) 228–230.
[13] D.V. Faget, Q. Ren, S.A. Stewart, Unmasking senescence: context-dependent effects of SASP in cancer, Nat. Rev. Cancer 19 (2019) 439–453.
[14] R. Lister, M. Pelizzola, Y.S. Kida, et al., Hotspots of aberrant epigenomic reprogramming in human induced pluripotent stem cells, Nature 471 (2011) 68–73.
[15] W. Xie, M.D. Schultz, R. Lister, et al., Epigenomic analysis of multilineage differentiation of human embryonic stem cells, Cell 153 (2013) 1134–1148.
[16] M.G. Roubelakis, K.I. Pappa, V. Bitsika, et al., Molecular and proteomic characterization of human mesenchymal stem cells derived from amniotic fluid: comparison to bone marrow mesenchymal stem cells, Stem Cells Dev. 16 (2007) 931–952.
[17] H. Baharvand, M. Hajheidari, S.K. Ashtiani, et al., Proteomic signature of human embryonic stem cells, Proteomics 6 (2006) 3544–3549.
[18] A.D. Panopoulos, O. Yanes, S. Ruiz, et al., The metabolome of induced pluripotent stem cells reveals metabolic changes occurring in somatic cell reprogramming, Cell Res. 22 (2012) 168–177.

[19] H. Sperber, J. Mathieu, Y. Wang, et al., The metabolome regulates the epigenetic landscape during naive-to-primed human embryonic stem cell transition, Nat. Cell Biol. 17 (2015) 1523–1535.

[20] R. Bao, L. Huang, J. Andrade, et al., Review of current methods, applications, and data management for the bioinformatics analysis of whole exome sequencing, Cancer Inform. 13 (2014), https://doi.org/10.4137/CIN.S13779.

[21] O. Sonmezler, I. Boga, A. Bisgin, Integration of liquid biopsies into clinical laboratory applications via NGS in cancer diagnostics, Clin. Lab. 66 (2020).

[22] A. Ivanov, J. Pawlikowski, I. Manoharan, et al., Lysosome-mediated processing of chromatin in senescence, J. Cell Biol. 202 (2013) 129–143.

[23] M. Kosar, J. Bartkova, S. Hubackova, et al., Senescence-associated heterochromatin foci are dispensable for cellular senescence, occur in a cell type- and insult-dependent manner and follow expression of p16ink4a, Cell Cycle 10 (2011) 457–468.

[24] P.P. Shah, G. Donahue, G.L. Otte, et al., Lamin B1 depletion in senescent cells triggers large-scale changes in gene expression and the chromatin landscape, Genes Dev. 27 (2013) 1787–1799.

[25] J.M. Van Deursen, The role of senescent cells in ageing, Nature 509 (2014) 439–446.

[26] A. Serra, Value of drug-coated balloons in small-vessel disease: have they come of age? JACC: Cardiovasc. Interv. (2018) 2393–2395.

[27] A.G. Georgakilas, O.A. Martin, W.M. Bonner, p21: a two-faced genome guardian, Trends Mol. Med. 23 (2017) 310–319.

[28] B.Y. Lee, J.A. Han, J.S. Im, et al., Senescence-associated β-galactosidase is lysosomal β-galactosidase, Aging Cell 5 (2006) 187–195.

[29] B.G. Childs, M. Durik, D.J. Baker, et al., Cellular senescence in aging and age-related disease: from mechanisms to therapy, Nat. Med. 21 (2015) 1424–1435.

[30] J.P. De Magalhães, J. Curado, G.M. Church, Meta-analysis of age-related gene expression profiles identifies common signatures of aging, Bioinformatics 25 (2009) 875–881.

[31] G. Casella, R. Munk, K.M. Kim, et al., Transcriptome signature of cellular senescence, Nucleic Acids Res. 47 (2019) 7294–7305.

[32] A.L. Santos, A.B. Lindner, Protein posttranslational modifications: roles in aging and age-related disease, Oxid. Med. Cell. Longev. 2017 (2017), 5716409.

[33] K.J. Brown, C.A. Formolo, H. Seol, et al., Advances in the proteomic investigation of the cell secretome, Expert Rev. Proteomics 9 (2012) 337–345.

[34] N. Basisty, A. Kale, O.H. Jeon, et al., A proteomic atlas of senescence-associated secretomes for aging biomarker development, PLoS Biol. 18 (2020), e3000599.

[35] C.D. Wiley, S. Davis, A. Ramanathan, Measurement of metabolite changes in senescent cells by mass spectrometry, Methods Mol. Biol. 1896 (2019) 139–147.

[36] J.N. Weinstein, E.A. Collisson, G.B. Mills, et al., The cancer genome atlas pan-cancer analysis project, Nat. Genet. 45 (2013) 1113–1120.

[37] R. Watanabe, S.-i. Kanno, A.M. Roushandeh, et al., Nucleosome remodelling, DNA repair and transcriptional regulation build negative feedback loops in cancer and cellular ageing, Philos. Trans. R. Soc. B 372 (2017) 20160473.

[38] N. Noren Hooten, M.K. Evans, Techniques to induce and quantify cellular senescence, J. Vis. Exp. (2017) 55533.

[39] W. Wagner, The link between epigenetic clocks for aging and senescence, Front. Genet. 10 (2019) 303.

[40] F. Galkin, B. Zhang, S.E. Dmitriev, et al., Reversibility of irreversible aging, Ageing Res. Rev. 49 (2019) 104–114.

[41] Y. Ma, M. Qi, Y. An, et al., Autophagy controls mesenchymal stem cell properties and senescence during bone aging, Aging Cell 17 (2018) e12709.

[42] J. Seok, H.S. Jung, S. Park, et al., Alteration of fatty acid oxidation by increased CPT1A on replicative senescence of placenta-derived mesenchymal stem cells, Stem Cell Res. Ther. 11 (2020) 1.

[43] J. Neves, P. Sousa-Victor, H. Jasper, Rejuvenating strategies for stem cell-based therapies in aging, Cell Stem Cell 20 (2017) 161–175.

[44] Y. Sun, Y. Zheng, C. Wang, et al., Glutathione depletion induces ferroptosis, autophagy, and premature cell senescence in retinal pigment epithelial cells, Cell Death Dis. 9 (2018) 1–15.

[45] D.K.W. Ocansey, B. Pei, Y. Yan, et al., Improved therapeutics of modified mesenchymal stem cells: an update, J. Transl. Med. 18 (2020) 42.

[46] X. Zhou, Y. Hong, H. Zhang, et al., Mesenchymal stem cell senescence and rejuvenation: current status and challenges, Front. Cell Dev. Biol. 8 (2020) 364.

Chapter 5

Signaling pathways influencing stem cell self-renewal and differentiation

Mahak Tiwari[a,b,*], Sinjini Bhattacharyya[a,b,*], and Deepa Subramanyam[a]
[a]*National Centre for Cell Science, Pune, India,* [b]*Savitribai Phule Pune University, Pune, India*

1 Introduction

Fertilization of an ovum by a sperm results in the formation of a single-celled embryo called the zygote. The zygote then undergoes a series of divisions to give rise to a structure called the blastocyst, consisting of an outer layer of cells forming the trophectoderm, and a handful of cells located toward the inner side of the embryo, called the inner cell mass. It is the cells of the inner cell mass that goes on to form the entire embryo. Isolation of the inner cell mass and maintenance of these cells in culture, henceforth called embryonic stem cells (ESCs), have helped establish a system to study early developmental changes and decisions [1, 2]. These embryonic stem cells have two distinct properties—the ability to self-renew, and to differentiate into cell types of all three lineages. This chapter focuses on four main signaling pathways involved in the maintenance of the pluripotent state of ESCs and their differentiation into specialized cells.

2 LIF and JAK/STAT3 signaling pathway

The leukemia inhibitory factor (LIF) belongs to the interleukin-6 (IL-6) cytokine family, which is involved in various biological processes including inflammation, immune responses, and embryonic development [3]. LIF was first identified as an essential cytokine for the pluripotency of mESCs during the late 1980s. During that period, different independent studies led to the identification of molecules that inhibited differentiation and promoted the proliferation of mESCs. These included DIA (differentiation inhibiting activity) produced by conditioned medium from buffalo rat liver cells, LIF from the conditioned medium of Krebs II ascite cells, and HILDA (human interleukin for DA cells) that facilitated the proliferation of the mouse leukemia cell line, DA-1a. All these molecules share sequence similarity, and function in a similar manner in the context of ESCs [4].

During blastocyst formation, LIF is highly expressed by the uterine endometrial glands in both humans and mice [5]. It is also expressed by the granulosa-lutein and ovarian stromal cells [6]. LIF expression is higher in the glandular epithelium than in the luminal epithelium. However, the expression of the LIF receptor (LIFR) is greater in the endometrial epithelium compared to the glandular epithelium. After blastocyst attachment, trophoblasts express both LIF and LIFR throughout pregnancy [7]. In addition to their production by ovarian cells, LIF is also produced by other cell types including endometrial cells, fibroblasts, monocytes, macrophages, and T cells [8–10].

The role of LIF during embryo implantation became apparent when it was observed that LIF knockout female mice were unable to attach the implanting blastocyst. However, when LIF was infused into the uterus of these mice, the blastocysts attached and grew [11]. *Lifr* mutant mice showed normal implantation. However, mutant neonates died within 24 h of birth due to multiple defects caused by abnormal placentation [12]. Several clinical studies have reported that lower expression of LIF in endometrial cells leads to infertility and repeated abortions [13, 14]. Diapause, a phenomenon of arrested embryonic development and delayed implantation of the mouse blastocyst, played a major role in the early derivation of mESCs [1]. During diapause, maternal estrogen induced the secretion of LIF from the trophectoderm to maintain the self-renewal of the ICM cells [15], contributing to the ease of ESC isolation from blastocysts [16].

The canonical LIF/JAK/STAT3 pathway begins with the binding of LIF to its receptor, LIFR that further recruits a membrane protein gp130. LIFR and gp130 form a heterodimer, which together result in the activation of Janus kinases (JAKs) by phosphorylation of a tyrosine residue at position 1022. There are four known JAK proteins: JAK1, JAK2, JAK3, and TYK2 that contain seven JAK homology (JH) domains, 1–7. The term Janus was given due to the presence of the JH1

---

* These authors contributed equally.

(catalytic domain for tyrosine kinase activity) and JH2 (kinase like-domain) domains of JAKs, which are adjacent to each other resembling the two-headed Roman God, Janus. JH3 and JH4 domains have homology with the Src-homology-2 (SH2) domains. JH4-JH7 domains together constitute a domain called FERM (band4.1, ezrin, radixin, and moesin), which is involved in the association of JAK with cytokine receptors. JAK1 and JAK2 are the kinases that are mainly involved in the LIF signaling pathway. It is reported that *Jak1* knockdown mESCs require a higher concentration of LIF for self-renewal than wild-type mESCs [17]. Activated JAKs function by further recruiting and phosphorylating STAT3 (signal transducer and activator of transcription 3) [18, 19].

STAT3 is a transcription factor belonging to the STAT family of proteins. STAT3 has six domains: a coiled-coiled domain, a DNA-binding domain, a dimerization domain, a linker domain, a transactivation domain, and an SH2 domain. Phosphorylation of STAT3 at tyrosine 705 by active JAK causes STAT3 dimerization via reciprocal SH2 interaction, followed by translocation to the nucleus to activate target gene transcription [20]. This is achieved by binding STAT3 to the consensus sequence TTCCSGGAA present in the enhancer region of its target genes [21]. STAT3 can also be phosphorylated at the Serine 727 position through the action of the LIF signaling pathway [22]. However, the implication of this particular phosphorylation event in the context of mESC self-renewal remains unclear. Other members of the IL-6 cytokine family including cardiotrophin 1, ciliary neurotrophic factor, and Oncostatin have also been shown to help maintain mESC self-renewal [23–25].

mESCs can be maintained independent of LIF, by the constitutive activation of STAT3 [26]. However, LIF failed to support the self-renewal of *Stat3*$^{-/-}$ mESCs [27], or in cells overexpressing a dominant negative mutant of *Stat3* [28]. It has also been reported that mutation of the alanine residue of STAT3 at position 661 and asparagine at position 663 to cysteine residues, could induce STAT3 dimerization without the phosphorylation of tyrosine 705. This mutant STAT3 can function as a constitutively active form and is called STAT3C [29].

Various studies have identified downstream targets of STAT3 that are involved in maintaining the pluripotency of mESCs. Genome-wide studies by various research groups have reported that STAT3 binding sites identified by chromatin immunoprecipitation (ChIP) sequencing technology are co-occupied/bound by pluripotency regulators such as Oct4, Sox2, and Nanog [21, 30]. The STAT3 target gene list comprises of both transcriptionally active and inactive genes. The active genes include the transcription factors Myc, Gbx2 (gastrulation brain homeobox 2), and Pim1/2, whose sustained expression maintains the self-renewal of mESCs even in the absence of LIF [31–33]. The list of inactive genes includes tissue-specific genes such as T-brachyury, Gata4, and Eomes, to name a few [30]. Bourillot et al. [34] have reported that at least 22 STAT3 target genes were required to maintain mESCs in an undifferentiated state, preventing the induction of the mesodermal and endodermal lineage. A large number of STAT3 targets were also co-bound and coregulated by Nanog. STAT3 thus plays a role in suppressing lineage-specific genes, thereby maintaining the self-renewal of mESCs.

The JAK/STAT signaling pathway is subject to regulation by numerous other proteins [3]. These include phosphatases such as SHPs, PTP-BL, and PTP1B that dephosphorylate tyrosine residues in the JAKs and STAT3. Additionally, a protein family called PIAS (protein inhibitor of activated STAT) can bind directly to STAT3 and inhibit its function [35]. Expression levels of members of the SOCS (suppressors of cytokine signaling) protein family are also upregulated upon LIF exposure. SOCS can inhibit JAK/STAT signaling, thereby forming a negative feedback loop. Upregulated SOCS1 and SOCS3 can bind to JAK1 and gp130 that can either result in their ubiquitination, or directly inhibit JAK's catalytic activity [36–38].

Several studies have shown that the overexpression of pluripotency genes such as *Nanog*, *Klf4*, and *c-Myc* can maintain the self-renewal of mESCs even in the absence of LIF signaling [27, 32, 39, 40]. A recent report in 2019 explained the role of Nanog in mESC self-renewal in the context of LIF signaling. NANOG, in the presence of LIF, promotes chromatin accessibility for pluripotency factors such as OCT4, SOX2, and ESRRB to enhancers in mESCs. However, in the absence of LIF, it can block differentiation by maintaining the H3K27me3-repressive mark on differentiation genes such as *Otx2* [41]. mESCs cultured without LIF can also be maintained in their self-renewing and pluripotent state through the inducible expression of *Oct4* (iOct4) in media containing the GSK3β inhibitor and FBS [42].

Overexpression of other factors including *Klf2*, *Klf5*, *Pramel7*, *Pim1/3*, *Tfcp2l1*, *MnSOD*, and *Gbx2* (gastrulation brain homeobox 2) in mESCs can also recapitulate the effects of LIF [27, 31, 33, 43–46]. A report from Wang et al. [47] shows that *Gbx2* can maintain the naïve pluripotent state of mESCs by inducing the expression of *Klf4*. Apart from maintaining the pluripotent state of mESCs, the LIF/JAK/STAT3 pathway is also important for the reprogramming of somatic cells to induced pluripotent stem cells (iPSCs). The overexpression of estrogen-related receptorβ (*Esrrb*) in partially reprogrammed cells (pre-iPSCs), which have repressed JAK/STAT3 signaling, allows resumption of complete reprogramming [48]. Elevated levels of LIF/STAT3 signaling were also sufficient to completely reprogram partially reprogrammed pre-iPS cells [49].

Along with LIF that inhibits endodermal and mesodermal differentiation, BMP2 and BMP4 proteins also help to maintain mESCs in an undifferentiated state by inhibiting neural differentiation [50], and mitogen-activated protein kinase (MAPK) pathways [51]. Other than LIF and BMP proteins, inhibitors of other signaling pathways can also be used to maintain the self-renewal of mESCs. These include CHIR99021, PD184352, and SU5402, which inhibit GSK3β, MEK, and FGF receptor tyrosine kinases, respectively [52].

Other than the JAK/STAT3 signaling pathway, LIF also triggers the PI3K (phosphoinositide 3-kinase)/AKT pathway and SHP2/MAPK (Src homology 2/mitogen-activated protein kinase) pathway via LIFR/gp130 receptor dimerization that further activates downstream signaling cascades [53, 54]. In the context of the PI3K/AKT pathway, LIF signaling leads to the phosphorylation of the regulatory subunit, p85, resulting in the activation of AKT serine/threonine kinases. Activation of AKT causes the inhibition of GSK3β (glycogen synthase kinase 3β) either by phosphorylating Serine at position 9 [55], or by facilitating its nuclear export [56]. Inhibition of GSK3β results in the upregulation of expression of NANOG and c-MYC, thereby maintaining ESC self-renewal. A report from Watanabe et al. has shown that constitutive expression of AKT can support the self-renewal of ESCs even in the absence of LIF [57].

As discussed previously, the phosphorylation of tyrosine at position 705 on STAT3 as a consequence of signaling downstream of LIF can activate STAT3, and other downstream targets, required for mESC self-renewal. Similarly, acetylation of the lysine residue at position 685 can activate STAT3 by promoting the formation of stable dimers, thereby driving active transcription of target genes without the need for tyrosine phosphorylation [58]. It should not be overlooked that the PI3K/AKT pathway can also be induced by various other factors including FGF4 [52], Insulin, insulin-like growth factor 1 (IGF1) [59], ESC-specific Ras-like protein (ERAS) [60], and retinol [61]. FGF4-induced activation of the PI3K/AKT pathway has been shown to result in the differentiation of mESCs. However, this pathway has not been studied as well as the MAPK/ERK pathway in the context of mESC differentiation [52,62].

The SH2/MAPK signaling pathway is also activated by the binding of LIF to LIFR, which results in the phosphorylation of tyrosine residues of gp130 by JAK. This activation leads to the recruitment and phosphorylation of SHP2 by JAKs, which then interacts with the growth factor receptor-bound protein 2 (Grb2)—Son of Sevenless (SOS) complex. This interaction leads to the activation of the Ras/Raf/MEK/ERK signaling cascade and thereby MAPK [63–66]. This signaling cascade is usually involved in the differentiation of mESCs by inhibiting the expression of Nanog [67] and T-box3 (Tbx3) [68]. However, this is balanced by the activation of other pathways such as JAK/STAT3, thus resulting in the maintenance of ESC pluripotency.

Despite possessing similar core pluripotency networks, mESCs and human ESCs (hESCs) are biologically very different. Human ESCs are not dependent on the LIF/JAK/STAT3 pathway for their self-renewal. Various studies have shown that inspite of STAT3 phosphorylation and its translocation to the nucleus, LIF fails to maintain the self-renewal of rat and human ESCs [22,69,70]. Instead, hESCs are dependent on FGF2 and Activin A for the maintenance of their pluripotency and self-renewal. However, it has also been recently reported that high levels of LIF/STAT3 activation were present in naïve human ESCs [71]. The reprogramming of human fibroblasts in the presence of LIF and five transcription factors *Oct4*, *Sox2*, *Nanog*, *Klf4*, and *c-Myc* can produce human-induced pluripotent stem cells that represent the naïve state of pluripotency and resemble mESCs [72].

An explanation for the differences between mESCs and hESCs became clear upon the isolation and establishment of epiblast stem cells (epiSCs) from the mouse epiblast [73,74]. Mouse epiSCs are very similar to hESCs in terms of colony morphology and their dependence on FGF2 and Activin A for self-renewal. A number of other studies have also pointed to the presence of at least two states of pluripotency—the "naïve" state represented by mESCs and the "primed" state that is represented by hESCs and mEpiSCs. While both naïve and primed mESCs can induce teratomas upon injection into immunocompromised mice, primed mESCs are unable to contribute to chimera formation. Additionally, *Stella* and *Rex1* are expressed only in naïve mESCs, whereas *Fgf5*, *T-brachyury*, and *Lefty* are expressed in mEpiSCs [62,73,74]. The naïve state is also referred to as the ground state and is achieved by culturing mESCs in media containing two inhibitors (2i), namely the GSK3β inhibitor, CHIR99021, and the MEK inhibitor, PD18352 [52]. It is reported that *Nanog* expression is inhibited by the action of GSK3β. Therefore, the inhibition of GSK3β relieves the repression on Nanog, thus maintaining mESC self-renewal. Similarly, inhibition of the MEK pathway, which is involved in cellular differentiation, helps maintain mESCs in a pluripotent state.

The naïve state of pluripotency can also be converted to the primed state by replacing LIF and 2i with FGF2 and Activin A [75]. Conversely, the dedifferentiation of the primed state to the naïve state was also possible by replacing FGF2 and Activin A with LIF and 2i, and overexpressing *Klf4* [75]. Enhanced activation of the JAK/STAT3 pathway was sufficient to convert cells to the primed state [71], even in the presence of FGF2 and Activin A [49]. Forced expression of STAT3 in combination with 2i/LIF could also reprogram primed hESCs to a naïve pluripotent state [76,77]. Human naïve or ground pluripotent stem cells (PSCs) can also be maintained by culturing them in LIF, TGF-β1, FGF2, 2i, JNKi, p38i, ROCKi, and PKCi called naïve human stem cell medium (NHSM) [78–82].

The LIF/JAK/STAT3 signaling has been studied extensively in maintaining the pluripotency of ESCs. Study of this signaling pathway is crucial in elucidating the differences between the naïve and primed states of pluripotency. Its function in combination with other signaling pathways such as LIF/PI3K/AKT, LIF/SHP2/MAPK, and the TGF-β signaling pathway may help us understand the distinction between the two states of pluripotency and the differentiation of ESCs.

3 TGF-β/Smad pathway

The transforming growth factor (TGF)-β superfamily is a group of proteins that are structurally related regulatory proteins that exist as a precursor protein and are converted to active ligands after cleavage at the N-terminus to form homo- or heterodimers linked by a single disulfide bond. This family involves a number of proteins including TGF-β, Activin, Nodal, Lefty, bone morphogenetic proteins (BMPs). The TGF-β superfamily can be divided into two distinct groups, one group comprising factors such as TGF-β, Activin, Nodal, and the other group including BMP, growth and differentiation factors (GDFs), and Mullerian inhibiting substance (MIS). The TGF-β superfamily of proteins has been involved in a variety of biological processes such as cell fate determination, morphogenesis, apoptosis, cell proliferation, and differentiation.

The TGF-β subfamily has three mammalian isoforms: TGF-β1, TGF-β2, and TGF-β3. The fourth member TGF-β4 has only been identified in birds, and TGF-β5 is found only in frogs [83]. The TGF-β isoforms share 70%–80% similarity in peptide sequence and are encoded as precursor proteins. The activated form of these isoforms is produced by proteolytic cleavage of the proregion of 20–30 amino acids present at N-terminal.

There are three different classes of receptors of the TGF-β family, which include type I (TGFβRI, also termed activin-like kinases (ALKs)), type II (TGFβRII), and type III receptors (TGFβRIII) [84,85]. Type I and type II receptors have serine/threonine kinase activity in their intracellular domain [86]. Activated TGF-β ligands can initiate a signaling cascade upon binding to TGF-β receptors I and II.

BMPs are a group of cytokines that were initially discovered for their role in bone and cartilage formation [87]. BMP signaling plays an important role in embryonic development, specifically embryonic patterning and skeletal development [88–90]. Till date, around 20 BMP proteins have been identified. With the exception of BMP1, which is a metalloprotease involved in cartilage development, all other BMPs belong to the TGF-β superfamily. Several BMPs are referred to as GDFs. BMP signaling initiates with the interaction of the ligand with BMP receptors (BMPRs) present on the cell surface. This interaction results in the mobilization of the Smad family of proteins [91].

Activins are protein hormones that exist as dimers of two β subunits, βA and βB, and are present in a wide range of tissues [92]. Primarily, they have a role in regulating follicular stimulating hormone (FSH) secretion [93]. They also initiate TGF-β signaling by binding with type I and type II TGF-β receptors [94].

Nodal is a secretory protein and is involved in cell differentiation during early embryogenesis [95]. The Nodal pathway plays a role in nervous system patterning, mesoendoderm induction, and determination of the dorsal-ventral axis [96]. Activation of Nodal signaling begins with the binding of Nodal to Activin resulting in the phosphorylation of Smad proteins and further activation of a downstream signaling cascade and transcription of genes such as Lefty, Cerberus [97].

The Lefty proteins (left-right determination factors) are antagonists of Nodal signaling. Humans and mice have two Lefty homologs—Lefty 1 and Lefty 2. As the name suggests, they play a role in determination of left-right asymmetry of organs during development. Expression of Lefty depends upon Nodal signaling, thus functioning as a feedback inhibitor for the Nodal pathway [97–99].

Activation of both TGF-β signaling pathways (BMP/GDF and TGF-β/Activin/Nodal) is initiated by the binding of the ligand to the transmembrane type I and type II TGF-β receptors on the cell surface. Binding of the ligand to the receptor results in phosphorylation of the SMAD proteins by activity of the serine/threonine kinase present in the intracellular domain of the receptors. Phosphorylated SMAD proteins form complexes with SMAD4 and translocate into the nucleus to activate transcription of target genes [100].

SMADs are the signal transducer proteins mainly involved in this pathway. SMAD is an abbreviation for the *C. elegans* SMA ("small" worm phenotype) and *D. melanogaster* MAD ("mothers" against decapentaplegic) gene family [101]. They are grouped into three categories: receptor-regulated SMADs (R-SMADs) SMAD1/SMAD2/SMAD3/SMAD5/SMAD8, the common partner SMAD4 (co-SMAD4), and the inhibitory SMADs (I-SMADs), SMAD6/SMAD7 [86,102].

TGF-β family ligands such as BMP, Nodal, and Activin are important for embryonic axis formation and tissue patterning during embryogenesis. In mice, Nodal is expressed throughout the epiblast and is required for anterior/posterior (AP) axis formation. Upon Nodal signaling, the distal visceral endoderm (DVE) is induced, which results in the secretion of Nodal antagonists such as Lefty, Cerberus from cells of visceral endoderm that further maintains the signaling gradient of Nodal along the proximal-distal axis. Nodal also induces DVE cells to migrate and form the anterior visceral endoderm (AVE) [103–105]. Various studies have reported that Nodal mutant embryos fail to form

AVE or primitive streak and therefore lack AP axis formation [96,106]. It is reported that conditional mutations in the Nodal gene in the VE result in incomplete DVE/AVE migration and decreased expression levels of Nodal in the epiblast [107]. It has also been reported that SMAD2 gets activated in the VE [96] and is required for DVE formation, which in turn expresses Nodal antagonists to inhibit Nodal signaling in the VE. Loss of SMAD2 results in the loss of expression of Nodal antagonists, leading to increased Nodal expression in the epiblast causing the induction of primitive streak formation [108,109].

Nodal also plays an important role in left/right (L/R) axis specification. Nodal is expressed symmetrically at the lateral edges of the mouse node at E7.0 but is expressed in the left lateral plate mesoderm (LPM) after E8.0. Mouse embryos lacking Nodal have shown multiple L/R patterning defects and also fail to induce molecular asymmetry in the left LPM. Nodal signaling induces the transcription of both Nodal and Lefty. Lefty acts as an antagonist for Nodal and hence mediates a negative feedback loop for regulating Nodal signaling [110].

BMP signaling is involved in the early stages of L/R patterning during node formation, with *BMP4* mutant mice showing compromised node formation [111]. It is also reported that ciliogenesis is malformed in mice lacking BMP type I receptor due to the cell cycle arrest of node cells [112]. Various reports have suggested that BMP signaling regulates asymmetrical Nodal expression in the LPM. Mouse embryos lacking *BMP4* fail to express Nodal in the left LPM [111]. BMP4 also inhibits Nodal expression in the right LPM, thereby allowing Nodal expression only in the left side of the embryo [113].

BMP signaling is also involved in limb development in mice. Mice lacking *BMP2* and *BMP4* have defects in posterior digit development [114]. The downstream transducers of this pathway also affect early mammalian development, with $Smad2^{+/-}$, $Smad3^{-/-}$ mutant embryos showing impaired anterior axial mesoendoderm during gastrulation. $Smad2^{-/-}$, $Smad3^{-/-}$ homozygous mutants do not develop mesoderm and also fail to undergo gastrulation [115].

BMP signaling initiates with the binding of BMP ligands to type I ALK2/ALK3/ALK6 receptors, resulting in the phosphorylation of SMAD1/SMAD5/SMAD8. Phosphorylated SMAD1/SMAD5/SMAD8 forms a complex with Co-SMAD4 that then translocates to the nucleus and activates transcription of target genes. In association with LIF/STAT3 signaling, the BMP pathway helps maintain the self-renewal of mESCs. BMP signaling inhibits neural differentiation of mESCs through SMAD1/SMAD5/SMAD8, while STAT3 signaling prevents mesoderm and endoderm differentiation [50,116]. It is reported that the binding sites of SMADs were identified in inhibitor of differentiation (Id) promoter regions, which are known to inhibit neural differentiation [50,117]. It has also been reported that BMP4, produced by mouse embryonic fibroblast (MEF) feeder cells maintains the pluripotency of mESCs by inhibiting extracellular receptor kinase (ERK) and p38-mitogen-activated protein kinase (MAPK) [51]. It has also been shown that BMP signaling can upregulate the expression of dual-specificity phosphatase 9 (DUSP9) via SMAD1/5 activation, resulting in reduced phosphorylation of ERK [118]. SMAD1 has been shown to have common targets with those of the pluripotency factors OCT4, SOX2, and NANOG, as reported by genome-wide chromatin occupancy analysis [21]. BMP signaling also plays a role in regulating the neural commitment of mouse ESCs. It is reported that neural differentiation of mESCs occurs in two stages: the first step requires the conversion of mESCs to mEpiSCs followed by a second step involving the conversion of mEpiSCs to neural precursor cells. BMP4 inhibits the conversion of ESCs into EpiSCs and thus inhibits the neural commitment of mouse EpiSCs, thereby promoting non-neural lineage differentiation [119]. It has been recently reported that $Smad2/3^{-/-}$ double-knockout mESCs show activated transcription of extra-embryonic genes (*Gata2*, *Fgfr2*, etc.), BMP target genes (*Id1/Id2/Id3/Id4*) and show disrupted cell fate allocation of the three germ layers [120].

Nodal signaling is required during the initial transition of ESCs from the naïve pluripotent state to acquire competence for multi-lineage differentiation [121]. It can induce SMAD7, which negatively regulates SMAD1/5 induction by BMP [122]. SMAD7 also plays a role in the activation of STAT3 by directly binding to the gp130 intracellular domain, thereby preventing the binding of SHP2 or SOCS to gp130. The maintenance of STAT3 activation results in maintaining the LIF-dependent self-renewal and pluripotency of mESCs [123]. Narayana et al. [124] have also reported increased levels of TGF-β and ERK/MAPK signaling upon knockdown of the clathrin heavy chain gene *CltC* in mESCs, resulting in a loss of pluripotency.

TGFβ/BMP signaling can also be inhibited by the action of small molecule inhibitors such as SB431542 and dorsomorphin. Inhibition of this pathway by these molecules and activation of Wnt signaling by CHIR990021 promote the differentiation of human pluripotent stem cells (hPSCs) into neurons with a chemical transitional embryoid-body-like state (CTras) [125]. It has been reported that TGFβ signaling is hyperactivated in differentiation-resistant-ESCs (DR-ESCs), which retain *Oct4* expression during the differentiation of mESCs to neural progenitor cells (NPCs) and therefore have higher tumorigenic potential. Inhibiting TGFβ signaling in DR-ESCs by SB431542 can induce complete differentiation of DR-ESCs [126]. Activation of the TGFβ pathway also regulates the activity of other signaling pathways. β-catenin activity, which is a downstream transducer of the Wnt pathway, increases upon the action of the TGF-β family ligands, Activin, and BMP [127].

As mentioned earlier, hESCs are similar to mouse EpiSCs as they both require FGF and Activin A for their self-renewal. Inhibition of TGF-β signaling in hESCs cultured on MEFs resulted in their differentiation, suggesting that TGF-β signaling may be required for hESC maintenance [128]. ChIP-seq analysis has shown that pluripotency factors such as OCT4, SOX2, and NANOG co-occupy promoter regions along with SMAD2/3 proteins in hESCs [129,130]. It is reported that PI3 kinase activity plays a dual role in regulating hESC self-renewal and differentiation, by acting through SMAD2/3 and Wnt3 signaling [131]. Noggin that is a known BMP antagonist maintains hESCs in a pluripotent state, with BMP treatment of hESCs resulting in mesoderm and trophoblast induction [132–134].

Nodal/Activin signaling maintains hESCs in a pluripotent and self-renewing state by inhibiting BMP-induced differentiation and maintaining *Oct4* and *Nanog* expression through the action of SMAD2/SMAD3 [135,136]. Inhibiting SMAD2 phosphorylation by the inhibitor SB431542 leads to the inhibition of the Activin/Nodal pathway resulting in the differentiation of hESCs [128,137]. Activation of the Activin/Nodal signaling pathway leads to binding of SMAD2/SMAD3 to the promoter region of *Nanog*, resulting in upregulation of *Nanog* in hESCs, which in turn prevents endodermal differentiation [137]. In contrast, mESCs lacking *Smad2/3* could self-renew. However, SMAD2/3 was required for accurate gene expression patterns during differentiation [120]. It has also been reported that activation of SMAD2/3 signaling leads to the expression of Lefty, which maintains hESCs in an undifferentiated state [138]. SMAD2/3 can also interact with the METTL3-METTL4-WTAP complex which is involved in the addition of N^6 methyladenosine (m^6A) to transcripts involved in cell fate decisions regulated by TGFβ signaling in hESCs [139].

As discussed previously, both TGF-β signaling and FGF signaling are required for the self-renewal of hESCs [128], with inhibition of either resulting in hESC differentiation. It is reported that inhibiting TGF-β signaling in the presence of continued FGF signaling can lead to neuroectoderm differentiation [4,140,141]. hESCs can be differentiated into mesoendoderm due to activated SMAD2/3 proteins, by activating Activin/Nodal signaling and inhibiting FGF signaling [142]. It has also been reported that activating BMP signaling and, inhibiting TGF-β and FGF signaling can promote ESC differentiation into primitive endoderm and extraembryonic trophoblast lineages [136,141]. TGFβ signaling has also been reported to play a role in the differentiation of hESCs to hepatocytes—a process involving sequential EMT-MET transitions [143].

Inhibition of Activin/Nodal signaling leads to an increased expression of SIP1/ZEB-2 that causes the repression of SMAD2/3 target genes resulting in neuroectoderm differentiation of mESCs [144,145]. Activation of Nodal signaling can also trigger differentiation of mESCs to mesendoderm lineage by the interaction of SMAD2/3 with TRIM33/TIF1gb upon Nodal signaling activation. This complex then binds to H3K9me3 and H3K18ac, resulting in the opening up of chromatin associated with mesodermal lineage genes [146].

Nodal signaling can also result in the removal of the repressive H3K27me3 mark and direct differentiation of hESCs toward definitive endoderm (DE) by recruiting demethylases such as JMJD3 [147,148]. Genome-wide expression profiling has revealed that JMJD3 co-localizes with SMAD3 at the promoters of TGF-β responsive genes in neural stem cells, and hence is required for SMAD3-dependent neuronal differentiation [149]. Apart from histone methylation, Nodal signaling can also be modulated by histone deacetylation. Histone deacetylase 1(HDAC1) can repress Nodal expression, thereby promoting neural induction in mouse embryos [150,151]. Several studies have reported that during differentiation, SMAD2/3 proteins interact with transcription factors such as FOXH1 and EOMES, as the expression of ESC transcription factors *Oct4* and *Nanog* are repressed, resulting in endodermal differentiation [129,135,148].

Previous studies have shown that inhibiting TGF-β signaling at specific stages improves the reprogramming efficiency of somatic cells, as TGF-β is a major inducer of the mesenchymal state [152,153]. Reprogramming of hESCs is dependent upon the acquisition of an epithelial state which is regulated by TGF-β signaling and EMT transcription factors [153–157]. During reprogramming, three different phases are observed; with the first phase involving the mesenchymal-to-epithelial transition (MET) phase with the help of BMP signaling and KLF4 [158]. The second phase involves the transition from MET to a pluripotent-competent state, followed by the third phase that is marked by the acquisition of full pluripotency [159,160]. *c-Jun* is a downstream target of TGF-β signaling, and similar to TGF-β, suppresses reprogramming by inducing the mesenchymal state via activation of mesenchymal-related genes [150,151,161].

The TGF-β/SMAD pathway is an important regulator of stem cell state and differentiation. The cell fate determination of both mouse and human ESCs largely depends on extracellular signals or ligands that activate different downstream signaling cascades in this pathway. Therefore, this leads to contradictory explanations for molecular mechanisms underlying embryonic stem cell fate choice. For instance, mESCs are dependent on BMP signaling for their self-renewal, whereas hESCs rely on Activin/Nodal/Smad2/3 pathway to remain in an undifferentiated state. Therefore, understanding the mechanisms of different TGF-β family signals in the context of pluripotency and differentiation of mESCs, hESCs, and iPSCs is important for understanding cell fate determination.

4 ERK1/2 signaling pathway

The role of the extracellular signal-regulated kinase (ERK)/mitogen-activated protein kinase (MAPK) signaling pathway in regulating the cell cycle, cell proliferation, and carcinogenesis has been known for decades [77,162,163]. Recent studies have shown that this signaling pathway is also required for the regulation of cell fate and for the normal development of organisms. The ERK/MAPK signal-transduction cascade is mediated when growth factors bind and activate specific receptor tyrosine kinases (RTKs) on the cell membrane. This activation leads to the binding of adaptor proteins such as GRB2 (growth factor receptor-bound protein 2), which in turn engages a guanine nucleotide exchange factor (GEF) such as SOS. Association of SOS promotes the activation of the GTPase RAS by replacing the bound GDP with GTP, which can now initiate a series of phosphorylations of downstream kinases, MAPKKK, MEK, and ERK [162–164]. Active ERK either phosphorylates cytoplasmic targets that translocate to the nucleus or undergoes nuclear translocation itself and regulates the activities of a number of transcription factors [162–164].

A number of experiments using pharmacological inhibitors or genetic manipulation approaches have shown that this signaling pathway is also involved in deciding the fate of cells during development. The activation of the ERK signaling cascade is associated with inducing differentiation in mESCs, with the most common upstream activators of this pathway being the fibroblast growth factors (FGFs) [165–167]. Austin Smith and his group demonstrated that upon depleting one of the fibroblast growth factors, FGF4, mESCs retained the expression of pluripotency markers such as OCT4. However, upon being subjected to differentiation, $Fgf4^{-/-}$ mESCs failed to upregulate primary neural markers such as *Sox1* or *Nes* [165]. As a consequence of this deletion, ERK signaling was significantly downregulated resulting in a blockade toward neural differentiation. The importance of ERK signaling was also studied in the context of mouse embryonic development. $Erk2^{-/-}$ mouse embryos were able to develop into blastocysts, implant, and give rise to the epiblast. However, they were unable to develop mesoderm and displayed severe defects in trophoblast formation, failing to survive postimplantation [165,168,169]. ERK2 is present in ES cells at higher levels than ERK1. Mouse embryos lacking *Erk1* developed into viable and fertile mice, but showed abnormal cell proliferation and compromised thymocyte maturation [170,171]. These findings indicate that the FGF-Ras-ERK1/2 signaling cascade is important for differentiation and regulates the commitment of ES cells to neural and non-neural lineages [165]. ERK signaling also initiates the differentiation of the ICM, as evidenced by the appearance of embryos with defective epiblast and hypoblast upon perturbation of any component of the pathway. Previous reports show that depleting the key adaptor protein of the pathway, *Grb2* results in defective formation of the hypoblast [172,173], while reducing ERK activity promotes the retention of OCT4-positive cells in the ICM [172,174,175]. When ERK signaling was suppressed by culturing mouse embryos in the presence of a pharmacological inhibitor against MEK, an upstream kinase required for the activation of ERK1/2, or against a further upstream target such as FGFR, the embryos exhibited a greater number of NANOG-positive cells (a marker specific for epiblast) and comparatively fewer GATA4-positive cells (a marker specific for the hypoblast) [172]. The FGF-ERK pathway is not only important for the differentiation of ICM cells but also for the differentiation of extra-embryonic tissues such as the trophoblast [172,176] and maintenance of the rate of proliferation of the diploid trophoblast [172,177–179]. Overall, these results in the context of mouse embryos support studies conducted on mESCs that the ERK signaling pathway is essential for the onset of differentiation.

Recently, Dhaliwal et al. showed that the activation of the MEK-ERK signaling pathway destabilized the core pluripotency maintenance circuit, setting the stage for the onset of differentiation. They further went on to show that phosphorylated (active) ERK localized to the nucleus where it caused nuclear export of KLF4 by phosphorylating it at residue Serine 132. This, in turn, resulted in the downregulation of *Klf4* and *Nanog* expression, which eventually led to decreased expression of OCT4 and SOX2 [180]. Additionally, KLF4 can also be phosphorylated at a different location (Serine 123) by ERK1/2, allowing βTrCP1 and βTrCP2 to recognize this phosphorylation mark and recruit E3 ubiquitin ligases to KLF4, targeting it for proteasomal degradation [77,181]. ERK signaling also stimulates the autopoly(ADP) ribosylation of PARP1 that can sequester SOX2, preventing it from binding to *Oct4/Sox2* enhancers, causing reduced expression of OCT4 and SOX2 [182].

Since NANOG is a substrate for ERK, this kinase also reduces the transactivation activity and stability of NANOG by phosphorylating it [77,183,184]. In 2018, Oscar Fernandez-Capetillo and his team studied the role of the RAS protein, which is an upstream activator of the ERK pathway in deciding the lineage commitment of mESCs. They found that RAS depletion rendered ESCs unable to differentiate, while retaining the expression of pluripotency markers. However, deletion of *Erf* (*E*-twenty-six 2 [Ets2]-repressive factor) restored the differentiation ability of RAS-deficient mESCs [185]. Apart from this, RSK (Ribosomal S6 kinase) has also been identified as a negative regulator of the ERK signaling cascade. Possibly, RSKs prevent the activation of ERK by phosphorylating two upstream adapters of the signaling cascade—phosphorylated SOS1 is sequestered by the 14–3-3 protein leading to its inactivation [186,187]; and phosphorylated GAB2 (Grb2-associated binder) prevents the recruitment of SHP2, thereby suppressing the MAPK/ERK signaling pathway [186,188]. This was shown by depleting ESCs of *Rsk* resulting in an accelerated exit from pluripotency, without compromised lineage commitment [186].

Besides promoting the onset of differentiation in mESCs, ERK signaling also controls the ESC cell cycle. Apart from inducing differentiation in mESCs, RAS deficiency also reduced the proliferation rate and growth of mESC colonies as the cells experienced hindrance at the G1/S boundary and failed to enter the mitotic phase. However, the deletion of *Erf* rescued the defects in the proliferation of RAS-deficient mESCs [185]. Recently, it has been reported that suppression of the ERK signaling cascade causes hypo-phosphorylated RB (a cell cycle checkpoint protein) resulting in an elongated G1 phase [189]. This suggests that the ERK signaling pathway abrogates the checkpoint and alters the unique cell cycle profile of ESCs, thereby implying a negative impact on the self-renewal of the ESCs. On the contrary, a number of contrasting pieces of evidence show that when other upstream activators of the ERK signaling pathway such as *Raf-A*, *Raf-B*, and *Raf-C* were knocked down, mouse ESCs failed to proliferate and survive [190]. Therefore, it still remains unclear as to exactly how the MAPK/ERK signaling pathway affects proliferation and self-renewal, especially in the context of mESCs.

Recently, two groups used different approaches to study the dynamics of the ERK signaling pathway in mESCs and in the adult organs of mice. The first group made an ESC line by knocking in the H2B Venus reporter into the *Sprouty4* locus, an early pathway target, and used the fluorescence intensity as a readout for the activity of the signaling pathway. They found heterogeneous activity in in vitro mESC culture. ERK signaling was active in the inner cell mass of the preimplantation embryo and also in the cells of the visceral endoderm of the early postimplantation embryo. During organogenesis in postimplantation embryos, ERK activity was enriched in the mesenchymal region including the limb bud, cephalic, olfactory, and genital tubercle and uterine mesenchyme, while the epithelium region including the apical ectodermal ridge showed lower levels of activity [191]. Deathridge et al. used a FRET-based sensor to study the dynamics of ERK signaling. Although ERK signaling is known to be associated with differentiation, they found a weak correlation between ERK activity and exit from the state of pluripotency. Rather, ERK showed heterogeneous spatiotemporal activity in ESCs undergoing differentiation, with the peripheral region of the colonies that had lower cell density exhibiting increased ERK activity. Owing to the single-cell resolution that could be achieved with this technique, it was observed that cells that were closely related showed a similar pattern of ERK activity compared to those that were unrelated or distantly related [192]. This suggests that ERK activity and signaling are linked to cell lineage and that this differential pattern of activity may determine the commitment of ESCs to different lineages. This is an underlying reason for the incorporation of the small molecule inhibitor of MEK as a component of the 2i media used to maintain ESCs in the pluripotent state for long-term culture. Suppression of ERK signaling by inhibiting the activity of MEK1/2 results in global hypomethylation of the DNA. This is achieved by downregulating the expression of DNA methyltransferases such as *Dnmt1*, *Dnmt3a*, *Dnmt3b* and associated cofactors such as *Dnmt3l*, *Uhrf1* both at the transcript and protein level. Additionally, inhibition of MEK activity upregulates the expression of transcription factors associated with naïve pluripotency. Thus, the inhibition of ERK signaling leads to the maintenance of the epigenetic status of ESCs in culture similar to that of the ICM [193]. However, the epigenetic state as well as the chromosomal stability of ESCs is determined by the degree of suppression of ERK signaling, because inhibiting MEK1/2 for longer time periods results in irreversible epigenetic aberrations in genes, particularly associated with germ layer formation, gastrulation or organogenesis, and ultimately impaired development [193]. In addition to these, ERK1/2 also maintains a permissive chromatin environment by binding to the promoter region of certain developmental genes along with the PRC2 complex. Thus, it prevents binding of TFIIH and also phosphorylates RNA polymerase II at Serine 5, to maintain a poised transcriptional state for these genes [77,194].

However, ERK signaling has a completely opposite role in hESCs compared to mESCs. Unlike mouse embryos, the formation of the epiblast and hypoblast in human embryos does not require FGF-ERK signaling [195]. Rather, FGF2 that is an upstream activator of ERK signaling is a necessary growth factor for pluripotency in hESCs [2,77]. This suggests a completely contrasting role of FGF-ERK signaling in mouse and human embryonic stem cells where FGF-ERK signaling is not essential for the differentiation of ICM in human embryos, but it is necessary for the maintenance of pluripotency in hESCs. Therefore, in contrast to mESCs, the inhibition of this pathway by knocking down *Erk2* compromises the pluripotency of hESCs [77,196]. This conflicting role of ERK signaling in mouse and human ESCs may have arisen due to the fact that hESCs are more closely related to mouse EpiSCs, which represent a primed state of pluripotency [73,74,77]. Therefore, existing literature suggests that ERK signaling is dispensable for the self-renewal of mouse ESCs but is necessary for their differentiation and that this signaling pathway has a very different effect on mouse and human ESCs.

5 Wnt signaling pathway

The evolutionarily conserved Wnt signaling pathway is one of the most studied signaling pathways. It is known to be responsible for vital biological processes involving cell division and proliferation, cell fate determination, and tissue homeostasis [197]. Scientific studies have established the association of this signaling cascade with the progression of cancer. Another important role of the Wnt signaling pathway is axis specification and progression to the gastrulation stage during embryogenesis [198,199]. Apart from this, mutations in the components of this pathway cause a number of developmental defects, with the loss of an important downstream component of this pathway, namely β-*catenin*, resulting in embryonic lethality [200–202]. The Wnt pathway is involved in regulating the self-renewal and pluripotency, and by extending the identity of ES cells.

WNTs are secreted ligands that bind to two receptor proteins on the membrane, namely frizzled (FZ), and the low-density lipoprotein (LDL) receptor-related proteins 5 and 6 (LRP5/6). The outcome of the canonical WNT pathway is the stabilization and nuclear translocation of β-CATENIN. When WNT binds to FZ and LRP5/6, Dishevelled (DVL) is recruited to the receptor complex on the membrane, which disassembles the "destruction complex." The destruction complex comprised of AXIN, glycogen synthase kinase 3β (GSK3β), casein kinase 1 (CK1), and adenomatous polyposis coli (APC). DVL sequesters AXIN, GSK3β, and CK1, resulting in their inability to phosphorylate β-CATENIN preventing its subsequent ubiquitination and degradation by the 26S proteasome [203–205]. The stabilized β-CATENIN can now translocate to the nucleus, and in association with TCF/LEF1 (lymphoid enhancer-binding factor 1) can regulate the expression of downstream target genes [204,206–208].

The exact role of WNT signaling in regulating the self-renewal and pluripotency of ESCs is still debatable. Moreover, the degree of WNT activity also determines whether ESCs maintain their pluripotency or differentiate. Studies have shown that mESCs exposed to high levels of WNT signaling promoted the expression of pluripotency markers, whereas mESCs underwent differentiation into neurons and cardiomyocytes in response to lower WNT stimulation [209]. In the context of differentiation, WNT signaling promotes mesodermal and endodermal lineage [199,210,211], but it attenuates neuroectodermal lineage [212]. Literature shows that the stabilization of β-CATENIN, either by the presence of mutations in APC or by the loss of both isoforms of GSK3β, is responsible for this effect. When both alleles of the *Apc* gene were mutated, mouse ESCs failed to give rise to teratomas, and even if teratomas formed under such conditions, they failed to differentiate into the neuroectodermal lineage [200,213,214]. Moreover, cells lacking both α and β isoforms of GSK3 failed to differentiate particularly toward the neural lineage and retained the expression of pluripotency markers such as NANOG, OCT4, and REX1 [55,200,215]. TCF3 is also an important player in regulating cell fate and is a member of the core pluripotency network, although its role is still controversial. A large amount of literature shows that the canonical WNT pathway helps in the maintenance of pluripotency by either converting repressive TCF3 complexes into activators or replacing them with other transcription factors belonging to the TCF family. These can interact with β-CATENIN and activate transcription of target genes [202,216]. Wray et al. [202] showed that the derepression of TCF3 caused by inhibition of GSK3β, and subsequent stabilization of β-CATENIN resulted in the transcription of its target genes that include components of the core pluripotency network [202]. Therefore, the downstream components of the WNT signaling pathway, such as GSK3β, β-CATENIN, and TCF3 work in harmony to maintain the pluripotency of mESCs by reducing the repressive impact on the core pluripotency network. However, the components of the WNT signaling pathway can also have a completely opposite effect on mESCs depending on environmental cues. In 2016, Austin Smith and his group studied the impact of GSK3β inhibition on differentiation. Upon inhibition of GSK3β, cMYC and β-CATENIN were stabilized, which in turn ensured commitment of naïve ESCs toward endodermal specification. cMYC bound to prebound MIZ1 at the *Tcf3* proximal promoter to repress its activity. As a result, the transcription of the pioneer factor for endoderm specification, *FoxA2* was derepressed. Subsequently, β-CATENIN activity could drive definitive endodermal lineage commitment by upregulating factors such as *Sox17* [217]. One of the targets of the TCF3 derepression and β-CATENIN stabilization is *Esrrb* which is an orphan nuclear receptor related to the estrogen receptor [218]. Owing to the inhibition of GSK3β, the expression of *Esrrb* increases, resulting in sustained self-renewal, maintenance of the core pluripotency network, and delayed onset of differentiation [219]. An RNAi-mediated knockdown of the *Tcf3* gene, stimulated self-renewal in mESCs, but this decrease in the endogenous expression of *Tcf3* inhibited differentiation of mESCs into three germ layers [200,220]. Under such conditions, ESCs failed to differentiate even in the presence of retinoic acid [200,221], or in the absence of exogenous LIF [200,216] (conditions that are reported to induce differentiation of ESCs). However, overexpression of *Tcf3* pushed ESCs toward differentiation [200,222] and inhibited the expression of stem cell-specific genes [200,223], whereas treatment with exogenously added WNT3a-inhibited differentiation and restored self-renewal [222]. Some articles demonstrate that TCF3 interacts with transcription factors that are associated with pluripotency such as OCT4, SOX2, and NANOG [216,221,224,225]. It is most likely that the combination of TCF3 and β-CATENIN serves as a transcriptional activator and supports pluripotency in mESCs [52,226].

More recently, scientists are of the opinion that TCF3 acts more as a repressor in ESCs than as an activator and that β-CATENIN derepresses TCF3 to promote the expression of pluripotency-associated genes [226]. Altogether, the inhibition of GSK3β and subsequent stabilization of β-CATENIN reduce the repressive effect of TCF3 and activate transcription of genes required to maintain the core pluripotency network [202]. Another aspect of regulating the fate of embryonic stem cells by WNT/β-CATENIN signaling that was unexplored until recent times is the epigenetic control. Recent studies have reported that the loss of WNT signaling and downregulation of β-CATENIN activity result in the global hypomethylation of ESC chromatin, causing impairment in their ability to differentiate [201]. Sustained WNT activity is essential to maintain the identity and genomic stability of ESCs, thereby underlining the significance of the inclusion of the GSK3β inhibitor as a component of 2i media required for the long-term in vitro culture of mESCs. The WNT signaling pathway is also known to influence different stages of embryo development, particularly during gastrulation [198,199,227]. Recent studies show that the excess of active β-CATENIN prevents the two-cell embryo from progressing to the 4-cell stage and subsequently to the morula and blastula stages. However, embryos lacking *β-catenin* exhibit no significant developmental defect; this may be due to the compensatory effect of maternal deposition [228]. However, it is likely that the key component of the pathway, β-CATENIN, regulates the developmental potential and is required mainly for the maternal to zygotic transition, and very early stages of embryonic development.

The WNT signaling pathway not only affects the balance between pluripotency and differentiation in mESCs but also exerts a positive effect on self-renewal. The GSK3β-mediated stabilization of β-CATENIN is also important to support and promote self-renewal [202]. Ying et al. [52] showed that the addition of a GSK3β inhibitor in culture media could support self-renewal of mESCs for a short period of time; but those grown in 2i (MEKi + GSK3i) could be maintained for a longer time [52,200,202]. While a combination of LIF and MEKi could support the self-renewal of mESCs, it was not as effective as a combination of GSK3i and MEKi [52,200,202,229]. This demonstrates that the activation of WNT signaling can sustain the self-renewal of mESCs but that cross talk with other signaling pathways makes it more efficient. Another view suggests that TCF3 can regulate self-renewal of mESCs. *Tcf3*$^{-/-}$ mESCs can continue to self-renew, whereas *Tcf3* overexpressing mESCs required WNT3a to restore their ability to self-renew [222]. This indicates that the stabilization of β-CATENIN by WNT suppresses the repressive activity of TCF3, revealing a pro-self-renewal effect of the WNT signaling pathway.

The WNT signaling pathway is also important in the context of hESCs. The addition of the WNT3a ligand activated the canonical WNT signaling pathway, resulting in their enhanced cell survival and cell proliferation. However, the hESCs gradually lost their property of pluripotency and started showing signs of differentiation under such conditions [230]. On the contrary, contrasting evidence showed that the active canonical WNT pathway sustains the expression of pluripotency markers such as *Oct-3/4*, *Rex1*, *Nanog* and maintains the undifferentiated phenotype of both mESCs and hESCs [231]. Murine fibroblasts can be efficiently reprogrammed into iPSCs using recombinant WNT3a even in the absence of *c-Myc* [224,225]. WNT signaling and β-CATENIN activity are required for the formation of neural crest cells from human pluripotent stem cells, where the magnitude of signaling determines the axial specificity. Lower WNT activity gives rise to anterior neural crest cells, whereas higher activity results in the formation of posterior ones [232]. In the case of human embryonic development, downregulation of β-*catenin* results in spontaneous abortion [228,233].

Therefore, one can summarize that the WNT signaling pathway can regulate cell fate decisions by favoring the retention of pluripotency or inducing the differentiation of mouse and human ESCs depending upon cues from the environment. Although controversy prevails about the exact role of WNT signaling in regulating the self-renewal of ESCs, literature supports that this pathway is critical and has a positive effect on self-renewal.

6 Conclusions and future perspectives

Based on the developments in stem cell research, various factors including cytokines, small molecule inhibitors, growth factors, hormones, and serum have been identified to be involved in embryonic stem cell maintenance and differentiation. These factors act through various signaling pathways that have been extensively studied in order to understand the molecular mechanisms underlying ESC fate determination. The pathways discussed earlier are indeed essential for the survival and maintenance of ESCs, but it should be noted that they do not operate independent of each other in the cellular context. It is rather the cross talk of these pathways that determine cell fate (Fig. 1). LIF/JAK/STAT3 signaling is indispensable for the maintenance of the pluripotent state of mESCs. However, hESCs rely on FGF2 and Activin A for their self-renewal. Similarly, the ERK pathway induces differentiation in mESCs, whereas it is important for maintaining the pluripotency of hESCs. This contradiction may be due to differences in their pluripotency state, with mESCs representing a naïve state of pluripotency and hESCs representing a primed state. Activin signaling cross-talks with the WNT signaling cascade by increasing the accessibility of β-CATENIN to the promoter regions of genes associated with meso/endodermal differentiation [234]. mESCs in addition to LIF signaling are also dependent on BMP signaling for their self-renewal [50,116]. A study reveals that a combination of the Nodal/SMAD pathway, the WNT/β-CATENIN, and the FGF/ERK pathway is required for neural differentiation. The inhibition of Nodal/SMAD signaling and WNT/β-CATENIN signaling accelerates neural induction in the early phase of differentiation. However, in the later period of differentiation, activation of the WNT/β-CATENIN pathway is required for neuronal differentiation, followed by steady activity of FGF/ERK signaling to maintain the neuronal population [235]. Therefore, these pathways not only act in combination but also follow a strict chronology to

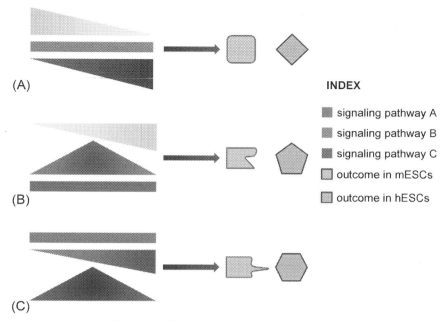

FIG. 1 Signaling pathways represent a complex interplay of signals dependent on concentration, timing, and location. (A), (B), and (C) represent three different conditions demonstrating the effect of variations in concentrations and activity of combinations of signaling pathways in a spatial and temporal manner. The schematic shows how a slight modification in the concentration or spatiotemporal activity of a signaling pathway can result in a different outcome in human and mouse embryonic stem cells. Additionally, signaling pathways affect the fate of mouse and human embryonic stem cells differentially.

modulate their activity for lineage commitment. Akin to mESCs, in human ESCs also, TGFβ signaling interacts with WNT signaling and modulates the β-CATENIN response [127].

Therefore, one can conclude that all these signaling pathways together form a regulatory network for determining the cell fate of both mESCs and hESCs. However, there is a need for studying the spatial, temporal, and concentration-dependent requirements of these pathways in the context of naïve/ground and primed states of pluripotency in order to gain a better understanding of the molecular underpinnings of pluripotency.

References

[1] M.J. Evans, M.H. Kaufman, Establishment in culture of pluripotential cells from mouse embryos, Nature 292 (5819) (1981) 154–156.

[2] J.A. Thomson, J. Itskovitz-Eldor, S.S. Shapiro, M.A. Waknitz, J.J. Swiergiel, V.S. Marshall, J.M. Jones, Embryonic stem cell lines derived from human blastocysts, Science 282 (5391) (1998) 1145–1147.

[3] P.C. Heinrich, I. Behrmann, S. Haan, H.M. Hermanns, G. Müller-Newen, F. Schaper, Principles of interleukin (IL)-6-type cytokine signalling and its regulation, Biochem. J. 374 (1) (2003) 1–20.

[4] A.G. Smith, J.K. Heath, D.D. Donaldson, G.G. Wong, J. Moreau, M. Stahl, D. Rogers, Inhibition of pluripotential embryonic stem cell differentiation by purified polypeptides, Nature 336 (6200) (1988) 688–690.

[5] D.S. Charnock-Jones, A.M. Sharkey, P. Fenwick, S.K. Smith, Leukaemia inhibitory factor mRNA concentration peaks in human endometrium at the time of implantation and the blastocyst contains mRNA for the receptor at this time, Reproduction 101 (2) (1994) 421–426.

[6] A. Arici, E. Oral, O. Bahtiyar, O. Engin, E. Seli, E.E. Jones, Leukaemia inhibitory factor expression in human follicular fluid and ovarian cells, Hum. Reprod. 12 (6) (1997) 1233–1239.

[7] A.M. Sharkey, A. King, D.E. Clark, T.D. Burrows, P.P. Jokhi, D.S. Charnock-Jones, Y.W. Loke, S.K. Smith, Localization of leukemia inhibitory factor and its receptor in human placenta throughout pregnancy, Biol. Reprod. 60 (2) (1999) 355–364.

[8] D. Metcalf, The unsolved enigmas of leukemia inhibitory factor, Stem Cells 21 (1) (2003) 5–14.

[9] P. Suman, S.S. Malhotra, S.K. Gupta, LIF-STAT signaling and trophoblast biology, JAKSTAT 2 (4) (2013) e25155.

[10] J.L. Taupin, V. Pitard, J. Dechanet, V. Miossec, N. Gualde, J.F. Moreau, Leukemia inhibitory factor: part of a large ingathering family, Int. Rev. Immunol. 16 (3–4) (1998) 397–426.

[11] C.L. Stewart, P. Kaspar, L.J. Brunet, H. Bhatt, I. Gadi, F. Köntgen, S.J. Abbondanzo, Blastocyst implantation depends on maternal expression of leukaemia inhibitory factor, Nature 359 (6390) (1992) 76–79.

[12] C.B. Ware, M.C. Horowitz, B.R. Renshaw, J.S. Hunt, D. Liggitt, S.A. Koblar, B.C. Gliniak, H.J. McKenna, T. Papayannopoulou, B. Thoma, Targeted disruption of the low-affinity leukemia inhibitory factor receptor gene causes placental, skeletal, neural and metabolic defects and results in perinatal death, Development 121 (5) (1995) 1283–1299.

[13] G. Delage, J.F. Moreau, J.L. Taupin, S. Frertas, E. Hambartsoumian, F. Olivennes, R. Fanchin, H. Letur-Könirsch, R. Frydman, G. Chaouat, In-vitro endometrial secretion of human interleukin for DA cells/leukaemia inhibitory factor by explant cultures from fertile and infertile women, MHR: Basic Sci. Reprod. Med. 1 (7) (1995) 335–340.

[14] E. Hambartsoumian, Endometrial leukemia inhibitory factor (LIF) as a possible cause of unexplained infertility and multiple failures of implantation, Am. J. Reprod. Immunol. 39 (2) (1998) 137–143.

[15] M.B. Renfree, G. Shaw, Diapause, Annu. Rev. Physiol. 62 (1) (2000) 353–375.

[16] F.A. Brook, R.L. Gardner, The origin and efficient derivation of embryonic stem cells in the mouse, Proc. Natl. Acad. Sci. 94 (11) (1997) 5709–5712.

[17] M. Ernst, A. Oates, A.R. Dunn, Gp130-mediated signal transduction in embryonic stem cells involves activation of Jak and Ras/mitogen-activated protein kinase pathways, J. Biol. Chem. 271 (47) (1996) 30136–30143.

[18] T. Kisseleva, S. Bhattacharya, J. Braunstein, C.W. Schindler, Signaling through the JAK/STAT pathway, recent advances and future challenges, Gene 285 (1–2) (2002) 1–24.

[19] W.J. Leonard, J.J. O'Shea, Jaks and STATs: biological implications, Annu. Rev. Immunol. 16 (1) (1998) 293–322.

[20] J. Sasse, U. Hemmann, C. Schwartz, U. Schniertshauer, B. Heesel, C. Landgraf, J. Schneider-Mergener, P.C. Heinrich, F. Horn, Mutational analysis of acute-phase response factor/Stat3 activation and dimerization, Mol. Cell. Biol. 17 (8) (1997) 4677–4686.

[21] X. Chen, H. Xu, P. Yuan, F. Fang, M. Huss, V.B. Vega, E. Wong, Y.L. Orlov, W. Zhang, J. Jiang, Y.H. Loh, Integration of external signaling pathways with the core transcriptional network in embryonic stem cells, Cell 133 (6) (2008) 1106–1117.

[22] L. Dahéron, S.L. Opitz, H. Zaehres, W.M. Lensch, P.W. Andrews, J. Itskovitz-Eldor, G.Q. Daley, LIF/STAT3 signaling fails to maintain self-renewal of human embryonic stem cells, Stem Cells 22 (5) (2004) 770–778.

[23] J.C. Conover, N.Y. Ip, W.T. Poueymirou, B. Bates, M.P. Goldfarb, T.M. DeChiara, G.D. Yancopoulos, Ciliary neurotrophic factor maintains the pluripotentiality of embryonic stem cells, Development 119 (3) (1993) 559–565.

[24] D. Pennica, K.L. King, K.J. SHAw, E. Luis, J. Rullamas, S.M. LuOH, W.C. Darbonne, D.S. Knutzon, R. Yen, K.R. Chien, Expression cloning of cardiotrophin 1, a cytokine that induces cardiac myocyte hypertrophy, Proc. Natl. Acad. Sci. 92 (4) (1995) 1142–1146.

[25] T.M. Rose, D.M. Weiford, N.L. Gunderson, A.G. Bruce, Oncostatin M (OSM) inhibits the differentiation of pluripotent embryonic stem cells in vitro, Cytokine 6 (1) (1994) 48–54.

[26] T. Matsuda, T. Nakamura, K. Nakao, T. Arai, M. Katsuki, T. Heike, T. Yokota, STAT3 activation is sufficient to maintain an undifferentiated state of mouse embryonic stem cells, EMBO J. 18 (15) (1999) 4261–4269.

[27] J. Hall, G. Guo, J. Wray, I. Eyres, J. Nichols, L. Grotewold, S. Morfopoulou, P. Humphreys, W. Mansfield, R. Walker, S. Tomlinson, Oct4 and LIF/Stat3 additively induce Krüppel factors to sustain embryonic stem cell self-renewal, Cell Stem Cell 5 (6) (2009) 597–609.

[28] H. Niwa, T. Burdon, I. Chambers, A. Smith, Self-renewal of pluripotent embryonic stem cells is mediated via activation of STAT3, Genes Dev. 12 (13) (1998) 2048–2060.

[29] J.F. Bromberg, M.H. Wrzeszczynska, G. Devgan, Y. Zhao, R.G. Pestell, C. Albanese, J.E. Darnell Jr., Stat3 as an oncogene, Cell 98 (3) (1999) 295–303.

[30] B.L. Kidder, J. Yang, S. Palmer, Stat3 and c-Myc genome-wide promoter occupancy in embryonic stem cells, PLoS One 3 (12) (2008) e3932.

[31] I. Aksoy, C. Sakabedoyan, P.Y. Bourillot, A.B. Malashicheva, J. Mancip, K. Knoblauch, M. Afanassieff, P. Savatier, Self-renewal of murine embryonic stem cells is supported by the serine/threonine kinases Pim-1 and Pim-3, Stem Cells 25 (12) (2007) 2996–3004.

[32] P. Cartwright, C. McLean, A. Sheppard, D. Rivett, K. Jones, S. Dalton, LIF/STAT3 controls ES cell self-renewal and pluripotency by a Myc-dependent mechanism, Development 132 (5) (2005) 885–896.

[33] C.I. Tai, Q.L. Ying, Gbx2, a LIF/Stat3 target, promotes reprogramming to and retention of the pluripotent ground state, J. Cell Sci. 126 (5) (2013) 1093–1098.

[34] P.Y. Bourillot, I. Aksoy, V. Schreiber, F. Wianny, H. Schulz, O. Hummel, N. Hubner, P. Savatier, Novel STAT3 target genes exert distinct roles in the inhibition of mesoderm and endoderm differentiation in cooperation with Nanog, Stem Cells 27 (8) (2009) 1760–1771.

[35] K. Shuai, B. Liu, Regulation of gene-activation pathways by PIAS proteins in the immune system, Nat. Rev. Immunol. 5 (8) (2005) 593–605.

[36] J.J. Babon, N.J. Kershaw, J.M. Murphy, L.N. Varghese, A. Laktyushin, S.N. Young, I.S. Lucet, R.S. Norton, N.A. Nicola, Suppression of cytokine signaling by SOCS3: characterization of the mode of inhibition and the basis of its specificity, Immunity 36 (2) (2012) 239–250.

[37] N.J. Kershaw, A. Laktyushin, N.A. Nicola, J.J. Babon, Reconstruction of an active SOCS3-based E3 ubiquitin ligase complex in vitro: identification of the active components and JAK2 and gp130 as substrates, Growth Factors 32 (1) (2014) 1–10.

[38] N.J. Kershaw, J.M. Murphy, I.S. Lucet, N.A. Nicola, J.J. Babon, Regulation of Janus kinases by SOCS proteins, Biochem. Soc. Trans. 41 (2013) 1042–1047.

[39] I. Chambers, D. Colby, M. Robertson, J. Nichols, S. Lee, S. Tweedie, A. Smith, Functional expression cloning of Nanog, a pluripotency sustaining factor in embryonic stem cells, Cell 113 (5) (2003) 643–655.

[40] K. Mitsui, Y. Tokuzawa, H. Itoh, K. Segawa, M. Murakami, K. Takahashi, M. Maruyama, M. Maeda, S. Yamanaka, The homeoprotein Nanog is required for maintenance of pluripotency in mouse epiblast and ES cells, Cell 113 (5) (2003) 631–642.

[41] V. Heurtier, N. Owens, I. Gonzalez, F. Mueller, C. Proux, D. Mornico, P. Clerc, A. Dubois, P. Navarro, The molecular logic of Nanog-induced self-renewal in mouse embryonic stem cells, Nat. Commun. 10 (1) (2019) 1–15.

[42] R. He, B. Xhabija, B. Al-Qanber, B.L. Kidder, OCT4 supports extended LIF-independent self-renewal and maintenance of transcriptional and epigenetic networks in embryonic stem cells, Sci. Rep. 7 (1) (2017) 1–19.

[43] E.A. Casanova, O. Shakhova, S.S. Patel, I.N. Asner, P. Pelczar, F.A. Weber, U. Graf, L. Sommer, K. Bürki, P. Cinelli, Pramel7 mediates LIF/STAT3-dependent self-renewal in embryonic stem cells, Stem Cells 29 (3) (2011) 474–485.

[44] S. Parisi, F. Passaro, L. Aloia, I. Manabe, R. Nagai, L. Pastore, T. Russo, Klf5 is involved in self-renewal of mouse embryonic stem cells, J. Cell Sci. 121 (16) (2008) 2629–2634.

[45] P. Sheshadri, A. Ashwini, S. Jahnavi, R. Bhonde, J. Prasanna, A. Kumar, Novel role of mitochondrial manganese superoxide dismutase in STAT3 dependent pluripotency of mouse embryonic stem cells, Sci. Rep. 5 (2015) 9516.

[46] S. Ye, P. Li, C. Tong, Q.L. Ying, Embryonic stem cell self-renewal pathways converge on the transcription factor Tfcp2l1, EMBO J. 32 (19) (2013) 2548–2560.

[47] M. Wang, L. Tang, D. Liu, Q.L. Ying, S. Ye, The transcription factor Gbx2 induces expression of Kruppel-like factor 4 to maintain and induce naïve pluripotency of embryonic stem cells, J. Biol. Chem. 292 (41) (2017) 17121–17128.

[48] D. Huang, L. Wang, J. Duan, C. Huang, X.C. Tian, M. Zhang, Y. Tang, LIF-activated Jak signaling determines Esrrb expression during late-stage reprogramming, Biol. Open 7 (1) (2018) bio029264.

[49] A.L. Van Oosten, Y. Costa, A. Smith, J.C. Silva, JAK/STAT3 signalling is sufficient and dominant over antagonistic cues for the establishment of naive pluripotency, Nat. Commun. 3 (1) (2012) 1–12.

[50] Q.L. Ying, J. Nichols, I. Chambers, A. Smith, BMP induction of Id proteins suppresses differentiation and sustains embryonic stem cell self-renewal in collaboration with STAT3, Cell 115 (3) (2003) 281–292.

[51] X. Qi, T.G. Li, J. Hao, J. Hu, J. Wang, H. Simmons, S. Miura, Y. Mishina, G.Q. Zhao, BMP4 supports self-renewal of embryonic stem cells by inhibiting mitogen-activated protein kinase pathways, Proc. Natl. Acad. Sci. 101 (16) (2004) 6027–6032.

[52] Q.L. Ying, J. Wray, J. Nichols, L. Batlle-Morera, B. Doble, J. Woodgett, P. Cohen, A. Smith, The ground state of embryonic stem cell self-renewal, Nature 453 (7194) (2008) 519–523.

[53] H. Hirai, P. Karian, N. Kikyo, Regulation of embryonic stem cell self-renewal and pluripotency by leukaemia inhibitory factor, Biochem. J. 438 (1) (2011) 11–23.

[54] T.S. Migone, S. Rodig, N.A. Cacalano, M. Berg, R.D. Schreiber, W.J. Leonard, Functional cooperation of the interleukin-2 receptor β chain and Jak1 in phosphatidylinositol 3-kinase recruitment and phosphorylation, Mol. Cell. Biol. 18 (11) (1998) 6416–6422.

[55] B.W. Doble, J.R. Woodgett, GSK-3: tricks of the trade for a multi-tasking kinase, J. Cell Sci. 116 (7) (2003) 1175–1186.

[56] M. Bechard, S. Dalton, Subcellular localization of glycogen synthase kinase 3β controls embryonic stem cell self-renewal, Mol. Cell. Biol. 29 (8) (2009) 2092–2104.

[57] S. Watanabe, H. Umehara, K. Murayama, M. Okabe, T. Kimura, T. Nakano, Activation of Akt signaling is sufficient to maintain pluripotency in mouse and primate embryonic stem cells, Oncogene 25 (19) (2006) 2697–2707.

[58] J. Braunstein, S. Brutsaert, R. Olson, C. Schindler, STATs dimerize in the absence of phosphorylation, J. Biol. Chem. 278 (36) (2003) 34133–34140.

[59] D.R. Alessi, M. Andjelkovic, B. Caudwell, P. Cron, N. Morrice, P. Cohen, B.A. Hemmings, Mechanism of activation of protein kinase B by insulin and IGF-1, EMBO J. 15 (23) (1996) 6541–6551.

[60] K. Takahashi, K. Mitsui, S. Yamanaka, Role of ERas in promoting tumour-like properties in mouse embryonic stem cells, Nature 423 (6939) (2003) 541–545.

[61] L. Chen, J.S. Khillan, A novel signaling by vitamin A/retinol promotes self renewal of mouse embryonic stem cells by activating PI3K/Akt signaling pathway via insulin-like growth factor-1 receptor, Stem Cells 28 (1) (2010) 57–63.

[62] J. Nichols, A. Smith, Naive and primed pluripotent states, Cell Stem Cell 4 (6) (2009) 487–492, https://doi.org/10.1016/j.stem.2009.05.015.

[63] T. Fukada, M. Hibi, Y. Yamanaka, M. Takahashi-Tezuka, Y. Fujitani, T. Yamaguchi, K. Nakajima, T. Hirano, Two signals are necessary for cell proliferation induced by a cytokine receptor gp130: involvement of STAT3 in anti-apoptosis, Immunity 5 (5) (1996) 449–460.

[64] F. Schaper, C. Gendo, M. Eck, J. Schmitz, C. Grimm, D. Anhuf, I.M. Kerr, P.C. Heinrich, Activation of the protein tyrosine phosphatase SHP2 via the interleukin-6 signal transducing receptor protein gp130 requires tyrosine kinase Jak1 and limits acute-phase protein expression, Biochem. J. 335 (3) (1998) 557–565.

[65] W.P. Schiemann, J.L. Bartoe, N.M. Nathanson, Box 3-independent signaling mechanisms are involved in leukemia inhibitory factor receptor α-and gp130-mediated stimulation of mitogen-activated protein kinase evidence for participation of multiple signaling pathways which converge at Ras, J. Biol. Chem. 272 (26) (1997) 16631–16636.

[66] H.M. Hermanns, S. Radtke, F. Schaper, P.C. Heinrich, I. Behrmann, Non-redundant signal transduction of interleukin-6-type cytokines. The adapter protein Shc is specifically recruited to rhe oncostatin M receptor, J. Biol. Chem. 275 (52) (2000) 40742–40748, https://doi.org/10.1074/jbc.M005408200.

[67] T. Hamazaki, S.M. Kehoe, T. Nakano, N. Terada, The Grb2/Mek pathway represses Nanog in murine embryonic stem cells, Mol. Cell. Biol. 26 (20) (2006) 7539–7549.

[68] H. Niwa, K. Ogawa, D. Shimosato, K. Adachi, A parallel circuit of LIF signalling pathways maintains pluripotency of mouse ES cells, Nature 460 (7251) (2009) 118–122.

[69] M. Buehr, S. Meek, K. Blair, J. Yang, J. Ure, J. Silva, R. McLay, J. Hall, Q.L. Ying, A. Smith, Capture of authentic embryonic stem cells from rat blastocysts, Cell 135 (7) (2008) 1287–1298.

[70] P. Li, C. Tong, R. Mehrian-Shai, L. Jia, N. Wu, Y. Yan, R.E. Maxson, E.N. Schulze, H. Song, C.L. Hsieh, M.F. Pera, Germline competent embryonic stem cells derived from rat blastocysts, Cell 135 (7) (2008) 1299–1310.

[71] J. Yang, A.L. Van Oosten, T.W. Theunissen, G. Guo, J.C. Silva, A. Smith, Stat3 activation is limiting for reprogramming to ground state pluripotency, Cell Stem Cell 7 (3) (2010) 319–328.

[72] C. Buecker, H.H. Chen, J.M. Polo, L. Daheron, L. Bu, T.S. Barakat, P. Okwieka, A. Porter, J. Gribnau, K. Hochedlinger, N. Geijsen, A murine ESC-like state facilitates transgenesis and homologous recombination in human pluripotent stem cells, Cell Stem Cell 6 (6) (2010) 535–546.

[73] I.G.M. Brons, L.E. Smithers, M.W. Trotter, P. Rugg-Gunn, B. Sun, S.M.C. de Sousa Lopes, S.K. Howlett, A. Clarkson, L. Ahrlund-Richter, R.A. Pedersen, L. Vallier, Derivation of pluripotent epiblast stem cells from mammalian embryos, Nature 448 (7150) (2007) 191–195.

[74] P.J. Tesar, J.G. Chenoweth, F.A. Brook, T.J. Davies, E.P. Evans, D.L. Mack, R.L. Gardner, R.D. McKay, New cell lines from mouse epiblast share defining features with human embryonic stem cells, Nature 448 (7150) (2007) 196–199.

[75] G. Guo, J. Yang, J. Nichols, J.S. Hall, I. Eyres, W. Mansfield, A. Smith, Klf4 reverts developmentally programmed restriction of ground state pluripotency, Development 136 (7) (2009) 1063–1069.

[76] H. Chen, I. Aksoy, F. Gonnot, P. Osteil, M. Aubry, C. Hamela, C. Rognard, A. Hochard, S. Voisin, E. Fontaine, M. Mure, Reinforcement of STAT3 activity reprogrammes human embryonic stem cells to naive-like pluripotency, Nat. Commun. 6 (2015) 7095.

[77] H. Chen, R. Guo, Q. Zhang, H. Guo, M. Yang, Z. Wu, S. Gao, L. Liu, L. Chen, Erk signaling is indispensable for genomic stability and self-renewal of mouse embryonic stem cells, Proc. Natl. Acad. Sci. 112 (44) (2015) E5936–E5943.

[78] Y.S. Chan, J. Göke, J.H. Ng, X. Lu, K.A.U. Gonzales, C.P. Tan, W.Q. Tng, Z.Z. Hong, Y.S. Lim, H.H. Ng, Induction of a human pluripotent state with distinct regulatory circuitry that resembles preimplantation epiblast, Cell Stem Cell 13 (6) (2013) 663–675.

[79] O. Gafni, L. Weinberger, A.A. Mansour, Y.S. Manor, E. Chomsky, D. Ben-Yosef, Y. Kalma, S. Viukov, I. Maza, A. Zviran, Y. Rais, Derivation of novel human ground state naive pluripotent stem cells, Nature 504 (7479) (2013) 282–286.

[80] Y. Takashima, G. Guo, R. Loos, J. Nichols, G. Ficz, F. Krueger, D. Oxley, F. Santos, J. Clarke, W. Mansfield, W. Reik, Resetting transcription factor control circuitry toward ground-state pluripotency in human, Cell 158 (6) (2014) 1254–1269.

[81] T.W. Theunissen, B.E. Powell, H. Wang, M. Mitalipova, D.A. Faddah, J. Reddy, Z.P. Fan, D. Maetzel, K. Ganz, L. Shi, T. Lungjangwa, Systematic identification of culture conditions for induction and maintenance of naive human pluripotency, Cell Stem Cell 15 (4) (2014) 471–487.

[82] C.B. Ware, A.M. Nelson, B. Mecham, J. Hesson, W. Zhou, E.C. Jonlin, A.J. Jimenez-Caliani, X. Deng, C. Cavanaugh, S. Cook, P.J. Tesar, Derivation of naive human embryonic stem cells, Proc. Natl. Acad. Sci. 111 (12) (2014) 4484–4489.

[83] A.B. Roberts, S.J. Kim, T. Noma, A.B. Glick, R. Lafyatis, R. Lechleider, S.B. Jaakowlew, A. Geiser, M.A. O'Reilly, D. Danielpour, M.B. Sporn, Multiple forms of TGF-: distinct promoters and differential expression, in: Clinical Applications of TGF, vol. 3, 1991, pp. 7–28.

[84] J. Massague, The transforming growth factor-beta family, Annu. Rev. Cell Biol. 6 (1) (1990) 597–641.

[85] J. Massagué, Receptors for the TGF-β family, Cell 69 (7) (1992) 1067–1070.

[86] J. Massagué, Y.G. Chen, Controlling TGF-β signaling, Genes Dev. 14 (6) (2000) 627–644.

[87] A. Hoffmann, G. Gross, BMP signaling pathways in cartilage and bone formation, Crit. Rev. Eukaryot. Gene Expr. 11 (1–3) (2001) 23–45.

[88] B.L. Hogan, Bone morphogenetic proteins in development, Curr. Opin. Genet. Dev. 6 (4) (1996) 432–438.

[89] S. Kishigami, Y. Mishina, BMP signaling and early embryonic patterning, Cytokine Growth Factor Rev. 16 (3) (2005) 265–278.

[90] M. Wan, X. Cao, BMP signaling in skeletal development, Biochem. Biophys. Res. Commun. 328 (3) (2005) 651–657.

[91] M.S. Rahman, N. Akhtar, H.M. Jamil, R.S. Banik, S.M. Asaduzzaman, TGF-β/BMP signaling and other molecular events: regulation of osteoblastogenesis and bone formation, Bone Res. 3 (2015) 15005.

[92] N. Ling, S.Y. Ying, N. Ueno, S. Shimasaki, F. Esch, M. Hotta, R. Guillemin, Pituitary FSH is released by a heterodimer of the β-subunits from the two forms of inhibin, Nature 321 (6072) (1986) 779–782.

[93] W. Vale, J. Rivier, J. Vaughan, R. McClintock, A. Corrigan, W. Woo, D. Karr, J. Spiess, Purification and characterization of an FSH releasing protein from porcine ovarian follicular fluid, Nature 321 (6072) (1986) 776–779.

[94] M. Namwanje, C.W. Brown, Activins and inhibins: roles in development, physiology, and disease, Cold Spring Harb. Perspect. Biol. 8 (7) (2016) a021881.

[95] M.M. Shen, Nodal signaling: developmental roles and regulation, Development 134 (6) (2007) 1023–1034.

[96] J. Brennan, C.C. Lu, D.P. Norris, T.A. Rodriguez, R.S. Beddington, E.J. Robertson, Nodal signalling in the epiblast patterns the early mouse embryo, Nature 411 (6840) (2001) 965–969.

[97] H. Hamada, C. Meno, D. Watanabe, Y. Saijoh, Establishment of vertebrate left–right asymmetry, Nat. Rev. Genet. 3 (2) (2002) 103–113.

[98] C. Meno, Y. Saijoh, H. Fujii, M. Ikeda, T. Yokoyama, M. Yokoyama, Y. Toyoda, H. Hamada, Left–right asymmetric expression of the TGFβ-family member lefty in mouse embryos, Nature 381 (6578) (1996) 151–155.

[99] C. Meno, A. Shimono, Y. Saijoh, K. Yashiro, K. Mochida, S. Ohishi, S. Noji, H. Kondoh, H. Hamada, lefty-1 is required for left-right determination as a regulator of lefty-2 and nodal, Cell 94 (3) (1998) 287–297.

[100] Y. Shi, J. Massagué, Mechanisms of TGF-β signaling from cell membrane to the nucleus, Cell 113 (6) (2003) 685–700.

[101] R. Derynck, W.M. Gelbart, R.M. Harland, C.H. Heldin, S.E. Kern, J. Massagué, D.A. Melton, M. Mlodzik, R.W. Padgett, A.B. Roberts, J. Smith, Nomenclature: vertebrate mediators of TGFβ family signals, Cell 87 (2) (1996) 173.

[102] J. Massagué, TGF-β signal transduction, Annu. Rev. Biochem. 67 (1998) 753–791.

[103] R.S. Beddington, E.J. Robertson, Axis development and early asymmetry in mammals, Cell 96 (2) (1999) 195–209.

[104] A.F. Schier, Nodal signaling in vertebrate development, Annu. Rev. Cell Dev. Biol. 19 (1) (2003) 589–621.

[105] M. Yamamoto, Y. Saijoh, A. Perea-Gomez, W. Shawlot, R.R. Behringer, S.L. Ang, H. Hamada, C. Meno, Nodal antagonists regulate formation of the anteroposterior axis of the mouse embryo, Nature 428 (6981) (2004) 387–392.

[106] F.L. Conlon, K.M. Lyons, N. Takaesu, K.S. Barth, A. Kispert, B. Herrmann, E.J. Robertson, A primary requirement for nodal in the formation and maintenance of the primitive streak in the mouse, Development 120 (7) (1994) 1919–1928.

[107] A. Kumar, M. Lualdi, G.T. Lyozin, P. Sharma, J. Loncarek, X.Y. Fu, M.R. Kuehn, Nodal signaling from the visceral endoderm is required to maintain Nodal gene expression in the epiblast and drive DVE/AVE migration, Dev. Biol. 400 (1) (2015) 1–9.

[108] A. Perea-Gomez, F.D. Vella, W. Shawlot, M. Oulad-Abdelghani, C. Chazaud, C. Meno, V. Pfister, L. Chen, E. Robertson, H. Hamada, R.R. Behringer, Nodal antagonists in the anterior visceral endoderm prevent the formation of multiple primitive streaks, Dev. Cell 3 (5) (2002) 745–756.

[109] W.R. Waldrip, E.K. Bikoff, P.A. Hoodless, J.L. Wrana, E.J. Robertson, Smad2 signaling in extraembryonic tissues determines anterior-posterior polarity of the early mouse embryo, Cell 92 (6) (1998) 797–808.

[110] J. Brennan, D.P. Norris, E.J. Robertson, Nodal activity in the node governs left-right asymmetry, Genes Dev. 16 (18) (2002) 2339–2344.

[111] T. Fujiwara, D.B. Dehart, K.K. Sulik, B.L. Hogan, Distinct requirements for extra-embryonic and embryonic bone morphogenetic protein 4 in the formation of the node and primitive streak and coordination of left-right asymmetry in the mouse, Development 129 (20) (2002) 4685–4696.

[112] Y. Komatsu, V. Kaartinen, Y. Mishina, Cell cycle arrest in node cells governs ciliogenesis at the node to break left-right symmetry, Development 138 (18) (2011) 3915–3920.

[113] N. Mine, R.M. Anderson, J. Klingensmith, BMP antagonism is required in both the node and lateral plate mesoderm for mammalian left-right axis establishment, Development 135 (14) (2008) 2425–2434.

[114] A. Bandyopadhyay, K. Tsuji, K. Cox, B.D. Harfe, V. Rosen, C.J. Tabin, Genetic analysis of the roles of BMP2, BMP4, and BMP7 in limb patterning and skeletogenesis, PLoS Genet. 2 (12) (2006) e216.

[115] N.R. Dunn, S.D. Vincent, L. Oxburgh, E.J. Robertson, E.K. Bikoff, Combinatorial activities of Smad2 and Smad3 regulate mesoderm formation and patterning in the mouse embryo, Development 131 (8) (2004) 1717–1728.

[116] P.A. Wilson, A. Hemmati-Brivanlou, Induction of epidermis and inhibition of neural fate by Bmp-4, Nature 376 (6538) (1995) 331–333.

[117] T. Fei, K. Xia, Z. Li, B. Zhou, S. Zhu, H. Chen, J. Zhang, Z. Chen, H. Xiao, J.D.J. Han, Y.G. Chen, Genome-wide mapping of SMAD target genes reveals the role of BMP signaling in embryonic stem cell fate determination, Genome Res. 20 (1) (2010) 36–44.

[118] Z. Li, T. Fei, J. Zhang, G. Zhu, L. Wang, D. Lu, X. Chi, Y. Teng, N. Hou, X. Yang, H. Zhang, BMP4 Signaling Acts via dual-specificity phosphatase 9 to control ERK activity in mouse embryonic stem cells, Cell Stem Cell 10 (2) (2012) 171–182.

[119] K. Zhang, L. Li, C. Huang, C. Shen, F. Tan, C. Xia, P. Liu, J. Rossant, N. Jing, Distinct functions of BMP4 during different stages of mouse ES cell neural commitment, Development 137 (13) (2010) 2095–2105.

[120] A.D. Senft, I. Costello, H.W. King, A.W. Mould, E.K. Bikoff, E.J. Robertson, Combinatorial Smad2/3 activities downstream of nodal signaling maintain embryonic/extra-embryonic cell identities during lineage priming, Cell Rep. 24 (8) (2018) 1977–1985.

[121] C. Mulas, T. Kalkan, A. Smith, NODAL secures pluripotency upon embryonic stem cell progression from the ground state, Stem Cell Rep. 9 (1) (2017) 77–91.

[122] K.E. Galvin, E.D. Travis, D. Yee, T. Magnuson, J.L. Vivian, Nodal signaling regulates the bone morphogenic protein pluripotency pathway in mouse embryonic stem cells, J. Biol. Chem. 285 (26) (2010) 19747–19756.

[123] Y. Yu, S. Gu, W. Li, C. Sun, F. Chen, M. Xiao, L. Wang, D. Xu, Y. Li, C. Ding, Z. Xia, Smad7 enables STAT3 activation and promotes pluripotency independent of TGF-β signaling, Proc. Natl. Acad. Sci. 114 (38) (2017) 10113–10118.

[124] Y.V. Narayana, C. Gadgil, R.D. Mote, R. Rajan, D. Subramanyam, Clathrin-mediated endocytosis regulates a balance between opposing signals to maintain the pluripotent state of embryonic stem cells, Stem Cell Rep. 12 (1) (2019) 152–164.

[125] K. Fujimori, T. Matsumoto, F. Kisa, N. Hattori, H. Okano, W. Akamatsu, Escape from pluripotency via inhibition of TGF-β/BMP and activation of Wnt signaling accelerates differentiation and aging in hPSC progeny cells, Stem Cell Rep. 9 (5) (2017) 1675–1691.

[126] X. Yang, R. Wang, X. Wang, G. Cai, Y. Qian, S. Feng, F. Tan, K. Chen, K. Tang, X. Huang, N. Jing, TGFβ signaling hyperactivation-induced tumorigenicity during the derivation of neural progenitors from mouse ESCs, J. Mol. Cell Biol. 10 (3) (2018) 216–228.

[127] J. Massey, Y. Liu, O. Alvarenga, T. Saez, M. Schmerer, A. Warmflash, Synergy with TGFβ ligands switches WNT pathway dynamics from transient to sustained during human pluripotent cell differentiation, Proc. Natl. Acad. Sci. 116 (11) (2019) 4989–4998.

[128] D. James, A.J. Levine, D. Besser, A. Hemmati-Brivanlou, TGFβ/activin/nodal signaling is necessary for the maintenance of pluripotency in human embryonic stem cells, Development 132 (6) (2005) 1273–1282.

[129] S. Brown, A. Teo, S. Pauklin, N. Hannan, C.H.H. Cho, B. Lim, L. Vardy, N.R. Dunn, M. Trotter, R. Pedersen, L. Vallier, Activin/Nodal signaling controls divergent transcriptional networks in human embryonic stem cells and in endoderm progenitors, Stem Cells 29 (8) (2011) 1176–1185.

[130] A.C. Mullen, D.A. Orlando, J.J. Newman, J. Lovén, R.M. Kumar, S. Bilodeau, J. Reddy, M.G. Guenther, R.P. DeKoter, R.A. Young, Master transcription factors determine cell-type-specific responses to TGF-β signaling, Cell 147 (3) (2011) 565–576.

[131] A.M. Singh, D. Reynolds, T. Cliff, S. Ohtsuka, A.L. Mattheyses, Y. Sun, L. Menendez, M. Kulik, S. Dalton, Signaling network crosstalk in human pluripotent cells: a Smad2/3-regulated switch that controls the balance between self-renewal and differentiation, Cell Stem Cell 10 (3) (2012) 312–326.

[132] R.H. Xu, X. Chen, D.S. Li, R. Li, G.C. Addicks, C. Glennon, T.P. Zwaka, J.A. Thomson, BMP4 initiates human embryonic stem cell differentiation to trophoblast, Nat. Biotechnol. 20 (12) (2002) 1261–1264.

[133] R.H. Xu, R.M. Peck, D.S. Li, X. Feng, T. Ludwig, J.A. Thomson, Basic FGF and suppression of BMP signaling sustain undifferentiated proliferation of human ES cells, Nat. Methods 2 (3) (2005) 185–190.

[134] P. Zhang, J. Li, Z. Tan, C. Wang, T. Liu, L. Chen, J. Yong, W. Jiang, X. Sun, L. Du, M. Ding, Short-term BMP-4 treatment initiates mesoderm induction in human embryonic stem cells, Blood 111 (4) (2008) 1933–1941.

[135] T.A. Beyer, A. Weiss, Y. Khomchuk, K. Huang, A.A. Ogunjimi, X. Varelas, J.L. Wrana, Switch enhancers interpret TGF-β and Hippo signaling to control cell fate in human embryonic stem cells, Cell Rep. 5 (6) (2013) 1611–1624.

[136] M. Sakaki-Yumoto, J. Liu, M. Ramalho-Santos, N. Yoshida, R. Derynck, Smad2 is essential for maintenance of the human and mouse primed pluripotent stem cell state, J. Biol. Chem. 288 (25) (2013) 18546–18560.

[137] L. Vallier, S. Mendjan, S. Brown, Z. Chng, A. Teo, L.E. Smithers, M.W. Trotter, C.H.H. Cho, A. Martinez, P. Rugg-Gunn, G. Brons, Activin/Nodal signalling maintains pluripotency by controlling Nanog expression, Development 136 (8) (2009) 1339–1349.

[138] L. Vallier, M. Alexander, R.A. Pedersen, Activin/Nodal and FGF pathways cooperate to maintain pluripotency of human embryonic stem cells, J. Cell Sci. 118 (19) (2005) 4495–4509.

[139] A. Bertero, S. Brown, P. Madrigal, A. Osnato, D. Ortmann, L. Yiangou, J. Kadiwala, N.C. Hubner, I.R. de Los Mozos, C. Sadée, A.S. Lenaerts, The SMAD2/3 interactome reveals that TGFβ controls m 6 A mRNA methylation in pluripotency, Nature 555 (7695) (2018) 256–259.

[140] L. Vallier, T. Touboul, S. Brown, C. Cho, B. Bilican, M. Alexander, J. Cedervall, S. Chandran, L. Ährlund-Richter, A. Weber, R.A. Pedersen, Signaling pathways controlling pluripotency and early cell fate decisions of human induced pluripotent stem cells, Stem Cells 27 (11) (2009) 2655–2666.

[141] L. Vallier, T. Touboul, Z. Chng, M. Brimpari, N. Hannan, E. Millan, L.E. Smithers, M. Trotter, P. Rugg-Gunn, A. Weber, R.A. Pedersen, Early cell fate decisions of human embryonic stem cells and mouse epiblast stem cells are controlled by the same signalling pathways, PLoS One 4 (6) (2009) e6082.

[142] K.A. D'Amour, A.D. Agulnick, S. Eliazer, O.G. Kelly, E. Kroon, E.E. Baetge, Efficient differentiation of human embryonic stem cells to definitive endoderm, Nat. Biotechnol. 23 (12) (2005) 1534.

[143] Q. Li, A.P. Hutchins, Y. Chen, S. Li, Y. Shan, B. Liao, D. Zheng, X. Shi, Y. Li, W.Y. Chan, G. Pan, A sequential EMT-MET mechanism drives the differentiation of human embryonic stem cells towards hepatocytes, Nat. Commun. 8 (1) (2017) 1–12.

[144] Z. Cheng, A. Teo, R.A. Pedersen, L. Vallier, SIP1 mediates cell-fate decisions between neuroectoderm and mesendoderm in human pluripotent stem cells, Cell Stem Cell 6 (1) (2010) 59–70.

[145] A.A. Postigo, J.L. Depp, J.J. Taylor, K.L. Kroll, Regulation of Smad signaling through a differential recruitment of coactivators and corepressors by ZEB proteins, EMBO J. 22 (10) (2003) 2453–2462.

[146] Q. Xi, Z. Wang, A.I. Zaromytidou, X.H.F. Zhang, L.F. Chow-Tsang, J.X. Liu, H. Kim, A. Barlas, K. Manova-Todorova, V. Kaartinen, L. Studer, A poised chromatin platform for TGF-β access to master regulators, Cell 147 (7) (2011) 1511–1524.

[147] Ø. Dahle, A. Kumar, M.R. Kuehn, Nodal signaling recruits the histone demethylase Jmjd3 to counteract polycomb-mediated repression at target genes, Sci. Signal. 3 (127) (2010) ra48.

[148] S.W. Kim, S.J. Yoon, E. Chuong, C. Oyolu, A.E. Wills, R. Gupta, J. Baker, Chromatin and transcriptional signatures for Nodal signaling during endoderm formation in hESCs, Dev. Biol. 357 (2) (2011) 492–504.

[149] C. Estarás, N. Akizu, A. García, S. Beltrán, X. de la Cruz, M.A. Martínez-Balbás, Genome-wide analysis reveals that Smad3 and JMJD3 HDM co-activate the neural developmental program, Development 139 (15) (2012) 2681–2691.

[150] J. Liu, Q. Han, T. Peng, M. Peng, B. Wei, D. Li, X. Wang, S. Yu, J. Yang, S. Cao, K. Huang, The oncogene c-Jun impedes somatic cell reprogramming, Nat. Cell Biol. 17 (7) (2015) 856–867.

[151] P. Liu, X. Dou, C. Liu, L. Wang, C. Xing, G. Peng, J. Chen, F. Yu, Y. Qiao, L. Song, Y. Wu, Histone deacetylation promotes mouse neural induction by restricting Nodal-dependent mesendoderm fate, Nat. Commun. 6 (1) (2015) 1–14.

[152] N. Maherali, K. Hochedlinger, Tgfβ signal inhibition cooperates in the induction of iPSCs and replaces Sox2 and cMyc, Curr. Biol. 19 (20) (2009) 1718–1723.

[153] D. Subramanyam, S. Lamouille, R.L. Judson, J.Y. Liu, N. Bucay, R. Derynck, R. Blelloch, Multiple targets of miR-302 and miR-372 promote reprogramming of human fibroblasts to induced pluripotent stem cells, Nat. Biotechnol. 29 (5) (2011) 443.

[154] J.A. Gingold, M. Fidalgo, D. Guallar, Z. Lau, Z. Sun, H. Zhou, F. Faiola, X. Huang, D.F. Lee, A. Waghray, C. Schaniel, A genome-wide RNAi screen identifies opposing functions of Snai1 and Snai2 on the Nanog dependency in reprogramming, Mol. Cell 56 (1) (2014) 140–152.

[155] R.A. Rao, N. Dhele, S. Cheemadan, A. Ketkar, G.R. Jayandharan, D. Palakodeti, S. Rampalli, Ezh2 mediated H3K27me3 activity facilitates somatic transition during human pluripotent reprogramming, Sci. Rep. 5 (2015) 8229.

[156] R. Teshigawara, K. Hirano, S. Nagata, J. Ainscough, T. Tada, OCT4 activity during conversion of human intermediately reprogrammed stem cells to iPSCs through mesenchymal-epithelial transition, Development 143 (1) (2016) 15–23.

[157] J.J. Unternaehrer, R. Zhao, K. Kim, M. Cesana, J.T. Powers, S. Ratanasirintrawoot, T. Onder, T. Shibue, R.A. Weinberg, G.Q. Daley, The epithelial-mesenchymal transition factor SNAIL paradoxically enhances reprogramming, Stem Cell Rep. 3 (5) (2014) 691–698, https://doi.org/10.1016/j.stemcr.2014.09.008.

[158] P. Samavarchi-Tehrani, A. Golipour, L. David, H.K. Sung, T.A. Beyer, A. Datti, K. Woltjen, A. Nagy, J.L. Wrana, Functional genomics reveals a BMP-driven mesenchymal-to-epithelial transition in the initiation of somatic cell reprogramming, Cell Stem Cell 7 (1) (2010) 64–77.

[159] A. Golipour, L. David, Y. Liu, G. Jayakumaran, C.L. Hirsch, D. Trcka, J.L. Wrana, A late transition in somatic cell reprogramming requires regulators distinct from the pluripotency network, Cell Stem Cell 11 (6) (2012) 769–782.

[160] J.M. Polo, E. Anderssen, R.M. Walsh, B.A. Schwarz, C.M. Nefzger, S.M. Lim, M. Borkent, E. Apostolou, S. Alaei, J. Cloutier, O. Bar-Nur, A molecular roadmap of reprogramming somatic cells into iPS cells, Cell 151 (7) (2012) 1617–1632.

[161] L. Pertovaara, L. Sistonen, T.J. Bos, P.K. Vogt, J. Keski-Oja, K. Alitalo, Enhanced jun gene expression is an early genomic response to transforming growth factor beta stimulation, Mol. Cell. Biol. 9 (3) (1989) 1255–1262.

[162] G. Pearson, F. Robinson, T. Beers Gibson, B.E. Xu, M. Karandikar, K. Berman, M.H. Cobb, Mitogen-activated protein (MAP) kinase pathways: regulation and physiological functions, Endocr. Rev. 22 (2) (2001) 153–183.

[163] Y.D. Shaul, R. Seger, The MEK/ERK cascade: from signaling specificity to diverse functions, Biochim. Biophys. Acta 1773 (8) (2007) 1213–1226.

[164] J. McCain, The MAPK (ERK) pathway: investigational combinations for the treatment of BRAF-mutated metastatic melanoma, Pharm. Ther. 38 (2) (2013) 96.

[165] T. Kunath, M.K. Saba-El-Leil, M. Almousailleakh, J. Wray, S. Meloche, A. Smith, FGF stimulation of the Erk1/2 signalling cascade triggers transition of pluripotent embryonic stem cells from self-renewal to lineage commitment, Development 134 (16) (2007) 2895–2902.

[166] P.P. Roux, J. Blenis, ERK and p38 MAPK-activated protein kinases: a family of protein kinases with diverse biological functions, Microbiol. Mol. Biol. Rev. 68 (2) (2004) 320–344.

[167] B. Thisse, C. Thisse, Functions and regulations of fibroblast growth factor signaling during embryonic development, Dev. Biol. 287 (2) (2005) 390–402.

[168] M.K. Saba-El-Leil, F.D. Vella, B. Vernay, L. Voisin, L. Chen, N. Labrecque, S.L. Ang, S. Meloche, An essential function of the mitogen-activated protein kinase Erk2 in mouse trophoblast development, EMBO Rep. 4 (10) (2003) 964–968.

[169] Y. Yao, W. Li, J. Wu, U.A. Germann, M.S. Su, K. Kuida, D.M. Boucher, Extracellular signal-regulated kinase 2 is necessary for mesoderm differentiation, Proc. Natl. Acad. Sci. 100 (22) (2003) 12759–12764.

[170] G. Pagès, S. Guérin, D. Grall, F. Bonino, A. Smith, F. Anjuere, P. Auberger, J. Pouysségur, Defective thymocyte maturation in p44 MAP kinase (Erk 1) knockout mice, Science 286 (5443) (1999) 1374–1377.

[171] J. Pouysségur, P. Lenormand, ERK1 and ERK2 map kinases: specific roles or functional redundancy? Front. Cell Dev. Biol. 4 (2016) 53.

[172] J. Nichols, J. Silva, M. Roode, A. Smith, Suppression of Erk signalling promotes ground state pluripotency in the mouse embryo, Development 136 (19) (2009) 3215–3222.

[173] A.M. Cheng, T.M. Saxton, R. Sakai, S. Kulkarni, G. Mbamalu, W. Vogel, C.G. Tortorice, R.D. Cardiff, J.C. Cross, W.J. Muller, T. Pawson, Mammalian Grb2 regulates multiple steps in embryonic development and malignant transformation, Cell 95 (6) (1998) 793–803.

[174] M. Buehr, A. Smith, Genesis of embryonic stem cells, Philos. Trans. R. Soc. Lond. Ser. B Biol. Sci. 358 (1436) (2003) 1397–1402.

[175] M. Buehr, J. Nichols, F. Stenhouse, P. Mountford, C.J. Greenhalgh, S. Kantachuvesiri, G. Brooker, J. Mullins, A.G. Smith, Rapid loss of Oct-4 and pluripotency in cultured rodent blastocysts and derivative cell lines, Biol. Reprod. 68 (1) (2003) 222–229.

[176] C.W. Lu, A. Yabuuchi, L. Chen, S. Viswanathan, K. Kim, G.Q. Daley, Ras-MAPK signaling promotes trophectoderm formation from embryonic stem cells and mouse embryos, Nat. Genet. 40 (7) (2008) 921.

[177] E. Arman, R. Haffner-Krausz, Y. Chen, J.K. Heath, P. Lonai, Targeted disruption of fibroblast growth factor (FGF) receptor 2 suggests a role for FGF signaling in pregastrulation mammalian development, Proc. Natl. Acad. Sci. 95 (9) (1998) 5082–5087.

[178] J. Nichols, B. Zevnik, K. Anastassiadis, H. Niwa, D. Klewe-Nebenius, I. Chambers, H. Schöler, A. Smith, Formation of pluripotent stem cells in the mammalian embryo depends on the POU transcription factor Oct4, Cell 95 (3) (1998) 379–391.

[179] S. Tanaka, T. Kunath, A.K. Hadjantonakis, A. Nagy, J. Rossant, Promotion of trophoblast stem cell proliferation by FGF4, Science 282 (5396) (1998) 2072–2075.

[180] N.K. Dhaliwal, K. Miri, S. Davidson, H.T. El Jarkass, J.A. Mitchell, KLF4 nuclear export requires ERK activation and initiates exit from naive pluripotency, Stem Cell Rep. 10 (4) (2018) 1308–1323.

[181] M.O. Kim, S.H. Kim, Y.Y. Cho, J. Nadas, C.H. Jeong, K. Yao, D.J. Kim, D.H. Yu, Y.S. Keum, K.Y. Lee, Z. Huang, ERK1 and ERK2 regulate embryonic stem cell self-renewal through phosphorylation of Klf4, Nat. Struct. Mol. Biol. 19 (3) (2012) 283.

[182] Y.S. Lai, C.W. Chang, K.M. Pawlik, D. Zhou, M.B. Renfrow, T.M. Townes, SRY (sex determining region Y)-box2 (Sox2)/poly ADP-ribose polymerase 1 (Parp1) complexes regulate pluripotency, Proc. Natl. Acad. Sci. 109 (10) (2012) 3772–3777.

[183] J. Brumbaugh, J.D. Russell, P. Yu, M.S. Westphall, J.J. Coon, J.A. Thomson, NANOG is multiply phosphorylated and directly modified by ERK2 and CDK1 in vitro, Stem Cell Rep. 2 (1) (2014) 18–25.

[184] S.H. Kim, M.O. Kim, Y.Y. Cho, K. Yao, D.J. Kim, C.H. Jeong, D.H. Yu, K.B. Bae, E.J. Cho, S.K. Jung, M.H. Lee, ERK1 phosphorylates Nanog to regulate protein stability and stem cell self-renewal, Stem Cell Res. 13 (1) (2014) 1–11.

[185] C. Mayor-Ruiz, T. Olbrich, M. Drosten, E. Lecona, M. Vega-Sendino, S. Ortega, O. Dominguez, M. Barbacid, S. Ruiz, O. Fernandez-Capetillo, ERF deletion rescues RAS deficiency in mouse embryonic stem cells, Genes Dev. 32 (7–8) (2018) 568–576.

[186] I.R. Nett, C. Mulas, L. Gatto, K.S. Lilley, A. Smith, Negative feedback via RSK modulates Erk-dependent progression from naïve pluripotency, EMBO Rep. 19 (8) (2018) e45642.

[187] M. Saha, A. Carriere, M. Cheerathodi, X. Zhang, G. Lavoie, J. Rush, P.P. Roux, B.A. Ballif, RSK phosphorylates SOS1 creating 14-3-3-docking sites and negatively regulating MAPK activation, Biochem. J. 447 (1) (2012) 159–166.

[188] X. Zhang, G. Lavoie, L. Fort, E.L. Huttlin, J. Tcherkezian, J.A. Galan, H. Gu, S.P. Gygi, S. Carreno, P.P. Roux, Gab2 phosphorylation by RSK inhibits Shp2 recruitment and cell motility, Mol. Cell. Biol. 33 (8) (2013) 1657–1670.

[189] M. Ter Huurne, J. Chappell, S. Dalton, H.G. Stunnenberg, Distinct cell-cycle control in two different states of mouse pluripotency, Cell Stem Cell 21 (4) (2017) 449–455.

[190] W. Guo, B. Hao, Q. Wang, Y. Lu, J. Yue, Requirement of B-Raf, C-Raf, and A-Raf for the growth and survival of mouse embryonic stem cells, Exp. Cell Res. 319 (18) (2013) 2801–2811.

[191] S.M. Morgani, N. Saiz, V. Garg, D. Raina, C.S. Simon, M. Kang, A.M. Arias, J. Nichols, C. Schröter, A.K. Hadjantonakis, A Sprouty4 reporter to monitor FGF/ERK signaling activity in ESCs and mice, Dev. Biol. 441 (1) (2018) 104–126.

[192] J. Deathridge, V. Antolović, M. Parsons, J.R. Chubb, Live imaging of ERK signalling dynamics in differentiating mouse embryonic stem cells, Development 146 (12) (2019) dev172940.

[193] J. Choi, A.J. Huebner, K. Clement, R.M. Walsh, A. Savol, K. Lin, H. Gu, B. Di Stefano, J. Brumbaugh, S.Y. Kim, J. Sharif, Prolonged Mek1/2 suppression impairs the developmental potential of embryonic stem cells, Nature 548 (7666) (2017) 219–223.

[194] W.W. Tee, S.S. Shen, O. Oksuz, V. Narendra, D. Reinberg, Erk1/2 activity promotes chromatin features and RNAPII phosphorylation at developmental promoters in mouse ESCs, Cell 156 (4) (2014) 678–690.

[195] M. Roode, K. Blair, P. Snell, K. Elder, S. Marchant, A. Smith, J. Nichols, Human hypoblast formation is not dependent on FGF signalling, Dev. Biol. 361 (2) (2012) 358–363.

[196] J. Göke, Y.S. Chan, J. Yan, M. Vingron, H.H. Ng, Genome-wide kinase-chromatin interactions reveal the regulatory network of ERK signaling in human embryonic stem cells, Mol. Cell 50 (6) (2013) 844–855.

[197] T.P. Rao, M. Kühl, An updated overview on Wnt signaling pathways: a prelude for more, Circ. Res. 106 (12) (2010) 1798–1806.

[198] H. Haegel, L. Larue, M. Ohsugi, L. Fedorov, K. Herrenknecht, R. Kemler, Lack of beta-catenin affects mouse development at gastrulation, Development 121 (11) (1995) 3529–3537.

[199] J. Huelsken, R. Vogel, V. Brinkmann, B. Erdmann, C. Birchmeier, W. Birchmeier, Requirement for β-catenin in anterior-posterior axis formation in mice, J. Cell Biol. 148 (3) (2000) 567–578.

[200] B.J. Merrill, Wnt pathway regulation of embryonic stem cell self-renewal, Cold Spring Harb. Perspect. Biol. 4 (9) (2012) a007971.

[201] I. Theka, F. Sottile, M. Cammisa, S. Bonnin, M. Sanchez-Delgado, U. Di Vicino, M.V. Neguembor, K. Arumugam, F. Aulicino, D. Monk, A. Riccio, Wnt/β-catenin signaling pathway safeguards epigenetic stability and homeostasis of mouse embryonic stem cells, Sci. Rep. 9 (1) (2019) 1–18.

[202] J. Wray, T. Kalkan, S. Gomez-Lopez, D. Eckardt, A. Cook, R. Kemler, A. Smith, Inhibition of glycogen synthase kinase-3 alleviates Tcf3 repression of the pluripotency network and increases embryonic stem cell resistance to differentiation, Nat. Cell Biol. 13 (7) (2011) 838–845.

[203] H. Aberle, A. Bauer, J. Stappert, A. Kispert, R. Kemler, β-catenin is a target for the ubiquitin–proteasome pathway, EMBO J. 16 (13) (1997) 3797–3804.

[204] H. Clevers, R. Nusse, Wnt/β-catenin signaling and disease, Cell 149 (6) (2012) 1192–1205.

[205] B. Rubinfeld, I. Albert, E. Porfiri, C. Fiol, S. Munemitsu, P. Polakis, Binding of GSK3β to the APC-β-catenin complex and regulation of complex assembly, Science 272 (5264) (1996) 1023–1026.

[206] J. Behrens, J.P. von Kries, M. Kühl, L. Bruhn, D. Wedlich, R. Grosschedl, W. Birchmeier, Functional interaction of β-catenin with the transcription factor LEF-1, Nature 382 (6592) (1996) 638–642.

[207] M. Molenaar, M. van de Wetering, M. Oosterwegel, J. Peterson-Maduro, S. Godsave, V. Korinek, J. Roose, O. Destrée, H. Clevers, XTcf-3 transcription factor mediates β-catenin-induced axis formation in Xenopus embryos, Cell 86 (3) (1996) 391–399.

[208] M. Van de Wetering, R. Cavallo, D. Dooijes, M. van Beest, J. van Es, J. Loureiro, A. Ypma, D. Hursh, T. Jones, A. Bejsovec, M. Peifer, Armadillo coactivates transcription driven by the product of the Drosophila segment polarity gene dTCF, Cell 88 (6) (1997) 789–799.

[209] J. Chen, C.M. Nefzger, F.J. Rossello, Y.B. Sun, S.M. Lim, X. Liu, S. De Boer, A.S. Knaupp, J. Li, K.C. Davidson, J.M. Polo, Fine tuning of canonical Wnt stimulation enhances differentiation of pluripotent stem cells independent of β-catenin-mediated T-cell factor signaling, Stem Cells 36 (6) (2018) 822–833.

[210] P. Liu, M. Wakamiya, M.J. Shea, U. Albrecht, R.R. Behringer, A. Bradley, Requirement for Wnt3 in vertebrate axis formation, Nat. Genet. 22 (4) (1999) 361–365.

[211] A. Tsakiridis, Y. Huang, G. Blin, S. Skylaki, F. Wymeersch, R. Osorno, C. Economou, E. Karagianni, S. Zhao, S. Lowell, V. Wilson, Distinct Wnt-driven primitive streak-like populations reflect in vivo lineage precursors, Development 141 (6) (2014) 1209–1221.

[212] Y. Atlasi, R. Noori, C. Gaspar, P. Franken, A. Sacchetti, H. Rafati, T. Mahmoudi, C. Decraene, G.A. Calin, B.J. Merrill, R. Fodde, Wnt signaling regulates the lineage differentiation potential of mouse embryonic stem cells through Tcf3 down-regulation, PLoS Genet. 9 (5) (2013) e1003424.

[213] N. Harada, Y. Tamai, T.O. Ishikawa, B. Sauer, K. Takaku, M. Oshima, M.M. Taketo, Intestinal polyposis in mice with a dominant stable mutation of the β-catenin gene, EMBO J. 18 (21) (1999) 5931–5942.

[214] M.F. Kielman, M. Rindapää, C. Gaspar, N. Van Poppel, C. Breukel, S. Van Leeuwen, M.M. Taketo, S. Roberts, R. Smits, R. Fodde, Apc modulates embryonic stem-cell differentiation by controlling the dosage of β-catenin signaling, Nat. Genet. 32 (4) (2002) 594–605.

[215] K.F. Kelly, D.Y. Ng, G. Jayakumaran, G.A. Wood, H. Koide, B.W. Doble, β-catenin enhances Oct-4 activity and reinforces pluripotency through a TCF-independent mechanism, Cell Stem Cell 8 (2) (2011) 214–227.

[216] M.F. Cole, S.E. Johnstone, J.J. Newman, M.H. Kagey, R.A. Young, Tcf3 is an integral component of the core regulatory circuitry of embryonic stem cells, Genes Dev. 22 (6) (2008) 746–755.

[217] G. Morrison, R. Scognamiglio, A. Trumpp, A. Smith, Convergence of cMyc and β-catenin on Tcf7l1 enables endoderm specification, EMBO J. 35 (3) (2016) 356–368.

[218] J. Luo, R. Sladek, J.A. Bader, A. Matthyssen, J. Rossant, V. Giguère, Placental abnormalities in mouse embryos lacking the orphan nuclear receptor ERR-β, Nature 388 (6644) (1997) 778–782.

[219] G. Martello, T. Sugimoto, E. Diamanti, A. Joshi, R. Hannah, S. Ohtsuka, B. Göttgens, H. Niwa, A. Smith, Esrrb is a pivotal target of the Gsk3/Tcf3 axis regulating embryonic stem cell self-renewal, Cell Stem Cell 11 (4) (2012) 491–504.

[220] N. Salomonis, C.R. Schlieve, L. Pereira, C. Wahlquist, A. Colas, A.C. Zambon, K. Vranizan, M.J. Spindler, A.R. Pico, M.S. Cline, T.A. Clark, Alternative splicing regulates mouse embryonic stem cell pluripotency and differentiation, Proc. Natl. Acad. Sci. 107 (23) (2010) 10514–10519.

[221] W.L. Tam, C.Y. Lim, J. Han, J. Zhang, Y.S. Ang, H.H. Ng, H. Yang, B. Lim, T-cell factor 3 regulates embryonic stem cell pluripotency and self-renewal by the transcriptional control of multiple lineage pathways, Stem Cells 26 (8) (2008) 2019–2031.

[222] F. Yi, L. Pereira, J.A. Hoffman, B.R. Shy, C.M. Yuen, D.R. Liu, B.J. Merrill, Opposing effects of Tcf3 and Tcf1 control Wnt stimulation of embryonic stem cell self-renewal, Nat. Cell Biol. 13 (7) (2011) 762–770.

[223] A. Nishiyama, L. Xin, A.A. Sharov, M. Thomas, G. Mowrer, E. Meyers, Y. Piao, S. Mehta, S. Yee, Y. Nakatake, C. Stagg, Uncovering early response of gene regulatory networks in ESCs by systematic induction of transcription factors, Cell Stem Cell 5 (4) (2009) 420–433.

[224] A. Marson, R. Foreman, B. Chevalier, S. Bilodeau, M. Kahn, R.A. Young, R. Jaenisch, Wnt signaling promotes reprogramming of somatic cells to pluripotency, Cell Stem Cell 3 (2) (2008) 132.

[225] A. Marson, S.S. Levine, M.F. Cole, G.M. Frampton, T. Brambrink, S. Johnstone, M.G. Guenther, W.K. Johnston, M. Wernig, J. Newman, J.M. Calabrese, Connecting microRNA genes to the core transcriptional regulatory circuitry of embryonic stem cells, Cell 134 (3) (2008) 521–533.

[226] H. Niwa, Wnt: what's needed to maintain pluripotency? Nat. Cell Biol. 13 (9) (2011) 1024–1026.

[227] S. Rudloff, R. Kemler, Differential requirements for β-catenin during mouse development, Development 139 (20) (2012) 3711–3721.

[228] J. Yu, X. Guo, X. Chen, C. Qiao, B. Chen, B. Xie, X. Luan, C. Shen, J. Zhu, J. Liu, Y. Yan, Activation of β-catenin causes defects in embryonic development during maternal-to-zygotic transition in mice, Int. J. Clin. Exp. Pathol. 11 (5) (2018) 2514.

[229] D. Ten Berge, D. Kurek, T. Blauwkamp, W. Koole, A. Maas, E. Eroglu, R.K. Siu, R. Nusse, Embryonic stem cells require Wnt proteins to prevent differentiation to epiblast stem cells, Nat. Cell Biol. 13 (9) (2011) 1070–1075.

[230] G. Dravid, Z. Ye, H. Hammond, G. Chen, A. Pyle, P. Donovan, X. Yu, L. Cheng, Defining the role of Wnt/β-catenin signaling in the survival, proliferation, and self-renewal of human embryonic stem cells, Stem Cells 23 (10) (2005) 1489–1501.

[231] N. Sato, L. Meijer, L. Skaltsounis, P. Greengard, A.H. Brivanlou, Maintenance of pluripotency in human and mouse embryonic stem cells through activation of Wnt signaling by a pharmacological GSK-3-specific inhibitor, Nat. Med. 10 (1) (2004) 55–63.

[232] G.A. Gomez, M.S. Prasad, M. Wong, R.M. Charney, P.B. Shelar, N. Sandhu, J.O. Hackland, J.C. Hernandez, A.W. Leung, M.I. García-Castro, WNT/β-catenin modulates the axial identity of embryonic stem cell-derived human neural crest, Development 146 (16) (2019) dev175604.

[233] S. Li, N. Li, P. Zhu, Y. Wang, Y. Tian, X. Wang, Decreased β-catenin expression in first-trimester villi and decidua of patients with recurrent spontaneous abortion, J. Obstet. Gynaecol. Res. 41 (6) (2015) 904–911.

[234] X. Xu, L. Wang, B. Liu, W. Xie, Y.G. Chen, Activin/Smad2 and Wnt/β-catenin up-regulate HAS2 and ALDH3A2 to facilitate mesendoderm differentiation of human embryonic stem cells, J. Biol. Chem. 293 (48) (2018) 18444–18453.

[235] Y. Song, S. Lee, E.H. Jho, Enhancement of neuronal differentiation by using small molecules modulating Nodal/Smad, Wnt/β-catenin, and FGF signaling, Biochem. Biophys. Res. Commun. 503 (1) (2018) 352–358.

Chapter 6

Immunity, stem cells, and aging

Ezhilarasan Devaraj[a,*], Muralidharan Anbalagan[b,*], R. Ileng Kumaran[c], and Natarajan Bhaskaran[d]

[a]Department of Pharmacology, Biomedical Research Unit and Laboratory Animal Center, Saveetha Dental College and Hospital, Saveetha Institute of Medical and Technical Sciences, Chennai, Tamil Nadu, India, [b]Department of Structural and Cellular Biology, Tulane University School of Medicine, New Orleans, LA, United States, [c]Biology Department, Farmingdale State College, Farmingdale, NY, United States, [d]Department of Biomedical Sciences, Sri Ramachandra Institute of Higher Education and Research, SRMC, Chennai, Tamil Nadu, India

1 Introduction

Aging and its related immunity is a complex process with various mechanisms in our normal physiological function that could lead to decreased body fitness and increased susceptibility to infections and various conditions such as metabolic, neurodegenerative, cardiovascular, and cancer diseases. Aging is an inevitable process that all of us have to undergo. Upon aging, all living organisms suffer a significant loss of tissue and organ function. This could be due to the significant loss of stem cells and their activity that occurs over a period. During homeostasis and under stress, stem cells regenerate body tissues. A significant decrease in T cell function without a decline in their number is observed with respect to the immune system and aging. This reduction in function weakens the immune system and in turn increases the risk of developing autoimmune diseases. Aging in stem cells deteriorates their functions and their ability to differentiate into various cell types, which eventually affects the repair mechanism of the cells.

A significant waning in tissue function observed during aging is attributed to the enhancement of proinflammatory mediators, decreased systemic inflammation, and impairment in wound repair [1]. A lack of communication between the immune and stem cells along with the accumulation of proinflammatory molecules in the tissue correlate with age-related defects. In a study, it was shown that the dendritic epidermal T cells (DETCs) of mouse skin facilitate wound repair, which demonstrates the interference of coordination between local immune cells and aged stem cells [2]. The wound bed of the aging skin is reepithelialized by the epidermal progenitors without the involvement of DETCs due to the breakdown of the communication network [3]. A similar miscommunication in the epidermis of aged individuals underlies chronic wounds. This communication interference between the immune and aged stem cells leads to the enhancement of proinflammatory markers, which in turn causes a significant decline in tissue function. Skin has different cells, including hair follicle stem cells that maintain hair growth and melanocyte stem cells (MSCs) that generate the melanin-making cells called melanocytes. In HFSCs, the growth undergoes three phases, that is, growth (anagen), regression (catagen), and rest (telogen). It is noted that aging does not decrease the frequency of hair follicle stem cells (HFSCs) [4], although they lose their function.

In skin, there is a significant reduction in melanocyte stem cells observed during the aging process by ectopic differentiation [5]. A similar effect on the MSCs was also observed when the cells were exposed to ionizing radiation, suggesting that hair graying associated with age [6] was due to genotoxic stress. By reducing this stress, the hair could retain its color. The rapid turnover of epithelial cells in the gut is provided by intestinal stem cells (ISCs). It was found that the aged macrophage in the intestine facilitates an increase in the expression of TNF alpha, diminishes ISC function, breaches the epithelial barrier, and increases the intestinal permeability. Drosophila studies demonstrated that a variety of factors related to the environment also play a vital role in ISC aging [7]. Normally, ISC aging occurs due to the activation of JNK, p38-mitogen-activated protein kinase [8], and vascular endothelial growth factor-linked signaling pathways [9]. In aging flies, the hemocyte interaction leads to intestinal dysplasia, whereas in younger flies, the same promotes ISC proliferation and infection resistance. The studies conducted resulted in the alteration, proliferation, and differentiation of the aged stem cells [10]. In the ISCs of mammals, two interconvertible markers exist, namely proliferative LGR5-expressing cells and quiescent label-retaining cells, present in the base and above the crypt, respectively [11].

Satellite stem cells (SCSs) are those that are responsible for the regeneration of skeletal muscle fibers during injury [12–14]. Aging plays a vital role in the decrease of satellite cells [15–17]. The satellite stem cells exhibited fibrogenic

---

[*] Equal contribution.

lineage due to changes in signaling pathways such as the Wnt pathway and the TGF-β pathway under skewed differentiation [18, 19]. The HSCs in the bone marrow were subjected to changes during aging. It was found that there was a significant decrease in HSC function with respect to age [20, 21]. Aged HSCs exhibited myeloid lineage under a skewed differentiation potential [22]. In aged individuals, anemia was also found to be linked to the aging of HSCs [23].

In germline stem cells (GSCs), aging shows an increase in GSCs that leads to cell cycle arrest [24]. In mammals, spermatogonial stem cells help to maintain the male germline and similarly decrease its number upon aging. Niche deterioration is said to be involved in germline cell aging [25, 26]. Insulin has a major role in the number and function of GSCs [27, 28].

The neural stem cells (NSCs) in the mammalian brain undergo neurogenesis in a specific region during adulthood. Upon aging, there is a decrease in the number of NSCs that leads to a decline in neurogenesis [29]. The senescence of NSCs during aging is one of the causes of the decline in the number of NSCs [30]. This decline affects memory, learning, and other cognitive functions in the aging process [31].

1.1 Epigenetics and aging stem cells

Throughout our lifetime, there are multiple epigenetic changes that occur in cells and tissues. Epigenetics usually involves changes in DNA methylation patterns, histone modifications, and chromatin remodeling. There are various enzymes that are important for the generation and maintenance of epigenetic signatures such as DNA methyl transferases, histone acetylases, histone deacetylases, histone methylases, and histone demethylases. The epigenetic changes that occur throughout our life affect stem cell function [32]. Studies have demonstrated that the DNA methylation pattern could foresee the ages in humans. In hematopoietic stem cells, mice studies showed that the DNA methylome comparison between young and old mice showed that all old mice had increased global DNA methylation. In contrast, the remaining adult stem cells exhibited a decline of DNA methylation. However, it should be noted that a strong methylation pattern was seen in the CPG island at specific loci in other adult stem cells. Histone repressive markers such as histone 3 lysine 9 trimethylation (H3K9me3) and histone 3 lysine 2 trimethylation (H3K2me3) were increased in aged HSC compared to the young. All these suggest that a process of increasing heterochromatinization occurs in a time-dependent manner, reducing stem cell plasticity during the aging process. A global methylomic profile by Hannum and his group using whole blood samples of 656 human individuals demonstrated that DNA methylomes age faster in men compared to women [33]. The age-related differences in the DNA methylome link directly to the functional changes in gene expression patterns [33].

1.2 Sex differences in aging

Regardless of sex, the age and immune system function differently. Old women are more vulnerable to autoimmune diseases compared to old men. Conversely, old men are more vulnerable to infections and they have less response toward vaccines compared to old women. This difference between men and women could be due to cell frequencies and cell intrinsic properties [34]. To differentiate the aging-related immune cell function, Marquez and coworkers isolated peripheral blood mononuclear cells (PBMCs) from 172 healthy individuals (age-matched female and male). Their study using ATAC-seq, RNA seq, and flow cytometry revealed that T cell function declines at a higher magnitude with age in men. Moreover, with aging there is an increase in cytotoxic cells and monocyte function in men compared to women [34]. With B cell aging, quite the opposite effect was observed in men and women. Upon aging, the B cell marker genes were upregulated only in women, but these genes are not activated in men [34]. This could be one of the possible reasons for sex differences in autoimmunity and humoral response.

2 Aging

Aging is the age-dependent biochemical change in tissues and the deterioration in the performance of various biological functions, including significant weakening in immunity, that leads to increased susceptibility to diseases and mortality [35]. Aging is one of the risk factors for several diseases such as dementia, cardiovascular disease, osteoporosis, cancer, diabetes, idiopathic pulmonary fibrosis, and glaucoma [36]. Aging is not caused by a single entity, but instead a variety of multifactorial processes are considered the hallmarks of aging, including DNA damage, accumulation of reactive oxygen species (ROS), inflammaging, disruption of circadian rhythms, change in metabolism, loss of telomeres, reduced stem cell function, disruption in mitochondrial dynamics, and senescence [37]. Senescence is a permanent cell-cycle arrest that limits the proliferation of aged stem cells and thereby has a vital role in aging [38]. Senescence occurs in response to cellular aging, inflammation, ROS activation, DNA damage, nucleotide depletion, and oncogenes. The human protein kinases such as (ataxia-telangiectasia, mutated) ATM and Rad3-related (ATR) are considered potential sensors of DNA damage [39]. DNA damage caused by telomeric attrition and stress (oncogenic or oxidative) activates ATM and ATR, which further activate the Chk2 and Chk1 pathways, respectively. They subsequently transactivate p53 and p21^{CIP1}. The activation of p21 induces the inhibition of cyclin-dependent kinase 4/6 activity, which contributes to the G1 phase cell cycle arrest or senescence [40]. Aging-related proinflammatory marker accumulation in tissues is the reason for aging, and this is known as inflammaging. The cells that developed a senescence-associated secretory phenotype have been implicated in the accumulation of inflammatory mediators. The aged cells acquire SASP due to DNA damage and cause the synthesis and liberation of proinflammatory markers such as interleukin-6 and others [41, 42].

Mechanistically, the aging-related protein aggregate accumulation in intracellular organelles such as mitochondria causes mitochondrial damage. Damaged mitochondria accumulate intracellular ROS and interfere with several cell signalings associated with stem cell functions [43]. Previously it has been shown that, intracellular ROS accumulation activates p38 mitogen-activated protein kinases (MAPK) and forkhead box protein O1 (FOXO) responsible for stress signaling-related impairment in stem cells proliferation and their differentiation [44–46]. Oxidative stress is also implicated in stem cell senescence and cytotoxicity via DNA damage [47]. These processes directly affect the regeneration potential and self-renewal of matured or aged stem cells, which subsequently depletes the stem cell pool [45]. From an epigenetic perspective, aging and autophagy impairment cause increased toxic protein aggregates in mitochondria. This disturbs homeostasis amid glycolysis and oxidative phosphorylation, triggering variations in the synthesis of epigenetic regulators and cofactors [45]. The disruption and imbalance between mitochondrial respiration processes such as glycolysis and oxidative phosphorylation subsequently cause epigenetic alterations via DNA methylation as well as histone methylation and acetylation that deplete the stem cell population [48, 49]. A variety of cells are involved in the aging process, including endothelial cells, stem cells, and vascular smooth muscle cells. Among these, stem cells are among the main drivers of the aging process.

3 Stem cells in aging

Embryonic stem cells (ESCs) originate from the blastocyst's inner cell mass. They are totipotent and have the variation potential to distinguish into any other cell type during embryonic development and growth [50]. Induced stem cells are pluripotent and are similar to ESCs. They can differentiate into any adult cell type by inserting four transcription factors into their genetic material such as octamer-binding transcription factor 4, sex-determining region Y-box 2, cMyc, and Krüppel-like factor-4 [51]. In a multicellular organism, somatic (adult) stem cells are found in almost all tissues in the body [47]. They have an essential role in aging via tissue regeneration and maintenance. These cells are in charge of the regeneration of tissues and the replenishment of dying cells due to their self-renewal capacity [45]. Hematopoietic stem cells (HSCs) are capable of differentiating into all types of blood cells, including immune cells, throughout adult life [52]. There are several processes such as oxidative stress as well as epigenetic, genomic, and proteomic changes that cause impairment in stem cell functions such as self-renewal and differentiation [53]. Upon aging, all tissues undergo a sequential regenerative decline. Decreases in stem cell number and their regenerative capability are linked with aging [54]. Also, a decreased self-renewal property and alterations in the cell cycles of stem cells are associated with aging. The proliferation ability of human HSCs is inversely correlated with age due to the shortening of telomeres [55,56]. Thus, aging reduces stem cell function, which in turn causes impairment in immune function and tissue maintenance as well as systemic side effects, senescence, and apoptosis [45].

HSCs constantly renew immune cells and the hematopoietic system function is also said to decline with age [57]. HSCs and their renewal capacity in bone marrow also decline with the aging process [58]. Changes in the HSC niche as well as alterations of hormone production and a variety of signaling disturb the self-renewal capability of immune cells and the lineage commitment of HSC. For instance, bone marrow niche TGF-β signaling controls the homeostasis of the immune system as well as the quiescence and self-renewal of HSCs [57]. Thus, the hematopoietic niche-derived cell-extrinsic factors are involved in HSC maintenance [59]. Once produced in the bone marrow and the thymus, the immature T and B cells wander to the secondary lymphoid organs such as the spleen. The primary lymphopoiesis is very active and robust in the young. Lymphocyte development is significantly diminished with aging due to changes in HSCs induced by intrinsic and extrinsic factors. Therefore, unsurprisingly, in the bone marrow and thymus, the B and T cell progenitor population significantly declines with age; this may be directly related to the reduced HSC function with aging [41]. The age-associated accumulation of cell-intrinsic factors such as the DNA damage response, enhanced ROS production, epigenetic changes, telomere attrition, and polarity changes induces HSC aging, and these processes indirectly diminish the immune cell function [59].

4 Immunity and immune cells

In an organism, the immune system is a vital element for the defense mechanism against battling the pathogenic disturbances. Age-related immune dysfunction, which is also known as immune senescence, reveals an enlarged vulnerability to infection and inflammation. It also intensifies the inception and development of autoimmune diseases and cancer [60]. In the immune system, the related organs are located all over the body and they are termed lymphoid organs. Lymphocytes, small white blood cells from lymphoid organs, are crucial in the immune system. The T cells and T lymphocytes mature in the lymphoid organ (thymus) and then travel to other tissues to fight pathogens while the B lymphocytes or B cells develop into activated and matured plasma cells, which in turn produce antibodies against diseases. All the immune cells start as immature cells in the bone marrow. By interacting with different cytokines and other molecules, they grow into specific types of cells such as T helper cells, natural killer cells, etc. These cells and their cellular factors such as cytokines, chemokines, interleukins, and antimicrobial peptides offer a front line of defense as the innate immune system and the adaptive immune system to fight various pathogens and their related ailments [61].

The cytokines and chemokines, which were produced by specific innate and adaptive immune cells, considerably modify along with aging. In particular, the inflammation causing cytokines like IL-6, Interleukin-1β, TNF-α, and TGF-β leads to inflammation and its associated inflammaging observed in the aging people [62]. The changes in the immune cells and their cytokines are shown in Fig. 1. These alterations in immune cells due to aging lead to various ailments.

5 Immunity in aging

A well-known symptom of aging is the deterioration in the immune system (immunosenescence). This leads to a marked reduction in T cell functions, predominantly T helper cells. This in turn upsets humoral immunity and impairs B-cell function. During the increase in age, regulatory T cells continue their role by escalating their numbers and importantly regulates the immune system in the old people. Natural killer cells, one of the important innate immune protection mechanisms, shows diminished cytotoxicity and cytokine production. This change in the innate immunity will also influence the purpose of the adaptive system and further affects the antiaging mechanism. Also, the changes in the foremost Th1 cells to the major Th2-type antigen responses and an alteration in their cytokines have been projected as machinery for age-oriented immune dysfunction; the same is shown in Fig. 1 and Table 1 [63]. During aging, there is a decline in the function

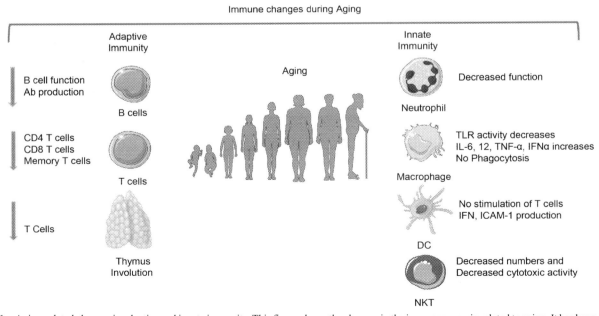

FIG. 1 Aging-related changes in adaptive and innate immunity. This figure shows the changes in the immune scenario related to aging. It has been clearly shown that as age increases, immunity decreases. During aging, most of the immune-related cells decrease in number and their function diminishes. The proinflammatory cytokine level goes up whereas the antibodies produced by B cells decline during aging. This picture is a schematic depiction of changes in immune system during aging. *AB*, antibody production; *DC*, dendritic cells; *NKT*, natural killer T cells; *ICAM1*, intercellular adhesion molecule 1; *IFN*, interferon; *IL*, interleukin; *TLR*, toll-like receptors; *TNFα*, tumor necrosis factor alpha.

TABLE 1 Innate and adaptive immunity during aging.

Innate immunity during aging		Adaptive immunity during aging	
Myeloid cell number	Increases	Number of naive cells	Decreases
		Number of memory cells	Increases
Cytokine production	Increases	Number of regulatory T cells	Increases
Free radical production	Increases	Function of regulatory T cells	Decreases
		Cell proliferation	Decreases
Phagocytosis	Decreases	Number of B cells	Decreases
		Function of B cells	Decreases
Chemotaxis	Decreases	Antibody production	Decreases
		Autoantibody production	Increases

This table represents the difference between innate and adaptive immunity in their functions during aging.

of various tissues due to the amplified levels of proinflammatory mediators, systemic inflammation, and weakened wound healing [1]. Studies suggest that a lack of communication between the immune and stem cells along with the accumulation of proinflammatory molecules in the tissue are correlated with age-related defects. During aging, there is a significant decrease in T cell functions and increases in cytotoxic and monocyte cell functions [34]. This signature matches age-associated changes in the immune system, including diminished adaptive responses and increased systemic inflammation with age [34, 64].

In younger individuals, the healthy production of B cells and T cells in the bone marrow and thymus is importantly noted. In HSCs, there are comparatively more lymphoid-based stem cells that generate lymphoid progenitors with high generative potential. However, in aged individuals, the number of lymphoid-biased HSCs drops considerably. During aging in HSC myeloid influenced stem cells that are predominate, leads to increased number of myeloid progenitors and reduced numbers of lymphoid progenitors [41]. B cell progenitors and T cell progenitors display proliferation compared with the young. Tumor suppressors such as p16Ink4a and ARF are elevated in the aging cells of different tissues [65]. An elevated level of p16Ink4a activates retinoblastoma, inducing cell cycle arrest. p19Arf activates p53, inducing either cell cycle arrest or apoptosis. The importance of lymphoid progenitors is condensed with age, and mice studies revealed that during aging, both Ink4a and Arf expressions increase in pro-B cells [66]. The amplified expression of Ink4a and Arf in pro-B cells and Ink4a in pro-T cells contributes to the decreased proliferation and increased apoptosis [41, 67]. Chronic low-grade inflammation that is mainly driven by endogenous signaling during aging is termed inflammaging [68]. One of the main features of inflammaging is the long-term activation of the innate immune system, where the macrophage plays a main role.

6 Protein damage, quality control mechanism, and aging

Stem cells use some mechanisms such as reduced metabolic activity and the accumulation of intracellular toxic metabolites to reduce the risk of the accumulation of damage during aging. Unless needed, they remain in a quiescence state so that replication-associated DNA damage can be avoided [69]. Stem cells do not have powerful DNA repair systems. Therefore, reducing the risk of DNA damage is critical, especially in these cells. With stem cells in the quiescence state, there is a significant decrease in RNA and protein synthesis, which provides the maintenance of a healthy proteome, as protein folding and the removal of misfolded and damaged proteins are processes that consume high amounts of energy that would bind the energy availability to other mechanisms [70]. Stem cells use various damage removal systems, including autophagy (an evolutionarily conserved process that is responsible for removing damaged proteins and organelles). A decline in the process of autophagy is linked with accelerated aging, where damage accumulation exceeds damage removal.

At a young age, proteostasis is sustained by protein quality-control mechanisms such as autophagy, the ubiquitin proteasome system (UPS), and the unfolded protein response in the mitochondria and the endoplasmic reticulum [45].

The productivity of these protein systems confirms the targeting of proteins and the exclusion of misfolded damage in the proteins, which declines with aging and compromises the cell function [71]. The mechanisms are moderately understood, but the compact action of regulatory transcription factors such as FOXO [72] is important in aging.

7 Molecules and mechanisms involved in stem cells and aging

Several intracellular signalings involve in the stem cell exhaustion-associated aging process. DNA damage accumulation is very common in aged stem cells. Stem cell DNA damage is responsible for the activation of several signaling cascades pertaining to cell-cycle arrest, apoptosis, senescence, or differentiation [73, 74]. The Arf/p53 pathway reportedly protects the stem cells against acute stress/DNA damage; this pathway also regulates the chronic type of stress associated with stem cells aging [75]. In mice, a modest increase in Arf/p53 activity was responsible for an antiaging effect while the downregulation of p53 promoted aging. Dysregulated Arf/p53 activity influences the lifespan of mice by a significant decrease in the function of stem cells as well as their regenerative capacity [76]. In contrast, p53 was also reported to promote aging and reduce tissue regeneration; it was also correlated to premature stem cell exhaustion. In mice, the loss of p53 function due to the hypomorphic Rad50 mutation led to a partial rescue and survival of stem cells, indicating that apoptotic signaling and double-strand breakthrough p53 results in stem cell death and the p53 knockout in these mice protects HSC exhaustion [77]. Therefore, the role of p53 in stem cells associated with aging is still subject to debate.

In the context of ROS, mitochondria act as both a source and target for ROS [78]. ROS are predominantly generated during OXPHOS, which interferes with the stem cell function and its destiny. In the stem cell niche of bone marrow, HSCs reside in a niche with low oxygen and low ROS conditions [55] has the long-term self-renewal activity [79]. Intracellular ROS accumulation in aged HSC causes oxidative stress, leading to the activation of the p53-associated DNA damage response pathways, the DNA damage sensing serine/threonine protein kinase, ATM, and FOXO3a, which in turn activate p16Ink4a/p19Arf and cell cycle inhibitors of HSC thereby inducing stem cell decline and senescence [80]. In aged HSCs, overwhelming intracellular ROS activates p38 MAPK signaling, thereby promoting HSC depletion and lineage skewing by upregulating inhibitors such as p16 and Arf [81].

Ample evidence suggests that mitochondrial function is crucial for HSC fate determination and function and gradual mitochondrial dysfunction is associated with aging [82]. Tissue stem cell mitochondria regulate aging by controlling the metabolic profile and respiration. Unlike aged stem cells, young stem cells contain metabolically inactive mitochondria [79, 83]. In addition to their metabolic function, mitochondria respond to a variety of stress signaling that can generate retrograde signals such as ROS that affect other intracellular organelles, which impair stem cell function and activity [84]. The glycolysis-OXPHOS balance is critical for stem cell homeostasis and its disturbance leads to the formation of insufficient levels of epigenetic regulators and cofactors [45, 85]. Mitochondrial damage causes excessive ROS, which in turn induces mitochondrial damage and metabolic imbalance that affects the homeostasis of glycolysis and OXPHOS. This disturbance causes histone and DNA acetylation, and methylation subsequently induces epigenetic alterations in stem cells [45]. Studies suggest that a mitochondrial homeostasis disturbance due to ROS is responsible for the inhibition of production and differentiation of stem cells and consequently the depletion of the stem cell pool. Further, the evolutionarily conserved FOXO transcription factors are identified as key stem cell regulators for the maintenance of homeostasis in different tissues [86]. The stem cell specific deletion of FOXO1, FOXO3a, and FOXO4 caused an increase in mitochondrial dysfunction and ROS accumulation, resulting in a loss of HSC quiescence and self-renewal capacity [79].

Autophagy is a vital proteostasis that involves homeostasis maintenance, immunity, and the development and prevention of disease [87, 88]. In quiescent stem cells, basal autophagy is responsible for the removal of metabolically active mitochondria. It thereby controls oxidative phosphorylation, glycolysis, stemness, and regenerative potential [89]. Accumulating evidence suggests that the impaired autophagy-related genes in aged stem cells result in a proteostasis imbalance and mitochondrial dysfunction [89]. The mammalian target of rapamycin (mTOR) is really important in stem cell propagation, self-renewal, differentiation, and progenitor cell maintenance [90]. mTOR hyperactivity is responsible for stem cell exhaustion via the modulation of autophagy and mitophagy in stem cells [91].

From the immune system perspective, all the above events are significantly modulated immune system homeostasis and its function. Over the aging process, stem cells are affected by several intrinsic changes. They are also responsive to extrinsic signaling pathways such as TGF-β, which decreases their differentiation, self-renewal, and homing and plasticity potential. In bone marrow, there was a remarkable reduction in HSC differentiation into immune cells as well as decreases in the functions of immune cells. Upon aging, the matured stem cells secrete several mediators that affect the proliferation, differentiation, and cytotoxic potential of the immune cells. The link among aging, stem cells, and the immune system is depicted in Fig. 2.

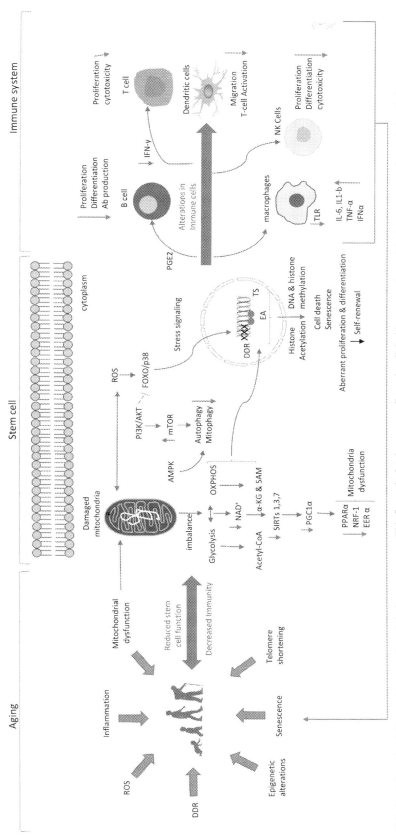

FIG. 2. Aging-related changes in stem cells and the immune system. This figure depicts how stem cells are interlinked between aging and the immune system by exploring the important signaling pathways and mechanisms. Aging directly affects the functions of stem cells while producing ROS-induced stress and mitochondrial dysfunction and stimulating autophagy and cell death. These processes in turn affect the immune system by increasing the proinflammatory cytokines, which causes cell senescence, epigenetic modifications, and stress. *DDR*, DNA damage response; *EA*, epigenetic alterations; *ERα*, estrogen receptor alpha; *FOXO*, Forkhead box protein O; *IFN*, interferon; *IL*, interleukin; *mTOR*, mammalian target of rapamycin; *NAD*, nicotinamide adenine dinucleotide; *NK*, natural killer cells; *NRF-1*, nuclear respiratory factor 1; *OXPHOS*, oxidative phosphorylation; *PGC-1α*, peroxisome proliferator-activated receptor gamma coactivator 1-alpha; *PGE2*, prostaglandin E2; *PI3K/AKT*, protein kinase B; *PPAR-γ*, peroxisome proliferator-activated receptor gamma; *ROS*, reactive oxygen species; *SAM*, S-adenosylmethionine; *SIRTs*, sirtuins; *TLR*, toll-like receptors; *TNFα*, tumor necrosis factor alpha; *TS*, telomeres shortening; *α-KG*, α-ketoglutarate.

8 Altered developmental pathways during aging

The major alterations and changes in the developing pathways are the major causes for stem cell aging in different tissues. Developmental pathways that are affected during aging include WNT, Notch, FGFs, TGF-β, p38, and cellular senescence signaling.

8.1 WNT signaling

Aging increases canonical WNT signaling in muscle stem cells (MuSCs), which in turn promotes the differentiation of MuSCs into fibrogenic lineages triggering muscle fibrosis. There is also a decline in muscle regeneration following injury [18]. A change in WNT signaling from canonical (WNT3A) to noncanonical (WNT5A) promotes HSC aging by causing harm to the stem cell polarity. This CDC42-dependent process is a key in modifying the interaction of HSCs with its niche, self-renewal, and differentiation [92, 93]. As WNT3A promotes lymphopoiesis and WNT5A augments myelopoiesis [94], the alteration from canonical to noncanonical WNT signaling might contribute to the age-associated skewing of hematopoiesis toward myelopoiesis [95].

8.2 Notch signaling

A Notch signaling decrease is associated with aging. The change in Notch signaling and an increase in TGF-β impairs the regeneration of MuSCs. TGF-β/SMAD3 and a Notch signaling imbalance have been shown to increase the expression of the cellular senescence markers p16 and p21, cyclin-dependent kinase CDK inhibitors [96]. Notch signaling is again suppressed in HSCs during aging, which leads to the impairment of activated HSC self-renewal. In ISCs, the suppression of Notch signaling leads to the impairment of ISC proliferation and differentiation.

The role of other developmental pathways in aging of stem cells remains to be studied. Some of the causes attributed to stem cell aging include: (a) Telomere attrition in which the telomere shortens with respect to age. It is suggested that the length of the telomere is involved in late-life survival [97, 98]. (b) Cellular senescence in which an irreversible cell cycle arrest that was induced by short telomeres is said to be a main cause of stem cell aging. (c) DNA damage and mutation. The accumulation of both DNA damage and mutation is attributed to stem cell aging whereas the enhancement of DNA repair is said to increase lifespan [99]. (d) Epigenetic alterations: Chromatin regulation is said to be one of the important factors responsible for stem cell function. During the aging process, there is a significant change in the gene expression and chromatin level. The changes in chromatin by DNA mutation are due to skewed lineage phenotypes that are exhibited by aged stem cells. (e) Nutrient sensing and metabolism: Caloric restriction plays a dynamic role in stem cell aging. It is said to decrease the insulin, IGF, and amino acid levels while increasing the NAD+ and AMP. This eventually results in improved DNA repair, epigenome stability, stress resistance, and oxidative metabolism, leading to longevity. (f) Cell polarity and proteostasis: Stem cells are said to maintain higher levels of autophagy and proteasome activity in order to repair protein damage. In HSCs and skin stem cells, autophagy is higher when compared to other differentiated cells [100]. (g) Functional decline and circulating factors: Apart from stem cell aging, the aging of the stem cell environment also brings about changes in stem cell function. These aged niche cells fail to transmit proper signals to stem cells, thereby affecting cell fate. The changes in the concentration of the circulating factors also influence stem cell aging. For example, insulin, IGF-1, and TGF-β, which increases during aging, are said to impair the function of both satellite cells and NSCs [19, 101]. The rejuvenation of stem cells can be obtained by reprogramming the stem cells, which includes DNA methylation and therapies that stimulate tissue stem cell regeneration and that target the accretion of inflammatory mediators during aging.

8.3 Cellular senescence

Cellular senescence is a state of permanent cell cycle arrest associated with changes in cell morphology, physiology, chromatin organization, and gene expression, accompanied by changes in the secretome [45, 102]. Cellular senescence is triggered by various types of stress such as telomere shortening, ROS, DNA replication stress, or signals such as oncogene activation or the overexpression of pluripotency factors. Cellular senescence contributes to the aging-associated deterioration in the function of hematopoietic stem cells, muscle stem cells, and intestinal stem cells. During the active aging process, senescence cells accumulate in tissues and are responsible for the synthesis of a variety of proinflammatory mediators called SASP [45, 102]. The removal of senescent cells in mice has improved the stem cell function and metabolic performance as well as tissue maintenance and lifespan [45].

9 Drugs and methods to prevent aging

In aging research, nordihydroguaiaretic acid and aspirin each increase the lifespan of male mice [103]. Another mice study demonstrated that acarbose, 17-α-estradiol, and nordihydroguaiaretic acid extend the mouse lifespan, specifically in males [104]. Studies of rapamycin (an mTOR inhibitor) have demonstrated very convincing evidence for targeting aging. Mice and yeast studies have shown that the pharmacological inhibition of mTOR extends the lifespan. Apart from increasing the lifespan, rapamycin treatment in mice has been shown to protect against various age-related pathologies such as cancer [105] and cardiovascular diseases [106] while recovering stem cell function [107] and augmenting muscle function [108]. When rapamycin is given in the latter part of life, it prolongs the lifespan [109, 110], slows aging in a dose-dependent manner, shows differential effects by sex [111], and is synergistic with metformin. Furthermore, in a clinical trial, 6 weeks of treatment of older people with a rapamycin derivative improved their immune response to the influenza vaccine [112].

Metformin is a first-line treatment for type 2 diabetes worldwide. Metformin has been used in humans for more than 60 years and has an excellent safety profile [113]. Data from multiple randomized clinical trials and other observational studies provide evidence for decreased incidence of age-related diseases in diabetic patients taking metformin to lower blood glucose [113]. Metformin has been shown to inhibit cytokine receptors, insulin, IGF-1, and adiponectin, all pathways that are activated with aging and, when modulated, are associated with longevity [113]. At the intracellular level, metformin has been shown to inhibit the inflammatory pathways, activate AMPK, and enhance the inhibition of mTOR, which is a well-known key target in aging. These mechanisms could help to reduce oxidative stress and remove senescent cells. Overall, these affect inflammation, cellular survival, stress defense, autophagy, and protein synthesis, which are major biological outcomes associated with aging [113].

From a stem cell perspective, the potential interventions to rejuvenate the aging immune system are to improve stem cell function; this is considered the primary cell type responsible for the differentiation and self-renewal of immune cells. In immune system perspective, increased B cells lymphopoiesis was acheived by stimulating B cell progenitors, which in turn induced thymus to stimulate naïve T cells production. From a therapeutic perspective, using antioxidants and antiinflammatory compounds reduces the accumulation of ROS and inflammatory mediators responsible for stem cell dysfunction, aging-associated immune system deterioration, and aging. Also, using an mTOR inhibitor, especially the mTOR complex 1 (mTORC1) inhibitor with rapamycin, may be useful in increasing the lifespan by increasing the aging-related autophagy in stem cells. The extracellular TGF-β signaling plays a crucial role in reduced stem cell function in aged tissue. Therefore, the specific inhibition of TGF-β signals is useful for the modulation of stem cell function and subsequent immune system dysfunction; this is elaborately shown in Fig. 2.

Apart from the aging inhibitors and synthetic drugs, there are more alternative methods available to prevent aging. In ancient as well as recent times, there has been interest in using Ayurveda and Siddha medicines for reversing the aging process. Plants have a variety of organic compounds and phytochemicals that fight against aging. Moreover, as mentioned in Fig. 3, it has been scientifically proven that highly nutritious food, yoga, physical exercises, music therapy, a reduction in stress, and immune-enrichment diets play a major role in fighting against aging.

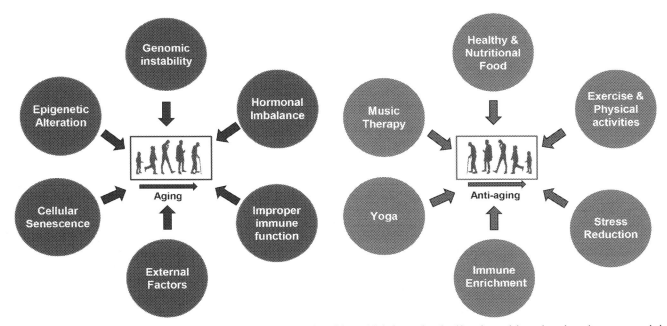

FIG. 3 Factors involved in aging and antiaging. Schematic representation of the multiple factors involved in aging and the various therapies recommended to reverse the aging process.

In this chapter, we tried to gather the most exciting evidence based on proofs about aging, its mechanism, how stem cells play a major role in aging, and how the immune system is important in aging.

10 Conclusion

Aging and immunity are closely related processes where there is a significant reduction in the immune function reported with aging. The B and T cell progenitor population is remarkably reduced during this aging process. The self-renewal and regeneration potential of stem cells decreases during the aging process. Mechanistically, DNA damage, enhanced ROS production, epigenetic changes, telomere attrition, and polarity changes have been severely associated with aging. Also, the reduction of the stem cell population and the regenerative potential are among the main causes of aging. Therefore, immune boosters, the prevention of stem cell aging, the modulation of stem cell-associated signaling pathways, the activation of the Arf/p53 pathway, and natural therapies might prevent aging due to various factors.

References

[1] H.M. Blau, B.D. Cosgrove, A.T. Ho, The central role of muscle stem cells in regenerative failure with aging, Nat. Med. 21 (8) (2015) 854–862.

[2] J. Jameson, K. Ugarte, N. Chen, P. Yachi, E. Fuchs, R. Boismenu, et al., A role for skin gamma delta T cells in wound repair, Science 296 (5568) (2002) 747–749.

[3] B.E. Keyes, S. Liu, A. Asare, S. Naik, J. Levorse, L. Polak, et al., Impaired epidermal to dendritic T cell signaling slows wound repair in aged skin, Cell 167 (5) (2016) 1323–1338.e14.

[4] L. Rittié, S.W. Stoll, S. Kang, J.J. Voorhees, G.J. Fisher, Hedgehog signaling maintains hair follicle stem cell phenotype in young and aged human skin, Aging Cell 8 (6) (2009) 738–751.

[5] E.K. Nishimura, Melanocyte stem cells: a melanocyte reservoir in hair follicles for hair and skin pigmentation, Pigment Cell Melanoma Res. 24 (3) (2011) 401–410.

[6] K. Inomata, T. Aoto, N.T. Binh, N. Okamoto, S. Tanimura, T. Wakayama, et al., Genotoxic stress abrogates renewal of melanocyte stem cells by triggering their differentiation, Cell 137 (6) (2009) 1088–1099.

[7] B. Biteau, C.E. Hochmuth, H. Jasper, JNK activity in somatic stem cells causes loss of tissue homeostasis in the aging Drosophila gut, Cell Stem Cell 3 (4) (2008) 442–455.

[8] J.S. Park, Y.S. Kim, M.A. Yoo, The role of p38b MAPK in age-related modulation of intestinal stem cell proliferation and differentiation in Drosophila, Aging 1 (7) (2009) 637–651.

[9] N.H. Choi, J.G. Kim, D.J. Yang, Y.S. Kim, M.A. Yoo, Age-related changes in Drosophila midgut are associated with PVF2, a PDGF/VEGF-like growth factor, Aging Cell 7 (3) (2008) 318–334.

[10] J. Oh, Y.D. Lee, A.J. Wagers, Stem cell aging: mechanisms, regulators and therapeutic opportunities, Nat. Med. 20 (8) (2014) 870–880.

[11] K. Takeda, I. Kinoshita, Y. Shimizu, Y. Matsuno, T. Shichinohe, H. Dosaka-Akita, Expression of LGR5, an intestinal stem cell marker, during each stage of colorectal tumorigenesis, Anticancer Res 31 (1) (2011) 263–270.

[12] J.R. Beauchamp, J.E. Morgan, C.N. Pagel, T.A. Partridge, Dynamics of myoblast transplantation reveal a discrete minority of precursors with stem cell-like properties as the myogenic source, J. Cell Biol. 144 (6) (1999) 1113–1122.

[13] A. Mauro, Satellite cell of skeletal muscle fibers, J. Biophys. Biochem. Cytol. 9 (2) (1961) 493–495.

[14] R.I. Sherwood, J.L. Christensen, I.M. Conboy, M.J. Conboy, T.A. Rando, I.L. Weissman, et al., Isolation of adult mouse myogenic progenitors: functional heterogeneity of cells within and engrafting skeletal muscle, Cell 119 (4) (2004) 543–554.

[15] A.S. Brack, H. Bildsoe, S.M. Hughes, Evidence that satellite cell decrement contributes to preferential decline in nuclear number from large fibres during murine age-related muscle atrophy, J. Cell Sci. 118 (20) (2005) 4813–4821.

[16] C.A. Collins, P.S. Zammit, A.P. Ruiz, J.E. Morgan, T.A. Partridge, A population of myogenic stem cells that survives skeletal muscle aging, Stem Cells 25 (4) (2007) 885–894.

[17] M.C. Gibson, E. Schultz, Age-related differences in absolute numbers of skeletal muscle satellite cells, Muscle Nerve 6 (8) (1983) 574–580.

[18] A.S. Brack, M.J. Conboy, S. Roy, M. Lee, C.J. Kuo, C. Keller, et al., Increased Wnt signaling during aging alters muscle stem cell fate and increases fibrosis, Science 317 (5839) (2007) 807–810.

[19] M.E. Carlson, M.J. Conboy, M. Hsu, L. Barchas, J. Jeong, A. Agrawal, et al., Relative roles of TGF-beta1 and Wnt in the systemic regulation and aging of satellite cell responses, Aging Cell 8 (6) (2009) 676–689.

[20] I. Beerman, D. Bhattacharya, S. Zandi, M. Sigvardsson, I.L. Weissman, D. Bryder, et al., Functionally distinct hematopoietic stem cells modulate hematopoietic lineage potential during aging by a mechanism of clonal expansion, Proc. Natl. Acad. Sci. U. S. A. 107 (12) (2010) 5465–5470.

[21] P. Genovese, G. Schiroli, G. Escobar, T.D. Tomaso, C. Firrito, A. Calabria, et al., Targeted genome editing in human repopulating haematopoietic stem cells, Nature 510 (7504) (2014) 235–240.

[22] D.J. Rossi, D. Bryder, J.M. Zahn, H. Ahlenius, R. Sonu, A.J. Wagers, et al., Cell intrinsic alterations underlie hematopoietic stem cell aging, Proc. Natl. Acad. Sci. U. S. A. 102 (26) (2005) 9194–9199.

[23] N. Berliner, Anemia in the elderly, Trans. Am. Clin. Climatol. Assoc. 124 (2013) 230–237.

[24] J. Cheng, N. Türkel, N. Hemati, M.T. Fuller, A.J. Hunt, Y.M. Yamashita, Centrosome misorientation reduces stem cell division during ageing, Nature 456 (7222) (2008) 599–604.

[25] M. Boyle, C. Wong, M. Rocha, D.L. Jones, Decline in self-renewal factors contributes to aging of the stem cell niche in the Drosophila testis, Cell Stem Cell 1 (4) (2007) 470–478.

[26] H. Toledano, C. D'Alterio, B. Czech, E. Levine, D.L. Jones, The let-7-Imp axis regulates ageing of the Drosophila testis stem-cell niche, Nature 485 (7400) (2012) 605–610.

[27] L. LaFever, D. Drummond-Barbosa, Direct control of germline stem cell division and cyst growth by neural insulin in Drosophila, Science 309 (5737) (2005) 1071–1073.

[28] W. Mair, C.J. McLeod, L. Wang, D.L. Jones, Dietary restriction enhances germline stem cell maintenance, Aging Cell 9 (5) (2010) 916–918.

[29] A.Y. Maslov, T.A. Barone, R.J. Plunkett, S.C. Pruitt, Neural stem cell detection, characterization, and age-related changes in the subventricular zone of mice, J. Neurosci. 24 (7) (2004) 1726–1733.

[30] J. Nishino, I. Kim, K. Chada, S.J. Morrison, Hmga2 promotes neural stem cell self-renewal in young but not old mice by reducing p16Ink4a and p19Arf expression, Cell 135 (2) (2008) 227–239.

[31] C. Zhao, W. Deng, F.H. Gage, Mechanisms and functional implications of adult neurogenesis, Cell 132 (4) (2008) 645–660.

[32] E.A. Pollina, A. Brunet, Epigenetic regulation of aging stem cells, Oncogene 30 (28) (2011) 3105–3126.

[33] G. Hannum, J. Guinney, L. Zhao, L. Zhang, G. Hughes, S. Sadda, B. Klotzle, et al., Genome-wide methylation profiles reveal quantitative views of human aging rates, Mol. Cell 49 (2) (2013) 359–367.

[34] E.J. Márquez, C.H. Chung, R. Marches, R.J. Rossi, D. Nehar-Belaid, A. Eroglu, et al., Sexual-dimorphism in human immune system aging, Nat. Commun. 11 (1) (2020) 751.

[35] A. Cellerino, E. Terzibasi Tozzini, Biology of aging: new models, new methods, Semin. Cell Dev. Biol. 70 (2017) 98.

[36] D. McHugh, J. Gil, Senescence and aging: causes, consequences, and therapeutic avenues, J. Cell Biol. 217 (1) (2018) 65–77.

[37] L. Harries, K. Goljanek-Whysall, The biology of ageing and the omics revolution, Biogerontology 19 (6) (2018) 435–436.

[38] C. Martínez-Cué, N. Rueda, Cellular senescence in neurodegenerative diseases, Front. Cell. Neurosci. 14 (2020) 16.

[39] E.Y. Min, I.H. Kim, J. Lee, E.Y. Kim, Y.H. Choi, T.J. Nam, The effects of fucodian on senescence are controlled by the p16INK4a-pRb and p14Arf-p53 pathways in hepatocellular carcinoma and hepatic cell lines, Int. J. Oncol. 45 (1) (2014) 47–56.

[40] M. Mijit, V. Caracciolo, A. Melillo, F. Amicarelli, A. Giordano, Role of p53 in the regulation of cellular senescence, Biomol. Ther. 10 (3) (2020) 420.

[41] E. Montecino-Rodriguez, B. Berent-Maoz, K. Dorshkind, Causes, consequences, and reversal of immune system aging, J. Clin. Invest. 123 (3) (2013) 958–965.

[42] F. Rodier, J.P. Coppé, C.K. Patil, W.A. Hoeijmakers, D.P. Muñoz, S.R. Raza, et al., Persistent DNA damage signalling triggers senescence-associated inflammatory cytokine secretion, Nat. Cell Biol. 11 (8) (2009) 973–979.

[43] F. Chen, Y. Liu, N.K. Wong, J. Xiao, K.F. So, Oxidative stress in stem cell aging, Cell Transplant. 26 (9) (2017) 1483–1495.

[44] L.B. Boyette, R.S. Tuan, Adult stem cells and diseases of aging, J. Clin. Med. 3 (1) (2014) 88–134.

[45] M. Ermolaeva, F. Neri, A. Ori, K.L. Rudolph, Cellular and epigenetic drivers of stem cell ageing, Nat. Rev. Mol. Cell Biol. 19 (9) (2018) 594–610.

[46] M. Khacho, R.S. Slack, Mitochondrial and reactive oxygen species signaling coordinate stem cell fate decisions and life long maintenance, Antioxid. Redox Signal. 28 (11) (2018) 1090–1101.

[47] K.J. Ahlqvist, A. Suomalainen, R.H. Hämäläinen, Stem cells, mitochondria and aging, Biochim. Biophys. Acta 1847 (11) (2015) 1380–1386.

[48] D. Cakouros, S. Gronthos, Epigenetic regulation of bone marrow stem cell aging: revealing epigenetic signatures associated with hematopoietic and mesenchymal stem cell aging, Aging Dis. 10 (1) (2019) 174–189.

[49] R. Ren, A. Ocampo, G.H. Liu, J.C. Izpisua Belmonte, Regulation of stem cell aging by metabolism and epigenetics, Cell Metab. 26 (3) (2017) 460–474.

[50] J. Dulak, K. Szade, A. Szade, W. Nowak, A. Józkowicz, Adult stem cells: hopes and hypes of regenerative medicine, Acta Biochim. Pol. 62 (3) (2015) 329–337.

[51] G. Koliakos, Stem cells and aging, Rejuvenation Res. 20 (1) (2017) 4–8.

[52] T. Sugiyama, Y. Omatsu, T. Nagasawa, Niches for hematopoietic stem cells and immune cell progenitors, Int. Immunol. 31 (1) (2019) 5–11.

[53] J. Lee, Y.S. Cho, H. Jung, I. Choi, Pharmacological regulation of oxidative stress in stem cells, Oxid. Med. Cell. Longev. 2018 (2018) 4081890.

[54] A.S. Ahmed, M.H. Sheng, S. Wasnik, D.J. Baylink, K.W. Lau, Effect of aging on stem cells, World J. Exp. Med. 7 (1) (2017) 1–10.

[55] Y. Liu, W. Weng, R. Gao, Y. Liu, New insights for cellular and molecular mechanisms of aging and aging-related diseases: herbal medicine as potential therapeutic approach, Oxid. Med. Cell. Longev. 2019 (2019) 4598167.

[56] D. Weiskopf, B. Weinberger, B. Grubeck-Loebenstein, The aging of the immune system, Transpl. Int. 22 (11) (2009) 1041–1050.

[57] U. Blank, S. Karlsson, TGF-β signaling in the control of hematopoietic stem cells, Blood 125 (23) (2015) 3542–3550.

[58] C.R. Keenan, R.S. Allan, Epigenomic drivers of immune dysfunction in aging, Aging Cell 18 (1) (2019) e12878.

[59] J. Lee, S.R. Yoon, I. Choi, H. Jung, Causes and mechanisms of hematopoietic stem cell aging, Int. J. Mol. Sci. 20 (6) (2019) 1272.

[60] P. Enck, K. Zimmermann, K. Rusch, A. Schwiertz, S. Klosterhalfen, J.S. Frick, The effects of ageing on the colonic bacterial microflora in adults, Z. Gastroenterol. 47 (7) (2009) 653–658.

[61] S. Ponnappan, U. Ponnappan, Aging and immune function: molecular mechanisms to interventions, Antioxid. Redox Signal. 14 (8) (2011) 1551–1585.

[62] L. O'Mahony, J. Holland, J. Jackson, C. Feighery, T.P. Hennessy, K. Mealy, Quantitative intracellular cytokine measurement: age-related changes in proinflammatory cytokine production, Clin. Exp. Immunol. 113 (2) (1998) 213–219.

[63] G. Pawelec, Y. Barnett, R. Forsey, D. Frasca, A. Globerson, J. McLeod, et al., T cells and aging, Front. Biosci. 7 (2002) d1056–d1183.
[64] T. Fulop, A. Larbi, G. Dupuis, A. Le Page, E.H. Frost, A.A. Cohen, et al., Immunosenescence and inflamm-aging as two sides of the same coin: friends or foes? Front. Immunol. 8 (2018) 1960.
[65] N.E. Sharpless, Ink4a/Arf links senescence and aging, Exp. Gerontol. 39 (11–12) (2004) 1751–1759.
[66] R.A. Signer, E. Montecino-Rodriguez, O.N. Witte, K. Dorshkind, Aging and cancer resistance in lymphoid progenitors are linked processes conferred by p16Ink4a and Arf, Genes Dev. 22 (22) (2008) 3115–3120.
[67] B. Berent-Maoz, E. Montecino-Rodriguez, R.A. Signer, K. Dorshkind, Fibroblast growth factor-7 partially reverses murine thymocyte progenitor aging by repression of Ink4a, Blood 119 (24) (2012) 5715–5721.
[68] C. Franceschi, P. Garagnani, P. Parini, C. Giuliani, A. Santoro, Inflammaging: a new immune-metabolic viewpoint for age-related diseases, Nat. Rev. Endocrinol. 14 (10) (2018) 576–590.
[69] A. Behrens, J.M. van Deursen, K.L. Rudolph, B. Schumacher, Impact of genomic damage and ageing on stem cell function, Nat. Cell Biol. 16 (2014) 201–207.
[70] A. Peth, J.A. Nathan, A.L. Goldberg, The ATP costs and time required to degrade ubiquitinated proteins by the 26 S proteasome, J. Biol. Chem. 288 (2013) 29215–29222.
[71] P.M. Douglas, A. Dillin, Protein homeostasis and aging in neurodegeneration, J. Cell Biol. 190 (5) (2010) 719–729.
[72] L.R. Lapierre, C. Kumsta, M. Sandri, A. Ballabio, M. Hansen, Transcriptional and epigenetic regulation of autophagy in aging, Autophagy 11 (2015) 867–880.
[73] S.P. Jackson, J. Bartek, The DNA-damage response in human biology and disease, Nature 461 (7267) (2009) 1071–1078.
[74] J. Kenyon, S.L. Gerson, The role of DNA damage repair in aging of adult stem cells, Nucleic Acids Res. 35 (22) (2007) 7557–7565.
[75] A. Matheu, A. Maraver, M. Serrano, The Arf/p53 pathway in cancer and aging, Cancer Res. 68 (15) (2008) 6031–6034.
[76] E. Carrasco-Garcia, M. Moreno, L. Moreno-Cugnon, A. Matheu, Increased Arf/p53 activity in stem cells, aging and cancer, Aging Cell 16 (2) (2017) 219–225.
[77] C.F. Bender, M.L. Sikes, R. Sullivan, L.E. Huye, M.M. Le Beau, D.B. Roth, et al., Cancer predisposition and hematopoietic failure in Rad50(S/S) mice, Genes Dev. 16 (17) (2002) 2237–2251.
[78] D. Ezhilarasan, Oxidative stress is bane in chronic liver diseases: clinical and experimental perspective, Arab J. Gastroenterol. 19 (2) (2018) 56–64.
[79] A.K. Singh, M.J. Althoff, J.A. Cancelas, Signaling pathways regulating hematopoietic stem cell and progenitor aging, Curr. Stem Cell Rep. 4 (2) (2018) 166–181.
[80] A. Mendelson, P.S. Frenette, Hematopoietic stem cell niche maintenance during homeostasis and regeneration, Nat. Med. 20 (8) (2014) 833–846.
[81] K. Ito, A. Hirao, F. Arai, K. Takubo, S. Matsuoka, K. Miyamoto, et al., Reactive oxygen species act through p38 MAPK to limit the lifespan of hematopoietic stem cells, Nat. Med. 12 (4) (2006) 446–451.
[82] H. Zhang, K.J. Menzies, J. Auwerx, The role of mitochondria in stem cell fate and aging, Development 145 (8) (2018) 143420.
[83] L. Kohli, E. Passegué, Surviving change: the metabolic journey of hematopoietic stem cells, Trends Cell Biol. 24 (8) (2014) 479–487.
[84] N. Sun, R.J. Youle, T. Finkel, The mitochondrial basis of aging, Mol. Cell 61 (5) (2016) 654–666.
[85] M.D. Filippi, S. Ghaffari, Mitochondria in the maintenance of hematopoietic stem cells: new perspectives and opportunities, Blood 133 (18) (2019) 1943–1952.
[86] R. Liang, S. Ghaffari, Stem cells seen through the FOXO lens: an evolving paradigm, Curr. Top. Dev. Biol. 127 (2018) 23–47.
[87] N.C. Chang, Autophagy and stem cells: self-eating for self-renewal, Front. Cell Dev. Biol. 8 (2020) 138.
[88] S.Q. Wong, A.V. Kumar, J. Mills, L.R. Lapierre, Autophagy in aging and longevity, Hum. Genet. 139 (3) (2020) 277–290.
[89] T.T. Ho, M.R. Warr, E.R. Adelman, O.M. Lansinger, J. Flach, E.V. Verovskaya, et al., Autophagy maintains the metabolism and function of young and old stem cells, Nature 543 (7644) (2017) 205–210.
[90] T. Weichhart, mTOR as regulator of lifespan, aging, and cellular senescence: a mini-review, Gerontology 64 (2) (2018) 127–134.
[91] D. Meng, A.R. Frank, J.L. Jewell, mTOR signaling in stem and progenitor cells, Development 145 (1) (2018) dev152595.
[92] M.C. Florian, K. Dörr, A. Niebel, D. Daria, H. Schrezenmeier, M. Rojewski, et al., Cdc42 activity regulates hematopoietic stem cell aging and rejuvenation, Cell Stem Cell 10 (5) (2012) 520–530.
[93] L. Yang, L. Wang, H. Geiger, J.A. Cancelas, J. Mo, Y. Zheng, Rho GTPase Cdc42 coordinates hematopoietic stem cell quiescence and niche interaction in the bone marrow, Proc. Natl. Acad. Sci. U. S. A. 104 (12) (2007) 5091–5096.
[94] F. Famili, B.A. Naber, S. Vloemans, E.F. de Haas, M.M. Tiemessen, F.J. Staal, Discrete roles of canonical and non-canonical Wnt signaling in hematopoiesis and lymphopoiesis, Cell Death Dis. 6 (11) (2015) e1981.
[95] M.C. Florian, K.J. Nattamai, K. Dörr, G. Marka, B. Uberle, V. Vas, et al., A canonical to non-canonical Wnt signalling switch in haematopoietic stem-cell ageing, Nature 503 (7476) (2013) 392–396.
[96] M.E. Carlson, M. Hsu, I.M. Conboy, Imbalance between pSmad3 and Notch induces CDK inhibitors in old muscle stem cells, Nature 454 (7203) (2008) 528–532.
[97] B.B. de Jesus, E. Vera, K. Schneeberger, A.M. Tejera, E. Ayuso, F. Bosch, M.A. Blasco, Telomerase gene therapy in adult and old mice delays aging and increases longevity without increasing cancer, EMBO Mol. Med. 4 (2012) 691–704.
[98] A. Tomás-Loba, I. Flores, P.J. Fernández-Marcos, M.L. Cayuela, A. Maraver, A. Tejera, et al., Telomerase reverse transcriptase delays aging in cancer-resistant mice, Cell 135 (2008) 609–622.
[99] Y. Kanfi, S. Naiman, G. Amir, V. Peshti, G. Zinman, L. Nahum, et al., The sirtuin SIRT6 regulates lifespan in male mice, Nature 483 (2012) 218–221.
[100] S. Salemi, S. Yousefi, M.A. Constantinescu, M.F. Fey, H.-U. Simon, Autophagy is required for self-renewal and differentiation of adult human stem cells, Cell Res. 22 (2012) 432–435.

[101] J. Pineda, M. Daynac, A. Chicheportiche, A. Cebrian-Silla, K. Sii Felice, J. Garcia-Verdugo, et al., Vascularderived TGF-β increases in the stem cell niche and perturbs neurogenesis during aging and following irradiation in the adult mouse brain, EMBO Mol. Med. 5 (2013) 548–562.

[102] N. Martin, D. Beach, J. Gil, Ageing as developmental decay: insights from p16(INK4a.), Trends Mol. Med. 20 (12) (2014) 667–674.

[103] R. Strong, R.A. Miller, C.M. Astle, R.A. Floyd, K. Flurkey, K.L. Hensley, et al., Nordihydroguaiaretic acid and aspirin increase lifespan of genetically heterogeneous male mice, Aging Cell 7 (5) (2008) 641–650.

[104] D.E. Harrison, R. Strong, D.B. Allison, B.N. Ames, C.M. Astle, H. Atamna, E. Fernandez, K. Flurkey, M.A. Javors, N.L. Nadon, et al., Acarbose, 17-a-estradiol, and nordihydroguaiaretic acid extend mouse lifespan preferentially in males, Aging Cell 13 (2014) 273–282.

[105] V.N. Anisimov, M.A. Zabezhinski, I.G. Popovich, T.S. Piskunova, A.V. Semenchenko, M.L. Tyndyk, et al., Rapamycin increases lifespan and inhibits spontaneous tumorigenesis in inbred female mice, Cell Cycle 10 (24) (2011) 4230–4236.

[106] J.M. Flynn, M.N. O'Leary, C.A. Zambataro, E.C. Academia, M.P. Presley, B.J. Garrett, et al., Late-life rapamycin treatment reverses age-related heart dysfunction, Aging Cell 12 (5) (2013) 851–862.

[107] C. Chen, Y. Liu, Y. Liu, P. Zheng, mTOR regulation and therapeutic rejuvenation of aging hematopoietic stem cells, Sci. Signal. 2 (98) (2009) ra75.

[108] A. Bitto, T.K. Ito, V.V. Pineda, N.J. LeTexier, H.Z. Huang, E. Sutlief, Transient rapamycin treatment can increase lifespan and healthspan in middle-aged mice, eLife 5 (2016) e16351.

[109] D.E. Harrison, R. Strong, Z.D. Sharp, J.F. Nelson, C.M. Astle, K. Flurkey, et al., Rapamycin fed late in life extends lifespan in genetically heterogeneous mice, Nature 460 (7253) (2009) 392–395.

[110] R.A. Miller, D.E. Harrison, C.M. Astle, J.A. Baur, A.R. Boyd, R. de Cabo, et al., Rapamycin, but not resveratrol or simvastatin, extends life span of genetically heterogeneous mice, J. Gerontol. A Biol. Sci. Med. Sci. 66 (2) (2011) 191–201.

[111] J.E. Wilkinson, L. Burmeister, S.V. Brooks, C.C. Chan, S. Friedline, D.E. Harrison, et al., Rapamycin slows aging in mice, Aging Cell 11 (4) (2012) 675–682.

[112] J.B. Mannick, G. Del Giudice, M. Lattanzi, N.M. Valiante, J. Praestgaard, B. Huang, et al., mTOR inhibition improves immune function in the elderly, Sci. Transl. Med. 6 (268) (2014) 268ra179.

[113] N. Barzilai, J.P. Crandall, S.B. Kritchevsky, M.A. Espeland, Metformin as a tool to target aging, Cell Metab. 23 (6) (2016) 1060–1065.

Chapter 7

Aging of hematopoietic stem cells: Insight into mechanisms and consequences

Bhaswati Chatterjee[a] and Suman S. Thakur[b]

[a]National Institute of Pharmaceutical Education and Research, Hyderabad, India, [b]Centre for Cellular and Molecular Biology, Hyderabad, India

1 Introduction

Hematopoietic stem cells (HSCs) form blood and immune cells. They are also known as blood stem cells and hematopoietic progenitor cells (HPCs). They are among the first-identified and best-studied stem cells (Fig. 1).[1] HSCs produce blood throughout their lives while maintaining self-renewal and differentiation for good health. Further, any discrepancies in HSCs cause diseases.[2] HSCs play an important role in the treatment of many cancers, including leukemia, lymphoma, and myeloma, as well as in noncancerous diseases related to blood disorders, immune system disorders, and inherited metabolic disorders such as sickle cell diseases and adrenoleukodystrophy. HSCs are successfully used for the replacement of blood and immune cells. There are some technical barriers of HSCs such as the inability to replicate and differentiate themselves *in vitro*. Another is the lack of a robust method to select HSCs from blood and bone marrow that hinders the widespread use of HSCs as cell replacement therapy in various diseases such as diabetes, neurological disorders, and spinal cord injuries. Healthy and young HSC cells are needed as compared to old HSC cells for use in clinics. Therefore, it is necessary to understand the aging of HSCs with the mechanism and consequences (Fig. 1).

Stem cells are among the longest-living cells within an organism and are responsible for aging and longevity.[3] The aging behavior of all stem cells is different, but theoretically there should be some similarity between the aging processes of stem cells. Notably, stem cell turnover is very low in HSCs while in the intestine, proliferation rates are very high. HSCs divide rarely and slowly and lose their efficiency of self-renewal with age. However, intestinal stem cells don't show functional declines during aging, even as they accumulate DNA damage.[4–7] With better lifestyles paired with advancements in science and therapeutics, the number of older people is increasing. Likewise, hematological disorders and leukemia are also increasing in aged persons.[8] Functional declines in innate and adaptive systems cause hematologic malignancies and other disorders, including autoimmunity with a higher chance of infection. The production of B cells and de novo T cells also decreases due to aging and thymic involution.[9]

Interestingly, the aging process cannot be reversed, but it may be slowed. Notably, the HSC regenerative capacity decreases with aging. HSC aging is related to hematological disorders, including myelodysplastic syndromes and acute myeloid leukemia. There is a decrease in the expression of apoptosis-promoting genes in old age side population stem cells compared to younger ages while the number of side population stem cells increases with age and is higher in the old compared to the young.[10] Moreover, there are several rejuvenating processes and agents that have the potential to partially recover or slow the aging process. Interestingly, rejuvenating agents and senolytic drugs may help in slowing the aging process by reprogramming and depleting senescence cells, respectively.[2]

2 Changes in hematopoietic stem cells (HSCs) due to aging

With age, the number of HSCs increases but their function decreases, especially their self-renewal properties. Dykstra et al. reported that old-age HSCs showed low self-renewal properties and produced small daughter clones during secondary transplants with respect to young HSCs.[11, 12] Further, HSCs from old-age mice showed lower cell cycle activity compared to the young. Notably, a transcription factor, Sox17, is required for fetal HSC hematopoiesis while in adults, HSCs become quiescent. Aged HSCs have a higher myeloid differentiation ability but lower T cell and B cell output compared to young HSCs after transplantation. In hematopoiesis, with age lymphopoiesis decreases while myelopoiesis becomes dominant with an increase in myeloid-related cells, progenitors, and biased HSCs as well as a decrease of B and T cells. Therefore, this results in an active innate immune system rather than an acquired one. This leads to inflammation, myeloid malignancies, and spontaneous anemia.[9]

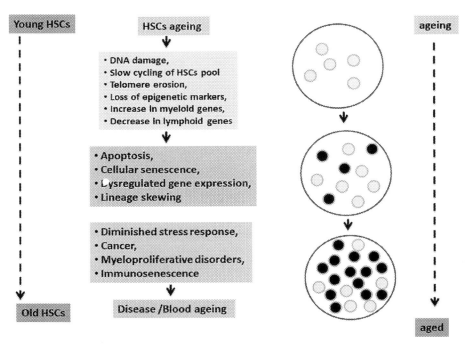

FIG. 1 Consequences of HSC aging lead to blood aging.

3 Differential expression of genes related to HSC aging

A microarray analysis has shown that lymphoid lineage-related genes are downregulated in HSC aging. This includes interleukin 7 receptor (IL7r), FMS-like tyrosine kinase 3 (Flt3), c-srk tyrosine kinase (Csk), phosphoinositide-3-kinase adaptor protein 1 (Pik3ap1/BCAP), tumor necrosis factor receptor superfamily 13c (Tnfrsf13c/Baff), B-cell leukemia/lymphoma 11B (Bcl11b), and SRY-box containing gene 4 (Sox4).[13] Further, the downregulation of epigenetic regulation genes are also reported. This includes the enhancer of zeste homolog 2 (Ezh2), high mobility group AT-hook 1 (Hmga1), special AT-rich sequence binding protein 1 (Sat1b), lymphoid gene interleukin 7 receptor (IL7r), and B-cell leukemia/lymphoma 11B (Bcl11b).

Interestingly, the upregulation of myeloid lineage-related genes is also reported. This includes friend leukemia integration site 1 (Fli-1), CCAAT/enhancer binding protein beta (Cebpb), CCAAT/enhancer binding protein delta (Cebpd), oncostatin M receptor (Osmr), homeo box B6 (Hoxb6), promyelocytic leukemia (Pml), runt related transcription factor 1 (AML1/Runx1), CBFA2T1 identified gene homolog (ETO/Cbfa2t1h), and fibroblast growth factor receptor 1 (Fgfr1) (Fig. 2). Notably, myeloid leukemia-related genes are upregulated in HSC aging. This includes promyelocytic leukemia (Pml), runt related transcription factor 1 (Runx1/AML1), CBFA2T1 identified gene homolog (Cbfa2t1h/ETO), fibroblast growth factor receptor 1 (Fgfr1), fibroblast growth factor receptor 3 (Fgfr3), FGFR1 Oncogene partner 2 (Fgfr1op2), Rho guanine nucleotide exchange factor (Arhgef12), and NCK interacting protein with SH# domain (Nckipsd) (Fig. 2)[13]. In addition, the upregulation of epigenetic gene SWI/SNF related, matrix associated, actin dependent regulator of chromatin subfamily a member 2 (Smarca2/Brm) were also reported.[14]

4 Mechanism of aging in HSCs

Primitive HSCs divide once in about 4 months in mice. With each division, they lose their long-term repopulating properties and developmental potential. The pool of stem cells is small in young mice but its potency is very high compared to aged mice. In mice, due to aging, the potency of stem cells is deeply decreased, even though the numbers are quite high compared to young mice. In the mouse lifetime, the most primitive HSCs divide only 4–5 times. Interestingly, HSCs have cellular memory and their aging phenomenon starts after the fifth division.[15,16] With increasing age, HSCs differentiate more toward myeloid lineage rather than lymphoid lineage.[1] The number of low regenerative potential cells is high in aged cells.[4] Murine genetic model studies shows that several pathways are important for longevity, including DNA repair, intracellular reactive oxygen species (ROS), telomere maintenance, and the stress response of the stem cells.

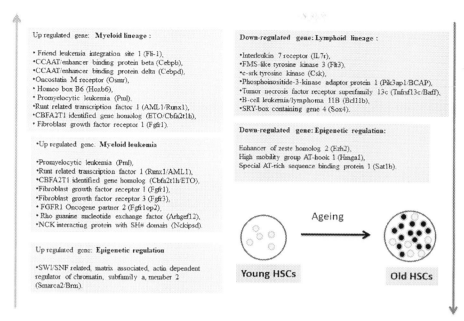

FIG. 2 Differential expression of age-related genes in HSCs.

All human body parts, tissues, blood, and HSCs need to go through the detrimental aging process. The mechanism of aging in HSCs is intrinsic, extrinsic, or both. One of the biggest challenges in the study of HSC molecular mechanisms is in the purity or removal of non-HSCs as the progenitor. Flow cytometry has helped in the isolation of single cells, but the purity was never more than 50% in a transplantation experiment.[4]

5 Cell-intrinsic mechanisms for HSC aging

There are several factors responsible for HSC aging, starting with DNA damage, slow cycling of the HSC pool leading to telomere erosion, the loss of epigenetic markers, an increase in myeloid genes, a decrease in lymphoid genes, apoptosis, cellular senescence, dysregulated gene expression, lineage skewing, and finally blood aging with phenomena such as diminished stress response, cancer, myeloproliferative disorders, and immunosenesscence (Fig. 3).[1]

It has been observed that the number of HSCs increases with age, as found in mice and humans, but this still cannot cover the damage and defects of HSC cells that occur due to aging.[11] Interestingly, young mouse HSCs are four times more functionally efficient compared to old mice.[17]

It has been observed that erythrocytes and platelets increase due to aging. Further, epigenetic regulators such as Amp3, Anxa7, Ap3b1, SELP, Egr1, Arhgef12, and Cbfa2t1h have also increased. Notably, cell surface markers such as CD28, CD38, CD41, CD47, CD62, CD 69, CD74, and CD81 are also upregulated due to aging. Extrinsic factors such as MSCs, neutrophils, megakaryocytes, and macrophages are upregulated with an increase in the cytokines IL-6 and IL-1B, the enzyme caspase-1, and β2-AR signaling. Important genes of the signaling pathway such as JAK, STAT, mTOR, NF-kB, p38, p38MAPK, and UPR^{mt} are also increased with aging (Fig. 4).[11]

Reports suggest that T cells; B cells; epigenetic regulators such as Flt3, Xab2, Rad52, Xrcc1, Sox17, Bcl11b, and Blnk; and cell surface markers such as CD27, CD34, CD37, CD44, CD48, CD52, CD63, CD79b, CD86, CD97, CD97b and CD160 have decreased with age. Notably, β3-AR signaling, TGF-β, TXNIP, sirtuin, and NAD+ are also downregulated with age (Fig. 4).[11]

6 DNA damage

The accumulation of random DNA damage is one of the irreversible causes of HSC aging. HSCs are majorly quiescent and divide rarely in a lifetime. The accumulation of DNA damage is mostly related to myeloid-biased HSCs. Widespread DNA damage was observed in aged HSCs. The erosion of telomeres was responsible for a specific type of DNA damage. In humans, telomere shortening was related with the functional decline of HSCs.[4] Aged HSCs have higher DNA damage. Unfortunately, the DNA repair pathway is also on the decline and is not effective. Age-associated DNA damage response

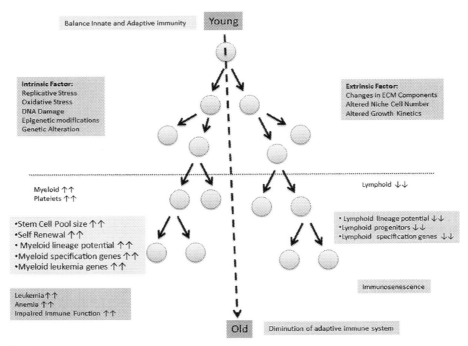

FIG. 3 Process of HSC aging.

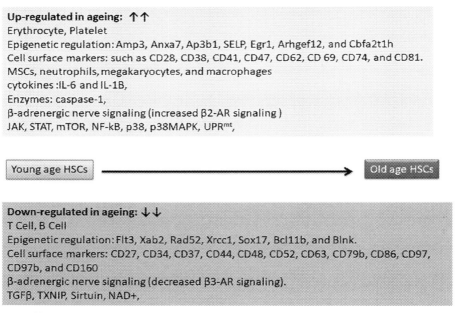

FIG. 4 Functional changes with HSC aging.

control is by several pathways, including nucleotide excision repair (NER) and nonhomologous end-joining (NHEJ). NER has an important role in maintaining HSCs, including self-renewal and preventing cell death in stress conditions. Xab2 associated with NER is downregulated in aged HSCs. Interestingly, hypomorphic mutations have been reported in important proteins such as murine ligase IV (Lig4y288c) of the NHEJ pathway. The expression of KU70 of the NHEJ pathway involved in HSCs having a role in self-renewal is negatively correlated with donor age.

7 Senescence and polarity

In the normal aging process, the cell accumulation of cycle-arrested senescent cells is observed in HSCs. The HSC potential can be regulated by the secretion of senescence cells from bone marrow. The expression of P16 is considered a marker or inducer of senescence.[4] The unequal distribution of specific proteins causes increases in polarity that are related with the overexpression of cdc42. The cdc42 inhibitor improves HSC functioning by restoring the polarity in aged HSCs. This property indicates that some parts of aging are reversible or aging can be slowed.[4]

8 Impaired autophagy and mitochondrial activity

Impaired autophagy has been observed in most aged HSCs, causing the accumulation of mitochondria and high levels of ROS that finally compromise the function of HSCs. The mitochondrial DNA mutation results in mitochondrial dysfunction, causing multiple hematopoietic defects in the elderly. Mitochondrial status, ROS, and mTOR signaling influence the cellular metabolism and are important to maintain HSCs.[18-21]

9 Epigenetics and aging

During the division of HSCs, all the genetic and epigenetic information propagates from parent to daughter cells properly while any error in transferring the information results in the compromised function of HSCs. Epigenetic genes such as DNMT3A, EZH2, TET2, and SETDB1 get mutated in aged persons with the oligoclonal hematopoietic system and patients of acute myeloid leukemia myelodysplastic syndromes.[22,23]

DNA methylation, histone modification, and noncoding RNA play an important role in the aging of HSCs, either independently or synergistically. DNMT1, DNMT3A, DNMT3B, TET1, TET2, 5-mC, and 5-hmC are downregulated with the alteration of age in HSCs. The loss of DNA methyltransferase 1 (DNMT1) is responsible for myeloid skewing.[24,25] Further, the downregulation of DNMT3A and DNMT3B is responsible for the arrest of HSC differentiation. Moreover, the decrease in the amount of TET2 attenuates the differentiation and leads to myeloid malignancies while the decrease in TET1 causes β cell malignancies.

Histone modifications such as methylation, acetylation, sumoylation, phosphorylation, and ubiquitination play important roles in the regulation of aging of HSCs, including self-renewal and differentiation. H3K4me3, H3K27me3, H3K23ac, H2BS14ph, and H3K9me2 are upregulated while H4K16ac, H3K27ac, H3K9me2, and H3K4me1 are downregulated in aged HSCs compared to the young. The increase of H3K4me3 and H3K27me3 alter the promoter usage. H3K4me3 is responsible for the upregulation of the genes Selp, Nupr1, and Sdpr while H3K27me3 is responsible for the downregulation of the Flt3 gene. The downregulation of H3K4me1 is associated with the differentiation and function of the myeloid and erythroid while the downregulation of H3K27ac is associated with leukocyte activation and apoptotic signaling.

Noncoding RNA, which is not able to translate into proteins, is also involved in the aging of HSCs by regulating gene expression and impacting epigenetic factors such as micro-RNA and ribosomal RNA. The miR-125b and microRNA-132/212 were increased during HSCs aging.[26,27]

10 miRNA

Several factors, including the transcription factor and miRNA (micro-RNA), decide the fate of hematopoietic cells.[28] Notably, miRNA is small noncoding RNA that has the potential to negatively regulate gene expression, thus playing a role in the survival and function of HSCs, including self-renewal, differentiation, and apoptosis. The miR-125b–let-7c–miR-99a–miR-100 cluster inhibits the overactivation of TGFβ and WNT1 signaling by targeting SMAD2, SMAD4, and anaphase-promoting complex subunit 2 (APC2) and hindering basic HSC properties such as self-renewal and differentiation.[28] The miR-132 helps in controlling the optimal forkhead box protein O3 (FOXO3) expression because overexpression causes apoptosis and hinders self-renewal and differentiation while lower expression causes proliferation.[28]

11 Clonal hematopoiesis

The expansion of the clonal population of blood cells by somatic mutation in hematopoietic stem cells (HSCs) is related to clonal hematopoiesis (CH). Different types of blood cells are made from the same genetic mutation in hematopoietic stem cells and have different genetic patterns from other blood cells. Mutations in Tet2, Dmnt3a, and JAK2 genes are responsible for CH. The upregulation of IL-1β, IL-6, NLRP3, and inflasome is responsible for the mutation in Tet2 while CLXCL1,2 and IL-6 cause the mutation in Dmnt3a. Further, the upregulation of IL-1β, IL-6, stat-1 signaling, and TNF-α causes the mutation in the JAK2 gene. CH is common in elderly people and is associated with a high risk of leukemia and cardiovascular disease.[4]

12 Hypercholesterolemia

It is responsible for speeding up HSC aging by decreasing Tet1 and H3k27me3 that upregulate the expression of p19 and p21, thereby leading to a loss of quiescence and an impaired reconstitution capacity of HSCs.[29] This causes an acceleration of HSC aging due to the changes inside the HSC compartment by the exhaustion of the long-term and side populations inside. Interestingly, this accelerated aging effect can be reversed by the restoration of Tet1 in HSCs.[29]

13 Signaling pathways and aging

There are several signaling pathways that play important roles in the loss of function of aging HSCs such as DNA damage and the JAK/STAT, NF-κB, mTOR, TGF-β, Wnt, ROS, and UPRmt pathways. The JAK/STAT signaling decreases in HSC aging with stem cell exhaustion, cell proliferation, and myeloid skewing.[30] NF-kB activity increases with the increasing age of HSCs, including a higher localization of the NF-kB subunit p65 protein (about 71%) in 22-month-old HSCs compared to 3% in young (2-month-old) HSCs.[31] In addition, phosphorylated mTOR and mTOR activity was reported to be very high in old-age HSC mice as compared to the young.[32]

Interestingly, the TGF-β signaling pathway with genes such as Smad4, Endoglin, Spectrin b2, Nr4a1, Cepba, Jun, and Junb decreases with aging HSCs.[33] Cdc42, a small RhoGTPase of the Wnt pathway, has higher activity in aged HSCs. This is related to polarity loss, the decline of self-renewal, and altered differentiation.[34]

Moreover, the hypoxic niche protects HSCs from apoptosis and the decline of their self-renewal properties. The thioredoxin-interacting protein (TXNIP) regulates intracellular ROS during oxidative stress and helps in maintaining HSCs.[35] The UPRmt regulator and Sirt7 decrease in aged HSCs. However, a higher level of Sirt7 helps in enhancing the regenerative ability of aged HSCs.[20] Notably, nicotinamide riboside, an UPRmt stimulator, rejuvenates muscle stem cells of aged mice by initiating the synthesis of prohibition proteins.[36]

14 Cell-extrinsic mechanism for HSCs aging

The surrounding cells of HSCs also play an important role in HSC aging and cause impaired lymphoid differentiation and myeloid expansion. The alteration of the HSC niche or microenvironment is also responsible for the functional loss of aging HSCs.[37] It contains a dynamic network of endothelial cells, mature hematopoietic cells, and other cell types (Fig. 3). Notably, HSC neighbor cells includes megakaryocytes, mesenchymal stromal cells (MSCs), osteoblasts, adipocytes, macrophages, neutrophils, cytokines, disrupted β-adrenergic nerve signaling, and enzymes including IL-6, IL-1B, and caspase-1. A higher level of cytokines IL-1B and Caspase 1 and senescent neutrophils is observed in aged mice.[9, 38–40] Notably, the replicative senescence and HSC homing are related with the significant increase of MSCs during aging. Similarly, HSC platelet bias is related with the dysfunction of aged marrow macrophages. It has been reported that during the aging process, senescent neutrophils, IL-1B, Caspase-1, platelet-biased HSCs, mesenchymal stromal cells, β2-AR signaling, IL-6, megakaryocytes, and Nos1 are upregulated while sympathetic nerve fibers and β3-AR signaling are downregulated.[9]

15 Single cell and HSCs

Several studies have been done on aging in mice and humans using a single cell. The single-cell RNA sequencing studies of bone marrow (BM) revealed that the chromatin marker has been increased in hematopoietic progenitors and terminally differentiated immune cells.[41]

Single-cell analysis helps to understand HSC aging, especially intrinsic molecular changes, as hundreds of genes are differentially regulated including a decrease in cycling HSCs and lymphoid-primed multipotent progenitors. Transcription studies have shown the differential expression of several genes: CD27, CD34, CD37, CD44, CD48, CD52, CD63, CD79b, CD86, CD97, CD97b, and CD160 were downregulated while CD28, CD38, CD41, CD61, CD47, CD62, CD69, CD74, and CD81 were upregulated with age. Single-cell epigenetic studies show that aged HSCs divided symmetrically while young HSCs divided asymmetrically.[42]

16 Consequence of aging HSCs

HSCs are able to maintain homeostasis in the entire mature blood cells due to their self-renewal and differentiation ability. Old or aged tissues do not have the ability to maintain tissue homeostasis and are unable to recover during stress and injury. Aging has a direct correlation with HSC fading and is responsible for disturbances in the adaptive immune system and increases of myeloproliferative disease (Fig. 1).[13] Understanding at the molecular and cellular levels can help to decipher the aging process of HSCs. The aging of the hematopoietic system causes a loss of immune function and myeloid leukemia. Rossi et al. have attempted to understand the mechanism of hematopoietic aging and studied long-term hematopoietic stem cells (LT-HSCs) from young and old mice. They observed the downregulation of lymphoid genes and the upregulation of myeloid genes related to fate and function during the aging process.[14] Notably, several genes involved in leukemic transformation were observed in LT-HSCs from old mice. The LT-HSC commitment toward the myeloid lineage is one of the reasons for higher incidences of myeloid leukemia in older people.[14]

17 Rejuvenating strategies for HSCs

There are several reports about old HSCs that can be rejuvenated to some extent by several ways, including fasting, genetic modulators, small molecules, and changes around bone marrow. It has been observed that young blood in aged animals can help in the rejuvenation of brain aging.

Prolonged fasting causes a reduction of circulating IGF-1 levels to protect mice from chemotoxicity.[43, 44] A reduction of IGF1 and protein kinase A (PKA) in cells due to fasting helps in lineage-balanced regeneration, self-renewal, and stress resistance (Fig. 5).[45] Satb1, Sirt3, and Sirt7 were found to be reduced in aged HSCs. Interestingly, Satb1 is related to a compromised lymphopoeitic while the overexpression of Satb1 helps in its restoration.[46] Sirt3 helps in the acetylation of mitochondrial protein and its upregulation helps to improve the regenerative capacity of aged HSCs.[47]

Pharmacological intervention can play an important role in the rejuvenation of aged HSCs. Inhibitors of mTOR, Cdc42, and p38 MAPK help in the rejuvenation of aged HSCs. mTOR signaling is increased in HSCs of aged mice and its inhibitor, rapamycin, helps to improve their immune response. The increase of Cdc42 in aged HSCs controls the regulation of cell division, transformation, and polarity while its inhibitor CASIN helps in the rejuvenation of aged HSCs by increasing polarized cells and the H4K16ac level while reducing myeloid lineage (Fig. 5).[48] During aging, the level of ROS was increased as HSCs prefer to stay in low-oxygen bone marrow niches; this is related with p38 MAPK. Notably, TN13 and SB203580 inhibit p38 MAPK, reduce the ROS level in cells, and help in the rejuvenation of aged HSC cells.[19, 49, 50]

Interestingly, ABT263 has the potential to selectively kill senescent cells that increased during aging and is an antiapoptotic inhibitor of BCL2 and BCL-xL. The oral treatment of ABT263 has successfully depleted senescent cells in aged mice and senescent BM HSCs.[51] Notably, the β3-AR agonist BRL37344 has the potential to decrease myeloid and rejuvenate aged HSCs in old mice.[52, 53] In newborn mice, a transfusion of HSC-derived monocytic cells helps in bone development in early onset autosomal recessive osteopetrosis.[54]

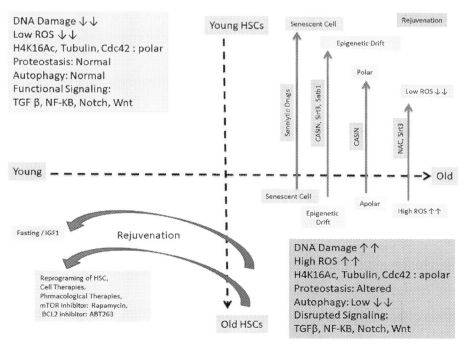

FIG. 5 Aging and rejuvenation factor.

Oncogenic mutations behave differently according to age. Notably, they are more harmful in old age compared to young. Reversing of aging can be achieved partially by reprogramming HSCs as well as dietary, pharmacological, and cell therapies.[4] Interestingly, aging in HSCs is the outcome of the cumulative effect of several processes and genes.

18 Future perspective

HSCs are active throughout their entire life, providing blood and immune cells in the body. Therefore, unearthing the mechanism of HSC aging is an important and challenging area. HSC aging happens by intrinsic or extrinsic systems or both. Understanding the comprehensive role of intrinsic and extrinsic factors in HSC aging, including DNA damage, epigenetics, and clonal hematopoiesis, may help to reduce aging and malignancies. Several complex signaling pathways are involved and interconnected in the aging process. Therefore, there is a need to understand their complete interaction network. JAK, STAT, mTOR, NF-kB, and the UPRmt signaling pathways are upregulated while TGF-β is downregulated. Single-cell studies of HSCs and overcoming the problem of purification of HSCs will help in a better understanding of HSCs. Understanding aging will help in finding ways of rejuvenation. A slowing of the aging process can be achieved by fasting, rejuvenating agents, reprogramming HSCs, cell therapies, and pharmacological therapies.

Conflict of interest

There is no conflict of interest.

References

[1] L.A. Warren, D.J. Rossi, Stem cells and aging in the hematopoietic system, Mech. Ageing Dev. 130 (2009) 46–53.

[2] S. Akunuru, H. Hartmut Geiger, Aging, clonality and rejuvenation of hematopoietic stem cells, Trends Mol. Med. 22 (8) (2016) 701–712.

[3] M.B. Schultz, D.A. Sinclair, When stem cells grow old: phenotypes and mechanisms of stem cell aging, Development 143 (2016) 3–14.

[4] G.D. Haan, S.S. Lazare, Aging of hematopoietic stem cells, Blood 131 (5) (2018) 479–487.

[5] H. Clevers, The intestinal crypt, a prototype stem cell compartment, Cell 154 (2) (2013) 274–284.

[6] F. Blokzijl, J. de Ligt, M. Jager, et al., Tissuespecific mutation accumulation in human adult stem cells during life, Nature 538 (7624) (2016) 260–264.

[7] N. Barker, M. Huch, P. Kujala, et al., Lgr5(1ve) stem cells drive self-renewal in the stomach and build long-lived gastric units in vitro, Cell Stem Cell 6 (1) (2010) 25–36.

[8] E. Mejia-Ramirez, M.C. Florian, Understanding intrinsic hematopoietic stem cell aging, Haematologica 105 (1) (2020) 22–37.

[9] L.V. Kovtonyuk, K. Kristin Fritsch, X. Feng, et al., Inflamm-aging of hematopoiesis, hematopoietic stem cells, and the bone marrow microenvironment, Front. Immunol. 7 (2016) 1–13. 502.

[10] D.J. Pearce, F. Anjos-Afonso, C.M. Ridler, A. Eddaoudi, D. Bonnet, Age-dependent increase in side population distribution within hematopoiesis: implications for our understanding of the mechanism of aging, Stem Cells 25 (4) (2007) 828–835.

[11] X. Li, X. Zeng, Y. Xu, et al., Mechanisms and rejuvenation strategies for aged hematopoietic stem cells, J. Hematol. Oncol. 13 (2020) 31, https://doi.org/10.1186/s13045-020-00864-8.

[12] B. Dykstra, S. Olthof, J. Schreuder, M. Ritsema, G. de Haan, Clonal analysis reveals multiple functional defects of aged murine hematopoietic stem cells, J. Exp. Med. 208 (13) (2011) 2691–2703.

[13] D.J. Rossi, D. Bryder, 2, Weissman IL, hematopoietic stem cell aging: mechanism and consequence, Exp. Gerontol. 42 (5) (2007) 385–390.

[14] D.J. Rossi, D. Bryder, J.M. Zahn, H. Ahlenius, R. Sonu, A.J. Wagers, I.L. Weissman, Cell intrinsic alterations underlie hematopoietic stem cell aging, version 2, Proc. Natl. Acad. Sci. U. S. A. 102 (26) (2005) 9194–9199.

[15] J.M. Bernitz, H.S. Kim, B. MacArthur, H. Sieburg, K. Moore, Hematopoietic stem cells count and remember self-renewal divisions, Cell 167 (5) (2016) 1296–1309.

[16] G. Van Zant, G. de Haan, I.N. Rich, Alternatives to stem cell renewal from a developmental viewpoint, Exp. Hematol. 25 (3) (1997) 187–192.

[17] S.J. Morrison, A.M. Wandycz, K. Akashi, A. Globerson, I.L. Weissman, The aging of hematopoietic stem cells, Nat. Med. 2 (9) (1996) 1011–1016.

[18] T.T. Ho, M.R. Warr, E.R. Adelman, et al., Autophagy maintains the metabolism and function of young and old stem cells, Nature 543 (7644) (2017) 205–210.

[19] K. Ito, A. Hirao, F. Arai, et al., Reactive oxygen species act through p38 MAPK to limit the lifespan of hematopoietic stem cells, Nat. Med. 12 (4) (2006) 446–451.

[20] M. Mohrin, J. Shin, Y. Liu, et al., Stem cell aging. A mitochondrial UPR-mediated metabolic checkpoint regulates hematopoietic stem cell aging, Science 347 (6228) (2015) 1374–1377.

[21] G.L. Norddahl, C.J. Pronk, M. Wahlestedt, et al., Accumulating mitochondrial DNA mutations drive premature hematopoietic aging phenotypes distinct from physiological stem cell aging, Cell Stem Cell 8 (5) (2011) 499–510.

[22] D.P. Steensma, R. Bejar, S. Jaiswal, et al., Clonal hematopoiesis of indeterminate potential and its distinction from myelodysplastic syndromes, Blood 126 (1) (2015) 9–16.

[23] F. Zink, S.N. Stacey, G.L. Norddahl, et al., Clonal hematopoiesis, with and without candidate driver mutations, is common in the elderly, Blood 130 (6) (2017) 742–752.

[24] A.M. Broske, L. Vockentanz, S. Kharazi, M.R. Huska, E. Mancini, M. Scheller, C. Kuhl, A. Enns, M. Prinz, R. Jaenisch, et al., DNA methylation protects hematopoietic stem cell multipotency from myeloerythroid restriction, Nat. Genet. 41 (11) (2009) 1207–1215.

[25] J.J. Trowbridge, J.W. Snow, J. Kim, S.H. Orkin, DNA methyltransferase 1 is essential for and uniquely regulates hematopoietic stem and progenitor cells, Cell Stem Cell 5 (4) (2009) 442–449.

[26] D. Djeghloul, K. Kuranda, I. Kuzniak, D. Barbieri, I. Naguibneva, C. Choisy, J.C. Bories, C. Dosquet, M. Pla, V. Vanneaux, et al., Age-associated decrease of the histone methyltransferase SUV39H1 in HSC perturbs heterochromatin and B lymphoid differentiation, Stem Cell Rep. 6 (6) (2016) 970–984.

[27] A. Mehta, J.L. Zhao, N. Sinha, G.K. Marinov, M. Mann, M.S. Kowalczyk, R.P. Galimidi, X. Du, E. Erikci, A. Regev, et al., The microRNA-132 and microRNA-212 cluster regulates hematopoietic stem cell maintenance and survival with age by buffering FOXO3 expression, Immunity 42 (6) (2015) 1021–1032.

[28] A. Mehta, D. Baltimore, MicroRNAs as regulatory elements in immune system logic, Nat. Rev. Immunol. 16 (5) (2016) 279–294.

[29] G. Tie, J. Yan, L. Khair, et al., Hypercholesterolemia accelerates the aging phenotypes of hematopoietic stem cells by a Tet1-dependent pathway, Sci. Rep. 10 (2020) 3567, https://doi.org/10.1038/s41598-020-60403-w.

[30] K. Kirschner, T. Chandra, V. Kiselev, D. Flores-Santa Cruz, I.C. Macaulay, H.J. Park, J. Li, D.G. Kent, R. Kumar, D.C. Pask, et al., Proliferation drives aging-related functional decline in a subpopulation of the hematopoietic stem cell compartment, Cell Rep. 19 (8) (2017) 1503–1511.

[31] S.M. Chambers, C.A. Shaw, C. Gatza, C.J. Fisk, L.A. Donehower, M.A. Goodell, Aging hematopoietic stem cells decline in function and exhibit epigenetic dysregulation, PLoS Biol. 5 (8) (2007), e201.

[32] C. Chen, Y. Liu, Y. Liu, P. Zheng, mTOR regulation and therapeutic rejuvenation of aging hematopoietic stem cells, Sci. Signal. 2 (98) (2009), ra75.

[33] D. Sun, M. Luo, M. Jeong, B. Rodriguez, Z. Xia, R. Hannah, H. Wang, T. Le, K.F. Faull, R. Chen, et al., Epigenomic profiling of young and aged HSCs reveals concerted changes during aging that reinforce self-renewal, Cell Stem Cell 14 (5) (2014) 673–688.

[34] S. Etienne-Manneville, Cdc42—the centre of polarity, J. Cell Sci. 117 (Pt 8) (2004) 1291–1300.

[35] H. Jung, M.J. Kim, D.O. Kim, W.S. Kim, S.-J. Yoon, Y.-J. Park, S.R. Yoon, T.-D. Kim, H.-W. Suh, S. Yun, et al., TXNIP maintains the hematopoietic cell pool by switching the function of p53 under oxidative stress, Cell Metab. 18 (1) (2013) 75–85.

[36] H. Zhang, D. Ryu, Y. Wu, K. Gariani, X. Wang, P. Luan, D. D'Amico, E.R. Ropelle, M.P. Lutolf, R. Aebersold, et al., NAD(+) repletion improves mitochondrial and stem cell function and enhances life span in mice, Science 352 (6292) (2016) 1436–1443.

[37] Y.H. Ho, S. Méndez-Ferrer, Microenvironmental contributions to hematopoietic stem cell aging, Haematologica 105 (1) (2020) 38–46.

[38] A. Chow, D. Lucas, A. Hidalgo, S. Méndez-Ferrer, D. Hashimoto, C. Scheiermann, et al., Bone marrow CD169 + macrophages promote the retention of hema-topoietic stem and progenitor cells in the mesenchymal stem cell niche, J. Exp. Med. 208 (2011) 261–271, https://doi.org/10.1084/jem.20101688.

[39] K. Kaushansky, Lineage-specific hematopoietic growth factors, N. Engl. J. Med. 354 (2006) 2034–2045, https://doi.org/10.1056/NEJMra052706.

[40] S.H. Orkin, L.I. Zon, Hematopoiesis: an evolving paradigm for stem cell biol-ogy, Cell 132 (2008) 631–644, https://doi.org/10.1016/j.cell.2008.01.025.

[41] K.A. Oetjen, K.E. Lindblad, M. Goswami, G. Gui, P.K. Dagur, C. Lai, L.W. Dillon, M.C. JP, C.S. Hourigan, Human bone marrow assessment by single-cell RNA sequencing, mass cytometry, and flow cytometry, JCI Insight 3 (23) (2018) e124928, https://doi.org/10.1172/jci.insight.124928.

[42] M.C. Florian, M. Klose, M. Sacma, J. Jablanovic, L. Knudson, K.J. Nattamai, G. Marka, A. Vollmer, K. Soller, V. Sakk, et al., Aging alters the epigenetic asymmetry of HSC division, PLoS Biol. 16 (9) (2018), e2003389.

[43] C. Lee, F.M. Safdie, L. Raffaghello, M. Wei, F. Madia, E. Parrella, D. Hwang, P. Cohen, G. Bianchi, V.D. Longo, Reduced levels of IGF-I mediate differential protection of normal and cancer cells in response to fasting and improve chemotherapeutic index, Cancer Res. 70 (4) (2010) 1564–1572.

[44] L. Raffaghello, C. Lee, F.M. Safdie, M. Wei, F. Madia, G. Bianchi, V.D. Longo, Starvation-dependent differential stress resistance protects normal but not cancer cells against high-dose chemotherapy, Proc. Natl. Acad. Sci. U. S. A. 105 (24) (2008) 8215–8220.

[45] C.W. Cheng, G.B. Adams, L. Perin, M. Wei, X. Zhou, B.S. Lam, S. Da Sacco, M. Mirisola, D.I. Quinn, T.B. Dorff, et al., Prolonged fasting reduces IGF-1/PKA to promote hematopoietic-stem-cell-based regeneration and reverse immunosuppression, Cell Stem Cell 14 (6) (2014) 810–823.

[46] Y. Satoh, T. Yokota, T. Sudo, M. Kondo, A. Lai, W. Kincade Paul, T. Kouro, R. Iida, K. Kokame, T. Miyata, et al., The Satb1 protein directs hematopoietic stem cell differentiation toward lymphoid lineages, Immunity 38 (6) (2013) 1105–1115.

[47] K. Brown, S. Xie, X. Qiu, M. Mohrin, J. Shin, Y. Liu, D. Zhang, D.T. Scadden, D. Chen, SIRT3 reverses aging-associated degeneration, Cell Rep. 3 (2) (2013) 319–327.

[48] H. Leins, M. Mulaw, K. Eiwen, V. Sakk, Y. Liang, M. Denkinger, H. Geiger, R. Schirmbeck, Aged murine hematopoietic stem cells drive aging associated immune remodeling, Blood 132 (6) (2018) 565–576.

[49] H. Jung, D.O. Kim, J.-E. Byun, W.S. Kim, M.J. Kim, H.Y. Song, Y.K. Kim, D.-K. Kang, Y.-J. Park, T.-D. Kim, et al., Thioredoxin-interacting protein regulates haematopoietic stem cell ageing and rejuvenation by inhibiting p38 kinase activity, Nat. Commun. 7 (1) (2016) 13674.

[50] M.L. Hamilton, H. Van Remmen, J.A. Drake, H. Yang, Z.M. Guo, K. Kewitt, C.A. Walter, A. Richardson, Does oxidative damage to DNA increase with age? Proc. Natl. Acad. Sci. U. S. A. 98 (18) (2001) 10469–10474.

[51] J. Chang, Y. Wang, L. Shao, R.M. Laberge, M. Demaria, J. Campisi, K. Janakiraman, N.E. Sharpless, S. Ding, W. Feng, et al., Clearance of senescent cells by ABT263 rejuvenates aged hematopoietic stem cells in mice, Nat. Med. 22 (1) (2016) 78–83.

[52] M. Maryanovich, A.H. Zahalka, H. Pierce, S. Pinho, F. Nakahara, N. Asada, Q. Wei, X. Wang, P. Ciero, J. Xu, et al., Adrenergic nerve degeneration in bone marrow drives aging of the hematopoietic stem cell niche, Nat. Med. 24 (6) (2018) 782–791.

[53] Y.H. Ho, R. Del Toro, J. Rivera-Torres, J. Rak, C. Korn, A. Garcia-Garcia, D. Macias, C. Gonzalez-Gomez, A. Del Monte, M. Wittner, et al., Remodeling of bone marrow hematopoietic stem cell niches promotes myeloid cell expansion during premature or physiological aging, Cell Stem Cell 25 (3) (2019) 407–418 (e406).

[54] C.E. Jacome-Galarza, G.I. Percin, J.T. Muller, E. Mass, T. Lazarov, J. Eitler, M. Rauner, V.K. Yadav, L. Crozet, M. Bohm, P.L. Loyher, G. Karsenty, C. Waskow, F. Geissmann, Developmental origin, functional maintenance and genetic rescue of osteoclasts, Nature 568 (7753) (2019) 541–545.

Chapter 8

Ocular stem cells and aging

Neethi Chandra Thathapudi[a] and Jaganmohan R. Jangamreddy[b]

[a]*L V Prasad Eye Institute, Hyderabad, India,* [b]*UR Advanced Therapeutics Private Limited, Aspire-BioNEST, University of Hyderabad, Hyderabad, India*

1 Introduction to stem cells

Stem cells are maintained throughout life and undergo self- renewal, as well as provide differentiated cells replacing the worn out/wounded adult cells through asymmetric division and thus are widely explored as treatment modalities for a variety of disease conditions leading to damaged tissues and organs [1]. Healthy and young persons are capable of quick (less scarred) recovery, partly due to the actively functioning stem cells; but in diseased and aging persons such a recovery is difficult and prolonged [2]. The key characteristics of stem cells include self- renewability, generation of single cells that can differentiate and functionally reorganization in a tissue. Stem cells are pluripotent, capable of regenerating/differentiating into any adult tissue but most adult stem cells are multipotent with limited differentiation capabilities. Altruistic, stem cells are assumed for an "unlimited" cell division capabilities that is crucial for their self renewal as well as differentiation [3]. However, in the biological system where aging is an important element, several factors, both intrinsic and extrinsic, influence the stem cells and their residing niche leading to their decreased potency [4]. Along with incapcitence of division due to aging and disease, stem cells also undergo gradual deterioration of function including decreased metabolic activity and thus leading to senescence [5,6]. The aging stem cells also morphologically appear more flatter and larger compared to their more potent precursor cells [5]. Both intrisic cellular mechanisms to recycle and protect cellular proteome as well as the external factors including alterations to their cellular niche and systemic cues affect the regenerative capacity as well as the homeostatic capability of the aging stem cell [6]. Due to the compromised cellular homeostasis mechanisms, the aging SC shows an increase in the accumulation of toxic metabolites leading to exessive Reactive Oxygen Species (ROS). Among aged stem cells, growth and metabolic pathways like Insulin growth factor (IGF-1 signaling pathway), that regulates FoxO1, FoxO3 and FoxO4 transcription factors along with PTEN /AKT/ mTOR pathway (which is upstream of the FoxO) causes increse in ROS formation through regulation of glycolytic and respiratory pathways [7]. Along preoteostasis and ROS generation, Longevity promoting pathways (such as insulin/IGF-1 signaling pathway), causes a downregulation of several genes (such as PSDM11) that are crucial for cell vitality and maintenance of pluripotency markers OCT4, Nanog, SOX2 and DPPA 2,4 among aged SC leading to their limited stemness charectoristics [8,9]. Embryonic stem cells (ESCs) have higher levels of glutathione/redoxin system enzymes as well as HSP proteins, than when they undergo differentiation. This ensures that the ESCs respond much better to misfolded proteins than older and differentiated cells. More over, Superoxide dismutase (Sod2), which is an antioxidant enzyme of the mitochondria, is observed to have more mutations in aged SC and and thus leading to its compromised function which further increases oxidative stress and ROS accumulation. Dysfunction in the autophagy/lysosomal pathway as well as the ubiquitin-proteosome pathway cause an impaired clearing mechanism within the cell.

The stem cell niche is regulated by external cues, which directly affect the functioning and maintenance of SC. The bone morphogenic protein (BMP) pathway is extremely important for maintenance of the SC niche and this is seen to be perturbed in the aging cell niche [8,10]. Reduction in levels of E-cadherin has also been associated to this. Fibro/adipogenic progenitors (FAPs) which are the key components of the stromal tissue, are the main ECM regulators; changes in the EGF, Wnt3A and Notch cause disturbances in the stem cell niche. There are also certain systemic cues involved leading to the process of aging and this is mostly due to an increase in TGF- β, along with the activation of the NF-κB pathway [8].

Telomere erosion along with oxidative damage by occumulation of ROS and misfolded proteins are also causes of the aging among SC and exhibit a senescence – like phenomenon which is reversivble with the ectopic expression telomerase reverse transcriptase (hTERT) that promote telomere extension and maintenance can extend the self- renewal ability of the stem cell [6]. Mutations in mammalian sirtuin, SirT1, is known to induce premature senecence among stem cells by the generation of excessive ROS [11,12]. Along with the metabolic and Telomere maintanance mechanisms, DNA damage repair (DDR) mechanisms and mitochondrial DNA proof reading mechanisms play a crucial role in stem cell vitality, quiescence and proliferative potential of stem cells. It was observed that γH2A.X, a variant of histone proteins that acts as a marker for breaks within the ds DNA, increased in older cells [8]. The mitochondrial proof-reading enzyme, mtDNA polymerase γ is known to cause premature aging of cells when it is dysfunctional; the importance of this enzyme in aging, is therefore critical [8]. Understanding the implications of the changes in stem cells during the process of aging would throw light on treating many age related disease conditions. Stem cells occur in almost all major regions of the eye but the most widely understood are the corneal limbal stem cells. The ability of the lens to terminally differentiate to fiber cells also hint at the presence of a stem cell niche [13]. More recent studies point to the presence of rare retinal stem cells but such studies are still novice. Lack of tissue localized adult stem cells population could be compensated with the circulating mesenchymal stem cells delivered through blood vessels [13]. Considering the vital characteristic of stem cells in disease and tissue repair, stem cell transplantation for various ocular injuries has been growing in the last few years. In this chapter, we focus both on the effects of aging on the resident ocular stem cells as well as the changes they undergo that affect their regenerative capacities.[a]

2 Stem cells of the cornea

2.1 Corneal stem cells

The outermost layer of the eye is called the cornea; it refracts light and focuses it onto the lens. The transparency of the cornea is maintained by the arrangement of cells within the corneal layers [14]. The epithelial layer is derived from the embryonic ectoderm, while the stromal and endothelium is derived from the neural crest [15]. The outer epithelial layer is continuously regenerated by the stem cells present at the margin of the cornea extending into the conjunctiva, called the limbal region (towards the periphery in the palisades of Vogt); this happens through a centripetal migration pattern [16]. The stroma, which is a collagen rich mesenchymal tissue (keratocytes), makes up the major bulk of the cornea; keratocytes are mostly quiescent in adults and rarely undergo cell cycle and mitotic renewal unlike the epithelial layer [17]. The inner endothelial layer does not undergo any cell divisions, maintaining a more or less same number of cells throughout adulthood, with gradual loss in number of cells due to aging [18]. Thus the stem cells of cornea mainly represent the epithelial and stromal layer with very few evidences of endothelial stem cells.

2.2 Limbal epithelial stem cells (LESC)

Limbal epithelial stem cells are somatic SC that divide asymmetrically to regenerate the corneal epithelium. Stem-like cells, whose loss results in impaired corneal repair during wound healing along with infiltration of conjunctival epithelial cells into the corneal layer, are localized in the corneal periphery region called Limbus which is marked with palisades of Vogt (corneo-scleral region) [19]. It is a pigmented region of the cornea towards its periphery and was first proposed by Davanger and Evensen in 1971 [20]. The location of the LESCs is key for stem cell survival; the basal cells are pigmented, providing an insulation against the solar rays, vasculature for nutrients as well as resistance to shear stress [21]. LESCs are currently identified using different markers that are not expressed in the differentiated corneal epithelial cells include, Bmi1 +ve, dNp63a +ve and C/EBPδ, along with ABCG2 +ve, Musashi1, integrina9 and N cadherin [22]. Connexin 43 is used as a negative marker for LESCs to differentiate them from corneal epithelial cells [23]. The basal layers of the epithelium does not express Cytokeratins 3 and12; these are expressed by the fully differentiated superficial epithelial cells. The basal layer of the epithelium has adult stem cells that divide and form transient amplifying cells (TACs) and move into the suprabasal layers; here they differentiate and move further on to the outer stratified layers [24]. The typical stem cell property of smaller cell size, high cytoplasm to nucleus ratio and increased euchromatin content, is observed in the basal limbal cells [25]. High expression of EGFR is also observed which gives these cells a proliferative potential signal. Based on the difference of expression of stem cell markers such as Oct4, Sox2, Nanog and Nestin, LESCs are divided into limbal niche cells (LNCs) and limbal epithelial progenitor/stem cells (LEPCs) [24]. LNCs show the suppression of the canonical Wnt pathway while the activation of non-canonical Wnt signaling pathway as well as the BMP signaling pathway is observed in them [26]. Also, the SDF1/CXCR4 pathway is modulated, where the LNCs secrete the cytokine CXCR4 [27]. It has been shown that Wnt and Pax6 play a role in regulation of the limbal niche. The basal layers of the limbus also show the presence of specific laminins and collagens that maintain the ECM structure, thereby providing the required microenvironment for the SC [28]. Apart from this, the physical and mechanical aspects are also involved which modulate the LESC maintenance. There is growing evidence of the presence of a "second niche" in the central corneal region, which functions irrespective of the limbus. The exact characterization of the cells and pathways involved has not yet been clearly understood [29].

---

a. Figure 1.

2.3 Corneal stromal stem cells (CSSC)

The presence of mesenchymal stem cells (MSCs) in the stroma has been shown by various studies. MSCs are defined by their characteristic features which are multipotency, asymmetric division and clonal growth [30]. These MSCs have been spotted in niches in the limbal region, in the anterior stromal layers. Markers used to identify these SC include ABCG2, BMI1, CD73, Notch1 and CD166 [31]. The CSSC have been shown to differentiate into keratocytes that ultimately differentiate into fibroblasts/myofibroblasts during wound healing [18,32]. Unlike the corneal epithelial layer, the stromal layer does not undergo continuous regeneration. These cells in culture show differentiating potential and secrete ECM proteins including collagen and aggrecan [33]. This property of the CSSCs is exploited in corneal stem cell transplantation techniques such as CLET [34]. There is evidence to show that the CSSCs help in maintaining the LESC population in the limbus in co-culture studies; STAT3 signaling pathway activation in LESCs happens through IL-6 production and this pathway is said to be a key component of the stromal population [35]. The origin of the CSSCs is the neural crest but there are certain theories suggesting a bone-marrow origin of these MSCs, however expression of neural crest proteins Snail, Slug and Sox9 markers in these cells provide a strong argument against it [36]. The development of keratocytes from the CSSCs has been shown to be regulated by its microenvironment; when cultured as a pellet, more keratocytes growth, with collagens similar to that of stroma *in vivo* is seen; when cultured on certain biomaterials, parallel formation of collagens was observed [37]. The topographical cues guide ECM formation and this is another aspect that has, and can be further exploited. CSSCs remodel and regenerate the stromal ECM and this has been shown in mice models. They help in ECM deposition by the stromal cells, that is regulated in a paracrine manner [38].

2.4 Effect of aging on the stem cells of the anterior cornea

As mentioned earlier, epithelial layer is derived from the embryonic ectoderm while the stromal and endothelium is derived from the neural crest as the CSSC express neural crest markers including Sox9, Slug and Snail [31]. Corneal stem cell aging, like other stem cells, has been studied and cytokine signaling pathway SDF1/CXCR4 has been shown to play an important role in maintenance of the limbal SC niche [39,40]. Limbal stem cells show a down regulation of b-catenin and Lef1, and an up regulation of Pax6, which play a key role in proliferation [41]. The "stemness" is maintained by Wnt1 and Wnt2 pathways, which are both up regulated in serum deprived stem cells [42]. The number of clones produced by the LESCs decrease with age and this could be due to the loss of stem cells or the gradual increase in quiescent stem cells. Physiologically as well, it has been observed that as the age increases, the palisades of Vogt become less visible along with decrease in the number of limbal crypts [43]. The size of the cells in the basal limbus that maintain the stem cell properties are also observed to increase with age marking the decreased regenerative capabilities [22]. However, the putative stem cell marker p63a showed no significant decrease in healthy older persons (60-75 years) and so did the telomerase activity [39]. However, telomerase reverse transcriptase (TERT) gene which helps in the entry of the quiescent stem cell into the cell cycle was also observed to be increased in the proliferative cells [44].

2.5 Corneal endothelial stem cells [45]

The endothelial layer of the cornea is widely known that it does not undergo any cell division and this could be possible due to TGFb2, p53 and TAp63, which prevent replication to happen by keeping them in G1 phase of the cell cycle [46]. Studies have shown that there is an increase in the density of endothelial cells towards the periphery of the cornea (almost near the limbal region); they also show telomerase activity and upregulated alkaline phosphatase activity similar to stem cells, building the hypothesis of "stem cell like cells" in the posterior endothelium [47]. Schwalbe's line cells, which are cells that found below the Schwalbe's ring, between the endothelium and the trabecular meshwork cells, have also been shown to express progenitor cell like properties [48]. It is hypothesized that the endothelial cells with stem cell like properties, at the periphery, move in a centripetal direction at a very slow rate [15]. If these cells can be isolated, they tend to show progenitor cell like properties, which can be exploited for various endothelial dysfunction diseases.

3 Stem cells of the conjunctiva

3.1 Conjunctival stem cells (CSC)

The conjunctiva is the protective mucous membrane that covers the outer portion of the eye and plays a very important role in protection and hydration of the eye. It is more prone to infections as it is the first layer of defense mechanism and can experience severe infections sometimes [49]. Conjunctival epithelial cells are of three types- bulbar epithelium, fornix epithelium and palpebral epithelium [50]. Conjunctival and corneal epithelial cells, though very distinct separate cells, initially were proposed to have an almost similar origin, but later studies have shown that there is presence of stem cells throughout the conjunctival epithelium but are more densely populated in certain parts [51]. During cases of corneal stem cell transplantation, there are varying result in its success rate and this has been hypothesized due to the conjunctival damage caused by the disease condition [50]. Therefore, identification of conjunctival stem cells and its further exploitation for transplantation is important. The inferior forniceal and medial cantal portions are rich in cells that are ABCG2 and p63 positive; physically, these regions resemble the limbal niches with rich vascularization and melanocytes [49]. These cells do express stem cell markers and also show proliferative potential [52]. It has been hypothesized that the basal layers of the conjunctival epithelium has "pockets" of stem cells and function similar to the interfollicular stem cells of the skin [53], but this has not yet been studied thoroughly. The CSC is bipotent and gives rise to goblet cells and keratinocytes [54]. The goblet cell region of the conjunctiva showed strong staining with ABCG2. Goblet cell population was identified among TACs indicating that it is a later stage process [55]. Meroclones from the stem cells are considered as TACs and they have a higher proliferative rate than other clones of SCs [56]. The pattern of the production of goblet cells from stem cell progenitors or TACs and their co-existence in the conjunctival stem cell niche is probably a necessary microenvironment [55]. "Conjunctival epithelial transdifferentiation" has been a widely discussed phenomenon where the conjunctival epithelial cells overlay the corneal epithelial cells, but these studies have not quite clearly separated out the role of the limbal stem cells in this [57]. Further studies have shown that the limbal lineage and the conjunctival lineage is quite different. Due to lack of specific CSC markers, these cells are identified on its slow cycling property *in vivo*; prolonged BrdU treatment has been employed for the same [53]. In rabbits, it was observed that the palpebral epithelial cells grew quicker and had more proliferative potential. It was proposed that the conjunctival stem cells occur in the palpebral region, more specifically in the mucocutaneous junction [53].

3.2 Aging in conjunctival stem cells

It has been observed in a variety of stem cell niche areas of various tissues, that the microenvironment deteriorates with aging – both in the stem cell number and their clonogenic potential. This has also been observed in the conjunctival stem cells [53]. As the donor age increased, the number of cells with clonogenic potential and stem cell marker expression decreased quite a bit [58]. There is a significant decrease in the number of cell differentiation cycles in the older meroclones as compared to the younger ones [53]. Conjunctival keratocytes give rise to goblet cells at a specific time in the cell cycle, which happens twice (specific to their doubling clock); once in the earlier stages and once just before senescence [59]. It is to be noted that the region of stem cell niche in the conjunctiva is more prone to diseases and therefore it is imperative that the regenerative capacity will reduce drastically as the process of aging occurs [60]. The goblet cells play the important role of mucin secretion, which decreases as aging occurs. The number of goblet cells remain the same as aging occurs but there is a loss of functionality; these cells undergo apoptosis more in old age persons. This is also one of the mechanism for the onset of Dry Eye Disease in older patients [61]. Also sometimes, in older patients, there is an increase in inflammatory cytokines, such as IFN-gamma, which leads to the degeneration of conjunctival cells [61].

4 Stem cells of the lens

4.1 Lens epithelial stem/progenitor cells (LECs)

Lens is a transparent structure of the eye present below the cornea in the anterior segment and is refractive and "accommodating" for the focus of vision. Its structure includes a protective capsule on all sides which is the basement membrane. Below this is an epithelial layer (LEC) [62]. The lens has a highly homogenous structure even at the cellular level with a monolayer of epithelial cells that have low levels of cell division and cell death [63]. The major bulk of the lens is made up of differentiated fiber cells that are formed by the cells in the central region, which are continuously formed from the equatorial segment of the epithelial layer [64]. The lens epithelium is formed opposite to the optic cup and exhibits proliferative capacity. Lens regeneration in amphibians has been found to occur due to transdifferentiation of the outer layers of the cornea [65]. More recent studies have discussed about the presence of "stem cell progenitors" due to the fact that mammalian lens (in rabbits) could be re-grown, to a certain extent, from the lens capsule [66]. Lens epithelial stem/progenitor cells are the endogenous stem cell-like population within the lens. LECs are present on the anterior surface of the lens and extend towards the equator as lens fiber cells. Pax6 and Sox2 expression has been identified in these cells along with P75 neurotrophin receptor which is one of the markers to identify LECs [67]. The lens undergoes cell divisions late into adulthood and that is a proof for the existence of stem cells within the lens. The germinative zone of the lens was thought to house the lens stem cells but this has been disproved; central lens also showed presence of a number of slow cycling cells [68]. The lens does not offer any typical stem cell niche area features such as vascularity, protection and pigmentation. The lens is avascular and without any pigment, which hints that there is an external niche area where the lens stem cells could reside [69]. The ciliary body, which is a structure in close proximity to the lens capsule, is both pigmented and vascularized and provides the ambience for stem cell niche area. It was observed that the ciliary body has the capacity to form lentoids, which are a cluster of cells that exhibit lens proteins and also few of its features [70]. In newts, lens was seen to regenerate from the pigmented iris, which also is along the lines of the existence of lens stem cells outside the lens [71]. There has been an interesting hypothesis that retinal stem cell could have the potential to differentiate into lens cells [68]. Cell migration would also be an important aspect of this, since the cells have to enter the lens, and this could be occurring with the help of zonular fibrils; they could be providing support for the cell migration process and also serve as the gate for the entry into the lens [71]. The continuous growth of lens (mitosis and DNA synthesis) happens in the anterior epithelial region while the remaining parts lose their proliferative capacity and form primary fibers [72].

4.2 Cell cycle regulation in lens epithelial stem cells

Cdk4-p57 complexes were found to be expressed in these epithelial stem cells hinting at a role of Cdk, CKI and cyclins. Cyclin A, B, D1, and D2 were seen to be expressed in these cells while decreased levels of cyclin E was observed. Cdk2 and cdk4 were also expressed in the epithelium, mostly in the peripheral regions and less in the central region; this could be due to a lesser proliferative capacity in the central epithelium [73]. Further, it was observed that c-myc is expressed very well in the epithelium but this was not the case with n-myc [74]. It is hypothesized that if the pocket proteins are required to regulate the cell cycle of these cells, then the functions of the cdk, cyclins and p53 needs to be altered [74]. pRb is activated in the cells that are just about to begin the process of differentiation while HPV oncoprotein E7 inactivates this in postnatal mice. The result is an increase in proliferative potential of the progenitor cells. Epithelial proliferative cells have been studied in vitro to understand the effect of different growth factors on these cells; FGF-1,2, PDGF, insulin, and HGF show a positive effect on their growth while TGFb shows a negative effect [75]. Of course, it has already been established that TGFb promotes the formation of fibroblasts but the effect on proliferative potential was looked into. The proliferating cells showed expression of receptor PDGFR –a, while PGDF-D was also seen to be expressed in the iris and ciliary body as well as the aqueous humor [76]. This could also be due to the housing of few lens stem cells in those regions. The involvement of bone morphogenic protein (BMP) was studied where it was seen that an Alk-3 mediated pathway could be enhancing the cell proliferative capacity [77]. Recent articles also discuss the role of PDZ proteins Dlg-1 and Scrib, which regulate cell adhesion, proliferation and differentiation. Zhou et al have reported the presence of LSC in center of the lens along with the germinative region [78] – this could mean that these cells also are affected in cataracts. Studying cataractous lens would throw light on the aging mechanisms of the lens epithelial stem cells.

4.3 Changes in the aging lens epithelial stem cells

The symmetry of the cells within the lens is maintained even as aging occurs but is disturbed in the cataractous lens. The process of aging and age related diseases have been suggested to be associated with oxidative stress through the formation of lipid peroxides (LPO) [79]. There has been a study which reported that there is an age related difference in size of telomeres in lens epithelial cells; it said that telomere length is regulated by oxidative stress and anti – oxidant mechanisms [80]. This has been studied in both aging lens and cataractous lens. An increase in LPO products was reported in aqueous humor samples of patients with cataract [79]. This directs us towards the implications of chronic oxidative stress on aging of the lens epithelial progenitor cells. Apoptosis induction is decreased by hTERT and the down regulation of p53 and Bcl; not due to telomerase synthesis. This means that irrespective of the endogenous telomerase activity, hTERT helps in the proliferation of the stem cells. The regeneration of potential of stem cells reduces as it ages due to "exhaustion" which involves the senescence mechanism. It was shown that the number of functional lens stem cells reduces with the increase in age [80]. The greater number of senescent cells in older persons also attributes to cataract formation and this occurs due to reduction in cell cycle activity in these cells. Advanced glycation end products (AGE) have been reported widely in the lens; the basement membrane secreted by the lens epithelial cells, have an accumulation of these AGE products that were analyzed and compared with younger lens [81]. These products hint towards EMT pathway with an upregulation of the TGF-b mediated EMT proteins. Data suggests that a TGF-b2 mediated fibrosis occurs in the lens epithelial cells [82]. Another AGE precursor, methylglyoxal (MGO) modifications (which are post synthetic modifications) have been reported to increase in aged lens and further enhanced in the cataractous lens; these modifications cause a decrease in the glyoxalase 1 (GLO1) activity in such lens [83]. Slow self – renewal properties are seen in aging stem cells and it has been observed that an elevated let-7 microRNA, an important regulatory factor in the process of cellular senescence, occurs [84]. It is proposed that this microRNA could be dysregulating genes associated to cell proliferation but the exact mechanism in the cataractous lens has not been understood [85]. In congenital cataract, LECs have been observed in lens regeneration post- surgery but this has not been the case in older patients [86]; the regenerative capacity of the stem cells has been lost due to physiological aging process. There is evidence to show that the number of LECs reduces in older patients, as compared to younger patients, when stained with BrdU, which is a marker for proliferative cells [86]. But upon injury, the same LECs seem to greatly increase its regenerative capacity [87]. Analysis of the aging lens has been done in vitro; the occurrence of 'retrodots', which are small opacities within the normal lens, has an increase in occurrence in people older than 40 years [70]. It was also noted that there was a much higher lens cell growth in patients younger than 40 years as those compared to greater than 60 years; serum stimulation helped older patients cells to grow better [88].

5 Stem cells of the retina

5.1 Retinal stem cells

The retina is the neural portion of the eye and is a part of the central nervous system (CNS). The optic cup is formed by the evagination of the diencephalon, which encompasses the retina on the inside and retinal pigment epithelial cells (RPE) on the exterior surface. Retinal cells are of different types – **vascular**, which include pericytes and endothelial cells that maintain the blood-brain barrier; **glial cells**, which includes Muller cells and astrocytes that regulate the retinal metabolism and integrate vascular and neuronal pathways in the retina; **neurons**, which include bipolar cells, ganglion cells, photoreceptor cells, horizontal cells and amacrine cells which subserve vision; **microglia**, which perform phagocytosis when called upon [89]. The cells of the mature retina arise from "founder cells" in the optic vesicle. The zone between the RPE layer and the retinal region is called the ciliary margin (CM). In vertebrates, this region differentiates into non-neural structures, while in fish and amphibians, they harbor SC population; in these species, the retina grows by cell multiplication throughout adulthood as well [67]. In vertebrates, until recently, retinal stem cells were assumed to not exist because in adult stage, the retina grows by cell stretching and not cell proliferation. But there is growing evidence on the presence of SC in the retina, though they mostly remain quiescent [89]. These post embryonic proliferating cells are found to be in the CM region. Studies have shown that these proliferating cells are multipotent; they developed into both neurons and glia. This has been completely established in fish and yet to be clearly understood in vertebrates. Studies also suggest that the earliest decision that a neural retinal cell takes is to decide if it will become a multipotent stem cell or not a stem cell [90].

5.2 Pathways involved in maintenance of retinal stem cells

There are studies which show that the multipotent cell population in the CM zone show sustained stem cell properties in adult rodents [63]; but they only form a very limited number of cells and therefore just using these cells for transplantation therapy is not an ideal option. This sustenance of stem cell like properties is attributed to the Wnt pathway induction in these cells; also this pathway is established to play a key role in eye development in vertebrates [89]. The canonical Wnt pathway is where the Wnt proteins bind to LRP and Fzd which inactivates Axin and GSK3; this inactivation stabilizes b– catenin which enters the nucleus and activates transcription of genes such as Sox and Myc; these are genes that critical for the G1 to S phase transition [91]. This pathway therefore leads to maintenance of stem cell like properties of these progenitor cells. Cells from the CM zone were cultured as spheres and induction of Wnt pathway in these neurospheres showed an increase in the number of BrdU stained cells as well as Ki-67 positive cells. The canonical Wnt pathway has also been demonstrated to play a key role in the maintenance of the progenitor cells, for proliferation and state of un-commitment to any lineage [92]. Inhibition of this pathway accelerates the differentiation of these retinal progenitor cells. This is achieved by the co-play of the Notch pathway, which regulates the expression of Wnt effector Lef1 and Wnt inhibitor sFRP2 [90]. There is also certain discussion regarding the influence of non-canonical pathways in this process but it is not yet clearly understood.

5.3 Pathway modifications in aged retinal stem cells

Glaucoma, Diabetic retinopathy and age related macular degeneration (AMD) are the most common age related diseases of the retina. The outer retina, RPE, ganglion cells and choriocapillaris undergo structural changes in these conditions [93]. AMD is caused by cellular changes in the RPE and outer retina including atrophy (in non-exudative AMD) and neovascularization (in exudative AMD) [94]. Gene expression profiling was done in these old retinal samples and compared with young ones; an increase in several genes was observed which include IFN – responsive TF sub unit (ISGF3G), chemokine ligand 2 (Ccl2) and Wnt inhibitory proteins DKK1, FZD10 and sFRP; these proteins are seen to be more expressed in the periphery. The inhibition of the Wnt pathway correlates with the previous understanding of its role in the maintenance of undifferentiated retinal progenitor cells [95]. The peripheral retina shows higher expression of these inhibitory proteins which would mean that the number of retinal stem cells would be much lesser due to differentiation. Aging has also shown to lessen the expression of retina "protection" genes such as X-linked inhibitor of apoptosis (XAF1), protein tyrosine kinase (PTK2) and cadherin CDH8 [96]. Genes involved in apoptosis have also been seen to be upregulated in these cells. The disturbed Wnt pathway might also enhance the degeneration of the retina in these disease conditions. It has been observed that an inflammatory- like response has been created in DR condition [97] and understanding how the stem cells are altered under these circumstances would be interesting. Stem cells under inflammatory conditions show that they induce generation of regulatory T cells and suppress T helper cells [96]. The cross talk between stem cells and inflammation cytokines is crucial in avoiding huge immune responses.[b]

6 Stem cells of the trabecular meshwork

6.1 Trabecular meshwork stem cells

The trabecular meshwork (TM) is responsible for the intra ocular pressure (IOP) regulation, wherein the outflow from the aqueous humor is maintained. It is present between the cornea and the sclera and is built up of collagenous lamellae on a basal lamina [98]. It comprises uveal meshwork, corneoscleral meshwork and juxtacanalicular tissue. The lamellar cells play a key phagocytic function in the aqueous humor [99]. There is a fourth region in the TM where the stem cells are said to be inhabiting called the "insert zone". The insert zone is the stem cell niche in the TM region which is almost near the corneal endothelium [100]. Various studies have shown that these cells express stem cell markers such as Sox2, Pax6 along with ABCG2, Notch1, AnkG, Nestin, Oct3/4 and LIF, which showed the presence of undifferentiated progenitor cells in high numbers [101]. Tay et al. identified TM-mesenchymal stem cells which are TM stem cells that expressed mesenchymal stem cell markers CD73, CD90, CD105 and vimentin [102]. Stem cells of the TM have also shown to be multipotent, giving rise to different types of cells expressing neural, keratocytes and adipocyte markers. Other studies have also shown that the TM stem cells could develop into mesenchymal and photoreceptor lineages, based on the microenvironmental cues provided [103]. iPS models have proved that fibroblast iPS can be used to generate TM cells, with the phagocytic function, expressing TM markers MGP, AQP1and CHI3L1 [103]. TMSCs developed into TM cells, with serum treatment, implying that this is their natural direction. Resident stem cells in the TM area exhibit the typical slow cycling and label retaining property of stem cells, which throws light upon the stem cell understanding in the TM region [104]. It has also been reported that TMSCs are multipotent; they differentiate into phagocytic TM cells and are primed in this direction by culturing them in the presence of aqueous humor or a high - serum condition [32]. TMSC have been shown to grow better in 3D spheres than as 2D cultures, and this could be due to the inherent influence of extracellular matrix on the stem cell behavior [32].

6.2 ER stress in aging TM stem cells

Primary open – angle glaucoma is the most common age related disease condition that occurs due to changes in the trabecular meshwork cells. The elevated IOP in this condition is caused due to blockage of the conventional outflow pathway, where the drainage of fluid happens though the TM cells [105]. It is established that the aging TM region has a decrease in its cellularity (maybe through increase in apoptosis) as well as an abnormal change in the ECM. The endoplasmic reticulum (ER) stress pathway is shown to be a key contributor to this type of glaucoma. Under ER stress, an increase in the expression of CHOP, GRP78, sXBP1 and GADD34, which are UPR (unfolded protein response) markers, was observed [106]. In hematopoietic SC, it has been reported that the UPR selectively induces apoptosis to the SC to stop the proliferation of diseased stem cells. The UPR and the PERK pathway have a similar effect on muscle stem cells and intestinal epithelial stem cells [107]. The stemness of a SC is reduced during diseased condition; this along with the reduction in cellularity, both could contribute to the dampened regeneration capacity of the progenitor cells.

6.3 Regeneration in the aging TM

TMSCs have been shown to differentiate to TM cells in post glaucoma cases, where they homed to the region of injury and actively divided. Other MSCs have also been shown to differentiate into TM cells. TM specific markers are expressed in both the cases which include AQP1 and CHI3L1 [99]. The involvement of chemokine CXCR4 in the homing of TMSCs

---

b. Figure 2.

to the site of wound has been studied, but the exact mechanism of the process yet to be understood. During the process of physiological aging, there is a decrease in the outflow in the uveoscleral canal which is compensated by the TM cells pathway [108]. It has very well been documented that as the age increases, the number of TM cells decreases and this is also seen in Glaucoma [109]. The cause suggested for this is senescence and increase induction of apoptosis, ECM abnormalities and blockage of the IOP regulation [110]. TM cells are phagocytic in function and clear ECM debris; this function is impaired in older patients.

7 Conclusion

Aging stem cells show a decline in function and this has been shown by various reports on different types of cells. Intrinsic changes that cause the microenvironment to change have a major effect on this phenomenon. Age-associated diseases are increasing, and the eye especially has many such conditions. The exploitation of stem cells to treat such conditions is not as effective as in younger patients, and therefore, there is a need to understand more in detail, the process of regeneration in aging stem cells. Induction of regenerative capacity in senescent stem cells holds a great deal of potential and should be explored.

References

[1] M. Boulton, J. Albon, Stem cells in the eye, Int. J. Biochem. Cell Biol. 36 (4) (2004) 643–657.
[2] C.M. Woolthuis, G. de Haan, G. Huls, Aging of hematopoietic stem cells: intrinsic changes or micro-environmental effects? Curr. Opin. Immunol. 23 (4) (2011) 512–517.
[3] J.W. Shay, W.E. Wright, Hayflick, his limit, and cellular ageing, Nat. Rev. Mol. Cell Biol. 1 (1) (2000) 72–76.
[4] A.S.I. Ahmed, M.H. Sheng, S. Wasnik, D.J. Baylink, K.-H.W. Lau, Effect of aging on stem cells, World J. Exp. Med. 7 (1) (2017) 1–10, https://doi.org/10.5493/wjem.v7.i1.1.
[5] A. Banito, et al., Senescence impairs successful reprogramming to pluripotent stem cells, Genes Dev. 23 (18) (2009) 2134–2139.
[6] R.C. Taylor, A. Dillin, Aging as an event of proteostasis collapse, Cold Spring Harb. Perspect. Biol. 3 (5) (2011), https://doi.org/10.1101/cshperspect.a004440.
[7] F. Chen, Y. Liu, N.-K. Wong, J. Xiao, K.-F. So, Oxidative stress in stem cell aging, Cell Transplant. 26 (9) (2017) 1483–1495, https://doi.org/10.1177/0963689717735407.
[8] J. Oh, Y.D. Lee, A.J. Wagers, Stem cell aging: mechanisms, regulators and therapeutic opportunities, Nat. Med. 20 (8) (2014) 870–880, https://doi.org/10.1038/nm.3651.
[9] A.D. Ho, W. Wagner, U. Mahlknecht, Stem cells and ageing. The potential of stem cells to overcome age-related deteriorations of the body in regenerative medicine, EMBO Rep. 6 Spec No (2005) S35–S38, https://doi.org/10.1038/sj.embor.7400436.
[10] H. Yousef, A. Morgenthaler, C. Schlesinger, L. Bugaj, I.M. Conboy, D.V. Schaffer, Age-associated increase in BMP signaling inhibits hippocampal neurogenesis, Stem Cells 33 (5) (2015) 1577–1588, https://doi.org/10.1002/stem.1943.
[11] G. Saretzki, T. Walter, S. Atkinson, J.F. Passos, B. Bareth, W.N. Keith, R. Stewart, S. Hoare, M. Stojkovic, L. Armstrong, T. von Zglinicki, M. Lako, Downregulation of multiple stress defense mechanisms during differentiation of human embryonic stem cells, Stem Cells 26 (2) (2008) 455–464, https://doi.org/10.1634/stemcells.2007-0628.
[12] Charlie Mantel, Hal E Broxmeyer, Sirtuin 1, stem cells, aging, and stem cell aging., Curr Opin Hematol 15 (4) (2008) 326–331, https://doi.org/10.1097/MOH.0b013e3283043819.
[13] Stefan Schreier, Wannapong Triampo, The Blood Circulating Rare Cell Population. What is it and What is it Good For? Cells 9 (4) (2020), https://doi.org/10.3390/cells9040790.
[14] Dawiyat Massoudi, Francois Malecaze, Stephane D Galiacy, Collagens and proteoglycans of the cornea: importance in transparency and visual disorders., Cell Tissue Res 363 (2) (2016) 337–349, https://doi.org/10.1007/s00441-015-2233-5.
[15] Z. He, et al., Revisited microanatomy of the corneal endothelial periphery: new evidence for continuous centripetal migration of endothelial cells in humans, Stem Cells 30 (11) (2012) 2523–2534.
[16] Takayuki Nagasaki, Jin Zhao, Centripetal movement of corneal epithelial cells in the normal adult mouse., Invest Ophthalmol Vis Sci 44 (2) (2003) 558–566, https://doi.org/10.1167/iovs.02-0705.
[17] K.M. Meek, C. Knupp, Corneal structure and transparency, Prog. Retin. Eye Res. 49 (2015) 1–16.
[18] M. Hovakimyan, et al., Morphological analysis of quiescent and activated keratocytes: a review of ex vivo and in vivo findings, Curr. Eye Res. 39 (12) (2014) 1129–1144.
[19] W M Townsend, The limbal palisades of Vogt., Trans Am Ophthalmol Soc 89 (1991) 721–756.
[20] M Davanger, A Evensen, Role of the pericorneal papillary structure in renewal of corneal epithelium., Nature 229 (5286) (1971) 560–561, https://doi.org/10.1038/229560a0.
[21] Marc A Dziasko, Julie T Daniels, Anatomical Features and Cell-Cell Interactions in the Human Limbal Epithelial Stem Cell Niche., Ocul Surf 14 (3) (2016) 322–330, https://doi.org/10.1016/j.jtos.2016.04.002.
[22] E.C. Figueira, et al., The phenotype of limbal epithelial stem cells, Invest. Ophthalmol. Vis. Sci. 48 (1) (2007) 144–156.

[23] Kevin Y H Chee, Anthony Kicic, Steven J Wiffen, Limbal stem cells: the search for a marker., Clin Exp Ophthalmol 34 (1) (n.d.) 64–73, doi:10.1111/j.1442-9071.2006.01147.x 16451261.

[24] H.S. Dua, et al., Limbal epithelial crypts: a novel anatomical structure and a putative limbal stem cell niche, Br. J. Ophthalmol. 89 (5) (2005) 529–532.

[25] Linheng Li, Ting Xie, Stem cell niche: structure and function., Annu Rev Cell Dev Biol 21 (2005) 605–631, https://doi.org/10.1146/annurev.cellbio.21.012704.131525.

[26] Bo Han, Szu-Yu Chen, Ying-Ting Zhu, Scheffer C G Tseng, Integration of BMP/Wnt signaling to control clonal growth of limbal epithelial progenitor cells by niche cells., Stem Cell Res 12 (2) (2014) 562–573, https://doi.org/10.1016/j.scr.2014.01.003.

[27] Szu-Yu Chen, Bo Han, Ying-Ting Zhu, Megha Mahabole, Jie Huang, David C Beebe, Scheffer C G Tseng, HC-HA/PTX3 Purified From Amniotic Membrane Promotes BMP Signaling in Limbal Niche Cells to Maintain Quiescence of Limbal Epithelial Progenitor/Stem Cells., Stem Cells 33 (11) (2015) 3341–3355, https://doi.org/10.1002/stem.2091.

[28] U Schlötzer-Schrehardt, T Dietrich, K Saito, L Sorokin, T Sasaki, M Paulsson, F E Kruse, Characterization of extracellular matrix components in the limbal epithelial stem cell compartment., Exp Eye Res 85 (6) (2007) 845–860, https://doi.org/10.1016/j.exer.2007.08.020.

[29] Tung-Tien Sun, Robert M Lavker, Corneal epithelial stem cells: past, present, and future., J Investig Dermatol Symp Proc 9 (3) (2004) 202–207, https://doi.org/10.1111/j.1087-0024.2004.09311.x.

[30] N. Pinnamaneni, J.L. Funderburgh, Concise review: stem cells in the corneal stroma, Stem Cells 30 (6) (2012) 1059–1063.

[31] Damien G Harkin, Leanne Foyn, Laura J Bray, Allison J Sutherland, Fiona J Li, Brendan G Cronin, Concise reviews: can mesenchymal stromal cells differentiate into corneal cells? A systematic review of published data., Stem Cells 33 (3) (2015) 785–791, https://doi.org/10.1002/stem.1895.

[32] Y. Du, et al., Multipotent stem cells from trabecular meshwork become phagocytic TM cells, Invest. Ophthalmol. Vis. Sci. 53 (3) (2012) 1566–1575.

[33] Yiqin Du, Martha L Funderburgh, Mary M Mann, Nirmala SundarRaj, James L Funderburgh, Multipotent stem cells in human corneal stroma., Stem Cells 23 (9) (2005) 1266–1275, https://doi.org/10.1634/stemcells.2004-0256.

[34] Pinnita Prabhasawat, Pattama Ekpo, Mongkol Uiprasertkul, Suksri Chotikavanich, Nattaporn Tesavibul, Efficacy of cultivated corneal epithelial stem cells for ocular surface reconstruction., Clin Ophthalmol 6 (2012) 1483–1492, https://doi.org/10.2147/OPTH.S33951.

[35] Nobuyuki Ebihara, Akira Matsuda, Shinji Nakamura, Hironori Matsuda, Akira Murakami, Role of the IL-6 classic- and trans-signaling pathways in corneal sterile inflammation and wound healing., Invest Ophthalmol Vis Sci 52 (12) (2011) 8549–8557, https://doi.org/10.1167/iovs.11-7956.

[36] Matthew James Branch, Khurram Hashmani, Permesh Dhillon, D Rhodri E Jones, Harminder Singh Dua, Andrew Hopkinson, Mesenchymal stem cells in the human corneal limbal stroma., Invest Ophthalmol Vis Sci 53 (9) (2012) 5109–5116, https://doi.org/10.1167/iovs.11-8673.

[37] Jian Wu, Yiqin Du, Mary M Mann, James L Funderburgh, William R Wagner, Corneal stromal stem cells versus corneal fibroblasts in generating structurally appropriate corneal stromal tissue., Exp Eye Res 120 (2014) 71–81, https://doi.org/10.1016/j.exer.2014.01.005.

[38] Jian Wu, Yiqin Du, Simon C Watkins, James L Funderburgh, William R Wagner, The engineering of organized human corneal tissue through the spatial guidance of corneal stromal stem cells., Biomaterials 33 (5) (2012) 1343–1352, https://doi.org/10.1016/j.biomaterials.2011.10.055.

[39] M. Notara, et al., The impact of age on the physical and cellular properties of the human limbal stem cell niche, Age (Dordr.) 35 (2) (2013) 289–300.

[40] Nada Sagga, Lucia Kuffová, Neil Vargesson, Lynda Erskine, J Martin Collinson, Limbal epithelial stem cell activity and corneal epithelial cell cycle parameters in adult and aging mice., Stem Cell Res 33 (2018) 185–198, https://doi.org/10.1016/j.scr.2018.11.001.

[41] Bina B Kulkarni, Patrick J Tighe, Imran Mohammed, Aaron M Yeung, Desmond G Powe, Andrew Hopkinson, Vijay A Shanmuganathan, Harminder S Dua, Comparative transcriptional profiling of the limbal epithelial crypt demonstrates its putative stem cell niche characteristics., BMC Genomics 11 (2010) 526, https://doi.org/10.1186/1471-2164-11-526.

[42] Sheyla González, Denise Oh, Elfren R Baclagon, Jie J Zheng, Sophie X Deng, Wnt Signaling Is Required for the Maintenance of Human Limbal Stem/Progenitor Cells In Vitro., Invest Ophthalmol Vis Sci 60 (1) (2019) 107–112, https://doi.org/10.1167/iovs.18-25740.

[43] R G Faragher, B Mulholland, S J Tuft, S Sandeman, P T Khaw, Aging and the cornea., Br J Ophthalmol 81 (10) (1997) 814–817, https://doi.org/10.1136/bjo.81.10.814.

[44] S.L. Piper, et al., Inducible immortality in hTERT-human mesenchymal stem cells, J. Orthop. Res. 30 (12) (2012) 1879–1885.

[45] R.G. Faragher, et al., Aging and the cornea, Br. J. Ophthalmol. 81 (10) (1997) 814–817.

[46] Nancy C Joyce, Proliferative capacity of the corneal endothelium., Prog Retin Eye Res 22 (3) (2003) 359–389, https://doi.org/10.1016/s1350-9462(02)00065-4.

[47] E.M. Espana, M. Sun, D.E. Birk, Existence of corneal endothelial slow-cycling cells, Invest. Ophthalmol. Vis. Sci. 56 (6) (2015) 3827–3837.

[48] Barbara M Braunger, Bahar Ademoglu, Sebastian E Koschade, Rudolf Fuchshofer, B'Ann T Gabelt, Julie A Kiland, Elizabeth A Hennes-Beann, Kevin G Brunner, Paul L Kaufman, Ernst R Tamm, Identification of adult stem cells in Schwalbe's line region of the primate eye., Invest Ophthalmol Vis Sci 55 (11) (2014) 7499–7507, https://doi.org/10.1167/iovs.14-14872.

[49] L P K Ang, D T H Tan, Ocular surface stem cells and disease: current concepts and clinical applications., Ann Acad Med Singap 33 (5) (2004) 576–580.

[50] Z G Wei, R L Wu, R M Lavker, T T Sun, In vitro growth and differentiation of rabbit bulbar, fornix, and palpebral conjunctival epithelia. Implications on conjunctival epithelial transdifferentiation and stem cells., Invest Ophthalmol Vis Sci 34 (5) (1993) 1814–1828.

[51] Rebecca C Taylor, Andrew Dillin, Aging as an event of proteostasis collapse., Cold Spring Harb Perspect Biol 3 (5) (2011), https://doi.org/10.1101/cshperspect.a004440.

[52] Rosalind M K Stewart, Carl M Sheridan, Paul S Hiscott, Gabriela Czanner, Stephen B Kaye, Human Conjunctival Stem Cells are Predominantly Located in the Medial Canthal and Inferior Forniceal Areas., Invest Ophthalmol Vis Sci 56 (3) (2015) 2021–2030, https://doi.org/10.1167/iovs.14-16266.

[53] W. Chen, et al., Wistar rat palpebral conjunctiva contains more slow-cycling stem cells that have larger proliferative capacity: implication for conjunctival epithelial homeostasis, Jpn. J. Ophthalmol. 47 (2) (2003) 119–128.
[54] Tiago Ramos, Deborah Scott, Sajjad Ahmad, An Update on Ocular Surface Epithelial Stem Cells: Cornea and Conjunctiva., Stem Cells Int 2015 (2015) 601731, https://doi.org/10.1155/2015/601731.
[55] Ilene K Gipson, Goblet cells of the conjunctiva: A review of recent findings., Prog Retin Eye Res 54 (2016) 49–63, https://doi.org/10.1016/j.preteyeres.2016.04.005.
[56] F. Majo, et al., Oligopotent stem cells are distributed throughout the mammalian ocular surface, Nature 456 (7219) (2008) 250–254.
[57] B J Cho, A R Djalilian, W F Obritsch, D M Matteson, C C Chan, E J Holland, Conjunctival epithelial cells cultured on human amniotic membrane fail to transdifferentiate into corneal epithelial-type cells., Cornea 18 (2) (1999) 216–224, https://doi.org/10.1097/00003226-199903000-00013.
[58] Ahdeah Pajoohesh-Ganji, Mary Ann Stepp, In search of markers for the stem cells of the corneal epithelium., Biol Cell 97 (4) (2005) 265–276, https://doi.org/10.1042/BC20040114.
[59] A.J. Huang, S.C. Tseng, K.R. Kenyon, Morphogenesis of rat conjunctival goblet cells, Invest. Ophthalmol. Vis. Sci. 29 (6) (1988) 969–975.
[60] Arianne J H van Velthoven, Marina Bertolin, Vanessa Barbaro, Mireille M J P E Sthijns, Rudy M M A Nuijts, Vanessa L S LaPointe, Mor M Dickman, Stefano Ferrari, Increased Cell Survival of Human Primary Conjunctival Stem Cells in Dimethyl Sulfoxide-Based Cryopreservation Media., Biopreserv Biobank (2020), https://doi.org/10.1089/bio.2020.0091.
[61] Takefumi Yamaguchi, Inflammatory Response in Dry Eye., Invest Ophthalmol Vis Sci 59 (14) (2018) DES192–DES199, https://doi.org/10.1167/iovs.17-23651.
[62] U P Andley, J S Rhim, L T Chylack Jr., T P Fleming, Propagation and immortalization of human lens epithelial cells in culture., Invest Ophthalmol Vis Sci 35 (7) (1994) 3094–3102.
[63] P.A. Tsonis, K. Del Rio-Tsonis, Lens and retina regeneration: transdifferentiation, stem cells and clinical applications, Exp. Eye Res. 78 (2) (2004) 161–172.
[64] J Fielding Hejtmancik, Alan Shiels, Overview of the Lens., Prog Mol Biol Transl Sci 134 (2015) 119–127, https://doi.org/10.1016/bs.pmbts.2015.04.006.
[65] Jonathan J Henry, Alvin G Thomas, Paul W Hamilton, Lisa Moore, Kimberly J Perry, Cell signaling pathways in vertebrate lens regeneration., Curr Top Microbiol Immunol 367 (2013) 75–98, https://doi.org/10.1007/82_2012_289.
[66] H. Lin, et al., Lens regeneration using endogenous stem cells with gain of visual function, Nature 531 (7594) (2016) 323–328.
[67] Mariko Hirano, Akitsugu Yamamoto, Naoko Yoshimura, Tomoyuki Tokunaga, Tsutomu Motohashi, Katsuhiko Ishizaki, Hisahiro Yoshida, Kenji Okazaki, Hidetoshi Yamazaki, Shin-Ichi Hayashi, Takahiro Kunisada, Generation of structures formed by lens and retinal cells differentiating from embryonic stem cells., Dev Dyn 228 (4) (2003) 664–671, https://doi.org/10.1002/dvdy.10425.
[68] Panagiotis A Tsonis, Katia Del Rio-Tsonis, Lens and retina regeneration: transdifferentiation, stem cells and clinical applications., Exp Eye Res 78 (2) (2004) 161–172, https://doi.org/10.1016/j.exer.2003.10.022.
[69] F M Watt, B L Hogan, Out of Eden: stem cells and their niches., Science 287 (5457) (2000) 1427–1430, https://doi.org/10.1126/science.287.5457.1427.
[70] C. Yang, et al., Efficient generation of lens progenitor cells and lentoid bodies from human embryonic stem cells in chemically defined conditions, FASEB J. 24 (9) (2010) 3274–3283.
[71] Susann G Remington, Rita A Meyer, Lens stem cells may reside outside the lens capsule: an hypothesis., Theor Biol Med Model 4 (2007) 22, https://doi.org/10.1186/1742-4682-4-22.
[72] E Hanssen, S Franc, R Garrone, Synthesis and structural organization of zonular fibers during development and aging., Matrix Biol 20 (2) (2001) 77–85, https://doi.org/10.1016/s0945-053x(01)00122-6.
[73] X Chen, W Xiao, W Chen, L Luo, S Ye, Y Liu, The epigenetic modifier trichostatin A, a histone deacetylase inhibitor, suppresses proliferation and epithelial-mesenchymal transition of lens epithelial cells., Cell Death Dis 4 (2013) e884, https://doi.org/10.1038/cddis.2013.416.
[74] Gabriel R Cavalheiro, Gabriel E Matos-Rodrigues, Anielle L Gomes, Paulo M G Rodrigues, Rodrigo A P Martins, c-Myc regulates cell proliferation during lens development., PLoS One 9 (2) (2014) e87182, https://doi.org/10.1371/journal.pone.0087182.
[75] N Ibaraki, L R Lin, V N Reddy, Effects of growth factors on proliferation and differentiation in human lens epithelial cells in early subculture., Invest Ophthalmol Vis Sci 36 (11) (1995) 2304–2312.
[76] J Wei, H Tang, Z Q Xu, B Li, L Q Xie, G X Xu, Expression and function of PDGF-α in columnar epithelial cells of age-related cataracts patients., Genet Mol Res 14 (4) (2015) 13320–13327, https://doi.org/10.4238/2015.October.26.28.
[77] Sonya C Faber, Michael L Robinson, Helen P Makarenkova, Richard A Lang, Bmp signaling is required for development of primary lens fiber cells., Development 129 (15) (2002) 3727–3737.
[78] Mingyuan Zhou, Joshua Leiberman, Jing Xu, Robert M Lavker, A hierarchy of proliferative cells exists in mouse lens epithelium: implications for lens maintenance., Invest Ophthalmol Vis Sci 47 (7) (2006) 2997–3003, https://doi.org/10.1167/iovs.06-0130.
[79] S Choudhary, W Zhang, F Zhou, G A Campbell, L L Chan, E B Thompson, N H Ansari, Cellular lipid peroxidation end-products induce apoptosis in human lens epithelial cells., Free Radic Biol Med 32 (4) (2002) 360–369, https://doi.org/10.1016/s0891-5849(01)00810-3.
[80] Xiao-Qin Huang, Juan Wang, Jin-Ping Liu, Hao Feng, Wen-Bin Liu, Qin Yan, Yan Liu, Shu-Ming Sun, Mi Deng, Lili Gong, Yun Liu, David Wan-Cheng Li, hTERT extends proliferative lifespan and prevents oxidative stress-induced apoptosis in human lens epithelial cells., Invest Ophthalmol Vis Sci 46 (7) (2005) 2503–2513, https://doi.org/10.1167/iovs.05-0154.
[81] A W Stitt, Advanced glycation: an important pathological event in diabetic and age related ocular disease., Br J Ophthalmol 85 (6) (2001) 746–753, https://doi.org/10.1136/bjo.85.6.746.
[82] F J Lovicu, E H Shin, J W McAvoy, Fibrosis in the lens. Sprouty regulation of TGFβ-signaling prevents lens EMT leading to cataract., Exp Eye Res 142 (2016) 92–101, https://doi.org/10.1016/j.exer.2015.02.004.

[83] Junghyun Kim, Ohn Soon Kim, Chan-Sik Kim, Eunjin Sohn, Kyuhyung Jo, Jin Sook Kim, Accumulation of argpyrimidine, a methylglyoxal-derived advanced glycation end product, increases apoptosis of lens epithelial cells both in vitro and in vivo., Exp Mol Med 44 (2) (2012) 167–175, https://doi.org/10.3858/emm.2012.44.2.012.

[84] J.J. Henry, et al., Cell signaling pathways in vertebrate lens regeneration, Curr. Top. Microbiol. Immunol. 367 (2013) 75–98.

[85] Chi-Hsien Peng, Jorn-Hon Liu, Lin-Chung Woung, Tzu-Jung Lin, Shih-Hwa Chiou, Po-Chen Tseng, Wen-Yuan Du, Cheng-Kuo Cheng, Chao-Chien Hu, Ke-Hung Chien, Shih-Jen Chen, MicroRNAs and cataracts: correlation among let-7 expression, age and the severity of lens opacity., Br J Ophthalmol 96 (5) (2012) 747–751, https://doi.org/10.1136/bjophthalmol-2011-300585.

[86] Chunbo Yang, Ying Yang, Lisa Brennan, Eric E Bouhassira, Marc Kantorow, Ales Cvekl, Efficient generation of lens progenitor cells and lentoid bodies from human embryonic stem cells in chemically defined conditions., FASEB J 24 (9) (2010) 3274–3283, https://doi.org/10.1096/fj.10-157255.

[87] A.E. Griep, Cell cycle regulation in the developing lens, Semin. Cell Dev. Biol. 17 (6) (2006) 686–697.

[88] Alan Shiels, J Fielding Hejtmancik, Mutations and mechanisms in congenital and age-related cataracts., Exp Eye Res 156 (2017) 95–102, https://doi.org/10.1016/j.exer.2016.06.011.

[89] V. Tropepe, et al., Retinal stem cells in the adult mammalian eye, Science 287 (5460) (2000) 2032–2036.

[90] A. Moshiri, J. Close, T.A. Reh, Retinal stem cells and regeneration, Int. J. Dev. Biol. 48 (8–9) (2004) 1003–1014.

[91] Chi-Hsien Peng, Yuh-Lih Chang, Chung-Lan Kao, Ling-Ming Tseng, Chih-Chia Wu, Yu-Chih Chen, Ching-Yao Tsai, Lin-Chung Woung, Jorn-Hon Liu, Shih-Hwa Chiou, Shih-Jen Chen, SirT1-a sensor for monitoring self-renewal and aging process in retinal stem cells., Sensors (Basel) 10 (6) (2010) 6172–6194, https://doi.org/10.3390/s100606172.

[92] G Astrid Limb, Julie T Daniels, Ocular regeneration by stem cells: present status and future prospects., Br Med Bull 85 (2008) 47–61, https://doi.org/10.1093/bmb/ldn008.

[93] L.P. Ang, D.T. Tan, Ocular surface stem cells and disease: current concepts and clinical applications, Ann. Acad. Med. Singapore 33 (5) (2004) 576–580.

[94] H. Gao, J.G. Hollyfield, Aging of the human retina. Differential loss of neurons and retinal pigment epithelial cells, Invest. Ophthalmol. Vis. Sci. 33 (1) (1992) 1–17.

[95] Leah P Foltz, Dennis O Clegg, Rapid, Directed Differentiation of Retinal Pigment Epithelial Cells from Human Embryonic or Induced Pluripotent Stem Cells., J Vis Exp (128) (2017), https://doi.org/10.3791/56274.

[96] Hassan Akrami, Zahra-Soheila Soheili, Keynoush Khalooghi, Hamid Ahmadieh, Mojgan Rezaie-Kanavi, Shahram Samiei, Malihe Davari, Shima Ghaderi, Fatemeh Sanie-Jahromi, Retinal pigment epithelium culture;a potential source of retinal stem cells., J Ophthalmic Vis Res 4 (3) (2009) 134–141.

[97] Agnese Fiori, Vincenzo Terlizzi, Heiner Kremer, Julian Gebauer, Hans-Peter Hammes, Martin C Harmsen, Karen Bieback, Mesenchymal stromal/stem cells as potential therapy in diabetic retinopathy., Immunobiology 223 (12) (2018) 729–743, https://doi.org/10.1016/j.imbio.2018.01.001.

[98] D.W Abu-Hassan, T.S. Acott, M.J. Kelley, The trabecular meshwork: a basic review of form and function, J. Ocul. Biol. 2 (1) (2014), https://doi.org/10.13188/2334-2838.1000017.

[99] A. Castro, Y. Du, Trabecular meshwork regeneration—a potential treatment for glaucoma, Curr. Ophthalmol. Rep. 7 (2) (2019) 80–88.

[100] H. Yun, Y. Zhou, A. Wills, Y. Du, Stem cells in the trabecular meshwork for regulating intraocular pressure, J. Ocul. Pharmacol. Ther. 32 (5) (2016) 253–260, https://doi.org/10.1089/jop.2016.0005.

[101] Y. Du, D.S. Roh, M.M. Mann, M.L. Funderburgh, J.L. Funderburgh, J.S. Schuman, Multipotent stem cells from trabecular meshwork become phagocytic TM cells, Invest. Ophthalmol. Vis. Sci. 53 (3) (2012) 1566–1575, https://doi.org/10.1167/iovs.11-9134.

[102] C.Y. Tay, P. Sathiyanathan, S.W.L. Chu, L.W. Stanton, T.T. Wong, Identification and characterization of mesenchymal stem cells derived from the trabecular meshwork of the human eye, Stem Cells Dev. 21 (9) (2012) 1381–1390, https://doi.org/10.1089/scd.2011.0655.

[103] E.J. Snider, R.T. Vannatta, L. Schildmeyer, W.D. Stamer, C.R. Ethier, Characterizing differences between MSCs and TM cells: toward autologous stem cell therapies for the glaucomatous trabecular meshwork, J. Tissue Eng. Regen. Med. 12 (3) (2018) 695–704, https://doi.org/10.1002/term.2488.

[104] M.J. Kelley, et al., Stem cells in the trabecular meshwork: present and future promises, Exp. Eye Res. 88 (4) (2009) 747–751.

[105] J. Zhao, S. Wang, W. Zhong, B. Yang, L. Sun, Y. Zheng, Oxidative stress in the trabecular meshwork (Review), Int. J. Mol. Med. 38 (4) (2016) 995–1002, https://doi.org/10.3892/ijmm.2016.2714.

[106] Y. Wang, D. Osakue, E. Yang, Y. Zhou, H. Gong, X. Xia, Y. Du, Endoplasmic reticulum stress response of trabecular meshwork stem cells and trabecular meshwork cells and protective effects of activated PERK pathway, Invest. Ophthalmol. Vis. Sci. 60 (1) (2019) 265–273, https://doi.org/10.1167/iovs.18-25477.

[107] P.P. Pattabiraman, P.V. Rao, Mechanistic basis of Rho GTPase-induced extracellular matrix synthesis in trabecular meshwork cells, Am. J. Physiol. Cell Physiol. 298 (3) (2010) C749–C763, https://doi.org/10.1152/ajpcell.00317.2009.

[108] J. Alvarado, et al., Age-related changes in trabecular meshwork cellularity, Invest. Ophthalmol. Vis. Sci. 21 (5) (1981) 714–727.

[109] A. Pulliero, et al., Oxidative damage and autophagy in the human trabecular meshwork as related with ageing, PLoS One 9 (6) (2014) e98106.

[110] H. Yun, et al., Human stem cells home to and repair laser-damaged trabecular meshwork in a mouse model, Commun. Biol. 1 (2018) 216.

Chapter 9

Skeletal muscle cell aging and stem cells

Shabana Thabassum Mohammed Rafi[a], Yuvaraj Sambandam[b], Sivanandane Sittadjody[c], Surajit Pathak[d], Ilangovan Ramachandran[a], and R. Ileng Kumaran[e]

[a]Department of Endocrinology, Dr. ALM PG Institute of Basic Medical Sciences, University of Madras, Taramani Campus, Chennai, Tamil Nadu, India. [b]Department of Surgery, Comprehensive Transplant Center, Northwestern University, Feinberg School of Medicine, Chicago, IL, United States, [c]Wake Forest Institute for Regenerative Medicine, Wake Forest University School of Medicine, Winston-Salem, NC, United States, [d]Faculty of Allied Health Sciences, Chettinad Hospital and Research Institute, Chettinad Academy of Research and Education, Chennai, Tamil Nadu, India, [e]Biology Department, Farmingdale State College, Farmingdale, NY, United States

1 Introduction

Aging is linked to a decline in the ability for cellular regeneration, and an age-related reduction in the number and function of adult stem cells. During age-related injury to skeletal muscles, their resident muscle stem cells undergo myogenic differentiation program to form myofibers to repair the damaged tissues, and also undergo proliferation, i.e., self-renewal to replenish the lost stem cells of the stem cell pool. However, the regeneration of the muscle cells becomes ineffective with aging, and the muscle tissue is replaced by fatty and fibrous tissue; and this type of muscular atrophy is known as sarcopenia. Therefore, sarcopenia can also decrease the muscle stem cell function, which in turn can additionally contribute to lowering of muscle regeneration capacity during aging [1]. The postmitotic skeletal muscle cells undergo lifelong maintenance, as they have a high regenerative potential that is mediated by the muscle stem cells, also called as satellite cells [2, 3]. The skeletal muscle satellite cells are located in a unique niche between the sarcolemma and basal lamina [4, 5]. Skeletal muscles are essential for locomotion; the tendons serve as the anchor between bone and muscle, giving life to the musculoskletal system [6]. When the muscle cell or myofiber suffers an injury, or exposed to oxidative stress, or undergoes an epigenetic alteration, the satellite cells that are residing in the muscle in quiescent state at G0 phase of cell cycle, gets activated and leaves the quiescent state to undergo myogenesis. to repair and regenerate the damaged muscle and also replenish the stem cell pool of satellite cells, resulting in the homeostasis of the adult muscle tissue [2, 7] (Fig. 1). The activated satellite cell enters into the G1 phase of cell cycle, leading to its division into a daughter satellite cell and a myoblast. While the daughter satellite cell replenishes the muscle stem cell pool, the myoblast undergoes differentiation via its myogenic-lineage program to generate myocytes, myotubes, and myofibers, which ultimately repairs and regenerates the damaged muscle [3]. In the quiescent satellite cells, though there are many transcription start sites for the myogenic lineage-specific genes, only the myogenic transcription factor paired box 3 (Pax3), which has the bivalent domains is expressed in premyoblasts [8, 9]. During the stage of de novo adult myogenesis, the satellite cells undergo asymmetric cell division in context to their cell-fate determination and the segregation of their DNA strands, for simultaneous self-renewal of the satellite cells to maintain the muscle stem cell pool and generation of the myogenic progeny for their differentiation into the muscle cells [1]. The protein markers, which are expressed in satellite cells are Pax3, Pax7, BARX homeobox 2 (BARX2), cluster of differentiation 34 (CD34), C-X-C motif chemokine receptor 4 (CXCR4), vascular cell adhesion molecule 1 (VCAM1), myogenic factor 5 (Myf5), etc. [2]. Accumulating evidence indicate that different molecular mechanisms, including epigenetic regulation, play an integral role in the transition through different stages of the myogenic program [5, 10]. This chapter discusses about skeletal muscle and its satellite cells or adult stem cells, and the altered regulation of gene expression, metabolism, extrinsic or intrinsic factors and epigenetics in satellite cells and adult myogenesis during skeletal muscle aging.

2 Skeletal muscle stem cells: Satellite cells

Skeletal muscles account for nearly 40% of the total body mass in humans, and are essential for locomotion, thermoregulation, respiration, and digestion [6, 11, 12]. This muscle tissue responds to physiological stimuli and has the capacity to regenerate after overt damage [13]. The muscle regenerative capacity and maintenance ability are mainly due to the population or pool of tissue-specific muscle stem cells, the satellite cells [5, 14]. In 1961, Alexander Mauro was the first to identify the satellite cells in frog skeletal muscles using an electron microscope. The satellite cells were observed to be wedged

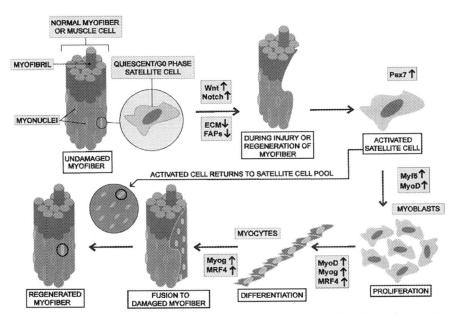

FIG. 1 **Role of satellite cells during injury to undamaged or normal muscle tissue.** Satellite cells in undamaged muscle tissue are in a quiescent state i.e., G0 phase, and this is maintained by the Notch and Wnt/β-catenin signaling pathways. There is no significant fibrosis in normal or young muscle tissue, and the extracellular matrix (ECM) level and fibro-adipogenic progenitors (FAPs) activity are downregulated. During muscle injury or regeneration, the satellite cells get activated from their quiescent state, and undergo differentiation to generate new myofibers. The paired box 7 (Pax7) transcription factor is highly expressed in satellite cells and participates in the initial stage of muscle development. The myoblast determination protein 1 (MyoD) [also known as myogenic differentiation 1 (MyoD1) or myogenic factor 3 (Myf3)] gene expression is upregulated, and the satellite cells become MyoD$^+$ during the proliferation stage, which enables their transition from proliferation to differentiation stage. MyoD also upregulates the expression of myogenic regulatory factor 4 (MRF4), which helps in the differentiation of myoblasts to myofibers. MRF4 along with MyoD upregulates the myogenin (Myog) expression, and upregulated Myog causes fusion and differentiation of myoblasts to myotubes. Some activated satellite cells return back to the satellite pool for self-renewal before proliferation, and this is enhanced by myogenic factor 5 (Myf5). The upward (↑) or downward (↓) arrow symbol indicates upregulation or downregulation, respectively.

between the sarcolemma and basement membrane of muscle fibers [14]. The microanatomical location of the satellite cells are in grooves, which are in intimate association with the muscle fibers. This precise location or niche of the satellite cells enable them to maintain the quiescent state, and also helps in the rapid activation in response to physiological or pathophysiological stimuli, for skeletal muscle regeneration or repair upon injury to muscle fibers in health or disease or aging [15].

2.1 Origin of satellite cells

During embryogenesis, all the skeletal muscles are derived from the dermomyotome, which is the precursor cell of the adult muscles [16]. They arise from a dorsal structure of the paraxial mesoderm called as somites. Epaxial and hypaxial are two subdomains of somites. The epaxial muscles develop from the dorsomedial dermomyotome to form the musculature of the back, and the hypaxial muscles develop from the ventrolateral dermomyotome and form the musculature of the abdomen, thorax, and limbs [17, 18]. During mouse development, between 16.5 and 18.5 embryonic days, a group of myogenic progenitors in the dermomyotome move to their forthcoming niche, which is positioned between the primitive basal lamina structure and the myotome [19]. However, during mouse postnatal development, the satellite cells begin to acquire their molecular characteristics in two steps. In the first step, muscle fibers are formed, and then, additional muscle fibers are added using the former fibers as the template [20]. During the second step, the muscle progenitors start to proliferate and maintain their nuclei percentage, which leads to the synthesis of myofibrillar proteins and the expression of specific surface markers [21]. After the muscle reaches its fully developed state in the young mouse, then the resident satellite cells switch to a quiescent state in muscle tissues [16].

2.2 Transcription factors-mediated gene regulation in satellite cells

The expression of cell-specific markers helps in the identification of satellite cells. The Pax7 protein, which belongs to the paired box (Pax) domain transcription factor family, is highly expressed in satellite cells and is considered to be responsible for the self-renewal of satellite cells [22, 23]. Pax3 is another transcription factor, and is involved in the early embryonic stages of muscle formation [24]. During embryogenesis, Pax3 and Pax7 are seen within the developing muscle in the central part of the dermomyotome [25]. Apart from paired domain transcription factors, the other markers that are expressed in satellite cells are myogenic factor 5 (Myf5), c-Met (a tyrosine protein kinase), myoblast determination protein 1 (MyoD) [also known as myogenic differentiation 1 (MyoD1) or myogenic factor 3 (Myf3)], cluster of differentiation 34 (CD34), neural cell adhesion molecule 1 (NCAM1), BARX homeobox 2 (BARX2), C-X-C motif chemokine receptor 4 (CXCR4), vascular cell adhesion molecule 1 (VCAM1), caveolin 1 (CAV1), calcitonin receptor (CALCR), epidermal growth factor receptor (EGFR), and integrin subunit alpha 7 (ITGA7) [2]. In addition, some of the key structural proteins of the nuclear envelope such as emerin and lamin A/C, which also play an important role in gene regulation, can be used as markers to identify the satellite cells [2,26–28]). Moreover, emerin and lamin A/C have been shown to be important for satellite cells or mouse embryonic stem cells (mESCs) to differentiate into muscle cells [29–31]. Myf5 and Barx2 are coexpressed along with Pax7, and are involved in the growth, maintenance, and regeneration of muscles. Pax3 is involved in the transcription of c-Met, for which the cell adhesion protein M-cadherin is coexpressed along with c-Met. Hence, these proteins are also used as markers to identify the satellite cells.

The highly conserved transcriptional factors, which are responsible for the activation, proliferation, and differentiation of muscle cells, are called myogenic regulatory factors (MRFs) (Fig. 1). MRFs are heterodimeric DNA binding transcription factors that can bind to the E box DNA motif of target genes and regulate the cell cycle process, mostly during terminal myogenic differentiation. Myf5, myogenic regulatory factor 4 [MRF4, also known as myogenic factor 6 (Myf6)], myogenin [Myog, also known as myogenic factor 4 (Myf4)], and MyoD are the four major MRFs that control myogenesis (Table 1) [32]. The activated satellite cells express Myf5 followed by MyoD expression. This results in myoblast proliferation, downregulation of Myf5, and upregulation of MRF4. The increase in the level of MRF4 leads to the upregulation of Myog expression. Then, Myog promotes the terminal differentiation and fusion of muscle fibers. When a group of myoblasts lose MyoD expression and exit the cell cycle, it contributes to repopulating the satellite cell pool for self-renewal [32–34].

TABLE 1 Role of myogenic regulatory factors.

Myogenic regulatory factors	Activation phase	Proliferation or differentiation phase	Functions
Myogenic factor 5 (Myf5)	Myf5 is the first gene, whose expression is upregulated during myogenesis in satellite cells.	Myf5 expression is upregulated in proliferation phase, but downregulated after adequate satellite cell proliferation.	Myf5 initiates and leads postnatal myogenesis. Myf5 also helps in the self-renewal of satellite cell pool.
Myoblast determination protein 1 (MyoD) [also known as myogenic differentiation 1 (MyoD1) or myogenic factor 3 (Myf3)]	MyoD gene expression is downregulated, but its expression is maintained in satellite cells.	MyoD expression is upregulated, and so the MyoD marker in cells change from MyoD$^-$ to MyoD$^+$.	MyoD is involved in the transition from proliferation to differentiation phase of myoblasts. MyoD upregulates MRF4, and the absence of MyoD leads to impairment in the differentiation of satellite cells.
Myogenic regulatory factor 4 (MRF4, also known as Myf6)	MRF4 is not expressed.	MRF4 expression is upregulated.	Upregulated MyoD and MRF4 regulate Myog expression. MRF4 coordinates myogenic differentiation at both transcription and epigenetic levels.
Myogenin (Myog, also known as Myf4)	Myog is not expressed.	Myog expression is upregulated.	Myog is involved in differentiation of myoblasts into myotubes. Myog is also involved in the terminal differentiation and fusion of muscle fibers.

The table lists the key myogenic regulatory factors, their regulation and functions in various phases of myogenesis.

2.3 Quiescent state of satellite cells

The process of activation of satellite cells from their quiescence or G0 phase, followed by their terminal differentiation into muscle cells is called myogenic determination or myogenic program [2]. The mesoderm-derived satellite cells are a heterogeneous population of adult stem cells of the skeletal muscle [2]. In the absence of muscle cell injury, the satellite cells remain dormant and rest in the G0 phase of the cell cycle, which is the quiescent state. However, when there is damage to muscle cells, the adjacent satellite cells in the G0 phase of the cell cycle get activated, and then proliferate and differentiate to form new myotubes and myofibers. Satellite cell activation is mediated in part by induced rapid expression of MyoD and Myf5. Also, a subgroup of proliferating satellite cells return to the quiescent state to replenish the depleted satellite cell pool [5].

3 Skeletal muscle stem cells in adult and aging

Satellite cells are skeletal muscle-resident adult stem cells, which are heterogeneous in nature and are actively involved in repair mechanisms during regeneration of damaged skeletal muscle. In normal tissue, the satellite cells reside in the quiescent state. However, when the muscle tissue exhibits any injury, these dormant satellite cells get activated by cell-intrinsic signals and their surrounding microenvironment, called the niche [1].

During aging, the function of satellite cells is impaired, and hence, there is a decline in the regenerative capacity of muscle tissues. Moreover, the support of the microenvironment is also lost when there is a reduction in satellite cell numbers and functions during aging [35]. Furthermore, when aged muscle cells are damaged, the differentiation of satellite cells is delayed, and the functions of both the cell-intrinsic and cell-extrinsic factors are diminished [36,37]. In addition, fibro-adipogenic progenitors (FAPs) continue to deposit the extracellular matrix (ECM) under the basal lamina and muscle niche, and higher levels of ECM make the myofibers become stiffer. Also, while factors such as transforming growth factor beta 1 (TGFβ1) and Delta-like canonical notch ligand 1 (DLL1) are downregulated, fibroblast growth factor 2 (FGF2) is upregulated, and the activity of these factors paves the way for the niche to become less supportive. Moreover, there is also a reduction in the levels of integrin β1, fibronectin, and focal adhesion kinase (FAK), which causes the niche to become less adhesive. The major signaling pathways that regulate the satellite cell function are p38 mitogen-activated protein kinase (p38 MAPK), Janus kinase (JAK)/signal transducer and activator of transcription (STAT), 16 kDa protein (p16), sprouty (Spry), NOTCH and Wingless/integrated (Wnt) molecules, and these pathways also show age-dependent changes. Therefore, age-dependent alterations in satellite cells and their microenvironment or niche have a major impact on muscle tissue regeneration.

3.1 Age-dependent changes in satellite cells

Aging affects the structure and function of all cells in the body, including the muscle cells. In 2015, the Sousa-Victor group [38] grafted the whole muscle tissue from geriatric mice into young mice, and the results indicated a defect in adult muscle tissue regeneration. In contrast, Lee et al. [39] observed a delay in regeneration rather than impairment in old muscle tissue. Carlson and Faulkner [40] transplanted the whole muscle grafts of a young and an old mouse to assess the regenerative ability of both. While the young host allowed the successful engraftment of both young and old muscles, the graft failed to regenerate in the old host. Aged muscle is more liable to fractal damage than the younger one [41]. From the above experiments, it is clear that aging is accompanied by an impairment in the response of satellite cells [42–44]. It is also impressive that even though there is a delayed regeneration rather than impairment in some cases, the tissues will have a serious health consequence after injury.

3.1.1 Function, behavior and heterogeneity of satellite cells

Chakkalakal and group [37] performed the label retention assay and showed that between 3 and 26 months of age, the satellite cell pool undergoes approximately 4–8 division. The satellite cells that undergo 4 division retain label [label retaining cells (LRCs)] while other cells which undergo 8 division lose label [nonlabel retaining cells (nLRCs). This suggests that the satellite cell pool undergoes turnover throughout the life. The transplantation assay reveals that LRCs function as bonafide stem cells and nLRCs as progenitors [37].

The satellite cells represent a heterogeneous population of muscle precursor cells and do not mark a unique type of cells although they express specific markers [45]. Muscle cells are present all over the body and satellite cells exhibit heterogeneity based on the location of the muscle and distinct origin [46,47]. Satellite cells can commit to muscle lineage differentiation to generate muscle cells or undergo self-renewal. The differences in the rate of proliferation of satellite cells can determine their role and function in muscle tissue. The fast dividing cells contribute to the myogenic lineage, whereas the slow dividing cells serve the long-term self-renewal role [48]..

During aging, LRCs are lost and nLRCs are retained (committed progenitors). Due to the loss of stem cell potential, there is a decline in the transplantation potential of aged satellite cells [37]. In support of this, Cosgrove et al. [49] observed that the number of functional satellite cells are diminished when aged muscle satellite cells are engrafted into an adult host. Taken together, there is a decline in the satellite cell function, behavior, and heterogeneity during aging.

4 Aging and metabolism of satellite cells

Metabolism includes both anabolism and catabolism, and it is very crucial for the regulation of cellular functions. To activate the quiescent satellite cells for myogenesis, active metabolism including mitochondrial activity [50] and autophagy are required [51]. Anabolism produces nucleotides, lipids, and proteins for the cells by the hydrolysis of adenosine triphosphate (ATP)/nicotinamide adenine dinucleotide phosphate (NADPH). Catabolism is a degradation process, which provides energy for the anabolism. Mitochondria generate their ATP molecules via oxidative phosphorylation (OXPHOS). OXPHOS is the key regulatory pathway for energy production during normal cellular respiration and metabolism, and it provides most of the energy that is required for cellular functions [10,52]. In contrast, there is a pathway of self-degradation of cellular components called as autophagy, and it is an evolutionarily conserved pathway or process. The autophagy pathway is carried out by lysosomes and autophagosomes, and they together control protein quality and prevent intracellular build up of toxic waste [53,54]. Thus, autophagy, which means self-eating, involves cellular and molecular mechanisms to clean up the damaged old cells to regenerate healthier new cells. So, autophagy has a major impact on satellite cell function in skeletal muscle regeneration during aging, and is discussed in more detail in the subsequent section.

4.1 Mitochondria in satellite cells

Mitochondria are considered as the powerhouse of the cells and they act as energy generators. During glycolysis, glucose is converted into pyruvate and two ATP molecules are generated per glucose molecule. Another effective way to generate ATP is by OXPHOS in mitochondria and it produces around 34 molecules of ATP per glucose molecule by the oxidation of pyruvate to acetyl-CoA (acetyl coenzyme A) in the tricarboxylic acid (TCA), also called as citric acid or Krebs cycle [55]. Dysfunction of this type of cellular metabolism is a common attribute in all the aged organisms as well as in muscle satellite cells. There is an age-dependent decline in the function of mitochondria [56]. Also, there is an increase in the activity of mitochondria in the quiescent satellite cells during aging, and it has an impact on the normal metabolic process and function of mitochondria. The satellite cells undergo a process called metabolic reprogramming to activate myogenesis. Therefore, the satellite cells switch from mitochondrial fatty acid oxidation to glycolysis. This shift also induces autophagy, which provides additional energy by increasing the catabolism [51]. The mitochondria also generate metabolites such as reactive oxygen species (ROS) and are involved in amino acid, nucleotide, and lipid metabolisms, which are important regulators of stem cell function [55]. Kyoto encyclopedia of genes and genomes (KEGG) analysis at the transcription level shows the downregulation of TCA and OXPHOS [57,58], consistent with diminished ATP and NAD^+; and therefore, the energy level is lower [58].

Mitochondria serve as a key producer of ROS. In normal young cells, a low level of ROS is produced; which is the chief regulator of the signal transduction pathway in different cellular functions inclusive of muscle [59,60]. Aging is related with the immoderate ROS level, which comes up with mitochondrial damage and mitochondrial-mediated apoptotic signaling [61]. The increased levels of ROS contribute to impairment of the satellite cell function and repairing during aging (Fig. 2). Beccafico and group [62] reported that there is a decline in the antioxidant levels and increase in ROS levels in the muscle stem cells with increasing age; and this process eventually diminishes the satellite cell function. Importantly, when the ROS level increases, it damages the quiescent satellite cells and affects the ability of the satellite cells during aging (Fig. 2).

The mitochondrial unfolded protein response (UPR^{mt}), which occurs in mitochondria is a conserved stress response pathway, and is important for longevity of organisms [63]. Restoration of NAD^+ level by nicotinamide riboside treatment reduced the DNA damage and its markers, and induced UPR^{mt}. Inducing or increasing the UPR^{mt} improved the quality and function of stem cells. Therefore, treating the aged satellite cells with nicotinamide riboside to increase their UPR^{mt} level, can potentially help them to regain their regenerative capacity [58]. Minet and Gaster [64] experimented by isolating the human muscle satellite cells from older humans in which the muscles have reduced mitochondria ATP production. However, when they were stimulated, there was normal production of ATP. This suggests that improving the mitochondrial function can increase the lifespan of the aged satellite cells.

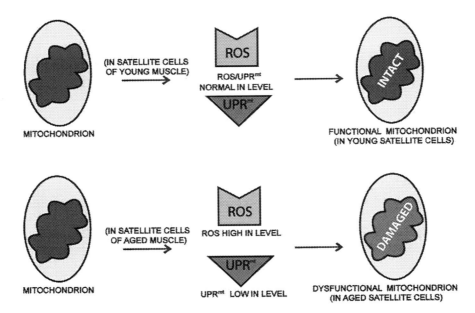

FIG. 2 **Metabolism in the mitochondrion of young versus aged satellite cells.** Mitochondria are key energy generators of cell and produce most of their ATP by oxidative phosphorylation. The dysfunctional mitochondrion in aged satellite cells lead to metabolic reprogramming of mitochondrial fatty acid oxidation to glycolysis to activate myogenesis. In this condition, mitochondrion produces high levels of reactive oxygen species (ROS) as metabolites. ROS serve important roles as a signal transducer and in the functions of muscle. Increased level of ROS damages the mitochondrion and impairs the function of satellite cells. The mitochondrial unfolded protein response (UPR^{mt}) pathway occurs in mitochondria, and is a conserved pathway involved in longevity of organisms. A reduction in UPR^{mt} pathway causes mitochondrial damage, lowers activity and quality of mitochondrion, and leads to a decrease in satellite cell number, function and regeneration. Dysfunction in mitochondrial metabolism is observed in aged satellite cells.

4.2 Autophagy in satellite cells

Autophagy is a physiological pathway or process that helps in the clearance of the damaged cell organelles, and it involves a self-degradation process [65]. The energy required for cell doubling and sustained proliferation is gained via autophagy [6]. The autophagy process or pathway is very important to maintain homeostasis needed for survival and stress response. The failure of autophagy pathway could lead to dysfunction of cellular organelles and result in senescence [66]. During aging, there is a decline in the self-degradation process of the stem cells [67], especially in satellite cells [68]. Generally, there are three types of autophagy: chaperone-mediated autophagy, microautophagy, and macroautophagy. Chaperone-mediated autophagy involves a specific transmembrane protein called lysosomal-associated membrane protein 2 (LAMP2) that facilitate the transport of misfolded cytosolic proteins into the lysosomes for clearance. Microautophagy is the direct engulfment of the cargo (organelles or cytosolic proteins) by the lysosomes. Macroautophagy involves the formation of double membrane vesicles called autophagosomes that sequester the cargo and deliver it to the lysosomes [69,70].

Usually, the quiescent satellite cells rest in a reversible G0 phase. However, when there is a damage, these satellite cells get activated and start to proliferate. This process is contradictory to aging, as there is dysfunction and reduction in the regeneration of satellite cells as they enter senescence that is, the quiescent satellite cells get into an irreversible G0 phase [68]. In skeletal muscles, satellite cell senescence has an important contribution for muscle aging [71]. In senescence, there is a cell cycle arrest that also increases the metabolic activity linked with the secretory phenotype [72]. The accumulation of senescence cells and other toxic waste in satellite cells affect the fitness and bonafide nature of the quiescent state, therefore, autophagy activity is disrupted, and this serves as the important factor for the aging process [68]. Active basal autophagy is responsible for the maintenance of organelles, however, in the old mice, the basal autophagic flux is impaired. Therefore, there is an accumulation of organelles, including the mitochondria [68,73].

Autophagy-related genes (ATGs) are highly conserved genes that serve as a marker for autophagy [54]. The satellite cells express the autophagy related 7 (Atg7) gene, and it plays a vital role in the formation of the autophagosome in young satellite cells (Fig. 3A). The expression of the Atg7 gene declines during aging, which reflects a decline in the autophagy activity (Fig. 3B). A reduction in autophagy activity negatively impacts mitochondrial function by increasing the level of ROS. Also, deletion of the Atg7 gene in the satellite cells caused a reduction in the number of satellite cells and led to the dysfunction of satellite cells. Hence, basal level of autophagy is required for lifelong maintenance of quiescent satellite cells. Importantly, autophagy is one of the important age-associated mechanisms [68].

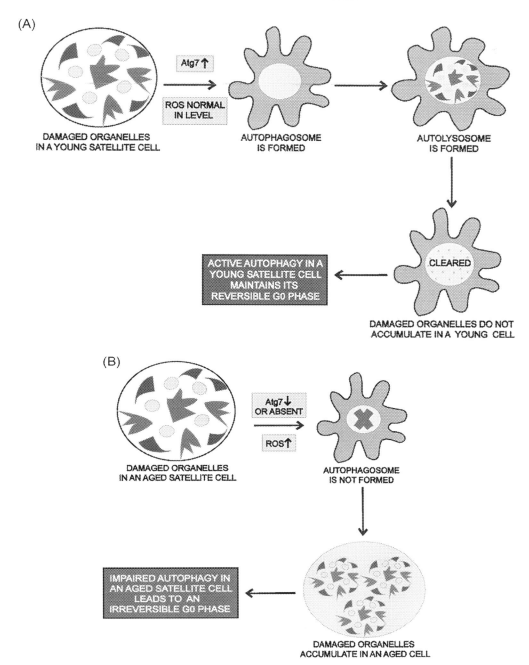

FIG. 3 **Role of autophagy in a young and aged satellite cells.** The self-degradation process of damaged organelles in a satellite cell occurs by macroautophagy. (A) In the normal young satellite cells, the expression of autophagy related 7 (Atg7) gene helps in autophagosome formation and ROS level is normal. The autophagosome carries cytosolic protein into the lysosome for clearance. No accumulation of toxic waste is observed; therefore, the G0 reversible quiescent state is maintained. (B) In the aged satellite cell, Atg7 gene expression is low or absent and the formation of the autophagosome does not take place. Diminished autophagy activity negatively acts on mitochondria and increases ROS, leading to cell cycle arrest due to senescence. The accumulation of toxic waste is observed; therefore, the G0 reversible quiescent state is not maintained. (A) or (B), the upward (↑) or downward (↓) arrow symbol indicates upregulation or downregulation, respectively.

Garcia-Prat et al. [68] demonstrated that antioxidant treatment upregulated the epigenetic factors $p16^{Ink4a}$ and Atg7 in the aged and the adult satellite cells, leading to the reduction in ROS in mitochondria. Interestingly, Cerletti et al. [74] reported that short term duration of caloric restriction showed proper homeostasis by maintaining the function of the satellite cells. Furthermore, it increased the number of mitochondria and their activity, and also improved the longevity of the satellite cells in the muscles of young and old mice. Caloric restriction appears to be independent of the developmental stage and it may have the same impact on the function of muscle satellite cells [74,75].

The mechanistic target of rapamycin (mTOR) is the master regulator of nutrient response pathway that influences the metabolism and growth of cells, and thus its inhibition could slow aging and extend the lifespan of the cells and organisms [76]. Therefore, downregulation of the mTOR pathway rejuvenates the autophagic flux in the adult satellite cells and increase the activity of the aged satellite cells. Available studies indicate that autophagy is essential for maintaining the G0 resting phase and stemness of cells, and it prevents the cellular senescence. However, a decline in autophagy leads to a complete alteration in the cellular metabolism and cell cycle arrest.

5 Aging and extrinsic factors of satellite cells

For effective muscle satellite cell function and regeneration, the local factors in the microenvironment are crucial [16]. The microenvironment involves muscle fibers, fibro-adipogenic progenitors (FAPs), growth factors, circulatory factors, macrophages, cytokines, etc. Myofibers are surrounded by the collagen-rich connective tissue known as endomysium, comprising ECM [77]. Linkage between the basal lamina and the endomysium arbitrates an adhesive molecule collagen IV that forms like a mesh. This collagen serves as a bridge and enhances the link between muscle regeneration-associated growth factors and their receptors [1]. During aging, the satellite cells are impaired due to the accumulation of toxic waste and dysfunction in the metabolisms. Several intrinsic and extrinsic factors play a crucial role in the aging of satellite cells. So, in the subsequent section, we discuss the microenvironmental changes in satellite cells during aging.

5.1 Extracellular matrix of satellite cells

The ECM contributes about 10% of the skeletal muscle. It acts as the scaffold between the cells and supports blood vessels and nerves. ECM plays an important role in the maintenance and repair of muscle fibers, and also in the force transmission of muscle fibers [78]. When quiescent satellite cells get activated, they undergo proliferation and differentiation with the production of ECM [79–82] and maintain myogenesis. In the skeletal muscle, there is a high amount of ECM, which is involved in the determination of satellite cell fate [83]. ECM contributes to fibrosis and hence increases the tissue stiffness [84,85]. Fibrosis is a characteristic of muscular dystrophy [86]. During injury, there is a formation of new temporary ECM [80,87] by fibrocytes, FAPs, and mesenchymal progenitors that express the platelet-derived growth factor receptor (PDGFRα). The deposition of ECM is under the control of TGF-β, connective tissue growth factor (CTGF), and the renin-angiotensin system (RAS) [88–92].

During aging, there is a loss of muscle mass, which is called sarcopenia. In this clinical condition, the fibrosis is higher [93]. When ECM declines, the FAPs get activated, resulting in a decrease in proliferation that leads to fibrosis [94,95]. Increased fibrosis impairs the myogenesis [84]. The conversion of myogenic to fibrogenic phenotype is controlled by Wnt/β-catenin signaling through the upregulation of TGF-β2 [84,96]. There is a cross-talk between Wnt/β-catenin and TGF-β signaling, and as a result, TGF-β can upregulate the Wnt/β-catenin signaling molecules and vice versa [97].

Myofibers are said to have increased stiffness during aging, which is due to the increased cross-link of the collagen in the aged satellite cells of the muscle [98,99]. Lacraz et al. [98] reported that adult primary myoblast cultured on hydrogel mimicked the stiffness of the aged muscle cells associated with differentiation. However, Gilbert et al. [83] reported that when adult stem cells are isolated and cultured in the hydrogel that mimics softness such as an ECM-coated plastic dish, accelerated proliferation and self-renewal occurred, which improved the engraftments. Proliferation is crucial for maintaining the homeostasis of the muscle regeneration. Considering PDGFRα as the target for FAP production, PDGFRα signaling inhibition can lead to the reduction of fibrotic deposition during regeneration of the aged muscles [100].

5.2 Circulatory factors and satellite cells

In aged muscle satellite cells, there is a decline in the regeneration capacity, and this can be effectively repaired by support and a healthy microenvironment. Parabiosis is usually used to test the relationship between the systemic factors and their difference from one animal to other [101]. Parabiosis and whole muscle transplantation show the importance of this extrinsic environment. Carlson and Faulkner [40] reported that grafting of whole muscle from young to old mice totally depends on the age. The young host showed successful engraftment in both the young and the old muscle, whereas in the aged host, the regeneration of the satellite cells failed in both. In another study, Lee et al. [39] observed that there is delayed regeneration rather than impairment of the satellite cells. The differences observed may be due to the age, gender, models, and types of injury.

During a short-term injury, the aging factors delayed the repair, whereas the young factors paved the way for the muscle repair. The age of the circulatory factors and microenvironment is directly proportional to the aging of muscle [102]. When satellite cells from aged mice were exposed to the young environment (serum), there was an increase in the proliferation of satellite cells in vitro. Furthermore, there was enhanced regeneration of aged satellite cells by the serum of young mice [84,102]. These results suggest that the circulating factors can regulate the aged satellite cells. Therefore, by manipulating the signaling pathways of receptor tyrosine kinase (RTK), Notch, Wnt, TGF-β, and oxytocin, the aged muscle impairment can be reduced [84,103–108]. Available studies support the role of growth factors and inflammatory cytokines in aging. During aging, there is an increase in inflammation. However, there is a delayed response to the injury in old mice [109,110]. Growth differentiation factor 11 (GDF11) is reduced in the circulation during aging. However, treatment with recombinant GDF11 brought it back and enhanced its role in aged muscle [108]. These findings reveal that the impaired muscle satellite cells during aging can be reversed by mimicking the young and healthy extrinsic microenvironment.

5.3 Growth factors and satellite cells

Growth factors play a crucial role in satellite cell function [82,111]. Hepatocyte growth factor (HGF) and fibroblast growth factors (FGFs) are involved in the activation and proliferation of aged satellite cells in vitro [112,113]. The FGF2 is highly expressed in aged muscle fibers, and it does not undergo any age-dependent changes in the satellite cells [37]. Moreover, the exposure of aged satellite cells to FGF2 affected the quiescent state. When the FGF2 was downregulated using an inhibitor of fibroblast growth factor receptor 1 (FGFR1), there was a decrease in the number and function of the aged muscle cells [37]. This suggests that FGFs play a crucial role in the regulation of muscle stem cells or satellite cells.

Insulin-like growth factors (IGFs) also play a crucial role in the proliferation and differentiation of muscle cells. IGF-I has an anabolic effect and can induce muscle hypertrophy; it can also increase the muscle-specific isoform of IGF-I [114]. Glucocorticoids interfere with the IGF-I signaling pathway and can cause muscle wasting [115,116]. Furthermore, treatment with corticosteroids lead to a decrease in IGF–I, causing muscle atrophy. Agents such as growth hormone (GH) and IGF regulate the muscle cells and maintain the muscle mass. The administration of IGF-I during growth and hypertrophy may prevent age-associated loss and function of myofibers [117].

Transforming growth factor (TGF-β) regulates both embryogenesis and adult tissue homeostasis. TGF-β can reduce the adult myofibers and increase the young satellite cells [118,119]. Aged muscles show an increased level of TGF-β1. TGFβ signaling leads to the phosphorylation of the SMAD complex that results in the delayed cell cycle. When TGF-β signaling is neutralized, the regeneration process is carried out in the aged muscles. From this, it is clear that the aging phenotype can be alleviated by inhibiting the TGF-β activity [120,121].

5.4 Cytokines and satellite cells

Several studies indicate that cytokines play a key role in cellular proliferation, protein synthesis, remodeling required for tissue repair, and angiogenesis [122]. During stress in muscles, an influx is initiated that serves as the aid for healing and releases inflammatory cytokines such as interleukin-6 (IL-6) and tumor necrosis factor alpha (TNF-α), and IGF-I [123,124]. IL-6 seems to be associated with muscle performance, and there is increased IL-6 during a decline in muscle function [125,126]. Moreover, in transgenic mice, an increased level of IL-6 causes muscle atrophy and is also involved in the breakdown of proteins [127].

The proinflammatory cytokine IL-6 can represent an isoform that is different from inflammatory cells; also, muscle-derived IL-6 inhibits TNF-α [126]. An increase in TNF-α can lead to impairment of the muscle regeneration in aged muscles, causing sarcopenia. Therefore, increased TNF-α is associated with lower appendicular skeletal muscle mass contributing to age-associated impairment [128].

6 Aging and intrinsic factors of satellite cells

6.1 Telomeres in satellite cells

Telomere attrition is one of the hallmarks of aging [129]. Telomeres are protective caps that cover the ends of the chromosomes. They are repetitive double-stranded DNA sequence TTA GGG consisting of about 2–20 kb and are surrounded by shelterin, a complex protein. There are six types of shelterin (Table 2). The shelterin and telomeres work together for the protection of the chromosome end from breaking the DNA [130–134].

Skeletal muscles are made of postmitotic multinucleated muscle fibers or cells, and understanding the structure and function of their telomeres have provided significant insights on aging and telomere biology. The telomere length has to be constant to avoid DNA damage [135]. The loss of telomere and its function leads to genomic instability, causing premature aging or aging and cancer. When the shelterin complex is perturbed, there is a loss in homeostasis of telomere length [136]. Satellite cells are crucial for muscle regeneration. When a single nucleated satellite cell divide asymmetrically, one daughter cell goes to

TABLE 2 Protein components of the shelterin complex.

List of shelterin proteins	
TRF1	Telomeric repeat binding factor 1
TRF2	Telomeric repeat binding factor 2
RAP1	Repressor/activator protein 1
TIN2	TRF1-interacting nuclear protein 2
TPP1	Tripeptidyl peptidase 1
POT1	Protection of telomeres 1
Recently discovered proteins of the telomeres	
HOT1	Homeobox telomere-binding protein 1
TZAP	Telomere zinc-finger associated protein

This table lists the different proteins of the shelterin complex that are involved in protecting the chromosome ends, that is, telomeres, from damage.

replenish the satellite cell pool and another leads to the regeneration of the muscle. Now, telomeres are shortened in new cells as they are incorporated from the precursor cells, therefore causing muscular dystrophies [82,137–140]. Daniali et al. [141] reported that skeletal muscle telomeres get shortened with age. These authors collected 87 samples such as skin, immune cells, and skeletal muscles within the age range of 19–77, and interestingly, they observed an age-dependent telomere shortening [141]. The telomere shortening was observed during oxidative stress [142] and 8-OXOG formation in DNA [143]. To remove oxidative stress induced damage from DNA, base excision repair (BER) pathway is very important. BER involves the removal of damaged bases, incision at basic site, cleaning of DNA terminal, insertion of correct nucleotide, and ligation of nicks in the DNA backbone [144]. Nucleotide excision repair (NER) is involved in the regulation and intergrity of telomeres [145]. There are transcription-coupled (TC)-NER and global-genome (GG)-NER where TC-NER is responsible for elimination of lesion in active gene and GG-NER recognizes and removes DNA lesion throughout the genome [146]. Furthermore, Xeroderma pigmentosum B (XPB/ERCC3/p89) plays a vital role in repair of DNA lesion and telomere dynamics [147].

The structure and length of the telomeres can be modified by various factors including epigenetic factors. Histone methylation is involved in human telomerase reverse transcriptase (hTERT) regulation, and the epigenetic status of *hTERT* locus is critical for its regulation in normal somatic cells and in tumorigenesis [148]. Moreover, the initial transcriptional repression of the hTERT promoter is independent of histone deacetylation and the nucleosomal deposition at the core promoter (nucleosomal remodeling) is responsible for hTERT transcriptional repression in human somatic cells [149]. Hence, the telomeres play a crucial role in the regulation of skeletal muscle aging.

Aging is negatively associated with satellite cell integrity over intrinsic changes [150]. Also, there was a decrease in Pax7 expression in the cultured aged satellite cells, and increased apoptosis in aged muscle cells, causing a reduction in the satellite cell pool [151–153]. The signaling pathways involved in myogenesis and their alterations during aging are discussed in the sections below.

6.2 Notch signaling pathway of satellite cells

In the Notch signaling pathway, the binding of Delta ligand with Notch receptor initiates proteolytic events resulting in the release of Notch intracellular domain (NICD), which translocates to the nucleus and activates the transcription factors that leads to the expression of Notch target genes (Hes and Hey) [102–104,154]. Interestingly, decreased Delta ligand expression was observed in isolated aged mice satellite cells when compared to young satellite cells showing less regenerative potential of the muscle. This suggests diminished Notch signaling pathway in the aged skeletal muscle [102].

Extrinsic factors in the circulation may increase the Notch signaling. Conboy et al. [102] reported that there was about five-fold increase in Delta ligand expression in injured aged mouse satellite cells, and when satellite cells from aged mouse were exposed to the young serum, that resulted in enhanced Notch signaling pathway. Thus, age-associated dysfunction of the satellite cells can be influenced by systemic factors [102].

The deterioration in the upstream molecules that activate the Notch signaling [121,155] and increased Notch signaling inhibitors [103,154] may be the reason for decreased Notch signaling in aged satellite cells. One of the most important factors that activates Notch is MAPK/PERK, which is decreased in the muscle cells. During aging, the treatment of the aged satellite cells with MAPK agonist resulted in the increase in Notch signaling [155]. TNFα and TGFβ are the two crucial inhibitors of Notch signaling. TNFα is increased in aging and it inhibits Notch signaling, contributing to the sarcopenia condition of the muscle cells [156]. TGFβ, another active inhibitor of Notch signaling is also increased, and therefore TGFβ negatively correlate with Notch signaling, and thus cause impairment in the regeneration of the satellite cells [103,154]. Furthermore, hormones also play a key role in regulating the Notch signaling. The testosterone level depleted during aging is directly associated with the decrease in the Notch signaling pathway [104,155].

6.3 Wnt/β-catenin signaling of satellite cells

The Wnt/β-catenin pathway is involved in important cellular events such as cell proliferation, invasion, cellular senescence, and angiogenesis [157,158]. When the Wnt ligand binds to the frizzled (FZD) and low-density lipoprotein receptor-related protein (LRP), the signaling cascade is activated by inhibition of GSK-3β phosphorylation of β-catenin. The accumulated β-catenin translocate into the nucleus where it binds with the transcription factors (TCF and LEF), leading to the expression of genes involved in skeletal muscle repair [159–161].

Wnt signaling plays a vital role during the early stage of myogenesis. Wnt1 and Wnt3a form the dorsal neural tube and Wnt7 forms the ectoderm that gives rise to myogenic precursor [162,163]. Aberrant Wnt signaling in the aged muscle results in the dysfunction of myogenesis [84,164]. Liu et al. [165] reported that the klotho mouse shows high Wnt signaling leading to the cellular senescence and reduced life span. Decreased levels of Wnt antagonists namely secreted frizzled-related protein 2 (SFRP2) and Wnt inhibitory factor 1 (WIF1) have been reported in aged muscles than the young [164], along with increased β-catenin and decreased GSK3β [84]. These findings suggest that there is increased Wnt/β-catenin signaling pathway in the muscle satellite cells that contribute to the dysfunction and impairment of aged satellite cells.

An increase in GSK3β and a reduction in Wnt signaling were observed when aged mice were paired with young mice [84]. GSK3β in the circulation help in the reduction of Wnt/β-catenin signaling pathway in the aged satellite cells, suggesting that upregulation of Wnt signaling may facilitate aging. Wnt also contributes to the formation of fibrosis by transition of aged skeletal muscle to fibrogenic tissue; and it promotes aging [84,165–167]. There is a high affinity to change from myogenesis to fibrosis [84]. Interestingly, injection of the Dickkopf Wnt signaling pathway inhibitor 1 (DKK1) in young mice demonstrated a decrease in fibrosis [168].

Taken together, Notch and Wnt signaling plays a crucial role in muscle cell function. The aggravation of Wnt signaling and the fading of Notch signaling results in dysfunctional repair of aged skeletal muscles [84,103,154,169].

7 Aging and epigenetic regulation of satellite cells

Chromatin (i.e., genomic DNA and histones) modification and chromatin remodeling are key epigenetic mechanisms that generate heritable chromatin signature/marks, and contribute to the epigenetic regulation of transcription and gene activation, repression/silencing or expression during stem cell maintenance and differentiation [170]. The epigenetic mechanisms also generate bivalent domains in stem cells. These bivalent domains are chromatin regions that possess both active and repressive chromatin marks. The bivalent domains are often observed on promoters of genes that show low level of expression. Intriguingly, more than 400 genes in the myoblasts were shown to exhibit bivalent domains [171]. Generally, the activated satellite cells leave the quiescent state, and undergo proliferation and differentiation during muscle regeneration. Many transcriptional factors such as Pax3, Pax7, and MRFs are involved in these processes. Therefore, it is important to understand the role of transcription factors and epigenetic regulators during muscle cell regeneration in the young and aged. This section discusses the mechanisms and role of epigenetic regulation in myogenesis.

7.1 Epigenetic events in quiescent state of satellite cells

Different types of epigenetic mechanism can regulate satellite cells gene expression or function during myogenesis. Cao et al. [172] reported in a genome-wide analysis that MyoD protein bound to promoter regions of several different genes in the cells of skeletal muscle. The level of MyoD binding correlated with the level of local histone hyperacetylation. This implicates a broader role for MyoD in epigenetic regulation of cellular reprogramming during skeletal muscle differentiation via recruitment of histone acetyltransferases (HATs).

The various mechanisms that are involved in the epigenetic regulation of satellite cell-dependent myogenesis are chromatin remodeling, DNA methylation, and covalent modification of histones and transcription factors. These mechanisms usually result in reversible activation or repression of genes [173,174]. H2A, H2B, H3, and H4 are the four core histone proteins that help in the formation of nucleosomes, over which the chromosomal/genomic DNA swaddles around to constitute chromatin. Generally, the amino-terminal tail of histones undergo posttranslational modifications. The H3K4me3 histone modification/mark indicate transcriptionally active/permissive chromatin state, while H3K9me2 and H3K27me3 histone modifications/marks indicate transcriptionally inactive/repressive chromatin state [170]. Histone marks can participate in rapid signal-dependent regulation of gene expression by responding to transcription cues in the form of differentiation signals during muscle regeneration [171].

Liu et al. [8] demonstrated that when the level of histones declines in satellite cells, there is also reduction in the level of transcription. This shows that chromatin, its modifications and epigenetic events can affect satellite cell gene expression and state during chronological aging. During aging, in quiescent satellite cells, the level of repressive H3K9me2 and H3K27me3 histone marks increase on histone genes, while the level of permissive H3K4me3 histone mark is maintained. Hence, this leads to a loss of transcription potential of histone genes and lower level of histone proteins, and an alteration in the epigenetic regulation of satellite cell function during aging [8,175,176]. Notably, only Pax3 gene has bivalent chromatin domains on it, whereas Pax7 gene has H3K4me3 or H3K27me3 mark on it, and the trithorax group (TrxG) protein complex is responsible for H3K4me3 mark, while the polycomb repressive complex (PRC) is responsible for H3K27me3 mark [8].

The activation of a satellite cell from its quiescent state, and its commitment to myogenic program during skeletal muscle repair and regeneration involves Pax7 protein, but requires Myf5 expression too. To activate Myf5 gene expression, the coactivator-associated arginine methyltransferase 1 (Carm 1) first interacts with and then methylates Pax7 protein. Then, methylated Pax7 protein recruits H3K4 histone methyltransferase (HMT) complex to promoter and enhancer regions of Myf5 gene, which leads to the epigenetic induction of its expression [177]. Enhancer of zeste homolog (Ezh2) is a HMT and the catalytic subunit of the polycomb repressive complex 2 (PRC2), which generates the H3K27me3 repressive histone mark. In a study, Woodhouse et al. [178] used the conditional Ezh2-null mice and reported that satellite cells derived from these mice lacked Ezh2 activity, and exhibited a decrease in muscle growth, repair and stem cell number. PR/SET domain 2 or PR domain 2 (PRDM2)/retinoblastoma interacting zinc-finger (RIZ) protein is a H3K9 HMT, and is in high level in quiescent satellite cells. In the G0 phase myoblasts, PRDM2/RIZ binds to several H3K9me2 marked gene promoters, including those involved in the regulation of myogenesis [e.g., myogenin (*Myog*) gene] and cell cycle [e.g., cyclin A2 (*Ccna2*) gene]. Interestingly, the *Ccna2* gene locus contains bivalent chromatin domains in the G0 phase of the cell cycle, and PRDM2 binds to Ezh2 protein and regulates its interaction with the bivalent domains in the *Ccna2* gene [179]. Together, Ezh2/PRC2 and PRDM2 contribute to the epigenetic regulation of quiescent state gene expression in satellite cells. Cell cycle phases can be regulated by several cyclin-dependent kinases or kinase inhibitors, including cyclin-dependent kinase inhibitor 2A (Cdkn2a). The *Cdkn2a* gene expresses several transcripts and alternatively spliced forms that encode distinct proteins. One the important transcripts expressed from *Cdkn2a* gene is the p16 protein. The p16 is a **16 kDa p**rotein, which is an **in**hibitor of the cyclin-dependent **k**inase **4** (CDK4), and hence called as p16^{Ink4a}. It functions as a tumor suppressor gene/protein and regulates cell cycle during cellular senescence. For more details on "Cellular Senescence", the readers can refer to our "Chapter 14: Cellular senescence and aging in bone". Notably, Sousa-Victor et al. [71] showed in aging mouse skeletal muscle satellite cells, a link between quiescence, epigenetic regulation of Cdkn2a/p16^{Ink4a} gene expression and cellular senescence.

7.1.1 Cdkn2a/p16^{Ink4a} and epigenetic regulation of satellite cells

In the young skeletal muscle, satellite cells rest in a reversible quiescent state with intrinsic self-renewal potential, and their Cdkn2a/p16^{Ink4a} gene expression is epigenetically repressed to maintain this state. However, Sousa-Victor et al. [71] showed that when the Cdkn2a/p16^{Ink4a} gene expression is derepressed via epigenetic regulation in the mouse geriatric satellite cells, they switched from their reversible quiescent to irreversible senescent state via geroconversion. This is a senescence program in which there is futile growth of cycle arrested cells. The geroconversion occurred in mouse geriatric satellite cells, even when their surrounding environment was youthful. Importantly, Cdkn2a/p16^{Ink4a} gene expression is also dysregulated in geriatric satellite cells in human muscle. The quiescent satellite cells in their reversible G0 phase is required for skeletal muscle regeneration, and senescent satellite cells in their irreversible G0 phase impairs this function [180]. Also, the increase in Cdkn2a/p16^{Ink4a} expressing positive cells, and their accumulation reduces the quality of the satellite cell pool, and thereby impairs skeletal muscle regeneration [181]. During aging, satellite cells produce high levels of ROS, and it induces cellular damage and senescence. Some of the senescence markers, which are used to identify the senescent satellite cells are, the senescence-associated β-galactosidase (SA-β-Gal) activity and Cdkn2a/p16^{Ink4a} [71]. Moreover, Sausa-Victor et al. [71] also reported that silencing the expression of Cdkn2a/p16^{Ink4a} in senescent satellite cells, can restore its normal function in skeletal muscle regeneration. Additionally, Baker et al. [180] demonstrated in the transgenic mice with the *INK-ATTAC* transgene containing a senescent cell active p16^{Ink4a} minimal promoter, and using the cell-permeable chemical drug

AP20187, that drug-induced and targeted-removal of senescence positive cells, extended the lifespan of mice. This study showed that senescent cells play a critical role in the aging of tissues and organs in an animal [180]. Together, these reports highlight the crucial contribution of quiescent sate and senescent state pathways, and their link for the normal and efficient function of satellite cells, especially in skeletal muscle regeneration during aging.

7.2 Epigenetic events in proliferative state of satellite cells

Epigenetic regulation of gene expression plays a key role in the proliferation state as in the quiescent state of satellite cells. DNA methylation is an important epigenetic mechanism and repressive system that regulates the gene loci, in proliferating satellite cells during adult myogenesis [182]. Moreover, in satellite cells undergoing proliferation, the promoter region of the genes involved in skeletal muscle differentiation contain hypoacetylated histones and H3K9me2, H3K9me3 and H3K27me3 histone marks, leading to gene repression. Histone hypoacetylation is catalyzed by histone deacetylases (HDACs) and methylation by HMTs from the Polycomb group (PcG) and suppressor of variegation 3-9 homolog (Suv39) families [174,183]. MyoD is an important myogenic regulatory factor, which is expressed in satellite cell-derived myoblasts and is involved in their proliferation. Notably, MyoD works in concert with other myogenic factors and epigenetic regulators during skeletal myogenesis [184]. In proliferating myoblasts, the expression of Ezh2 prevented myoblast differentiation via its association with the transcriptional regulator Ying Yang 1 (YY1), which in turn recruited the HDAC1 to hypoacetylate histones in promoters of muscle specific genes [185]. Additionally, Ezh2 generated its repressive H3K27me3 histone mark directly, and repressive DNA methylation mark indirectly via the recruitment of DNA methyltransferases 3A and 3B (DNMT3A and DMNT3B), on promoters of muscle specific genes in proliferating myoblasts [186]. Together, these epigenetic mechanisms contribute to repress the expression of muscle-specific genes during myoblast proliferation. The Pax7 transcription factor plays a significant role in satellite cell proliferation, and its expression enables the satellite cells to retain their differentiation potential. Interestingly, Palacios et al. [187] showed a novel link between inflammatory signaling by TNF-α via p38α kinase, PRC2/Ezh2 and YY1 in regulating the Pax7 gene expression in satellite cell-mediated myogenesis during muscle regeneration. As discussed in an earlier section in this book chapter, Notch signaling is crucial for satellite cell function. Nevertheless, Terragni et al. [188] reported that at the DNA/gene level, i.e., in the intergenic or intragenic regions of the NOTCH1/2 receptor and its ligand DLL or JAG2, they were hydroxymethylated and hypomethylated in myoblasts. This finding suggests that components of the Notch signaling can be under epigenetic regulation during myogenesis. Several inhibitors of DNA methyltransferase are known, and 5-azacytidine is one of the main inhibitor. Interestingly, the inhibition of DNA methyltransferase using 5-azacytidine leads to the transdifferentiation of fibroblast into myoblast [189]. Together, these studies show that epigenetic regulation contributes significantly, to repress the expression of muscle-specific genes during myoblast proliferation, and thereby helps to prevent the premature differentiation of myoblasts.

7.3 Epigenetic events in differentiation state of satellite cells

The differentiation of the satellite cell-derived myoblasts into myocytes starts at the end of their proliferation, and after they exit their cell cycle. However, the complete details of the molecular mechanisms involved in this process still remains to be known.

The transcription factor Pax7 has an important function in satellite cell proliferation, and the expression of Pax7 helps it to preserve its capacity to undergo skeletal muscle differentiation. Notably, an inflammatory signaling pathway involving TNF-α/p38α kinase, PRC2/Ezh2 and YY1 axis was shown to regulate the expression of the Pax7 gene during muscle differentiation [187]. Importantly, the retinoblastoma protein (pRb) that plays a vital role in mitotic cell cycle exit, enables the permanent cell cycle arrest in myoblasts by maintaining the H3K27me3 repressive histone marks on the cell cycle genes, and promotes myoblast differentiation [190]. Additionally, the small chromatin associated protein p8 [also called as nuclear protein 1, transcriptional regulator (Nupr1)], which is related to high mobility group (HMG) proteins, was shown to bind to the histone acetyltransferase (HAT), p300 [also called as CREB binding protein (CREBP/CBP)], and promote cell cycle arrest and myogenic differentiation [191].

During myoblast differentiation, and prior to transcriptional activation of the MyoD-target muscle genes, the 60 KDa Brg-1/Brm-associated factor subunit C (BAF60C) interacts with the MyoD protein, and they are associated as MyoD-BAF60C complex on MyoD-target genes. BAF60C is a subunit of the switch/sucrose nonfermentable (SWI/SNF) chromatin-remodeling complex. In myoblasts, upon activation of p38α kinase signaling during differentiation, the assembled MyoD-BAF60C complex on MyoD-target genes, recruit the core-complex of the SWI/SNF chromatin remodeler and activate the transcription of muscle-specific genes [192]. Therefore, to promote myogenesis, chromatin remodeling is crucial, and it depends on the collaboration between the activated p38 signaling cascade and SWI/SNF complex, which acts on the muscle-specific gene loci [193].

7.3.1 p³⁸MAPK and epigenetic regulation of satellite cells

SNF2-related CBP activator protein (SRCAP)/H2A.Z-mediated chromatin remodeling is an early event in muscle-specific gene expression, and it helps in histone replacement. SRCAP deposits H2A.Z on chromatin, and this replacement by SRCAP occurs in a p38 MAPK-dependent manner [194]. Mitogen-activated protein kinase (MAPK), a superfamily of intracellular serine/threonine protein kinase is regulated in response to stress, especially by p38 subgroup [195]. Moreover, p38 is also activated during nonstress conditions [196]. Treatment with the p38α/β inhibitors prevented the fusion of the myoblasts into myotubes, and the p38 MAPK pathway plays a vital role in the skeletal muscle differentiation [197]. Brien et al. [198] demonstrated that p38α promotes myoblast differentiation and regulates proliferation. These reports highlight the significance of p38 MAPK role in myogenesis.

Aging leads to a decline in skeletal muscle satellite cells or stem cells function, differentiation and regeneration capacity. This decline correlated with higher levels and incidences of senescence markers, and elevated activity of p38 MAPK in aged satellite cells. Moreover, the decline in stem cell function was independent of stem cell niche in the young versus aged mice. Importantly, the p38 protein and its direct target MAPK activated protein kinase 2 (MAPKAPK2, also indicated as MK2), showed elevated phosphorylation in satellite cells of aged mice as compared to young mice. Additionally, aged satellite cells took longer time to get activated as compared to younger satellite cells [49,57]. Together, these findings highlight the critical role of p38 MAPK signaling pathway in stem cell maintenance and myogenic differentiation during aging.

Jones et al. [199] showed that FGF2, which is increased in aged satellite cells, can directly activate p38 via the FGFR1-p38 MAPK signaling pathway [57]. Moreover, administration of FGFR1 inhibitor to aged mice was able to reduce the age-related phenotypes in vivo [37]. Furthermore, p38 activity in aged satellite cells can be targeted by using small interfering RNAs (SiRNAs) knockdown approach, which reduces the protein level of p38. Together, these studies show that targeting p38 can help to ameliorate the age-related problems caused by the increased activation of p38 MAPK signaling [49,57]. Of significance, Liu et al. [8] observed that there is a change in the distribution and quality of the H3K27me3 chromatin. Also, in geriatric satellite cells, Cdkn2a/p16^{Ink4a} expression was increased, and this lead to the conversion of satellite cells from their quiescence to senescence [71]. p38 is regulated during stress conditions, and it gets recruited to genes that are stress-induced. This binding induces the expression of stress-response genes in a p38-dependent manner [200]. The p38 MAPK signaling can enable the myoblasts, to determine their choice between proliferation versus differentiation. It is involved epigenetic regulation of Pax7 gene expression via the TNF-α/p38 MAPK/YY1/PRC2/Ezh2 axis. This lead to the generation of repressive histone mark on the Pax7 promoter region during muscle regeneration. The Pax7 locus has bivalent chromatin domains, which get resolved into repressive chromatin domains upon p38 activation. Moreover, during differentiation, chromatin-remodeling complex SRCAP was recruited to the myogenin promoter in a p38-dependent manner. The recruitment of SRCAP to myogenin promoter initiated histone exchange, during which the histone variant H2A.Z replaced the core histone H2A. These events lead to H2A.Z accumulation on the chromatin of the myogenin promoter. Subsequently, H2A.Z/SRCAP mediated remodeling of chromatin lead to muscle-specific myogenin gene expression. Thus, p38 MAPK signaling leads to structural alteration in the promoter region, and this epigenetic mechanism facilitates the myogenin gene expression, during terminal differentiation of satellite cell-derived myoblasts into myocytes [187,194,201]. The MyoD protein, its target muscle-specific genes, and BAF60C, a SWI/SNF chromatin-remodeler subunit participate in myoblast differentiation. Notably, the BAF60C phosphorylation by p38 kinase is the cue that helps the incorporation of BAF60C-MyoD into SWI/SNF complex, to remodel chromatin and activate transcription of MyoD-target genes. Additionally, p38 MAPK-mediated phosphorylation of the E47 protein [also called as transcription factor 3 (TCF3) induces E47/MyoD interaction, and thereby regulates formation of the functional E47/MyoD heterodimers, which activate muscle-specific gene expression. Also, inflammatory signaling by mitogen-activated protein kinase kinase 6 (MKK6, also called as MAP2K6, and an important component of p38 MAPK signaling pathway), and insulin-like growth factor 1 (IGF-I)-induced by phosphoinositide 3-kinase (PI3K)/RAC-alpha serine/threonine-protein kinase (AKT) are functionally codependent, and converge on muscle-specific genes to regulate their expression. Interestingly, the myocyte enhancer factor 2C (MEF2C) is not only a substrate for p38 MAPK, but also a co-activator of MyoD. Importantly, p38 MAPK signaling pathway was shown to regulate MyoD activity at least partly by activating MEF2C, and thereby promote skeletal muscle differentiation [192,202–204]. Taken together, these studies and their findings emphasize the importance of p38 MAPK signaling and its role in the diverse mechanisms of epigenetic regulation during myogenic differentiation. Therefore, a decline or impairment in the function of any of the key molecules involved in the epigenetic regulation of satellite cell function, will have a major impact on the myogenic program and skeletal muscle repair or regeneration during aging.

8 Conclusions

The high regenerative capacity of skeletal muscles is attributed to their resident tissue-specific muscle stem cells, which account for both the muscle cell lineage and self-renewal potential. This is achieved via an appropriate balance in the number of skeletal muscle satellite or stem cells, which is in the quiescent/G0 phase versus committed activated differentiating satellite cells participating in the repair of damaged/injured muscle. A visible change that occurs universally during the aging process is a decline in the sensory perception and motor abilities of the muscle as a result of alteration or impairment in the complex metabolisms of muscle cells. As organisms or individuals age, the mitochondrial stress induces or stimulates ROS production/accumulation and the UPR pathway. These lead to a decrease in autophagy activity, which promotes cell cycle arrest-induced cellular senescence, thereby resulting in an impaired ability to repair muscle and maintain muscle mass. In the repair and maintenance of muscles, extrinsic environmental factors are crucial as collagen IV serves as the link between growth factors, circulatory factors, and cytokines as well as their corresponding receptors. There are age-dependent alterations in the levels of FGF2, TGF-β, EMC, FAPs, IGF, TNF-α, and IL-6. A decrease in ECM has a direct effect on increasing the FAPs, which can cause the age-related disease, sarcopenia. However, there is evidence that shows that the restoration of a systemic environment to a youthful environment can maintain the niche and also regain its ability to regenerate. But, intrinsic factors are also important and have to be considered. Some of the crucial intrinsic factors such as Wnt, Notch, p16, and p38, which are involved in myogenesis, are altered by circulating extrinsic factors during aging. Both intrinsic and extrinsic microenvironment factors are interlinked, and the interactive signaling pathways appear to be altered during aging. Transcription factors that are involved in myogenesis can also contribute to epigenetic alterations/changes via the recruitment of HATs. p38 and its signaling serves as a critical factor and pathway in satellite cells during their differentiation to muscle cells via myogenesis or self-renewal. Both p38 and these mechanisms (differentiation or self-renewal) are dramatically impaired or altered in aging. Therefore, future research could explore in detail the link between extrinsic and intrinsic factors, metabolic activity, and epigenetic alteration during aging in satellite versus muscle cells to better understand the basic biology of skeletal muscle aging, and also to focus on developing or discovering novel therapeutic interventions to treat or alleviate aging toward improving healthy longevity or the lifespan of an individual or organism.

Acknowledgements

We sincerely thank Dr. Satish Ramalingam (Department of Genetic Engineering, School of Bioengineering, SRM Institute of Science and Technology, Kanchipuram, Tamil Nadu, India) for critically reading this manuscript and providing his valuable comments. Financial support provided by the Science and Engineering Research Board (SERB) [ECR/2015/000277], Government of India, to Dr. Ilangovan Ramachandran is sincerely acknowledged.

Conflict of interest

The authors declare no conflict of interest.

References

[1] H. Silva, I.M. Conboy, Aging and Stem Cell Renewal, StemBook, Cambridge (MA), 2008.
[2] C.F. Almeida, S.A. Fernandes, A.F. Ribeiro Junior, et al., Muscle satellite cells: exploring the basic biology to rule them, Stem Cells Int. 2016 (2016) 1078686.
[3] T.J. Hawke, D.J. Garry, Myogenic satellite cells: physiology to molecular biology, J. Appl. Physiol. 91 (2001) 534–551.
[4] S. Kuang, K. Kuroda, F. Le Grand, et al., Asymmetric self-renewal and commitment of satellite stem cells in muscle, Cell 129 (2007) 999–1010.
[5] J. Segales, E. Perdiguero, P. Munoz-Canoves, Regulation of muscle stem cell functions: a focus on the p38 MAPK signaling pathway, Front. Cell Dev. Biol. 4 (2016) 91.
[6] G. Purohit, J. Dhawan, Adult muscle stem cells: exploring the links between systemic and cellular metabolism, Front. Cell Dev. Biol. 7 (2019) 312.
[7] A.S. Brack, T.A. Rando, Tissue-specific stem cells: lessons from the skeletal muscle satellite cell, Cell Stem Cell 10 (2012) 504–514.
[8] L. Liu, T.H. Cheung, G.W. Charville, et al., Chromatin modifications as determinants of muscle stem cell quiescence and chronological aging, Cell Rep. 4 (2013) 189–204.
[9] T.S. Mikkelsen, M. Ku, D.B. Jaffe, et al., Genome-wide maps of chromatin state in pluripotent and lineage-committed cells, Nature 448 (2007) 553–560.
[10] J.G. Ryall, T. Cliff, S. Dalton, et al., Metabolic reprogramming of stem cell epigenetics, Cell Stem Cell 17 (2015) 651–662.
[11] W.R. Frontera, J. Ochala, Skeletal muscle: a brief review of structure and function, Calcif. Tissue Int. 96 (2015) 183–195.
[12] I. Janssen, S.B. Heymsfield, Z.M. Wang, et al., Skeletal muscle mass and distribution in 468 men and women aged 18-88 yr, J. Appl. Physiol. 89 (2000) 81–88.

[13] I. Irrcher, P.J. Adhihetty, A.M. Joseph, et al., Regulation of mitochondrial biogenesis in muscle by endurance exercise, Sports Med. 33 (2003) 783–793.
[14] A. Mauro, Satellite cell of skeletal muscle fibers, J. Biophys. Biochem. Cytol. 9 (1961) 493–495.
[15] J. Dhawan, T.A. Rando, Stem cells in postnatal myogenesis: molecular mechanisms of satellite cell quiescence, activation and replenishment, Trends Cell Biol. 15 (2005) 666–673.
[16] C.F. Bentzinger, Y.X. Wang, M.A. Rudnicki, Building muscle: molecular regulation of myogenesis, Cold Spring Harb. Perspect. Biol. 4 (2012), a008342.
[17] M. Buckingham, F. Relaix, The role of Pax genes in the development of tissues and organs: Pax3 and Pax7 regulate muscle progenitor cell functions, Annu. Rev. Cell Dev. Biol. 23 (2007) 645–673.
[18] H.P. Shih, M.K. Gross, C. Kioussi, Muscle development: forming the head and trunk muscles, Acta Histochem. 110 (2008) 97–108.
[19] F. Relaix, D. Rocancourt, A. Mansouri, et al., A Pax3/Pax7-dependent population of skeletal muscle progenitor cells, Nature 435 (2005) 948–953.
[20] R. Sambasivan, S. Tajbakhsh, Skeletal muscle stem cell birth and properties, Semin. Cell Dev. Biol. 18 (2007) 870–882.
[21] T.A. Davis, M.L. Fiorotto, Regulation of muscle growth in neonates, Curr. Opin. Clin. Nutr. Metab. Care 12 (2009) 78–85.
[22] H.C. Olguin, B.B. Olwin, Pax-7 up-regulation inhibits myogenesis and cell cycle progression in satellite cells: a potential mechanism for self-renewal, Dev. Biol. 275 (2004) 375–388.
[23] P. Seale, L.A. Sabourin, A. Girgis-Gabardo, et al., Pax7 is required for the specification of myogenic satellite cells, Cell 102 (2000) 777–786.
[24] M. Buckingham, L. Bajard, T. Chang, et al., The formation of skeletal muscle: from somite to limb, J. Anat. 202 (2003) 59–68.
[25] J. Gros, M. Manceau, V. Thome, et al., A common somitic origin for embryonic muscle progenitors and satellite cells, Nature 435 (2005) 954–958.
[26] R.I. Kumaran, B. Muralikrishna, V.K. Parnaik, Lamin A/C speckles mediate spatial organization of splicing factor compartments and RNA polymerase II transcription, J. Cell. Biol. 159 (5) (2002) 783–793.
[27] R.I. Kumaran, D.L. Spector, A genetic locus targeted to the nuclear periphery in living cells maintains its transcriptional competence, J. Cell. Biol. 180 (1) (2008) 51–65.
[28] V.F. Gnocchi, R.B. White, Y. Ono, et al., Further characterisation of the molecular signature of quiescent and activated mouse muscle satellite cells, PLoS One 4 (4) (2009) e5205.
[29] B. Muralikrishna, J. Dhawan, N. Rangaraj, et al., Distinct changes in intranuclear lamin A/C organization during myoblast differentiation, J. Cell. Sci. 114 (Pt 22) (2001) 4001–4011.
[30] R.L. Frock, B.A. Kudlow, A.M. Evans, et al., Lamin A/C and emerin are critical for skeletal muscle satellite cell differentiation, Genes Dev. 20 (4) (2006) 486–500.
[31] P. Sehgal, P. Chaturvedi, R.I. Kumaran, et al., Lamin A/C haploinsufficiency modulates the differentiation potential of mouse embryonic stem cells, PLoS One 8 (2) (2013) e57891.
[32] D.D. Cornelison, B.B. Olwin, M.A. Rudnicki, et al., MyoD(−/−) satellite cells in single-fiber culture are differentiation defective and MRF4 deficient, Dev. Biol. 224 (2000) 122–137.
[33] R.N. Cooper, S. Tajbakhsh, V. Mouly, et al., In vivo satellite cell activation via Myf5 and MyoD in regenerating mouse skeletal muscle, J. Cell Sci. 112 (Pt 17) (1999) 2895–2901.
[34] D.D. Cornelison, B.J. Wold, Single-cell analysis of regulatory gene expression in quiescent and activated mouse skeletal muscle satellite cells, Dev. Biol. 191 (1997) 270–283.
[35] H.M. Blau, B.D. Cosgrove, A.T. Ho, The central role of muscle stem cells in regenerative failure with aging, Nat. Med. 21 (2015) 854–862.
[36] A.S. Brack, P. Munoz-Canoves, The ins and outs of muscle stem cell aging, Skelet. Muscle 6 (2016) 1.
[37] J.V. Chakkalakal, K.M. Jones, M.A. Basson, et al., The aged niche disrupts muscle stem cell quiescence, Nature 490 (2012) 355–360.
[38] P. Sousa-Victor, L. Garcia-Prat, A.L. Serrano, et al., Muscle stem cell aging: regulation and rejuvenation, Trends Endocrinol. Metab. 26 (2015) 287–296.
[39] A.S. Lee, J.E. Anderson, J.E. Joya, et al., Aged skeletal muscle retains the ability to fully regenerate functional architecture, BioArchitecture 3 (2013) 25–37.
[40] B.M. Carlson, J.A. Faulkner, Muscle transplantation between young and old rats: age of host determines recovery, Am. J. Physiol. 256 (1989) C1262–C1266.
[41] M.D. Grounds, Age-associated changes in the response of skeletal muscle cells to exercise and regeneration, Ann. N. Y. Acad. Sci. 854 (1998) 78–91.
[42] H.C. Dreyer, C.E. Blanco, F.R. Sattler, et al., Satellite cell numbers in young and older men 24 hours after eccentric exercise, Muscle Nerve 33 (2006) 242–253.
[43] B.R. McKay, D.I. Ogborn, J.M. Baker, et al., Elevated SOCS3 and altered IL-6 signaling is associated with age-related human muscle stem cell dysfunction, Am. J. Physiol. Cell Physiol. 304 (2013) C717–C728.
[44] B.R. McKay, D.I. Ogborn, L.M. Bellamy, et al., Myostatin is associated with age-related human muscle stem cell dysfunction, FASEB J. 26 (2012) 2509–2521.
[45] S. Oustanina, G. Hause, T. Braun, Pax7 directs postnatal renewal and propagation of myogenic satellite cells but not their specification, EMBO J. 23 (2004) 3430–3439.
[46] S. Biressi, T.A. Rando, Heterogeneity in the muscle satellite cell population, Semin. Cell Dev. Biol. 21 (2010) 845–854.
[47] I. Harel, E. Nathan, L. Tirosh-Finkel, et al., Distinct origins and genetic programs of head muscle satellite cells, Dev. Cell 16 (2009) 822–832.
[48] Y. Ono, S. Masuda, H.S. Nam, et al., Slow-dividing satellite cells retain long-term self-renewal ability in adult muscle, J. Cell Sci. 125 (2012) 1309–1317.

[49] B.D. Cosgrove, P.M. Gilbert, E. Porpiglia, et al., Rejuvenation of the muscle stem cell population restores strength to injured aged muscles, Nat. Med. 20 (2014) 255–264.

[50] J.T. Rodgers, K.Y. King, J.O. Brett, et al., mTORC1 controls the adaptive transition of quiescent stem cells from G0 to G(alert), Nature 510 (2014) 393–396.

[51] A.H. Tang, T.A. Rando, Induction of autophagy supports the bioenergetic demands of quiescent muscle stem cell activation, EMBO J. 33 (2014) 2782–2797.

[52] N.S. Chandel, H. Jasper, T.T. Ho, et al., Metabolic regulation of stem cell function in tissue homeostasis and organismal ageing, Nat. Cell Biol. 18 (2016) 823–832.

[53] A.M. Cuervo, E. Bergamini, U.T. Brunk, et al., Autophagy and aging: the importance of maintaining "clean" cells, Autophagy 1 (2005) 131–140.

[54] D.J. Klionsky, A.M. Cuervo, P.O. Seglen, Methods for monitoring autophagy from yeast to human, Autophagy 3 (2007) 181–206.

[55] J.G. Ryall, S. Dell'Orso, A. Derfoul, et al., The NAD(+)-dependent SIRT1 deacetylase translates a metabolic switch into regulatory epigenetics in skeletal muscle stem cells, Cell Stem Cell 16 (2015) 171–183.

[56] L. Garcia-Prat, P. Munoz-Canoves, Aging, metabolism and stem cells: spotlight on muscle stem cells, Mol. Cell. Endocrinol. 445 (2017) 109–117.

[57] J.D. Bernet, J.D. Doles, J.K. Hall, et al., p38 MAPK signaling underlies a cell-autonomous loss of stem cell self-renewal in skeletal muscle of aged mice, Nat. Med. 20 (2014) 265–271.

[58] J. Zhang, J.C. Grindley, T. Yin, et al., PTEN maintains haematopoietic stem cells and acts in lineage choice and leukaemia prevention, Nature 441 (2006) 518–522.

[59] R.S. Frey, X. Gao, K. Javaid, et al., Phosphatidylinositol 3-kinase gamma signaling through protein kinase Czeta induces NADPH oxidase-mediated oxidant generation and NF-kappaB activation in endothelial cells, J. Biol. Chem. 281 (2006) 16128–16138.

[60] S.K. Powers, L.L. Ji, A.N. Kavazis, et al., Reactive oxygen species: impact on skeletal muscle, Compr. Physiol. 1 (2011) 941–969.

[61] P. Sousa-Victor, L. Garcia-Prat, P. Munoz-Canoves, New mechanisms driving muscle stem cell regenerative decline with aging, Int. J. Dev. Biol. 62 (2018) 583–590.

[62] S. Beccafico, C. Puglielli, T. Pietrangelo, et al., Age-dependent effects on functional aspects in human satellite cells, Ann. N. Y. Acad. Sci. 1100 (2007) 345–352.

[63] N. Sun, R.J. Youle, T. Finkel, The mitochondrial basis of aging, Mol. Cell 61 (2016) 654–666.

[64] A.D. Minet, M. Gaster, Cultured senescent myoblasts derived from human vastus lateralis exhibit normal mitochondrial ATP synthesis capacities with correlating concomitant ROS production while whole cell ATP production is decreased, Biogerontology 13 (2012) 277–285.

[65] C. He, D.J. Klionsky, Regulation mechanisms and signaling pathways of autophagy, Annu. Rev. Genet. 43 (2009) 67–93.

[66] D. Glick, S. Barth, K.F. Macleod, Autophagy: cellular and molecular mechanisms, J. Pathol. 221 (2010) 3–12.

[67] T.T. Ho, M.R. Warr, E.R. Adelman, et al., Autophagy maintains the metabolism and function of young and old stem cells, Nature 543 (2017) 205–210.

[68] L. Garcia-Prat, M. Martinez-Vicente, E. Perdiguero, et al., Autophagy maintains stemness by preventing senescence, Nature 529 (2016) 37–42.

[69] N. Mizushima, B. Levine, A.M. Cuervo, et al., Autophagy fights disease through cellular self-digestion, Nature 451 (2008) 1069–1075.

[70] Z. Yang, D.J. Klionsky, Mammalian autophagy: core molecular machinery and signaling regulation, Curr. Opin. Cell Biol. 22 (2010) 124–131.

[71] P. Sousa-Victor, S. Gutarra, L. Garcia-Prat, et al., Geriatric muscle stem cells switch reversible quiescence into senescence, Nature 506 (2014) 316–321.

[72] D. Munoz-Espin, M. Serrano, Cellular senescence: from physiology to pathology, Nat. Rev. Mol. Cell Biol. 15 (2014) 482–496.

[73] P. Sousa-Victor, E. Perdiguero, P. Munoz-Canoves, Geroconversion of aged muscle stem cells under regenerative pressure, Cell Cycle 13 (2014) 3183–3190.

[74] M. Cerletti, Y.C. Jang, L.W. Finley, et al., Short-term calorie restriction enhances skeletal muscle stem cell function, Cell Stem Cell 10 (2012) 515–519.

[75] E.D. Smith, T.L. Kaeberlein, B.T. Lydum, et al., Age- and calorie-independent life span extension from dietary restriction by bacterial deprivation in Caenorhabditis elegans, BMC Dev. Biol. 8 (2008) 49.

[76] S.C. Johnson, P.S. Rabinovitch, M. Kaeberlein, mTOR is a key modulator of ageing and age-related disease, Nature 493 (2013) 338–345.

[77] M.D. Grounds, L. Sorokin, J. White, Strength at the extracellular matrix-muscle interface, Scand. J. Med. Sci. Sports 15 (2005) 381–391.

[78] A.R. Gillies, R.L. Lieber, Structure and function of the skeletal muscle extracellular matrix, Muscle Nerve 44 (2011) 318–331.

[79] T. Laumonier, J. Menetrey, Muscle injuries and strategies for improving their repair, J. Exp. Orthop. 3 (2016) 15.

[80] C.J. Mann, E. Perdiguero, Y. Kharraz, et al., Aberrant repair and fibrosis development in skeletal muscle, Skelet. Muscle 1 (2011) 21.

[81] M. Saclier, S. Cuvellier, M. Magnan, et al., Monocyte/macrophage interactions with myogenic precursor cells during skeletal muscle regeneration, FEBS J. 280 (2013) 4118–4130.

[82] H. Yin, F. Price, M.A. Rudnicki, Satellite cells and the muscle stem cell niche, Physiol. Rev. 93 (2013) 23–67.

[83] P.M. Gilbert, K.L. Havenstrite, K.E. Magnusson, et al., Substrate elasticity regulates skeletal muscle stem cell self-renewal in culture, Science 329 (2010) 1078–1081.

[84] A.S. Brack, M.J. Conboy, S. Roy, et al., Increased Wnt signaling during aging alters muscle stem cell fate and increases fibrosis, Science 317 (2007) 807–810.

[85] L. Lukjanenko, M.J. Jung, N. Hegde, et al., Loss of fibronectin from the aged stem cell niche affects the regenerative capacity of skeletal muscle in mice, Nat. Med. 22 (2016) 897–905.

[86] P. Pessina, D. Cabrera, M.G. Morales, et al., Novel and optimized strategies for inducing fibrosis in vivo: focus on Duchenne muscular dystrophy, Skelet. Muscle 4 (2014) 7.

[87] A.L. Serrano, C.J. Mann, B. Vidal, et al., Cellular and molecular mechanisms regulating fibrosis in skeletal muscle repair and disease, Curr. Top. Dev. Biol. 96 (2011) 167–201.
[88] A.W. Hahn, F. Kern, F.R. Buhler, et al., The renin-angiotensin system and extracellular matrix, Clin. Investig. 71 (1993) S7–12.
[89] G. Juban, B. Chazaud, Metabolic regulation of macrophages during tissue repair: insights from skeletal muscle regeneration, FEBS Lett. 591 (2017) 3007–3021.
[90] D.R. Lemos, F. Babaeijandaghi, M. Low, et al., Nilotinib reduces muscle fibrosis in chronic muscle injury by promoting TNF-mediated apoptosis of fibro/adipogenic progenitors, Nat. Med. 21 (2015) 786–794.
[91] P. Munoz-Canoves, A.L. Serrano, Macrophages decide between regeneration and fibrosis in muscle, Trends Endocrinol. Metab. 26 (2015) 449–450.
[92] X. Wang, W. Zhao, R.M. Ransohoff, et al., Identification and function of fibrocytes in skeletal muscle injury repair and muscular dystrophy, J. Immunol. 197 (2016) 4750–4761.
[93] S. Ciciliot, S. Schiaffino, Regeneration of mammalian skeletal muscle. Basic mechanisms and clinical implications, Curr. Pharm. Des. 16 (2010) 906–914.
[94] D. Fiore, R.N. Judson, M. Low, et al., Pharmacological blockage of fibro/adipogenic progenitor expansion and suppression of regenerative fibrogenesis is associated with impaired skeletal muscle regeneration, Stem Cell Res. 17 (2016) 161–169.
[95] A.W. Joe, L. Yi, A. Natarajan, et al., Muscle injury activates resident fibro/adipogenic progenitors that facilitate myogenesis, Nat. Cell Biol. 12 (2010) 153–163.
[96] S. Biressi, E.H. Miyabara, S.D. Gopinath, et al., A Wnt-TGFbeta2 axis induces a fibrogenic program in muscle stem cells from dystrophic mice, Sci. Transl. Med. 6 (2014), 267ra176.
[97] Y. Guo, L. Xiao, L. Sun, et al., Wnt/beta-catenin signaling: a promising new target for fibrosis diseases, Physiol. Res. 61 (2012) 337–346.
[98] G. Lacraz, A.J. Rouleau, V. Couture, et al., Increased stiffness in aged skeletal muscle impairs muscle progenitor cell proliferative activity, PLoS One 10 (2015), e0136217.
[99] L.K. Wood, E. Kayupov, J.P. Gumucio, et al., Intrinsic stiffness of extracellular matrix increases with age in skeletal muscles of mice, J. Appl. Physiol. (1985) 117 (2014) 363–369.
[100] A.A. Mueller, C.T. van Velthoven, K.D. Fukumoto, et al., Intronic polyadenylation of PDGFRalpha in resident stem cells attenuates muscle fibrosis, Nature 540 (2016) 276–279.
[101] M.J. Conboy, I.M. Conboy, T.A. Rando, Heterochronic parabiosis: historical perspective and methodological considerations for studies of aging and longevity, Aging Cell 12 (2013) 525–530.
[102] I.M. Conboy, M.J. Conboy, A.J. Wagers, et al., Rejuvenation of aged progenitor cells by exposure to a young systemic environment, Nature 433 (2005) 760–764.
[103] M.E. Carlson, M. Hsu, I.M. Conboy, Imbalance between pSmad3 and notch induces CDK inhibitors in old muscle stem cells, Nature 454 (2008) 528–532.
[104] I.M. Conboy, M.J. Conboy, G.M. Smythe, et al., Notch-mediated restoration of regenerative potential to aged muscle, Science 302 (2003) 1575–1577.
[105] M.A. Egerman, S.M. Cadena, J.A. Gilbert, et al., GDF11 increases with age and inhibits skeletal muscle regeneration, Cell Metab. 22 (2015) 164–174.
[106] C. Elabd, W. Cousin, P. Upadhyayula, et al., Oxytocin is an age-specific circulating hormone that is necessary for muscle maintenance and regeneration, Nat. Commun. 5 (2014) 4082.
[107] P. Paliwal, N. Pishesha, D. Wijaya, et al., Age dependent increase in the levels of osteopontin inhibits skeletal muscle regeneration, Aging (Albany NY) 4 (2012) 553–566.
[108] M. Sinha, Y.C. Jang, J. Oh, et al., Restoring systemic GDF11 levels reverses age-related dysfunction in mouse skeletal muscle, Science 344 (2014) 649–652.
[109] A. Salminen, K. Kaarniranta, A. Kauppinen, Inflammaging: disturbed interplay between autophagy and inflammasomes, Aging (Albany NY) 4 (2012) 166–175.
[110] T. Shavlakadze, J. McGeachie, M.D. Grounds, Delayed but excellent myogenic stem cell response of regenerating geriatric skeletal muscles in mice, Biogerontology 11 (2010) 363–376.
[111] F. Gattazzo, A. Urciuolo, P. Bonaldo, Extracellular matrix: a dynamic microenvironment for stem cell niche, Biochim. Biophys. Acta 1840 (2014) 2506–2519.
[112] R. Bischoff, A satellite cell mitogen from crushed adult muscle, Dev. Biol. 115 (1986) 140–147.
[113] S.M. Sheehan, R. Tatsumi, C.J. Temm-Grove, et al., HGF is an autocrine growth factor for skeletal muscle satellite cells in vitro, Muscle Nerve 23 (2000) 239–245.
[114] G. McKoy, W. Ashley, J. Mander, et al., Expression of insulin growth factor-1 splice variants and structural genes in rabbit skeletal muscle induced by stretch and stimulation, J. Physiol. 516 (Pt 2) (1999) 583–592.
[115] G. Gayan-Ramirez, F. Vanderhoydonc, G. Verhoeven, et al., Acute treatment with corticosteroids decreases IGF-1 and IGF-2 expression in the rat diaphragm and gastrocnemius, Am. J. Respir. Crit. Care Med. 159 (1999) 283–289.
[116] J.R. Singleton, B.L. Baker, A. Thorburn, Dexamethasone inhibits insulin-like growth factor signaling and potentiates myoblast apoptosis, Endocrinology 141 (2000) 2945–2950.
[117] E.R. Barton-Davis, D.I. Shoturma, A. Musaro, et al., Viral mediated expression of insulin-like growth factor I blocks the aging-related loss of skeletal muscle function, Proc. Natl. Acad. Sci. U. S. A. 95 (1998) 15603–15607.
[118] N. Murakami, I.S. McLennan, I. Nonaka, et al., Transforming growth factor-beta2 is elevated in skeletal muscle disorders, Muscle Nerve 22 (1999) 889–898.

[119] P. Noirez, S. Torres, J. Cebrian, et al., TGF-beta1 favors the development of fast type identity during soleus muscle regeneration, J. Muscle Res. Cell Motil. 27 (2006) 1–8.

[120] M.L. Beggs, R. Nagarajan, J.M. Taylor-Jones, et al., Alterations in the TGFbeta signaling pathway in myogenic progenitors with age, Aging Cell 3 (2004) 353–361.

[121] M.E. Carlson, H.S. Silva, I.M. Conboy, Aging of signal transduction pathways, and pathology, Exp. Cell Res. 314 (2008) 1951–1961.

[122] J.G. Cannon, Cytokines in muscle homeostasis and disease, in: J.J. Oppenheim, J.L. Rossio, A.J.H. Gearing (Eds.), Clinical Applications of Cytokines, Oxford University Press, New York, 1993, pp. 329–336.

[123] R. Roubenoff, Catabolism of aging: is it an inflammatory process? Curr. Opin. Clin. Nutr. Metab. Care 6 (2003) 295–299.

[124] M. Visser, M. Pahor, D.R. Taaffe, et al., Relationship of interleukin-6 and tumor necrosis factor-alpha with muscle mass and muscle strength in elderly men and women: the health ABC study, J. Gerontol. A Biol. Sci. Med. Sci. 57 (2002) M326–M332.

[125] M.A. Febbraio, B.K. Pedersen, Muscle-derived interleukin-6: mechanisms for activation and possible biological roles, FASEB J. 16 (2002) 1335–1347.

[126] B.K. Pedersen, A. Steensberg, C. Fischer, et al., Exercise and cytokines with particular focus on muscle-derived IL-6, Exerc. Immunol. Rev. 7 (2001) 18–31.

[127] T. Tsujinaka, C. Ebisui, J. Fujita, et al., Muscle undergoes atrophy in association with increase of lysosomal cathepsin activity in interleukin-6 transgenic mouse, Biochem. Biophys. Res. Commun. 207 (1995) 168–174.

[128] M. Pedersen, H. Bruunsgaard, N. Weis, et al., Circulating levels of TNF-alpha and IL-6-relation to truncal fat mass and muscle mass in healthy elderly individuals and in patients with type-2 diabetes, Mech. Ageing Dev. 124 (2003) 495–502.

[129] C. Lopez-Otin, M.A. Blasco, L. Partridge, et al., The hallmarks of aging, Cell 153 (2013) 1194–1217.

[130] E.H. Blackburn, C.W. Greider, J.W. Szostak, Telomeres and telomerase: the path from maize, Tetrahymena and yeast to human cancer and aging, Nat. Med. 12 (2006) 1133–1138.

[131] T. de Lange, Shelterin: the protein complex that shapes and safeguards human telomeres, Genes Dev. 19 (2005) 2100–2110.

[132] R. Diotti, D. Loayza, Shelterin complex and associated factors at human telomeres, Nucleus 2 (2011) 119–135.

[133] C.W. Greider, Telomerase is processive, Mol. Cell. Biol. 11 (1991) 4572–4580.

[134] J.D. Watson, Origin of concatemeric T7 DNA, Nat. New Biol. 239 (1972) 197–201.

[135] F. Kadi, E. Ponsot, The biology of satellite cells and telomeres in human skeletal muscle: effects of aging and physical activity, Scand. J. Med. Sci. Sports 20 (2010) 39–48.

[136] J.R. Walker, X.D. Zhu, Post-translational modifications of TRF1 and TRF2 and their roles in telomere maintenance, Mech. Ageing Dev. 133 (2012) 421–434.

[137] S. Decary, C.B. Hamida, V. Mouly, et al., Shorter telomeres in dystrophic muscle consistent with extensive regeneration in young children, Neuromuscul. Disord. 10 (2000) 113–120.

[138] S. Decary, V. Mouly, C.B. Hamida, et al., Replicative potential and telomere length in human skeletal muscle: implications for satellite cell-mediated gene therapy, Hum. Gene Ther. 8 (1997) 1429–1438.

[139] V. Mouly, A. Aamiri, A. Bigot, et al., The mitotic clock in skeletal muscle regeneration, disease and cell mediated gene therapy, Acta Physiol. Scand. 184 (2005) 3–15.

[140] V. Renault, L.E. Thornell, G. Butler-Browne, et al., Human skeletal muscle satellite cells: aging, oxidative stress and the mitotic clock, Exp. Gerontol. 37 (2002) 1229–1236.

[141] L. Daniali, A. Benetos, E. Susser, et al., Telomeres shorten at equivalent rates in somatic tissues of adults, Nat. Commun. 4 (2013) 1597.

[142] A.T. Ludlow, E.E. Spangenburg, E.R. Chin, et al., Telomeres shorten in response to oxidative stress in mouse skeletal muscle fibers, J. Gerontol. A Biol. Sci. Med. Sci. 69 (2014) 821–830.

[143] R. Tan, L. Lan, Guarding chromosomes from oxidative DNA damage to the very end, Acta Biochim. Biophys. Sin. Shanghai 48 (2016) 617–622.

[144] J.L. Parsons, G.L. Dianov, Co-ordination of base excision repair and genome stability, DNA Repair (Amst) 12 (2013) 326–333.

[145] X.D. Zhu, L. Niedernhofer, B. Kuster, et al., ERCC1/XPF removes the 3' overhang from uncapped telomeres and represses formation of telomeric DNA-containing double minute chromosomes, Mol. Cell 12 (2003) 1489–1498.

[146] J.P. Melis, H. van Steeg, M. Luijten, Oxidative DNA damage and nucleotide excision repair, Antioxid. Redox Signal. 18 (2013) 2409–2419.

[147] A.P. Ting, G.K. Low, K. Gopalakrishnan, et al., Telomere attrition and genomic instability in xeroderma pigmentosum type-b deficient fibroblasts under oxidative stress, J. Cell. Mol. Med. 14 (2010) 403–416.

[148] J. Zhu, Y. Zhao, S. Wang, Chromatin and epigenetic regulation of the telomerase reverse transcriptase gene, Protein Cell 1 (2010) 22–32.

[149] S. Wang, C. Hu, J. Zhu, Distinct and temporal roles of nucleosomal remodeling and histone deacetylation in the repression of the hTERT gene, Mol. Biol. Cell 21 (2010) 821–832.

[150] C.E. Lee, A. McArdle, R.D. Griffiths, The role of hormones, cytokines and heat shock proteins during age-related muscle loss, Clin. Nutr. 26 (2007) 524–534.

[151] S.E. Alway, P.M. Siu, Nuclear apoptosis contributes to sarcopenia, Exerc. Sport Sci. Rev. 36 (2008) 51–57.

[152] C.A. Collins, P.S. Zammit, A.P. Ruiz, et al., A population of myogenic stem cells that survives skeletal muscle aging, Stem Cells 25 (2007) 885–894.

[153] S.S. Jejurikar, E.A. Henkelman, P.S. Cederna, et al., Aging increases the susceptibility of skeletal muscle derived satellite cells to apoptosis, Exp. Gerontol. 41 (2006) 828–836.

[154] M.F. Buas, T. Kadesch, Regulation of skeletal myogenesis by notch, Exp. Cell Res. 316 (2010) 3028–3033.

[155] M.E. Carlson, C. Suetta, M.J. Conboy, et al., Molecular aging and rejuvenation of human muscle stem cells, EMBO Mol. Med. 1 (2009) 381–391.

[156] S. Acharyya, S.M. Sharma, A.S. Cheng, et al., TNF inhibits Notch-1 in skeletal muscle cells by Ezh2 and DNA methylation mediated repression: implications in duchenne muscular dystrophy, PLoS One 5 (2010), e12479.
[157] I. Ramachandran, V. Ganapathy, E. Gillies, et al., Wnt inhibitory factor 1 suppresses cancer stemness and induces cellular senescence, Cell Death Dis. 5 (2014), e1246.
[158] I. Ramachandran, E. Thavathiru, S. Ramalingam, et al., Wnt inhibitory factor 1 induces apoptosis and inhibits cervical cancer growth, invasion and angiogenesis in vivo, Oncogene 31 (2012) 2725–2737.
[159] T.P. Rao, M. Kuhl, An updated overview on Wnt signaling pathways: a prelude for more, Circ. Res. 106 (2010) 1798–1806.
[160] S. Tanaka, K. Terada, T. Nohno, Canonical Wnt signaling is involved in switching from cell proliferation to myogenic differentiation of mouse myoblast cells, J. Mol. Signal. 6 (2011) 12.
[161] S. Tsivitse, Notch and Wnt signaling, physiological stimuli and postnatal myogenesis, Int. J. Biol. Sci. 6 (2010) 268–281.
[162] H.M. Stern, A.M. Brown, S.D. Hauschka, Myogenesis in paraxial mesoderm: preferential induction by dorsal neural tube and by cells expressing Wnt-1, Development 121 (1995) 3675–3686.
[163] J. Wagner, C. Schmidt, W. Nikowits Jr., et al., Compartmentalization of the somite and myogenesis in chick embryos are influenced by wnt expression, Dev. Biol. 228 (2000) 86–94.
[164] A. Scime, J. Desrosiers, F. Trensz, et al., Transcriptional profiling of skeletal muscle reveals factors that are necessary to maintain satellite cell integrity during ageing, Mech. Ageing Dev. 131 (2010) 9–20.
[165] H. Liu, M.M. Fergusson, R.M. Castilho, et al., Augmented Wnt signaling in a mammalian model of accelerated aging, Science 317 (2007) 803–806.
[166] E. Edstrom, M. Altun, E. Bergman, et al., Factors contributing to neuromuscular impairment and sarcopenia during aging, Physiol. Behav. 92 (2007) 129–135.
[167] A.L. Serrano, P. Munoz-Canoves, Regulation and dysregulation of fibrosis in skeletal muscle, Exp. Cell Res. 316 (2010) 3050–3058.
[168] F. Trensz, S. Haroun, A. Cloutier, et al., A muscle resident cell population promotes fibrosis in hindlimb skeletal muscles of mdx mice through the Wnt canonical pathway, Am. J. Physiol. Cell Physiol. 299 (2010) C939–C947.
[169] B.D. White, N.K. Nguyen, R.T. Moon, Wnt signaling: it gets more humorous with age, Curr. Biol. 17 (2007) R923–R925.
[170] A. Avgustinova, S.A. Benitah, Epigenetic control of adult stem cell function, Nat. Rev. Mol. Cell Biol. 17 (2016) 643–658.
[171] D. Puri, H. Gala, R. Mishra, et al., High-wire act: the poised genome and cellular memory, FEBS J. 282 (2015) 1675–1691.
[172] Y. Cao, Z. Yao, D. Sarkar, et al., Genome-wide MyoD binding in skeletal muscle cells: a potential for broad cellular reprogramming, Dev. Cell 18 (2010) 662–674.
[173] Y. Bergman, H. Cedar, DNA methylation dynamics in health and disease, Nat. Struct. Mol. Biol. 20 (2013) 274–281.
[174] J. Segales, E. Perdiguero, P. Munoz-Canoves, Epigenetic control of adult skeletal muscle stem cell functions, FEBS J. 282 (2015) 1571–1588.
[175] B.E. Bernstein, T.S. Mikkelsen, X. Xie, et al., A bivalent chromatin structure marks key developmental genes in embryonic stem cells, Cell 125 (2006) 315–326.
[176] H. Marks, T. Kalkan, R. Menafra, et al., The transcriptional and epigenomic foundations of ground state pluripotency, Cell 149 (2012) 590–604.
[177] Y. Kawabe, Y.X. Wang, I.W. McKinnell, et al., Carm1 regulates Pax7 transcriptional activity through MLL1/2 recruitment during asymmetric satellite stem cell divisions, Cell Stem Cell 11 (2012) 333–345.
[178] S. Woodhouse, D. Pugazhendhi, P. Brien, et al., Ezh2 maintains a key phase of muscle satellite cell expansion but does not regulate terminal differentiation, J. Cell Sci. 126 (2013) 565–579.
[179] S. Cheedipudi, D. Puri, A. Saleh, et al., A fine balance: epigenetic control of cellular quiescence by the tumor suppressor PRDM2/RIZ at a bivalent domain in the cyclin a gene, Nucleic Acids Res. 43 (2015) 6236–6256.
[180] D.J. Baker, B.G. Childs, M. Durik, et al., Naturally occurring p16(Ink4a)-positive cells shorten healthy lifespan, Nature 530 (2016) 184–189.
[181] J. Campisi, Aging, cellular senescence, and cancer, Annu. Rev. Physiol. 75 (2013) 685–705.
[182] R.C. Laker, J.G. Ryall, DNA methylation in skeletal muscle stem cell specification, proliferation, and differentiation, Stem Cells Int. 2016 (2016) 5725927.
[183] D. Palacios, P.L. Puri, The epigenetic network regulating muscle development and regeneration, J. Cell. Physiol. 207 (2006) 1–11.
[184] B.M. Ling, N. Bharathy, T.K. Chung, et al., Lysine methyltransferase G9a methylates the transcription factor MyoD and regulates skeletal muscle differentiation, Proc. Natl. Acad. Sci. U. S. A. 109 (2012) 841–846.
[185] G. Caretti, M. Di Padova, B. Micales, et al., The Polycomb Ezh2 methyltransferase regulates muscle gene expression and skeletal muscle differentiation, Genes Dev. 18 (2004) 2627–2638.
[186] E. Vire, C. Brenner, R. Deplus, et al., The Polycomb group protein EZH2 directly controls DNA methylation, Nature 439 (2006) 871–874.
[187] D. Palacios, C. Mozzetta, S. Consalvi, et al., TNF/p38alpha/polycomb signaling to Pax7 locus in satellite cells links inflammation to the epigenetic control of muscle regeneration, Cell Stem Cell 7 (2010) 455–469.
[188] J. Terragni, G. Zhang, Z. Sun, et al., Notch signaling genes: myogenic DNA hypomethylation and 5-hydroxymethylcytosine, Epigenetics 9 (2014) 842–850.
[189] S.M. Taylor, P.A. Jones, Multiple new phenotypes induced in 10T1/2 and 3T3 cells treated with 5-azacytidine, Cell 17 (1979) 771–779.
[190] A. Blais, C.J. van Oevelen, R. Margueron, et al., Retinoblastoma tumor suppressor protein-dependent methylation of histone H3 lysine 27 is associated with irreversible cell cycle exit, J. Cell Biol. 179 (2007) 1399–1412.
[191] R. Sambasivan, S. Cheedipudi, N. Pasupuleti, et al., The small chromatin-binding protein p8 coordinates the association of anti-proliferative and pro-myogenic proteins at the myogenin promoter, J. Cell Sci. 122 (2009) 3481–3491.

[192] S.V. Forcales, S. Albini, L. Giordani, et al., Signal-dependent incorporation of MyoD-BAF60c into Brg1-based SWI/SNF chromatin-remodelling complex, EMBO J. 31 (2012) 301–316.
[193] C. Simone, S.V. Forcales, D.A. Hill, et al., p38 pathway targets SWI-SNF chromatin-remodeling complex to muscle-specific loci, Nat. Genet. 36 (2004) 738–743.
[194] A. Cuadrado, N. Corrado, E. Perdiguero, et al., Essential role of p18Hamlet/SRCAP-mediated histone H2A.Z chromatin incorporation in muscle differentiation, EMBO J. 29 (2010) 2014–2025.
[195] M. Cargnello, P.P. Roux, Activation and function of the MAPKs and their substrates, the MAPK-activated protein kinases, Microbiol. Mol. Biol. Rev. 75 (2011) 50–83.
[196] A. Cuenda, S. Rousseau, p38 MAP-kinases pathway regulation, function and role in human diseases, Biochim. Biophys. Acta 1773 (2007) 1358–1375.
[197] F. Lluis, E. Perdiguero, A.R. Nebreda, et al., Regulation of skeletal muscle gene expression by p38 MAP kinases, Trends Cell Biol. 16 (2006) 36–44.
[198] P. Brien, D. Pugazhendhi, S. Woodhouse, et al., p38alpha MAPK regulates adult muscle stem cell fate by restricting progenitor proliferation during postnatal growth and repair, Stem Cells 31 (2013) 1597–1610.
[199] N.C. Jones, K.J. Tyner, L. Nibarger, et al., The p38alpha/beta MAPK functions as a molecular switch to activate the quiescent satellite cell, J. Cell Biol. 169 (2005) 105–116.
[200] I. Ferreiro, M. Barragan, A. Gubern, et al., The p38 SAPK is recruited to chromatin via its interaction with transcription factors, J. Biol. Chem. 285 (2010) 31819–31828.
[201] C. Mozzetta, S. Consalvi, V. Saccone, et al., Selective control of Pax7 expression by TNF-activated p38alpha/polycomb repressive complex 2 (PRC2) signaling during muscle satellite cell differentiation, Cell Cycle 10 (2011) 191–198.
[202] F. Lluis, E. Ballestar, M. Suelves, et al., E47 phosphorylation by p38 MAPK promotes MyoD/E47 association and muscle-specific gene transcription, EMBO J. 24 (2005) 974–984.
[203] C. Serra, D. Palacios, C. Mozzetta, et al., Functional interdependence at the chromatin level between the MKK6/p38 and IGF1/PI3K/AKT pathways during muscle differentiation, Mol. Cell 28 (2007) 200–213.
[204] A. Zetser, E. Gredinger, E. Bengal, p38 mitogen-activated protein kinase pathway promotes skeletal muscle differentiation. Participation of the Mef2c transcription factor, J. Biol. Chem. 274 (1999) 5193–5200.

Chapter 10

Aging and stability of cardiomyocytes

Shouvik Chakravarty[a], Johnson Rajasingh[b], and Satish Ramalingam[a]
[a]Department of Genetic Engineering, School of Bio-Engineering, SRM Institute of Science and Technology, Kanchipuram, Tamil Nadu, India.
[b]Bioscience Research, Medicine-Cardiology, University of Tennessee Health Science Center, Memphis, TN, United States

1 Introduction

Age is often found to be in strong correlation with a higher incidence of complex or neurodegenerative ailments, such as cancers, Parkinson's disease, Alzheimer's disease, and dementia, to name a few. Age is also a major risk factor associated with cardiovascular disease-related mortality [1, 2]. Heart diseases remain the principal health threat in several developed countries, including the United States. The elderly population are the worst affected more often than the rest of the age group. Centers for Disease Control and Prevention (CDC), a leading US healthcare and research organization, claims that 12% of adults in US population (≥ 18 years of age) suffer from either hypertension, stroke, cardiomyopathies, or heart failure (HF) [3]. Cardiomyocytes generate contractile force in the heart, which is essential for normal functionality. They also enable the heart to beat rhythmically [4]. Altered CM function is the main cause of reduced heart function and gradually less tolerance to stress. These changes often manifest in the form of decrease in the total number of CMs due to necrosis and apoptosis, ultimately leading to disease condition. Inability to meet the chronic demand for increased contractile force produces insufficient cardiac output, less than what organism needs (also known as heart failure), a major cause of mortality in western countries of the world. After birth, myocytes increase in size without substantial increases in cell number. Molecular mechanisms behind this process of hypertrophy are not very well understood. However, several intracellular signaling pathways that might play a role have been identified in numerous studies [5].

First changes occurring in cardiac gene expression are often found to be associated with age. It mainly involves proteins that are essential for cardiac functionalities such as contraction and relaxation [1, 6, 7]. Research initiated in this aspect involved gene-based analyses mainly focusing on myosin heavy chain (MHC) proteins. They are known to provide structural integrity to cardiomyocytes. Several studies showed an isoform shift in MHC transcript/protein abundance in case of aging myocardium in rodents and in the atria of humans. Also, there are reports of altered myosin ATPase activity [8–10]. The human heart contains mainly the β-MHC isoform, and also α-MHC to a lesser extent. β-MHC gene activity is found to be upregulated in disease or aging conditions in case of rodents [11]. One of the major mutations found in the β-MHC gene is R403Q, and these mutations are known to cause 45% of familial hypertrophic cardiomyopathies (FHC) cases in humans [12].

A number of changes (mostly functional) occur in SERCA2 [S(sarcoplasmic reticulum)/ER (endoplasmic reticulum) Ca^{2+}-ATPase 2] pump with the progress of age. Regulatory protein phospholamban (PLB) is found to repress SERCA2 activity, when SERCA2 is in unphosphorylated state. This situation is reversed when SERCA2 is phosphorylated by either one of the two proteins: protein kinase A (PKA) or CaMKII (Ca^{2+}/calmodulin-dependent protein kinase II), at amino acid position Ser16 and Thr17, respectively. The said protein is responsible for pumping cytosolic calcium ions into the SR [13–15], and the changes might lead to slowed myocardial relaxation. The changes of expression in NCX1 protein (which is actually a Na^+/Ca^{2+} exchanger) can also have an effect on cardiac relaxation with aging, as evidenced by the study by Koban et al. [16]. These studies provide a molecular explanation to the biological aging of the heart.

A number of other cardiac factors were found to be altered with age, in a number of studies [17]. In rats, mRNA level increase in case of IP3R (inositol 1,4,5-trisphosphate receptor) and transient receptor potential canonical (TRPC) channels was found to be associated with age, this incident being reported in case of both atria and ventricles [18]. Studies have found that the expression levels of angiotensin II receptor subtypes 1 and 2, COX-2 (cyclooxygenase 2) and NF-κB (nuclear factor κB) increase with age [19]. A number of studies found inversely proportional relations between expression of M2-cholinoreceptor, β1-adrenergic receptor, adenylate cyclases and estrogen receptor, and age [20–23]. Two markers have been associated with aging-related hypertrophy, namely atrial natriuretic peptide (ANP) and brain natriuretic peptide (BNP) [24]. It is also shown that mitochondrial-encoded transcripts are also reduced, which can often lead to deteriorated mitochondrial function with age [25, 26].

2 Cardiomyocytes and aging

Cardiac aging involves alterations on numerous levels. The changes can be on a cellular level, might involve certain macromolecules, can be related to mitochondria and energy, and so on. Cell renewal (stem cell function) might also be involved. Alterations in the extracellular matrix (ECM) are often suspected to be instrumental for the development of structural changes that ultimately have a devastating effect in aged individuals, promoting the breakdown of cardiac machinery [27]. Cellular senescence is absolutely critical for the stem cell pool homeostasis, signifying its pivotal role in tissue remodeling (in normal and pathological conditions) [28].

2.1 Stability of cardiomyocytes and potential regeneration

In the recent past, researchers have adopted different approaches for the regeneration of functional cardiomyocytes (CMs). A group of researchers proposed to leverage intrinsic proliferative capacity of cardiomyocytes, while the others sought to tackle the question by enhancing resident/nonresident progenitor cell differentiation [29–32]. In their seminal 2006 paper, Takahashi and Yamanaka described the discovery of induced pluripotent stem cells (iPSCs), which revolutionized the regenerative medicine of the heart. It broke new ground as regards cardiac regeneration [33]. After that discovery, different mechanisms have been put forth, such as cardiac differentiation of iPSCs and conversion between differentiated cell types bypassing a pluripotent intermediate (dubbed as "direct reprogramming"). Studies have reported evidence in favor of these hypotheses for a number of cell types, which includes CM [34].

2.2 Mechanism of aging in cardiomyocytes

Researchers have identified aging to be a major barrier to cell reprogramming. Aging cells need to adjust for some age-associated characteristics, to help keep the adverse effects in check and to significantly increase the efficiency of reprogramming [35, 36]. Senescence can occur via some of the mechanisms described below:

(a) As a tissue remodeling mechanism

Cell senescence has evolved as a mechanism that prevents the replication of damaged DNA and subsequently prevents it from getting transmitted in the future generations of cells. Therefore, it has a vital role in mediating tissue remodeling and repair. It is also said to act as a tumor suppressor mechanism. Evidences suggest that embryonic development is associated with cellular senescence [37]. A peculiar pro-inflammatory behavior is often observed in senescent cells, called the senescence-associated secretory phenotype (SASP), which can enable the cells to induce changes in surrounding normal tissues [38]. SASP exerts its function by numerous mechanisms: chemokine-mediated, pro-inflammatory cytokine-mediated, or interleukin-mediated inflammation [39, 40]. SASP can also utilize a reactive oxygen species (ROS) generation-mediated mechanism and subsequent DNA damage [41]. Some other characteristics of senescent cells include the appearance of senescence-associated heterochromatin foci (SAHF) and senescence-associated β-galactosidase (SA-β-gal) expression.

Epigenomic change and genomic damage coupled with the activation of DNA damage response (DDR) are the salient features of senescence-induced stimuli. Such stimuli mediate initiation as well as maintenance of senescence growth arrest in mice and humans (in vitro and in vivo) [42]. Erroneous replication origins and replication fork collapse can be caused by mitogenic signals driven by oncogenes or overexpression of genes that promote proliferation. This situation can give rise to a situation called "oncogene-induced senescence" (OIS), which is basically a kind of senescence growth arrest [43]. Overall, it can be stated that senescence works toward maintaining and sustaining in vivo reprogramming, modulating the tissue microenvironment accordingly [44].

(b) Reprogramming-induced senescence (RIS), A cell-autonomous approach

The reprogramming process can itself initiate and trigger senescence. It is called reprogramming-induced senescence (RIS) (Fig. 1). It has a pivotal role in preventing aging cell-fate decisions and manipulations, using a p16/p21-dependent senescence response [45]. RIS is also demonstrated to be a cause of the reversal of a number of features associated with age, such as arrest of cell cycle, DNA methylation/histone modifications [46], length of telomere [47], and expression of pro-inflammatory factors [48].

(c) In vivo reprogramming

In 2016, it was shown for the first time that the Yamanaka factors (Oct3/4, Sox2, Klf4, c-myc) impose DNA damage on many other cells, in addition for being directly responsible for starting cell reprogramming process. As a result, they can drive the cells to reach a senescent stage [49]. The resultant senescent cells (reprogramming-induced) are shown to exert their reprogramming functionality in vivo, in conjunction with the paracrine action of certain SASP components

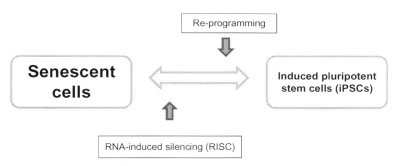

FIG. 1 Effect of cellular senescence on the reprogramming process.

(such as IL-6, an interleukin) [50]. As it has been shown that SASP may have the capability to initiate stemness or progenitor characteristics in a number of tissues, characterization of a signal able to induce tissue regeneration to ultimately repair tissues can be proved to be of great significance [51].

This process is mediated by the genes **Oct4, Sox2, Klf4,** and **c-myc**, as well as **senescence-associated secretory phenotype (SASP)**.

Cardiac senescence can be defined as the decrease in the number and increase in size of cardiomyocytes (often correlated with age and cardiovascular diseases). Along with increase in amount of the extracellular matrix, it is one of the basic features of cardiac remodeling [52]. Telomere length is also an important feature of aging. It has often been cited as one of the reasons of cardiovascular disease development and associated mortality in case of the elderly sufferers [53]. Two of the major aspects associated with cardiomyocyte senescence are defective adrenergic signaling (dependent on age) and handling of calcium. Defective (reduced) plasma clearance and excessive spillover from the affected tissues can result in increased norepinephrine levels in circulation. Impaired calcium handling can result from a dysfunctional myocardial sarcoplasmic reticulum calcium adenosine triphosphatase (SERCA2a). Cardiomyocyte aging is thought to be the reason for this condition [54]. Inability of mitochondria in maintaining homeostasis of reactive oxygen species causes these highly reactive, often toxic products to accumulate. Sirtuins (which is a family of nicotinamide adenine dinucleotide-dependent deacetylases) have been implicated in aging, having a demonstrated role [55]. Aging also can occur when several ECM proteins are overexpressed, such as collagen, fibronectin, and alpha 1 and alpha 5 integrin [56].

3 Regeneration of cardiomyocytes—A mechanistic overview

There exist numerous possible mechanisms for the potential regeneration of cardiomyocytes.

3.1 Cardiomyocyte survival and protection

If the heart has to function normally, it must preserve its tissue and cellular function. Cardiomyocyte survival is essential for myocardium protection. Cardiomyocyte survival can help to maintain hemodynamic performance. A number of signaling pathways are involved in cardiomyocyte survival [57–59]. These pathways function through a combined mechanism: inhibition of senescence and reduction of inflammation. Akt, a Ser/Thr kinase pathway (Fig. 2), is widely studied and reported in the case of survival signaling [60]. Akt signaling pathways have been demonstrated to promote cardiomyocyte survival in human [61] and other models [62, 63]. In several studies, the downstream targets of Akt in cardiomyocytes are found to be the following: phosphoinositol 3-kinase by Ras/ERK1/2 (extracellular signal-related kinase-1/2) [64], mTOR [65], Raf1 [66], CREB [67], FoxO [68], and PIM-1 [69], among others. Therefore, Akt can act as a key regulator for different biological activities, including survival.

Autophagy can be a mechanism for cardiomyocyte survival after injury. BNIP-3 (BCL2-interacting protein 3) activation can help to achieve this goal [70]. Antiapoptotic protein Bcl-2 is an inhibitor of Beclin-I. Beclin-I helps in the regulation of cell survival after initial injury has occurred [71]. Antiapoptotic signaling also has the potential to promote the survival of the cardiomyocytes. If ischemia occurs, subsequently apoptosis also peaks. Both extrinsic and intrinsic pathways have the ability to promote apoptosis [72]. Congestive heart failure can increase the amount of tumor necrosis factor-α (TNF-α) [73]. When the stress protein high-mobility group box 1 is released, TNF-α activity increases [74]. It has a significant influence on apoptotic effects in cardiomyocytes. In the intrinsic pathway, BAX (Bcl-2-associated X protein) and BH-3 (both are proapoptotic proteins) are found to have the ability to destabilize the outer mitochondrial membrane by promoting apoptosis. Subsequently, caspases are activated [75].

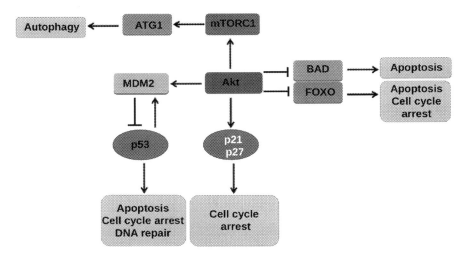

FIG. 2 Akt mediates cell survival process.

3.2 By means of inflammation reduction

Cardiac healing involves multiple types of cells. Mononuclear cells and mast cells are infiltrated in response to different cytokine and growth factors. These are C5a, TGF (transforming growth factor)-β1, MCP-1 (methyl-accepting chemotaxis protein), IL (interleukin)-8, histamine, TNF-α, IL-6, among others [76]. In some cases, mast cells are accumulated in the region where an injury or infarction has occurred. The process is mediated by chemoattractants, mainly stem cell factor (SCF), platelet-derived growth factor (PDGF), vascular endothelial growth factor (VEGF), and basic fibroblast growth factor (bFGF) [77–81]. Myofibroblasts are found to express a frizzled 2 homolog gene. Apoptosis can occur in the scar. Thereby, cardiac healing (reduction of inflammation) can also be spatially controlled [82]. There may be other inflammatory response-mediated mechanisms, such as infiltration of neutrophils, neutrophil-endothelial interactions, selectin activation, leukocyte β2 integrins, chemotaxis-mediated mechanisms, and chemokines. Additional cells are recruited by neutrophils to the required site via the release of oxidases, proteases, etc. [83]. Studies have provided strong evidence in favor of both humoral inflammatory responses and antibody-mediated immune responses, hence coordinating inflammation [84].

3.3 Cell-to-cell communication

Myocardial mechanical force has to be adjusted to preserve proper functionalities of the heart. Signaling between cells and extracellular matrix can mediate this. It can preserve proper cardiac functionalities in the case of a sudden myocardial injury. Fibroblast-induced cell-extracellular matrix signaling is influenced by multiple factors, including integrins, extracellular matrix proteins (collagen, laminin, fibronectin, etc.), and metalloproteinases. Stiffness of tissue can trigger focal adhesion kinases (FAKs) and yes-associated protein (YAP)/TAZ (transcriptional coactivator with PDZ-binding motif). Together, they activate the interstitial cells [85]. Communication between cardiomyocytes is necessary, and this happens by a number of mechanisms such as gap junctions (Macula communicans) and adhesion complexes. Secretome factors might also play a role in this. Endothelium can have a role in the cardiomyocyte response via a neuregulin (Nrg)-regulated mechanism, which can ultimately provide the protection from ischemic damages. It is an example of paracrine signaling [86].

3.4 Through angiogenesis and vascularization

Angiogenic response disruption and angiocrine signaling (which is mediated by endothelial cells) can potentially cause heart failure in mice [87]. Endothelial progenitor cells (EPCs) get incorporated into capillaries that are newly formed. EPCs play an important role in vascular reorganization and repair after a myocardial injury, mainly by moving from bone marrow to the site of injury [88]. EPCs are known to have pro-angiogenic activity. This is mainly mediated by paracrine signaling. Different types of inflammatory cells may act together with EPCs to achieve the same goal, namely adipose tissue-derived hematopoietic progenitor cells, monocytes, T cell subpopulations, and mesenchymal cells derived from bone marrow and tissue [89].

3.5 Cardiomyogenesis

New cardiomyocyte formation can occur mainly by two ways: (a) from differentiation of resident stem and progenitor cell pool and (b) dedifferentiation of preexisting cardiomyocytes. This second mechanism is thought to be functionally insignificant, as evidenced in a number of studies [90, 91]. In lower vertebrate animals, cardiomyocyte dedifferentiation and proliferation are the major mechanisms of heart regeneration [92]. Considerable plasticity is observed in the case of mammalian postnatal cardiomyocytes. Epigenomic reprogramming (e.g., methylation) is the chief mechanism responsible for cardiomyocyte dedifferentiation. Genes maintaining cardiac structure and function may lose their expression, therefore adapting a mechanism different from cell cycle reentry [93].

3.6 Molecular mechanisms behind proliferation and cell cycle

Identification of genetic factors that mediate cardiomyocyte mitosis in the adult heart is of utmost importance to researchers and clinicians alike, because these factors can help to pinpoint targets for cardiac regeneration (Fig. 3). Hippo-YAP signaling pathway is a major pathway for the regulation of cardiomyocyte proliferation (Fig. 4). Hippo-deficient mouse embryos are demonstrated to develop a condition known as cardiomegaly (intense cardiomyocyte proliferation). Canonical Wnt/β-catenin signaling also plays a role in this [94, 95]. Activation of YAP enhances cardiac regeneration and cardiac function. If an injury or infarction is induced, YAP promotes cardiomyocyte proliferation and renewal [96]. Other factors that regulate

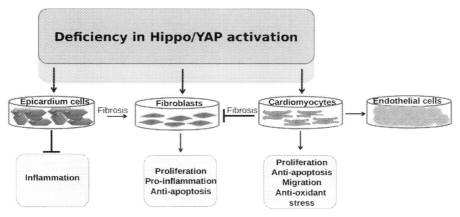

FIG. 3 Hippo/YAP pathway in the heart.

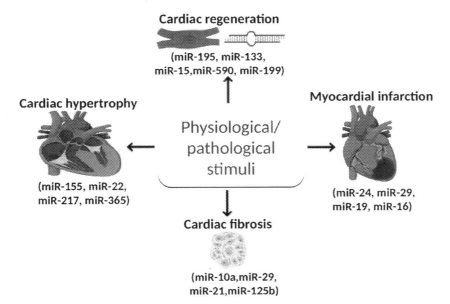

FIG. 4 Different microRNAs in cardiac regeneration and other cardiac conditions.

cardiomyocyte proliferation are Meis-1 (Meis homeobox 1, a homeodomain transcription factor, which is crucial for embryonic hematopoiesis) and Nrg-1. Deletion of Meis-1 in adult hearts produces increased cytokinesis. Simultaneously, cardiomyocytes can enter into the cell cycle [97]. In the case of neonates, overexpression of Meis-1 causes premature arrest of the cell cycle. Recombinant Nrg-1 has the capability to promote new myocardium generation postinjury in neonates. Therefore, the administration of recombinant Nrg-1 in infants suffering from congenital heart diseases can be a possible alternative to surgery [98]. Myocardial repair can be mediated by different paracrine factors and corresponding signaling pathways. FGF is shown to be essential for regenerating muscles, which mediates the said regeneration by mainly orchestrating a crosstalk between epicardial and myocardial incidents. It thus promotes the survival of myocytes in mammalian hearts that have suffered injury as well as promotes angiogenesis [99]. Fibroblasts are also major players in this aspect. They promote a WNT signaling pathway-mediated transdifferentiation process. During this fibroblast transdifferentiation, contractile protein expression and stress fiber formation occur, promoting myocardial repair [100].

3.7 miRNA-mediated regeneration

miRNAs are often described as so-called molecular switches (that have the ability to either activate or inhibit mechanisms that regulate the development of cells and tissues, in simpler terms promote on/off situations) [101]. miRNAs are also found to exert their effects in heart development (Fig. 5) [102]. In a miRNA expression profiling study using mouse cardiac ventricles, the investigators found that 71 miRNAs are either up- or downregulated in two developmental stages: within and beyond a specific regenerative window [103]. They found miRNA-195 to be the most highly upregulated miRNA. In a study by Eulalio et al. [104], the investigators sought to identify miRNAs that can promote the proliferation of cardiomyocytes. Almost 40 miRNAs were found to increase the DNA synthesis and cytokinesis process in the case of both neonatal mouse and rat cardiomyocytes. They also found that two miRNAs miRNA-590 and miRNA-199 have the ability to promote cardiomyocyte reentry into the cell cycle. Studies have implicated that miRNA-34a also has the ability for regulating cell cycle activity and cardiomyocyte death. miRNA-34a mediates its function in both neonatal and adult heart. Overexpression of the miRNA-34a is found to prevent cardiac regeneration in the neonatal mouse heart, while inhibition of miRNA-34a improves cardiac function and repair in the adult heart that has suffered injury or infarction. A number of cell cycle and survival genes play key roles in this process, for example, Bcl2, Cyclin D1, and Sirt1, to name a few [105].

3.8 Cardiac-resident stem cell-mediated regeneration

Adult mouse heart contains a population of c-kit$^+$Lin$^-$ cells. These cells are known to express markers of cardiomyocyte progenitors. They possess self-renewal potential and can also differentiate into cardiomyocytes [106]. Studies have investigated the cardiomyocyte progenitor activity of these cells in adult mouse model for heart injury [107]. It is reported that the human heart also has a c-kit$^+$ stem cell population having the capability of dividing and differentiating into myocytes

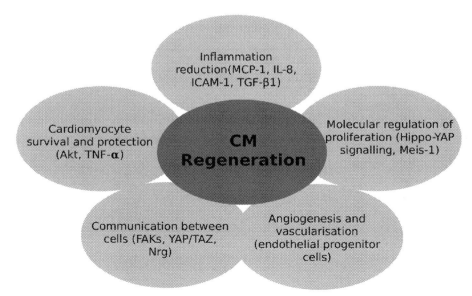

FIG. 5 Major mechanisms of cardiomyocyte regeneration and associated factors.

[108]. It is often considered to be associated with a decrease in telomerase activity and/or cellular senescence. Although the research community has long believed that regeneration of the neonatal heart happens through dedifferentiation of cardiomyocytes, there is scant evidence available experimentally. In particular, the precise role which the cardiac progenitor/stem cells play is yet to be clearly understood.

4 Conclusion and future perspectives

A considerable amount of progress has been made in the field of cardiac research since the path-breaking idea was put forward that the human heart is capable of regeneration and repair. Cardiac research has been focused on studying the characteristics of early adult myocardial structures that set it apart from its neonatal counterpart. Molecular control of cellular behavior is thought to be crucial in this respect. However, researchers faced significant challenge to actually characterize the regulation of cellular regenerative potential in animal systems and validating their hypotheses in vivo. A vast amount of theoretical and practical challenges still remain unaddressed. Limitations of current approaches should be recognized and acknowledged, which can ultimately help to build on what knowledge is already there, and to formulate and design effective treatment strategies.

References

[1] E.G. Lakatta, Cardiovascular regulatory mechanisms in advanced age, Physiol. Rev. 73 (1993) 413–467.
[2] S.S. Najjar, A. Scuteri, E.G. Lakatta, Arterial aging: is it an immutable cardiovascular risk factor? Hypertension 46 (2005) 454–462.
[3] W. Rosamond, K. Flegal, K. Furie, A. Go, K. Greenlund, N. Haase, S.M. Hailpern, M. Ho, V. Howard, B. Kissela, et al., Heart disease and stroke statistics - 2008 update: a report from the American Heart Association Statistics Committee and Stroke Statistics Subcommittee, Circulation 117 (2008) E25–E146.
[4] E.A. Woodcock, S.J. Matkovich, Cardiomyocytes structure, function and associated pathologies, Int. J. Biochem. Cell Biol. 37 (2005) 1746–1751.
[5] G.W. Dorn II, T. Force, Protein kinase cascades in the regulation of cardiac hypertrophy, J. Clin. Investig. 115 (2005) 527–537.
[6] B. Swynghedauw, Developmental and functional adaptation of contractile proteins in cardiac and skeletal muscles, Physiol. Rev. 66 (1986) 710–771.
[7] B. Swynghedauw, S. Besse, P. Assayag, F. Carre, B. Chevalier, D. Charlemagne, C. Delcayre, S. Hardouin, C. Heymes, J.M. Moalic, Molecular and cellular biology of the senescent hypertrophied and failing heart, Am. J. Cardiol. 76 (1995) D2–D7.
[8] X. Long, M.O. Boluyt, L. O'Neill, J.S. Zheng, G. Wu, Y.K. Nitta, M.T. Crow, E.G. Lakatta, Myocardial retinoid X receptor, thyroid hormone receptor, and myosin heavy chain gene expression in the rat during adult aging, J. Gerontol. A Biol. Sci. Med. Sci. 54 (1999) B23–B27.
[9] A.-M. Lompré, J.J. Mercadier, C. Wisnewsky, P. Bouveret, C. Pantaloni, A. D'Albis, K. Schwartz, Species and age-dependent changes in the relative amounts of cardiac myosin isoenzymes in mammals, Dev. Biol. 84 (1981) 286–290.
[10] J.J. Mercadier, D. de la Bastie, P. Menasche, A.N. Van Cao, P. Bouveret, P. Lorente, A. Piwnica, R. Slama, K. Schwartz, α-Myosin heavy chain isoform and atrial size in patients with various types of mitral valve dysfunction: a quantitative study, J. Am. Coll. Cardiol. 9 (1987) 1024–1030.
[11] B. Swynghedauw, C. Delcayre, J.M. Moalic, Y. Lecarpentier, A. Ray, J.J. Mercadier, A.M. Lompre, M.C. Aumont, K. Schwartz, Isoenzymic changes in myosin and hypertrophy; adaptation during chronic mechanical overload, Eur. Heart J. 3 (Suppl. A) (1982) 75–82.
[12] Q. Xu, S. Dewey, S. Nguyen, A.V. Gomes, Malignant and benign mutations in familial cardiomyopathies: insights into mutations linked to complex cardiovascular phenotypes, J. Mol. Cell. Cardiol. 48 (2010) 899–909.
[13] P. Assayag, D. Charlemagne, J. deLeiris, F. Boucher, P.E. Valere, S. Lortet, B. Swynghedauw, S. Besse, Senescent heart compared with pressure overload-induced hypertrophy, Hypertension 29 (1997) 15–21.
[14] M.T. Jiang, M.P. Moffat, N. Narayanan, Age-related alterations in the phosphorylation of sarcoplasmic reticulum and myofibrillar proteins and diminished contractile response to isoproterenol in intact rat ventricle, Circ. Res. (1993).
[15] B.S. Cain, D.R. Meldrum, K.S. Joo, J.E. Wang, X.Z. Meng, J.C. Cleveland, A. Banerjee, A.H. Harken, Human SERCA2a levels correlate inversely with age in senescent human myocardium, J. Am. Coll. Cardiol. 32 (1998) 458–467.
[16] M.U. Koban, S.A. Brugh, D.R. Riordon, K.A. Dellow, H.T. Yang, D. Tweedie, K.R. Boheler, A distant upstream region of the rat multipartite Na^+-Ca^{2+} exchanger NCX1 gene promoter is sufficient to confer cardiac-specific expression, Mech. Dev. 109 (2001) 267–279.
[17] E.G. Lakatta, S.J. Sollott, Perspectives on mammalian cardiovascular aging: humans to molecules, Comp. Biochem. Physiol. A Mol. Integr. Physiol. 132 (2002) 699–721.
[18] P. Kaplan, D. Jurkovicova, E. Babusikova, S. Hudecova, P. Racay, M. Sirova, J. Lehotsky, A. Drgova, D. Dobrota, O. Krizanova, Effect of aging on the expression of intracellular Ca 2+ transport proteins in a rat heart, Mol. Cell. Biochem. 301 (2007) 219–226.
[19] S. Del Ry, M. Maltinti, D. Giannessi, G. Cavallini, E. Bergamini, Age-related changes in endothelin-1 receptor subtypes in rat heart, Exp. Aging Res. 34 (2008) 251–266.
[20] M. Volkova, R. Garg, S. Dick, K.R. Boheler, Aging-associated changes in cardiac gene expression, Cardiovasc. Res. 66 (2005) 194–204.
[21] S. Hardouin, P. Mansier, B. Bertin, T. Dakhly, B. Swynghedauw, J.M. Moalic, β-Adrenergic and muscarinic receptor expression are regulated in opposite ways during senescence in rat left ventricle, J. Mol. Cell. Cardiol. 29 (1997) 309–319.
[22] S. Hardouin, E. Bourgeois, S. Besse, C.A. Machida, B. Swynghedauw, J.M. Moalic, Decreased accumulation of β 1-adrenergic receptor, G-α-S and total myosin heavy-chain messenger-rnas in the left-ventricle of senescent rat-heart, Mech. Ageing Dev. 71 (1993) 169–188.

[23] J.G. Dobson, J. Fray, J.L. Leonard, R.E. Pratt, Molecular mechanisms of reduced β-adrenergic signaling in the aged heart as revealed by genomic profiling, Physiol. Genomics 15 (2003) 142–147.

[24] A. Younes, M.O. Boluyt, L. O'Neill, A.L. Meredith, M.T. Crow, E.G. Lakatta, Age-associated increase in rat ventricular ANP gene expression correlates with cardiac hypertrophy, Am. J. Phys. 269 (1995) H1003–H1008.

[25] D.C. Wallace, A mitochondrial paradigm for degenerative diseases and ageing, Novartis Found. Symp. 235 (2001) 247–263.

[26] E.J. Lesnefsky, S. Moghaddas, B. Tandler, J. Kerner, C.L. Hoppel, Mitochondrial dysfunction in cardiac disease: ischemia-reperfusion, aging, and heart failure, J. Mol. Cell. Cardiol. 33 (2001) 1065–1089.

[27] L. Iop, E. dal Sasso, L. Schirone, M. Forte, M. Peruzzi, E. Cavarretta, et al., The light and shadow of senescence and inflammation in cardiovascular pathology and regenerative medicine, Mediat. Inflamm. 2017 (2017), 7953486.

[28] D. Muñoz-Espín, M. Serrano, Cellular senescence: from physiology to pathology, Nat. Rev. Mol. Cell Biol. 15 (7) (2014) 482–496.

[29] Y.L. Tang, Y.J. Wang, L.J. Chen, Y.H. Pan, L. Zhang, N.L. Weintraub, Cardiac-derived stem cell-based therapy for heart failure: progress and clinical applications, Exp. Biol. Med. 238 (3) (2013) 294–300.

[30] S.J. Jansen of Lorkeers, J.E. Eding, H.M. Vesterinen, T.I. van der Spoel, E.S. Sena, H.J. Duckers, et al., Similar effect of autologous and allogeneic cell therapy for ischemic heart disease: systematic review and meta-analysis of large animal studies, Circ. Res. 116 (1) (2015) 80–86.

[31] J.J.H. Chong, X. Yang, C.W. don, E. Minami, Y.-W. Liu, J.J. Weyers, et al., Human embryonic-stem-cell-derived cardiomyocytes regenerate non-human primate hearts, Nature 510 (7504) (2014) 273–277.

[32] S.E. Senyo, R.T. Lee, B. Kühn, Cardiac regeneration based on mechanisms of cardiomyocyte proliferation and differentiation, Stem Cell Res. 13 (3) (2014) 532–541.

[33] K. Takahashi, S. Yamanaka, Induction of pluripotent stem cells from mouse embryonic and adult fibroblast cultures by defined factors, Cell 126 (4) (2006) 663–676.

[34] B. Ebrahimi, In vivo reprogramming for heart regeneration: a glance at efficiency, environmental impacts, challenges and future directions, J. Mol. Cell. Cardiol. 108 (2017) 61–72.

[35] S. Mahmoudi, A. Brunet, Aging and reprogramming: a two-way street, Curr. Opin. Cell Biol. 24 (6) (2012) 744–756.

[36] L. Lapasset, O. Milhavet, A. Prieur, E. Besnard, N. Ait-Hamou, et al., Rejuvenating senescent and centenarian human cells by reprogramming through the pluripotent state, Genes Dev. 25 (21) (2011) 2248–2253.

[37] H. Davaapil, J.P. Brockes, M.H. Yun, Conserved and novel functions of programmed cellular senescence during vertebrate development, Development 144 (1) (2017) 106–114.

[38] A. Freund, A.V. Orjalo, P.-Y. Desprez, J. Campisi, Inflammatory networks during cellular senescence: causes and consequences, Trends Mol. Med. 16 (5) (2010) 238–246.

[39] W. Xue, L. Zender, C. Miething, R.A. Dickins, E. Hernando, V. Krizhanovsky, et al., Senescence and tumour clearance is triggered by p53 restoration in murine liver carcinomas, Nature 445 (7128) (2007) 656–660.

[40] L. Hoenicke, L. Zender, Immune surveillance of senescent cells—biological significance in cancer- and non-cancer pathologies, Carcinogenesis 33 (6) (2012) 1123–1126.

[41] J.C. Acosta, A. Banito, T. Wuestefeld, A. Georgilis, P. Janich, J.P. Morton, et al., A complex secretory program orchestrated by the inflammasome controls paracrine senescence, Nat. Cell Biol. 15 (8) (2013) 978–990.

[42] F. Rodier, J.-P. Coppé, C.K. Patil, W.A.M. Hoeijmakers, D.P. Muñoz, S.R. Raza, et al., Persistent DNA damage signalling triggers senescence-associated inflammatory cytokine secretion, Nat. Cell Biol. 11 (8) (2009) 973–979.

[43] V.G. Gorgoulis, T.D. Halazonetis, Oncogene-induced senescence: the bright and dark side of the response, Curr. Opin. Cell Biol. 22 (6) (2010) 816–827.

[44] J.A. Menendez, T. Alarcón, Senescence-inflammatory regulation of reparative cellular reprogramming in aging and Cancer, Front. Cell Dev. Biol. 5 (2017) 49.

[45] A. Banito, S.T. Rashid, J.C. Acosta, S. Li, C.F. Pereira, I. Geti, et al., Senescence impairs successful reprogramming to pluripotent stem cells, Genes Dev. 23 (18) (2009) 2134–2139.

[46] G.A. Garinis, G.T.J. van der Horst, J. Vijg, J.H.J. Hoeijmakers, DNA damage and ageing: new-age ideas for an age-old problem, Nat. Cell Biol. 10 (11) (2008) 1241–1247.

[47] M.A. Blasco, Telomere length, stem cells and aging, Nat. Chem. Biol. 3 (10) (2007) 640–649.

[48] R.M. Marion, K. Strati, H. Li, A. Tejera, S. Schoeftner, S. Ortega, et al., Telomeres acquire embryonic stem cell characteristics in induced pluripotent stem cells, Cell Stem Cell 4 (2) (2009) 141–154.

[49] L. Mosteiro, C. Pantoja, A. de Martino, M. Serrano, Senescence promotes in vivo reprogramming through p16(INK)(4a) and IL-6, Aging Cell (2017) e12711.

[50] A. Chiche, I. Le Roux, M. von Joest, H. Sakai, S.B. Aguín, C. Cazin, et al., Injury-induced senescence enables in reprogramming in skeletal muscle, Cell Stem Cell 20 (3) (2017) 407–414.

[51] B. Ritschka, M. Storer, A. Mas, F. Heinzmann, M.C. Ortells, J.P. Morton, et al., The senescence-associated secretory phenotype induces cellular plasticity and tissue regeneration, Genes Dev. 31 (2) (2017) 172–183.

[52] M. Serrano, A.W. Lin, M.E. Mccurrach, D. Beach, S.W. Lowe, Oncogenic ras provokes premature cell senescence associated with accumulation of p53 and p16INK4a, Cell 88 (5) (1997) 593–602.

[53] A. Laina, K. Stellos, K. Stamatelopoulos, Vascular ageing: underlying mechanisms and clinical implications, Exp. Gerontol. 5565 (17) (2017) 30273–30275.

[54] D.J. Kurz, et al., Degenerative aortic valve stenosis, but not coronary disease, is associated with shorter telomere length in the elderly, Arterioscler. Thromb. Vasc. Biol. 26 (6) (2006) e114–e117.

[55] E. Migliaccio, M. Giorgio, S. Mele, G. Pelicci, P. Reboldi, P.P. Pandolfi, et al., The p66shc adaptor protein controls oxidative stress response and life span in mammals, Nature 402 (1999) 309–313.
[56] M.L. Burgess, J.C. Mccrea, H.L. Hedrick, Age-associated changes in cardiac matrix and integrins, Mech. Ageing Dev. 122 (15) (2001) 1739–1756.
[57] D.J. Hausenloy, D.M. Yellon, Survival kinases in ischemic preconditioning and postconditioning, Cardiovasc. Res. 70 (2006) 240–253.
[58] S.C. Kolwicz Jr., S. Purohit, R. Tian, Cardiac metabolism and its interactions with contraction, growth, and survival of cardiomyocytes, Circ. Res. 113 (2013) 603–616.
[59] W. O'Neal, W. Griffin, S. Kent, J. Virag, Cellular pathways of death and survival in acute myocardial infarction, J. Clin. Exp. Cardiol. S6 (2013) 003.
[60] T. Matsui, J. Tao, F. del Monte, K.H. Lee, L. Li, M. Picard, T.L. Force, T.F. Franke, R.J. Hajjar, A. Rosenzweig, Akt activation preserves cardiac function and prevents injury after transient cardiac ischemia in vivo, Circulation 104 (2001) 330–335.
[61] S. Mohsin, M. Khan, H. Toko, et al., Human cardiac progenitor cells engineered with Pim-I kinase enhance myocardial repair, J. Am. Coll. Cardiol. 60 (2012) 1278–1287.
[62] G. Bendig, M. Grimmler, I.G. Huttner, G. Wessels, T. Dahme, S. Just, N. Trano, H.A. Katus, M.C. Fishman, W. Rottbauer, Integrin-linked kinase, a novel component of the cardiac mechanical stretch sensor, controls contractility in the zebrafish heart, Genes Dev. 20 (2006) 2361–2372.
[63] Y. Fujio, T. Nguyen, D. Wencker, R.N. Kitsis, K. Walsh, Akt promotes survival of cardiomyocytes in vitro and protects against ischemia-reperfusion injury in mouse heart, Circulation 101 (2000) 660–667.
[64] X.M. Yang, T. Krieg, L. Cui, J.M. Downey, M.V. Cohen, NECA and bradykinin at reperfusion reduce infarction in rabbit hearts by signaling through PI3K, ERK, and NO, J. Mol. Cell. Cardiol. 36 (2004) 411–421.
[65] O.J. Kemi, M. Ceci, U. Wisloff, S. Grimaldi, P. Gallo, G.L. Smith, G. Condorelli, O. Ellingsen, Activation or inactivation of cardiac Akt/mTOR signaling diverges physiological from pathological hypertrophy, J. Cell. Physiol. 214 (2008) 316–321.
[66] E. O'Neill, L. Rushworth, M. Baccarini, W. Kolch, Role of the kinase MST2 in suppression of apoptosis by the proto-oncogene product Raf-1, Science 306 (2004) 2267–2270.
[67] B. Li, M.A. Kaetzel, J.R. Dedman, Signaling pathways regulating murine cardiac CREB phosphorylation, Biochem. Biophys. Res. Commun. 350 (2006) 179–184.
[68] C. Skurk, Y. Izumiya, H. Maatz, P. Razeghi, I. Shiojima, M. Sandri, K. Sato, L. Zeng, S. Schiekofer, D. Pimentel, S. Lecker, H. Taegtmeyer, A.L. Goldberg, K. Walsh, The FOXO3a transcription factor regulates cardiac myocyte size downstream of AKT signaling, J. Biol. Chem. 280 (2005) 20814–20823.
[69] J.A. Muraski, M. Rota, Y. Misao, et al., Pim-1 regulates cardiomyocyte survival downstream of Akt, Nat. Med. 13 (2007) 1467–1475.
[70] Y. Dong, V.V. Undyala, R.A. Gottlieb, R.M. Mentzer Jr., K.A. Przyklenk, Definition, molecular machinery, and potential role in myocardial ischemia-reperfusion injury, J. Cardiovasc. Pharmacol. Ther. 15 (2010) 220–230.
[71] S. Pattingre, A. Tassa, X. Qu, R. Garuti, X.H. Liang, N. Mizushima, M. Packer, M.D. Schneider, B. Levine, Bcl-2 antiapoptotic proteins inhibit beclin 1-dependent autophagy, Cell 122 (2005) 927–939.
[72] P. Anversa, W. Cheng, Y. Liu, A. Leri, G. Redaelli, J. Kajstura, Apoptosis and myocardial infarction, Basic Res. Cardiol. 93 (Suppl. 3) (1998) 8–12.
[73] D.P. Dutka, J.S. Elborn, F. Delamere, D.J. Shale, G.K. Morris, Tumour necrosis factor alpha in severe congestive cardiac failure, Br. Heart J. 70 (1993) 141–143.
[74] H. Xu, Y. Yao, Z. Su, Y. Yang, R. Kao, C.M. Martin, T. Rui, Endogenous HMGB1 contributes to ischemia-reperfusion-induced myocardial apoptosis by potentiating the effect of TNF-α/JNK, Am. J. Physiol. Heart Circ. Physiol. 300 (2011) H913–H921.
[75] L.A. Kubasiak, O.M. Hernandez, N.H. Bishopric, K.A. Webster, Hypoxia and acidosis activate cardiac myocyte death through the Bcl-2 family protein BNIP3, Proc. Natl. Acad. Sci. U. S. A. 99 (2002) 12825–12830.
[76] G.L. Kukielka, C.W. Smith, A.M. Manning, K.A. Youker, L.H. Michael, M.L. Entman, Induction of interleukin-6 synthesis in the myocardium. Potential role in postreperfusion inflammatory injury, Circulation 92 (1995) 1866–1875.
[77] S.P. Levick, J.L. McLarty, D.B. Murray, R.M. Freeman, W.E. Carver, G.L. Brower, Cardiac mast cells mediate left ventricular fibrosis in the hypertensive rat heart, Hypertension 53 (2009) 1041–1047.
[78] Y. Okayama, T. Kawakami, Development, migration, and survival of mast cells, Immunol. Res. 34 (2006) 97–115.
[79] T.C. Theoharides, D.E. Cochrane, Critical role of mast cells in inflammatory diseases and the effect of acute stress, J. Neuroimmunol. 146 (2004) 1–12.
[80] N.G. Frangogiannis, J.L. Perrard, L.H. Mendoza, A.R. Burns, M.L. Lindsey, C.M. Ballantyne, L.H. Michael, C.W. Smith, M.L. Entman, Stem cell factor induction is associated with mast cell accumulation after canine myocardial ischemia and reperfusion, Circulation 98 (1998) 687–698.
[81] V. Patella, G. de Crescenzo, B. Lamparter-Schummert, G. De Rosa, M. Adt, G. Marone, Increased cardiac mast cell density and mediator release in patients with dilated cardiomyopathy, Inflamm. Res. 46 (Suppl. 1) (1997) S31–S32.
[82] Y. Sun, M.F. Kiani, A.E. Postlethwaite, K.T. Weber, Infarct scar as living tissue, Basic Res. Cardiol. 97 (2002) 343–347.
[83] D. Bell, M. Jackson, J.J. Nicoll, A. Millar, J. Dawes, A.L. Muir, Inflammatory response, neutrophil activation, and free radical production after acute myocardial infarction: effect of thrombolytic treatment, Br. Heart J. 63 (1990) 82–87.
[84] R.N. Pinckard, M.S. Olson, P.C. Giclas, R. Terry, J.T. Boyer, R.A. O'Rourke, Consumption of classical complement components by heart subcellular membranes in vitro and in patients after acute myocardial infarction, J. Clin. Invest. 56 (1975) 740–750.
[85] D. Mosqueira, S. Pagliari, K. Uto, M. Ebara, S. Romanazzo, C. Escobedo-Lucea, J. Nakanishi, A. Taniguchi, O. Franzese, P. Di Nardo, M.J. Goumans, E. Traversa, P. Pinto-do-Ó, T. Aoyagi, G. Forte, Hippo pathway effectors control cardiac progenitor cell fate by acting as dynamic sensors of substrate mechanics and nanostructure, ACS Nano 8 (2014) 2033–2047.
[86] N. Hedhli, Q. Huang, A. Kalinowski, M. Palmeri, X. Hu, R.R. Russell, K.S. Russell, Endothelium-derived neuregulin protects the heart against ischemic injury, Circulation 123 (2011) 2254–2262.

[87] I. Shiojima, K. Sato, Y. Izumiya, S. Schiekofer, M. Ito, R. Liao, W.S. Colucci, K. Walsh, Disruption of coordinated cardiac hypertrophy and angiogenesis contributes to the transition to heart failure, J. Clin. Invest. 115 (2005) 2108–2118.

[88] H. Iwasaki, A. Kawamoto, M. Ishikawa, A. Oyamada, S. Nakamori, H. Nishimura, K. Sadamoto, M. Horii, T. Matsumoto, S. Murasawa, T. Shibata, S. Suehiro, T. Asahara, Dose-dependent contribution of CD34-positive cell transplantation to concurrent vasculogenesis and cardiomyogenesis for functional regenerative recovery after myocardial infarction, Circulation 113 (2006) 1311–1325.

[89] J.S. Burchfield, S. Dimmeler, Role of paracrine factors in stem and progenitor cell mediated cardiac repair and tissue fibrosis, Fibrogenesis Tissue Repair 1 (2008) 4.

[90] K.E. Hatzistergos, J.M. Hare, Murine models demonstrate distinct vasculogenic and cardiomyogenic cKit+ lineages in the heart, Circ. Res. 118 (2016) 382–387.

[91] K. Alkass, J. Panula, M. Westman, T.D. Wu, J.L. Guerquin-Kern, O. Bergmann, No evidence for cardiomyocyte number expansion in preadolescent mice, Cell 163 (2015) 1026–1036.

[92] C. Jopling, E. Sleep, M. Raya, M. Martí, A. Raya, J.C. Izpisúa Belmonte, Zebrafish heart regeneration occurs by cardiomyocyte dedifferentiation and proliferation, Nature 464 (2010) 606–609.

[93] Y. Zhang, J.F. Zhong, H. Qiu, W.R. MacLellan, E. Marbán, C. Wang, Epigenomic reprogramming of adult cardiomyocyte-derived cardiac progenitor cells, Sci. Rep. 5 (2015) 17686.

[94] T. Heallen, M. Zhang, J. Wang, M. Bonilla-Claudio, E. Klysik, R.L. Johnson, J.F. Martin, Hippo pathway inhibits Wnt signaling to restrain cardiomyocyte proliferation and heart size, Science 332 (2011) 458–461.

[95] Q. Zhou, L. Li, B. Zhao, K.L. Guan, The hippo pathway in heart development, regeneration, and diseases, Circ. Res. 116 (2015) 1431–1447.

[96] Z. Lin, A. von Gise, P. Zhou, F. Gu, Q. Ma, J. Jiang, A.L. Yau, J.N. Buck, K.A. Gouin, P.R. van Gorp, B. Zhou, J. Chen, J.G. Seidman, D.Z. Wang, W.T. Pu, Cardiac-specific YAP activation improves cardiac function and survival in an experimental murine MI model, Circ. Res. 115 (2014) 354–363.

[97] A.I. Mahmoud, F. Kocabas, S.A. Muralidhar, W. Kimura, A.S. Koura, S. Thet, E.R. Porrello, H.A. Sadek, Meis1 regulates postnatal cardiomyocyte cell cycle arrest, Nature 497 (2013) 249–253.

[98] B.D. Polizzotti, B. Ganapathy, S. Walsh, S. Choudhury, N. Ammanamanchi, D.G. Bennett, C.G. dos Remedios, B.J. Haubner, J.M. Penninger, B. Kühn, Neuregulin stimulation of cardiomyocyte regeneration in mice and human myocardium reveals a therapeutic window, Sci. Transl. Med. 7 (2015) 281ra45.

[99] F.B. Engel, P.C. Hsieh, R.T. Lee, M.T. Keating, FGF1/p38 MAP kinase inhibitor therapy induces cardiomyocyte mitosis, reduces scarring, and rescues function after myocardial infarction, Proc. Natl. Acad. Sci. U. S. A. 103 (2006) 15546–15551.

[100] W. Chen, N.G. Frangogiannis, Fibroblasts in post-infarction inflammation and cardiac repair, Biochim. Biophys. Acta 1833 (2013) 945–953.

[101] A.H. Williams, N. Liu, E. van Rooij, E.N. Olson, MicroRNA control of muscle development and disease, Curr. Opin. Cell Biol. 21 (2009) 461–469.

[102] R.A. Espinoza-Lewis, D.Z. Wang, MicroRNAs in heart development, Curr. Top. Dev. Biol. 100 (2012) 279–317.

[103] E.R. Porrello, B.A. Johnson, A.B. Aurora, E. Simpson, Y.J. Nam, S.J. Matkovich, et al., miR-15 family regulates postnatal mitotic arrest of cardiomyocytes, Circ. Res. 109 (2011) 670–679.

[104] A. Eulalio, M. Mano, M. Dal Ferro, L. Zentilin, G. Sinagra, S. Zacchigna, et al., Functional screening identifies miRNAs inducing cardiac regeneration, Nature 492 (2012) 376–381.

[105] Y. Yang, H.W. Cheng, Y. Qiu, D. Dupee, M. Noonan, Y.D. Lin, et al., MicroRNA-34a plays a key role in cardiac repair and regeneration following myocardial infarction, Circ. Res. 117 (2015) 450–459.

[106] A.P. Beltrami, L. Barlucchi, D. Torella, M. Baker, F. Limana, S. Chimenti, et al., Adult cardiac stem cells are multipotent and support myocardial regeneration, Cell 114 (2003) 763–776.

[107] G.M. Ellison, C. Vicinanza, A.J. Smith, I. Aquila, A. Leone, C.D. Waring, et al., Adult c-kit(pos) cardiac stem cells are necessary and sufficient for functional cardiac regeneration and repair, Cell 154 (2013) 827–842.

[108] K. Urbanek, D. Torella, F. Sheikh, A. De Angelis, D. Nurzynska, F. Silvestri, et al., Myocardial regeneration by activation of multipotent cardiac stem cells in ischemic heart failure, Proc. Natl. Acad. Sci. U. S. A. 102 (2005) 8692–8697.

Chapter 11

Signaling pathways influencing stem cell self-renewal and differentiation—Special emphasis on cardiomyocytes

Selvaraj Jayaraman[a], Ponnulakshmi Rajagopal[b], Vijayalakshmi Periyasamy[c], Kanagaraj Palaniyandi[d], R. Ileng Kumaran[e], Sakamuri V. Reddy[g], Sundaravadivel Balasubramanian[f], and Yuvaraj Sambandam[h]

[a]Department of Biochemistry, Saveetha Dental College & Hospitals, Saveetha Institute of Medical and Technical Sciences (SIMATS), Chennai, Tamil Nadu, India, [b]Central Research Laboratory, Meenakshi Academy of Higher Education and Research, Chennai, Tamil Nadu, India, [c]Department of Biotechnology & Bioinformatics, Holy Cross College (Autonomous), Trichy, Tamil Nadu, India, [d]Department of Biotechnology, School of Bioengineering, Cancer Science Laboratory, SRM Institute of Science and Technology, Chennai, Tamil Nadu, India, [e]Biology Department, Farmingdale State College, Farmingdale, NY, United States, [f]Department of Radiation Oncology, Hollings Cancer Center, Medical University of South Carolina, Charleston, SC, United States, [g]Darby Children's Research Institute, Department of Pediatrics, Medical University of South Carolina, Charleston, SC, United States, [h]Department of Surgery, Comprehensive Transplant Center, Northwestern University, Feinberg School of Medicine, Chicago, IL, United States

1 Introduction

The human body is a structural framework that encompasses cells, tissues, and organs. Among them, the human heart plays a vital role in the supply of oxygen and nutrients to the cells/tissues/organs via the bloodstream. The heart rhythmically beats at the rate of 72 beats per minute to pump blood throughout the body. Heart is a muscular organ that consists of various types of cells such as cardiomyocytes, cardiac conduction system cells, smooth muscle cells, endothelial cells, cardiac fibroblasts, and cardiac progenitor cells. The cardiomyocytes contribute up to 40% of the cells in the heart but occupy approximately 70–85% tissue volume of the total mammalian heart [1–3].

The heart is one of the initial organs that become functional around 21 days in the vertebrate embryo [4]. The tissues of the heart are derived from the mesodermal layer, and the induction of cardiomyogenic phenotype largely depends on the signal obtained from the neighboring layers of endoderm and ectoderm [5]. Though its development is early, cardiogenesis is regarded as a highly regulated process that combines the differentiation, specialization of cellular types along with spatial integration, and most importantly coordination of various signaling pathways.

Cardiovascular diseases are common noncommunicable diseases with an estimate of 17·9 million deaths during the year 2017, globally [6]. Cardiovascular problems increase with age, and the mortality rates have been a significant factor indicating a necessary treatment. Cellular-based therapies have been considered as an alternative to conventional medicine. Depletion of stem cell pool and loss of regenerative population of stem cells are the prominent factors responsible for aging accompanied by loss of organ function and homeostasis of tissues. Aging and chronic conditions like cardiovascular diseases have more complications accompanied by diabetes, hypertension, obesity, and increase or decrease in levels of various biological substances associated with the functioning of the heart [7, 8].

The ability of stem and progenitor-like cells to differentiate into cardiomyocytes has been identified as novel cardiovascular repair-oriented therapeutics. It is a well-known fact that heart regeneration is possible with the help of stem cells as supported by many research data. Recent focus has been on the studies of stem cells and its' signaling mechanism that are vital for differentiation and subsequent regeneration of impaired tissue/organ. This has led to the development of cellular-based treatment approaches to overcome the drawbacks and increase the effectiveness of cardiovascular repair related to aging. It has also been disseminated that deeper understanding and identification of the molecular mechanism of stem cell-based therapy may provide treatment for the repair and damage of the heart and are crucial for age-associated cardiac issues [7, 8].

Previous studies suggest that cellular senescence through an innate response to cell damage also leads to disruption of normal organ function and an increase in age-related illness. Cardiac stem cells undergo apoptosis, and a decrease in telomere length is observed with age. There are many signaling factors involved in cardiovascular repair mechanisms including stromal cell-derived factor 1 (SDF-1), vascular endothelial growth factor (VEGF), granulocyte colony-stimulating factor (G-CSF), and tenascin C [9–12]. Additional evidence suggests that platelet-derived growth factor (PDGF) pathway plays a role in the maintenance of Oct-3/4 + and cardiomyocyte differentiation from bone marrow cells that decrease with age [13].

Despite the advent of technology in medical research, heart diseases are the leading cause of death in the global arena [14]. Heart diseases have been known as the "World's biggest killer" [15,16]. To overcome this major threat, the scientific fraternity needs to work in a coordinated manner to reduce the risk of heart problems in humans. Clinical trials have demonstrated that stem cell therapy can be an effective approach to treat cardiovascular diseases [17]. At this juncture, understanding the signaling pathways that are important for the self-renewal and differentiation of the stem cells would facilitate an effective strategy for stem cell therapy [18].

In this chapter, we provide information about the signaling pathways that influence the stem cells, highlighting the potential of these signals in cardiomyocyte self-renewal and differentiation.

2 Stem cells

Stem cells are capable of prolonged self-renewal as well as possess the ability to differentiate into specific cell types, including cardiomyocytes [19]. They have the potential to repair and regenerate the damaged tissues of the heart. These are broadly classified as totipotent (able to differentiate into any sort of cells, including placenta), pluripotent (capable of differentiating into several different types of cells), and multipotent (differentiate into a limited number of cell types in a particular lineage) stem cells. Unlike normal cells, stem cells are capable of many rounds of replication [20].

Stem cell therapy is considered as a valuable approach for conditions that affect the heart such as myocardial infarction (MI), coronary artery disease, peripheral artery disease, stroke, and heart failure [21]. The sources of stem cells are embryonic stem cells (ESCs), mesenchymal stem cells (MSCs), bone marrow, umbilical cord blood, placenta, hematopoietic stem cells, and cardiac stem cells. ESCs belong to the pluripotent stem cells (PSCs) category, and other cells are characterized into multipotent stem cells [22].

ESCs are derived from the blastocyst that develops approximately 5 days after fertilization. ESC lines were established in 1981 by different groups of researchers from mouse blastocysts [23,24]. Subsequently, the first human ESCs were established in the year 1998 [25]. Somatic stem cells or adult stem cells (ASCs) have a low self-renewal capacity with restricted differentiation capabilities [26]. All multipotent type cells belong to this category, where cells are replenished within an organ owing to physiological or pathological constraints.

3 Stem cell culture and therapy

Researchers have been focusing on the use of stem cells for cardiac diseases by two promising methods. First is by developing heart muscles from stem cells in the laboratory. This helps to identify new drugs based on the genetic origin of heart disease. Second, they are being used to develop therapies to repair and replace damaged heart tissues. The treatment strategies that are put to practice include stem cell transplantation, tissue engineering, and direct reprogramming of stem cells [27]. These approaches are known to improve the function of cells that have been transplanted and also able to stimulate preexisting cells into cardiomyocytes. Human pluripotent ESCs were first isolated from fertilized oocytes/primitive embryos (2n) during the blastocyst stage. Due to ethical and legal issues, alternative approaches are also utilized to obtain ESCs or ESC-like cells such as artificially activated oocytes (1n) using chemical or physical substances known as parthenotes (2n), and reprogrammed oocytes (2n) by employing somatic cell nuclear transfer (SCNT) method using enucleated oocytes. Further, this technology is also applied to generate the induced pluripotent stem cells (iPSCs) from somatic cells alone, which mimic ESCs and develop into any specific cell types, including cardiomyocytes (Fig. 1) [28].

Cellular therapy offers efficient treatment for heart deformities compared to other treatments [29]. Scientists around the world report the use of many stem cell types to repair damaged heart tissues [30–32]. The major focus of cellular therapy is targeted toward cardiomyocytes, beating muscle cells that constitute the atria, chambers of the heart, and ventricles. These cells could be cultured in the laboratory from ESCs and iPSCs [33].

ESCs are derived from the embryo of an organism that is capable of developing into any type of cells in the body. Cultured ESCs show peculiar characteristics of pluripotency, enormous differentiation capability, and immortality. They remain undifferentiated and maintain normal chromosomal composition. ESCs are known to express an abundance of cell surface and other markers such as CD9, CD24, CD324, CD90, CD49f/CD29, TRA-1-60, alkaline phosphatase (ALP), and genes that are linked to pluripotency, including Oct4, Rex1, Sox2, Klf4, LN28, DCT4, NANOG, c-Myc, Cripto-1, SSEA-3/4,

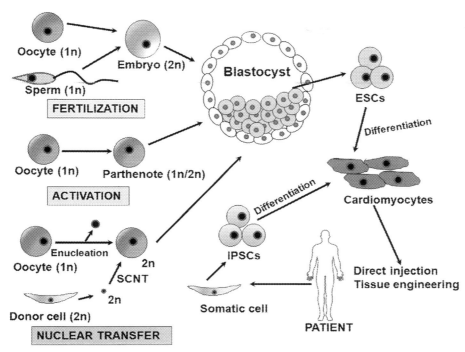

FIG. 1 Generation of cardiomyocytes in vitro. The best sources of pluripotent stem cells (PSCs) are human embryonic stem cells (hESCs) and induced pluripotent stem cells (iPSCs). Human ESCs were first separated from early human embryos in the blastocyst stage with low efficiency. However, the technical advances improved the yield by about 90%. These PSCs can be obtained from the fertilized embryo (fertilization of the oocyte by sperm), chemically or artificially activated oocyte (parthenotes), and reprogramming the cells by somatic cell nuclear transfer (SCNT) to an enucleated oocyte. In contrast, the iPSCs are generated from somatic cells; therefore, they are conceivable, are free of ethical issues, and have an advantage over immunological complications.

and Thy-1 [34–38]. Immortality in culture is associated with the increased expression of telomerase (TERT) in vitro. Though the methods to maintain ESCs seem to be technically tedious, its impact on treating life-threatening diseases makes it important and essential to improve its efficiency. There is a need to develop strategies for the maintenance of ESC in long-term culture without the use of animal products such as mouse fibroblasts as feeder cells when it comes to human ESCs (hESCs). Also, there is a definite need to allow cells to differentiate into specific cell types based on the nature of treatment offered for a particular disease.

iPSCs are cells that are reprogrammed *in vitro* to mimic ESCs bestowed with the ability to develop into any specific cell types, including cardiomyocytes. This type of stem cell-derived cardiomyocytes behaves in a similar manner as that of a live heart muscle cells. The cells are known to beat in unison in the culture dish. Cardiac cells derived from iPSCs serve as human models of heart disease. This helps to study abnormalities associated with the heart and for testing the efficacy of drugs or treatments. It is possible to become a sought after remedy in place of transplantation that is being currently performed in patients suffering from heart diseases [39].

Efforts are being made to develop strategies to rebuild the damaged heart muscles. One such approach is the use of ASCs since it is a reliable source for heart cell regeneration. It has been demonstrated that these cells express cardiomyocyte-specific markers and gain functional ability. Initial investigations carried out with ASCs are known to provide positive outcomes related to cardiac functioning though they are succumbed to death in a short span of time [40]. Hematopoietic stem cells (HSCs) have been successfully used clinically for almost four decades [41]. MSCs of stromal origin can be easily isolated from any tissue for wide use in clinical practice. Large trials based on ASCs have proved their safety and feasibility but are less efficient in terms of regaining the lost health [42]. The major advantage of deploying ESCs instead of ASCs is mainly attributed to their pluripotent nature and vast diversity. At the same time, ethical and safety concerns have limited the use of hESCs as opposed to ASCs. Also, it is too early to predict the functional benefits of stem cell therapy.

In order to reap the beneficial aspects of stem cell therapy, we need to have a clear understanding of mode of actions and the factors responsible for the proper regulation of its function. The vital role of TGFβ and NODAL signaling in the self-renewal of hESCs has been demonstrated [43,44]. SMAD4 contributes extensively to enhance the modulation of TGFβ signaling and to achieve self-renewal capacity of stem cells [45]. Evidently, the targeted deletion of *SMAD4* decreased the self-renewal capacity of HSCs in vivo [46].

Takahashi and Yamanaka [47] have devised a protocol to reprogram somatic cells into PSCs like ESCs. These cells were then called iPSCs, and they were also established in human cells, followed by murine [48]. This has gained momentum in recent years wherein forced expressions of genes that are master regulators of pluripotency are developed. The methodology requires the treatment of mature cells from donors or patients with necessary genes and other supplements that facilitate to replenish healthy cells. The discovery of iPSCs has contributed to the technical advancement of direct reprogramming and enabled the growth of being a trendsetter in the field of regenerative medicine. This technology has given rise to the development of sufficient and diverse nature of cell types from fibroblasts, neurons, cardiomyocytes, endothelial cells, and hepatocytes [49].

Research has envisaged the fact that iPSC-derived cardiomyocytes are capable of developing into heart muscle cells that release signals. These cells can be utilized to replace the muscles that have been damaged due to heart attack [50]. Such transplantation studies have also proved to be successful in animal models [51–53]. After some positive outcome, recently researchers from Osaka University started the clinical trials using iPSCs and it is underway to restore cardiac function in humans [54,55].

4 Signaling pathways and factors involved in the self-renewal and differentiation of cardiomyocytes

Pluripotent stem cells (PSCs) have been fruitful in producing cardiomyocytes and hence serve as a model to predict the process of cardiomyogenic differentiation. It has been possible to culture and study the nature of the heart cell development using certain regulatory factors and/or altering the proportion of key elements in signaling pathways. Some of the such regulatory factors involved in the developmental stages are bone morphogenetic proteins (BMPs), activin and Nodal, fibroblast growth factors (FGFs), and Wingless/integrated (Wnt) pathway [56].

Primarily, the uptake of cardiac cells into the heart necessitates positive-acting signals from adjacent germ layers. The major signaling molecules are BMPs and FGFs. BMP-2 and 4 play a pivotal role in initiating the synthesis of cardiac cells from non-precardiac mesoderm cells. BMP-2 signaling is the major factor contributing to cardiogenic induction, maintenance of cells into the lineage, and their further differentiation into cardiomyocyte-like cells [57]. The activation of the signal transduction pathway of BMP depends on type 1 and type II BMP receptors. Once BMP binds to these receptors, they undergo phosphorylation by kinase receptors. This initiates the activation of the transforming growth factor β-activated kinase 1, mitogen-activated protein kinase kinase 3/6,p38 mitogen-activated protein kinases (TAK1-MKK3/6-p38), c-Jun N-terminal kinases (JNK) pathway as well as mothers against decapentaplegic (Mad) and the *C. elegans* gene Sma (SMAD) pathway. TAK1 belongs to the mitogen-activated protein kinase (MAPK) superfamily of proteins. Subsequent phosphorylation leads to the activation of a cascade of events that stimulate the expression of nuclear activating transcription factor 2 (ATF2) with the upregulation of downstream genes. In the SMAD pathway, BMPs activate type 1 receptor with SMAD 1 proteins allowing the phosphorylation and further recruitment of BMP ligand-specific SMAD to coordinate with SMAD4 leading to the formation of a transcription complex. This complex gets translocated from the cytoplasm into the nucleus leading to the activation of ATF2 enabling the transcription of genes associated with cardiomyogenesis. SMAD4 signaling pathway is found to be crucial in the formation of human cardiac mesoderm and its further differentiation [58–60].

Fibroblast growth factor (FGF) possesses four receptors that belong to tyrosine kinases capable of being activated by ligands. Once the FGF molecule binds to these receptors, it causes dimerization and autophosphorylation of residual tyrosine present within the cell. This acts as a signal for the formation and induction of signaling complexes present downstream. FGF signals promote differentiation by three major pathways such as phospholipase (C-γ/Ca^{+2}) pathway, Ras/MAPK pathway, and the phosphatidylinositol 3-kinase (PI3K)/AKT pathway. The main pathway involved in the signaling of FGF is the Ras/MAPK pathway [61]. FGF promotes the suppression of autophagy and prevents premature differentiation of cardiac progenitor cells [62].

5 Wnt/β-catenin signaling pathway

Another major signaling molecule that necessitates the induction of cardiogenic cells of mesoderm belongs to the Wnt family. This group comprises of both positive and negative regulators of cardiogenesis. These are secreted glycoproteins that are involved in developmental processes like organization, differentiation, proliferation, and migration of a variety of cells. Based on the studies carried out in Xenopus and chicken, cardiogenic induction of progenitor cells of mesoderm needs Wnt/ca^{+2} and Wnt/polarity to be activated along with the repression of Wnt/β-catenin pathway. It has been observed that Wnt/ca^{+2} activates protein kinase C (PKC), whereas the polarity/Wnt family triggers the JNK pathway culminating in the transcription of nuclear genes. Cell surface receptors such as frizzled (Fzd) and low-density lipoprotein receptor-related proteins (LRPs) are the essential elements of Wnt signaling [63]. Inhibition of Wnt/β-catenin signaling in anterior lateral mesoderm is facilitated by Dickkopf 1 (DKK1) and Crescent, while activation of Wnt (Ca^{+2}/polarity) pathways requires Wnt2 in the precardiac mesoderm. β-catenin is a dual function protein that contributes to the regulation of transcription as well as adhesion important for stem cell differentiation. ESCs possess the capacity to self-renew in medium containing serum/BMP, leukemia inhibiting factor (LIF), or glycogen synthase kinase 3 (GSK3) and MEK1/2 inhibitors also known as "2i" inhibitors. In the presence of GSK3β inhibitor (CHIR99021), Tcf3 gets degraded, which allows β-catenin to induce estrogen-related receptor beta (ESRRB) expression. Subsequently, binding of β-catenin in the promoter region along with lymphoid enhancer factor 1 (LEF1) and Kruppel-like factor 4 (KLF4) molecules induces TERT expression to maintain the self-renewal status of the newly formed naïve cells. On the other hand, inhibition of MEK/ERK (PD0325901) and its downstream E26 transformation-specific or E-twenty-six (ETS) molecule prevents the β-catenin, lymphoid enhancer factor/T-cell factor (LEF/TCF) complex to induce the differentiation-associated genes. In the case of primed pluripotent cells, Axin2 along with tankyrase inhibitor XAV (XAV939, and stabilizer of Axin) retains β-catenin in the cytoplasm and thus facilitates the cells to undergo self-renewal even in basal medium. Low quantities of Wnt/β-catenin signaling also favor the pluripotency of these cells to some extent by the expression of octamer-binding transcription factor 4 (OCT4) and NANOG. During differentiation, β-catenin, LEF/TCF complex bind with ETS that actively induce the mesendoderm-associated genes responsible for lineage-specific differentiation. Also, binding of the β-catenin complex with SMAD2/3 stimulates mix paired-like homeobox 1 (MIXL1) expression and induces differentiation (Fig. 2) [64].

6 Differentiation of stem cells

ESCs have been established from early mouse embryos as permanent lines of undifferentiated pluripotent cells. Mouse ESCs have the capacity to self-renew in medium containing serum/BMP and leukemia inhibitory factor (LIF) or glycogen synthase kinase 3 (GSK3) and MEK1/2 also referred to as "2i" inhibitors [65].

ESCs are derived from inner cell mass (ICM), embryonic ectoderm, and primordial germ cells of fetal genital ridge. They are pluripotent bestowed with the innate ability of self-renewal and differentiation. ESC requires feeder layers like mouse embryonic fibroblasts or induction of differentiation inhibitory factor (LIF) to maintain a stable karyotype [66]. hESCs have also been isolated and cultured for investigations pertaining to therapeutic needs. Ethical issues remain a serious concern over the exploitation of embryos [67].

The differentiation of ESCs in vitro begins with the initial aggregation termed as "embryoid bodies" (EBs). Certain parameters influence this process of producing different specialized cells like cardiomyocytes with regard to EB. They are the initial count of cells in EB, media, FBS, growth factor supplements (FGFs, EGFs, activin, etc.), cell lines, and time of culture [68].

Inside the EBs, cardiomyocytes are found intermediate to the epithelial layer and basal layer of mesenchymal cells. These could be identified precisely within 1–4 days after plating as they contract spontaneously. During differentiation, the number of spontaneously beating foci increases, creating a localized area of beating cells in EBs. The rate of contraction in each beating area rapidly increases due to differentiation, followed by a plunge in the average beating rate on reaching maturation [69]. Thus, developmental stages of cardiomyocytes could be well-correlated with time span and divided into three stages of differentiation as early (pacemaker-like or primary myocardial like), intermediate, and terminal (atrial, ventricular, nodal, His, Purkinje-like cells). During the early stages of differentiation, cardiomyocytes present in EBs are small and round. After maturation, they become elongated with well-developed myofibrils and sarcomeres [70].

ESC-derived cardiomyocytes express GATA binding protein 4 (GATA4) nuclear transcription factors, that regulate cardiac specific genes such as atrial natriuretic factor (ANF), myosin light chain (MLC)-2v, α-myosin heavy chain (αMHC), sodium, calcium exchangers. Sarcomeric proteins such as titin, myomesin, α-actins, cardiac troponin T and M proteins are also found to be expressed in cultured cardiomyocytes. Expression of these factors in cardiomyocytes in vitro culture system help us to gain a better understanding of the developmental process as they mimic normal myocardial development [71].

The increased mortality owing to heart failure has led to the development of regenerative cardiac repair using stem cells. Some of the molecules known to exhibit the rejuvenation of cardiac progenitor cells are Pim-1 kinase, Notch 1 signaling, and TERT. Rejuvenation of myogenic response in muscle cells was observed after the activation of the Notch signaling pathway [72].

7 Notch signaling pathway

The Notch signaling system is highly conserved in most of the organisms. There were four different notch receptors, which have been identified, namely NOTCH1, NOTCH2, NOTCH3, and NOTCH4 in mammals. The Notch receptor is a

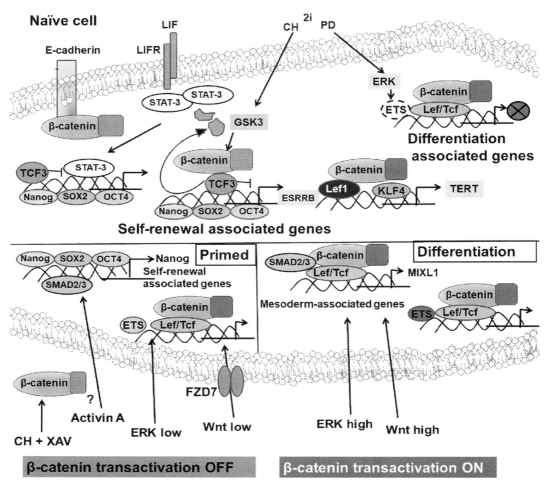

FIG. 2 LIF activation of the STAT3 pathway stimulates the self-renewal gene expression and maintains the self-renewal status in naïve cells, which are mediated by β-catenin/E-cadherin signaling pathway. During 2i conditions (selective 2 inhibitors, PD0325901 and CHIR99021 for MEK 1/2 GSK3β), Tcf3 is degraded, and therefore, β-catenin favors estrogen-related receptor beta (ESRRB) expression; also along with LEF1 and KLF4 binding, it induces TERT expression, which maintains the self-renewal status of the naïve cells. Due to the inhibition of MEK/ERK and its downstream E26 transformation-specific or E-twenty-six (ETS) molecule, the β-catenin complexes with LEF/TCF are not able to induce the differentiation. In primed pluripotent cells, Axin2 with tankyrase inhibitor XAV (XAV939, and stabilizer of Axin) retains β-catenin in the cytoplasm and allows the cells to self-renew even in basal medium. Nevertheless, the mechanisms involved in this process are unknown. Low levels of Wnt/β-catenin signaling also contribute to pluripotency to some extent with OCT4 and NANOG expression. Activation of β-catenin and ETS factors along with LEF/TCF induces the mesendoderm-associated genes and lineage-specific differentiation. Furthermore, the β-catenin complex binds with SMAD2/3 to stimulate MIXL1 and induce differentiation.

mono-transmembrane receptor protein that consists of a large extracellular portion with a Notch protein-binding domain and a small intracellular region [73]. This signaling pathway plays an important role in cardiovascular development/differentiation. Also, it exerts positive and adverse effects on cardiac stem cells during cardiac injury.

Notch 1 signaling plays a significant role in cardiomyocyte differentiation from human bone marrow-derived MSCs [74]. It is involved in fibrotic and regenerative repair; hence, the Notch signaling is considered vital throughout the healing process of injured hearts in adults. The Notch is known to regulate the MSC population by modulating major cellular mechanisms that specifically balance fibrotic and regenerative repair in the matured heart [75]. Also, MSCs are shown to attenuate doxorubicin-induced cellular senescence through the VEGF/Notch/TGF-β pathway in H9c2 cardiomyocytes. Hence, MSCs with active Notch signaling possess therapeutic benefits in the suppression of cardiotoxicity [76]. Furthermore, activation of HIF-1 α/Jagged1/Notch 1 signaling promotes early cardiac-specific differentiation under hypoxia-stressed conditions [77].

Notch signaling cascades are also essential for repair and the lineage commitment of cardiac progenitor cells [78]. Activation of Notch signaling as a heart's injury repair-response generates multipotent cell population from adult epicardium and thus repairs fibrosis. Knock-out of the Notch 1 gene demonstrated that Notch signaling pathways facilitate regeneration and cardiac repair after MI. Besides, it augments the early cardiac differentiation of dedifferentiated fat cells by dimethyloxalylglycine [79,80].

Tung et al. [81] have developed tunable microparticle-based Notch signaling biomaterials, which enhance cardiomyocyte population during specific developmental stages. Additionally, this nanotechnology provides a way for researchers to generate large quantities of cardiomyocytes. They also observed a biphasic effect of Notch signals: Activation of Notch in undifferentiated hESCs induces ectodermal differentiation, whereas activation of these signals in the specific cardiovascular progenitor cells enhances cardiac differentiation.

Yan and colleagues [82] have studied the role of thymosin β4 (Tβ4), which is a small G-actin sequestering peptide involved in cell proliferation, migration, and angiogenesis. Transplantation of Tβ4 overexpressed ESCs enhances the ability of these cells to differentiate into cardiomyocytes in both in vitro and in vivo conditions and thus promotes cardiac protection and repairs the cardiac function in an infarcted heart. To demonstrate these, stable cell lines were established using a mouse ES cell line with a red fluorescent protein (RFP) such as RFP-ESCs, Tβ4-ESCs, and RFP-Tβ4 fusion protein. A significant population of spontaneously beating EBs was comparatively elevated in Tβ4 expressed ESCs than in RPF alone expressing ESCs in vitro. Overexpression of cardiac transcriptional factors such as GATA-4, MEF 2c, and Txb6 was observed in Tβ4-ESCs with a high level of EBs, which indicates the increased number of functional cardiomyocytes in the culture. Furthermore, Tβ4-ESCs increase newly formed cardiomyocytes linked to the activation of the Notch pathway. Also, it decreases apoptotic cells by elevating Akt and suppressing the PTEN level in myocardial infarcted (MI) mice. Reduction of cardiac fibrosis and function has been observed in the left ventricles of Tβ4-ESC transplanted mice. Taken together, genetically modified Tβ4-ESC cells could develop into cardiomyocytes in both in vitro and in vivo [82].

Likewise, Notch 1 signaling modulates cardiosphere-derived cell (CDC) differentiation. In a MI mouse model, the effect of CDCs enhanced by Notch signals protects the heart function, suggesting the therapeutic potential of CDC protein in cardiac repair. Notch 1 promotes smooth muscle cell differentiation of CDCs through the recombination signal binding protein for immunoglobulin kappa J region (RBPJ)-dependent signaling pathway in vitro, which in turn suggests its implications in progenitor cell-mediated angiogenesis [83].

Croquelois et al. [84] have identified the role of Notch 1 signaling that controls pathophysiological cardiac remodeling in the TG1306/1R transgenic mouse model. It was observed that the absence of Notch 1 induced changes such as exacerbation of cardiac hypertrophy, development of fibrosis, altered function, and increase in cell death. Evidence indicates that cell cycle entry of mouse ESC-derived cardiomyocytes developed rapidly owing to the activation of the Notch pathway, which in turn maintains the expression and nuclear localization of cyclin D1. Transcriptional activation of cyclin D1 genes is mediated by the Notch pathway by employing transcription factor RBPJ.

Extracellular vesicles from activated Notch signaling in cardiac MSCs enhance neovasculogenesis and proliferation of myocytes. Cardiac MSCs derived from cardiac tissue have been found to be responsible for heart cell regeneration. Proper cardiomyocyte development depends on mitochondrial fusion. Any ablation of mitochondrial fusion proteins Mitofusin 1 and 2 in the mouse embryonic heart, or gene-trapping of Mitofusin 2 and/or optic atrophy 1 in mouse ESCs, blocks the development of the heart and also impedes the differentiation of ESCs into cardiomyocytes.

Furthermore, expression of transcription factors such as a nuclear factor of activated T cells (NFAT1), GATA4, and myocyte enhancer factor 2c (MEF2C) positively correlated with elevated Ca(2 +)-dependent calcineurin activity in ischemic cardiomyopathy [85]. Further, it has been reported that mitochondrial fusion controls cardiomyocyte differentiation through the activation of calcineurin and Notch signaling pathways. In this study, they revealed that increased Ca(2 +)-dependent calcineurin activity and Notch 1 signaling regulate GATA4 and MEF2 and subsequently suppress the ESC differentiation. Cardiomyocyte differentiation by mitochondrial fusion signifies the interaction between mitochondria, Ca(2 +), and calcineurin with the Notch 1 signaling pathway [86].

Apart from these beneficial effects, the Notch pathway also plays an adverse role in cardiomyocyte rejuvenation via stem cells. A classical Notch inhibitor known as DAPT favors mouse fibroblast differentiation into induced cardiac-like myocytes by activating transcription factors GATA4, HAND2, TBX5, and MEF2C [87]. In addition, coordination of DAPT along with AKT kinase efficiently increases these fibroblast differentiation into functional cardiomyocytes process up to 70%.

Also, DAPT significantly enhances the calcium flux, the structure of sarcomere, and the population of spontaneous rhythmically beating cells; importantly, it is involved in differentiating cells to gain cardiomyocyte-specific features.

Inhibition of Notch signaling by DAPT increases the reprogramming of mouse fibroblasts into induced cardiac-like myocytes by increasing the binding of transcription factor MEF2C in the gene promoter region. Also, DAPT alters the genetic programs related to the development, differentiation and contraction-excitation coupling of muscles [16]. Numb family proteins (NFPs) comprising of Numb and Numb-like (NumbL) are considered as determinants of cell fate for numerous progenitor cell types [88].

In order to study the function of NFP in the late stages of cardiac development, both Numb and NumbL were deleted to generate myocardial double-knockout (MDKO) mice. This led to embryonic lethality and portrayed various defects in the differentiation of cardiac progenitor cells, proliferation, outflow tract (OFT) alignment, and atrioventricular septation. Ablation of NFPs in numerous types of the cardiac population followed by lineage tracing proved the requirement of NFPs in the second heart field (SHF) for atrioventricular septation and OFT alignment. A defect in the differentiation of the SHF progenitor cells was observed in MDKOs by various methods, including mRNA deep sequencing.

Numb-associated regulation of cardiac progenitor differentiation depends on the endocytosis process. The use of a transgenic Notch reporter cell line exhibited the upregulation of Notch signaling in MDKO. The investigations also showed that the suppression of Notch 1 signaling in MDKOs rescued defects in the expression of p57, proliferation, and thickness of trabeculae. Further, Numb proteins are known to inhibit Notch 1 signaling by enhancing the degradation of the intracellular domain of Notch 1 in cardiomyocytes. Collectively, the research signifies that NFPs are responsible for modulating trabecular thickness and cardiac morphogenesis by inhibiting Notch 1 signaling pathways. Additionally, NFPs are shown to control cardiac progenitor cell differentiation through regulating endocytosis. Thus, the function of NFPs in the differentiation of cardiac progenitor cells and morphogenesis indicates the therapeutic potential of NFPs for heart muscle regeneration as well as congenital heart diseases. [89,90].

8 Mechanotransduction pathways

Integrins are cell surface heterodimeric (alpha/beta) transmembrane proteins that transduce both mechanical and chemical signals bidirectionally [91]. Inside-out and outside-in signaling by integrin subtypes and their dimeric combination (over 7 alpha and 17 beta subunits) mediate cellular processes, including matrix adhesion, spreading, cytoskeletal rearrangement, production of extracellular matrix (ECM) proteins, adhesion-coupled transcriptional and translational activation, and cell cycle progression. Upon binding to extracellular ligands such as RGD motif-containing ECM molecules such as fibronectin and vitronectin, integrins induce cytoskeletal rearrangement via actin-myosin contractile machinery. Integrins are activated in both physiological and pathological cardiac growth leading to increased protein synthesis, cardiomyocyte hypertrophy, PDGF-induced cardiac fibroblast proliferation, and myocyte beta-actin dynamics at the Z-disks and play a crucial role in the overall structural maintenance of the myocardium [92,93]. Recent evidence indicates that integrins and their downstream mediators such as focal adhesion kinases (FAKs) have a strong role in mESC and hESC differentiation [94]. In murine, studies have demonstrated that the activation of MAPK signaling by beta1 integrin is required for the maintenance of stemness in the neural stem cells. The presence of integrins alpha(5)beta(1), alpha(v)beta(5), alpha(6)beta(1), and alpha(9)beta(1) in mESC plays a role in maintaining the stem cell pluripotency and promoting self-renewal capability [95,96]. The self-renewal of human ESCs through the activation of alpha(v)beta(5) integrin expression has also been reported. Further, it has been shown that integrin expression strongly supports human iPSC growth in vitro [97,98].

Many studies have revealed that integrin downstream signaling molecules play important role in ESC pluripotency. Also, integrin-activating substrates such as vitronectin, laminin, and fibronectin maintain the self-renewal capacity of undifferentiated hESCs [97,99]. Interaction of integrin α6β1 with laminin 1 (LN1) induces the differentiation of ESCs by employing CD151 protein to activate FAK and Akt signaling pathway and lineage-specific transcription factor Er71 (ETS-related 71) [100]. It has been reported that fibronectin (FN) binding to integrin β1, phosphorylates Src, caveolin-1, and FAK stimulates mESC proliferation by activating the RhoA-PI3K/Akt-ERK1/2 signaling pathway [101]. Collectively, ECM remarkably influences hESC fate and function via integrin/FAK signaling mechanisms that contribute to the pluripotency of the stem cells [94].

9 Conclusion

Stem cell therapy holds great promise in the field of medicine and provides a strategy to treat various diseases, including the infracted heart. Despite advancements in technology, challenges and constraints still emerge pertaining to the safety and efficacy of differentiation and maturation of cardiomyocytes with further implications on their therapeutic applications. However, a deeper understanding of the signaling pathway insights could improve the development of cellular therapies and tissue engineering using stem cells to address the problem efficiently for the millions of people suffering from heart diseases worldwide.

References

[1] P. Zhou, W.T. Pu, Recounting cardiac cellular composition, Circ. Res. 118 (3) (2016) 368–370.
[2] Y. Tang, J.R. Nyengaard, J.B. Andersen, U. Baandrup, H.J. Gundersen, The application of stereological methods for estimating structural parameters in the human heart, Anat. Rec. (Hoboken) 292 (10) (2009) 1630–1647.

[3] V. Talman, R. Kivelä, Cardiomyocyte-endothelial cell interactions in cardiac remodeling and regeneration, Front. Cardiovasc. Med. 5 (2018) 101.
[4] A. Moorman, S. Webb, N.A. Brown, W. Lamers, R.H. Anderson, Development of the heart: (1) formation of the cardiac chambers and arterial trunks, Heart 89 (7) (2003) 806–814.
[5] A. Leitolis, A.W. Robert, I.T. Pereira, A. Correa, M.A. Stimamiglio, Cardiomyogenesis modeling using pluripotent stem cells: the role of microenvironmental signaling, Front. Cell. Dev. Biol. 7 (2019) 164.
[6] G.A. Roth, C. Johnson, A. Abajobir, F. Abd-Allah, et al., Global, regional, and national burden of cardiovascular diseases for 10 causes, 1990 to 2015, J. Am. Coll. Cardiol. 70 (1) (2017) 1–25.
[7] N. Hariharan, M.A. Sussman, Cardiac aging - getting to the stem of the problem, J. Mol. Cell. Cardiol. 83 (2015) 32–36.
[8] V.L. Ballard, J.M. Edelberg, Stem cells and the regeneration of the aging cardiovascular system, Circ. Res. 100 (8) (2007) 1116–1127.
[9] Y. Tan, H. Shao, D. Eton, Z. Yang, L. Alonso-Diaz, H. Zhang, A. Schulick, A.S. Livingstone, H. Yu, Stromal cell-derived factor-1 enhances pro-angiogenic effect of granulocyte-colony stimulating factor, Cardiovasc. Res. 73 (4) (2007) 823–832.
[10] J. Honold, R. Lehmann, C. Heeschen, D.H. Walter, B. Assmus, K. Sasaki, H. Martin, J. Haendeler, A.M. Zeiher, S. Dimmeler, Effects of granulocyte colony simulating factor on functional activities of endothelial progenitor cells in patients with chronic ischemic heart disease, Arterioscler. Thromb. Vasc. Biol. 26 (10) (2006) 2238–2243.
[11] T. Asahara, T. Takahashi, H. Masuda, C. Kalka, D. Chen, H. Iwaguro, Y. Inai, M. Silver, J.M. Isner, VEGF contributes to postnatal neovascularization by mobilizing bone marrow-derived endothelial progenitor cells, EMBO J. 18 (14) (1999) 3964–3972.
[12] K. Imanaka-Yoshida, M. Hiroe, T. Nishikawa, S. Ishiyama, T. Shimojo, Y. Ohta, T. Sakakura, T. Yoshida, Tenascin-C modulates adhesion of cardiomyocytes to extracellular matrix during tissue remodeling after myocardial infarction, Lab. Invest. 81 (7) (2001) 1015–1024.
[13] B.A. Pallante, I. Duignan, D. Okin, A. Chin, M.C. Bressan, T. Mikawa, J.M. Edelberg, Bone marrow Oct3/4+ cells differentiate into cardiac myocytes via age-dependent paracrine mechanisms, Circ. Res. 100 (1) (2007) e1–11.
[14] J.A. Leopold, J. Loscalzo, Emerging role of precision medicine in cardiovascular disease, Circ. Res. 122 (9) (2018) 1302–1315.
[15] J.E. Cole, N. Astola, A.P. Cribbs, M.E. Goddard, I. Park, P. Green, A.H. Davies, R.O. Williams, M. Feldmann, C. Monaco, Indoleamine 2,3-dioxygenase-1 is protective in atherosclerosis and its metabolites provide new opportunities for drug development, Proc. Natl. Acad. Sci. U. S. A. 112 (42) (2015) 13033–13038.
[16] M. Abad, H. Hashimoto, H. Zhou, M.G. Morales, B. Chen, R. Bassel-Duby, E.N. Olson, Notch inhibition enhances cardiac reprogramming by increasing MEF2C transcriptional activity, Stem Cell Rep. 8 (3) (2017) 548–560.
[17] P. Muller, H. Lemcke, R. David, Stem cell therapy in heart diseases - cell types, mechanisms and improvement strategies, Cell. Physiol. Biochem. 48 (6) (2018) 2607–2655.
[18] J.C. Bilgimol, S. Ragupathi, L. Vengadassalapathy, N.S. Senthil, K. Selvakumar, M. Ganesan, S.R. Manjunath, Stem cells: an eventual treatment option for heart diseases, World J. Stem Cells 7 (8) (2015) 1118–1126.
[19] J.K. Biehl, B. Russell, Introduction to stem cell therapy, J. Cardiovasc. Nurs. 24 (2) (2009) 98–103 (quiz 104-105).
[20] R. Jaenisch, R. Young, Stem cells, the molecular circuitry of pluripotency and nuclear reprogramming, Cell 132 (4) (2008) 567–582.
[21] R. Hajar, Risk factors for coronary artery disease: historical perspectives, Heart Views 18 (3) (2017) 109–114.
[22] M.L. Weiss, D.L. Troyer, Stem cells in the umbilical cord, Stem Cell Rev. 2 (2) (2006) 155–162.
[23] G.R. Martin, Isolation of a pluripotent cell line from early mouse embryos cultured in medium conditioned by teratocarcinoma stem cells, Proc. Natl. Acad. Sci. U. S. A. 78 (12) (1981) 7634–7638.
[24] M.J. Evans, M.H. Kaufman, Establishment in culture of pluripotential cells from mouse embryos, Nature 292 (5819) (1981) 154–156.
[25] J.A. Thomson, J. Itskovitz-Eldor, S.S. Shapiro, M.A. Waknitz, J.J. Swiergiel, V.S. Marshall, J.M. Jones, Embryonic stem cell lines derived from human blastocysts, Science 282 (5391) (1998) 1145–1147.
[26] P.A. Redondo, M. Pavlou, M. Loizidou, U. Cheema, Elements of the niche for adult stem cell expansion, J .Tissue Eng. 8 (2017). 2041731417725464.
[27] M.T. Alrefai, D. Murali, A. Paul, K.M. Ridwan, J.M. Connell, D. Shum-Tim, Cardiac tissue engineering and regeneration using cell-based therapy, Stem Cells Cloning 8 (2015) 81–101.
[28] K. Breckwoldt, F. Weinberger, T. Eschenhagen, Heart regeneration, Biochim. Biophys. Acta 1863 (7 Pt B) (2016) 1749–1759.
[29] A.N. Raval, T.J. Kamp, L.F. Hogle, Cellular therapies for heart disease: unveiling the ethical and public policy challenges, J. Mol. Cell. Cardiol. 45 (4) (2008) 593–601.
[30] K. Lunde, S. Solheim, S. Aakhus, H. Arnesen, M. Abdelnoor, T. Egeland, K. Endresen, A. Ilebekk, A. Mangschau, J.G. Fjeld, H.J. Smith, E. Taraldsrud, H.K. Grogaard, R. Bjornerheim, M. Brekke, C. Muller, E. Hopp, A. Ragnarsson, J.E. Brinchmann, K. Forfang, Intracoronary injection of mononuclear bone marrow cells in acute myocardial infarction, N. Engl. J. Med. 355 (12) (2006) 1199–1209.
[31] B. Assmus, A. Rolf, S. Erbs, A. Elsasser, W. Haberbosch, R. Hambrecht, H. Tillmanns, J. Yu, R. Corti, D.G. Mathey, C.W. Hamm, T. Suselbeck, T. Tonn, S. Dimmeler, T. Dill, A.M. Zeiher, V. Schachinger, R.-A. Investigators, Clinical outcome 2 years after intracoronary administration of bone marrow-derived progenitor cells in acute myocardial infarction, Circ. Heart Failure 3 (1) (2010) 89–96.
[32] V. Schachinger, S. Erbs, A. Elsasser, W. Haberbosch, R. Hambrecht, H. Holschermann, J. Yu, R. Corti, D.G. Mathey, C.W. Hamm, T. Suselbeck, B. Assmus, T. Tonn, S. Dimmeler, A.M. Zeiher, R.-A. Investigators, Intracoronary bone marrow-derived progenitor cells in acute myocardial infarction, N. Engl. J. Med. 355 (12) (2006) 1210–1221.
[33] I. Batalov, A.W. Feinberg, Differentiation of cardiomyocytes from human pluripotent stem cells using monolayer culture, Biomark. Insights 10 (Suppl 1) (2015) 71–76.
[34] L.A. Boyer, T.I. Lee, M.F. Cole, S.E. Johnstone, S.S. Levine, J.P. Zucker, M.G. Guenther, R.M. Kumar, H.L. Murray, R.G. Jenner, D.K. Gifford, D.A. Melton, R. Jaenisch, R.A. Young, Core transcriptional regulatory circuitry in human embryonic stem cells, Cell 122 (6) (2005) 947–956.
[35] A.M. Ghaleb, M.O. Nandan, S. Chanchevalap, W.B. Dalton, I.M. Hisamuddin, V.W. Yang, Kruppel-like factors 4 and 5: the yin and yang regulators of cellular proliferation, Cell Res. 15 (2) (2005) 92–96.

[36] S.H. Orkin, J. Wang, J. Kim, J. Chu, S. Rao, T.W. Theunissen, X. Shen, D.N. Levasseur, The transcriptional network controlling pluripotency in ES cells, Cold Spring Harb. Symp. Quant. Biol. 73 (2008) 195–202.

[37] C. Hadjimichael, K. Chanoumidou, N. Papadopoulou, P. Arampatzi, J. Papamatheakis, A. Kretsovali, Common stemness regulators of embryonic and cancer stem cells, World J. Stem Cells 7 (9) (2015) 1150–1184.

[38] W. Zhao, X. Ji, F. Zhang, L. Li, L. Ma, Embryonic stem cell markers, Molecules 17 (6) (2012) 6196–6236.

[39] E. Matsa, P.W. Burridge, J.C. Wu, Human stem cells for modeling heart disease and for drug discovery, Sci. Transl. Med. 6 (239) (2014) 239ps6.

[40] J.H. van Berlo, O. Kanisicak, M. Maillet, R.J. Vagnozzi, J. Karch, S.C. Lin, R.C. Middleton, E. Marban, J.D. Molkentin, C-kit+ cells minimally contribute cardiomyocytes to the heart, Nature 509 (7500) (2014) 337–341.

[41] A. Gratwohl, M.C. Pasquini, M. Aljurf, Y. Atsuta, H. Baldomero, L. Foeken, M. Gratwohl, L.F. Bouzas, D. Confer, K. Frauendorfer, E. Gluckman, H. Greinix, M. Horowitz, M. Iida, J. Lipton, A. Madrigal, M. Mohty, L. Noel, N. Novitzky, J. Nunez, M. Oudshoorn, J. Passweg, J. van Rood, J. Szer, K. Blume, F.R. Appelbaum, Y. Kodera, D. Niederwieser, Worldwide Network for Blood and Marrow Transplantation (WBMT), One million haemopoietic stem-cell transplants: a retrospective observational study, Lancet Haematol. 2 (3) (2015) e91–100.

[42] M.F. Pittenger, D.E. Discher, B.M. Péault, D.G. Phinney, J.M. Hare, A.I. Caplan, Mesenchymal stem cell perspective: cell biology to clinical progress, NPJ Regen. Med. 4 (2019) 22.

[43] G.M. Beattie, A.D. Lopez, N. Bucay, A. Hinton, M.T. Firpo, C.C. King, A. Hayek, Activin a maintains pluripotency of human embryonic stem cells in the absence of feeder layers, Stem Cells 23 (4) (2005) 489–495.

[44] L. Vallier, M. Alexander, R.A. Pedersen, Activin/nodal and FGF pathways cooperate to maintain pluripotency of human embryonic stem cells, J. Cell Sci. 118 (Pt 19) (2005) 4495–4509.

[45] A.C. Mullen, J.L. Wrana, TGF-β family signaling in embryonic and somatic stem-cell renewal and differentiation, Cold Spring Harbor Perspect. Biol. 9 (7) (2017) a022186.

[46] G. Karlsson, U. Blank, J.L. Moody, M. Ehinger, S. Singbrant, C.X. Deng, S. Karlsson, Smad4 is critical for self-renewal of hematopoietic stem cells, J. Exp. Med. 204 (3) (2007) 467–474.

[47] K. Takahashi, S. Yamanaka, Induction of pluripotent stem cells from mouse embryonic and adult fibroblast cultures by defined factors, Cell 126 (4) (2006) 663–676.

[48] K. Takahashi, K. Tanabe, M. Ohnuki, M. Narita, T. Ichisaka, K. Tomoda, S. Yamanaka, Induction of pluripotent stem cells from adult human fibroblasts by defined factors, Cell 131 (5) (2007) 861–872.

[49] T. Sadahiro, S. Yamanaka, M. Ieda, Direct cardiac reprogramming: progress and challenges in basic biology and clinical applications, Circ. Res. 116 (8) (2015) 1378–1391.

[50] J.J. Chong, C.E. Murry, Cardiac regeneration using pluripotent stem cells-progression to large animal models, Stem Cell Res. 13 (3 Pt B) (2014) 654–665.

[51] R. Bolli, X.L. Tang, S.K. Sanganalmath, O. Rimoldi, F. Mosna, A. Abdel-Latif, H. Jneid, M. Rota, A. Leri, J. Kajstura, Intracoronary delivery of autologous cardiac stem cells improves cardiac function in a porcine model of chronic ischemic cardiomyopathy, Circulation 128 (2) (2013) 122–131.

[52] K.E. Hatzistergos, H. Quevedo, B.N. Oskouei, Q. Hu, G.S. Feigenbaum, I.S. Margitich, R. Mazhari, A.J. Boyle, J.P. Zambrano, J.E. Rodriguez, R. Dulce, P.M. Pattany, D. Valdes, C. Revilla, A.W. Heldman, I. McNiece, J.M. Hare, Bone marrow mesenchymal stem cells stimulate cardiac stem cell proliferation and differentiation, Circ. Res. 107 (7) (2010) 913–922.

[53] P.V. Johnston, T. Sasano, K. Mills, R. Evers, S.T. Lee, R.R. Smith, A.C. Lardo, S. Lai, C. Steenbergen, G. Gerstenblith, R. Lange, E. Marban, Engraftment, differentiation, and functional benefits of autologous cardiosphere-derived cells in porcine ischemic cardiomyopathy, Circulation 120 (12) (2009) 1075–1083.

[54] D. Cyranoski, 'Reprogrammed' stem cells approved to mend human hearts for the first time, Nature 557 (7707) (2018) 619–620.

[55] S. Yla-Herttuala, iPSC-derived cardiomyocytes taken to rescue infarcted heart muscle in coronary heart disease patients, Mol. Ther. 26 (9) (2018) 2077.

[56] A. Parikh, J. Wu, R.M. Blanton, E.S. Tzanakakis, Signaling pathways and gene regulatory networks in cardiomyocyte differentiation, Tissue Eng., Part B 21 (4) (2015) 377–392.

[57] M.J. Walters, G.A. Wayman, J.L. Christian, Bone morphogenetic protein function is required for terminal differentiation of the heart but not for early expression of cardiac marker genes, Mech. Dev. 100 (2) (2001) 263–273.

[58] J. Xu, P.J. Gruber, K.R. Chien, SMAD4 is essential for human cardiac mesodermal precursor cell formation, Stem Cells 37 (2) (2019) 216–225.

[59] K. Monzen, Y. Hiroi, S. Kudoh, H. Akazawa, T. Oka, E. Takimoto, D. Hayashi, T. Hosoda, M. Kawabata, K. Miyazono, S. Ishii, Y. Yazaki, R. Nagai, I. Komuro, Smads, TAK1, and their common target ATF-2 play a critical role in cardiomyocyte differentiation, J. Cell Biol. 153 (4) (2001) 687–698.

[60] Y. Sano, J. Harada, S. Tashiro, R. Gotoh-Mandeville, T. Maekawa, S. Ishii, ATF-2 is a common nuclear target of Smad and TAK1 pathways in transforming growth factor-beta signaling, J. Biol. Chem. 274 (13) (1999) 8949–8957.

[61] D.M. Ornitz, N. Itoh, The fibroblast growth factor signaling pathway, Wiley Interdiscip. Rev. Dev. Biol. 4 (3) (2015) 215–266.

[62] J. Zhang, J. Liu, L. Liu, W.L. McKeehan, F. Wang, The fibroblast growth factor signaling axis controls cardiac stem cell differentiation through regulating autophagy, Autophagy 8 (4) (2012) 690–691.

[63] B.T. MacDonald, K. Tamai, X. He, Wnt/beta-catenin signaling: components, mechanisms, and diseases, Dev. Cell 17 (1) (2009) 9–26.

[64] G.S. Sineva, V.A. Pospelov, β-catenin in pluripotency: adhering to self-renewal or Wnting to differentiate? Int. Rev. Cell Mol. Biol. 312 (2014) 53–78.

[65] A. Romito, G. Cobellis, Pluripotent stem cells: current understanding and future directions, Stem Cells Int. 2016 (2016) 9451492.

[66] S. Dakhore, B. Nayer, K. Hasegawa, Human pluripotent stem cell culture: current status, challenges, and advancement, Stem Cells Int. 2018 (2018) 7396905.

[67] T.S. Park, Z. Galic, A.E. Conway, A. Lindgren, B.J. van Handel, M. Magnusson, L. Richter, M.A. Teitell, H.K. Mikkola, W.E. Lowry, K. Plath, A.T. Clark, Derivation of primordial germ cells from human embryonic and induced pluripotent stem cells is significantly improved by coculture with human fetal gonadal cells, Stem Cells 27 (4) (2009) 783–795.

[68] C.L. Mummery, J. Zhang, E.S. Ng, D.A. Elliott, A.G. Elefanty, T.J. Kamp, Differentiation of human embryonic stem cells and induced pluripotent stem cells to cardiomyocytes: a methods overview, Circ. Res. 111 (3) (2012) 344–358.

[69] M. Simunovic, A.H. Brivanlou, Embryoids, organoids and gastruloids: new approaches to understanding embryogenesis, Development 144 (6) (2017) 976–985.

[70] Y. Jiang, P. Park, S.M. Hong, K. Ban, Maturation of cardiomyocytes derived from human pluripotent stem cells: current strategies and limitations, Mol. Cells 41 (7) (2018) 613–621.

[71] J.D. Molkentin, D.V. Kalvakolanu, B.E. Markham, Transcription factor GATA-4 regulates cardiac muscle-specific expression of the alpha-myosin heavy-chain gene, Mol. Cell. Biol. 14 (7) (1994) 4947–4957.

[72] K.M. Broughton, B.J. Wang, F. Firouzi, F. Khalafalla, S. Dimmeler, F. Fernandez-Aviles, M.A. Sussman, Mechanisms of cardiac repair and regeneration, Circ. Res. 122 (8) (2018) 1151–1163.

[73] R. Kopan, Notch signaling, Cold Spring Harb Perspect. Biol. 4 (10) (2012) a011213.

[74] Z. Yu, Y. Zou, J. Fan, C. Li, L. Ma, Notch1 is associated with the differentiation of human bone marrow-derived mesenchymal stem cells to cardiomyocytes, Mol. Med. Rep. 14 (6) (2016) 5065–5071.

[75] M. Nemir, M. Metrich, I. Plaisance, M. Lepore, S. Cruchet, C. Berthonneche, A. Sarre, F. Radtke, T. Pedrazzini, The notch pathway controls fibrotic and regenerative repair in the adult heart, Eur. Heart J. 35 (32) (2014) 2174–2185.

[76] L. Chen, W. Xia, M. Hou, Mesenchymal stem cells attenuate doxorubicin-induced cellular senescence through the VEGF/Notch/TGF-β signaling pathway in H9c2 cardiomyocytes, Int. J. Mol. Med. 42 (1) (2018) 674–684.

[77] K. Wang, R. Ding, Y. Ha, Y. Jia, X. Liao, S. Wang, R. Li, Z. Shen, H. Xiong, J. Guo, W. Jie, Hypoxia-stressed cardiomyocytes promote early cardiac differentiation of cardiac stem cells through HIF-1alpha/Jagged1/Notch1 signaling, Acta Pharm. Sin. B 8 (5) (2018) 795–804.

[78] N. Gude, E. Joyo, H. Toko, P. Quijada, M. Villanueva, N. Hariharan, V. Sacchi, S. Truffa, A. Joyo, M. Voelkers, R. Alvarez, M.A. Sussman, Notch activation enhances lineage commitment and protective signaling in cardiac progenitor cells, Basic Res. Cardiol. 110 (3) (2015) 29.

[79] J.L. Russell, S.C. Goetsch, N.R. Gaiano, J.A. Hill, E.N. Olson, J.W. Schneider, A dynamic notch injury response activates epicardium and contributes to fibrosis repair, Circ. Res. 108 (1) (2011) 51–59.

[80] Y. Li, Y. Hiroi, J.K. Liao, Notch signaling as an important mediator of cardiac repair and regeneration after myocardial infarction, Trends Cardiovasc. Med. 20 (7) (2010) 228–231.

[81] J.C. Tung, S.L. Paige, B.D. Ratner, C.E. Murry, C.M. Giachelli, Engineered biomaterials control differentiation and proliferation of human-embryonic-stem-cell-derived cardiomyocytes via timed notch activation, Stem Cell Rep. 2 (3) (2014) 271–281.

[82] B. Yan, R.D. Singla, L.S. Abdelli, P.K. Singal, D.K. Singla, Regulation of PTEN/Akt pathway enhances cardiomyogenesis and attenuates adverse left ventricular remodeling following thymosin β4 overexpressing embryonic stem cell transplantation in the infarcted heart, PLoS One 8 (9) (2013) e75580.

[83] L. Chen, M. Ashraf, Y. Wang, M. Zhou, J. Zhang, G. Qin, J. Rubinstein, N.L. Weintraub, Y. Tang, The role of notch 1 activation in cardiosphere derived cell differentiation, Stem Cells Dev. 21 (12) (2012) 2122–2129.

[84] A. Croquelois, A.A. Domenighetti, M. Nemir, M. Lepore, N. Rosenblatt-Velin, F. Radtke, T. Pedrazzini, Control of the adaptive response of the heart to stress via the Notch1 receptor pathway, J. Exp. Med. 205 (13) (2008) 3173–3185.

[85] R. Cortés, M. Rivera, E. Roselló-Lletí, L. Martínez-Dolz, L. Almenar, I. Azorín, F. Lago, J.R. González-Juanatey, M. Portolés, Differences in MEF2 and NFAT transcriptional pathways according to human heart failure aetiology, PLoS One 7 (2) (2012) e30915.

[86] A. Kasahara, S. Cipolat, Y. Chen, G.W. Dorn 2nd, L. Scorrano, Mitochondrial fusion directs cardiomyocyte differentiation via calcineurin and notch signaling, Science 342 (6159) (2013) 734–737.

[87] K. Song, Y.J. Nam, X. Luo, X. Qi, W. Tan, G.N. Huang, A. Acharya, C.L. Smith, M.D. Tallquist, E.G. Neilson, J.A. Hill, R. Bassel-Duby, E.N. Olson, Heart repair by reprogramming non-myocytes with cardiac transcription factors, Nature 485 (7400) (2012) 599–604.

[88] J.M. Verdi, R. Schmandt, A. Bashirullah, S. Jacob, R. Salvino, C.G. Craig, A.E. Program, H.D. Lipshitz, C.J. McGlade, Mammalian NUMB is an evolutionarily conserved signaling adapter protein that specifies cell fate, Curr. Biol. 6 (9) (1996) 1134–1145.

[89] C. Zhao, H. Guo, J. Li, T. Myint, W. Pittman, L. Yang, W. Zhong, R.J. Schwartz, J.J. Schwarz, H.A. Singer, M.D. Tallquist, M. Wu, Numb family proteins are essential for cardiac morphogenesis and progenitor differentiation, Development 141 (2) (2014) 281–295.

[90] J. Grego-Bessa, L. Luna-Zurita, G. del Monte, V. Bolos, P. Melgar, A. Arandilla, A.N. Garratt, H. Zang, Y.S. Mukouyama, H. Chen, W. Shou, E. Ballestar, M. Esteller, A. Rojas, J.M. Perez-Pomares, J.L. de la Pompa, Notch signaling is essential for ventricular chamber development, Dev. Cell 12 (3) (2007) 415–429.

[91] R.K. Johnston, S. Balasubramanian, H. Kasiganesan, C.F. Baicu, M.R. Zile, D. Kuppuswamy, Beta3 integrin-mediated ubiquitination activates survival signaling during myocardial hypertrophy, FASEB J. 23 (8) (2009) 2759–2771.

[92] S. Balasubramanian, D. Kuppuswamy, RGD-containing peptides activate S6K1 through beta3 integrin in adult cardiac muscle cells, J. Biol. Chem. 278 (43) (2003) 42214–42224.

[93] S. Balasubramanian, L. Quinones, H. Kasiganesan, Y. Zhang, D.L. Pleasant, K.P. Sundararaj, M.R. Zile, A.D. Bradshaw, D. Kuppuswamy, Beta3 integrin in cardiac fibroblast is critical for extracellular matrix accumulation during pressure overload hypertrophy in mouse, PLoS One 7 (9) (2012) e45076.

[94] L. Vitillo, S.J. Kimber, Integrin and FAK regulation of human pluripotent stem cells, Curr. Stem Cell Rep. 3 (4) (2017) 358–365.

[95] L.S. Campos, D.P. Leone, J.B. Relvas, C. Brakebusch, R. Fassler, U. Suter, C. ffrench-Constant, Beta1 integrins activate a MAPK signalling pathway in neural stem cells that contributes to their maintenance, Development 131 (14) (2004) 3433–3444.

[96] S.T. Lee, J.I. Yun, Y.S. Jo, M. Mochizuki, A.J. van der Vlies, S. Kontos, J.E. Ihm, J.M. Lim, J.A. Hubbell, Engineering integrin signaling for promoting embryonic stem cell self-renewal in a precisely defined niche, Biomaterials 31 (6) (2010) 1219–1226.

[97] S.R. Braam, L. Zeinstra, S. Litjens, D. Ward-van Oostwaard, S. van den Brink, L. van Laake, F. Lebrin, P. Kats, R. Hochstenbach, R. Passier, A. Sonnenberg, C.L. Mummery, Recombinant vitronectin is a functionally defined substrate that supports human embryonic stem cell self-renewal via alphavbeta5 integrin, Stem Cells 26 (9) (2008) 2257–2265.

[98] T.J. Rowland, L.M. Miller, A.J. Blaschke, E.L. Doss, A.J. Bonham, S.T. Hikita, L.V. Johnson, D.O. Clegg, Roles of integrins in human induced pluripotent stem cell growth on Matrigel and vitronectin, Stem Cells Dev. 19 (8) (2010) 1231–1240.

[99] S. Rodin, A. Domogatskaya, S. Strom, E.M. Hansson, K.R. Chien, J. Inzunza, O. Hovatta, K. Tryggvason, Long-term self-renewal of human pluripotent stem cells on human recombinant laminin-511, Nat. Biotechnol. 28 (6) (2010) 611–615.

[100] S.P. Toya, K.K. Wary, M. Mittal, F. Li, P.T. Toth, C. Park, J. Rehman, A.B. Malik, Integrin alpha6beta1 expressed in ESCs instructs the differentiation to endothelial cells, Stem Cells 33 (6) (2015) 1719–1729.

[101] J.H. Park, J.M. Ryu, H.J. Han, Involvement of caveolin-1 in fibronectin-induced mouse embryonic stem cell proliferation: role of FAK, RhoA, PI3K/Akt, and ERK 1/2 pathways, J. Cell. Physiol. 226 (1) (2011) 267–275.

Chapter 12

Angiogenesis in aging hearts—Cardiac stem cell therapy

Vinu Ramachandran and Anandan Balakrishnan
Department of Genetics, Dr. ALM PG Institute of Basic Medical Sciences, University of Madras, Taramani Campus, Chennai, Tamil Nadu, India

1 Introduction

Cardiovascular Disease (CVD) is a major cause of death and a global burden. CVD is regarded as an age-dependent condition with gradually increasing incidence in the aging population. Age-related risk factors for CVD such as increased oxidative stress (causing electrical and functional abnormalities similar to arrhythmias, atrial fibrillation and heart failure) due to production of ROS and inflammatory signals which along with additional (co-morbid) risk factors like obesity, diabetes and frailty contribute to increased prevalence of CVD among the elderly (> 70 years of age) population [1–3]. Visible age-related signs are indicative of poor cardiovascular health [4]. Understanding the mechanism behind aged hearts enables the development of therapies to prevent heart failure [5]. Impaired angiogenesis and endothelial dysfunction are associated with aging and induction of angiogenesis is a promising therapeutic intervention for ischemic diseases [6]. The ischemic CVDs such as myocardial infarction are dependent on the growth of blood vessels and are associated with worst outcomes in elderly patients. To extend the lifespan of the elderly and individuals with CVD, reversal of cardiac aging or antiaging intervention is crucial. Cell-based therapies have entered clinical trials (summarized by Psaltis et al. [7]) and have been encouraging to define treatment modalities for aging and ischemic diseases [8]. Stem cells provide cardioprotective effects via several mechanisms including neoangiogenesis [9].

2 Cardiac aging

The structure and function of the adult human heart declines progressively with age. The phenotypic changes of an aging heart are observed as cardiac hypertrophy (left ventricle), diastolic dysfunction, myocardial fibrosis and atrial fibrillation, considered as intrinsic cardiac aging [10]. The clinical issues met with CVD in the aging population for treatment are systolic hypertension, vascular aging, heart failure with preserved and reduced ejection fraction, calcification of the valvular and cardiac skeleton, frailty and sarcopenia (revised in detail by Paneni et al. [11]). Understanding the cellular and molecular mechanisms behind cardiac aging is essential for the design of better prevention and treatment strategies.

2.1 Factors influencing cardiac aging

Cellular and molecular signals together interact to influence the overall function of the heart. Intrinsic cellular processes encompassing immune response (inflammation, senescence-associated secretory phenotype), impaired calcium homeostasis, metabolic imbalance (impaired autophagy, oxidative stress-induced mitochondrial dysfunction, metabolic processes—mTOR signaling), adverse extracellular matrix (ECM) remodeling and increased cellular senescence are reported to impact cardiac aging. Molecular changes that include altered growth signaling (mTOR and IGF-1), chronic activation in neurohormonal signaling (renin-angiotensin aldosterone system (RAAS)), genomic instability (aggravated telomere shortening, Single nucleotide polymorphisms, DNA damage) and epigenetic changes (limited chromatin remodeling, limited methylation, histone preservation, age-related dysregulation of microRNAs expression) and cardiac stem-/progenitor-cell aging influence the functional capacity of heart (summarized by Chiao and Rabinovitch [12], Gude et al. [13]). The heart dysfunction is mediated by the cardiac cells in a cell-specific and tissue-specific manner. At the cellular level, the vascular cell senescence is associated with atherosclerosis, fibroblasts and mesenchymal cells (interstitial cells) exit the cell cycle and secrete senescence-associated secretory phenotype (SASP) factors, the cardiomyocytes are hypertrophied and show decreased contractility function, and cardiac progenitor cells (CPCs) lose the self-renewal and regeneration capacity.

The extrinsic factors such as lifestyle (alterations in the levels of nutrients, physical inactivity, psychological stress), behavior (social integration) and environment (pollution, chemical exposure) also modulate cardiac aging at the cellular level (described in detail by Gude et al. [13]). Insight on exogenous and endogenous mechanisms underlying cardiac aging is essential to prompt innovative strategies for improvements on replicative and reparative processes of cardiomyocytes and noncardiomyocytes.

2.1.1 Cellular senescence in cardiac aging

Cellular senescence is an irreversible cell cycle arrest that is progressive with age. The accumulation of these poorly functional senescent cells results in impaired intercellular communications and compromise tissue function promoting inflammation, consequently induce cell death and loss of cardiomyocytes.

Senescence also occurs in nonmyocytes of the heart such as endothelial cells, fibroblasts, CPCs, hematopoietic stem cells, bone marrow-derived mononuclear cells. They show several indicators of cellular senescence such as shortened-telomeres and reduced differentiation capacity, resulting in cardiac aging [14–16]. The build-up of senescent cells and calcification are characteristic of atherosclerotic plaques in the vasculature [17, 18]. A significant approach to mitigate cardiac aging has been to antagonize cardiac cellular senescence [19, 20].

2.2 Vascular changes (impaired angiogenesis) in cardiac aging

2.2.1 Angiogenesis

Angiogenesis is the sprouting of new blood capillaries from preexisting vascular structures. This mechanism is initiated by the migration and proliferation of endothelial cells. The new capillaries form as a network, consisting of endothelial cells as tubes, but lack smooth muscle cells and stabilizing cell structures [21]. An age-associated effect on cardiovascular function includes dysregulated angiogenic repair mechanism, which is responsible to restore blood flow after ischemia [22].

2.2.2 Impaired angiogenesis in aged hearts

Among the several cell types of the noncardiomyocyte population, the chief blood vessel types comprise vascular endothelial cells, vascular smooth muscle cells and pericytes. One of the clinical features in an aging elderly human heart is vascular aging characterized by endothelial dysfunction and increased stiffness of central arteries [23]. Irrespective of the proliferative capacity held by vascular cells in vitro and following injury in vivo, age-associated plaque deposits in the blood vessels result in stiffness of walls, inflammation, myocardial infarction and vascular cell death. Subsequently, resulting in impaired vascular structure and function [17, 24].

Evidences for age-associated impairment of angiogenesis are presented as decreased capillary density, defective functionality of eNOS, impaired sensitivity to insulin, reduced proliferative capacity of senescent endothelial cells, impaired telomerase activity, reduction in the production of angiogenic growth factors like VEGF-A, reduced endothelial migration, in hypoxia-inducible factor 1α (HIF1α) and Peroxisome proliferator-activated receptor gamma coactivator 1-alpha (PGC-1α) activity, and deterioration in amount and function of the stem and progenitor cells [6, 25–27]. Chronic psychological stress in mice caused a decline in angiogenic action (impaired aortic endothelial sprouting) and accelerated vascular (aortic) senescence [28].

3 Therapy for aging heart

The therapeutic approaches for cardiac aging include calorie restriction, pharmacological intervention (e.g., rapamycin), dietary supplement (e.g., Telomerase activator TA-65) recombinant protein therapy (e.g., ACE inhibitors), gene therapy (antagomirs/anti-miRs) and cell therapy (cardiac stem-/progenitor cells) has been discussed [12, 13]. The use of epigenetic modifiers and gene editing in iPSC generated cardiac muscle cells for the treatment of age-associated CVD are new potential therapeutic approaches but is in early stages of development for clinical application, has been reviewed [13]. To restore the youthful state of the aged heart, the aged somatic bone marrow cells reprogrammed into induced pluripotent stem cells (iPSCs) to develop into cardiac tissue were investigated to demonstrate a partial rejuvenation in mice [29].

3.1 Cardiac stem cells for therapy

Cardiac stem cells exhibit self-renewal and clonogenic capacity. Transplanting sufficient cells is key to retain the maximum number of cells in the heart for repair. Current cell delivery strategies include transvascular approach (intracoronary, intravenous, stem cell mobilization by cytokines) and direct injection into left ventricular wall of heart (transepicardial, transendocardial, transcoronary vein). Application of cardiac stem cells (CSCs) for cardiac repair appear safe [30].

3.1.1 Cardiac stem cell population

The mesoderm rising from pluripotent stem cells differentiate to cardiac mesoderm and cardiac progenitors to functional cardiomyocytes [31]. The cardiac stem cell lineage is presented in Fig. 1. The aging mammalian heart was traditionally considered a postmitotic organ with reduced cell renewal capacity. This has been confirmed with the transcriptional profiles of cardiomyocytes and nonmyocytes (fibroblasts, leukocytes, endothelial cells) from infarcted and noninfarcted neonatal against adult mouse hearts [32]. However, a c-Kit$^+$ CSC from the adult heart has self-renewal capability and can differentiate into cardiomyocytes and nonmyocytes (endothelial and smooth muscle cells), supporting the regeneration of injured heart [33–35]. The c-Kit$^+$ cells are multipotent which expresses the tyrosine kinase receptor c-kit.

CPCs, a subset of cardiac interstitial cells facilitates a limited reparative process in the heart during injury. The biological relevance of the progenitor cells in the heart has earned interest to supplement the inadequate proliferative potential of the adult mammalian myocytes to enhance cardiac repair. Besides the c-Kit$^+$ cells, subpopulations of cardiac progenitors have been identified in the adult mammalian heart. Several CSC and CPC populations have been discovered including, cardiosphere-derived cells (CDCs) [36, 37], stem cell antigen (Sca)-1$^+$ cells [38, 39], insulin gene enhancer protein (Isl)-1$^+$ cells [40, 41], cardiac side population cells [42] and Cardiac colony-forming unit-fibroblasts (cCFU-F) [43]. These cells belong to the CPC population and exhibit several markers [44]. The cardiac stem cells and the progenitor cells are essential for cardiac repair and replacement of cardiomyocytes in myocardial infarction. Persistent adult progenitor population has been located in the subepicardium and myocardial interstitium [45].

Transplantation of Sca-1$^+$ cells antagonized left ventricular remodeling in mouse heart with myocardial infarction [46]. Cell-based therapy approach has utilized autologous KIT$^+$ CPC for patients with heart failure, for which phase II clinical trial is in preparation [47]. However, their diminished reparative capacity in stem cells from the aged and diseased individuals has been recorded [48]. Alternatively, combinatorial stem cell therapy, renewal of the aged CPCs via genetic engineering and preconditional hypoxia are suggested approaches for betterment of treatment [49–51]. Hypoxia preconditioning of the bone marrow mesenchymal stem cells has shown improved survival and function of stem cells through autophagy

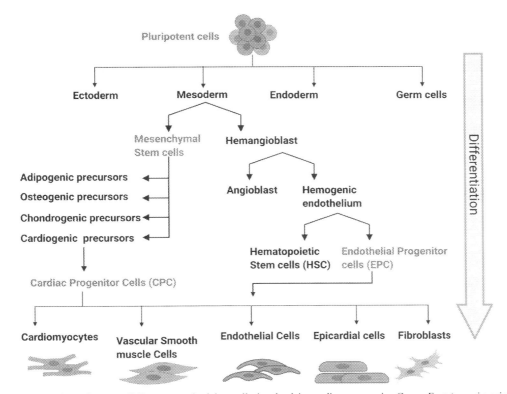

FIG. 1 Schematic presentation of stem cell lineage emphasizing cells involved in cardiomyogenesis. *Green Font* (gray in print version)—Cells potentially employed in therapies for cardiac repair.

regulation [52]. Apart from direct cell delivery methods, stimulation of the CPCs using exosomes, growth factors, drugs, paracrine-secreting exogenous cells is also another stem cell-based strategy for the treatment of cardiac aging [53].

3.2 Targeting angiogenesis using cardiac stem cells for therapy

Regenerative stem cell therapy aims to substitute the loss of myocardium with healthy cells [34]. Several types of stem cells potentiate cardiac function postmyocardial infarction [54–56]. They mediate beneficial effects via paracrine action (secretion of cytokines and growth factors) for cardio protection and angiogenesis. However, the effects are moderate in clinical trials which could be owing to limited differentiation capacity [57]. Additionally, the stem cells transplanted create an ischemic environment. Hence along with cardiogenesis, the angiogenic process also should be promoted for optimal cardiac repair. The pro-angiogenic reparative strategy is illustrated in Fig. 2.

3.2.1 Pro-angiogenic potential of cardiac stem cells

Cardiac stem cells that reside in the adult human heart have the potential for both cardiogenesis and angiogenesis. Advanced age is said to affect the regenerative potential of tissue-specific stem cells [58]. The angiogenic potential of stem cells has also been reported to be impaired [59]. However, the influence of age over CSCs is not completely understood. Nakamura and colleagues investigated the effect of age over CDCs. The cultured CDCs using the right atria from patients (younger <65 years and elder >65 years of age) showed that the impact of aging over the quantity and quality of the cells was limited. While the markers of cellular senescence (β-galactosidase and DNA damage) were higher in aged patients, the expression of beneficial factors that mediate angiogenesis, antiapoptosis, recruitment of stem cells via paracrine effects, such as VEGF, HGF, IGF-1, SDF-1 and TGF-β did not reduce with age. Alongside, in vitro tube formation angiogenic assay and migration assay that showed no impairment of angiogenic potency, indicate possible implication of

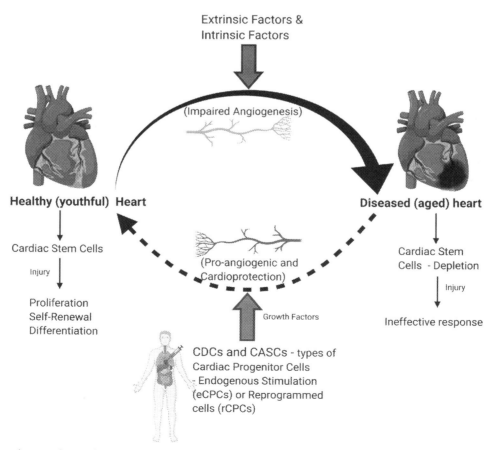

FIG. 2 Pro-angiogenic reparative mechanism for cardiac aging. (Abbreviations: *CASCs*, cardiac atrial appendage stem cells; *CDCs*, cardiosphere-derived cells; *eCPCs*, endogenous cardiac progenitor cells; *rCPCs*, reprogrammed cardiac progenitor cells).

autologous CSC transplantation therapy in patients belonging to the elderly population [60]. Correspondingly, previous reports imply CDCs to be characteristically similar to other stem cells, and endothelial cells identified using antigenic markers such as c-Kit, Sca-1, CD34, CD105, KDR and showed cardiac regeneration potential to reduce scar size in myocardial infarction [61–63]. Autologous transplantation of CDCs isolated from cardiac specimens has reversed ventricular dysfunction in the human trials [64].

Exosomes also exhibit cardioprotective functions by augmentation of angiogenesis, reducing fibrotic tissue formation or decreasing apoptosis [65]. Human CPC derived exosomes under physoxic conditions demonstrated tube formation capability and is a recommended therapeutic strategy for myocardial repair [66].

Cardiac atrial appendage stem cells (CASCs), a type of cardiac stem cell, exhibit myocardial differentiation and regeneration potential. CASCs cultured from patients with myocardial infarction has shown to promote cardiac angiogenesis via paracrine mechanisms such as increased production of growth factors (VEGF, IGFBP3, ET-1), promotion of endothelial cell proliferation, migration, tube formation which are important steps of the angiogenic process in vitro and blood vessel formation in vivo. These signs indicate the angiogenic potential of CASCs and appear to be an optimal stem cell for ischemic heart disease therapy [67].

3.2.2 Cardiomyogenic/angiogenic stimulants

Potential therapeutic agents for enhancing endogenous stem-cell or progenitor-cell function include stromal-cell-derived factor (SDF)-1, platelet-derived growth factor (PDGF), vascular endothelial growth factor (VEGF), and Tenascin-C. These agents are known to promote cardiac regeneration, induce angiogenesis and function [68]. The effects of transplantation of CSC along with VEGF were greater to the effect of CSC alone, toward improved cardiomyogenesis and angiogenesis [69]. The combined therapy of CSC and SDF-1 enhanced angiogenesis and heart function while scarring was reduced [70]. Prior treatment with combined angiogenic growth factors such as PDGF and VEGF preceding cardiac allograft transplantation promoted local angiogenesis and EPC-mediated vasculogenesis, that are usually impaired in aging mice. This aided the successful allografting for vascularization [71].

3.2.3 Pro-angiogenic potential of stem cells other than CPCs

The stem cells other than cardiac-derived stem cells, also enhance angiogenesis to combat cardiovascular dysfunction. The bone marrow-derived progenitor cells secrete chemokines (angiopoietin-1 and VEGF) via paracrine mechanisms at the site of injury. EPCs promote neovascularization as well as angiogenesis via angiogenic factors [72]. The endothelial progenitor cells and mesenchymal stem cells are proven to promote angiogenesis within ischemic CVDs (reviewed in Hou et al. [73]). Hypoxia preconditioning also increased the expression of prosurvival and proangiogenic factors in mesenchymal stem cells in vitro and transplantation in Cynomolgus monkey increased cardiomyocyte proliferation, vascular density, glucose uptake and reduced apoptosis without arrhythmogenic complications [74].

4 Conclusion

Cardiac stem cell therapy targeting the promotion of angiogenesis in aged hearts serves an important attribute to confront the rise in CVD-prone elderly population and ameliorate the negative effects of aging. However, more research studies are required to overcome the following existing challenges. The beneficial mechanism underlying angiogenesis promoted by stem cells in an aged heart is poorly understood. The therapeutic effects of stem cells are limited by the proliferation, engraftment, survival, and persistence of the transplanted cells. Stem-cell or progenitor-cell delivery systems need improvement. The success of this therapy is dependent on clinical efficacy and safety to recover the quality of life for the aging population.

References

[1] J.L. Rodgers, J. Jones, S.I. Bolleddu, S. Vanthenapalli, L.E. Rodgers, K. Shah, K. Karia, S.K. Panguluri, Cardiovascular risks associated with gender and aging, J. Cardiovasc. Dev. Dis. 6 (2) (2019) 19, https://doi.org/10.3390/jcdd6020019.

[2] M. Steenman, G. Lande, Cardiac aging and heart disease in humans, Biophys. Rev. 9 (2) (2017) 131–137, https://doi.org/10.1007/s12551-017-0255-9.

[3] A. Yazdanyar, A.B. Newman, The burden of cardiovascular disease in the elderly: morbidity, mortality, and costs, Clin. Geriatr. Med. 25 (4) (2009) 563–vii, https://doi.org/10.1016/j.cger.2009.07.007.

[4] M. Christoffersen, R. Frikke-Schmidt, P. Schnohr, G.B. Jensen, B.G. Nordestgaard, A. Tybjærg-Hansen, Visible age-related signs and risk of ischemic heart disease in the general population: a prospective cohort study, Circulation 129 (9) (2014) 990–998, https://doi.org/10.1161/CIRCULATIONAHA.113.001696.

[5] J.B. Strait, E.G. Lakatta, Aging-associated cardiovascular changes and their relationship to heart failure, Heart Fail. Clin. 8 (1) (2012) 143–164, https://doi.org/10.1016/j.hfc.2011.08.011.

[6] J. Lähteenvuo, A. Rosenzweig, Effects of aging on angiogenesis, Circ. Res. 110 (9) (2012) 1252–1264, https://doi.org/10.1161/CIRCRESAHA.111.246116.

[7] P.J. Psaltis, N. Schwarz, D. Toledo-Flores, S.J. Nicholls, Cellular therapy for heart failure, Curr. Cardiol. Rev. 12 (3) (2016) 195–215, https://doi.org/10.2174/1573403x12666160606121858.

[8] S. Dimmeler, A. Leri, Aging and disease as modifiers of efficacy of cell therapy, Circ. Res. 102 (11) (2008) 1319–1330, https://doi.org/10.1161/CIRCRESAHA.108.175943.

[9] B. Wernly, M. Mirna, R. Rezar, C. Prodinger, C. Jung, B.K. Podesser, A. Kiss, U.C. Hoppe, M. Lichtenauer, Regenerative cardiovascular therapies: stem cells and beyond, Int. J. Mol. Sci. 20 (6) (2019) 1420, https://doi.org/10.3390/ijms20061420.

[10] E.G. Lakatta, D. Levy, Arterial and cardiac aging: major shareholders in cardiovascular disease enterprises: part II: the aging heart in health: links to heart disease, Circulation 107 (2) (2003) 346–354, https://doi.org/10.1161/01.cir.0000048893.62841.f7.

[11] F. Paneni, C. Diaz Cañestro, P. Libby, T.F. Lüscher, G.G. Camici, The aging cardiovascular system: understanding it at the cellular and clinical levels, J. Am. Coll. Cardiol. 69 (15) (2017) 1952–1967, https://doi.org/10.1016/j.jacc.2017.01.064.

[12] Y.A. Chiao, P.S. Rabinovitch, The aging heart, Cold Spring Harb. Perspect. Med. 5 (9) (2015) a025148, https://doi.org/10.1101/cshperspect.a025148.

[13] N.A. Gude, K.M. Broughton, F. Firouzi, M.A. Sussman, Cardiac ageing: extrinsic and intrinsic factors in cellular renewal and senescence, Nat. Rev. Cardiol. 15 (9) (2018) 523–542, https://doi.org/10.1038/s41569-018-0061-5.

[14] A. Biernacka, N.G. Frangogiannis, Aging and cardiac fibrosis, Aging Dis. 2 (2) (2011) 158–173.

[15] N. Hariharan, M.A. Sussman, Cardiac aging—getting to the stem of the problem, J. Mol. Cell. Cardiol. 83 (2015) 32–36, https://doi.org/10.1016/j.yjmcc.2015.04.008.

[16] E. Nollet, V.Y. Hoymans, I.R. Rodrigus, D. De Bock, M. Dom, V. Van Hoof, C.J. Vrints, E.M. Van Craenenbroeck, Accelerated cellular senescence as underlying mechanism for functionally impaired bone marrow-derived progenitor cells in ischemic heart disease, Atherosclerosis 260 (2017) 138–146, https://doi.org/10.1016/j.atherosclerosis.2017.03.023.

[17] J.C. Kovacic, P. Moreno, E.G. Nabel, V. Hachinski, V. Fuster, Cellular senescence, vascular disease, and aging: part 2 of a 2-part review: clinical vascular disease in the elderly, Circulation 123 (17) (2011) 1900–1910, https://doi.org/10.1161/CIRCULATIONAHA.110.009118.

[18] T. Minamino, H. Miyauchi, T. Yoshida, Y. Ishida, H. Yoshida, I. Komuro, Endothelial cell senescence in human atherosclerosis: role of telomere in endothelial dysfunction, Circulation 105 (13) (2002) 1541–1544, https://doi.org/10.1161/01.cir.0000013836.85741.17.

[19] D. McHugh, J. Gil, Senescence and aging: causes, consequences, and therapeutic avenues, J. Cell Biol. 217 (1) (2018) 65–77, https://doi.org/10.1083/jcb.201708092.

[20] I. Shimizu, T. Minamino, Cellular senescence in cardiac diseases, J. Cardiol. 74 (4) (2019) 313–319, https://doi.org/10.1016/j.jjcc.2019.05.002.

[21] M. Heil, I. Eitenmüller, T. Schmitz-Rixen, W. Schaper, Arteriogenesis versus angiogenesis: similarities and differences, J. Cell. Mol. Med. 10 (1) (2006) 45–55, https://doi.org/10.1111/j.1582-4934.2006.tb00290.x.

[22] M.J. Reed, J.M. Edelberg, Impaired angiogenesis in the aged, Sci. Aging Knowl. Environ. 2004 (7) (2004) pe7, https://doi.org/10.1126/sageke.2004.7.pe7.

[23] M.L. Meyer, H. Tanaka, P. Palta, S. Cheng, N. Gouskova, D. Aguilar, G. Heiss, Correlates of segmental pulse wave velocity in older adults: the Atherosclerosis Risk in Communities (ARIC) Study, Am. J. Hypertens. 29 (1) (2016) 114–122, https://doi.org/10.1093/ajh/hpv079.

[24] J.C. Kovacic, P. Moreno, V. Hachinski, E.G. Nabel, V. Fuster, Cellular senescence, vascular disease, and aging: part 1 of a 2-part review, Circulation 123 (15) (2011) 1650–1660, https://doi.org/10.1161/CIRCULATIONAHA.110.007021.

[25] M.A. Creager, J.W. Olin, J.J. Belch, G.L. Moneta, T.D. Henry, S. Rajagopalan, B.H. Annex, W.R. Hiatt, Effect of hypoxia-inducible factor-1alpha gene therapy on walking performance in patients with intermittent claudication, Circulation 124 (16) (2011) 1765–1773, https://doi.org/10.1161/CIRCULATIONAHA.110.009407.

[26] B.B. Groen, H.M. Hamer, T. Snijders, J. van Kranenburg, D. Frijns, H. Vink, L.J. van Loon, Skeletal muscle capillary density and microvascular function are compromised with aging and type 2 diabetes, J. Appl. Physiol. (1985) 116 (8) (2014) 998–1005, https://doi.org/10.1152/japplphysiol.00919.2013.

[27] S. Rajagopalan, J. Olin, S. Deitcher, A. Pieczek, J. Laird, P.M. Grossman, C.K. Goldman, K. McEllin, R. Kelly, N. Chronos, Use of a constitutively active hypoxia-inducible factor-1alpha transgene as a therapeutic strategy in no-option critical limb ischemia patients: phase I dose-escalation experience, Circulation 115 (10) (2007) 1234–1243, https://doi.org/10.1161/CIRCULATIONAHA.106.607994.

[28] L. Piao, G. Zhao, E. Zhu, A. Inoue, R. Shibata, Y. Lei, L. Hu, C. Yu, G. Yang, H. Wu, W. Xu, K. Okumura, N. Ouchi, T. Murohara, M. Kuzuya, X.W. Cheng, Chronic psychological stress accelerates vascular senescence and impairs ischemia-induced neovascularization: the role of dipeptidyl peptidase-4/glucagon-like peptide-1-adiponectin axis, J. Am. Heart Assoc. 6 (10) (2017) e006421, https://doi.org/10.1161/JAHA.117.006421.

[29] Z. Cheng, H.L. Peng, R. Zhang, X.M. Fu, G.S. Zhang, Rejuvenation of cardiac tissue developed from reprogrammed aged somatic cells, Rejuvenation Res. 20 (5) (2017) 389–400, https://doi.org/10.1089/rej.2017.1930.

[30] R. Sun, X. Li, M. Liu, Y. Zeng, S. Chen, P. Zhang, Advances in stem cell therapy for cardiovascular disease (Review), Int. J. Mol. Med. 38 (1) (2016) 23–29, https://doi.org/10.3892/ijmm.2016.2607.

[31] K. Rajala, M. Pekkanen-Mattila, K. Aalto-Setälä, Cardiac differentiation of pluripotent stem cells, Stem Cells Int. 2011 (2011), https://doi.org/10.4061/2011/383709, 383709.

[32] G.A. Quaife-Ryan, C.B. Sim, M. Ziemann, A. Kaspi, H. Rafehi, M. Ramialison, A. El-Osta, J.E. Hudson, E.R. Porrello, Multicellular transcriptional analysis of mammalian heart regeneration, Circulation 136 (12) (2017) 1123–1139, https://doi.org/10.1161/CIRCULATIONAHA.117.028252.

[33] C. Bearzi, M. Rota, T. Hosoda, J. Tillmanns, A. Nascimbene, A. De Angelis, S. Yasuzawa-Amano, I. Trofimova, R.W. Siggins, N. Lecapitaine, S. Cascapera, A.P. Beltrami, D.A. D'Alessandro, E. Zias, F. Quaini, K. Urbanek, R.E. Michler, R. Bolli, J. Kajstura, A. Leri, et al., Human cardiac stem cells, Proc. Natl. Acad. Sci. U. S. A. 104 (35) (2007) 14068–14073, https://doi.org/10.1073/pnas.0706760104.

[34] A.P. Beltrami, L. Barlucchi, D. Torella, M. Baker, F. Limana, S. Chimenti, H. Kasahara, M. Rota, E. Musso, K. Urbanek, A. Leri, J. Kajstura, B. Nadal-Ginard, P. Anversa, Adult cardiac stem cells are multipotent and support myocardial regeneration, Cell 114 (6) (2003) 763–776, https://doi.org/10.1016/s0092-8674(03)00687-1.

[35] G.M. Ellison, C. Vicinanza, A.J. Smith, I. Aquila, A. Leone, C.D. Waring, B.J. Henning, G.G. Stirparo, R. Papait, M. Scarfò, V. Agosti, G. Viglietto, G. Condorelli, C. Indolfi, S. Ottolenghi, D. Torella, B. Nadal-Ginard, Adult c-kit(pos) cardiac stem cells are necessary and sufficient for functional cardiac regeneration and repair, Cell 154 (4) (2013) 827–842, https://doi.org/10.1016/j.cell.2013.07.039.

[36] K. Malliaras, T.S. Li, D. Luthringer, J. Terrovitis, K. Cheng, T. Chakravarty, G. Galang, Y. Zhang, F. Schoenhoff, J. Van Eyk, L. Marbán, E. Marbán, Safety and efficacy of allogeneic cell therapy in infarcted rats transplanted with mismatched cardiosphere-derived cells, Circulation 125 (1) (2012) 100–112, https://doi.org/10.1161/CIRCULATIONAHA.111.042598.

[37] A.J. White, R.R. Smith, S. Matsushita, T. Chakravarty, L.S. Czer, K. Burton, E.R. Schwarz, D.R. Davis, Q. Wang, N.L. Reinsmoen, J.S. Forrester, E. Marbán, R. Makkar, Intrinsic cardiac origin of human cardiosphere-derived cells, Eur. Heart J. 34 (1) (2013) 68–75, https://doi.org/10.1093/eurheartj/ehr172.

[38] S. Uchida, P. De Gaspari, S. Kostin, K. Jenniches, A. Kilic, Y. Izumiya, I. Shiojima, K. Grosse Kreymborg, H. Renz, K. Walsh, T. Braun, Sca1-derived cells are a source of myocardial renewal in the murine adult heart, Stem Cell Rep. 1 (5) (2013) 397–410, https://doi.org/10.1016/j.stemcr.2013.09.004.

[39] X. Wang, Q. Hu, Y. Nakamura, J. Lee, G. Zhang, A.H. From, J. Zhang, The role of the sca-1+/CD31- cardiac progenitor cell population in postinfarction left ventricular remodeling, Stem Cells 24 (7) (2006) 1779–1788, https://doi.org/10.1634/stemcells.2005-0386.

[40] R. Genead, C. Danielsson, A.B. Andersson, M. Corbascio, A. Franco-Cereceda, C. Sylvén, K.H. Grinnemo, Islet-1 cells are cardiac progenitors present during the entire lifespan: from the embryonic stage to adulthood, Stem Cells Dev. 19 (10) (2010) 1601–1615, https://doi.org/10.1089/scd.2009.0483.

[41] K.L. Laugwitz, A. Moretti, J. Lam, P. Gruber, Y. Chen, S. Woodard, L.Z. Lin, C.L. Cai, M.M. Lu, M. Reth, O. Platoshyn, J.X. Yuan, S. Evans, K.R. Chien, Postnatal isl1+ cardioblasts enter fully differentiated cardiomyocyte lineages, Nature 433 (7026) (2005) 647–653, https://doi.org/10.1038/nature03215.

[42] T. Oyama, T. Nagai, H. Wada, A.T. Naito, K. Matsuura, K. Iwanaga, T. Takahashi, M. Goto, Y. Mikami, N. Yasuda, H. Akazawa, A. Uezumi, S. Takeda, I. Komuro, Cardiac side population cells have a potential to migrate and differentiate into cardiomyocytes in vitro and in vivo, J. Cell Biol. 176 (3) (2007) 329–341, https://doi.org/10.1083/jcb.200603014.

[43] J.A. Cornwell, R.E. Nordon, R.P. Harvey, Analysis of cardiac stem cell self-renewal dynamics in serum-free medium by single cell lineage tracking, Stem Cell Res. 28 (2018) 115–124, https://doi.org/10.1016/j.scr.2018.02.004.

[44] T. Le, J. Chong, Cardiac progenitor cells for heart repair, Cell Death Discov. 2 (2016) 16052, https://doi.org/10.1038/cddiscovery.2016.52.

[45] Z.E. Clayton, R.D. Hume, D. Selvakumar, J.J.H. Chong, The Cardiac Stem Cell Niche During Aging, 2020, https://doi.org/10.1016/bs.asn.2020.05.004.

[46] H. Oh, S.B. Bradfute, T.D. Gallardo, T. Nakamura, V. Gaussin, Y. Mishina, J. Pocius, L.H. Michael, R.R. Behringer, D.J. Garry, M.L. Entman, M.D. Schneider, Cardiac progenitor cells from adult myocardium: homing, differentiation, and fusion after infarction, Proc. Natl. Acad. Sci. U. S. A. 100 (21) (2003) 12313–12318, https://doi.org/10.1073/pnas.2132126100.

[47] A.R. Chugh, G.M. Beache, J.H. Loughran, N. Mewton, J.B. Elmore, J. Kajstura, A. Pappas, A. Tatooles, M.F. Stoddard, J.A. Lima, M.S. Slaughter, P. Anversa, R. Bolli, Administration of cardiac stem cells in patients with ischemic cardiomyopathy: the SCIPIO trial: surgical aspects and interim analysis of myocardial function and viability by magnetic resonance, Circulation 126 (11 Suppl. 1) (2012) S54–S64, https://doi.org/10.1161/CIRCULATIONAHA.112.092627.

[48] R. Ren, A. Ocampo, G.H. Liu, J.C. Izpisua Belmonte, Regulation of stem cell aging by metabolism and epigenetics, Cell Metab. 26 (3) (2017) 460–474, https://doi.org/10.1016/j.cmet.2017.07.019.

[49] K.M. Fischer, C.T. Cottage, W. Wu, S. Din, N.A. Gude, D. Avitabile, P. Quijada, B.L. Collins, J. Fransioli, M.A. Sussman, Enhancement of myocardial regeneration through genetic engineering of cardiac progenitor cells expressing Pim-1 kinase, Circulation 120 (21) (2009) 2077–2087, https://doi.org/10.1161/CIRCULATIONAHA.109.884403.

[50] S. Mohsin, M. Khan, H. Toko, B. Bailey, C.T. Cottage, K. Wallach, D. Nag, A. Lee, S. Siddiqi, F. Lan, K.M. Fischer, N. Gude, P. Quijada, D. Avitabile, S. Truffa, B. Collins, W. Dembitsky, J.C. Wu, M.A. Sussman, Human cardiac progenitor cells engineered with Pim-I kinase enhance myocardial repair, J. Am. Coll. Cardiol. 60 (14) (2012) 1278–1287, https://doi.org/10.1016/j.jacc.2012.04.047.

[51] M. Natsumeda, V. Florea, A.C. Rieger, B.A. Tompkins, M.N. Banerjee, S. Golpanian, J. Fritsch, A.M. Landin, N.D. Kashikar, V. Karantalis, V.Y. Loescher, K.E. Hatzistergos, L. Bagno, C. Sanina, M. Mushtaq, J. Rodriguez, M. Rosado, A. Wolf, K. Collon, L. Vincent, et al., A combination of allogeneic stem cells promotes cardiac regeneration, J. Am. Coll. Cardiol. 70 (20) (2017) 2504–2515, https://doi.org/10.1016/j.jacc.2017.09.036.

[52] Z. Zhang, C. Yang, M. Shen, M. Yang, Z. Jin, L. Ding, W. Jiang, J. Yang, H. Chen, F. Cao, T. Hu, Autophagy mediates the beneficial effect of hypoxic preconditioning on bone marrow mesenchymal stem cells for the therapy of myocardial infarction, Stem Cell Res Ther 8 (1) (2017) 89, https://doi.org/10.1186/s13287-017-0543-0.

[53] A. Finan, S. Richard, Stimulating endogenous cardiac repair, Front. Cell Dev. Biol. 3 (2015) 57, https://doi.org/10.3389/fcell.2015.00057.

[54] M. Hendrikx, K. Hensen, C. Clijsters, H. Jongen, R. Koninckx, E. Bijnens, M. Ingels, A. Jacobs, R. Geukens, P. Dendale, J. Vijgen, D. Dilling, P. Steels, U. Mees, J.L. Rummens, Recovery of regional but not global contractile function by the direct intramyocardial autologous bone marrow transplantation: results from a randomized controlled clinical trial, Circulation 114 (1 Suppl) (2006) I101–I107, https://doi.org/10.1161/CIRCULATIONAHA.105.000505.

[55] G. Song, X. Li, Y. Shen, L. Qian, X. Kong, M. Chen, K. Cao, F. Zhang, Transplantation of iPSc restores cardiac function by promoting angiogenesis and ameliorating cardiac remodeling in a post-infarcted swine model, Cell Biochem. Biophys. 71 (3) (2015) 1463–1473, https://doi.org/10.1007/s12013-014-0369-7.

[56] X.L. Tang, Q. Li, G. Rokosh, S.K. Sanganalmath, N. Chen, Q. Ou, H. Stowers, G. Hunt, R. Bolli, Long-term outcome of administration of c-kit(POS) cardiac progenitor cells after acute myocardial infarction: transplanted cells do not become cardiomyocytes, but structural and functional improvement and proliferation of endogenous cells persist for at least one year, Circ. Res. 118 (7) (2016) 1091–1105, https://doi.org/10.1161/CIRCRESAHA.115.307647.

[57] M. Gnecchi, Z. Zhang, A. Ni, V.J. Dzau, Paracrine mechanisms in adult stem cell signaling and therapy, Circ. Res. 103 (11) (2008) 1204–1219, https://doi.org/10.1161/CIRCRESAHA.108.176826.

[58] J. Oh, Y.D. Lee, A.J. Wagers, Stem cell aging: mechanisms, regulators and therapeutic opportunities, Nat. Med. 20 (8) (2014) 870–880, https://doi.org/10.1038/nm.3651.

[59] T.S. Li, M. Kubo, K. Ueda, M. Murakami, A. Mikamo, K. Hamano, Impaired angiogenic potency of bone marrow cells from patients with advanced age, anemia, and renal failure, J. Thorac. Cardiovasc. Surg. 139 (2) (2010) 459–465, https://doi.org/10.1016/j.jtcvs.2009.07.053.

[60] T. Nakamura, T. Hosoyama, D. Kawamura, Y. Takeuchi, Y. Tanaka, M. Samura, K. Ueno, A. Nishimoto, H. Kurazumi, R. Suzuki, H. Ito, K. Sakata, A. Mikamo, T.S. Li, K. Hamano, Influence of aging on the quantity and quality of human cardiac stem cells, Sci. Rep. 6 (2016) 22781, https://doi.org/10.1038/srep22781.

[61] M. Bonios, C.Y. Chang, A. Pinheiro, V.L. Dimaano, T. Higuchi, C. Melexopoulou, F. Bengel, J. Terrovitis, T.P. Abraham, M.R. Abraham, Cardiac resynchronization by cardiosphere-derived stem cell transplantation in an experimental model of myocardial infarction, J. Am. Soc. Echocardiogr. 24 (7) (2011) 808–814, https://doi.org/10.1016/j.echo.2011.03.003.

[62] R.R. Makkar, R.R. Smith, K. Cheng, K. Malliaras, L.E. Thomson, D. Berman, L.S. Czer, L. Marbán, A. Mendizabal, P.V. Johnston, S.D. Russell, K.H. Schuleri, A.C. Lardo, G. Gerstenblith, E. Marbán, Intracoronary cardiosphere-derived cells for heart regeneration after myocardial infarction (CADUCEUS): a prospective, randomised phase 1 trial, Lancet 379 (9819) (2012) 895–904, https://doi.org/10.1016/S0140-6736(12)60195-0.

[63] R.R. Smith, E. Marbán, L. Marbán, Enhancing retention and efficacy of cardiosphere-derived cells administered after myocardial infarction using a hyaluronan-gelatin hydrogel, Biomatter 3 (1) (2013) e24490, https://doi.org/10.4161/biom.24490.

[64] K. Malliaras, R.R. Makkar, R.R. Smith, K. Cheng, E. Wu, R.O. Bonow, L. Marbán, A. Mendizabal, E. Cingolani, P.V. Johnston, G. Gerstenblith, K.H. Schuleri, A.C. Lardo, E. Marbán, Intracoronary cardiosphere-derived cells after myocardial infarction: evidence of therapeutic regeneration in the final 1-year results of the CADUCEUS trial (CArdiosphere-Derived aUtologous stem CElls to reverse ventricUlar dySfunction), J. Am. Coll. Cardiol. 63 (2) (2014) 110–122, https://doi.org/10.1016/j.jacc.2013.08.724.

[65] A.G. Ibrahim, K. Cheng, E. Marbán, Exosomes as critical agents of cardiac regeneration triggered by cell therapy, Stem Cell Rep. 2 (5) (2014) 606–619, https://doi.org/10.1016/j.stemcr.2014.04.006.

[66] J.A. Dougherty, N. Patel, N. Kumar, S.G. Rao, M.G. Angelos, H. Singh, C. Cai, M. Khan, Human cardiac progenitor cells enhance exosome release and promote angiogenesis under physoxia, Front. Cell Dev. Biol. 8 (2020) 130, https://doi.org/10.3389/fcell.2020.00130.

[67] Y. Fanton, C. Houbrechts, L. Willems, A. Daniëls, L. Linsen, J. Ratajczak, A. Bronckaers, I. Lambrichts, J. Declercq, J.L. Rummens, M. Hendrikx, K. Hensen, Cardiac atrial appendage stem cells promote angiogenesis in vitro and in vivo, J. Mol. Cell. Cardiol. 97 (2016) 235–244, https://doi.org/10.1016/j.yjmcc.2016.06.005.

[68] V.L. Ballard, J.M. Edelberg, Stem cells and the regeneration of the aging cardiovascular system, Circ. Res. 100 (8) (2007) 1116–1127, https://doi.org/10.1161/01.RES.0000261964.19115.e3.

[69] H.J. Chung, J.T. Kim, H.J. Kim, H.W. Kyung, P. Katila, J.H. Lee, T.H. Yang, Y.I. Yang, S.J. Lee, Epicardial delivery of VEGF and cardiac stem cells guided by 3-dimensional PLLA mat enhancing cardiac regeneration and angiogenesis in acute myocardial infarction, J. Control. Release 205 (2015) 218–230, https://doi.org/10.1016/j.jconrel.2015.02.013.

[70] E.L. Tilokee, N. Latham, R. Jackson, A.E. Mayfield, B. Ye, S. Mount, B.K. Lam, E.J. Suuronen, M. Ruel, D.J. Stewart, D.R. Davis, Paracrine engineering of human explant-derived cardiac stem cells to over-express stromal-cell derived factor 1α enhances myocardial repair, Stem Cells 34 (7) (2016) 1826–1835, https://doi.org/10.1002/stem.2373.

[71] M. Xaymardan, J. Zheng, I. Duignan, A. Chin, J.M. Holm, V.L. Ballard, J.M. Edelberg, Senescent impairment in synergistic cytokine pathways that provide rapid cardioprotection in the rat heart, J. Exp. Med. 199 (6) (2004) 797–804, https://doi.org/10.1084/jem.20031639.

[72] C. Urbich, A. Aicher, C. Heeschen, E. Dernbach, W.K. Hofmann, A.M. Zeiher, S. Dimmeler, Soluble factors released by endothelial progenitor cells promote migration of endothelial cells and cardiac resident progenitor cells, J. Mol. Cell. Cardiol. 39 (5) (2005) 733–742, https://doi.org/10.1016/j.yjmcc.2005.07.003.

[73] L. Hou, J.J. Kim, Y.J. Woo, N.F. Huang, Stem cell-based therapies to promote angiogenesis in ischemic cardiovascular disease, Am. J. Physiol. Heart Circ. Physiol. 310 (4) (2016) H455–H465, https://doi.org/10.1152/ajpheart.00726.2015.

[74] X. Hu, Y. Xu, Z. Zhong, Y. Wu, J. Zhao, Y. Wang, H. Cheng, M. Kong, F. Zhang, Q. Chen, J. Sun, Q. Li, J. Jin, Q. Li, L. Chen, C. Wang, Y. Zhan, Y. Fan, Q. Yang, L. Yu, et al., A large-scale investigation of hypoxia-preconditioned allogeneic mesenchymal stem cells for myocardial repair in nonhuman primates: paracrine activity without remuscularization, Circ. Res. 118 (6) (2016) 970–983, https://doi.org/10.1161/CIRCRESAHA.115.307516.

Chapter 13

Gut stem cells: Interplay with immune system, microbiota, and aging

Francesco Marotta[a], Baskar Balakrishnan[b], Azam Yazdani[c], Antonio Ayala[d], Fang He[e], and Roberto Catanzaro[f]

[a]ReGenera R&D International for Aging Intervention and Vitality & Longevity Medical Science Commission, Femtec, Milano, Italy, [b]Department of Immunology, Mayo Clinic, Rochester, MN, United States, [c]Department of Anesthesiology, Perioperative and Pain Medicine, Brigham and Women's Hospital, Harvard Medical School, Boston, MA, United States, [d]Department of Molecular Biochemistry and Biology, University of Seville, Seville, Spain, [e]Department of Nutrition, Food Safety and Toxicology, West China School of Public Health, Sichuan University, Chengdu, People's Republic of China, [f]Department of Clinical and Experimental Medicine, Section of Gastroenterology, University of Catania, Catania, Italy

1 Fundamental background

The intestinal tract offers a unique opportunity to quali-quantitatively study epithelial homeostasis and, in particular, intestinal stem cell (ISC) dynamics. Indeed, the intestinal epithelial cells are daily replaced in a stochastic manner (neutral drift), and this occurs on a weekly basis or less. This astounding turnover, by moving along its tract, permeating the epithelial lining surface, and shedding into the lumen, also helps providing a morphofunctional protection against the accumulation of injured epithelial cells (EC) due to chemical and biological xenobiotics. On the contrary, the ISCs compartment is lifetime preserved. This takes place under a fine interplay between ISCs, moving through transit-amplifying (TA) cells and niche cells (Paneth cells), intimately placed at the pit of the crypts of Lieberkühn [1]. This is due to their asymmetrical generation of clonal ribbons of progeny regulating ISC activity, which further differentiates into enteroblast (EB) and then into enteroendocrine cells (EECs) or enterocytes and also self-renewing cells that protect the characteristics of stem cell pool [2]. Overall, roughly 80% of the small intestinal epithelium is represented by enterocytes, whereas goblet cells account for up to 10%, Paneth cells for 5%, and EECs for 1%. Active stem cells are also classified as actively cycling crypt base columnar cells (CBCs) that relentlessly contribute to the whole crypt-villus axis kinetics and quiescent stem cells marking the +4 position. The former reacts to niche signals that provide critical factors for ISC homeostasis and proliferation and supply R-spondins and potent Wnt signaling agonists, while the latter, i.e., the quiescent +4 label-retaining cells (LRCs), then may generate other cells in case of injury of any kind [3].

In this context, leucine-rich-repeat-containing G-protein-coupled receptor 5-expressing (Lgr5$^+$) cells, located in between the Paneth cells, have been found to represent a crucial niche, thus gaining attention as EECs and robust marker for CBCs. Nonetheless, however, long-lived and behaving as ISCs in homeostatic condition, Lgr5$^+$ stem cells per se are not vital to preserve the integrity.

Overall, nutritional factors are highly involved in the regulation of multiple morphofunctional aspects of gastrointestinal tract. Although the dietary modulation/ISC interplay remains to be fully explained, glutamate, amino acids, and methionine exert a more defined stimulus to ISCs [4]. The same research group has shown that also, dietary *S*-adenosylmethionine and methionine metabolism modulates protein synthesis in ISCs and Upd3 cytokine signaling from ECs to maintain ISC division. Accordingly, dietary lipids, by involving Notch signaling mechanisms, regulate EE cell number at an experimental level [5], and the hexosamine biosynthetic pathway (HBP) promotes the proliferation by the insulin signaling mechanisms on ISCs [6].

To get deeper insights into such an area, Mattila et al. [5] reported that Drosophila ISCs avail themselves of a fine cellular nutrient-sensing system residing in a regulatory HBP. This bridges the inner metabolic signal with environmental proliferation pathways via a Warburg effect-like metabolic modulatory switch promoting ISC proliferation but also affecting their insulin receptor sensitivity. Indeed, this group also found out that under *N*-acetyl-D-glucosamine feeding, an HBP metabolite, Drosophila displays a significant ISC proliferation.

Recently, a deeper understanding of intestinal physiopathology has been achieved by observing crypt organoids in health and disease [7]. These consist of a tissue culture system encompassing ISCs and differentiated ISCs within an organotypic 3-D structure including crypt-, villus-, and lumen-centered domains. Moreover, only in the last few years, some light has been shed on the role of extracellular cell matrix (ECM) in the functional niche regulation. You et al. [8] have shown in Drosophila that the ECM regulates ISC-ECM adhesion, with anchorage to the basement membrane, and integrin signaling that maintains ISC's identity and proliferation potential.

2 Regulatory mechanisms

We mentioned earlier that, when for any physiological or pathological reason, it occurs a decrement of $Lgr5^+$ cellular pool; two cell types are generated to preserve the epithelial homeostasis. One is a slow-cycling, dormant cellular units located at the $+4$ location within crypts, and the second type is the absorptive/secretory progenitors [9]. Nonetheless, as alluded before, "$+4$" cells and $Lgr5^+$ cells share a number of markers such as SPARC-related modular calcium binding-2 (Smoc2), Kruppel-like factor 5, B lymphoma Mo-MLV insertion region 1 homolog (Bmi1), achaete-scute complex homolog 2 (a master regulatory gene for $Lgr5^+$ISCs stemness program), HOP homeobox, Hedgehog, murine telomerase reverse-transcriptase (mTert), leucine-rich repeats/immunoglobulin-like domains 1 (Lrig1), SRY-box 9 (Sox9), prominin 1 (Prom1), the oncoprotectors ring-finger-43, zinc-ring-finger-3, and more colon-targeted RNA-binding protein Musashi1 and Olfactomedin 4 [10].

By using in vivo lineage tracing technique, it has been shown that tissue damage triggers $mTert+$cells, expressing at $+4$ position similarly to label-retaining cell (LRC), to generate a full array of intestinal cell lineages and where slow-dividing, reserve stem cell "$+4$" LRCs have reduced EGFR expression [11]. Genetic factors may play a role in ICSs differentiation into mature intestinal cells and few years ago, Zhai et al. [12] have indeed reported that the gene Sox21a, encoding a transcription factor, is definitively involved through JAK/STAT signaling mechanisms. This is likely to end up in preferential enterocyte lineages rather than in EECs.

Moreover, following tissue damage, the pool of Paneth precursor LRC placed at the "$+4$" position can attain stem cell features, although this may be less efficient in aging [13]. Briefly, although the typology of injury model may variably affect this process, it has been shown that Paneth cells, goblet cells, enterocytes, and EECs can display stem cell potentiality. Nonetheless, it seems that even during homeostatic situations, some secretory progenitor clusters may acquire clonal/stem cell capacity in a stochastic manner [14]. It was of interest to note that, while displaying dissimilar lineage outcome, a subpopulation of $Lgr5^+$ cells and LRCs share similar transcriptomic signatures. This suggests a rather blurred boundary between "1–3" and "$+4$" crypt cell locations [11]. Thus, despite the location-dependent differences among functional marker expression of CBCs, it seems they are all able to display multipotent capacity if properly activated. Such bilateral transition from ISCs to differentiated cells may be advocated for the inner dynamic of chromatin remodeling [15]. On the contrary, no significant epigenetic modification has been observed during (de)differentiation. One more explanation can come from niche signaling leading to a reversible ISCs functional phenotype transformation [16]. Indeed, ISCs intimately depend on signals from the niche microenvironment, particularly from Paneth cells, which secrete epithelial growth factor, Wnt3, the surface-bound Notch ligand Delta-like 4, and transforming growth factor-α, [17]. As a matter of fact, among a number of pathways involved in installing and maintaining the ISC phenotype, such as also EGFR/MAPK, Notch, and ErbB, Wnt signaling represents the most relevant one, and it can be also abundantly retrieved by the nearby mesenchymal CBCs cellular pool. Concurrently, Wnt ligands bind to their cell surface cognate receptors frizzled and low-density lipoprotein receptor, controlling the nuclear translocation of cytoplasmic β-catenin while acting together with DNA-binding transcription factors to yield a transactivation of target genes. In this context, Ascl2 holds a key role as a Wnt-responsive master transcription regulator by which $Lgr5^+$ ISC gene expression is controlled [18]. Wnt and EGFR/MAPK activity fluctuating interrelationship also seems to play a key role in modulating the balance between active and dormant "$+4$" $Lgr5^+$ cells. In all this, ISCs move on from Wnt-rich milieu signaling pathways, such as Notch by switching on by paracrine pathways, across the secretion of Delta-like 1 and 4 ligands, to determine an absorptive cellular lineage. On the contrary, the repression of Notch signals leads ISCs toward the secretory lineage via Atonal Homolog 1 (Atoh1) mediation [19]. Parallel to the previously mentioned, the inhibitory bone morphogenetic protein (BMP) signaling and its counterbalance at a niche-located Noggin exert an overall modulatory effect on ISC proliferation [20]. Indeed, the switching off of the BMP pathway and inactivation of Wnt signals trigger significant proliferation stimuli for ISC niche toward differential lineages. BMP and Ephrin-B signaling are expressed at gradually higher intensity from the crypts upward to populate the postmitotic villus compartment, hence fostering the EC differentiation along the crypt-villus axis [16]. After CBC loss, there is a vital recovery, among others, of niche factors provided by newly located Paneth cells, to regain ISC activity. Thus, it seems that intestinal EC are endowed with pluripotent potentiality given that even fully differentiated Paneth cells and EECs may still exhibit the ability to rewind the cellular status back to an ISC stage.

This finding is of great therapeutic relevance suggesting that there may not be a systematic predefined irreversible clear-cut maturation state unable to switching on dedifferentiation. At the same time, this brings to take into account that extrinsic inflammatory signals conjuring up with promoting factors, such as antiapoptotic protein BCL-2, may equally place tumorigenetic-prone differentiated cells [21]. Indeed, the intrinsically highly proliferative nature of fast-cycling Lgr5$^+$ cells makes them potentially liable to DNA damage and it has been shown by Schwitalla et al. [22] that uncontrolled heightened NF-κB signaling can lead a dedifferentiation process up to a tumor-initiating Lgr5$^+$ status in otherwise well-differentiated enterocytes expressing stem cell features. In this regard the peculiar capacity of EC to display wide plasticity along the process of tissue-damage-related regeneration turnover, the associated extrinsic inflammatory signals may install differentiated cells with oncogenic features as observed in ISCs. The same applies also to the differentiated tuft cells that normally do not play any proactive role in tissue turnover, but clearly express ISC activity during gut inflammatory injury.

Colon epithelium. Although morphologically lacking the villa, colonic epithelium shares many aspects with small intestine lining, including stem cell-endowed crypt cells. The crypt bottom cells such as Reg4$^+$ and cKIT$^+$ cells have Lgr5$^+$ cellular pool expressing growth factors, whereas no "+4" population (Paneth cells) or Bmi1$^+$ cells can be retrieved. Here ISCs express the R-spondin binding to receptors of the Lgr4–6 and Znrf3/Rnf43 with significantly enhancement of Wnt signaling. As a proof of concept when Drosophila is treated with R-spondin, it occurs an overt crypt hyperplasia and gut epithelium overgrowth either in vitro or in vivo [23]. Overall, R-spondin3 plays a crucial role for epithelial repair via induction of Wnt signaling and, in its absence, even upon mild-intensity injury would alter normal crypt regeneration.

Novel lineage tracing techniques have allowed to investigate stem cell stochastic dynamics in the human adult colon generating further clonal lineages. This has helped understanding that at the colon level, the mesenchymal cellular pool located in the crypts promotes through Wnt signaling, the self-replacement of stem cells identified as Lgr5$^+$ and EphrB2^{high}. Highly regenerative Lgr5$^+$/Axin2$^+$ cells and secretory Lgr5$^-$/Axin2$^+$ cells represent two distinct colonic cell clusters with marked Wnt signaling features to be quickly recruited during crypt regeneration.

Moreover, also colonic stem cells present cell cycle dissimilarities with the high Notch and Lrig1 expression identifying the slower cycling cellular pool [24]. Peculiarly, colonic human stem cells show a significantly slower niche fixation speed as compared to mice since colonic stem cells are replaced within the crypt over a year unlike every 3 days as in murine colon.

Drosophila melanogaster has been used as a model to investigate the key stress response pathway in ECs against cellular injury by keeping adequate ISC proliferative turnover and differentiation. The complex phenomenology involves cytokine secretion after JNK signaling and consequent activation of JAK/STAT at ISC level [25]. This pathway is likely to move from Domeless receptor regulating the activation of JAK/STAT target gene Socs36E and of Sox21, a transcription factor that fosters their proliferation and differentiation [8]. Moreover, a decrease in LDH activity may affect ISCs by altering redox status not only through NAD-level variations but also through hexosamine biosynthesis [15]. It has also to be noted that Nrf2 is constitutively active in ISCs and its repression by its counterregulator Keap1 has a key role in controlling the ISC proliferation and overall intracellular redox balance.

The role of the intestinal stromal pool has not been fully clarified as yet, as well as the functional regulation in health and disease with gut microflora or the potential supporting effect played by epithelial/endothelial-mesenchymal transition mechanisms [26].

3 Biorhythm and gut stem cells

The mechanisms underlying the molecular pacemaker system have attained a great deal of scientific interest, given the complex synchronology and neuromodulatory effect. This is guided by pineal gland at a systemic level through circadian changes in gene transcription as well as protein expression. Studies in mammals have shown that the circadian regulation is represented by the transactivators CLK and BMAL1 and their counterregulator cryptochrome *CRY1-2* together with PER1-3 [27]. The circadian clock in the gastrointestinal tract follows a rhythmic production of TNF, myeloid cells, and EC that initiate the JNK stress response pathway. Following tissue injury, *BMAL1*, which regulates the JNK stress cytokine modification daily profile, gets activated leading to a regeneration process.

In this regard, Drosophila is representing a robust model since it shares remarkable similarities of the gastrointestinal tract morphofunctional aspects with the human ones. It has been suggested that clock activity is present in ISCs and EBs, as well as differentiated ECs, but is switched off during EE differentiation. Indeed, clock function is needed by nondividing differentiated ECs to beget circadian rhythms in ISC proliferation. Indeed, by disrupting clock function in specific cell lines either in differentiated EC or in undifferentiated precursors (ISCs + EBs), it became clear that it exists a circadian pathway system interrelating the two populations.

3.1 Mitochondrial function in the gut and stem cell interplay

Mitochondrial signaling and related specific changes in pyruvate metabolism in ECs have been reported to play a relevant role in gut physiology and ISCs by regulating EC dedifferentiation-differentiation mechanisms and intestinal function in health and disease. Indeed, mitochondrial dysfunction and its triggering of unfolded protein response represent a fundamental abnormality found in all gastrointestinal diseases. A deranged mitochondrial pyruvate metabolism in differentiated ECs, via activation of JAK/STAT signaling, is effective enough to impair ISC proliferation. This has been confirmed by using myo1A-GAL4 (wholegut marker) to target RNAi for a gene encoding the multisubunit pyruvate dehydrogenase (PDH) complex. PDH acts downstream from the MPC in the mitochondrial matrix by triggering the pyruvate-acetyl-CoA transformation, thus supplying the first metabolite in the tricarboxylic acid cycle. A very recent study from Wisidagama [28] quantifying phosphohistone-H3-staining cells in myo1A > PDH RNAi intestinal tract has shown a substantial ISC proliferation.

Accordingly, by adding LDHRNAi to PDHRNAi in ECs, Wisidagama et al. [28] demonstrated that ISC proliferation would be markedly suppressed. This brings to infer that following a reduced mitochondrial pyruvate metabolism, their adequate lactate levels generated by LDH in ECs are needed to yield an upregulation of Upd3 cytokine. This is preceded by JNK activation of JAK/STAT pathway and represents a mechanism by which dysfunctional intestinal cells are replaced by differentiated ones.

However, although a pyruvate oxidation decrement-induced mitochondrial dysfunction may alter redox status, this specific aspect may not represent a relevant nonautonomous factor of EC metabolism on ISC proliferation. As a matter of fact, when animals with EC-specific defective mitochondrial pyruvate uptake and silencing of *MPC1* (*d*MPC1RNA), i.e., of a key carrier present also in humans [29], were supplemented with N-acetylcysteine, significant variation in the rate of ISC proliferation was noted.

Taken together, the aforementioned data reconcile with the concept that mitochondrial function plays a relevant role in regulating intestinal homeostasis in response to stress [29] but also during physiological aging process.

4 Aging gut, microbiota, immune system, and stem cells

Experimental animal studies have supported the hypothesis that aging is associated with reduced height and quantity of the crypts and of proliferating ISCs leading to a compromised absorption of nutrients and overall exhaustion of intestinal functional capacity [9]. Very recently, He et al. [30] have further unveiled a loss of the number of Ki67+ progenitor cell population although this was compatible with the villus size and the numbers of apoptotic, senescent cells. In particular, Mihaylova et al. [31] observed that, against a well-maintained stable enterocyte differentiated pool, during aging process, it occurs a significant decline of the number and potency of ISCs and TA progenitors. Indeed, during aging, it occurs an increase of EC and ISC mTORC1 activation with the upregulation of mitogen-activated protein kinase kinase-6/(p38-MAPK) and p53, causing cell overgrowth and trophism with age-related morphofunctional decay of ISCs [32]. Accordingly, He et al. [30] experimentally demonstrated that during aging per se, mTORC1 signaling is robustly activated in ISCs and TA cells, whereas its inhibition can partially restore the morphofunctional characteristics of villa in the same setting. In agreement with this, the shutdown in Lgr5+ ISCs of Tuberous sclerosis 1 (TSC1), which encodes a protein forming a complex that inhibits signal transduction to the mTOR, causes premature crypt aging phenotypes in the colon. However, increased cell stress response pathways too can be advocated for to further explain this phenomenon.

The age-associated species diversity of gut microbiota may modify the intraluminal and systemic metabolomics, thus affecting, besides mitochondria dysregulation, also ISCs. The two age-related damaged compartments are interlinked one another given that a protracted glycolysis/oxidative phosphorylation invariably causes a ROS-induced decrease of ISCs.

The aging process determines complex and factor-dense inflammatory phenotype changes of ISCs, increasing the percentage of their positivity for IL-6 and pNF-κB when compared to young ISC control. Overall, it also appeared that the competence of the DNA damage response pathway was curbed in old ISCs. The same research group had also previously shown that old mesenchymal stem cells would also exhibit chemokines called senescence-associated secretory phenotype (SASP) as a hallmark of aging process. However, a direct unidirectional role of SASP cannot be taken for granted since such large secretome (pooling PDGF, IGF1-binding proteins, EGF, transforming growth factor beta-TGF-β, HGF, receptor modulators, cytokines, chemokines, extracellular-matrix-remodeling proteins, and others) plays also a role in tissue remodeling and proliferation pathways. Overall, as very recently shown [33], with physiological aging, Paneth cells, through higher mTORC1 activity, secrete an increased amount of Notum that acts as an extracellular Wnt inhibitor that associates with a detrimental PPAR-α inhibition as well.

5 Gut microbiome

Recent advancements in gut microbiome studies demonstrated that there is a strong relationship between gut microbial diversity and gut stem cell aging [34]. Gut microbial diversity abnormalities favor adverse microbial members that can disturb regular cellular mechanisms including immune cells and end up in forced epigenetic modulation of ISCs. Also, there are many microbial metabolites reported to induce stress responses not involving stem cells [34] directly, thus affecting its aging and regeneration process.2 In particular, host microbiome is significantly involved in aging-related reduced regenerative capacity of ISCs by involving p38-MAPK, Wnt, Notch pathway suppression, TGF-β, and JNK signaling through their metabolites, ending up in a recruitment of p16 among others cyclin-dependent kinase inhibitors. A further gut flora aging-associated phenotype is represented by a decrease in Faecalibacterium prausnitzii and an increase of Firmicutes, namely ethanol-producing proteobacteria that exerts significant TGF beta-activated kinase 1 (TAK1)-MKK-p38/MAPK-hampering effect on tissue regeneration (35xxxx). This associates with a promotion of gut permeability by disrupting epithelial tight junctions and ultimately leading to a depletion of ISCs, besides causing also an ethanol-related inhibition of hippocampal stem cells.

As a matter of fact, during aging, there is an overall decrease of gut *Bacteroides* and of its induction of TGF-β. Moreover, short-chain fatty acids (SCFAs) such as butyrate propionate and acetate, common metabolites of resident gut flora, beneficially act on animal life span. Among these, SCFA may inhibit histone deacetylase and enhance Forkhead box protein (FOXO) proliferation-promoting activity on ISCs. Moreover, SCFAs control redox balance to regulate p38- and JNK-mediated signaling for SC differentiation. In the residing gut flora of Drosophila, it can be commonly retrieved *Acetobacter* pomorum that is involved in the host metabolic homeostasis via insulin-like growth factor pathway, thus increasing the basal ISC numbers.

However, with aging progression, G-protein-coupling receptor-SCFA binding inhibits insulin signaling causing mitochondrial dysfunction associated with dysfunction of NAD-dependent deacetylase sirtuin-1(SIRT1)/peroxisome proliferator-activated receptor gamma co-activator 1 alphaPGC1α) mechanism.

Taken as a whole, this may trigger a cascade phenomenon with increasingly larger redox imbalance and mitochondrial damage, autophagy inhibition, and abnormal β-catenin accumulation as a Wnt-signaling substrate. It is conceivable that long-standing gut dysbiosis may favor the pooling of T cells within "leaky" foci with massive delivery of inflammatory cytokines, exhaustion of ISCs, decrease of their differentiation capability, and accelerating senescence if not even fostering the clonal proliferation of cancer stem cells in the worst scenario. Although the mechanisms have not been unfolded as yet, it is known that fecal transplantation from old Drosophila to young ones shortens the life span.

Collectively, microbiota is involved in the regulation of age-related gene expressions, namely STAT and interferon regulatory factors, via transcription factor promotions in the maturation of ISCs.

5.1 Microbiome on the immunity of stem cell aging

Intestinal T cell homeostasis is essential in keeping equilibrium between gut microbial measures and host intestinal cell functions. Disturbances in this integrity activate humoral immune response results in inflammation [35]. The primary immune mechanism affecting ISC aging is T cell regulation, where regulated T cell mechanisms can keep ISC self-regeneration. Indeed, following specific gut ecosystem events, T helper cells 1 (Th1)-IFN-gamma foster ISC differentiation to Paneth cell whereas T helper cells 2 (Th2)-released IL-13 triggers tuft cell. Several studies proved that reduced Treg cells and their cytokines increase the susceptibility of ISCs leading to lowered tissue regeneration [34]. The specific probiotic treatment proved to restore the eubiosis and increase the immune regulatory mechanism (Fig. 1). Some bacterial taxa have been reported to increase the production of regulatory T cells by inducing antiinflammatory cytokine production such as IL-10, IL-25, IL-33, and TSLP. IL-10 is produced by monocytes, Th2 cells, B cells, Tregs, DCs, and keratinocytes, whereas recently discovered cytokines IL-25, TSLP, and IL-33 produced by intestinal EC [36]. Studies showed that changes in particular gut-dominant taxa, for example, reduction in Faecalibacterium prausnitzii, are associated with aging-related inflammatory diseases [34]. *Faecalibacterium prausnitzii* treatment in animal models proved to increase antiinflammatory cytokine IL-10 [37]. This represents that an essential immune-regulatory cytokine also regulates ISC expression by influencing IL-10-dependent IFN-β, which induces TLR-3 expression in stem cells [38]. Intestinal *Salmonella* demonstrated to induce the pericryptal fibroblasts to release IL-33, which is involved in ISC regulation [36]. Moreover, also parasitic helminths can influence the tuft cells to secrete IL-25 and induce innate lymphoid cells to produce ISC-stimulating factor IL-13. LAB plays a vital role in maintaining gut homeostasis; its broad spectrum of activities in the gut can keep immune regulatory mechanisms in control. Aubry et al. [39] proved that the LAB gives a protective effect to the colitis by stimulating Treg cell production via inducing the EC-generated TSLP. Therefore, the ISC aging and its immunity chiefly rely on antiinflammatory functions of the immune system.

5.2 Microbiome on epigenetics of stem cell aging

Stem cell aging is connected to epigenetic changes via the DNA methylation process, alongside augmentation of histone repressive markers such as H3K9me3 and H3K27me3. Microbial metabolites that influence epigenetic modulations like

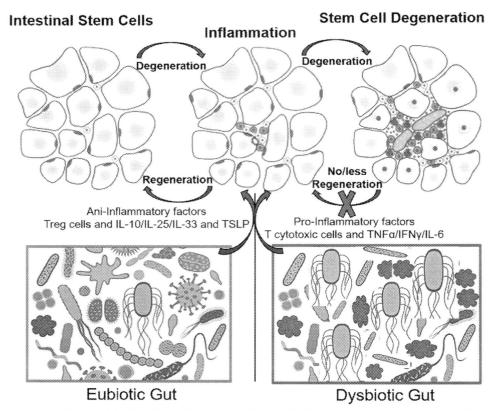

FIG. 1 Influence of gut microbiome on intestinal stem cell regulation. Healthy eubiotic gut induces the production of antiinflammatory factors to reverse the degeneration of stem cells in an inflammatory condition, whereas dysbiotic gut increases the production of proinflammatory substances and degeneration of ISCs.

SCFAs can tend to enhance the histone repressive markers. Butyrate is one such a metabolite that can able to increase the production of H3K9me3 in gut stem cells downregulating glycolysis and NADH/NAD$^+$ ratio [34]. This epigenetic mechanism helps in maintaining equilibrium between oxidative phosphorylation and glycolysis, which is responsible for keeping ISCs inactive in terms of differentiation, hence keeping activation of the antioxidizing system toward regulated antiaging mechanisms [34]. SCFAs also impair the differentiation ability of ISCs as butyrate does in colonic epithelial stem and progenitor cells by switching on stress response for FOXO3. In fact, microbiota metabolites such as SCFA may interfere with the pro-/antiinflammatory balance system by affecting the regulatory T cell modulation via Forkhead box P3 (FOXP3) mechanism.

Also gut dysbiosis causing changes in the production of gut flora metabolites may result in epigenetics changes of the ISCs leading to their degeneration. By doing so, quiescent ISCs, given their vulnerability to ROS, cause an overreactive inflammatory responses triggering overdifferentiation and a positive feedback loop leading to ISC exhaustion and accelerated aging.

In addition to modulation in histone gene expressions, microbes are also involved in the modulation of a specific transcription factor such as STATs and interferon regulatory factors. Overexpression of these regulatory factors weakens the self-regeneration of stem cells. Likewise, DNA damaging is a significant factor in the aging of stem cells. Bacterial metabolites, such as toxins like colibactin from *Escherichia coli, have been* reported to damage DNA of EC and to increase senescent-associated microRNA (miR-20a-5p) inhibits p53 degradation and causing growth arrest [40].

Finally, extracellular vesicles (EV) are membrane-surrounded vesicles that appear to be a novel mode of intercellular networking by carrying biologically relevant molecules under a protected structure (Oszvald et al. [41]). By using 3-D murine and human intestinal organoids, it has been shown that intestinal fibroblast-derived EVs are novel carriers of Wnt and epidermal growth factor (EGF) in forming the ISC niche. In agreement, it also appeared that fibroblast-derived EVs could protect ISCs in the lack of EGF.

6 Tentative interventional perspectives

From the conceptual viewpoint, the remodulation of intrinsic and extrinsic factors leading to restoration of stem cell functionality represents a fascinating therapeutic avenue. Caloric restriction (CR) is known to help expanding ISCs and protein synthesis by inhibiting mTORC1 signaling while upregulating SIRT1 activity [42]. This mechanism is so robust that Paneth cell signaling overrides also the nutrient sensing in ISCs to privilege such feedback to CR. This induces a mechanism via bone stromal antigen 1 and the generation of the paracrine factor cyclic ADP ribose (cADPR) secretion self-renewal and growth of the ISC pool [43].

Of interest is also the finding that during CR and associated intestinal atrophy, when HBP activity is limiting ISC divisions, *N*-acetyl-D-glucosamine is equally able to enhance ISC proliferation [5]. Linked upstream to mTORC1, the suppression of p38 MAPK has been demonstrated by Park et al. [44] to improve the morphofunction of aged villi, thus representing a target for a potential prevention of ISCs and villus aging. Interestingly, it has been found in Drosophila that when the CR diet is enriched with GlcNAc, the clone size remains at the of noncalorie-restricted.

Animal studies have shown that also growth hormone (GH) and glutamine (Gln exerts a positive effect on the EC proliferation differentiation and on ISCs with increased the stemness marker expression as well. These data have been confirmed in in vitro (human and murine) and in vivo (murine) studies with GH treatment alone as for Paneth cells and EC with Ki67 upregulation in the crypt organoid formation and ISC stemness and differentiate into goblet cells. The aforesaid effect seems to be age-related in that GH effect in aged animals needs a high protein intake.

A recent new interesting area of potential intervention is represented by food-derived exosome-like nanoparticles. Such particles, as in the case of grape exosome-like nanoparticles (GELN), namely GELN lipids, specifically affect Lgr5$^+$ ISCs by Wnt-mediated activation of the Tcf4 transcription machinery in the crypts while playing a role in gut tissue remodeling during injury [45]. Also, pectin has been shown to protect ISCs and increase crypt survival in high-dose irradiation (IR) mice model by upregulating Msi1 and Notch1 expression (which plays significant roles in putative stem cells and differentiation) against IR injury [46]. By using an impaired autophagy mouse model lacking Atg16l1, Jones et al. [47] showed a dysregulation of Paneth cells due to a decreased protein breakdown and downregulation of exocytosis. This aspect prompts further research in this area as a potential interventional target.

Igarashi et al. [48] have very recently observed that treating old mice with a 6-week supplementation with the NAD(+) precursor nicotinamide riboside (500 mg/kg body weight) would rejuvenate ISCs, reversing an impaired ability to repair gut damage.

Another interventional avenue is suggested to be a microbiome profile remodeling leading to systemic antiinflammatory mechanism. Fecal microbiota transplantation, ideally restoring the intracellular redox balance, reducing the intrinsic transduction mechanisms such as STATs, and affecting gut and stem cell-associated aging process, is under investigation. Indeed, it is not fully clarified yet whether fecal transplantation from young animals to old receivers may reestablish the self-renewal, differentiation, and regenerative capacity of ISCs till to prolong the health span. On the contrary, the aforementioned Gln supplementation has been found to decrease the Firmicutes/Bacteroidetes ratio, thus favoring ISC trophism and EC proliferation. Genetic or chemical inhibition of Paneth cell-produced Notum (such as LP-922056 using 1-chloro-1,2-benziodoxol-3-one or 2-phenoxyacetamides and the neutralizing antibody 2.78.33) or Wnt-mimicking compounds are amenable avenues to restored function of aged intestinal organoids [49].

Finally, we have recently reported that a specific bioactive fraction of Rhodiola complexed with a patented marine lipoprotein extract from *Trachurus* sp. could significantly stimulate cell proliferation rate and stem cell stemness in normal cells (Fig. 2) and also in cells under oxidative stress, mimicking what observed in aging process. This phytomarine complex with intriguing senolytic activity has also shown to significantly upregulate "vitagenes" such as SIRT-1 and MMP2 while downregulating MMP9 and modulating Serpina6 genes. From our ongoing studies, it appears also that while this is effective on the normal lung cells (L132), it does not ignite proliferation in H522 cancer cells [50] while improving melatonin biorhythm in a preliminary clinical pilot study whose implications with gut is a field of growing interest (preliminary data by S. Pathak selected for presentation at the World Antiaging Congress, Monte Carlo, 2017, full paper accepted by Current Aging Science, in press).

Dedication

This work is dedicated to Braveheart little giant Carole "Titty" for reflecting to me with her glowing grace, the endurance and willpower she proved herself in successfully tackling much harder hurdles.

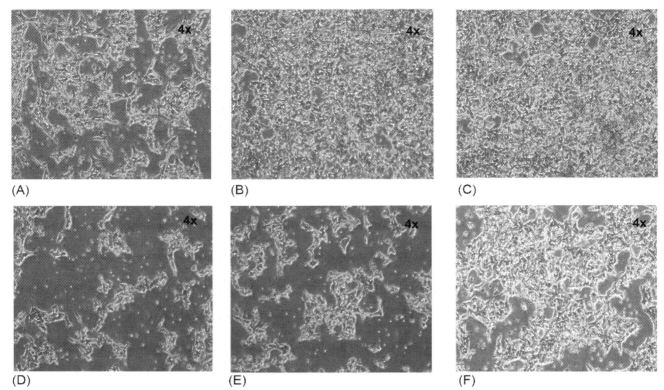

FIG. 2 FL cells (human amniotic stem cell-like cells) treated with 0.05% (R-L1) and 0.1% (R-L2) concentrations of R-L senolytic compound extract, respectively, along with H_2O_2 (H) and R-L+H_2O_2 compound extract for a period of 3 days and increased cell proliferation was observed on day 3. About 0.5% and 0.05% R-L senolytic compound extract treatment showed increased cell proliferation when compared to control cells (A–C). Stress-induced cells when treated with R-L compound 0.5% and 0.05% showed visible increase in cell growth and proliferation when compared to stress-induced control cells (H_2O_2-treated cells) (D–F). Increased proliferation after the treatment with R+L senolytic compound (0.05% and 0.1%) (A) FL cells; (B) and (C) cells treated with 0.05% and 0.1% of R+L senolytic compound, respectively; (D) cells treated with H_2O_2; (E) and (F) cells treated with H_2O_2, in combination with 0.05% and 0.1% of R+L senolytic compound, respectively. Human amniotic cells treated with RL compound day 0 and day 2. Magnification 10×.

References

[1] L. Ritsma, S.I.J. Ellenbroek, A. Zomer, H.J. Snippert, F.J. de Sauvage, B.D. Simons, H. Clevers, J. van Rheenen, Intestinal crypt homeostasis revealed at single-stem-cell level by in vivo live imaging, Nature 507 (7492) (2014) 362–365, https://doi.org/10.1038/nature12972.

[2] B. Ohlstein, T. Kai, E. Decotto, A. Spradling, The stem cell niche: theme and variations, Curr. Opin. Cell Biol. 16 (2004) 693–699, https://doi.org/10.1016/j.ceb.2004.09.003.

[3] M. Buszczak, H. Krämer, Autophagy keeps the balance in tissue homeostasis, Dev. Cell 49 (2019) 499–500, https://doi.org/10.1016/j.devcel.2019.05.005.

[4] F. Obata, K. Tsuda-Sakurai, T. Yamazaki, R. Nishio, K. Nishimura, et al., Nutritional control of stem cell division through S-adenosylmethionine in drosophila intestine, Dev. Cell 44 (2018) 741–751 e743, https://doi.org/10.1016/j.devcel.2018.02.017.

[5] J. Mattila, K. Kokki, V. Hietakangas, M. Boutros, Stem cell intrinsic hexosamine metabolism regulates intestinal adaptation to nutrient content, Dev. Cell 47 (2018) 112–121.e3, https://doi.org/10.1016/j.devcel.2018.08.011.

[6] D.R. Jones, W.J. Keune, K.E. Anderson, L.R. Stephens, P.T. Hawkins, N. Divecha, The hexosamine biosynthesis pathway and O-GlcNAcylation maintain insulin-stimulated PI3K-PKB phosphorylation and tumour cell growth after short-term glucose deprivation, FEBS J. 281 (16) (2014) 3591–3608, https://doi.org/10.1111/febs.12879.

[7] W. Dieterich, M.F. Neurath, Y. Zopf, Intestinal ex vivo organoid culture reveals altered programmed crypt stem cells in patients with celiac disease, Sci. Rep. 10 (1) (2020) 3535, https://doi.org/10.1038/s41598-020-60521-5.

[8] J. You, Y. Zhang, Z. Li, Z. Lou, J. Lin, X. Lin, Drosophila perlecan regulates intestinal stem cell activity via cell-matrix attachment, Stem Cell Rep. 2 (2014) 761–769, https://doi.org/10.1016/j.stemcr.2014.04.007.

[9] N. Li, A. Nakauka-Ddamba, J. Tobias, S.T. Jensen, C.J. Lengner, Mouse label-retaining cells are molecularly and functionally distinct from reserve intestinal stem cells, Gastroenterology 151 (2) (2016) 298–310.e7, https://doi.org/10.1053/j.gastro.2016.04.049.

[10] D.T. Breault, I.M. Min, D.L. Carlone, L.G. Farilla, D.M. Ambruzs, D.E. Henderson, S. Algra, R.K. Montgomery, A.J. Wagers, N. Hole, Generation of mTert-GFP mice as a model to identify and study tissue progenitor cells, Proc. Natl. Acad. Sci. U. S. A. 105 (30) (2008) 10420–10425, https://doi.org/10.1073/pnas.0804800105.

[11] A.J.M. Santos, Y.H. Lo, A.T. Mah, C.J. Kuo, The intestinal stem cell niche: homeostasis and adaptations, Trends Cell Biol. 28 (12) (2018) 1062–1078, https://doi.org/10.1016/j.tcb.2018.08.001.

[12] Z. Zhai, J.P. Boquete, B. Lemaitre, A genetic framework controlling the differentiation of intestinal stem cells during regeneration in Drosophila, PLoS Genet. 13 (2017) 1006854.

[13] J. Choi, N. Rakhilin, P. Gadamsetty, D.J. Joe, T. Tabrizian, S.M. Lipkin, D.M. Huffman, X. Shen, N. Nishimura, Intestinal crypts recover rapidly from focal damage with coordinated motion of stem cells that is impaired by aging, Sci. Rep. 8 (1) (2018) 10989, https://doi.org/10.1038/s41598-018-29230-y.

[14] A. Sada, T. Tumbar, New insights into mechanisms of stem cell daughter fate determination in regenerative tissues, Int. Rev. Cell Mol. Biol. 300 (2013) 1–50, https://doi.org/10.1016/B978-0-12-405210-9.00001-1.

[15] J. Kazakevych, S. Sayols, B. Messner, C. Krienke, N. Soshnikova, Dynamic changes in chromatin states during specification and differentiation of adult intestinal stem cells, Nucleic Acids Res. 45 (10) (2017) 5770–5784, https://doi.org/10.1093/nar/gkx167.

[16] Z. Qi, Y. Li, B. Zhao, C. Xu, Y. Liu, H. Li, B. Zhang, X. Wang, X. Yang, W. Xie, B. Li, J.J. Han, Y.G. Chen, BMP restricts stemness of intestinal Lgr5$^+$ stem cells by directly suppressing their signature genes, Nat. Commun. 8 (2017) 13824, https://doi.org/10.1038/ncomms13824.

[17] A. Pastuła, J. Marcinkiewicz, Cellular interactions in the intestinal stem cell niche, Arch. Immunol. Ther. Exp. (Warsz.) 67 (1) (2019) 19–26, https://doi.org/10.1007/s00005-018-0524-8.

[18] J. Li, M.R. Dedloff, K. Stevens, L. Maney, M. Prochaska, C.F. Hongay, K.N. Wallace, A novel group of secretory cells regulates development of the immature intestinal stem cell niche through repression of the main signaling pathways driving proliferation, Dev. Biol. 456 (1) (2019) 47–62, https://doi.org/10.1016/j.ydbio.2019.08.005.

[19] F. Ishibashi, H. Shimizu, T. Nakata, S. Fujii, K. Suzuki, A. Kawamoto, S. Anzai, R. Kuno, S. Nagata, G. Ito, T. Murano, T. Mizutani, S. Oshima, K. Tsuchiya, T. Nakamura, M. Watanabe, R. Okamoto, Contribution of ATOH1+ cells to the homeostasis, repair, and tumorigenesis of the colonic epithelium, Stem Cell Rep. 10 (1) (2018) 27–42, https://doi.org/10.1016/j.stemcr.2017.11.006.

[20] A. Guezguez, F. Paré, Y.D. Benoit, N. Basora, J.F. Beaulieu, Modulation of stemness in a human normal intestinal epithelial crypt cell line by activation of the WNT signaling pathway, Exp. Cell Res. 322 (2) (2014) 355–364, https://doi.org/10.1016/j.yexcr.2014.02.009.

[21] M. van der Heijden, C.D. Zimberlin, A.M. Nicholson, S. Colak, R. Kemp, S.L. Meijer, J.P. Medema, F.R. Greten, M. Jansen, D.J. Winton, L. Vermeulen, Bcl-2 is a critical mediator of intestinal transformation, Nat. Commun. 7 (2016) 10916, https://doi.org/10.1038/ncomms10916.

[22] S. Schwitalla, A.A. Fingerle, P. Cammareri, T. Nebelsiek, S.I. Göktuna, P.K. Ziegler, O. Canli, J. Heijmans, D.J. Huels, G. Moreaux, R.A. Rupec, M. Gerhard, R. Schmid, N. Barker, H. Clevers, R. Lang, J. Neumann, T. Kirchner, M.M. Taketo, G.R. van den Brink, O.J. Sansom, M.C. Arkan, F.R. Greten, Intestinal tumorigenesis initiated by dedifferentiation and acquisition of stem-cell-like properties, Cell 152 (1–2) (2013) 25–38, https://doi.org/10.1016/j.cell.2012.12.012.

[23] C. Harnack, H. Berger, A. Antanaviciute, R. Vidal, S. Sauer, A. Simmons, T.F. Meyer, M. Sigal, R-spondin 3 promotes stem cell recovery and epithelial regeneration in the colon, Nat. Commun. 10 (1) (2019) 4368, https://doi.org/10.1038/s41467-019-12349-5.

[24] A. Hirata, J. Utikal, S. Yamashita, H. Aoki, A. Watanabe, T. Yamamoto, H. Okano, N. Bardeesy, T. Kunisada, T. Ushijima, A. Hara, R. Jaenisch, K. Hochedlinger, Y. Yamada, Dose-dependent roles for canonical Wnt signalling in de novo crypt formation and cell cycle properties of the colonic epithelium, Development 140 (1) (2013) 66–75, https://doi.org/10.1242/dev.084103.

[25] K. Beebe, W.C. Lee, C.A. Micchelli, JAK/STAT signaling coordinates stem cell proliferation and multilineage differentiation in the drosophila intestinal stem cell lineage, Dev. Biol. 338 (2010) 28–37.

[26] D.W. Powell, I.V. Pinchuk, J.I. Saada, X. Chen, R.C. Mifflin, Mesenchymal cells of the intestinal lamina propria, Annu. Rev. Physiol. 73 (2011) 213–237, https://doi.org/10.1146/annurev.physiol.70.113006.100646.

[27] I. Iwashina, K. Mochizuki, Y. Inamochi, T. Goda, Clock genes regulate the feeding schedule-dependent diurnal rhythm changes in hexose transporter gene expressions through the binding of BMAL1 to the promoter/enhancer and transcribed regions, J. Nutr. Biochem. 22 (4) (2011) 334–343, https://doi.org/10.1016/j.jnutbio.2010.02.012.

[28] D.R. Wisidagama, C.S. Thummel, Regulation of drosophila intestinal stem cell proliferation by enterocyte mitochondrial pyruvate metabolism, G3 (Bethesda). 9 (2019) 3623–3630.

[29] D.K. Bricker, E.B. Taylor, J.C. Schell, T. Orsak, A. Boutron, Y.C. Chen, J.E. Cox, C.M. Cardon, J.G. Van Vranken, N. Dephoure, C. Redin, S. Boudina, S.P. Gygi, M. Brivet, C.S. Thummel, J. Rutter, Science 337 (6090) (2012) 96–100, https://doi.org/10.1126/science.1218099.

[30] D. He, H. Wu, J. Xiang, X. Ruan, P. Peng, Y. Ruan, Y.G. Chen, Y. Wang, Q. Yu, H. Zhang, S.L. Habib, R.A. De Pinho, H. Liu, B. Li, Gut stem cell aging is driven by mTORC1 via a p38 MAPK-p53 pathway, Nat. Commun. 11 (2020) 37, https://doi.org/10.1038/s41467-019-13911-x.

[31] M.M. Mihaylova, C.W. Cheng, A.Q. Cao, S. Tripathi, M.D. Mana, K.E. Bauer-Rowe, M. Abu-Remaileh, L. Clavain, A. Erdemir, C.A. Lewis, E. Freinkman, A.S. Dickey, A.R. La Spada, Y. Huang, G.W. Bell, V. Deshpande, P. Carmeliet, P. Katajisto, D.M. Sabatini, Ö.H. Yilmaz, Fasting activates fatty acid oxidation to enhance intestinal stem cell function during homeostasis and aging, Cell Stem Cell 22 (2018) 769–778.

[32] D. Witte, H. Otterbein, M. Förster, K. Giehl, R. Zeiser, H. Lehnert, H. Ungefroren, Negative regulation of TGF-β1-induced MKK6-p38 and MEK-ERK signalling and epithelial-mesenchymal transition by Rac1b, Sci. Rep. 7 (1) (2017) 17313, https://doi.org/10.1038/s41598-017-15170-6.

[33] N. Pentinmikko, S. Iqbal, M. Mana, S. Andersson, A.B. Cognetta III, R.M. Suciu, J. Roper, K. Luopajärvi, E. Markelin, S. Gopalakrishnan, O.P. Smolander, S. Naranjo, T. Saarinen, A. Juuti, K. Pietiläinen, P. Auvinen, A. Ristimäki, N. Gupta, T. Tammela, T. Jacks, D.M. Sabatini, B.F. Cravatt, Ö.H. Yilmaz, P. Katajisto, Notum produced by paneth cells attenuates regeneration of aged intestinal epithelium, Nature 571 (7765) (2019) 398–402, https://doi.org/10.1038/s41586-019-1383-0.

[34] Y. Tan, Z.K. Wei, J.L. Chen, J.L. An, M.L. Li, L.Y. Zhou, et al., Save your gut save your age: the role of the microbiome in stem cell ageing, J. Cell. Mol. Med. 23 (8) (2019) 4866–4875.

[35] B. Balakrishnan, V. Taneja, Microbial modulation of the gut microbiome for treating autoimmune diseases, Expert Rev. Gastroenterol. Hepatol. 12 (10) (2018) 985–996, https://doi.org/10.1080/17474124.2018.1517044.

[36] M.R. Howitt, S. Lavoie, M. Michaud, et al., Tuft cells, taste-chemosensory cells, orchestrate parasite type 2 immunity in the gut, Science 351 (6279) (2016) 1329–1333, https://doi.org/10.1126/science.aaf1648.

[37] D.K.W. Ocansey, L. Wang, J. Wang, Y. Yan, H. Qian, X. Zhang, et al., Mesenchymal stem cell-gut microbiota interaction in the repair of inflammatory bowel disease: an enhanced therapeutic effect, Clin. Transl. Med. 8 (1) (2019) 31, https://doi.org/10.1186/s40169-019-0251-8.

[38] J.Y. Lim, K.I. Im, E.S. Lee, et al., Enhanced immunoregulation of mesenchymal stem cells by IL-10-producing type 1 regulatory T cells in collagen-induced arthritis, Sci. Rep. 6 (2016) 26851, https://doi.org/10.1038/srep26851.

[39] C. Aubry, C. Michon, F. Chain, Y. Chvatchenko, L. Goffin, S.C. Zimmerli, Protective effect of TSLP delivered at the gut mucosa level by recombinant lactic acid bacteria in DSS-induced colitis mouse model, Microb. Cell Fact. 14 (2015), https://doi.org/10.1186/s12934-015-0367-5.

[40] K. Watanabe, Y. Ikuno, Y. Kakeya, S. Ikeno, H. Taniura, M. Kurono, K. Minemori, Y. Katsuyama, H. Naka-Kaneda, Age-related dysfunction of the DNA damage response in intestinal stem cells, Inflamm. Regen. 39 (2019) 8, https://doi.org/10.1186/s41232-019-0096-y.

[41] Á. Oszvald, Z. Szvicsek, G.O. Sándor, et al., Extracellular vesicles transmit epithelial growth factor activity in the intestinal stem cell niche, Stem Cells 38 (2020) 291–293, https://doi.org/10.1002/stem.3113.

[42] K.L. Tinkum, K.M. Stemler, L.S. White, A.J. Loza, S. Jeter-Jones, B.M. Michalski, C. Kuzmicki, R. Pless, T.S. Stappenbeck, D. Piwnica-Worms, H. Piwnica-Worms, Fasting protects mice from lethal DNA damage by promoting small intestinal epithelial stem cell survival, Proc. Natl. Acad. Sci. U. S. A. 112 (51) (2015) E7148–E7154, https://doi.org/10.1073/pnas.1509249112.

[43] Ö.H. Yilmaz, P. Katajisto, D.W. Lamming, Y. Gültekin, K.E. Bauer-Rowe, S. Sengupta, K. Birsoy, A. Dursun, V.O. Yilmaz, M. Selig, G.P. Nielsen, M. Mino-Kenudson, L.R. Zukerberg, A.K. Bhan, V. Deshpande, D.M. Sabatini, mTORC1 in the Paneth cell niche couples intestinal stem-cell function to calorie intake, Nature 486 (7404) (2012) 490–495, https://doi.org/10.1038/nature11163.

[44] J.S. Park, Y.S. Kim, M.A. Yoo, The role of p38bMAPK in age-related modulation of intestinal stem cell proliferation and differentiation in Drosophila, Aging (Albany NY) 1 (2009) 637–651, https://doi.org/10.18632/aging.100054.

[45] S. Ju, J. Mu, T. Dokland, X. Zhuang, Q. Wang, H. Jiang, X. Xiang, Z.B. Deng, B. Wang, L. Zhang, M. Roth, R. Welti, J. Mobley, Y. Jun, D. Miller, H.G. Zhang, Grape exosome-like nanoparticles induce intestinal stem cells and protect mice from DSS-induced colitis, Mol. Ther. 21 (2013) 1345–1357, https://doi.org/10.1038/mt.2013.64.

[46] S.M. Sureban, R. May, D. Qu, P. Chandrakesan, N. Weygant, N. Ali, et al., Dietary pectin increases intestinal crypt stem cell survival following radiation injury, PLoS One 10 (8) (2015), https://doi.org/10.1371/journal.pone.0135561, e0135561.

[47] E.J. Jones, Z.J. Matthews, L. Gul, P. Sudhakar, A. Treveil, D. Divekar, J. Buck, T. Wrzesinski, M. Jefferson, S.D. Armstrong, L.J. Hall, A.J.M. Watson, S.R. Carding, W. Haerty, Integrative analysis of Paneth cell proteomic and transcriptomic data from intestinal organoids reveals functional processes dependent on autophagy, Dis. Model. Mech. 12 (3) (2019), https://doi.org/10.1242/dmm.037069, dmm037069.

[48] M. Igarashi, M. Miura, E. Williams, F. Jaksch, T. Kadowaki, T. Yamauchi, L. Guarente, NAD_+ supplementation rejuvenates aged gut adult stem cells, Aging Cell 18 (3) (2019) e12935, https://doi.org/10.1111/acel.12935.

[49] N. Pentinmikko, S. Iqbal, M. Mana, S. Andersson, A.B. Cognetta 3rd, R.M. Suciu, J. Roper, K. Luopajärvi, E. Markelin, S. Gopalakrishnan, O.P. Smolander, S. Naranjo, T. Saarinen, Notum produced by Paneth cells attenuates regeneration of aged intestinal epithelium, Nature 571 (7765) (2019) 398–402, https://doi.org/10.1038/s41586-019-1383-0.

[50] F. Marotta, J. Arunachalam, A. Banerjee, R. Catanzaro, S. Adalti, A. Das, A. Kolyada, S. Pathak, Oxidative stress and smoke-related lung diseases: a tentative approach through the blood, lungs, and gut, in: S. Chakraborti, T. Chakraborti, S.K. Das, D. Chattopadhyay (Eds.), Oxidative Stress in Lung Diseases, Springer Nature Singapore Pte Ltd, 2019, pp. 27–50.

Chapter 14

Cellular senescence and aging in bone

Manju Mohan[a], Sridhar Muthusami[b,c], Nagarajan Selvamurugan[d], Srinivasan Narasimhan[e], R. Ileng Kumaran[f], and Ilangovan Ramachandran[a]

[a]Department of Endocrinology, Dr. ALM PG Institute of Basic Medical Sciences, University of Madras, Taramani Campus, Chennai, Tamil Nadu, India, [b]Department of Biochemistry, Karpagam Academy of Higher Education, Coimbatore, Tamil Nadu, India, [c]Karpagam Cancer Research Centre, Karpagam Academy of Higher Education, Coimbatore, Tamil Nadu, India, [d]Department of Biotechnology, College of Engineering and Technology, SRM Institute of Science and Technology, Kattankulathur, Tamil Nadu, India, [e]Faculty of Allied Health Sciences, Chettinad Hospital and Research Institute, Chettinad Academy of Research and Education, Chennai, Tamil Nadu, India, [f]Biology Department, Farmingdale State College, Farmingdale, NY, United States

1 Introduction

Aging population is increasing globally with advancements in medicine and healthcare procedures. This positive feature has also inherited an increase in age-related morbidities and mortality with a soaring medical expenditure on one side and enhanced spending on geriatric research, on the other side. Aging is the major risk factor for many diseases including neurodegenerative diseases, cardiovascular diseases, diabetes, cancer, and bone diseases. The hallmarks of this complex aging process are cellular senescence, mitochondrial dysfunction, genomic instability, epigenetic alterations, telomere attrition, stem cell exhaustion, altered intercellular communication, loss of proteostasis, and deregulated nutrient sensing (Fig. 1) [1]. The rate of aging is distinct from each individual, and it is determined by one or a few critical cell populations. Cellular health is regulated at distinct areas in the cell, initially from the structural organization of chromosome, transcriptional regulation, protein translation, quality check, autophagic recycling of cellular components, and finally maintenance of the cytoskeletal integrity, extracellular matrix (ECM), and its signaling. Each cellular system passes and receives signals from other systems, and these interplays regulate the aging of cells.

Cellular senescence was first reported in 1961 by Hayflick and Moorhead [2], wherein cells in culture entered a state of irreversible growth arrest, and that state of the cells were described as "replicative senescence." Senescence plays an important role in the physiological process during tissue homeostasis, embryonic development, repair mechanism, and aging. In the last decade, our knowledge in understanding the senescence's detrimental impact on aging and age-associated diseases has improved significantly. Of importance, the senescent cell accumulation drives aging and age-related diseases that include atherosclerosis, inflammation, sarcopenia, osteoporosis, and osteoarthritis, which have impact on normal physiological function causing a progressive tissue deterioration. Cellular senescence can be accelerated by certain stimuli such as DNA damage, telomere shortening, oncogenic activation, chemotherapeutic drugs, reactive oxygen species (ROS), and genotoxic stress that ultimately limits the premalignant cell expansion (Fig. 2). Additionally, the oncogenic stress is induced by the overexpression of oncogenes or loss of functional mutation of tumor suppressor genes in premalignant cells, which can also induce cellular senescence [3]. Most importantly, because of antiproliferative effects, the activated cellular senescence arrests tumorigenesis *in vivo*, highlighting the senescence-induced treatment for tumor progression and cancer therapy.

A healthy life of an adult generally depends on functional homeostatic mechanisms. During aging, mesenchymal stromal/stem cells (MSCs), satellite cells (skeletal muscle stem cells), osteoblasts, osteoclasts, chondrocytes, and adipocytes exhibit reduced self-renewal capacity with age, which is associated with elevated expression of senescent markers [4, 5]. Furthermore, the senescent MSCs secrete various factors including proinflammatory factors such as cytokines (interleukins), growth factors, and metalloproteinases, known as senescence-associated secretory phenotype (SASP) or senescence-messaging secretome (SMS) in order to distinguish the senescent cells from other nonsenescent and cell cycle-arrested cells, thus favoring the reparative capacity of tissues [3]. Therefore, targeting SASP provides an effective antiaging strategy to enhance the life span. The general aging features of bone include alterations in the structural bone constituents, increased bone turnover and bone marrow fat deposition, and ultimately lower bone mineral density (BMD) [6]. Microarchitecture of bone is altered by the trabecular strut rearrangement, enlarged medullary cavity, and subperiosteal expansion with the

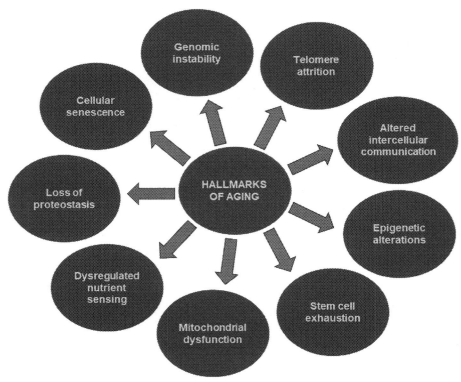

FIG. 1 Hallmarks of aging. The schematic illustration represents nine hallmarks of aging such as genomic instability, telomere attrition, altered intercellular communication, epigenetic alterations, stem cell exhaustion, mitochondrial dysfunction, dysregulated nutrient sensing, loss of proteostasis and cellular senescence.

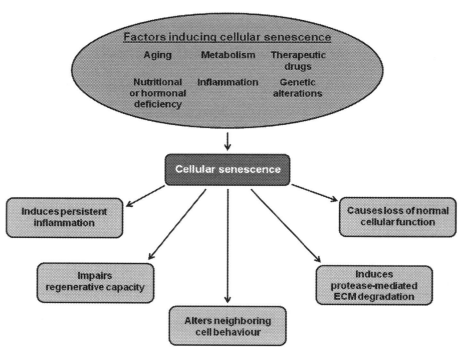

FIG. 2 Factors inducing cellular senescence. Cellular senescence is induced by several intrinsic and extrinsic factors. The senescence state or senescent cells have the ability to induce biological dysfunction through (1) loss of normal cellular function, (2) impairment in regenerative capacity, (3) induction of persistent inflammation by secretion of proinflammatory factors, (4) alteration in the behavior of neighboring cells, and (5) protease-mediated degradation of extracellular components. These various factors contribute to age-related pathologies.

alterations in the organic matrix and crystalline properties of calcium deposition. The age-related imbalance in skeletal homeostasis occurs when these factors are associated with changes in hormonal levels, reduced physical activity, and malnutrition that ultimately leads to osteoporosis as well as reduced healing capacity after injury [7]. Data from mouse and human samples suggest that a subset of cell lineages in bone microenvironment become senescent and synthesize a heterogeneous SASP with aging [8]. However, the mechanisms by which senescent cells and SASP possibly alter bone remodeling are poorly understood.

Bone remodeling is affected primarily due to decreased levels of sex steroids, namely androgens and estrogens, with aging. Estrogen has a stimulatory role on osteoblasts and an inhibitory effect on osteoclasts, and loss of this inhibition causes higher bone turnover [9]. Estrogen deficiency is the major cause of bone loss in postmenopausal women and also contributes to age-associated bone loss and pathogenesis of osteoporosis in both men and women. Also, estrogen deficiency leads to the elevated levels of tumor necrosis factor-alpha (TNF-α), which stimulate receptor activator of nuclear factor kappa B ligand (RANKL)-induced osteoclast resorption. In addition to sex steroid deficiency, several other factors contribute to age-related bone loss that includes hypercalciuria, excess glucocorticoids, hyperthyroidism, gastrointestinal disorders, alcoholism, cigarette smoking, and some types of malignancy [10]. Major chronic endocrine diseases such as osteoporosis and diabetes mellitus become more prevalent with aging. Formation of advanced glycation end products (AGEs) due to high circulating glucose in diabetic condition causes senescence of MSCs and inhibits bone formation leading to secondary osteoporosis in a higher population of diabetic patients [11, 12].

Among the several cell signaling pathways, the canonical Wnt signaling pathway plays an essential role in bone health. Dysregulation of Wnt pathway results in reduced bone formation and alterations in this signaling activates osteoclast formation. Alterations in bone morphogenetic protein (BMP) and transforming growth factor-beta (TGF-β) signaling were associated with decreased bone formation and increased adipogenesis with aging [13]. Another important factor that regulates bone is insulin-like growth factor (IGF) system components, which are important for the proliferation of bone cells and skeletal development. We and other groups have observed that a reduction in IGFs and IGF binding proteins (IGFBPs) altered the bone formation and osteocyte function, ultimately leading to bone loss and reduced BMD due to imbalance in bone remodeling (Fig. 3) [14–17]. Moreover, the authors and others have reported an age-related decrease in serum levels of IGFs and IGFBP-3, but an increase in IGFBP-4, which correlated with a reduction in BMD and higher risk fractures, implicating the role of IGFs in skeletal aging [18, 19]. Taken together, available reports suggest that changes in tissue-intrinsic growth factors and cytokines contribute to bone aging.

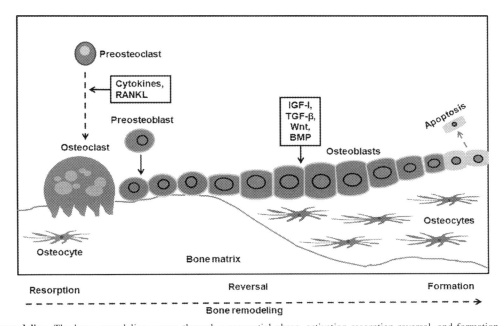

FIG. 3 **Bone remodeling.** The bone remodeling occurs through a sequential phase, activation-resorption-reversal, and formation. The first stage is activated in response to appropriate stimuli, thereby recruiting osteoclast precursors into remodeling site. After the resorption, macrophage-like uncharacterized cells remove debris produced after matrix degradation during reversal phase. The final formation phase is triggered by several growth factors including insulin-like growth factors (IGFs), transforming growth factor-β (TGF-β), bone morphogenetic proteins (BMPs), and Wnt signaling molecules that are responsible for the recruitment of osteoblasts into resorption site and promote mineralization, thus completing the cycle of bone remodeling.

As the authors have noted earlier, enormous geriatric research activities have recognized that targeting cellular senescence can bring in therapeutic remedies to most of the aging-associated morbidities. In this chapter, the authors attempted to present and discuss the recent findings on the role of cellular senescence and aging on bone diseases.

2 Skeletal aging: An insight into bone cellular senescence

Bone is a highly specialized and dynamic tissue that serves several functions including protection of vital organs, providing a rigid framework for locomotion, reservoir of minerals, and endocrine regulation. It undergoes a continuous remodeling, which is executed by the bone-forming osteoblasts and bone-resorbing osteoclasts as well as differentiated osteocytes. The bone cells are influenced by various systemic hormones, growth factors, cytokines, and other signaling molecules through different mechanisms [20–25]. In bone, the sequential cascade of osteogenesis is initiated by the recruitment of MSCs to remodeling sites followed by cell proliferation, lineage commitment, expression of lineage-specific markers, collagen secretion, and matrix mineralization [26].

With aging, bone is affected not only by increased resorption but also by functional impairment of osteoblasts as well as MSCs and a concomitant decline in their ability to self-renew. Over an extended time period, these events cause an imbalance in bone tissue homeostasis that finally leads to dramatic bone loss, resulting in osteoporosis. Another age-related feature of human bone may be attributed to a reduction in osteoblast number, thereby contributing to a decline in osteocyte density. Micropetrosis is a phenomenon of osteocyte lacuna mineralization, which may also contribute to decreased osteocyte density associated with age. The major cause of micropetrosis is unknown, but this condition could be one of the potential outcomes of osteocyte death [27].

All normal cells, including bone cells such as osteoblasts, osteoclasts, and osteocytes, have a defined life span that is controlled mainly by external factors and the number of replication cycles. Normal human cells can only undergo a limited number of cell divisions, indicated by Hayflick limit replication theory and with aging those cell cycle number decreases [28]. Telomere has a repetitive sequence of DNA and associated proteins that confer stability to the end of chromosomes. Telomere length at the end of genes is the determinant of cell replication number. Cellular senescence and apoptosis are initiated when cell replication is hindered and telomere length is shortened in each cell division. Similarly, telomere damage caused by oxidative stress and UV radiation possibly accelerates telomere shortening, thereby curtailment of cells [29]. Bone aging is accelerated with knockdown of telomerase reverse transcriptase (TERT)-specific gene in mice that exhibited a reduction in osteoblast number with an increase in osteoclast number, which resulted from a proinflammatory cytokine environment, though the source of this inflammatory response is unknown [30, 31]. Telomere shortening impairs the differentiation of osteoblasts and possibly attributes to skeletal aging due to elevated senescence of skeletal MSCs.

Skeletal aging at the tissue level is characterized by bone loss and bone marrow adipose tissue (BMAT) accumulation. In rodents, loss of bone strength and mass is associated with increased prevalence of apoptotic osteocytes and osteoblasts, and a comparable reduction in the number of osteoblasts and bone formation rate with advancing age [32]. The bone formation rate generally depends upon MSC progenitor pool availability and adequate MSCs recruitment as well as individual activity of osteoblasts on the bone surface (Fig. 4).

In men, bone formation markers in the circulation decline steadily with age but increase in the case of older women because estrogen level declines during menopause, which results in the basic multicellular unit (BMU) activation that leads to a high bone turnover state and causes bone imbalance due to reduced bone formation at the cellular level [33]. Thus, aging exerts defects in bone homeostasis in both men and women. These age-related changes appear as a sequel effect of intrinsic and extrinsic factors. Intrinsic factors, namely oxidative stress, epigenetic changes, telomere shortening, and autophagy, regulate the intercellular communication and repair mechanisms, which impact the skeletal stem cell renewal and differentiation and modulate the bone matrix composition and production. Extrinsic factors such as sex hormone deficiency, excessive glucocorticoid production, and physical activity influence the skeletal integrity and bone functions associated with aging [34].

3 Key factors regulating senescence in bone cells

3.1 Intrinsic factors

3.1.1 Oxidative stress

Cellular damage is induced by ROS and this is considered as a key component associated with aging. Metabolism and its regulation generates ROS, and enzymes like lipoxygenase (LOX) and nicotinamide adenine dinucleotide phosphate

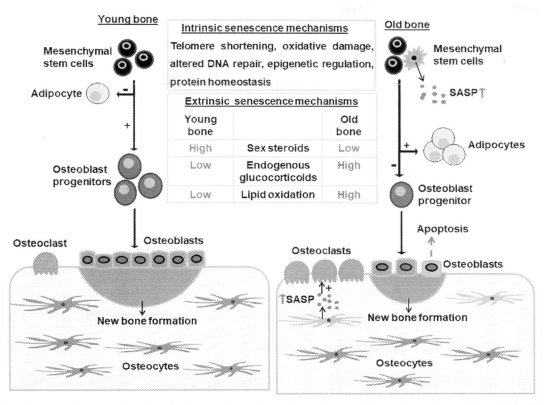

FIG. 4 **Mechanisms regulating age-related cellular changes in the bone microenvironment.** Age-related changes are associated with alterations in intrinsic and extrinsic mechanisms, resulting in an increased level of adipogenic lineage commitment, reduced osteoblast formation, and increased apoptosis of osteoblasts and osteocytes. Accumulation of senescent cells in bone microenvironment promotes the release of senescence-associated secretory phenotype (SASP) and elevates bone resorption, leading to altered skeletal integrity and function.

(NADPH) oxidase can produce ROS in response to signaling by growth factors and cytokines [35]. During oxidative phosphorylation, the mitochondrial electron transport chain generates most of the cellular ROS. In general, mitochondrial-derived ROS and oxidative stress can accelerate telomere shortening and dysfunction, which is considered as one of the characteristics of cellular senescence and aging. Accumulation of oxidative stress causes tissue and cellular damage that leads to the adverse effects of skeletal aging. The amount of ROS production increases in bone with age and sex steroid deficiency, and therefore, the increased level of oxidative stress plays a crucial role in the development of skeletal fragility and pathological disorders [32, 36].

The balance between oxidant production and quenching of oxidants in normal tissues is maintained by various antioxidants. The most important antioxidant enzyme is superoxide dismutase (SOD) that converts superoxide anion free radical to hydrogen peroxide (H_2O_2) and then to water and oxygen. Also, thiol-containing oligopeptides can detoxify the excess ROS production, of which most abundant are thioredoxin and glutathione. The diminishing effect of oxidative stress on bone formation has been demonstrated in mice treated with buthionine sulfoximine (BSO), a pro-oxidant [37]. The skeletal features of normal aged mouse are similar to SOD knockout mouse, which exhibited decreased bone mass, lower osteoclast and osteoblast numbers with decreased expression of RANKL in bone [38]. This emphasizes the bone-sparing effects of antioxidants in aged animal models. Wnt/β-catenin pathway plays a pivotal role in cell proliferation and is implicated in various cellular events including senescence [39, 40]. Oxidative stress mitigated the osteoblastogenic effects of Wnt/β-catenin signaling pathway *in vitro* wherein ROS inhibited the Wnt signaling by dephosphorylation and activation of GSK-3β [41]. Thus, ROS has a major role in bone tissue homeostasis and oxidative stress exerts detrimental effects on skeletal mass and strength during aging.

3.1.2 Epigenetic modifications

Epigenetic modifications such as DNA methylation and histone modifications show pattern change throughout the life span of a human. The overall loss in methylation pattern with cellular senescence and aging is termed as "epigenetic drift" [42].

In this phenomenon, age-related decline in DNA methylation occurs but does not necessarily have the same pattern for all individuals. In addition, the "epigenetic clock" identifies DNA methylation levels at a specific region of the human genome that are highly related to age in specific tissues. During cellular senescence and aging, the expression of DNA methyltransferases (DNMTs) such as DNMT1 and DNMT3B was substantially downregulated in MSCs, suggesting an apparent hypomethylation in aged MSCs. Also, the inhibition of DNA methyltransferases promotes cellular senescence, thereby emphasizing the role of epigenetic regulators in aging [43].

MSCs can differentiate into osteocytes, chondrocytes, and adipocytes. During long-term culture, the globally distributed DNA methylation CpG pattern of MSCs is retained. Conversely, upon culture expansion, some CpG sites showed significant modifications in DNA methylation. Few of these alterations are strongly reproducible in culture *in vitro*, which exhibit a linear correlation to total doubling population, time duration, and even passage number [44]. Thus, senescence-related DNA methylation alterations of specific sites of CpGs may be used to evaluate the senescence state of MSCs [45, 46]. Also, senescence-associated DNA methylation modifications in MSCs are correlated with histone mark repression than with H3K4me3 and H3K27me3 modifications. Epigenetic modifications to replicative senescence in cultured cells and *in vivo* aging tend to relate but diverse regulatory pathways can be expected to vary in these two systems. Senescence-associated hypo- and hypermethylations in H3K27me3 and H3K4me3 regions and targets of enhancer of zeste homolog 2 (EZH2) were reported [47]. These authors also suggested that replicative senescence is regulated by polycomb repressive complex 2 (PRC2) protein complex.

Numerous transcription factors are associated with stem cell growth and development, which have an enriched binding domain in differentially methylated sites and differentially expressed promoter gene during senescence. The basic helix-loop-helix transcription factor Twist-1 is reported to maintain MSC stemness and prevent MSC senescence by promoting EZH2 recruitment to p16^{Ink4a} locus, triggering H3K27me3 and suppressing the expression [48].

Gene transcription during osteoblastogenesis and osteoclastogenesis is influenced by several epigenetic mechanisms. Particularly, DNA methylation regulates osteocalcin (OCN), osteoprotegerin (OPG), RANKL, sclerostin, CCAAT/enhancer-binding protein alpha (C/EBP-α), and BMP-2 expression. Extensive studies on the epigenetic state of gene coding sclerostin (*SOST*) have been performed. Delgado-Calle et al. [49] reported 2 CpG-rich regions, namely region 1 located in the proximal promoter and region 2 located in the exon 1 of *SOST*. While the CpG-rich promoter region 1 was hypomethylated in human osteocytes, it is hypermethylated in osteoblasts and extraskeletal cells. Treatment of osteoblastic cells with the demethylating agent 5-aza-2′-deoxycytidine (AzadC) caused a significant increase in gene expression of SOST [49]. Thus, these authors suggested that methylation status of CpG dinucleotides influences the regulation of *SOST* gene expression in osteocytes and also during osteoblast–osteocyte transition. In another study, Reppe et al. [50] demonstrated that genetic and epigenetic alterations in *SOST* influence the sclerostin mRNA expression in bone and serum levels of sclerostin in postmenopausal women. Furthermore, these authors demonstrated a higher level of promoter hypermethylation of *SOST* in osteoporotic women that is correlated with decreased mRNA expression of sclerostin in bone and reduced levels of serum sclerostin in osteoporotic postmenopausal women, compared with healthy subjects. Interestingly, the hypermethylation of *SOST* promoter in osteoporotic subjects appears to be a compensatory mechanism that reduces the circulating levels of sclerostin, which in turn diminishes the inhibition of Wnt signaling pathway in order to enhance the bone formation [50].

Histone methylation regulates peroxisome proliferator-activated receptor gamma (PPARγ) target gene promoters, runt-related transcription factor 2 (Runx2) [also known as core-binding factor subunit alpha-1 (Cbfa1)], and osterix (Osx) [also called Sp7 transcription factor] expression. The disturbances in epigenetic regulators may contribute to age-related bone loss. Additionally, epigenetic dysregulation results in silencing of genes or inappropriate gene expression could contribute to the development of rheumatic and musculoskeletal diseases, such as osteoarthritis, rheumatoid arthritis, and systemic lupus erythematosus (SLE) [51].

3.1.3 Telomere shortening

The age-related bone diseases such as osteoporosis and osteoarthritis are associated with excessive telomere shortening and TERT activity dysfunction [52]. Telomere shortening and irreversible cell cycle arrest are the major mechanisms responsible for the age-related intrinsic cellular dysfunction. The length of telomere is preserved by TERT, and other protein complex telomerase RNA component (TERC). Shortening of telomere occurs during replication and DNA damage at telomeres causes dysfunction, uncapping, and eventually cellular senescence. The dysfunctional telomere is identified as telomere dysfunction-induced foci (TIFs), which is used as a senescent marker for osteoblasts, osteocytes, MSCs as well as other types of cells [8, 53].

The senescence mechanism in bone cells is characterized by the upregulation of cell cycle inhibitors, namely cyclin-dependent kinase inhibitor 1A (Cdkn1a) also referred to as p21$^{Cip1/Waf1}$, Cdkn2a (p16^{Ink4a}), and CDK4/6 in osteoprogenitors, osteoblasts, osteocytes, and myeloid cells [54]. In the osteoprogenitor cells, age-related telomere shortening contributes to limited stem cell pool and suppressed differentiation into osteoblasts. The role of telomeres in bone aging is strongly supported by the findings from Wrn−/− Terc−/− mutant mice that exhibited a decline in bone mass [30, 55]. The telomere-mediated defects in osteoblast differentiation are associated with increased expression of p53/p21, suggesting the role of telomere dysfunction-induced senescence in age-related osteoporosis [55]. In another study, Terc−/− mice exhibited enhanced osteoclastogenesis and increased bone resorption, due to the creation of proinflammatory osteoclast-activating microenvironment [31]. Additionally, MSC culture from Terc−/− mice displayed high levels of DNA damage and senescent β-galactosidase-positive cells [31]. In long-term culture, MSCs lack telomerase activity and exhibit telomere shortening with replicative senescence. In human bone marrow stromal cells (BMSCs), ectopic expression of human TERT (hTERT) led to telomere elongation, stemness maintenance, life span extension of cells, and enhanced bone formation *in vivo* and *in vitro*, suggesting that TERT expression could prevent the replicative senescence and telomere shortening in osteoprogenitor cells. Thus, telomere dysfunction limits the differentiation of osteoblasts from osteoprogenitor cells and contributes to age-related bone loss [56].

3.1.4 Autophagy

Autophagy is a basic mechanism of degrading dysfunctional cellular components or damaged proteins accumulated due to various stress conditions, through lysosome-dependent degradation pathway. A decline in autophagy is marked as a hallmark of cellular aging and effector of cellular senescence. All the bone cells in the skeleton exhibit autophagy, where long-lived postmitotic osteocytes can be replaced by bone turnover, thus autophagy plays a major role in maintenance, survival, and pathophysiology of bone [57].

Autophagy helps osteocytes to survive ROS-generated stress. In bone-resorbing cells, autophagy induces osteoclast precursor cell differentiation in response to oxidative stress and also participates in the secretion of lysosomal contents into ECM. P66^{shc}, an adaptor protein, amplifies mitochondrial ROS production in osteocytes of young adult mice lacking autophagy. Therefore, autophagy suppression in osteocytes of young adult mice caused skeletal changes and is similar to wild-type aged mice, contributing to bone aging [58]. Deletion of focal adhesion kinase family interacting protein of 200 kDa (FIP200), a component of mammalian autophagy complex, resulted in impaired osteoblast differentiation, decreased bone formation, and osteopenia in mice [59]. Furthermore, Atg7 (a regulator of autophagy) deletion decreased bone formation and resulted in low bone mass associated with elevated oxidative stress, highlighting the role of autophagy in osteoblastogenesis and bone health [59]. Thus, autophagy serves as a cell surveillance process and protects the cells from senescence and associated morbidities.

3.2 Extrinsic factors

3.2.1 Sex hormone deficiency

Sex hormones have various effects on osteoblastic lineage. Estrogens and androgens are essential for the acquisition of bone mass during the skeletal growth and development. In osteoblasts, estrogens act through estrogen receptors (ERs) (ERα and ERβ). Androgens act directly through the androgen receptor (AR) or indirectly via aromatization to estradiol. Bone loss occurs at a higher rate during the onset of menopause and the immediate subsequent years, indicating the adverse impact of estrogen deficiency on bone mass and its acceleration on bone involution. Estrogen deficiency is an important determinant of age-related bone loss in both genders [60] through increased production of cytokines and a conducive microenvironment for osteoclast differentiation [61]. Additionally, we have shown that nonaromatizable androgen dihydrotestosterone (DHT) plays an important role in bone health and is a determinant of BMD in men [14].

Alterations in the circulating sex hormone levels may have a dramatic change in IGF-1 and TGF-β, which regulate the bone formation and bone resorption rate. Furthermore, sex hormone deficiency accelerates osteoblast apoptosis and age-related oxidative stress via stimulation of cytoplasmic kinases [62]. Nevertheless, in a recent study, both gonadectomized female and male wild-type mice had no evidence for increased cellular senescence in bone compared with sham-operated controls. Further, a similar finding has been reported in human bone biopsies obtained from older healthy postmenopausal women who received either no treatment or estrogen therapy for 3 weeks [63]. These authors have concluded that estrogen deficiency and cellular senescence are independent mechanisms in the pathogenesis of osteoporosis. Although the authors cannot rule out the role of estrogen deficiency in bone cell senescence, previous studies in mice have implicated proinflammatory cytokines, some of which are part of the SASP, in the bone microenvironment in response to ovariectomy [64–67]. Even though the accumulation of senescent cells has a definite role in age-related bone loss in preclinical models and humans, how estrogen deficiency can be attributed to cellular senescence still remains a mute question. So there is enough scope for future investigations focusing on this aspect.

3.2.2 Excess glucocorticoids

Glucocorticoid-induced osteoporosis is a serious and devastating consequence of either long-term exogenous glucocorticoid therapy or endogenous hypercortisolism. Glucocorticoids affect bone formation by influencing the function of osteoprogenitors, osteoblasts, and osteocytes. About 50% of the patients treated with glucocorticoids orally for 6 months or longer were affected by glucocorticoid-induced osteoporosis due to diminished glucocorticoid feedback inhibition of ACTH and also due to increased expression of 11β-hydroxysteroid dehydrogenase (HSD) type 1 (an enzyme that activates glucocorticoids) in bone with aging. Studies carried out in animal models also demonstrated the detrimental effects of glucocorticoids on the bone. An increase in osteoclastic bone resorption is generally associated with postmenopausal osteoporosis but suppression of osteoblastic activity and bone formation is the predominant feature of glucocorticoid-induced osteoporosis. Additionally, osteoclast numbers are slightly increased, suggesting that glucocorticoids increase the mature osteoclast's life span. In bone, osteoblasts and osteocytes are the primary targets of glucocorticoid actions. The loss of function of osteoblasts and osteocytes, induction of apoptosis, and autophagy are associated with glucocorticoid-induced bone loss as observed in both clinical and experimental studies [68].

Glucocorticoids inhibit osteoblastogenesis and stimulate osteoclast formation. They increase bone resorption as a result of secondary hyperparathyroidism and decreased absorption of calcium [69]. The inhibitory effects of glucocorticoids on bone formation include a decrease in osteoblast number, diminished bone matrix formation, and enhanced adipogenesis in bone marrow. In osteoblasts, the generation of ROS and activation of PKCβ/p66shc/JNK signaling are responsible for the proapoptotic effects of glucocorticoids [70]. Moreover, higher level of the circulating cortisol was associated with reduced BMD and increased bone loss rate in healthy elderly women and men even after adjustment for age, sex hormone levels, parathyroid hormone (PTH), alcohol consumption, smoking, adiposity, and dietary calcium level, suggesting that endogenous glucocorticoids contribute to involutional bone loss [71]. Similar to humans, age-associated decline in bone mass and strength is also exhibited in mice. The reduced bone mass is associated with an elevated level of adrenal glucocorticoids and an expression of 11β-HSD1 in bone cells. In a transgenic mouse model [osteocalcin gene 2 (OG2)-11βHSD2], where the osteoblasts and osteocytes were shielded from glucocorticoids through the cell-specific gene expression of 11β-HSD type 2 (enzyme that inactivates glucocorticoids), the adverse effects of aging on osteoblast and osteocyte apoptosis, rate of bone formation, interstitial fluid, crystallinity, microarchitecture, bone vasculature, and strength are prevented, demonstrating that endogenous glucocorticoids enhance the fragility of bones in old age [72].

Dexamethasone, a synthetic glucocorticoid, which is used in the differentiation of MSCs to osteoblasts *in vitro*, is shown to inhibit the function of osteoblast by inducing cellular senescence and stimulating some factors of SASP. Furthermore, Wnt regulatory proteins are downregulated by glucocorticoids, and Wnt pathway suppression is associated with reduced bone formation and impaired osteoblastic activity [73]. Importantly, glucocorticoids also induce telomere dysfunction, suggesting that they act as senescence inducers. Thus, elevation in endogenous glucocorticoids highlights another mechanism of age-related involutional osteoporosis.

3.2.3 Physical activity

Aging is associated with gradual loss of muscle quantity and quality that leads to loss of muscle strength and potentially mediates some negative impact on the skeleton. Therefore, exercise is advised to improve and maintain bone health. Age-related mitochondrial changes such as loss of mitophagy, increased mitochondrial uncoupling, superoxide synthesis, and changes in fission and fusion could occur in the muscle as well as osteocytes. Physical activity is a nonpharmacological strategy that could conserve mitochondrial function and number, thereby preventing the cellular senescence. In fact, skeletal unloading results in extensive bone loss with aging and immobilization due to long-term bed rest among aged individuals is an important factor for low BMD and associated fractures. Age-associated osteoporosis is correlated with the lack of cell sensitivity toward mechanical loading induced by weight bearing. Importantly, senescent MSCs failed to respond to the low magnitude mechanical signal due to alterations in the secretome [74]. Developing resistance to apoptosis might induce senescence in osteocytes. The osteocyte apoptosis can be triggered by both damage-inducing loading and unloading, but appropriate mechanical stimulation prevents apoptosis of osteocytes. In addition, mechanical loading stimulates the activation of Src/ERK via integrin and cytoskeleton in osteocytes, thereby inhibiting senescent-related and apoptotic pathways and enhancing the survival of osteocytes [75]. A gap junction protein connexin43 is involved in mechanotransduction and protects the osteocytes from apoptosis; however, it declines with aging. Senescence can be induced by unloading through the inactivity and reduction in myokines, which cause severe alterations in whole-body nutrient trafficking, and results in the lipid accumulation in osteoblasts and osteocytes [76]. Several types of myokines have the ability to influence bone turnover. Exercise-induced myokines such as irisin and β-aminoisobutyric acid have the ability to slow down age-related osteocyte senescence by maintaining the integrity of mitochondria [76]. Srinivasan et al. [77] presented an agent-based model of real-time Ca^{2+}/nuclear factor of activated T cells (NFAT) signaling in bone cells and demonstrated that various loading stimuli-induced periosteal bone formation in young and aged female mice. The stimulations in this model indicated that restoring age-related deficits in DNA binding capacity and NFAT dephosphorylation or translocation effectively improved the skeletal sensitivity toward mechanical loading during senescence. Furthermore, these authors showed that supplementation of low dose of cyclosporin A with mechanical stimuli can completely rescue the loading-induced bone formation in the aged bone [77].

A decline in physical activity leads to impaired osteoblastogenesis and loss of bone with age. Among many factors that contribute to bone imbalance, the fate of MSCs plays a major role. Age-associated bone degeneration is due to decreased MSC proliferation and also increased commitment of MSCs into an adipogenic lineage. For bone health, higher intensity or quicker loading exercises like jumping and resistance training induce stimuli for existing osteocytes and activate the MSC differentiation into osteoblast lineage and further differentiate to form more osteocytes. Interestingly, in an animal study where mice trained over a period of 5 weeks on a treadmill at progressive speeds exhibited an increased MSC number and their osteogenic commitment along with decreased differentiation into adipocytes [78]. Moreover, these authors demonstrated that physical exercise increased the bone formation markers [alkaline phosphatase (ALP), OCN, and osteopontin (OPN)] and decreased the amount of fat in bone marrow cavity of mice. Similarly in rats, skeletal unloading resulted in decreased bone formation due to the recruitment of impaired osteoblasts and their differentiation [79]. Importantly, skeletal unloading caused alterations in the expression of growth factors and thus decreased bone formation in rats [80].

Aerobic exercise is not often suggested as an important part of promoting bone strength and mass. However, aerobic exercise could maintain the viability of osteocytes in a manner that is independent of mechanical loading. On the contrary, unloading may induce greater senescence due to a lack of stimulus for promoting cell viability [81]. In osteoblasts and osteocytes, several signaling pathways such as IGF signaling, prostaglandins (PGs), nitric oxide (NO), adenosine triphosphate (ATP) signaling, and membrane ion channels were shown to promote mechanotransduction *in vitro* [82]. Similarly, another important signaling that responds to mechanical loading on bone is Wnt/β-catenin signaling, which regulates MSC differentiation upon mechanical loading. Physical exercise improves the quality of life in young and aged people by inducing stem cell commitment and differentiation. Therefore, physical activity may be considered as a key factor for regenerative medicine.

4 Role of senescence-associated secretory phenotype and senescent markers on skeletal aging

Cellular senescence is a process that causes irreversible growth arrest in cells exposed to various forms of stress response [83]. Senescence occurs throughout the life span but particularly with aging and senescent cells accumulate in several tissues. The metabolically active senescent cells secrete a wide array of soluble proteins, which include proinflammatory cytokines, chemokines, growth factors, proteases, and ECM degrading factors. Senescent cells are defined on the occurrence of these proinflammatory secretome, often referred to as senescence-associated secretory phenotype (SASP). This SASP type of secretome distinguishes senescent cells from nonsenescent cells or cell cycle-arrested cells, such as quiescent cells and terminally differentiated cells, on the basis of SASP-related factors such as p21, p53, and interleukins (ILs), namely IL-1α, IL-1β, IL-6, and IL-8 [8]. In the tissue microenvironment, the SASP functions physiologically through the production of these autocrine and paracrine factors, which can transmit the stress response and interact with the neighboring cells. Most importantly, the intensity and the composition of SASP-secreted factors differ depending upon the environmental factors, cell types, triggers of cellular senescence, and elapsed time since the initiation of senescence. SASP is thought to be responsible and detrimental for chronic inflammation, which leads to several age-related phenotypes. Indeed, senescent cell accumulation contributes to the onset of aging and several age-related diseases, such as sarcopenia, pulmonary fibrosis, osteoarthritis, cataract, tumorigenesis, frailty, and osteoporosis [84]. *In vivo* age-related senescence impact on bone is not explored completely. Furthermore, the mechanisms by which senescent cells and bone cell SASP alters bone remodeling and tissue homeostasis are not well understood.

In the bone microenvironment, various cell types undergo senescence with aging, where a significant population of myeloid cells and senescent osteocytes are the major sources driving bone loss through SASP with the increased levels of interleukins (IL-1α, IL-8, and IL-6), NF-κB, and other key SASP markers including plasminogen activator inhibitor type 1 (PAI-1) and PAI-2, regulated upon activation, normal T cell expressed and presumably secreted (RANTES), macrophage colony stimulating factor (M-CSF), tumor necrosis factor-alpha (TNF-α), and matrix metalloproteinases (MMPs) (Table 1). In old mice, osteocytes exhibited telomere dysfunction and satellite distension; also, the osteoprogenitor cells showed an elevated level of DNA damage markers, increased expression of GATA4, and NF-κB activation suggesting SASP initiation [8, 85]. SASP-related Janus kinase (JAK) pathway inhibition downregulated the SASP circulating proteins

TABLE 1 Cellular senescence markers.

Criteria	Key markers
Senescence-associated secretory phenotype (SASP)	*Cytokines*: IL-1α, IL-1β, IL-6, IL-8, RANTES, PAI-1 and 2, M-CSF, and TNF-α
	Growth factors: TGF-β, VEGF
	ECM proteins: matrix metalloproteinases (MMPs)
DNA double-stranded break (DSB) damage	γH2AX, p53, TAF
Lysosomal damage	Lysosomal β-galactosidase enzyme activity
Cell cycle arrest	$p16^{Ink4a}$, $p21^{Cip1}$, and p53
Other markers	Lamin B1, HMGB1

The table shows the list of key biomarkers of cellular senescence. These biomarkers can be used to detect the senescent or aging cells. Abbreviations used are: IL: interleukin, RANTES: regulated upon activation, normal T cell expressed and presumably secreted, PAI-1: plasminogen activator inhibitor type 1, M-CSF: macrophage colony stimulating factor, TNF-α: tumor necrosis factor-alpha, TGF-β: transforming growth factor-beta, VEGF: vascular endothelial growth factor, and TAF: telomere-associated foci.

PAI-1, IL-6, and IL-8 in old mice, improved the bone strength and microarchitecture of both cortical and trabecular bones *in vivo*, and attenuated pro-osteoclastogenic effects *in vitro*. Taken together, these findings indicate that senescent cells with SASP contribute to skeletal aging.

Gorissen et al. [86] reported that osteoclast acquired SASP-like phenotype during osteoclastogenesis. In both genders of mice, the primary senescence marker $p16^{Ink4a}$ is elevated in myeloid cells, T cells, B cells, osteoprogenitor cells, osteoblasts, and osteocytes with aging. Additionally, a relatively few SASP factors increased in T cells and B cells of hematopoietic cell population with aging, but in old myeloid cells, 26 genes significantly upregulated among a panel of 36 genes evaluated. Clinical findings obtained from needle biopsies of young and old-aged bones suggested that the expression of $p16^{Ink4a}$ and p21 was elevated and 12 SASP factors were upregulated in an old female bone. Taken together, these data demonstrate that in bone microenvironment, a subset of cells become senescent; however, the SASP upregulation occurs primarily in myeloid cells and osteocytes with aging in both mice and humans. In turn, these myeloid cell lineages become senescent and produce signals, which result in excessive synthesis and secretion of proinflammatory cytokines, forming a toxic microenvironment that leads to senescence of bone cells and ultimately contributes to age-related bone loss. Similarly, senescent markers such as γH2AX, $p16^{Ink4a}$, and several other SASP markers were increased in the osteocytes of old mice. These changes were associated with enhanced RANKL expression by osteocytes and increased porosity in cortical bone, revealing that osteocyte senescence may lead to increased resorption of endocortical and intracortical bone. Studies on osteoprogenitor cells showed a decrease in osterix-expressing osteoprogenitor cells that were associated with increased γH2AX expression and other senescence markers in aged mice. The cellular senescence and SASP secretion of bone marrow MSCs and osteoblasts enhanced the osteoclastogenesis and osteoporotic development in $Ercc^{-/-}$ and $Ercc1^{-/\Delta}$ mice [87–89]. The identification of senescent osteoblasts or osteocytes during aging, together with the rescue role of eliminating dysfunctional senescent cells under osteoporosis, could reverse SASP, suggesting the important role of senescent osteocytes or osteoblasts for age-related bone loss and fragility. However, more studies are warranted to elucidate the mechanisms involved in the SASP development and to explore the characteristics of SASP in the senescent cells of bone microenvironment.

5 Key pathways regulating the skeletal aging

The molecular mechanisms that are associated with bone aging and age-related diseases are not completely understood. Studies performed on the critical signaling pathways that regulate senescence in aging and age-associated pathologies revealed that Wnt and p53/p21 signaling pathways play an essential role in the regulation of age-related bone loss.

5.1 Wnt signaling pathway

Wnt signaling is the major regulator of bone tissue homeostasis and its impairment might be associated with age-related bone loss. This pathway is essential particularly during the commitment of MSCs into osteoblast differentiation. On activation of Wnt, adipogenic transcription factors are inhibited and preadipocytes are maintained in the undifferentiated state. Defect in Wnt signaling could impair bone tissue response to mechanical demand and thus increase the risk of fracture.

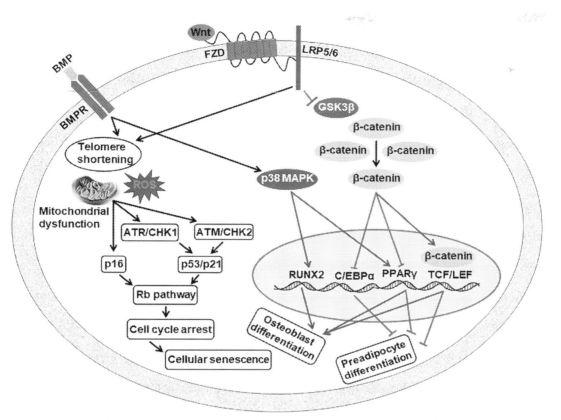

FIG. 5 Crucial signaling pathways involved in bone cell senescence. Schematic representation of Wnt and p53/p21 mediated signaling in cellular senescence of bone. Wnt and BMP signaling pathways regulate the differentiation of mesenchymal stem cells into osteoblasts or adipocytes by either promoting or inhibiting the respective transcription factors. A variety of stress stimuli like ROS accumulation, mitochondrial damage, and telomere shortening can activate p53/p21 and p16/Rb signaling that induces cellular senescence.

β-Catenin, a key transcriptional regulator in Wnt signaling pathway, plays a critical role in bone remodeling and microdamage repair and acts as a critical factor determining the bone mass (Fig. 5). The differentiation of osteoblasts and adipocytes is controlled by Wnt3a, Wnt5a, and Wnt10b, and their expressions are markedly decreased with aging in mice. Consistent loss of Wnt10b leads to an osteopenic condition in mice, which is attributed to multipotent progenitor loss. Therefore, reduction in Wnt signaling molecules contributed to altered bone formation in aging mice. In the noncanonical pathway, elevated Wnt5a expression in hematopoietic stem cells resulted in stem cell aging in aged mice [90]. *In vivo* study suggested an increase in Wnt10b expression in old bone, whereas reduced expression of Wnt16a resulted in decreased bone mass and spontaneous fractures of cortical bone in mice. Apart from this, β-catenin acts as a coactivator of forkhead box O (FoxO). Oxidative stress induces the activation of FoxO and association with β-catenin and is essential for the FoxO gene activation in bone cells and other cell types. The ability of FoxO to sequester β-catenin might limit the proliferation and differentiation of osteoprogenitor cells, which results in reduced bone formation and low bone mass with aging. β-Catenin increases the OPG expression and reduces the RANKL expression, thereby inhibiting the osteoclastogenesis. In addition to the inhibitory effect of oxidative stress on β-catenin-mediated TCF transcription, the expression of Wnt-targeted genes and OPG in murine bone decreases with age. Wnt signaling suppression by ROS/FoxO might be the mechanism by which lipid oxidation results in the reduction of osteoblast formation and low osteoblast number that occurs during aging [32]. These studies clearly show that Wnt/β-catenin pathway plays a critical function in skeletal homeostasis and age-related bone loss.

5.2 p53/p21 signaling pathway

p53/p21 and p16/Rb are the two key pathways involved in the growth arrest of senescent cells. This pathway is regulated in response to telomere shortening, DNA damage, accumulation of ROS, and other stimuli such as oncogene activation. p53/p21 and p16/Rb signaling pathways are based on antiproliferative mechanisms that cease the cell division and are further allowed to repair. DNA damage accumulation leads to senescence and accumulation of ROS-triggered DNA damage response and activated p53/p21 and p16/Rb pathways ultimately result in cellular senescence [91]. p53 phosphorylation is increased with aging or sex hormone deficiency in both male and female mice. Also, genetically engineered p53 mutant mouse models displayed aging phenotypes including osteoporosis. On the contrary, bone formation rate and osteoblast number increased with high bone mass phenotype in p53 null mice. This contrast effect of p53 expression on osteoblast formation has been attributed to the suppression of osterix and Runx2 expression [92]. Several studies revealed the role of p53/p21 pathway in the MSC senescence. Generally, p53 regulates the MSC proliferation and in a study with old rhesus monkeys showed upregulation of p53 and p21 in bone marrow MSCs when compared with young ones. Also, the late passage of MSCs displayed an increase in the expression of p21 and decreased proliferative capacity. Interestingly, accelerated proliferation of MSCs can be achieved by downregulating p21. In another study, bone marrow MSC senescence is induced by TGF-β with an increased expression of p53 and p21 and decrease in pRb expression. Moreover, cell growth arrest was suppressed on treatment with cell growth factors via repression of p53 and p21 and increased level of pRb expression [93].

In osteoblast, p53 activity mitigated the formation of osteoclast by decreasing the M-CSF expression. Most importantly, the downstream of p53 activation generates ROS and is associated with decreased proliferation of osteoblast, showing that p53 activation in aged bone might reduce osteoblast and bone mass in FoxO1-deficient murine model [94]. Additionally, targeting p21 using small interfering RNA in MSCs showed a faster proliferation rate while p21 knockdown enhanced the stemness marker expression, osteogenic potential, and proliferation. In p21$^{-/-}$ mice, a fourfold higher rate of proliferation and about sevenfold lower senescence in normal MSCs were observed. Interestingly, telomere shortened on activation of p53 resulted in mitochondrial alteration, which increased the ROS production mediated by p53-induced PPAR coactivator 1α and PPAR coactivator 1 repression, suggesting that mitochondrial dysfunction arises with age. Interestingly, failure to maintain telomere impairs differentiation of osteoblasts and induces osteoporosis similar to p53 activation, showing that p53/p21, ROS, and telomere dysfunction play a pivotal role in the aging of osteoprogenitor cells [32, 95].

6 Cellular senescence as a therapeutic target to improve bone frailty

Age-related bone loss starts as early as the third decade of life, after attaining peak bone mass. Age-associated cellular senescence is also considered to play an important role in the pathogenesis of osteoporosis. MSCs of elderly patients display senescence characteristics, including senescence-associated β-galactosidase (SA-β-Gal) activity, telomere shortening, and reduced differentiation capacity of osteoblastic lineage [96]. Moreover, these authors reported that MSCs from osteoporotic patients displayed an increased level of p21 and p16^{ink4a} than MSCs from normal individuals. Thus, senescence is considered to contribute pathophysiologically to the disease.

In the context of aging, intensive research in animal models provided more understanding of the characteristic features of skeletal aging and changes in function of bone cells. A study using old INK-ATTAC (apoptosis through targeted activation of caspase 8) transgenic mice targeting senescent cells by genetic knockdown allowed systemic elimination of p16^{Ink4a}-expressing cells upon AP20187 drug administration. This analysis was compared with systemic senescent cell elimination by the administration of senolytics or along with SASP inhibitor; both approaches improved the bone microarchitecture and increased the strength in old mice. Reduction or elimination of senescent cells resulted in suppressed osteoclastogenesis in cortical and trabecular bone along with simultaneously increased bone formation in endocortical region and maintained the bone formation in the trabecular region [97]. These findings indicate that partial elimination of senescent osteocytes favored the bone remodeling and metabolism. Similarly, Klotho-deficient mice showed accelerated senescence and osteoporosis was attenuated by crossing these mice with p16^{ink4a} knockdown mice, indicating that bone deterioration can be prevented by senescent cell ablation [98]. Contrary to the impact of SASP on the inhibition of adipogenesis in visceral and peripheral fat depots, SASP induced adipocyte differentiation from BMSCs, suggesting that SASP factors secreted from senescent cells contributed to the alteration in MSC lineage commitment to adipocytes than osteoblastic lineage. Consistent with this, the naturally aged INK-ATTAC mice eliminated senescent cells, which resulted in the prevention of excessive bone marrow adipose tissue accumulation, the characteristic of skeletal aging.

Pharmacological interventions in naturally aged mice with the combination of senolytic cocktail and quercetin or with JAK inhibitor, which can suppress SASP, improved the microarchitecture of cortical and trabecular bone and thereby maintained the whole bone microarchitecture and bone strength. These collective data indicate that senescence plays a fundamental role in age-related bone loss and osteoporosis [97, 99, 100]. In addition, an old SAMP6 (senescence-accelerated mouse prone 6) mouse received normal allogeneic BMSC's intrabone marrow injection and another similar approach induced locally into bone marrow showed an increase in trabecular bone mass and reduced loss of BMD, thus preventing age-related osteoporosis. Likewise, ovariectomized osteoporotic female rats injected with BMSCs isolated from healthy individuals, into the bone marrow of femur bone resulted in increased bone mass. Interestingly, an increase in bone formation was observed in senile osteoporotic mouse injected systemically with normal allogeneic BMSCs [101]. Senolytic drug such as ABT263, an inhibitor of antiapoptotic protein, can reduce senescence-associated factors and clear senescent hematopoietic stem cells of aged mice. Similarly, treatment with antiapoptotic inhibitors, namely A1331852 and A1155463, cleared the senescent cells *in vitro* [102]. Taken together, these potential agents could be considered for protecting against age-related bone loss.

A recent study has identified the systemic and local proinflammatory environment as the cause for cellular senescence and to the decline in skeletal stem/progenitor cell (SSPC) number and function [103]. Furthermore, this study demonstrated that by inhibiting NF-κB activation in aged mice, a reversal of aging phenotype, functional activation of aged SSPCs with decreased senescence, increased SSPC number, and increased osteogenic function as well as bone healing were observed in an ectopic model [103]. Thus, targeting the senescent cells may provide a therapeutic approach to prevent bone loss in osteoporotic patients.

7 Conclusions

Recent advancements in science have revolutionized the field of medicine, leading to a demographic shift toward an increase in aging populations. Somatic cells have a finite life span as a result of senescence, in which proliferating cells slow down, and enter a nondividing state after a limited number of cell divisions. Cellular senescence is one of the hallmarks of aging. Senescence is a complex process and is accelerated by a combination of intrinsic and extrinsic factors, which may have beneficial or deleterious effects based on their tissue microenvironment. In childhood, bone growth and development are regulated and maintained by programmed cellular senescence. But with aging, senescent cells accumulate and induce chronic inflammation in the bone microenvironment. This causes the release of SASP factors, which damage the bone tissues, and results in age-associated bone loss. Exploring various factors and signaling pathways that regulate cellular senescence can enhance the current cell-based therapy or potentiate the development of novel therapeutic approaches to prevent or lower the rate of age-related comorbidities. Therefore, targeting SASP factors may have clinical implications in protecting against senescence-induced osteoporosis and other related diseases, to enhance the quality of life and healthy life span of the aged/aging individuals.

Acknowledgments

The authors express sincere thanks to Dr. Yuvaraj Sambandam (Department of Surgery, Comprehensive Transplant Center, Northwestern University, Feinberg School of Medicine, Chicago, IL 60611, USA) for critically reading this manuscript and providing his valuable comments. The financial support provided by Science and Engineering Research Board (SERB) [ECR/2015/000277], Government of India, to Dr. Ilangovan Ramachandran, is gratefully acknowledged.

Conflict of interest

The authors declare no conflict of interest.

References

[1] C. Lopez-Otin, M.A. Blasco, L. Partridge, M. Serrano, G. Kroemer, The hallmarks of aging, Cell 153 (6) (2013) 1194–1217.
[2] L. Hayflick, P.S. Moorhead, The serial cultivation of human diploid cell strains, Exp. Cell Res. 25 (1961) 585–621.
[3] X.L. Liu, J.L. Liu, Y.C. Xu, X. Zhang, Y.X. Wang, L.H. Qing, W. Guo, J. Ding, L.H. Meng, Membrane metallo-endopeptidase mediates cellular senescence induced by oncogenic PIK3CA(H1047R) accompanied with pro-tumorigenic secretome, Int. J. Cancer 145 (3) (2019) 817–829.
[4] D. McHugh, J. Gil, Senescence and aging: causes, consequences, and therapeutic avenues, J. Cell Biol. 217 (1) (2018) 65–77.
[5] A.V. Molofsky, S.G. Slutsky, N.M. Joseph, S. He, R. Pardal, J. Krishnamurthy, N.E. Sharpless, S.J. Morrison, Increasing p16INK4a expression decreases forebrain progenitors and neurogenesis during ageing, Nature 443 (7110) (2006) 448–452.
[6] K. Boros, T. Freemont, Physiology of ageing of the musculoskeletal system, Best Pract. Res. Clin. Rheumatol. 31 (2) (2017) 203–217.
[7] O. Demontiero, C. Vidal, G. Duque, Aging and bone loss: new insights for the clinician, Ther. Adv. Musculoskelet Dis. 4 (2) (2012) 61–76.
[8] J.N. Farr, D.G. Fraser, H. Wang, K. Jaehn, M.B. Ogrodnik, M.M. Weivoda, M.T. Drake, T. Tchkonia, N.K. LeBrasseur, J.L. Kirkland, L.F. Bonewald, R.J. Pignolo, D.G. Monroe, S. Khosla, Identification of senescent cells in the bone microenvironment, J. Bone Miner. Res. 31 (11) (2016) 1920–1929.
[9] S. Khosla, Pathogenesis of age-related bone loss in humans, J. Gerontol. A Biol. Sci. Med. Sci. 68 (10) (2013) 1226–1235.
[10] S.R. Cummings, M.C. Nevitt, W.S. Browner, K. Stone, K.M. Fox, K.E. Ensrud, J. Cauley, D. Black, T.M. Vogt, Risk factors for hip fracture in white women. Study of osteoporotic fractures research group, N. Engl. J. Med. 332 (12) (1995) 767–773.

[11] K. Wongdee, N. Charoenphandhu, Osteoporosis in diabetes mellitus: possible cellular and molecular mechanisms, World J. Diabetes 2 (3) (2011) 41–48.

[12] V. Carnevale, E. Romagnoli, L. D'Erasmo, E. D'Erasmo, Bone damage in type 2 diabetes mellitus, Nutr. Metab. Cardiovasc. Dis. 24 (11) (2014) 1151–1157.

[13] E.J. Moerman, K. Teng, D.A. Lipschitz, B. Lecka-Czernik, Aging activates adipogenic and suppresses osteogenic programs in mesenchymal marrow stroma/stem cells: the role of PPAR-gamma2 transcription factor and TGF-beta/BMP signaling pathways, Aging Cell 3 (6) (2004) 379–389.

[14] R. Ilangovan, S. Sittadjody, M. Balaganesh, R. Sivakumar, B. Ravi Sankar, K. Balasubramanian, S. Srinivasan, C. Subramanian, D.M. Thompson, L. Queimado, N. Srinivasan, Dihydrotestosterone is a determinant of calcaneal bone mineral density in men, J. Steroid Biochem. Mol. Biol. 117 (4–5) (2009) 132–138.

[15] S. Mohan, C. Kesavan, Role of insulin-like growth factor-1 in the regulation of skeletal growth, Curr. Osteoporos. Rep. 10 (2) (2012) 178–186.

[16] A.R. Guntur, C.J. Rosen, IGF-1 regulation of key signaling pathways in bone, Bonekey Rep. 2 (2013) 437.

[17] S. Roberts, P. Colombier, A. Sowman, C. Mennan, J.H. Rolfing, J. Guicheux, J.R. Edwards, Ageing in the musculoskeletal system, Acta Orthop. 87 (sup363) (2016) 15–25.

[18] S. Sittadjody, R. Ilangovan, T. Thangasamy, R.C. Vignesh, S. Veni, A.G. Bertoni, S. Srinivasan, C. Subramanian, N. Srinivasan, Age-related changes in serum levels of insulin-like growth factor-II and its binding proteins correlate with calcaneal bone mineral density among post-menopausal south-Indian women, Clin. Chim. Acta 414 (2012) 281–288.

[19] S. Yakar, H. Werner, C.J. Rosen, Insulin-like growth factors: actions on the skeleton, J. Mol. Endocrinol. 61 (1) (2018) T115–T137.

[20] S. Saiganesh, R. Saathvika, B. Arumugam, M. Vishal, V. Udhaya, R. Ilangovan, N. Selvamurugan, TGF-beta1-stimulation of matrix metalloproteinase-13 expression by down-regulation of miR-203a-5p in rat osteoblasts, Int. J. Biol. Macromol. 132 (2019) 541–549.

[21] L.I. Plotkin, A. Bruzzaniti, Molecular signaling in bone cells: regulation of cell differentiation and survival, Adv. Protein Chem. Struct. Biol. 116 (2019) 237–281.

[22] V. Mohanakrishnan, A. Balasubramanian, G. Mahalingam, N.C. Partridge, I. Ramachandran, N. Selvamurugan, Parathyroid hormone-induced down-regulation of miR-532-5p for matrix metalloproteinase-13 expression in rat osteoblasts, J. Cell. Biochem. 119 (7) (2018) 6181–6193.

[23] Y. Han, X. You, W. Xing, Z. Zhang, W. Zou, Paracrine and endocrine actions of bone-the functions of secretory proteins from osteoblasts, osteocytes, and osteoclasts, Bone Res. 6 (2018) 16.

[24] J.A. Siddiqui, N.C. Partridge, Physiological bone Remodeling: systemic regulation and growth factor involvement, Physiology (Bethesda) 31 (3) (2016) 233–245.

[25] D.J. Baylink, R.D. Finkelman, S. Mohan, Growth factors to stimulate bone formation, J. Bone Miner. Res. 8 (Suppl 2) (1993) S565–S572.

[26] A. Infante, C.I. Rodriguez, Osteogenesis and aging: lessons from mesenchymal stem cells, Stem Cell Res. Ther. 9 (1) (2018) 244.

[27] B. Busse, D. Djonic, P. Milovanovic, M. Hahn, K. Puschel, R.O. Ritchie, M. Djuric, M. Amling, Decrease in the osteocyte lacunar density accompanied by hypermineralized lacunar occlusion reveals failure and delay of remodeling in aged human bone, Aging Cell 9 (6) (2010) 1065–1075.

[28] G.M. Martin, C.A. Sprague, C.J. Epstein, Replicative life-span of cultivated human cells. Effects of donor's age, tissue, and genotype, Lab. Invest. 23 (1) (1970) 86–92.

[29] M. Muller, Cellular senescence: molecular mechanisms, in vivo significance, and redox considerations, Antioxid. Redox Signal. 11 (1) (2009) 59–98.

[30] R.J. Pignolo, R.K. Suda, E.A. McMillan, J. Shen, S.H. Lee, Y. Choi, A.C. Wright, F.B. Johnson, Defects in telomere maintenance molecules impair osteoblast differentiation and promote osteoporosis, Aging Cell 7 (1) (2008) 23–31.

[31] H. Saeed, B.M. Abdallah, N. Ditzel, P. Catala-Lehnen, W. Qiu, M. Amling, M. Kassem, Telomerase-deficient mice exhibit bone loss owing to defects in osteoblasts and increased osteoclastogenesis by inflammatory microenvironment, J. Bone Miner. Res. 26 (7) (2011) 1494–1505.

[32] M. Almeida, Aging mechanisms in bone, Bonekey Rep. 1 (2012) 102.

[33] B.L. Riggs, S. Khosla, L.J. Melton 3rd., Sex steroids and the construction and conservation of the adult skeleton, Endocr. Rev. 23 (3) (2002) 279–302.

[34] P.J. Marie, Bone cell senescence: mechanisms and perspectives, J. Bone Miner. Res. 29 (6) (2014) 1311–1321.

[35] Y.M. Janssen-Heininger, B.T. Mossman, N.H. Heintz, H.J. Forman, B. Kalyanaraman, T. Finkel, J.S. Stamler, S.G. Rhee, A. van der Vliet, Redox-based regulation of signal transduction: principles, pitfalls, and promises, Free Radic. Biol. Med. 45 (1) (2008) 1–17.

[36] R.L. Jilka, M. Almeida, E. Ambrogini, L. Han, P.K. Roberson, R.S. Weinstein, S.C. Manolagas, Decreased oxidative stress and greater bone anabolism in the aged, when compared to the young, murine skeleton with parathyroid hormone administration, Aging Cell 9 (5) (2010) 851–867.

[37] C.J. Jagger, J.M. Lean, J.T. Davies, T.J. Chambers, Tumor necrosis factor-alpha mediates osteopenia caused by depletion of antioxidants, Endocrinology 146 (1) (2005) 113–118.

[38] H. Nojiri, Y. Saita, D. Morikawa, K. Kobayashi, C. Tsuda, T. Miyazaki, M. Saito, K. Marumo, I. Yonezawa, K. Kaneko, T. Shirasawa, T. Shimizu, Cytoplasmic superoxide causes bone fragility owing to low-turnover osteoporosis and impaired collagen cross-linking, J. Bone Miner. Res. 26 (11) (2011) 2682–2694.

[39] I. Ramachandran, E. Thavathiru, S. Ramalingam, G. Natarajan, W.K. Mills, D.M. Benbrook, R. Zuna, S. Lightfoot, A. Reis, S. Anant, L. Queimado, Wnt inhibitory factor 1 induces apoptosis and inhibits cervical cancer growth, invasion and angiogenesis in vivo, Oncogene 31 (22) (2012) 2725–2737.

[40] I. Ramachandran, V. Ganapathy, E. Gillies, I. Fonseca, S.M. Sureban, C.W. Houchen, A. Reis, L. Queimado, Wnt inhibitory factor 1 suppresses cancer stemness and induces cellular senescence, Cell Death Dis. 5 (2014) e1246.

[41] S.Y. Shin, B.R. Chin, Y.H. Lee, J.H. Kim, Involvement of glycogen synthase kinase-3beta in hydrogen peroxide-induced suppression of Tcf/Lef-dependent transcriptional activity, Cell. Signal. 18 (5) (2006) 601–607.

[42] G.M. Martin, Epigenetic drift in aging identical twins, Proc. Natl. Acad. Sci. U. S. A. 102 (30) (2005) 10413–10414.

[43] D. Cakouros, S. Gronthos, Epigenetic regulation of bone marrow stem cell aging: revealing epigenetic signatures associated with hematopoietic and mesenchymal stem cell aging, Aging Dis. 10 (1) (2019) 174–189.

[44] J. Franzen, W. Wagner, E. Fernandez-Rebollo, Epigenetic modifications upon senescence of mesenchymal stem cells, Curr. Stem Cell Rep. 2 (2016) 248–254.

[45] A. Schellenberg, S. Mauen, C.M. Koch, R. Jans, P. de Waele, W. Wagner, Proof of principle: quality control of therapeutic cell preparations using senescence-associated DNA-methylation changes, BMC. Res. Notes 7 (2014) 254.

[46] C.M. Koch, S. Joussen, A. Schellenberg, Q. Lin, M. Zenke, W. Wagner, Monitoring of cellular senescence by DNA-methylation at specific CpG sites, Aging Cell 11 (2) (2012) 366–369.

[47] A. Schellenberg, Q. Lin, H. Schuler, C.M. Koch, S. Joussen, B. Denecke, G. Walenda, N. Pallua, C.V. Suschek, M. Zenke, W. Wagner, Replicative senescence of mesenchymal stem cells causes DNA-methylation changes which correlate with repressive histone marks, Aging (Albany NY) 3 (9) (2011) 873–888.

[48] D. Cakouros, S. Isenmann, L. Cooper, A. Zannettino, P. Anderson, C. Glackin, S. Gronthos, Twist-1 induces Ezh2 recruitment regulating histone methylation along the Ink4A/Arf locus in mesenchymal stem cells, Mol. Cell. Biol. 32 (8) (2012) 1433–1441.

[49] J. Delgado-Calle, C. Sanudo, A. Bolado, A.F. Fernandez, J. Arozamena, M.A. Pascual-Carra, J.C. Rodriguez-Rey, M.F. Fraga, L. Bonewald, J.A. Riancho, DNA methylation contributes to the regulation of sclerostin expression in human osteocytes, J. Bone Miner. Res. 27 (4) (2012) 926–937.

[50] S. Reppe, A. Noer, R.M. Grimholt, B.V. Halldorsson, C. Medina-Gomez, V.T. Gautvik, O.K. Olstad, J.P. Berg, H. Datta, K. Estrada, A. Hofman, A.G. Uitterlinden, F. Rivadeneira, R. Lyle, P. Collas, K.M. Gautvik, Methylation of bone SOST, its mRNA, and serum sclerostin levels correlate strongly with fracture risk in postmenopausal women, J. Bone Miner. Res. 30 (2) (2015) 249–256.

[51] J. Loughlin, L.N. Reynard, Osteoarthritis: epigenetics of articular cartilage in knee and hip OA, Nat. Rev. Rheumatol. 11 (1) (2015) 6–7.

[52] P. Fragkiadaki, D. Nikitovic, K. Kalliantasi, E. Sarandi, M. Thanasoula, P.D. Stivaktakis, C. Nepka, D.A. Spandidos, T. Tosounidis, A. Tsatsakis, Telomere length and telomerase activity in osteoporosis and osteoarthritis, Exp. Ther. Med. 19 (3) (2020) 1626–1632.

[53] V. Raz, B.J. Vermolen, Y. Garini, J.J. Onderwater, M.A. Mommaas-Kienhuis, A.J. Koster, I.T. Young, H. Tanke, R.W. Dirks, The nuclear lamina promotes telomere aggregation and centromere peripheral localization during senescence of human mesenchymal stem cells, J. Cell Sci. 121 (Pt. 24) (2008) 4018–4028.

[54] R.J. Pignolo, R.M. Samsonraj, S.F. Law, H. Wang, A. Chandra, Targeting cell senescence for the treatment of age-related bone loss, Curr. Osteoporos. Rep. 17 (2) (2019) 70–85.

[55] J. Wang, C.L. Clauson, P.D. Robbins, L.J. Niedernhofer, Y. Wang, The oxidative DNA lesions 8,5′-cyclopurines accumulate with aging in a tissue-specific manner, Aging Cell 11 (4) (2012) 714–716.

[56] J.L. Simonsen, C. Rosada, N. Serakinci, J. Justesen, K. Stenderup, S.I. Rattan, T.G. Jensen, M. Kassem, Telomerase expression extends the proliferative life-span and maintains the osteogenic potential of human bone marrow stromal cells, Nat. Biotechnol. 20 (6) (2002) 592–596.

[57] X. Yin, C. Zhou, J. Li, R. Liu, B. Shi, Q. Yuan, S. Zou, Autophagy in bone homeostasis and the onset of osteoporosis, Bone Res. 7 (2019) 28.

[58] M. Almeida, C.A. O'Brien, Basic biology of skeletal aging: role of stress response pathways, J. Gerontol. A Biol. Sci. Med. Sci. 68 (10) (2013) 1197–1208.

[59] F. Liu, F. Fang, H. Yuan, D. Yang, Y. Chen, L. Williams, S.A. Goldstein, P.H. Krebsbach, J.L. Guan, Suppression of autophagy by FIP200 deletion leads to osteopenia in mice through the inhibition of osteoblast terminal differentiation, J. Bone Miner. Res. 28 (11) (2013) 2414–2430.

[60] J. Glowacki, J.P. Bilezikian, C.J. Rosen, The Aging Skeleton, first ed., Academic Press, 1999, pp. 1–642.

[61] A. Zallone, Direct and indirect estrogen actions on osteoblasts and osteoclasts, Ann. N. Y. Acad. Sci. 1068 (2006) 173–179.

[62] M. Almeida, M. Martin-Millan, E. Ambrogini, R. Bradsher 3rd, L. Han, X.D. Chen, P.K. Roberson, R.S. Weinstein, C.A. O'Brien, R.L. Jilka, S.C. Manolagas, Estrogens attenuate oxidative stress and the differentiation and apoptosis of osteoblasts by DNA-binding-independent actions of the ERalpha, J. Bone Miner. Res. 25 (4) (2010) 769–781.

[63] J.N. Farr, J.L. Rowsey, B.A. Eckhardt, B.S. Thicke, D.G. Fraser, T. Tchkonia, J.L. Kirkland, D.G. Monroe, S. Khosla, Independent roles of Estrogen deficiency and cellular senescence in the pathogenesis of osteoporosis: evidence in Young adult mice and older humans, J. Bone Miner. Res. 34 (8) (2019) 1407–1418.

[64] R.L. Jilka, G. Hangoc, G. Girasole, G. Passeri, D.C. Williams, J.S. Abrams, B. Boyce, H. Broxmeyer, S.C. Manolagas, Increased osteoclast development after estrogen loss: mediation by interleukin-6, Science 257 (5066) (1992) 88–91.

[65] S. Cenci, G. Toraldo, M.N. Weitzmann, C. Roggia, Y. Gao, W.P. Qian, O. Sierra, R. Pacifici, Estrogen deficiency induces bone loss by increasing T cell proliferation and lifespan through IFN-gamma-induced class II transactivator, Proc. Natl. Acad. Sci. U. S. A. 100 (18) (2003) 10405–10410.

[66] S. Cenci, M.N. Weitzmann, C. Roggia, N. Namba, D. Novack, J. Woodring, R. Pacifici, Estrogen deficiency induces bone loss by enhancing T-cell production of TNF-alpha, J. Clin. Invest. 106 (10) (2000) 1229–1237.

[67] R.H. Straub, The complex role of estrogens in inflammation, Endocr. Rev. 28 (5) (2007) 521–574.

[68] H. Zhou, M.S. Cooper, M.J. Seibel, Endogenous glucocorticoids and bone, Bone Res. 1 (2) (2013) 107–119.

[69] M.S. Cooper, 11beta-Hydroxysteroid dehydrogenase: a regulator of glucocorticoid response in osteoporosis, J. Endocrinol. Invest. 31 (7 Suppl) (2008) 16–21.

[70] M. Almeida, L. Han, E. Ambrogini, R.S. Weinstein, S.C. Manolagas, Glucocorticoids and tumor necrosis factor alpha increase oxidative stress and suppress Wnt protein signaling in osteoblasts, J. Biol. Chem. 286 (52) (2011) 44326–44335.

[71] R.M. Reynolds, E.M. Dennison, B.R. Walker, H.E. Syddall, P.J. Wood, R. Andrew, D.I. Phillips, C. Cooper, Cortisol secretion and rate of bone loss in a population-based cohort of elderly men and women, Calcif. Tissue Int. 77 (3) (2005) 134–138.

[72] R.S. Weinstein, C. Wan, Q. Liu, Y. Wang, M. Almeida, C.A. O'Brien, J. Thostenson, P.K. Roberson, A.L. Boskey, T.L. Clemens, S.C. Manolagas, Endogenous glucocorticoids decrease skeletal angiogenesis, vascularity, hydration, and strength in aged mice, Aging Cell 9 (2) (2010) 147–161.

[73] F.S. Wang, C.L. Lin, Y.J. Chen, C.J. Wang, K.D. Yang, Y.T. Huang, Y.C. Sun, H.C. Huang, Secreted frizzled-related protein 1 modulates glucocorticoid attenuation of osteogenic activities and bone mass, Endocrinology 146 (5) (2005) 2415–2423.

[74] H. Wang, T.A. Brennan, E. Russell, J.H. Kim, K.P. Egan, Q. Chen, C. Israelite, D.C. Schultz, F.B. Johnson, R.J. Pignolo, R-Spondin 1 promotes vibration-induced bone formation in mouse models of osteoporosis, J. Mol. Med. (Berl) 91 (12) (2013) 1421–1429.

[75] L. Qin, W. Liu, H. Cao, G. Xiao, Molecular mechanosensors in osteocytes, Bone Res. 8 (2020) 23.

[76] V.D. Sherk, C.J. Rosen, Senescent and apoptotic osteocytes and aging: exercise to the rescue? Bone 121 (2019) 255–258.

[77] S. Srinivasan, B.J. Ausk, J. Prasad, D. Threet, S.D. Bain, T.S. Richardson, T.S. Gross, Rescuing loading induced bone formation at senescence, PLoS Comput. Biol. 6 (9) (2010) e1000924.

[78] M. Maredziak, A. Smieszek, K. Chrzastek, K. Basinska, K. Marycz, Physical activity increases the total number of bone-marrow-derived mesenchymal stem cells, enhances their Osteogenic potential, and inhibits their adipogenic properties, Stem Cells Int. 2015 (2015), 379093.

[79] M. Machwate, E. Zerath, X. Holy, M. Hott, D. Modrowski, A. Malouvier, P.J. Marie, Skeletal unloading in rat decreases proliferation of rat bone and marrow-derived osteoblastic cells, Am. J. Physiol. 264 (5 Pt. 1) (1993) E790–E799.

[80] P.J. Marie, E. Zerath, Role of growth factors in osteoblast alterations induced by skeletal unloading in rats, Growth Factors 18 (1) (2000) 1–10.

[81] M.T. Valenti, L. Dalle Carbonare, G. Dorelli, M. Mottes, Effects of physical exercise on the prevention of stem cells senescence, Stem Cell Rev. Rep. 16 (1) (2020) 33–40.

[82] C.H. Turner, S.J. Warden, T. Bellido, L.I. Plotkin, N. Kumar, I. Jasiuk, J. Danzig, A.G. Robling, Mechanobiology of the skeleton, Sci. Signal. 2 (68) (2009) pt3.

[83] B.G. Childs, M. Durik, D.J. Baker, J.M. van Deursen, Cellular senescence in aging and age-related disease: from mechanisms to therapy, Nat. Med. 21 (12) (2015) 1424–1435.

[84] X. Liu, M. Wan, A tale of the good and bad: cell senescence in bone homeostasis and disease, Int. Rev. Cell Mol. Biol. 346 (2019) 97–128.

[85] H.N. Kim, J. Chang, L. Shao, L. Han, S. Iyer, S.C. Manolagas, C.A. O'Brien, R.L. Jilka, D. Zhou, M. Almeida, DNA damage and senescence in osteoprogenitors expressing Osx1 may cause their decrease with age, Aging Cell 16 (4) (2017) 693–703.

[86] B. Gorissen, A. de Bruin, A. Miranda-Bedate, N. Korthagen, C. Wolschrijn, T.J. de Vries, R. van Weeren, M.A. Tryfonidou, Hypoxia negatively affects senescence in osteoclasts and delays osteoclastogenesis, J. Cell. Physiol. 234 (1) (2018) 414–426.

[87] Q. Chen, K. Liu, A.R. Robinson, C.L. Clauson, H.C. Blair, P.D. Robbins, L.J. Niedernhofer, H. Ouyang, DNA damage drives accelerated bone aging via an NF-kappaB-dependent mechanism, J. Bone Miner. Res. 28 (5) (2013) 1214–1228.

[88] M. Piemontese, M. Almeida, A.G. Robling, H.N. Kim, J. Xiong, J.D. Thostenson, R.S. Weinstein, S.C. Manolagas, C.A. O'Brien, R.L. Jilka, Old age causes de novo intracortical bone remodeling and porosity in mice, JCI Insight 2 (17) (2017) e93771.

[89] S. Khosla, J.N. Farr, J.L. Kirkland, Inhibiting cellular senescence: a new therapeutic paradigm for age-related osteoporosis, J. Clin. Endocrinol. Metab. 103 (4) (2018) 1282–1290.

[90] M. Rauner, W. Sipos, P. Pietschmann, Age-dependent Wnt gene expression in bone and during the course of osteoblast differentiation, Age (Dordr.) 30 (4) (2008) 273–282.

[91] A. Qadir, S. Liang, Z. Wu, Z. Chen, L. Hu, A. Qian, Senile osteoporosis: the involvement of differentiation and senescence of bone marrow stromal cells, Int. J. Mol. Sci. 21 (1) (2020) 349.

[92] X. Wang, H.Y. Kua, Y. Hu, K. Guo, Q. Zeng, Q. Wu, H.H. Ng, G. Karsenty, B. de Crombrugghe, J. Yeh, B. Li, p53 functions as a negative regulator of osteoblastogenesis, osteoblast-dependent osteoclastogenesis, and bone remodeling, J. Cell Biol. 172 (1) (2006) 115–125.

[93] T. Ito, R. Sawada, Y. Fujiwara, Y. Seyama, T. Tsuchiya, FGF-2 suppresses cellular senescence of human mesenchymal stem cells by down-regulation of TGF-beta2, Biochem. Biophys. Res. Commun. 359 (1) (2007) 108–114.

[94] M.T. Rached, A. Kode, L. Xu, Y. Yoshikawa, J.H. Paik, R.A. Depinho, S. Kousteni, FoxO1 is a positive regulator of bone formation by favoring protein synthesis and resistance to oxidative stress in osteoblasts, Cell Metab. 11 (2) (2010) 147–160.

[95] B.J. Mehrara, T. Avraham, M. Soares, J.G. Fernandez, A. Yan, J.C. Zampell, V.P. Andrade, A.P. Cordeiro, C.M. Sorrento, p21cip/WAF is a key regulator of long-term radiation damage in mesenchyme-derived tissues, FASEB J. 24 (12) (2010) 4877–4888.

[96] S. Zhou, J.S. Greenberger, M.W. Epperly, J.P. Goff, C. Adler, M.S. Leboff, J. Glowacki, Age-related intrinsic changes in human bone-marrow-derived mesenchymal stem cells and their differentiation to osteoblasts, Aging Cell 7 (3) (2008) 335–343.

[97] J.N. Farr, M. Xu, M.M. Weivoda, D.G. Monroe, D.G. Fraser, J.L. Onken, B.A. Negley, J.G. Sfeir, M.B. Ogrodnik, C.M. Hachfeld, N.K. LeBrasseur, M.T. Drake, R.J. Pignolo, T. Pirtskhalava, T. Tchkonia, M.J. Oursler, J.L. Kirkland, S. Khosla, Targeting cellular senescence prevents age-related bone loss in mice, Nat. Med. 23 (9) (2017) 1072–1079.

[98] M.P. Baar, E. Perdiguero, P. Munoz-Canoves, P.L. de Keizer, Musculoskeletal senescence: a moving target ready to be eliminated, Curr. Opin. Pharmacol. 40 (2018) 147–155.

[99] M. Xu, A.K. Palmer, H. Ding, M.M. Weivoda, T. Pirtskhalava, T.A. White, A. Sepe, K.O. Johnson, M.B. Stout, N. Giorgadze, M.D. Jensen, N.K. LeBrasseur, T. Tchkonia, J.L. Kirkland, Targeting senescent cells enhances adipogenesis and metabolic function in old age, Elife 4 (2015), e12997.

[100] J.N. Farr, S. Khosla, Cellular senescence in bone, Bone 121 (2019) 121–133.

[101] J. Kiernan, S. Hu, M.D. Grynpas, J.E. Davies, W.L. Stanford, Systemic mesenchymal stromal cell transplantation prevents functional bone loss in a mouse model of age-related osteoporosis, Stem Cells Transl. Med. 5 (5) (2016) 683–693.

[102] J. Chang, Y. Wang, L. Shao, R.M. Laberge, M. Demaria, J. Campisi, K. Janakiraman, N.E. Sharpless, S. Ding, W. Feng, Y. Luo, X. Wang, N. Aykin-Burns, K. Krager, U. Ponnappan, M. Hauer-Jensen, A. Meng, D. Zhou, Clearance of senescent cells by ABT263 rejuvenates aged hematopoietic stem cells in mice, Nat. Med. 22 (1) (2016) 78–83.

[103] A.M. Josephson, V. Bradaschia-Correa, S. Lee, K. Leclerc, K.S. Patel, E. Muinos Lopez, H.P. Litwa, S.S. Neibart, M. Kadiyala, M.Z. Wong, M.M. Mizrahi, N.L. Yim, A.J. Ramme, K.A. Egol, P. Leucht, Age-related inflammation triggers skeletal stem/progenitor cell dysfunction, Proc. Natl. Acad. Sci. U. S. A. 116 (14) (2019) 6995–7004.

Chapter 15

Aging-induced stem cell dysfunction: Molecular mechanisms and potential therapeutic avenues

Yander Grajeda[a], Nataly Arias[a], Albert Barrios[a], Shehla Pervin[b,c,d,e], and Rajan Singh[c,d,e,f]

[a]California State University Dominguez Hills, Los Angeles, CA, United States, [b]Department of Biology, California State University, Los Angeles, CA, United States, [c]Division of Endocrinology and Metabolism, Charles R. Drew University of Medicine and Science, Los Angeles, CA, United States, [d]Department of Obstetrics and Gynecology, UCLA School of Medicine, Los Angeles, CA, United States, [e]Johnson Comprehensive Cancer Center, UCLA School of Medicine, Los Angeles, CA, United States, [f]Research Program in Men's Health: Aging and Metsabolism, Brigham and Women's Hospital, Harvard Medical School, Boston, MA, United States

1 Introduction

Epidemiological and experimental evidence gathered over the years suggests that aging is a predominant risk factor for many chronic age-related diseases including neurodegenerative diseases such as Alzheimer's and Parkinson's, as well as frailty, sarcopenia, cancer, heart disease, and chronic obstructive pulmonary diseases that limit health span. In simple terms, aging is the process of becoming older that involves a number of biological mechanisms associated with diminishing tissues and organs which lead to deterioration of both cognitive and physical health over time [1]. These changes are nearly universal and are evident at both microscopic and macroscopic levels. According to projections, by 2050, the number of adults older than 80 will have tripled compared to the numbers estimated in 2015 [2]. Seniors are usually more susceptible to conditions like loss of muscle mass and strength that lead to all-cause mortality [3]. As the aging population increases, the cost of social programs like Medicare, Medicaid, and Social Security that the elderly population relies on will also increase [4].

The ability of any living organism to maintain normal healthy function during aging exclusively relies on mechanisms that regulate tissue homeostasis and regeneration [5]. Maintenance of tissue homeostasis and regenerative capacity in response to injury depends on tissue-specific stem cells with the ability to self-renew and differentiate. The regenerative response of these stem cells depends on their ability to balance quiescence by supporting proliferative activity, which is critical for survival [6]. Resident stem cells present in many organs of mature vertebrates play a key role during such tissue homeostasis and regeneration [7, 8]. It has become increasingly evident that this complex biological process has an impact on stem cells that lead to their loss of regenerative potential, apoptosis, or senescence [9]. Many of the regulatory mechanisms responsible for maintaining their role in stem cells are also diminished due to their inability to maintain homeostasis, self-renewal, and proliferation [10].

The effects of aging on stem cells remain a widely understudied area of research. Organisms composed of multicellular structures undergo a decline in body functions as they age. This observed decline in body functions and degradation of stem cell function is a result of accumulated damage and changes over time [11]. Without functional stem cells, tissues are unable to grow or regenerate under normal conditions or respond toward injuries. With the loss of stem cell function, the degradation and dysfunction of old tissues will result in the decreased longevity of organisms. During aging, this loss of function becomes exponential until it reaches a terminal state of dysfunction resulting from accumulated damage and changes to major stem cells [12]. The characteristics and effects of aging on various stem groups will be discussed along with how they affect the longevity of the organism. Combining our understanding of pluripotent stem cells' role in the body and interactions they undergo within an aging body can open more opportunities for pharmacological therapies and cellular reprogramming to help alleviate the effects of aging.

2 Hematopoietic cells

Hematopoietic stem cells (HSCs) in the bone marrow of adult mammals are capable of blood formation and immune system replenishment for long periods, but as mammals begin to age, the HSC's ability decreases due to the decline of active HSCs [13]. The degradation of HSCs does not occur during normal aging; it occurs when the concentration of aged HSCs increases [14]. Studies have shown that the concentration of aged HSCs has a significantly reduced overall function, which results from DNA damage accumulated during aging and the induction of tumor suppressor pathways that together decreased the regenerating abilities of HSCs [15]. The aging of HSCs is related to the decline of lymphoid lineage commitment that differentiates into hematopoietic cells, which are responsible for replenishing the concentration of functional HSCs. The decline in lineage contribution results from the downregulation of genes involved in the repair and preservation of DNA and the upregulation of genes related to inflammation and responses to stress [16]. When HSCs lack the ability to self-renew due to decline in the overall function, the demands for blood formation and immune system replenishment become difficult to meet. The results of inadequate response from aged HSCs for the demand of blood formation and immune system replenishment are identified by the decline in immune competence, increased autoimmunity, diminished stress response, anemia, and increased risk of diseases, such as myelodysplastic syndrome and myeloid leukemia [15]. In summary, HSCs are susceptible to aging due to the loss of epigenetic regulation and the lack of self-renewal that results in the decline of blood formation and immune system replenishment.

3 Intestinal stem cells

Intestinal stem cells (ISCs) have been studied using a Drosophila intestine model due to its lack of immunity, naturally occurring age-related intestinal dysplasia, and short life span [17]. These studies have shown that the age-related decline in intestinal structure and functionality is caused by environmental factors that result in the activation of stress response pathways such as platelet-derived growth factor (PDGF), c-Jun N-terminal kinase (JNK), and P38 mitogen-activated protein kinases (p38-MAPK) [17–19]. ISCs are used to regenerate the gut epithelium after experiencing damage from environmental sources or oxidative stress. JNK, one of the major signaling pathways, has a strong influence on the aging of the intestinal cells, which is released as a response to stress exposure from environmental factors. JNK promotes cytoprotective gene expression to prevent gastric mucosal injury from ulcerogenic and necrotizing agents without inhibiting the secretion of gastric acid or neutralizing intragastric acidity [20]. The intestinal system declines in function and structure due to age-related progressive loss of the mitochondrial functionality and generation of ROS [21]. In response to the damage from oxidative stress to the intestine, the proliferation of ISCs increases but with the downregulation of regenerative and protective genes. Such ICS atypical differentiation has been shown to contribute to colorectal cancer and increased number of dysfunctional intestinal cells [18, 22].

4 Germline stem cells

Germline stem cells (GSCs) differentiate into different germ cells depending on the organism's biological gender [23]. The germ cell production is maintained differently in each biological gender during the progression of aging. In males, the germ cell population is maintained by spermatogonial stem cells (SSCs), which decline as a result of accumulated DNA damage and mutations [24]. Such accumulated DNA damage causes an increase of undifferentiated cells during spermatogenesis. Even with the accumulation of damage and accumulation of undifferentiated cells, males remain fertile throughout their entire life. Females do not remain fertile throughout their life; females undergo oogenesis from oogonia in the developing ovary that stops before birth [18]. There are mammals such as bats that generate oocytes called oogonial stem cells (OSCs) even after birth. OSCs have been found in adult human ovaries that can form oocyte-like cells when placed into ovarian tissues [25, 26]. GSCs differentiate differently in males and females, but the cause of aging in GSC is cell-extrinsic. Hormonal signaling factors like insulin have a large influence on the maintenance of the GSC population and function [27]. Decreased insulin secretion and its signaling during aging result in a decline in the GSC population resulting from atypical or abnormal differentiation. It was demonstrated that GSC could reset their aging when placed in younger organisms as a result of higher population and functional GSC, suggesting that the decline of GSCs may be the result of decline in insulin-like external hormonal factors [27].

5 Skeletal muscle stem cells

Skeletal muscle stem cells, also called satellite cells (SCs), function by responding to injuries by regenerating the skeletal muscle fibers. As aging progresses, skeletal muscle fiber regeneration declines along with decreased muscle mass, muscle strength, and protein synthesis due to the decreased levels of hormones such as growth hormones and insulin [28]. These changes are a result of SCs frequency decreasing with age. Large muscle fibers that contain a large population of SC decline as aging occurs and similar outcomes with the smaller muscle fibers have been reported [29]. With the low frequency of SCs due to decreased proliferation in vitro, the regenerative and engraftment ability of SCs declines during aging. The causes of SCs aging may also be partially due to the nature of modern-day diets that are high in fats, which increase the rates of oxidative stress and increased DNA damage [10]. With increased DNA damage during aging, the expression of SCs genes associated with differentiation such as MyoD, Wnt, and TGF-β decreases, resulting in the decreased regeneration and maintenance of muscles [18, 30]. With increased DNA damage accumulation during advanced aging, the SCs also undergo cellular senescence. In summary, the downregulation of SC differentiation genes causes decreased muscle regeneration and maintenance as aging progresses.

6 Neural stem cells

Neural stem cells (NSCs) are post-mitotic cells that sustain neurogenesis in portions of the adult brain. NSCs are produced in the dentate gyrus of the hippocampus, and a number of environmental and cell-intrinsic factors are used to adapt the regulation of neurogenesis to environmental changes [31]. The environmental changes cause the NSCs to self-renew and generate neural progenitors that differentiate into neuroblasts and mature into dentate granules cells. There is a decrease in neurogenesis during aging, caused by the decline in immunoreactivity with polysialylated neural cell adhesion molecules involved in the migration and elongation of developing neurons [32]. The decline of neurogenesis in aging adults leads to a reduction in the functional plasticity of neurons. The decline in cognitive abilities during Alzheimer's and Parkinson's diseases is the result of the dysregulation of neurogenesis [33]. The effect of aging of NSCs is caused by the depletion of the adult neural cell pool. Without any method of self-renewal, the number of NSCs declines along with the ability to regenerate neural progenitors [31, 33]. The lack of self-renewal is influenced by the increase of chemokines such as CCL11, which inhibit the proliferation of neural progenitors, along with increased BMP levels that promote proliferation and reduce neurogenesis [33]. Overall, NSC aging results in the dysregulation of neurogenesis due to the lack of self-renewal and regenerative abilities.

7 Mesenchymal stem cells

Mesenchymal stem cells (MSCs) have the ability for multilineage differentiation and self-renewal. MSCs are found in several tissues like bone marrow, adipose tissue, and muscle tissues. MSCs contribute to the regeneration of organs such as bone, skin, liver, and muscle tissue by differentiating into the cell type the organ is composed of [34]. As aging progresses, there is a significant decline of MSCs in older adults, which decreases the regenerative ability of tissues. This aging-associated decline of MSCs limits the quality of regenerated tissues and delays regeneration and in vivo bone formation efficiency [34, 35]. DNA damage accumulation from oxidative stress and telomere shortening during aging is caused by the intrinsic changes in MSCs. The shortening of telomeres causes a decrease in expression of MSCs-specific markers such as CD73, CD90, CD105, and CD11b. Once telomere declines to the threshold of 10 kb, it enters senescence [36]. MSCs entering senescence result in the high expression of miR-55-5p along with MSCs increasing in adipogenic differentiation potential [37, 38]. The progenitors of three types of adipose tissues, namely white, brown, and beige adipocytes, which perform different functions from storing excess energy to dissipate energy, are also derived from MSCs. Aging has been reported to negatively affect the formation of both brown and beige adipocytes. Examination of anatomical locations has displayed an age-dependent loss of brown adipocytes [39]. Significant functional decline in both brown and beige adipocytes is accompanied by reduced expression of uncoupling protein-1 (UCP1), a key mitochondrial protein that plays a significant role in dissipating energy in the form of heat [40]. A progressive decline the in browning capacity of white adipocytes is also observed during aging [41]. In addition, impairment of regenerative capacity of mesenchymal progenitor cells responsible for the self-renewal, proliferation, and differentiation into mature adipocytes is significantly increased with age [42]. It is also suggested that the pool of progenitors present in the adipose tissues are highly heterogeneous in nature and aging negatively impacts the ability of such stem/progenitor cells to undergo maturation and regeneration. Age-related changes in endocrine signals and trophic factors that regulate proliferation and differentiation also play a significant role in the loss of this progenitor population [43]. A recent report demonstrated that aging impairs beige adipocyte differentiation of MSC via reduced expression of sirtuin 1 (SirT1) [44]. In adipose-derived MSC (AT-MSC) isolated from elderly human subjects, impaired beige adipocyte differentiation and significantly reduced expression of key beige adipogenic transcription factors, including UCP1, Cox8b, and Cidea, were observed [44]. On the other hand, AT-MSC isolated from an infant group showed higher expression of these factors compared to the elderly group [44], suggesting that aging impairs beige adipocyte differentiation ability of AT-MSC. Activation of senescence pathway has previously been reported to inhibit the differentiation potential of beige progenitor cells [45]. Furthermore, AT-MSCs isolated from the elderly showed upregulation of senescence-associated phenotype by overexpressing p16, p21, IL6, and IL8 and reduced expression of SirT1.

8 Models used to study stem cell dysfunction

The aging of an organism is a progressive change in bodily functions and signaling that results in the overall dysfunction of the body due to the lack of tissue renewal and functional repair mechanisms. The complex mechanism of aging that occurs in both stem cells and post-mitotic cells can be better understood using different models. Many of the modern models used in the genetic and signaling pathway analysis in aging do not have a longevity that matches the span of the human aging process, but do provide a crucial view into the aging mechanism in a short span of time [7]. Due to its intensive replication and ability to retain the full potential of its daughter cells made by the mother cell, *S. cerevisiae* has been widely used as a model to study aging-associated changes in stem cell self-renewal and function [7, 46, 47]. *C. elegans*, composed of post-mitotic cells, has also been used to study GSC proliferation and delineate the relationship between GSCs and the longevity [48, 49]. Similar to the two models discussed, the Drosophila model offers a tractable genetic system that is useful in investigating the relationship between stem cell regenerative ability and longevity [17]. The Drosophila model has several different tissues that are maintained and renewed by resident stem cells, such as the hindgut which provides an understanding to the mechanisms used by the model organism to repair stress-induced tissues in the gut following injury [50, 51]. The Drosophila model also showed that the regenerative and self-renewal system that extends longevity is dependent on the expression of repair genes [52]. Overall, these models provide a useful opportunity to understand the complex mechanism of aging-related stem cell dysfunction during the maintenance and repair of tissues.

9 Factors responsible for stem cell dysfunction

9.1 Stem cell exhaustion

Stem cell exhaustion is one of the predominant hallmarks of aging that play a crucial role in promoting the aging phenotypes. Stem cell exhaustion of any type leads to many health complications and threatens overall longevity. For example, HSC exhaustion leads to anemia and myelodysplasia, whereas MSC exhaustion leads to osteoporosis and to decreased function in bone fracture repair [53, 54]. Although cellular senescence is detrimental for homeostasis, it was conversely found that a high proliferative rate could lead to accelerated stem cell exhaustion in Drosophila [53]. This is also driven by the imbalance of stem cell quiescence and proliferation due to the excessive proliferation and deregulation of cell cycle checkpoints [55]. Stem cells need to remain in quiescence for long-term self-renewal and tissue homeostasis [53, 55, 56]. ROS induction was shown to lead stem cells to proliferation and differentiation, whereas low levels of ROS are required for stem cell maintenance and quiescence [57]. Autophagy, a highly conserved process to extend stem cell survival, functions to maintain cellular homeostasis under both normal and stress conditions. Aging-induced decrease in autophagy leads to decreased regenerative ability of MSC and HSC resulting from several changes such as impaired self-renewal capacity, loss of proteostasis, increased mitochondrial activity, and oxidative stress and an activated metabolic rate. However, the causes of a dysfunctional stem cell are not strictly limited to the intracellular interactions of a cell, but to the various extracellular interactions, an individual stem cell undergoes in its microenvironment as it tries to increase its longevity.

9.2 Microenvironment

Many external factors are reported to cause the dysregulation of stem cells during aging. The stem cell's microenvironment may be one of the most prominent influences that affect the function and number of stem cells [58]. Under homeostatic conditions, especially for skeletal muscle stem cells, it is normal to have pro-inflammatory and anti-inflammatory cytokines to promote bone healing [59]. In an aging individual, the microenvironment tends to be more pro-inflammatory than anti-inflammatory due to the overabundant quantity of cytokines produced from innate immune systems such as IL-6, TNF-α, and IL-1β. However, IL-10, an anti-inflammatory cytokine, is also upregulated simultaneously as part of an anti-inflammatory compensatory system found in centenarians, indicating that it promotes a longer life span [60]. Inflamm-aging, a chronic inflammatory condition, is one of the main drivers known to influence the immune system and cause stem cell dysfunction due to abnormal elevation of pro-inflammatory cytokines that negatively affect the regenerative capabilities of stem cells and lead to cell senescence [18, 58, 59, 61, 62]. Many factors including the deregulation of inflammatory pathways like NF-kB and the hypersensitivity of aged macrophages may help us understand this universal phenomenon [59, 63]. Activation of the NF-kB pathway is mainly involved in aging by promoting the chronic inflammatory phenotype in the microenvironment, which is promoted by overexpression of key transcriptional factors in the NF-kB pathway, resulting in stem cell dysfunction by primarily targeting pro-inflammatory genes such as TNF-α and IL-1 [55, 64]. Maintenance of the adaptive immune system declines as an organism ages, leading the organism to overstimulate the innate immune system, which results in inflamm-aging and immunosenescence. Oxidative stress also promotes the switch from adaptive to innate immune systems during aging by ROS-mediated damage on hematopoietic stem cells, responsible for the production of naïve T cells to conduct antigen surveillance. This ROS-mediated inflammatory process on stem cells, which increases the expression of pro-inflammatory mediators, is known as oxy-inflamm-aging [61, 62]. The immune system starts to have more memory T cells than naïve cells in aging population, in correlation with an increase of pro-inflammatory cytokines such as IFN-γ, IL-2, and TNF-a, whereas other factors like IL-4, IL-6, and IL-10 are increased in CD8+ cell population. The expression of CD8+CD28- cells, a subset of CD8+ cells that have shortened telomeres, is increased in adults and is an indicator of immunosenescence [64]. These CD8+CD28- cells tend to have an increased expression of NF-kB that can promote inflamm-aging and immunosenescence. This leads to decreased naïve T cell population, growing memory, and effector T cell population, resulting in reduced immune responsiveness [64]. Immune efficiency is reduced in a state of inflammation and leads to immune dysfunction that impedes immune responsiveness to diverse antigens introduced during aging. CD4+ and CD8+ memory and effector cells are found to be increased in older organisms, and these subsets were reported to secrete pro-inflammatory cytokines like IFN-γ, IL-2, and TNF-α [64]. The number of functional stem cells declined during aging as well due to the presence of senescent cells. Cellular senescence occurs due to several factors, including ROS-mediated damage, acquisition of senescence-associated secretory phenotype (SASP), and/or chronic inflammation [55, 62]. Elevated levels of ROS in the microenvironment are another major factor that causes stem cell dysregulation. Elevated ROS generated by aged tissues can lead to oxidative damage to proteins, lipids, and DNA that can eventually lead to mitochondrial dysfunction. Several factors including the decreasing activity of ROS-scavenging enzymes, accumulated mutations of mitochondrial DNA, and decline of mitochondrial function are among several factors that are suggested to contribute to increased ROS production during aging [9, 65]. As shown in Fig. 1, the stem cell's microenvironment consists of neighboring cells and blood vessels whose interaction in an aging system via endocrine, paracrine, or autocrine cell signaling may promote problematic consequences to the longevity of stem cells and its host.

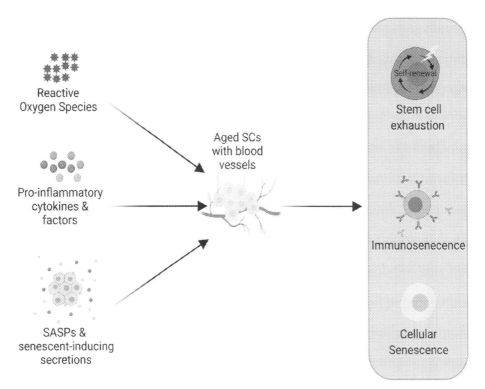

FIG. 1 Aging-related changes in microenvironmental factors and stem cell dysfunction. Aged stem cells (SCs) interact with many factors from the microenvironment such as reactive oxygen species (ROS) accumulation from neighboring cells, elevated pro-inflammatory cytokines, and factors and interactions from neighboring senescent cells. Breakage from its quiescence state from either ROS-accumulating damage or pro-inflammatory pathways leads to stem cell exhaustion. SCs undergo cellular senescence due to ROS, senescence-associated secretory phenotypes (SASPs) from neighboring senescent cells, and pro-inflammatory cytokines and factors that activate pathways like the NF-kB pathway. Immunosenescence occurs when the immune cells' ability for antigen surveillance is impaired with decreased naïve T cell population, increased memory, and effector T cell population.

9.3 DNA damage

A commonality between all stem cells, independent of their niche, is their ability to remain in their dormant state during their life span. Consequently, during an extended period, stem cells can become exposed to many external factors such as radiation and environmental toxins or internal factors such as the accumulation of ROS and errors in DNA replication that could lead to DNA damage and contribute to the dysfunction of these cells [58]. The stem cell viability is dependent on its ability to perform its function; however, if these cells are not tightly regulated through replication checkpoints, they have the potential to be predisposed to oncogenic transformation. If these extensive modifications remain unchecked, they can threaten the genomic integrity of stem cells and can negatively affect their function by altering the transcriptional pathways associated with their proliferative state. Many of these DNA lesions often arise during DNA replication where DNA mismatches may be introduced during aberrant topoisomerase activity, causing single- or double-stranded breaks. Some lesions occur during hydrolytic reactions and methylations that result in variations in base and sugar residues [66]. In addition, lesions can be produced from the generation of ROS and reactive nitrogen species (RNS), as well as various reactions involving metals and antioxidants. These types of reactions can lead to the formation of DNA adducts that can hinder base pairing and DNA replication and transcription, the loss of base, or the introduction of DNA single-stranded breaks [67]. Furthermore, subsequent proximity of two single-stranded breaks can result in the formation of double-stranded breaks. Although this type of DNA damage is less common, it is more difficult to forego DNA repair mechanisms because the DNA backbone is physically cleaved, resulting in compromised genomic and cellular integrity [68]. Due to the common occurrence of mutations, cells have integrated DNA repair mechanisms to ensure their survival although such repair mechanisms themselves are subject to mutations [69]. Despite the presence of these repair mechanism pathways, recent studies suggest that DNA damage plays a role in stem cell aging and dysfunction. Several studies have noted increased staining of phosphorylated form of variant histone H2AX in nuclear foci of HSC and muscle stem cells, also known as satellite cells [70–72]. This measure of DNA damage was found to increase with age in both HSC and satellite cell populations. Studies have shown that errors in DNA repair mechanisms are more strongly associated with premature aging [73]. Contrastingly, overexpression of SIRT6, a conserved family of deacetylases shown to regulate life span by regulating DNA repair mechanisms in simpler animal models, results in longer life span of male mice [11]. Furthermore, many studies have implicated faulty DNA repair mechanisms in the acceleration of the aging process that lead to age-related diseases in humans such as Seckel syndrome [74]. Overexpression of BUBR1, a gene encoding a mitotic regulator involved in mitotic checkpoint fidelity, leads to protection against aneuploidy and cancer and overall extended life span when compared to wild-type mice [75]. Mutations in the nuclear lamina as causative factors in age-related diseases like the Hutchinson-Gilford progeria syndrome and the Nestor-Guillermo progeria syndrome have been widely studied [76, 77]. These contributing DNA damage-induced factors and their resulting pathologically altered mechanisms during stem cell dysfunction are highlighted in Fig. 2. Although these findings strongly support the hypothesis that errors in the nuclear structure lead to aging-related dysfunction, the mechanistic applications relating to stem cells are not yet completely understood.

9.4 Mitochondrial dysfunction

The mitochondria produce the energy necessary to run the large number of reactions that occur in the body, but along with producing energy, it also produces reactive oxygen species (ROS) as a by-product. ROS is extremely harmful to the body causing damage to DNA and disrupting bodily functions. The mitochondria are capable of dealing with ROS, but due to environmental factors and accumulated damage, the mitochondria undergo mitochondrial dysfunction (MD) [78]. MD overall is the reduction of mitochondrial function as a result of loss of maintenance and repair capabilities, the decline of critical metabolites entering the mitochondria, and the increase of ROS [79]. MD is caused by excessive exposure to many environmental pollutants such as heavy metals that result in increased mitochondrial oxidative stress and damage to the mitochondrial DNA [80]. The oxidative stress within the mitochondria leads to the accumulation of point mutations in mitochondrial DNA and altered copy number leading to defects in mitochondrial enzymes [81, 82]. The changes accelerate mitochondrial DNA mutagenesis and further induce MD that can be measured by different indicators such as reduction of tetrazolium salts or water-soluble tetrazolium derivatives or the increased production of lactate by cells that results in increased compensatory glycolytic rate [78].

MD causes stem cell dysfunction such as epigenetic changes, stem cell exhaustion, and cellular senescence that progresses the aging process. Although MD leads to decline in energy production, the major consequence of MD is the increased production of ROS [83]. The increased frequency of ROS leads to significantly increased damage in the nucleic acid, proteins, and lipids that disrupt cellular functions such as cell regeneration and self-renewal. DNA damage by ROS causes stem cells to cease proliferation, resulting in diseases such as cancer, diabetes, fatigue, and age-related neurological

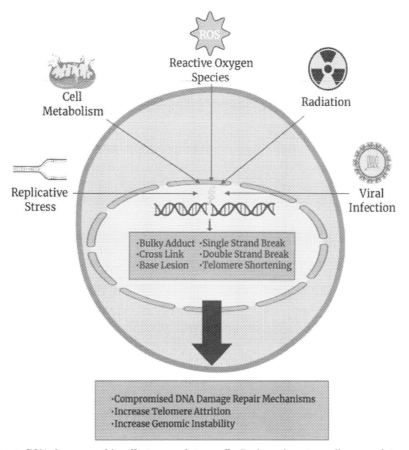

FIG. 2 Factors contributing to DNA damage and its effect on aged stem cells. During aging, stem cells accumulate nuclear DNA damage through common genomic stressors like replicative stress, accumulation of toxic metabolites like ROS, exposure to radiation, and viral infections. Exposure to such stressors results in compromised DNA damage repair mechanisms, an increase in telomere attrition, and ultimate genomic instability.

disorders [84, 85]. MD can also cause cell death. ROS released by interfibrillar mitochondria (IFM) with dysfunction complex 3 increases the stress on cardiovascular muscle fibers that result in increased susceptibility for infractions that lead to cell death [86]. MD also affects chemical signaling; in animal models, the increased ROS contributes to insulin resistance due to the activation of pro-inflammatory molecules like c-Jun N-terminal kinase (JNK) and inhibitor of nuclear factor kappa-B kinase subunit beta (IKK-β) that block the signaling ability of insulin receptor substrate 1 (IRS-1) [87, 88].

9.5 Proteostasis dysfunction

Similar to genome stability, the integrity of the stem cell proteome is equally important to maintain stem cell survival and function [89, 90]. This process collectively known as proteostasis comprises the pathways responsible for regulating the control of protein synthesis, folding, trafficking, aggregation, and degradation of proteins [89, 90]. Proteostasis is crucial for cellular functions such as genomic replication, maintenance of cellular structure, catalysis of metabolic reactions, and signaling in immune response pathways [90]. The integrity of proteins is maintained through a complex network that regulates the concentration, subcellular location, and folding of proteins. Proteins are also accompanied by molecular chaperones and folding enzymes to ensure proper protein folding, and when needed, protein degradation drives dysregulated proteins through pathways involved in lysosome and autophagy pathways [89–91]. Previous studies have reported the progressive deterioration of proteome during aging, emphasizing that the accumulation of misfolded or altered protein is a hallmark of the aging process [92]. Furthermore, the accumulation of these misregulated proteomes is associated with age-related pathologies such as Alzheimer's disease, Parkinson's disease, and Huntington's disease [90]. Some of the quality control mechanisms that have evolved over time are the heat shock family of proteins and the correction mechanisms responsible for proper elimination of misfolded proteins [90, 93]. There are also highly conserved regulators of age-related proteotoxicity, such as MOAG-4, that act independently of the known proteotoxicity regulatory pathways [94]. These pathways work in a systematic manner to restore proteome functionality or remove them entirely, to prevent the accumulation of these defective proteins. Several studies also suggest loss of function of chaperone-mediated folding during the progression of aging. The use of transgenic mouse, worm, and fly models overexpressing chaperones has shown a longer life span for these organisms [95, 96]. Marks of accelerated aging in mutant mice deficient in a co-chaperone heat shock protein were observed [97]. In *C. elegans*, longer life span was observed after the activation of the heat shock transcription factor HSP-1, while amyloid-binding compounds are observed to maintain protein homeostasis in vivo and in vitro [98, 99]. Similarly, the pharmacological induction of HSP72 was found to preserve muscle function and slow severe muscular dystrophy in mice [100]. It is important to note that these studies will vary greatly in complexity from organism to organism.

9.6 Epigenetics

Epigenetic modifiers and metabolism regulators have a significant impact on life span [53]. Among various epigenetic factors, chromatins and core histones (H2A, H2B, H3, and H4) regulate the genetic expression of age-associated genes. These factors can be depleted and remodeled with age that leads to an imbalance of histone modifications found commonly in all aging models [101]. Deletion of components from some of these histone methylation complexes, such as H3K4 and H3K27, was found to increase the life span of nematodes and flies [53, 102, 103]. For example, mutations on PRC2 subunits E(z) and esc in Drosophila led to increased resistance to oxidative stress and starvation. These mutants were also shown to reduce H3K27me3 levels, while trithorax (trx), on the other hand, was found to elevate histone 3 lysine 27 trimethylation (H3K27me3) and suppress their life span [103]. In an in vitro study, inhibition of histone methyltransferase G9A increased the self-renewal and division of HSPCs [104]. Other factors, such as cellular NAD+ and co-substrate of sirtuin deacetylases, elevate H4K16 acetylation to induce myogenic program in muscle stem cells when exiting quiescence [105]. In satellite cells during quiescence state, the SIRT1 deacetylase activity and NAD+ levels are elevated compared to activated skeletal muscle stem cells [106]. Metabolic and epigenetic modifications therefore contribute to stem cell aging by changing the expression of genes crucial for stem cell renewal. Although nucleosomes are constant structures established during early life, studies in transgenic mouse models of premature aging show epigenetic effects can subsequently cause disarray of global and local DNA methylation changes triggered by reduced histone 3 lysine 9 trimethylation (H3K9me3) levels and heterochromatin-associated proteins like HP1 [107–109]. Moreover, epigenetic changes in human adult stem cells have been attributed to alterations such as H3K9me3 global loss, centromeric heterochromatin decondensation, physical attrition of telomeres, and alterations in nucleolus organization [109–113]. Epigenetic profiling of young and old HSCs has detected variations in H3K4me3 levels across vital self-renewal genes as well as an increase in DNA methylation on differentiation genes. These changes enforce self-renewal capabilities and antagonize differentiation paralleling aging phenotype [114]. Another effect of aging on stem cell dysfunction can be noted when observing the effects of epigenetic regulation during chromatin condensation. A key subset of genes involved in the heterochromatin formation facilitates differentiation; however, disruption in these sites causes delayed differentiation and subsequent changes to cellular functionality [104]. A similar epigenetic signature was observed in human multipotent stem cells where a decreased expression of histone deacetylases (HDACs) coupled with subsequent downregulation of polycomb group genes like BMI1, EZH2, and SUZ12 and upregulation of JMJD3 was observed in senescent cells in vitro. These findings suggest that HDAC activity regulates genomic expression of genes governing cellular senescence [115]. The induction of senescence was also observed in MSCs during the dysregulation of methylation using 5-azacytidine or small interfering RNA [116]. In satellite cells, mapping of protein-DNA interactions using CHIP sequencing observed an increase in H3K27me3 and a decrease in H3K4me3 as aged satellite cells become quiescent [117]. The upregulation of H3K27me3 activity has been widely implicated in the downregulation of genes involved with stem cell function and maintenance of satellite cells. Thus, an enzymatic component, demethylase UTX, is shown to mediate muscle regeneration [118]. However, there is contradictory evidence that negates the significance of H3K27me3 regulators including JMJD3 and UTX in dictating the life span of organisms [119–121]. It is therefore important to keep in mind that model organisms vary in complexity and more studies are needed to determine their scientific relevance to human therapeutic application.

Epigenetic erosion is the decline of epigenetic regulation that results in stem cells losing the ability to differentiate and self-renew. Epigenetic erosion is at the epicenter of stem cell aging with both intrinsic and extrinsic factors further progressing epigenetic erosion in adult male stem cells [122]. Studies have shown that old animals become more epigenetically impaired, resulting in quiescent stem cells, which do not respond to activation signals efficiently [16, 122]. The effect of deregulation by epigenetic erosion can be seen in HSCs, where the genes for the self-renewal were downregulated by an increase in DNA methylation that occurs during aging. When DNMT3A, TET2, and ASXL1 genes are mutated in HSC, they contribute to mutation on the coding strand, as there is an overall loss of transcriptional regulation in older HSCs [16]. Similarly, muscle tissue is unable to regenerate due to overexpression of H4k16ac that causes the depletion of the stem cell pool that decreases the regenerative abilities in aging muscle [122]. In conclusion, epigenetic erosion affects stem cells through deregulation, causing stem cells to cease proliferating or increasing the frequency of dysfunction.

9.7 Changes in stem cell metabolic intake

Stem cells undergo many hurdles as they age that lead to the deterioration of tissues and reduce their regenerative capabilities. As stem cells age, they experience a decrease in metabolic intake, which is characterized by insulin resistance, changes in physiological decline in growth hormone and insulin-like growth factor (IGF), and sex steroids [123]. One of the hallmarks of aging is characterized by changes to stem cell metabolic intake due to deregulated nutrient sensing and mitochondrial dysfunction. The main drivers of insulin interference are cytokines secreted from visceral fat and pro-inflammatory cytokines from senescent cells, which are caused by fat tissue inflammation. In many in vitro studies, it was reported that differentiated 3T3-L1 adipocytes exhibited oxidative stress and expressed pro-inflammatory cytokines like monocyte chemoattractant protein-1 (MCP1), TNFα, and IL6 [124–128]. Many co-factors such as NAD+ are involved in epigenetics and play a role in metabolic intake in stem cells. Sirt1 plays a central role in regulating metabolism of various stem cells. In muscle stem cells, Sirt1 regulates the stem cell fate by afferent responses to the NAD+ levels of the cell [106]. More studies are needed to develop any therapeutic intervention that would not only promote redox homeostasis to deter mitochondrial dysfunction but will also promote metabolic intake.

9.8 Stem cell aging and gender

A careful glance at the life span of human supercentenarians reveals that over 95% of them are females [129]. This striking gender difference in predicting longevity was long known to the scientific community without any clear explanation as to why females live longer than males. The role of key sex steroids, namely estrogen and testosterone, which are predominant female and male hormones respectively, has been suggested to contribute to this sexual dimorphism of longevity. Estrogen supplementation to male mice led to significant increase in their life span [130]. On the other hand, studies with human eunuchs (castrated males) have shown that their average life span is about 14 years longer than non-castrated males [131]. Recent studies have also provided strong evidence related to the gender of an organism in regulating the stem cell behavior. Implantation of female stem cells displays better skeletal muscle regeneration than those taken from male mice [132]. Furthermore, HSCs are more abundant and proliferative in female mice compared to male mice [133]. A further increase in HSC population was observed during pregnancy [133], suggesting a role for estrogen-dependent mechanisms that may influence the ability of stem cells to regenerate. Estrogen also increases the proliferation of neuronal stem cells [134]. Muscle satellite cells obtained from females exhibit increased self-renewal and regeneration capacity compared to the male counterparts [132]. In addition, female stem cell population exhibited increased capacity for rapid wound healing and liver regeneration [132]. Bone-marrow stem cells isolated from non-human primates produce more neurogenic cells with increased capacity to contribute to reinnervation of damaged tissues [135]. Collectively, these evidences highlight the importance of gender differences in stem cell self-renewal, regeneration, and proliferation. Several potential mechanisms linking stem cell aging and gender differences have been suggested. Gender-specific regulation of DNA damage and ROS accumulation is one fascinating hypothesis to explain this link between stem cell aging and sex. Estrogen is known to induce antioxidant genes and reduce ROS, whereas testosterone is known to increase oxidative stress. Involvement of FOXO3, a transcription factor involved in stem cell maintenance of both HSC and NSCs by preventing premature stem cell depletion [136], is proposed to be another potential contributor. FOXO3 is known to interact with estrogen receptor alpha (ER-α) in an estrogen-dependent manner [137]. Another potential mechanism suggests the involvement of telomerases that are important for regenerative potential. Estrogen activates telomerase directly through ER-dependent transcription of TERT [138], suggesting that estrogen-enriched microenvironment could slow the aging process in females through the maintenance of telomerase. In conclusion, gender-associated differences in stem cell aging could be linked to sexual dimorphism in defining longevity and predisposition to age-related complications.

10 Common therapeutic approaches for the treatment of aging-associated stem cell dysfunction

10.1 Parabiosis

As the understanding of the effects of aging on stem cells becomes more developed, therapeutic interventions are being explored to counteract the effects of aging on stem cells. One possible therapeutic intervention is the introduction of major factors to aged systematic environments in a mouse model system known as parabiosis [139]. Parabiosis results in the exchange of cells and soluble factors through the shared vascular system from the younger organism to the older organism. The exchange revitalizes the older organism causing an increase in tissue regeneration and stem cell proliferation [140]. The results from this shared vascular system showed that factors from the younger systemic environment transferred to the older systemic environment had reactivated the molecular signaling pathways in hepatic, muscle, and neural stem cells. As a result, increased tissue regeneration and enhanced proliferation of hepatocytes and beta cells were observed in the older systemic environment [140, 141]. Soluble factors from the older systemic environment were transferred to the younger systemic environment, resulting in decreased synaptic plasticity and impaired spatial memory characterized by identifying the transferred factors that are circulating that have either pro-aging or de-aging effects on the organism [141, 142].

Parabiosis has been studied to develop therapeutic approaches for aging in muscle stem cells and has shown positive results in the rejuvenation of muscle stem cells. Satellite cells introduced from the youthful systemic environment also restored the genomic integrity in satellite cells from aged mice, indicating that parabiosis introduces factors from younger organisms that can reverse the effects of aging on muscle stem cells [72]. Introduction of GDF11, a member of the activin/TGF β superfamily of growth and differentiation factors, from the younger mice increased the number of satellite cells with intact DNA, and the number of satellite cells with damaged DNA decreased in aged mice [72, 143]. With GDF11 restored to youthful levels, the rate of muscle regeneration in aged mice was on a par with that of youthful mice in both rate of regeneration and muscle fiber density along with increased average exercise endurance and improved mitochondrial function [72, 144]. As shown in Fig. 3, parabiosis can introduce other factors that provide insight to creating therapeutic approaches toward the aging of other stem cells such as neural stem cells. By introducing factors from younger animals, it is possible to reactivate remyelination in aged organisms via the regenerative process that will restore saltatory conduction, prevent axonal degeneration, and promote functional recovery of the CNS in aged animals [145]. Introducing these factors reversed the effect of aging on CNS and restored the CNS to its youthful state capable of remyelination. Further experimentation with parabiosis will continue to show how aging can be reversed in aged organisms, making it a powerful tool toward the goal of discovering more therapeutic approaches for aging.

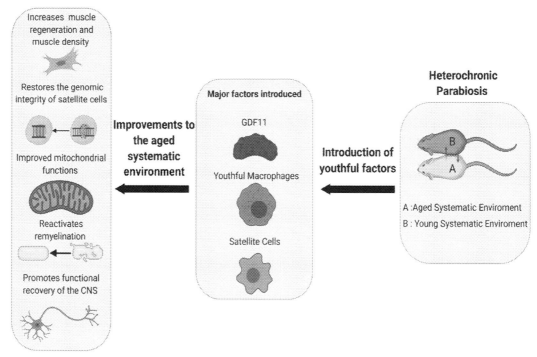

FIG. 3 Systemic rejuvenation by heterochronic parabiosis. The aged and young systematic environments share a joint vascular system in heterochronic parabiosis, resulting in the exchange of cells and soluble factors. Major factors like GDF11, youthful macrophages, and satellite cells from the young systematic environment reactivate cellular regeneration and improve system functions when introduced to the aged systematic environment.

10.2 Retrotransposons

Retrotransposons are mobile DNA elements that are commonly found in eukaryotic genomes and are thought to be activated as organisms age [146]. These DNA elements use reverse transcription to replicate their RNA transcripts and introduce their newly synthesized DNA copies in the genome. The genetic variation is increased using this mechanism to increase genomic copy numbers, which is restricted to the specific cell in question. The well-documented unrestricted mobility of these DNA elements is believed to cause genomic instability that leads to stem cell dysfunction [146]. These DNA elements make up 40% of the human genome and are referred to as transposable elements, which are viewed as the molecular parasites of the genome [146, 147]. Non-long terminal repeat (LTR) retrotransposons comprise approximately 20% of the human genome [148]. L1 retrotransposons, also known as long interspersed nuclear elements (LINE-1) gene products, were found to promote the risk of genomic instability, eroding the double-strand break repair mechanism and DNA damage [9, 148–150]. It was reported that L1 activities were suppressed by the SIRT6 expression [9, 149]. A mouse model study on premature aging usually found in Cockayne syndrome led to the identification of SIRT6 as one of the prominent candidates for treating accelerated aging. In this study, SIRT6 expression was upregulated following activation of PARP1, which is usually associated with DNA damage [151].

Retrotransposons remain inactive in the adult life of the organism but are active in embryonic stem cells (ESCs) and have a role in the development of embryos. Retrotransposons are expressed in distinct patterns in both mouse and human ESCs cultures and have a key impact on the reprogramming of cells [152]. There are three methods in which retrotransposons reprogram the cell and minimize the risk of damaging the genome integrity of the cell. The first method of programming retrotransposon binds to the host ESCs and provides a pool of regulatory sequences that can act as novel enhancers and promoters [152, 153]. The second method involves the retrotransposon long noncoding RNA (lncRNA) in the production of noncoding RNA that controls the pluripotency in early development. The third method involves the role of retroviral proteins in reproduction and development by providing genome defense [152]. The activation of retrotransposons in ESCs makes it possible to change the genome, leading to the evolution of organisms without risking damage to genome integrity.

As activated retrotransposons may play a role in gene expression, retrotransposon repression may also play a role in shaping gene regulation in critical events of development in embryonic stem cells [152]. There are various mechanisms utilizing RNAi and histone methyltransferases where the epigenetic modification of certain retrotransposon targets like KRAB-ZFP and KAP1 that are repressed may result in an increase of differentiation in ESCs [152, 154]. However, their progressive expression during aging could be a by-product of general decline of cellular and mechanistic function of regulatory mechanisms behind the repression of retrotransposons. In fact, they may further contribute to age-related cellular dysfunction. While an inverse relationship between aging and increased expression of retrotransposons is clear, the direct relationship between retrotransposons and specific aging phenotypes like increased DNA damage is not clearly understood. Analyses of these DNA elements may serve as a useful marker for aging, but exact mechanisms that lead to its expression and subsequent exacerbation of aging are yet to be understood [146]. Retrotransposons often cause genomic instability and mutations to affect many cellular functions during aging that subsequently contribute to stem cell dysfunction. They are found to be associated with the senescence of human adipose-derived mesenchymal stem cells (hADMSCs) via specific activation of Alu retrotransposons that cause continual DNA damage [155].

10.3 Cellular reprogramming of induced pluripotent stem cells (iPSCs)

Stem cell dysfunction is caused by numerous factors that lead to cellular senescence and exhaustion. One therapeutic method to rejuvenate stem cells for the treatment against the effects of aging is cellular reprogramming toward induced pluripotent stem cells (iPSCs). iPSCs are a subset of pluripotent stem cells that are derived from somatic adult cells that are rejuvenated into a stem cell. It is used against senescent cells because numerous studies have reported that telomere elongation and reduced oxidative stress were reduced [9]. However, the pathway to reprogram centenarian and senescent cells to be rejuvenated into a functional pluripotent state is a huge impediment [156]. One of the small hurdles with reprogramming somatic adult cells to rejuvenation by iPSCs is the promotion of replicative stress. However, it is usually overcome by the physiological changes the cell undergoes during the rejuvenation process [157]. A generation of iPSCs derived from centenarian senescent fibroblasts was accomplished utilizing a gene cocktail therapy [158]. The senescent fibroblasts extracted from centenarian subjects were treated with a six-gene combination therapy via lentivirus to successfully generate iPSCs and reinstate self-renewal capacity, pluripotency, mitochondria metabolism, increased telomere length, and rejuvenated cell physiology [158]. These improved functions suggest iPSCs to be of similar physiological state akin to young stem cells [159]. Increased telomere length found in IPSCs and rejuvenated adult stem cells helps explore any therapeutic opportunities in lengthening telomeres to treat stem cell exhaustion and senescent cells to increase the longevity of stem cells.

10.4 Telomere lengthening

Telomeres are special heterochromatic structures located at the end of chromosomes composed of tandem TTAGGG nucleotide repeats and an associated protein complex called shelterin [160]. The shelterin complex is characterized by the presence of a G strand, which rearranges to form a T-loop structure thought to be essential for chromosomal stability by working to protect chromosomes from end-to-end fusions, degradation, and DNA repair and telomerase activities [161, 162]. During early fetal development telomerase, a cellular reverse transcriptase makes de novo additions of the previously mentioned nucleotide repeats onto the telomeres at the end of chromosomes and is silenced in most tissues at about 3–4 months gestation [163]. With each cell division, these TTAGGG repeats shorten due to the incomplete replication of DNA polymerases, in a phenomenon known as the end-replication problem [164, 165]. Telomerase is known to be active in adult stem cell compartments; however, it is not enough to counter telomere attrition and thus results in telomere shortening as organisms age [166, 167]. The consequent shortening of telomeres can lead to the impaired tissue and self-renewal capacity of these stem cell compartments and is proposed as one of the molecular hallmarks of disease [53]. This shortening has been observed in various mouse stem cell compartments independent of tissue proliferative rate [167]. Once a cell has reached a critically shortened state, it can proceed to growth arrest where DNA damage mechanisms will first attempt to repair the damage and trigger replicative senescence [168, 169]. Over time, cells can remain in this senescent state and have the potential to secrete factors that influence age-associated diseases. In fact, it is believed that senescence serves as a tumor suppressor mechanism by limiting the rate of division that leads to oncogenic mutations as aging progresses in organisms with longer life spans [170]. To study the effects of telomere lengthening during aging, many pharmacological tools have been employed. For example, TA-65 is a small molecule used as a telomerase activator, which has shown the potential to lengthen telomeres and improve aging outcomes in mice and humans [171, 172]. Most interesting is the observation of telomerase reactivation in aged mice deficient in telomerase activity. The studies indicated that late-life reactivation of mTERC did not only lead to arrest of the degenerative phenotype, but reversed it in these genetically modified mice. Analysis of various cell compartments showed reduced DNA damage-associated signaling and even proliferative capacity of quiescent cells. The most noteworthy observation is seen in the reversal of neurodegeneration where an increase in proliferation of SOX2+ neural progenitors, DCX+ newborn neurons, and OLIG2+ oligodendrocyte populations was seen, suggesting the rejuvenation potential of telomere reactivation [173]. Although telomere reactivation has the potential to rescue telomere attrition in stem cell compartments, it remains independent of genetic or epigenetic alterations that could produce malignant neoplastic cell types. Overexpression of telomerase is used by cancer cells to overcome replicative senescence and proceed to unchecked replication [174–176].

10.5 Caloric restriction

Caloric restriction (CR) was shown to be one of the most longevity-extending interventions in several species by efficiently delaying age-related pathologies. It has been demonstrated that CR preserves stem cell number, function, or both in a variety of tissues. CR-induced physiological cues are reported to influence stem cell biology and function through both intrinsic and extrinsic regulatory mechanisms [133, 177]. Short-term CR improved stem cell frequency and led to improved muscle regeneration after injury as measured by FACS analysis of Pax7-expressing satellite cells (SCs) [177]. Significant increase in mitochondrial biogenesis and oxygen consumption rate (OCR) in CR-treated SC suggests alterations in mitochondrial biogenesis as a possible mechanism for the observed increase in myogenic function. Furthermore, increased expression of longevity and metabolic regulators SirT1 and FOXO3a was upregulated in young and aged-CR treated mice [177]. The same group further demonstrated that transplantation of GFP-labeled SC from ad libitum-fed mice to CR-treated mice displayed significantly greater engraftment of myofibers, possibly due to improved survival of transplanted cells in CR muscles compared to the control muscle group. CR improves the function of several stem cell populations, including HSCs and GSCs in mice [13, 178]. Rad2/cohesin, a critical mediator of NF-kB signaling, is known to limit HSC self-renewal in aging [13]. The use of long-term caloric restriction as an intervention to slow down aging-related HSC dysfunction was expected to have beneficial effects on the improvement of the HSC population; however, it also conversely also resulted in the impairment of lymphoid differentiation capacity and immune function [179]. Concurrent with extending life span, CR attenuates age-related decline in male GSC numbers in Drosophila [178]. CR also promotes ISC self-renewal by the induction of bone stromal antigen 1 (Bst1) in Paneth cells, the key constituents of mammalian ISC niche [180]. Although it is not evident why CR would increase ISC numbers, it is possible that CR favorably shifts the balance toward self-renewal of ISC by reducing the proliferative pool.

11 Conclusions

Stem cells are often considered as "fountain of youth" that typically self-renew and differentiate into multiple tissues within a growing adult. Although stem cells remain active into old age, changes in the stem cells and their microenvironments inhibit their regenerative potential. This self-renewal capacity of stem cells is susceptible to age-related functional damages that not only contribute to degeneration and dysfunction of aging tissues but also affect the life span. A better understanding of both the cell-intrinsic stem cell changes and concomitant changes to the stem cell niche and the systemic environment is crucial for the development of regenerative medicine strategies that can significantly reverse degenerative changes during aging. Several potential mechanisms are believed to contribute to the aging-associated dysfunction of stem cells. Common pathways associated with these stem cell dysfunctions include DNA damage, mitochondrial dysfunction, depletion and senescence of stem cells, accumulation of ROS in aged stem cells, defects in proteostasis that is fundamental to the maintenance of stem cell, and aging-associated changes in systemic factors among others. Over recent decades, significant advancements have been made to develop therapeutic interventions to prevent or delay aging and age-associated pathology. Several stem cell-based therapies are shown to be effective in animal models. As senescence and age-related status of the grafted stem cells are important biological factors for these stem cell-based therapies, it is crucial to use careful and advanced analysis of the stem cell grafts. Supplementation of recombinant growth and differentiation factor 11 (GDF11) in aged mice led to significant reversal of satellite cell dysfunction and restored its regenerative function [72]. Development of such rejuvenating interventions, which manipulate blood-borne factors to reverse stem cell aging and extend healthy life, has generated significant excitement. With a significant increase in aging population and aging-associated diseases worldwide, the need for effective regenerative medicine strategies is more important than ever. Recent technical advancements in stem cell and regenerative medicine continue to provide a better understanding of aging-associated stem cell dysfunction and allow developing more effective therapies and diagnostic technologies to treat aged patients. Further research on better understanding of the cellular mechanisms contributing to the aging of stem cells will provide key information related to cell-based therapies and pave the way to improve healthy aging.

Acknowledgments

The authors acknowledge the financial supports from NIH grants SC1AG049682 (RS), TRDRP T31IP1551 (RS), SC1CA232319 (SP), NIMHD S21 MD000103 and AXIS-U54MD007598 (Charles R Drew University), and NIGMS RISE R25GM62252 (California State University, Dominguez Hills).

References

[1] S. Amarya, K. Singh, M. Sabharwal, Ageing process and physiological changes, in: G. D'Onofrio, A. Greco, D. Sancarlo (Eds.), Gerontology, InTech, 2018, https://doi.org/10.5772/intechopen.76249.

[2] E. Jaul, J. Barron, Age-related diseases and clinical and public health implications for the 85 years old and over population, Front. Public Health 5 (2017) 335, https://doi.org/10.3389/fpubh.2017.00335.

[3] Q.-L. Xue, B.A. Beamer, P.H.M. Chaves, J.M. Guralnik, L.P. Fried, Heterogeneity in rate of decline in grip, hip, and knee strength and the risk of all-cause mortality: the Women's health and aging study II, J. Am. Geriatr. Soc. 58 (2010) 2076–2084, https://doi.org/10.1111/j.1532-5415.2010.03154.x.

[4] H. Gleckman, The Federal Government Will Spend Half Its Budget On Older Adults In Ten Years, Forbes, 2019. https://www.forbes.com/sites/howardgleckman/2019/02/01/the-federal-government-will-spend-half-its-budget-on-older-adults-in-ten-years/#109e289b56b6.

[5] M.A. Goodell, T.A. Rando, Stem cells and healthy aging, Science 350 (2015) 1199–1204, https://doi.org/10.1126/science.aab3388.

[6] T.H. Cheung, T.A. Rando, Molecular regulation of stem cell quiescence, Nat. Rev. Mol. Cell Biol. 14 (2013) 329–340, https://doi.org/10.1038/nrm3591.

[7] D.L. Jones, T.A. Rando, Emerging models and paradigms for stem cell ageing, Nat. Cell Biol. 13 (2011) 506–512, https://doi.org/10.1038/ncb0511-506.

[8] D.R. Bell, G. Van Zant, Stem cells, aging, and cancer: inevitabilities and outcomes, Oncogene 23 (2004) 7290–7296, https://doi.org/10.1038/sj.onc.1207949.

[9] A.S.I. Ahmed, M.H. Sheng, S. Wasnik, D.J. Baylink, K.-H.W. Lau, Effect of aging on stem cells, World J. Exp. Med. 7 (2017) 1–10, https://doi.org/10.5493/wjem.v7.i1.1.

[10] D. Lee, A. Bareja, D. Bartlett, J. White, Autophagy as a therapeutic target to enhance aged muscle regeneration, Cell 8 (2019) 183, https://doi.org/10.3390/cells8020183.

[11] Y. Kanfi, S. Naiman, G. Amir, V. Peshti, G. Zinman, L. Nahum, Z. Bar-Joseph, H.Y. Cohen, The sirtuin SIRT6 regulates lifespan in male mice, Nature 483 (2012) 218–221, https://doi.org/10.1038/nature10815.

[12] L. García-Prat, P. Muñoz-Cánoves, Aging, metabolism and stem cells: spotlight on muscle stem cells, Mol. Cell. Endocrinol. 445 (2017) 109–117, https://doi.org/10.1016/j.mce.2016.08.021.

[13] J. Chen, C.M. Astle, D.E. Harrison, Hematopoietic senescence is postponed and hematopoietic stem cell function is enhanced by dietary restriction, Exp. Hematol. 31 (2003) 1097–1103, https://doi.org/10.1016/s0301-472x(03)00238-8.

[14] S.J. Morrison, A.M. Wandycz, K. Akashi, A. Globerson, I.L. Weissman, The aging of hematopoietic stem cells, Nat. Med. 2 (1996) 1011–1016, https://doi.org/10.1038/nm0996-1011.

[15] I. Beerman, D. Bhattacharya, S. Zandi, M. Sigvardsson, I.L. Weissman, D. Bryder, D.J. Rossi, Functionally distinct hematopoietic stem cells modulate hematopoietic lineage potential during aging by a mechanism of clonal expansion, Proc. Natl. Acad. Sci. U. S. A. 107 (2010) 5465–5470, https://doi.org/10.1073/pnas.1000834107.

[16] S.M. Chambers, C.A. Shaw, C. Gatza, C.J. Fisk, L.A. Donehower, M.A. Goodell, Aging hematopoietic stem cells decline in function and exhibit epigenetic dysregulation, PLoS Biol. 5 (2007), https://doi.org/10.1371/journal.pbio.0050201, e201.

[17] H. Jasper, Exploring the physiology and pathology of aging in the intestine of Drosophila melanogaster, Invertebr. Reprod. Dev. 59 (2015) 51–58, https://doi.org/10.1080/07924259.2014.963713.

[18] M.B. Schultz, D.A. Sinclair, When stem cells grow old: phenotypes and mechanisms of stem cell aging, Development 143 (2016) 3–14, https://doi.org/10.1242/dev.130633.

[19] B. Biteau, C.E. Hochmuth, H. Jasper, JNK activity in somatic stem cells causes loss of tissue homeostasis in the aging Drosophila gut, Cell Stem Cell 3 (2008) 442–455, https://doi.org/10.1016/j.stem.2008.07.024.

[20] A. Tarnawski, D. Hollander, H. Cergely, Cytoprotective drugs. Focus on essential fatty acids and sucralfate, Scand. J. Gastroenterol. Suppl. 127 (1987) 39–43, https://doi.org/10.3109/00365528709090949.

[21] L. Guarente, Mitochondria—a Nexus for aging, calorie restriction, and sirtuins? Cell 132 (2008) 171–176, https://doi.org/10.1016/j.cell.2008.01.007.

[22] A. Merlos-Suárez, F.M. Barriga, P. Jung, M. Iglesias, M.V. Céspedes, D. Rossell, M. Sevillano, X. Hernando-Momblona, V. da Silva-Diz, P. Muñoz, H. Clevers, E. Sancho, R. Mangues, E. Batlle, The intestinal stem cell signature identifies colorectal Cancer stem cells and predicts disease relapse, Cell Stem Cell 8 (2011) 511–524, https://doi.org/10.1016/j.stem.2011.02.020.

[23] R. Lehmann, Germline stem cells: origin and destiny, Cell Stem Cell 10 (2012) 729–739, https://doi.org/10.1016/j.stem.2012.05.016.

[24] C. Paul, M. Nagano, B. Robaire, Aging results in molecular changes in an enriched population of undifferentiated rat spermatogonia, Biol. Reprod. 89 (2013), https://doi.org/10.1095/biolreprod.113.112995.

[25] C.E. Dunlop, E.E. Telfer, R.A. Anderson, Ovarian stem cells—potential roles in infertility treatment and fertility preservation, Maturitas 76 (2013) 279–283, https://doi.org/10.1016/j.maturitas.2013.04.017.

[26] Y.A.R. White, D.C. Woods, Y. Takai, O. Ishihara, H. Seki, J.L. Tilly, Oocyte formation by mitotically active germ cells purified from ovaries of reproductive-age women, Nat. Med. 18 (2012) 413–421, https://doi.org/10.1038/nm.2669.

[27] H.-J. Hsu, D. Drummond-Barbosa, Insulin levels control female germline stem cell maintenance via the niche in Drosophila, Proc. Natl. Acad. Sci. U. S. A. 106 (2009) 1117–1121, https://doi.org/10.1073/pnas.0809144106.

[28] S. Bortoli, V. Renault, E. Eveno, C. Auffray, G. Butler-Browne, G. Piétu, Gene expression profiling of human satellite cells during muscular aging using cDNA arrays, Gene 321 (2003) 145–154, https://doi.org/10.1016/j.gene.2003.08.025.

[29] A.S. Brack, P. Muñoz-Cánoves, The ins and outs of muscle stem cell aging, Skelet. Muscle 6 (2016), https://doi.org/10.1186/s13395-016-0072-z.

[30] S.J. Tapscott, The circuitry of a master switch: Myod and the regulation of skeletal muscle gene transcription, Development 132 (2005) 2685–2695, https://doi.org/10.1242/dev.01874.

[31] E. Horgusluoglu, K. Nudelman, K. Nho, A.J. Saykin, Adult neurogenesis and neurodegenerative diseases: a systems biology perspective, Am. J. Med. Genet. B Neuropsychiatr. Genet. 174 (2017) 93–112, https://doi.org/10.1002/ajmg.b.32429.

[32] H. Kuhn, H. Dickinson-Anson, F. Gage, Neurogenesis in the dentate gyrus of the adult rat: age-related decrease of neuronal progenitor proliferation, J. Neurosci. 16 (1996) 2027–2033, https://doi.org/10.1523/JNEUROSCI.16-06-02027.1996.

[33] T. Toda, S. Parylak, S.B. Linker, F.H. Gage, The role of adult hippocampal neurogenesis in brain health and disease, Mol. Psychiatry 24 (2019) 67–87, https://doi.org/10.1038/s41380-018-0036-2.

[34] G. Kasper, L. Mao, S. Geissler, A. Draycheva, J. Trippens, J. Kühnisch, M. Tschirschmann, K. Kaspar, C. Perka, G.N. Duda, J. Klose, Insights into mesenchymal stem cell aging: involvement of antioxidant defense and actin cytoskeleton, Stem Cells 27 (2009) 1288–1297, https://doi.org/10.1002/stem.49.

[35] M.A. Baxter, R.F. Wynn, S.N. Jowitt, J.E. Wraith, L.J. Fairbairn, I. Bellantuono, Study of telomere length reveals rapid aging of human marrow stromal cells following in vitro expansion, Stem Cells 22 (2004) 675–682, https://doi.org/10.1634/stemcells.22-5-675.

[36] J. Boulestreau, M. Maumus, P. Rozier, C. Jorgensen, D. Noël, Mesenchymal stem cell derived extracellular vesicles in aging, Front. Cell Dev. Biol. 8 (2020), https://doi.org/10.3389/fcell.2020.00107.

[37] Y. Hong, H. He, G. Jiang, H. Zhang, W. Tao, Y. Ding, D. Yuan, J. Liu, H. Fan, F. Lin, X. Liang, X. Li, Y. Zhang, miR-155-5p inhibition rejuvenates aged mesenchymal stem cells and enhances cardioprotection following infarction, Aging Cell (2020), e13128. n/a https://doi.org/10.1111/acel.13128.

[38] J.-S. Park, H.-Y. Kim, H.-W. Kim, G.-N. Chae, H.-T. Oh, J.-Y. Park, H. Shim, M. Seo, E.-Y. Shin, E.-G. Kim, S.C. Park, S.-J. Kwak, Increased caveolin-1, a cause for the declined adipogenic potential of senescent human mesenchymal stem cells, Mech. Ageing Dev. 126 (2005) 551–559, https://doi.org/10.1016/j.mad.2004.11.014.

[39] J.M. Heaton, The distribution of brown adipose tissue in the human, J. Anat. 112 (1972) 35–39. https://www.ncbi.nlm.nih.gov/pmc/articles/PMC1271341/. (Accessed 29 March 2020).

[40] R.B. McDonald, B.A. Horwitz, Brown adipose tissue thermogenesis during aging and senescence, J. Bioenerg. Biomembr. 31 (1999) 507–516, https://doi.org/10.1023/a:1005404708710.

[41] N.H. Rogers, A. Landa, S. Park, R.G. Smith, Aging leads to a programmed loss of brown adipocytes in murine subcutaneous white adipose tissue, Aging Cell 11 (2012) 1074–1083, https://doi.org/10.1111/acel.12010.

[42] J.L. Kirkland, T. Tchkonia, T. Pirtskhalava, J. Han, I. Karagiannides, Adipogenesis and aging: does aging make fat go MAD? Exp. Gerontol. 37 (2002) 757–767, https://doi.org/10.1016/s0531-5565(02)00014-1.

[43] M. Florez-Duquet, R.B. McDonald, Cold-induced thermoregulation and biological aging, Physiol. Rev. 78 (1998) 339–358, https://doi.org/10.1152/physrev.1998.78.2.339.

[44] V.C. Khanh, A.F. Zulkifli, C. Tokunaga, T. Yamashita, Y. Hiramatsu, O. Ohneda, Aging impairs beige adipocyte differentiation of mesenchymal stem cells via the reduced expression of Sirtuin 1, Biochem. Biophys. Res. Commun. 500 (2018) 682–690, https://doi.org/10.1016/j.bbrc.2018.04.136.

[45] D.C. Berry, Y. Jiang, R.W. Arpke, E.L. Close, A. Uchida, D. Reading, E.D. Berglund, M. Kyba, J.M. Graff, Cellular aging contributes to failure of cold-induced beige adipocyte formation in old mice and humans, Cell Metab. 25 (2017) 166–181, https://doi.org/10.1016/j.cmet.2016.10.023.

[46] K.A. Steinkraus, M. Kaeberlein, B.K. Kennedy, Replicative aging in yeast, Annu. Rev. Cell Dev. Biol. 24 (2008) 29–54, https://doi.org/10.1146/annurev.cellbio.23.090506.123509.

[47] P. Fabrizio, V.D. Longo, The chronological life span of Saccharomyces cerevisiae, Aging Cell 2 (2003) 73–81, https://doi.org/10.1046/j.1474-9728.2003.00033.x.

[48] K.A. Waters, V. Reinke, Extrinsic and intrinsic control of germ cell proliferation in Caenorhabditis elegans, Mol. Reprod. Dev. 78 (2011) 151–160, https://doi.org/10.1002/mrd.21289.

[49] N. Arantes-Oliveira, J. Apfeld, A. Dillin, C. Kenyon, Regulation of life-span by germ-line stem cells in *Caenorhabditis elegans*. (Reports), Science 295 (2002) 502. http://link.gale.com/apps/doc/A82554065/AONE?u=csudh&sid=zotero&xid=2c2b8035. (Accessed 29 March 2020).

[50] D.T. Fox, A.C. Spradling, The Drosophila hindgut lacks constitutively active adult stem cells but proliferates in response to tissue damage, Cell Stem Cell 5 (2009) 290–297, https://doi.org/10.1016/j.stem.2009.06.003.

[51] Y. Zhou, T. Lu, T. Xie, A PGC-1 tale: healthier intestinal stem cells, longer life, Cell Metab. 14 (2011) 571–572, https://doi.org/10.1016/j.cmet.2011.10.005.

[52] M. Nászai, L.R. Carroll, J.B. Cordero, Intestinal stem cell proliferation and epithelial homeostasis in the adult Drosophila midgut, Insect Biochem. Mol. Biol. 67 (2015) 9–14, https://doi.org/10.1016/j.ibmb.2015.05.016.

[53] C. López-Otín, M.A. Blasco, L. Partridge, M. Serrano, G. Kroemer, The hallmarks of aging, Cell 153 (2013) 1194–1217, https://doi.org/10.1016/j.cell.2013.05.039.

[54] R. Gruber, H. Koch, B.A. Doll, F. Tegtmeier, T.A. Einhorn, J.O. Hollinger, Fracture healing in the elderly patient, Exp. Gerontol. 41 (2006) 1080–1093, https://doi.org/10.1016/j.exger.2006.09.008.

[55] J. Oh, Y.D. Lee, A.J. Wagers, Stem cell aging: mechanisms, regulators and therapeutic opportunities, Nat. Med. 20 (2014) 870–880, https://doi.org/10.1038/nm.3651.

[56] J.V. Chakkalakal, K.M. Jones, M.A. Basson, A.S. Brack, The aged niche disrupts muscle stem cell quiescence, Nature 490 (2012) 355–360, https://doi.org/10.1038/nature11438.

[57] E. Owusu-Ansah, U. Banerjee, Reactive oxygen species prime Drosophila haematopoietic progenitors for differentiation, Nature 461 (2009) 537–541, https://doi.org/10.1038/nature08313.

[58] P. Rožman, K. Jazbec, M. Jež, Stem cell aging, in: R. Sharma (Ed.), Stem Cells in Clinical Practice and Tissue Engineering, InTech, 2018, https://doi.org/10.5772/intechopen.71764.

[59] A.M. Josephson, V. Bradaschia-Correa, S. Lee, K. Leclerc, K.S. Patel, E. Muinos Lopez, H.P. Litwa, S.S. Neibart, M. Kadiyala, M.Z. Wong, M.M. Mizrahi, N.L. Yim, A.J. Ramme, K.A. Egol, P. Leucht, Age-related inflammation triggers skeletal stem/progenitor cell dysfunction, Proc. Natl. Acad. Sci. U. S. A. 116 (2019) 6995–7004, https://doi.org/10.1073/pnas.1810692116.

[60] C. Franceschi, S. Salvioli, P. Garagnani, M. de Eguileor, D. Monti, M. Capri, Immunobiography and the heterogeneity of immune responses in the elderly: a focus on Inflammaging and trained immunity, Front. Immunol. 8 (2017) 982, https://doi.org/10.3389/fimmu.2017.00982.

[61] T. Fulop, A. Larbi, G. Dupuis, A. Le Page, E.H. Frost, A.A. Cohen, J.M. Witkowski, C. Franceschi, Immunosenescence and Inflamm-aging as two sides of the same coin: friends or foes? Front. Immunol. 8 (2018) 1960, https://doi.org/10.3389/fimmu.2017.01960.

[62] H.-O. Byun, Y.-K. Lee, J.-M. Kim, G. Yoon, From cell senescence to age-related diseases: differential mechanisms of action of senescence-associated secretory phenotypes, BMB Rep. 48 (2015) 549–558, https://doi.org/10.5483/BMBRep.2015.48.10.122.

[63] E. Gibon, L. Lu, S.B. Goodman, Aging, inflammation, stem cells, and bone healing, Stem Cell Res. Ther. 7 (2016) 44, https://doi.org/10.1186/s13287-016-0300-9.

[64] A. Bektas, S.H. Schurman, R. Sen, L. Ferrucci, Human T cell immunosenescence and inflammation in aging, J. Leukoc. Biol. 102 (2017) 977–988, https://doi.org/10.1189/jlb.3RI0716-335R.

[65] A. Bratic, N.-G. Larsson, The role of mitochondria in aging, J. Clin. Invest. 123 (2013) 951–957, https://doi.org/10.1172/JCI64125.

[66] S.P. Jackson, J. Bartek, The DNA-damage response in human biology and disease, Nature 461 (2009) 1071–1078, https://doi.org/10.1038/nature08467.

[67] M. Valko, C.J. Rhodes, J. Moncol, M. Izakovic, M. Mazur, Free radicals, metals and antioxidants in oxidative stress-induced cancer, Chem. Biol. Interact. 160 (2006) 1–40, https://doi.org/10.1016/j.cbi.2005.12.009.

[68] K.K. Khanna, S.P. Jackson, DNA double-strand breaks: signaling, repair and the cancer connection, Nat. Genet. 27 (2001) 247–254, https://doi.org/10.1038/85798.

[69] V. Gorbunova, A. Seluanov, Z. Mao, C. Hine, Changes in DNA repair during aging, Nucleic Acids Res. 35 (2007) 7466–7474, https://doi.org/10.1093/nar/gkm756.

[70] I. Beerman, J. Seita, M.A. Inlay, I.L. Weissman, D.J. Rossi, Quiescent hematopoietic stem cells accumulate DNA damage during aging that is repaired upon entry into cell cycle, Cell Stem Cell 15 (2014) 37–50, https://doi.org/10.1016/j.stem.2014.04.016.

[71] C.E. Rübe, A. Fricke, T.A. Widmann, T. Fürst, H. Madry, M. Pfreundschuh, C. Rübe, Accumulation of DNA damage in hematopoietic stem and progenitor cells during human aging, PLoS One 6 (2011), https://doi.org/10.1371/journal.pone.0017487, e17487.

[72] M. Sinha, Y.C. Jang, J. Oh, D. Khong, E.Y. Wu, R. Manohar, C. Miller, S.G. Regalado, F.S. Loffredo, J.R. Pancoast, M.F. Hirshman, J. Lebowitz, J.L. Shadrach, M. Cerletti, M.-J. Kim, T. Serwold, L.J. Goodyear, B. Rosner, R.T. Lee, A.J. Wagers, Restoring systemic GDF11 levels reverses age-related dysfunction in mouse skeletal muscle, Science 344 (2014) 649–652, https://doi.org/10.1126/science.1251152.

[73] A.A. Freitas, J.P. de Magalhães, A review and appraisal of the DNA damage theory of ageing, Mutat. Res. 728 (2011) 12–22, https://doi.org/10.1016/j.mrrev.2011.05.001.

[74] M. Murga, S. Bunting, M.F. Montaña, R. Soria, F. Mulero, M. Cañamero, Y. Lee, P.J. McKinnon, A. Nussenzweig, O. Fernandez-Capetillo, A mouse model of ATR-Seckel shows embryonic replicative stress and accelerated aging, Nat. Genet. 41 (2009) 891–898, https://doi.org/10.1038/ng.420.

[75] D.J. Baker, M.M. Dawlaty, T. Wijshake, K.B. Jeganathan, L. Malureanu, J.H. van Ree, J. Crespo-Diaz, S. Reyes, L. Seaburg, V. Shapiro, A. Behfar, A. Terzic, B. van de Sluis, J.M. van Deursen, Increased expression of BubR1 protects against aneuploidy and cancer and extends healthy lifespan, Nat. Cell Biol. 15 (2013) 96–102, https://doi.org/10.1038/ncb2643.

[76] M. Eriksson, W.T. Brown, L.B. Gordon, M.W. Glynn, J. Singer, L. Scott, M.R. Erdos, C.M. Robbins, T.Y. Moses, P. Berglund, A. Dutra, E. Pak, S. Durkin, A.B. Csoka, M. Boehnke, T.W. Glover, F.S. Collins, Recurrent de novo point mutations in lamin A cause Hutchinson-Gilford progeria syndrome, Nature 423 (2003) 293–298, https://doi.org/10.1038/nature01629.

[77] R. Cabanillas, J. Cadiñanos, J.A.F. Villameytide, M. Pérez, J. Longo, J.M. Richard, R. Álvarez, N.S. Durán, R. Illán, D.J. González, C. López-Otín, Néstor-Guillermo progeria syndrome: a novel premature aging condition with early onset and chronic development caused by BANF1 mutations, Am. J. Med. Genet. 155 (2011) 2617–2625, https://doi.org/10.1002/ajmg.a.34249.

[78] M.D. Brand, D.G. Nicholls, Assessing mitochondrial dysfunction in cells, Biochem. J. 435 (2011) 297–312, https://doi.org/10.1042/BJ20110162.

[79] G.L. Nicolson, Mitochondrial dysfunction and chronic disease: treatment with natural supplements, Integr. Med. (Encinitas) 13 (2014) 35–43. https://www.ncbi.nlm.nih.gov/pmc/articles/PMC4566449/. (Accessed 23 May 2020).

[80] G. Jia, A.R. Aroor, L.A. Martinez-Lemus, J.R. Sowers, Mitochondrial functional impairment in response to environmental toxins in the cardiorenal metabolic syndrome, Arch. Toxicol. 89 (2015) 147–153, https://doi.org/10.1007/s00204-014-1431-3.

[81] K.J. Ahlqvist, A. Suomalainen, R.H. Hämäläinen, Stem cells, mitochondria and aging, Biochim. Biophys. Acta 1847 (2015) 1380–1386, https://doi.org/10.1016/j.bbabio.2015.05.014.

[82] C.-C. Hsu, L.-M. Tseng, H.-C. Lee, Role of mitochondrial dysfunction in cancer progression, Exp. Biol. Med. (Maywood) 241 (2016) 1281–1295, https://doi.org/10.1177/1535370216641787.

[83] R.K. Lane, T. Hilsabeck, S.L. Rea, The role of mitochondrial dysfunction in age-related diseases, Biochim. Biophys. Acta 1847 (2015) 1387–1400, https://doi.org/10.1016/j.bbabio.2015.05.021.

[84] M. Khacho, A. Clark, D.S. Svoboda, J.G. MacLaurin, D.C. Lagace, D.S. Park, R.S. Slack, Mitochondrial dysfunction underlies cognitive defects as a result of neural stem cell depletion and impaired neurogenesis, Hum. Mol. Genet. 26 (2017) 3327–3341, https://doi.org/10.1093/hmg/ddx217.

[85] M. Modanloo, M. Shokrzadeh, Analyzing mitochondrial dysfunction, oxidative stress, and apoptosis: potential role of L-carnitine, Iran. J. Kidney Dis. 13 (2019) 74–86.

[86] C.L. Hoppel, E.J. Lesnefsky, Q. Chen, B. Tandler, Mitochondrial dysfunction in cardiovascular aging, in: G. Santulli (Ed.), Mitochondrial Dynamics in Cardiovascular Medicine, Springer International Publishing, Cham, 2017, pp. 451–464, https://doi.org/10.1007/978-3-319-55330-6_24.

[87] C. de Luca, J.M. Olefsky, Inflammation and insulin resistance, FEBS Lett. 582 (2008) 97–105, https://doi.org/10.1016/j.febslet.2007.11.057.

[88] A. Gonzalez-Franquesa, M.-E. Patti, Insulin resistance and mitochondrial dysfunction, in: G. Santulli (Ed.), Mitochondrial Dynamics in Cardiovascular Medicine, Springer International Publishing, Cham, 2017, pp. 465–520, https://doi.org/10.1007/978-3-319-55330-6_25.

[89] W.E. Balch, R.I. Morimoto, A. Dillin, J.W. Kelly, Adapting Proteostasis for disease intervention, Science 319 (2008) 916–919, https://doi.org/10.1126/science.1141448.

[90] E.T. Powers, R.I. Morimoto, A. Dillin, J.W. Kelly, W.E. Balch, Biological and chemical approaches to diseases of proteostasis deficiency, Annu. Rev. Biochem. 78 (2009) 959–991, https://doi.org/10.1146/annurev.biochem.052308.114844.

[91] M. Bucciantini, E. Giannoni, F. Chiti, F. Baroni, L. Formigli, J. Zurdo, N. Taddei, G. Ramponi, C.M. Dobson, M. Stefani, Inherent toxicity of aggregates implies a common mechanism for protein misfolding diseases, Nature 416 (2002) 507–511, https://doi.org/10.1038/416507a.

[92] H. Koga, S. Kaushik, A.M. Cuervo, Protein homeostasis and aging: the importance of exquisite quality control, Ageing Res. Rev. 10 (2011) 205–215, https://doi.org/10.1016/j.arr.2010.02.001.

[93] F.U. Hartl, A. Bracher, M. Hayer-Hartl, Molecular chaperones in protein folding and proteostasis, Nature 475 (2011) 324–332, https://doi.org/10.1038/nature10317.

[94] T.J. van Ham, M.A. Holmberg, A.T. van der Goot, E. Teuling, M. Garcia-Arencibia, H. Kim, D. Du, K.L. Thijssen, M. Wiersma, R. Burggraaff, P. van Bergeijk, J. van Rheenen, G. Jerre van Veluw, R.M.W. Hofstra, D.C. Rubinsztein, E.A.A. Nollen, Identification of MOAG-4/SERF as a regulator of age-related Proteotoxicity, Cell 142 (2010) 601–612, https://doi.org/10.1016/j.cell.2010.07.020.

[95] G.A. Walker, G.J. Lithgow, Lifespan extension in *C. elegans* by a molecular chaperone dependent upon insulin-like signals, Aging Cell 2 (2003) 131–139, https://doi.org/10.1046/j.1474-9728.2003.00045.x.

[96] G. Morrow, M. Samson, S. Michaud, R.M. Tanguay, Overexpression of the small mitochondrial Hsp22 extends *Drosophila* life span and increases resistance to oxidative stress, FASEB J. 18 (2004) 598–599, https://doi.org/10.1096/fj.03-0860fje.

[97] W.R. Swindell, M.M. Masternak, J.J. Kopchick, C.A. Conover, A. Bartke, R.A. Miller, Endocrine regulation of heat shock protein mRNA levels in long-lived dwarf mice, Mech. Ageing Dev. 130 (2009) 393–400, https://doi.org/10.1016/j.mad.2009.03.004.

[98] W.-C. Chiang, T.-T. Ching, H.C. Lee, C. Mousigian, A.-L. Hsu, HSF-1 regulators DDL-1/2 link insulin-like signaling to heat-shock responses and modulation of longevity, Cell 148 (2012) 322–334, https://doi.org/10.1016/j.cell.2011.12.019.

[99] S. Alavez, M.C. Vantipalli, D.J.S. Zucker, I.M. Klang, G.J. Lithgow, Amyloid-binding compounds maintain protein homeostasis during ageing and extend lifespan, Nature 472 (2011) 226–229, https://doi.org/10.1038/nature09873.

[100] S.M. Gehrig, C. van der Poel, T.A. Sayer, J.D. Schertzer, D.C. Henstridge, J.E. Church, S. Lamon, A.P. Russell, K.E. Davies, M.A. Febbraio, G.S. Lynch, Hsp72 preserves muscle function and slows progression of severe muscular dystrophy, Nature 484 (2012) 394–398, https://doi.org/10.1038/nature10980.

[101] P. Sen, P.P. Shah, R. Nativio, S.L. Berger, Epigenetic mechanisms of longevity and aging, Cell 166 (2016) 822–839, https://doi.org/10.1016/j.cell.2016.07.050.

[102] E.L. Greer, T.J. Maures, A.G. Hauswirth, E.M. Green, D.S. Leeman, G.S. Maro, S. Han, M.R. Banko, O. Gozani, A. Brunet, Members of the H3K4 trimethylation complex regulate lifespan in a germline-dependent manner in *C. elegans*, Nature 466 (2010) 383–387, https://doi.org/10.1038/nature09195.

[103] A.P. Siebold, R. Banerjee, F. Tie, D.L. Kiss, J. Moskowitz, P.J. Harte, Polycomb repressive complex 2 and Trithorax modulate Drosophila longevity and stress resistance, Proc. Natl. Acad. Sci. U. S. A. 107 (2010) 169–174, https://doi.org/10.1073/pnas.0907739107.

[104] F. Ugarte, R. Sousae, B. Cinquin, E.W. Martin, J. Krietsch, G. Sanchez, M. Inman, H. Tsang, M. Warr, E. Passegué, C.A. Larabell, E.C. Forsberg, Progressive chromatin condensation and H3K9 methylation regulate the differentiation of embryonic and hematopoietic stem cells, Stem Cell Rep. 5 (2015) 728–740, https://doi.org/10.1016/j.stemcr.2015.09.009.

[105] A. Brunet, T.A. Rando, Interaction between epigenetic and metabolism in aging stem cells, Curr. Opin. Cell Biol. 45 (2017) 1–7, https://doi.org/10.1016/j.ceb.2016.12.009.

[106] J.G. Ryall, S. Dell'Orso, A. Derfoul, A. Juan, H. Zare, X. Feng, D. Clermont, M. Koulnis, G. Gutierrez-Cruz, M. Fulco, V. Sartorelli, The NAD(+)-dependent SIRT1 deacetylase translates a metabolic switch into regulatory epigenetics in skeletal muscle stem cells, Cell Stem Cell 16 (2015) 171–183, https://doi.org/10.1016/j.stem.2014.12.004.

[107] P. Scaffidi, T. Misteli, Reversal of the cellular phenotype in the premature aging disease Hutchinson-Gilford progeria syndrome, Nat. Med. 11 (2005) 440–445, https://doi.org/10.1038/nm1204.

[108] P. Scaffidi, Lamin A-dependent nuclear defects in human aging, Science 312 (2006) 1059–1063, https://doi.org/10.1126/science.1127168.

[109] W. Zhang, J. Li, K. Suzuki, J. Qu, P. Wang, J. Zhou, X. Liu, R. Ren, X. Xu, A. Ocampo, T. Yuan, J. Yang, Y. Li, L. Shi, D. Guan, H. Pan, S. Duan, Z. Ding, M. Li, F. Yi, R. Bai, Y. Wang, C. Chen, F. Yang, X. Li, Z. Wang, E. Aizawa, A. Goebl, R.D. Soligalla, P. Reddy, C.R. Esteban, F. Tang, G.-H. Liu, J.C.I. Belmonte, A Werner syndrome stem cell model unveils heterochromatin alterations as a driver of human aging, Science 348 (2015) 1160–1163, https://doi.org/10.1126/science.aaa1356.

[110] R. Ren, A. Ocampo, G.-H. Liu, J.C. Izpisua Belmonte, Regulation of stem cell aging by metabolism and epigenetics, Cell Metab. 26 (2017) 460–474, https://doi.org/10.1016/j.cmet.2017.07.019.

[111] G.-H. Liu, B.Z. Barkho, S. Ruiz, D. Diep, J. Qu, S.-L. Yang, A.D. Panopoulos, K. Suzuki, L. Kurian, C. Walsh, J. Thompson, S. Boue, H.L. Fung, I. Sancho-Martinez, K. Zhang, J.Y. Iii, J.C.I. Belmonte, Recapitulation of premature ageing with iPSCs from Hutchinson–Gilford progeria syndrome, Nature 472 (2011) 221–225, https://doi.org/10.1038/nature09879.

[112] N. Kubben, W. Zhang, L. Wang, T.C. Voss, J. Yang, J. Qu, G.-H. Liu, T. Misteli, Repression of the antioxidant NRF2 pathway in premature aging, Cell 165 (2016) 1361–1374, https://doi.org/10.1016/j.cell.2016.05.017.

[113] L. Wang, F. Yi, L. Fu, J. Yang, S. Wang, Z. Wang, K. Suzuki, L. Sun, X. Xu, Y. Yu, J. Qiao, J.C.I. Belmonte, Z. Yang, Y. Yuan, J. Qu, G.-H. Liu, CRISPR/Cas9-mediated targeted gene correction in amyotrophic lateral sclerosis patient iPSCs, Protein Cell 8 (2017) 365–378, https://doi.org/10.1007/s13238-017-0397-3.

[114] D. Sun, M. Luo, M. Jeong, B. Rodriguez, Z. Xia, R. Hannah, H. Wang, T. Le, K.F. Faull, R. Chen, H. Gu, C. Bock, A. Meissner, B. Göttgens, G.J. Darlington, W. Li, M.A. Goodell, Epigenomic profiling of young and aged HSCs reveals concerted changes during aging that reinforce self-renewal, Cell Stem Cell 14 (2014) 673–688, https://doi.org/10.1016/j.stem.2014.03.002.

[115] J.-W. Jung, S. Lee, M.-S. Seo, S.-B. Park, A. Kurtz, S.-K. Kang, K.-S. Kang, Histone deacetylase controls adult stem cell aging by balancing the expression of polycomb genes and jumonji domain containing 3, Cell. Mol. Life Sci. 67 (2010) 1165–1176, https://doi.org/10.1007/s00018-009-0242-9.

[116] A.-Y. So, J.-W. Jung, S. Lee, H.-S. Kim, K.-S. Kang, DNA methyltransferase controls stem cell aging by regulating BMI1 and EZH2 through MicroRNAs, PLoS One 6 (2011), https://doi.org/10.1371/journal.pone.0019503, e19503.

[117] R. Liu, H. Zhang, M. Yuan, J. Zhou, Q. Tu, J.-J. Liu, J. Wang, Synthesis and biological evaluation of Apigenin derivatives as antibacterial and Antiproliferative agents, Molecules 18 (2013) 11496–11511, https://doi.org/10.3390/molecules180911496.

[118] H. Faralli, C. Wang, K. Nakka, A. Benyoucef, S. Sebastian, L. Zhuang, A. Chu, C.G. Palii, C. Liu, B. Camellato, M. Brand, K. Ge, F.J. Dilworth, UTX demethylase activity is required for satellite cell-mediated muscle regeneration, J. Clin. Invest. 126 (2016) 1555–1565, https://doi.org/10.1172/JCI83239.

[119] J. Labbadia, R.I. Morimoto, Repression of the heat shock response is a programmed event at the onset of reproduction, Mol. Cell 59 (2015) 639–650, https://doi.org/10.1016/j.molcel.2015.06.027.

[120] T.J. Maures, E.L. Greer, A.G. Hauswirth, A. Brunet, The H3K27 demethylase UTX-1 regulates *C. elegans* lifespan in a germline-independent, insulin-dependent manner: The H3K27me3 demethylase UTX-1 regulates worm lifespan, Aging Cell 10 (2011) 980–990, https://doi.org/10.1111/j.1474-9726.2011.00738.x.

[121] C. Merkwirth, V. Jovaisaite, J. Durieux, O. Matilainen, S.D. Jordan, P.M. Quiros, K.K. Steffen, E.G. Williams, L. Mouchiroud, S.U. Tronnes, V. Murillo, S.C. Wolff, R.J. Shaw, J. Auwerx, A. Dillin, Two conserved histone demethylases regulate mitochondrial stress-induced longevity, Cell 165 (2016) 1209–1223, https://doi.org/10.1016/j.cell.2016.04.012.

[122] C. Kosan, F.H. Heidel, M. Godmann, H. Bierhoff, Epigenetic erosion in adult stem cells: drivers and passengers of aging, Cell 7 (2018), https://doi.org/10.3390/cells7120237.

[123] N. Barzilai, D.M. Huffman, R.H. Muzumdar, A. Bartke, The critical role of metabolic pathways in aging, Diabetes 61 (2012) 1315–1322, https://doi.org/10.2337/db11-1300.

[124] A. Sepe, T. Tchkonia, T. Thomou, M. Zamboni, J.L. Kirkland, Aging and regional differences in fat cell progenitors—a mini-review, Gerontology 57 (2011) 66–75, https://doi.org/10.1159/000279755.

[125] P.A. Permana, C. Menge, P.D. Reaven, Macrophage-secreted factors induce adipocyte inflammation and insulin resistance, Biochem. Biophys. Res. Commun. 341 (2006) 507–514, https://doi.org/10.1016/j.bbrc.2006.01.012.

[126] K. Takahashi, S. Yamaguchi, T. Shimoyama, H. Seki, K. Miyokawa, H. Katsuta, T. Tanaka, K. Yoshimoto, H. Ohno, S. Nagamatsu, H. Ishida, JNK- and IκB-dependent pathways regulate MCP-1 but not adiponectin release from artificially hypertrophied 3T3-L1 adipocytes preloaded with palmitate in vitro, Am. J. Physiol. Endocrinol. Metab. 294 (2008) E898–E909, https://doi.org/10.1152/ajpendo.00131.2007.

[127] T. Suganami, J. Nishida, Y. Ogawa, A paracrine loop between adipocytes and macrophages aggravates inflammatory changes: role of free fatty acids and tumor necrosis factor α, Arterioscler. Thromb. Vasc. Biol. 25 (2005) 2062–2068, https://doi.org/10.1161/01.ATV.0000183883.72263.13.

[128] T. Suganami, K. Tanimoto-Koyama, J. Nishida, M. Itoh, X. Yuan, S. Mizuarai, H. Kotani, S. Yamaoka, K. Miyake, S. Aoe, Y. Kamei, Y. Ogawa, Role of the toll-like receptor 4/NF-κB pathway in saturated fatty acid–induced inflammatory changes in the interaction between adipocytes and macrophages, Arterioscler. Thromb. Vasc. Biol. 27 (2007) 84–91, https://doi.org/10.1161/01.ATV.0000251608.09329.9a.

[129] B. Dulken, A. Brunet, Stem cell aging and sex: are we missing something? Cell Stem Cell 16 (2015) 588–590, https://doi.org/10.1016/j.stem.2015.05.006.

[130] L. Fontana, L. Partridge, Promoting health and longevity through diet: from model organisms to humans, Cell 161 (2015) 106–118, https://doi.org/10.1016/j.cell.2015.02.020.

[131] K.-J. Min, C.-K. Lee, H.-N. Park, The lifespan of Korean eunuchs, Curr. Biol. 22 (2012) R792–R793, https://doi.org/10.1016/j.cub.2012.06.036.

[132] B.M. Deasy, A. Lu, J.C. Tebbets, J.M. Feduska, R.C. Schugar, J.B. Pollett, B. Sun, K.L. Urish, B.M. Gharaibeh, B. Cao, R.T. Rubin, J. Huard, A role for cell sex in stem cell-mediated skeletal muscle regeneration: female cells have higher muscle regeneration efficiency, J. Cell Biol. 177 (2007) 73–86, https://doi.org/10.1083/jcb.200612094.

[133] D. Nakada, H. Oguro, B.P. Levi, N. Ryan, A. Kitano, Y. Saitoh, M. Takeichi, G.R. Wendt, S.J. Morrison, Oestrogen increases haematopoietic stem-cell self-renewal in females and during pregnancy, Nature 505 (2014) 555–558, https://doi.org/10.1038/nature12932.

[134] J.L. Pawluski, S. Brummelte, C.K. Barha, T.M. Crozier, L.A.M. Galea, Effects of steroid hormones on neurogenesis in the hippocampus of the adult female rodent during the estrous cycle, pregnancy, lactation and aging, Front. Neuroendocrinol. 30 (2009) 343–357, https://doi.org/10.1016/j.yfrne.2009.03.007.

[135] J. Yuan, J. Yu, J. Ge, Sexual dimorphism on the neurogenic potential of rhesus monkeys mesenchymal stem cells, Biochem. Biophys. Res. Commun. 396 (2010) 394–400, https://doi.org/10.1016/j.bbrc.2010.04.103.

[136] R.A.J. Signer, S.J. Morrison, Mechanisms that regulate stem cell aging and life span, Cell Stem Cell 12 (2013) 152–165, https://doi.org/10.1016/j.stem.2013.01.001.

[137] D. Sisci, P. Maris, M.G. Cesario, W. Anselmo, R. Coroniti, G.E. Trombino, F. Romeo, A. Ferraro, M. Lanzino, S. Aquila, M. Maggiolini, L. Mauro, C. Morelli, S. Andò, The estrogen receptor α is the key regulator of the bifunctional role of FoxO3a transcription factor in breast cancer motility and invasiveness, Cell Cycle 12 (2013) 3405–3420, https://doi.org/10.4161/cc.26421.

[138] S. Kyo, M. Takakura, T. Kanaya, W. Zhuo, K. Fujimoto, Y. Nishio, A. Orimo, M. Inoue, Estrogen activates telomerase, Cancer Res. 59 (1999) 5917–5921.

[139] A. Eggel, T. Wyss-Coray, A revival of parabiosis in biomedical research, Swiss Med. Wkly. 144 (2014) w13914, https://doi.org/10.4414/smw.2014.13914.

[140] M. Conese, A. Carbone, E. Beccia, A. Angiolillo, The fountain of youth: a tale of parabiosis, stem cells, and rejuvenation, Open Med. 12 (2017), https://doi.org/10.1515/med-2017-0053.

[141] I.M. Conboy, T.A. Rando, Heterochronic parabiosis for the study of the effects of aging on stem cells and their niches, Cell Cycle 11 (2012) 2260–2267, https://doi.org/10.4161/cc.20437.

[142] M.J. Conboy, I.M. Conboy, T.A. Rando, Heterochronic parabiosis: historical perspective and methodological considerations for studies of aging and longevity, Aging Cell 12 (2013) 525–530, https://doi.org/10.1111/acel.12065.

[143] F.S. Loffredo, M.L. Steinhauser, S.M. Jay, J. Gannon, J.R. Pancoast, P. Yalamanchi, M. Sinha, C. Dall'Osso, D. Khong, J.L. Shadrach, C.M. Miller, B.S. Singer, A. Stewart, N. Psychogios, R.E. Gerszten, A.J. Hartigan, M.-J. Kim, T. Serwold, A.J. Wagers, R.T. Lee, Growth differentiation factor 11 is a circulating factor that reverses age-related cardiac hypertrophy, Cell 153 (2013) 828–839, https://doi.org/10.1016/j.cell.2013.04.015.

[144] R.I. Sherwood, J.L. Christensen, I.L. Weissman, A.J. Wagers, Determinants of skeletal muscle contributions from circulating cells, bone marrow cells, and hematopoietic stem cells, Stem Cells 22 (2004) 1292–1304, https://doi.org/10.1634/stemcells.2004-0090.

[145] I.D. Duncan, A. Brower, Y. Kondo, J.F. Curlee, R.D. Schultz, Extensive remyelination of the CNS leads to functional recovery, Proc. Natl. Acad. Sci. U. S. A. 106 (2009) 6832–6836, https://doi.org/10.1073/pnas.0812500106.

[146] P.H. Maxwell, What might retrotransposons teach us about aging? Curr. Genet. 62 (2016) 277–282, https://doi.org/10.1007/s00294-015-0538-2.

[147] M. De Cecco, S.W. Criscione, A.L. Peterson, N. Neretti, J.M. Sedivy, J.A. Kreiling, Transposable elements become active and mobile in the genomes of aging mammalian somatic tissues, Aging (Albany NY) 5 (2013) 867–883, https://doi.org/10.18632/aging.100621.

[148] G.S. Laurent, N. Hammell, T.A. McCaffrey, A LINE-1 component to human aging: do LINE elements exact a longevity cost for evolutionary advantage? Mech. Ageing Dev. 131 (2010) 299–305, https://doi.org/10.1016/j.mad.2010.03.008.

[149] M. Van Meter, M. Kashyap, S. Rezazadeh, A.J. Geneva, T.D. Morello, A. Seluanov, V. Gorbunova, SIRT6 represses LINE1 retrotransposons by ribosylating KAP1 but this repression fails with stress and age, Nat. Commun. 5 (2014) 5011, https://doi.org/10.1038/ncomms6011.

[150] J.L. Goodier, Restricting retrotransposons: a review, Mob. DNA 7 (2016) 16, https://doi.org/10.1186/s13100-016-0070-z.

[151] M. Scheibye-Knudsen, S.J. Mitchell, E.F. Fang, T. Iyama, T. Ward, J. Wang, C.A. Dunn, N. Singh, S. Veith, M.M. Hasan-Olive, A. Mangerich, M.A. Wilson, M.P. Mattson, L.H. Bergersen, V.C. Cogger, A. Warren, D.G. Le Couteur, R. Moaddel, D.M. Wilson, D.L. Croteau, R. de Cabo, V.A. Bohr, A high-fat diet and NAD + activate Sirt1 to rescue premature aging in Cockayne syndrome, Cell Metab. 20 (2014) 840–855, https://doi.org/10.1016/j.cmet.2014.10.005.

[152] L. Robbez-Masson, H.M. Rowe, Retrotransposons shape species-specific embryonic stem cell gene expression, Retrovirology 12 (2015) 45, https://doi.org/10.1186/s12977-015-0173-5.

[153] P.-É. Jacques, J. Jeyakani, G. Bourque, The majority of primate-specific regulatory sequences are derived from transposable elements, PLoS Genet. 9 (2013), https://doi.org/10.1371/journal.pgen.1003504, e1003504.

[154] R.K. Slotkin, R. Martienssen, Transposable elements and the epigenetic regulation of the genome, Nat. Rev. Genet. 8 (2007) 272–285, https://doi.org/10.1038/nrg2072.

[155] J. Wang, G.J. Geesman, S.L. Hostikka, M. Atallah, B. Blackwell, E. Lee, P.J. Cook, B. Pasaniuc, G. Shariat, E. Halperin, M. Dobke, M.G. Rosenfeld, I.K. Jordan, V.V. Lunyak, Inhibition of activated pericentromeric SINE/Alu repeat transcription in senescent human adult stem cells reinstates self-renewal, Cell Cycle 10 (2011) 3016–3030, https://doi.org/10.4161/cc.10.17.17543.

[156] J.M. Freije, C. López-Otín, Reprogramming aging and progeria, Curr. Opin. Cell Biol. 24 (2012) 757–764, https://doi.org/10.1016/j.ceb.2012.08.009.

[157] C. Soria-Valles, C. López-Otín, iPSCs: on the road to reprogramming aging, Trends Mol. Med. 22 (2016) 713–724, https://doi.org/10.1016/j.molmed.2016.05.010.

[158] L. Lapasset, O. Milhavet, A. Prieur, E. Besnard, A. Babled, N. Ait-Hamou, J. Leschik, F. Pellestor, J.-M. Ramirez, J. De Vos, S. Lehmann, J.-M. Lemaitre, Rejuvenating senescent and centenarian human cells by reprogramming through the pluripotent state, Genes Dev. 25 (2011) 2248–2253, https://doi.org/10.1101/gad.173922.111.

[159] M. Wahlestedt, G.L. Norddahl, G. Sten, A. Ugale, M.-A.M. Frisk, R. Mattsson, T. Deierborg, M. Sigvardsson, D. Bryder, An epigenetic component of hematopoietic stem cell aging amenable to reprogramming into a young state, Blood 121 (2013) 4257–4264, https://doi.org/10.1182/blood-2012-11-469080.

[160] M.A. Blasco, Telomeres and human disease: ageing, cancer and beyond, Nat. Rev. Genet. 6 (2005) 611–622, https://doi.org/10.1038/nrg1656.

[161] J.D. Griffith, L. Comeau, S. Rosenfield, R.M. Stansel, A. Bianchi, H. Moss, T. de Lange, Mammalian telomeres end in a large duplex loop, Cell 97 (1999) 503–514, https://doi.org/10.1016/S0092-8674(00)80760-6.

[162] Y. Doksani, J.Y. Wu, T. de Lange, X. Zhuang, Super-resolution fluorescence imaging of telomeres reveals TRF2-dependent T-loop formation, Cell 155 (2013) 345–356, https://doi.org/10.1016/j.cell.2013.09.048.

[163] W.E. Wright, M.A. Piatyszek, W.E. Rainey, W. Byrd, J.W. Shay, Telomerase activity in human germline and embryonic tissues and cells, Dev. Genet. 18 (1996) 173–179, https://doi.org/10.1002/(SICI)1520-6408(1996)18:2<173::AID-DVG10>3.0.CO;2-3.

[164] J.D. Watson, Origin of concatemeric T7 DNA, Nat. New Biol. 239 (1972) 197–201, https://doi.org/10.1038/newbio239197a0.

[165] A.M. Olovnikov, A theory of marginotomy, J. Theor. Biol. 41 (1973) 181–190, https://doi.org/10.1016/0022-5193(73)90198-7.

[166] J. Lindsey, N.I. McGill, L.A. Lindsey, D.K. Green, H.J. Cooke, In vivo loss of telomeric repeats with age in humans, Mutat. Res. 256 (1991) 45–48, https://doi.org/10.1016/0921-8734(91)90032-7.

[167] I. Flores, A. Canela, E. Vera, A. Tejera, G. Cotsarelis, M.A. Blasco, The longest telomeres: a general signature of adult stem cell compartments, Genes Dev. 22 (2008) 654–667, https://doi.org/10.1101/gad.451008.

[168] Y. Zou, A. Sfeir, S.M. Gryaznov, J.W. Shay, W.E. Wright, Does a sentinel or a subset of short telomeres determine replicative senescence? MBoC. 15 (2004) 3709–3718, https://doi.org/10.1091/mbc.e04-03-0207.

[169] M. Fumagalli, F. Rossiello, M. Clerici, S. Barozzi, D. Cittaro, J.M. Kaplunov, G. Bucci, M. Dobreva, V. Matti, C.M. Beausejour, U. Herbig, M.P. Longhese, F. d'Adda di Fagagna, Telomeric DNA damage is irreparable and causes persistent DNA-damage-response activation, Nat. Cell Biol. 14 (2012) 355–365, https://doi.org/10.1038/ncb2466.

[170] T. Tchkonia, Y. Zhu, J. van Deursen, J. Campisi, J.L. Kirkland, Cellular senescence and the senescent secretory phenotype: therapeutic opportunities, J. Clin. Invest. 123 (2013) 966–972, https://doi.org/10.1172/JCI64098.

[171] B.B. de Jesus, K. Schneeberger, E. Vera, A. Tejera, C.B. Harley, M.A. Blasco, The telomerase activator TA-65 elongates short telomeres and increases health span of adult/old mice without increasing cancer incidence: TA-65 elongates short telomeres and increases health span of adult/old mice, Aging Cell 10 (2011) 604–621, https://doi.org/10.1111/j.1474-9726.2011.00700.x.

[172] C.B. Harley, W. Liu, P.L. Flom, J.M. Raffaele, A natural product telomerase activator as part of a health maintenance program: metabolic and cardiovascular response, Rejuvenation Res. 16 (2013) 386–395, https://doi.org/10.1089/rej.2013.1430.

[173] M. Jaskelioff, F.L. Muller, J.-H. Paik, E. Thomas, S. Jiang, A.C. Adams, E. Sahin, M. Kost-Alimova, A. Protopopov, J. Cadiñanos, J.W. Horner, E. Maratos-Flier, R.A. DePinho, Telomerase reactivation reverses tissue degeneration in aged telomerase-deficient mice, Nature 469 (2011) 102–106, https://doi.org/10.1038/nature09603.

[174] A. Canela, J. Martín-Caballero, J.M. Flores, M.A. Blasco, Constitutive expression of tert in thymocytes leads to increased incidence and dissemination of T-cell lymphoma in Lck-Tert mice, Mol. Cell. Biol. 24 (2004) 4275–4293, https://doi.org/10.1128/mcb.24.10.4275-4293.2004.

[175] S.E. Artandi, S. Alson, M.K. Tietze, N.E. Sharpless, S. Ye, R.A. Greenberg, D.H. Castrillon, J.W. Horner, S.R. Weiler, R.D. Carrasco, R.A. DePinho, Constitutive telomerase expression promotes mammary carcinomas in aging mice, Proc. Natl. Acad. Sci. U. S. A. 99 (2002) 8191–8196, https://doi.org/10.1073/pnas.112515399.

[176] E. González-Suárez, E. Samper, A. Ramírez, J.M. Flores, J. Martín-Caballero, J.L. Jorcano, M.A. Blasco, Increased epidermal tumors and increased skin wound healing in transgenic mice overexpressing the catalytic subunit of telomerase, mTERT, in basal keratinocytes, EMBO J. 20 (2001) 2619–2630, https://doi.org/10.1093/emboj/20.11.2619.

[177] M. Cerletti, Y.C. Jang, L.W.S. Finley, M.C. Haigis, A.J. Wagers, Short-term calorie restriction enhances skeletal muscle stem cell function, Cell Stem Cell 10 (2012) 515–519, https://doi.org/10.1016/j.stem.2012.04.002.

[178] W. Mair, C.J. McLeod, L. Wang, D.L. Jones, Dietary restriction enhances germline stem cell maintenance, Aging Cell 9 (2010) 916–918, https://doi.org/10.1111/j.1474-9726.2010.00602.x.

[179] D. Tang, S. Tao, Z. Chen, I.O. Koliesnik, P.G. Calmes, V. Hoerr, B. Han, N. Gebert, M. Zörnig, B. Löffler, Y. Morita, K.L. Rudolph, Dietary restriction improves repopulation but impairs lymphoid differentiation capacity of hematopoietic stem cells in early aging, J. Exp. Med. 213 (2016) 535–553, https://doi.org/10.1084/jem.20151100.

[180] Ö.H. Yilmaz, P. Katajisto, D.W. Lamming, Y. Gültekin, K.E. Bauer-Rowe, S. Sengupta, K. Birsoy, A. Dursun, V.O. Yilmaz, M. Selig, G.P. Nielsen, M. Mino-Kenudson, L.R. Zukerberg, A.K. Bhan, V. Deshpande, D.M. Sabatini, mTORC1 in the Paneth cell niche couples intestinal stem-cell function to calorie intake, Nature 486 (2012) 490–495, https://doi.org/10.1038/nature11163.

Chapter 16

Therapeutic approaches for the treatment of aging-induced stem cell dysfunction

Debora Bizzaro, Francesco Paolo Russo, and Patrizia Burra
Department of Surgery, Oncology and Gastroenterology, Gastroenterology/Multivisceral Transplant Section, University Hospital Padova, Padova, Italy

1 Introduction

Aging is an inevitable physiological process in all living animals that is mediated by complex pathways and driven by various acquired and genetic factors [1]. Since ancient times, humans have continuously sought to find a way to avoid aging and to live forever, such as the superstitious search for the fountain of youth, sleep with virgins, bathing, or drinking blood. Although stem cell-based regenerative medicine offers promising new effective treatment to restore or rejuvenate tissues using endogenous or exogenous stem/progenitor cells [2, 3], we must be aware that also adult stem cells are intended for aging. With aging, the stem cells reduce their regenerative potential and functional ability [4, 5], which ultimately can lead to cell death, senescence, or loss of regenerative potential [6].

The stem cell theory of aging assumes that the loss of regenerative ability of stem cells is in large part responsible for the aging process [1]. Moreover, alterations in cell homeostasis in aging tissues might further induce microenvironmental alteration, promoting the selection of potential neoplastic stem cells [7].

Therefore, understanding the reciprocal connection of the aging process and the deterioration of stem cell functions is crucial, not only to understand the pathophysiology of the disorders associated with aging, but also for the future development of new potential therapies. Such therapeutic interventions should be aimed to slow down, and possibly reverse, the age-related degenerative processes, to improve the repair processes, and to maintain healthy function in aging tissues.

Cumulative cellular alterations, such as DNA damage, oxidative stress, increased expression of cell cycle inhibitors, and mitochondrial dysfunction, have been considered the likely principal factors responsible for tissue decline observed with aging. However, numerous studies of cellular rejuvenation revealed that a fundamental role is played by extracellular signals, both systemic and local. In the present chapter, we aimed to give an overview on the principal cellular and extracellular mechanisms responsible for stem cell aging and the potential therapeutic approaches for the treatment of aging-induced stem cell dysfunction.

2 The past of rejuvenation research

The ancient and superstitious idea that blood could have rejuvenating effects was not entirely unfounded. Indeed, some laboratories have applied in aging researches a surgical technique developed in the mid-1800s, the parabiosis (from the Greek para, meaning "alongside," and bios, meaning "life"), that reciprocally joins the vasculature of two living animals. The first experiment of parabiosis aimed to investigate the possibility of animal rejuvenation was performed by McCay et al. [8] by connecting the circulatory system of an old rat to that of a young one (heterochronic parabiosis), observing a qualitative amelioration of the aspect of cartilage of the older rat. Later, in 1972, Ludwig et al. [9] performed more quantitative experiments demonstrating life extension in older animals, who benefitted from sharing the blood supply of younger animals. Later studies have demonstrated that the blood of young mice seems to rejuvenate brain [10], muscles [11], and liver tissues in the aged animals [12], making old mice stronger, smarter, and healthier. Conversely, the younger partner exhibited decreased neurogenesis and cognitive abilities, consistent with that of an older animal [13].

Other approaches in rejuvenation research of tissue and organs involved cross-age transplantation. The first study of such type performed demonstrated that old muscle grafted into young hosts recovers the same mass and maximum force of young muscle [14]. Similarly, young muscle grafted into old hosts shows the characteristics of the old muscle. Hence, the authors concluded that the regeneration potential of muscle is dependent on the environment provided by the host and not on the tissue chronological age alone. In the same way, a thymus transplantation study demonstrated that senescent, involuted thymus gland becomes fully functional upon transplantation into young hosts [15].

These findings strongly suggest that aging-related cellular dysfunctions seem to be controlled also in an "extrinsic manner" by extracellular factors and therefore can be repaired successfully by modulating the microenvironment of the tissue in addition to cell-intrinsic aging factors [16].

3 Therapeutic approaches targeting extracellular aging factors

3.1 Rejuvenation of stem cell niche

The peculiar functions of stem cells—quiescence, proliferation, multipotency, and differentiation—are all potentially influenced by extracellular signals, which can be derived either from the systemic environment, reaching the stem cells via the bloodstream, or from the local microenvironment—the so-called niche.

The niche components are peculiar for each tissue and include somatic and stromal cells, immune cells, extracellular matrix, innervating neuronal fibers, and the vasculature [17]. The significance of niche cell support has been demonstrated for different tissue stem cells, including germline stem cells in the testes and ovaries [18, 19], intestinal stem cells [20], and skeletal muscle satellite cells [21].

The microenvironment of the niche is fundamental to stem cell maintenance and strictly regulates their functions. For this reason, for many tissues, the age of stem cells appears to be determined by the age of their niche or environment, rather than the age of the stem cell itself [22]. Consequently, alterations in the components of niches induced by aging can cause abnormality in stem cell functions and vice versa, and young local and/or systemic environment can promote effective regeneration of old stem cells. Considering these points, rejuvenating strategies should be performed targeting both stem cells and their niches, or more likely, rejuvenating of niches could directly rejuvenate the stem cells within them. Plasma exchange, from a young body, to an older body would be a potential means to rejuvenate stem cell niches. Indeed, as described above, in parabiosis studies, it was demonstrated that aged stem cells were rejuvenated by young blood and young stem cells were aged by exposure to blood from old animals, highlighting the importance of the systemic environment in stem cell aging [12]. Thus, the presence of negative factors that promote aging phenotypes in old plasma is conceivable. On the contrary, in young plasma, positive factors that promote young phenotypes and/or factors that inhibit or neutralize "negative" or "aging" factors may be present. Numerous soluble molecules secreted from different organs and tissues and carried by the bloodstream could affect the functions of stem cells, including hormones, growth factors, and any other signaling molecule of immune origin secreted by infiltrating immune cells.

Improving the quality of interactions with the niche and the systemic environment could be an efficient strategy to positively modulate the regenerative potential of stem cells, reducing the aging-induced dysfunction [23–25]. Such approaches include (i) modulation of inflammation mediators and immune cells; (ii) integration of molecular signals that weaken with age; and (iii) elimination of detrimental local or systemic signals that accumulate with age or elimination of their cellular sources, such as senescent cells.

However, the effectiveness and safety of these procedures have not yet been examined in humans, so extremely careful analysis of risks and benefits should be performed. In addition, the current challenge is to identify which components of the blood are responsible for these rejuvenating effects.

3.1.1 Modulation of "inflammaging" and "senolysis"

The deregulation of the immune response is one of the principal changes that occur during aging, leading to a chronic systemic inflammatory state.

The chronic, low-grade inflammation observed during aging is denominated "inflammaging" and has been proposed to be associated with the pathology of most age-related diseases and with the loss of regenerative capacity of many tissues [26, 27].

Several key intra- or intercellular signaling pathways are closely associated with chronic inflammatory changes during aging. Among these, interleukin (IL)-6, tumor necrosis factor (TNF)-α, and their receptors, chemokines, and C-reactive protein have been found to be involved in age-related pathogenesis [28]. In particular, the role of NF-κB pathway as mediator of chronic inflammation is clearly established in aged tissues such as skeletal muscle, skin, bone, and nervous system [29]; however, its direct effect on stem function is still under investigation.

Together, the scientific evidence available today suggests that limiting inflammaging is crucial for improving the effectiveness of regenerative processes and stem cell-based therapies. Since inflammatory signals regulate and are regulated by immune cells, direct targeting of these cells is probably a useful approach to reduce age-related inflammation and improve regeneration. Ultimately, it could be considered an effective approach in the treatment of stem cell aging.

Furthermore, recent studies suggest that the targeting of senescent cells and their products in aging tissues, the so-called senolysis, could be a possible strategy to restoring stem cell function. Clearance of senescent cells, using an inducible genetic model for senescent cell ablation, performed in young animals, is able to delay the onset of pathologies in multiple aging tissues, including fat, muscle, and eye. When performed in old animals, senolysis does not recover established age-related pathologies but its action is limited to attenuate their progression [30]. These results have prompted further studies investigating pharmacological strategies that could similarly induce the depletion of senescent cells from aged tissues. These so-called senolytic molecules, if identified, could potentially be useful in a number of age-associated pathologies, including metabolic disease and sarcopenia.

3.1.2 "Aging" and "rejuvenating" circulating factors

As previously described, it was demonstrated that aged stem cells were rejuvenated by young plasma, and on the contrary, young stem cells were aged by exposure to plasma from old animals [31]. Thus, two hypotheses were formulated: (i) the presence of positive factors in young plasma that promote young phenotypes and negative factors in old plasma that promote aging phenotypes and (ii) the presence of factors inhibiting or neutralizing "negative" factors in young plasma.

Numerous studies were directed to this direction and found that higher levels of circulating growth differentiation factor 11 (GDF11), oxytocin, and the IL-15 appear related to the "rejuvenating" effect of a young systemic environment [32–34]. The IL-15 naturally decreases during aging causing sarcopenia and obesity, suggesting that this cytokine could be a "positive" factor for young phenotypes [33]. Similarly, oxytocin normally decreases with the age and when administrated is able to improve the function of aged mesenchymal and muscle satellite stem cells through the activation of the MAPK/ERK signaling pathway [34, 35]. Interestingly, it was demonstrated that the treatment of old mice with either recombinant GDF11 or oxytocin reduces dysfunction of aged satellite cells and restores regenerative function in aged mice [11, 34]. GDF11 supplementation in old mice is further able to reverse age-related heart hypertrophy [32], enhance neural stem cell and neuronal function [36], and increase strength and endurance exercise capacity [11].

On the other hand, examples of aging factors include CCL11 (also known as eotaxin) and lamin A (specifically the truncated form progerin). The plasma levels of CCL11 correlate with dysfunction in neurogenesis, brain aging, and loss of cognitive function, and its levels are increased in the plasma and cerebrospinal fluid of healthy aging humans [13]. Progerin reduces the regenerative capacity of stem cells by disrupting the self-renewal markers, in part by deregulating Oct1, which perturbs both mTOR and autophagy pathways [37, 38]. These studies evidence that the management of blood-borne factors could provide an attractive strategy for the treatment of age-related disease. It is still under study whether an increase in positive factors, a decrease in negative factors (possibly by dilution in young plasma), or their combination represents the most effective strategy.

Hopefully, further knowledge on the molecular mechanisms underlying aging and rejuvenation will make more concrete the possibility to use molecules and agents as therapy for aging stem cell dysfunction. Of course, stem cell rejuvenation by targeting these molecular pathways by synthetic agents or natural sources will be more feasible than plasma exchange, since some agonists or antagonists of specific signaling pathways have already been developed and approved, even if not yet used for rejuvenation purposes.

4 Intrinsic aging factors and therapeutic approaches

4.1 Improving protein homeostasis and regulation of autophagy

The cellular processes responsible for the synthesis, folding, and turnover of proteins, called protein homeostasis or proteostasis, are essential for development and for most of the cellular functions [39]. Defects in this process can cause cellular damage and tissue dysfunction due to aberrant folding, toxic aggregation, or accumulation of altered proteins [40]. Aged cells are more affected by alteration in proteostasis and more inclined to accumulate misfolded protein aggregates [41, 42], and consequently, age is one of the main risk factors for most diseases associated with protein misfolding, (such as Alzheimer's disease, Parkinson's disease, and Huntington's disease) [43]. Specifically to stem cell functions, proteostasis is an important determinant of stem cell preservation. Several studies have demonstrated that the modulation of molecules involved in the control of autophagy pathway [autophagy-related gene 7 (*Atg7*), FoxO3A, mTOR] has been shown to regulate oxidative and metabolic stress, the self-renewal, and differentiation of hematopoietic stem cells (HSCs) and intestinal stem cells [44–47].

Conversely, the potential role of stimulation of autophagy or proteasome activity in aged stem cells is still under investigation, although few preliminary studies provide intriguing evidences on potential therapeutic interventions for age-related dysfunction in stem cell aging. For example, an interesting study demonstrated that the inhibition of mTOR by the administration of rapamycin is able to restore the self-renewal and hematopoietic potential of aged HSCs [48]. However, despite the encouraging results, this aspect remains an intriguing subject for future research.

4.2 Regulation of oxidative metabolism and improving mitochondrial function

The reactive oxygen species (ROS) are produced principally during mitochondrial oxidative phosphorylation and are hypothesized to be a putative character of stem cell dysfunction in aging [49, 50]. According to this hypothesis, a vicious circle leads to cellular decay in aged cells. Cell damage induced by aging and the decline of mitochondrial integrity stimulate high ROS production, which in turn leads to damage to cell macromolecules and disrupts mitochondrial oxidative phosphorylation, leading to further cell damage [49]. The role of ROS generation in promoting stem cell aging has been demonstrated in aged human mesenchymal stem cells (MSCs), where elevated levels of ROS were found [51], in murine HSCs and neural stem cells, in which excessive cellular concentrations of ROS lead to abnormal proliferation and compromised self-renewal of stem cells [52, 53].

Studies of the sirtuin (SIRT) family of NAD-dependent protein deacetylases further link the regulation of oxidative metabolism to stem cell aging. A role in maintaining MSCs growth and differentiation was recently suggested for SIRT1, which has been shown to decline with age [54] and SIRT3 that is essential in the maintenance of mitochondrial function and in control of ROS production during HSCs aging [55]. Furthermore, induction of expression of SIRT3 enhances the antioxidant activity of SOD2, an important regulator of oxidative stress, improving the function of aged HSCs [56]. Together, these studies suggest that the modulation of stem cell metabolic and redox states, which in turn can influence intracellular accumulation of ROS, could be a potential therapeutic approach.

Numerous antioxidants were tested as potential rejuvenating agents. One of these is the antioxidant *N*-acetyl-L-cysteine (NAC), a precursor of glutathione that acts as a direct ROS scavenger. NAC has been used as a therapeutic agent to ameliorate the damaging effects of ROS, restoring the quiescence, survival, and repopulating capacity in HSCs [57] and improving survival of myogenic stem cells in skeletal muscle, both in vitro and in vivo models [58, 59]. Although the potential of antioxidants as therapeutic agents is promising, whether they have a direct or indirect effect on age-dependent deficits in stem cells or stem cell function is still unclear and remains an open area of research. In addition, a better understanding of the endogenous sources of ROS and the cellular compartments in which they act is also likely to be important for clarifying the stem cell regulatory actions and potential therapeutic value of ROS-modulating agents.

5 Nuclear aging as potential therapeutic targets

5.1 Targeting DNA damage repair to restore stem cell function

One of the earliest theories of the aging process is the "mutational theory," linked to the concept that the accumulation of damaged DNA may in part be responsible for the cell aging process [60–62]. Indeed, spontaneous and extrinsic mutational events occur on DNA on daily basis in mammal cells. While normal DNA repair mechanisms are able to repair most of the DNA damages, some of them escape from the surveillance and accumulate over time. Therefore, there would be a significant accumulation of mutated or damaged DNA in aging somatic cells compared to young cells [61]. The accumulation of DNA damages is a common finding also in aged stem cells impairing their function [11, 63–65]. Indeed, stem cells are able to remain in a state of quiescence in mammalian tissues and organs for a long time, expositing to a greater risk of possible genotoxic damage. A high damage rate could result from an accumulation of damage over time, an increase in the damage rate itself, a decrease in the DNA repair rate, or a combination of these possibilities. Therefore, a reasonable therapeutic approach aimed at increasing the activity of DNA repair pathways could slow down or even prevent the accumulation of age-related defects in stem cells, promoting healthy tissue functioning.

In order to obtain a powerful and global DNA repair system, it is necessary to identify the main master regulatory genes that respond to multiple DNA damage response pathways, in order to exercise a more efficient action. This approach has been illustrated in the case of telomerase reactivation. Telomeres are the regions of repetitive nucleotide sequences at each end of the chromosome and have the function of protecting the genome from nucleolytic degradation, unnecessary recombination, repair, or fusion with neighboring chromosomes [66]. Following the shortening of telomeres, the cell undergoes senescence and proliferative blockage and can undergo apoptosis. Many approaches to increase telomere length protecting cells from chromosome shortening included telomerase activation drugs and telomerase gene therapy [67, 68] have been developed in recent years to protect the cells from premature aging. However, great caution must be accounted, due to the risk of malignant transformation [69, 70]. Whether it will be possible to translate these findings to design effective therapeutic strategies, mitigating risks of tumorigenesis remains to be established. Additional assessment of further DNA damage repair/response pathways will be crucial to assess the full rejuvenation potential of these pathways.

5.2 Epigenetic reprogramming

Epigenome refers to changes in gene expression without affecting the DNA sequence, including but not limited to changes in DNA methylation and histone modifications [71]. Many studies demonstrated that the epigenetic regulation is an important determinant of stem cell function and is transmitted heritably to their daughter cells [72]. Alterations in the epigenome occurring with aging can interfere with cellular processes, contributing to age-associated dysfunction of stem cells, and predispose to the risk of cancers [73]. Indeed, the acetylation and methylation status of DNA modifies with age, and alteration of enzymes responsible for these variations can alter organism longevity [74]. Therefore, there is a great interest in understanding these regulatory mechanisms that contribute to the decline of stem cell and tissue function with age. Due to their intrinsic reversibility, the epigenetic modification could be considered good therapeutic targets for potential therapies for stem cell rejuvenation.

5.2.1 Chemicals against epigenetic alteration

Positive effect on longevity was reached in studies in which alterations of histone modification were pharmacologically induced in aging and progeroid mouse models, suggesting that future therapeutic interventions targeting epigenetic regulators may be feasible [75, 76].

Among these compounds tested for treating age-related conditions and for delaying aging, metformin and rapamycin hold great promise [77, 78]. Metformin is involved in the modulation of AMP-activated protein kinase (AMPK), a direct activity regulator of several epigenetic enzymes, such as histone acetyltransferases (HATs), histone deacetylases (HDACs), and DNA methyltransferases (DNMTs) [79]. While metformin has an indirect effect, rapamycin treatment is demonstrated to directly reduce the accumulation of epigenetic aging signatures in mouse liver cells [80].

Specific chemicals active in delaying cellular senescence in stem cells are acetylsalicylate acid and ascorbate. As observed in mice and nematodes, aspirin metabolite salicylate is able to competitively inhibit the HAT p300 and trigger the cardioprotective effect [81]. Ascorbate acts potentially in an epigenetic-dependent manner on different fronts: resets gene expression profiles, restores heterochromatin structure, and alleviates aging defects [82]. Additionally, it was demonstrated that ascorbate maintains high levels of HSCs in humans and mice and suppresses leukemogenesis in vivo [83].

5.2.2 MicroRNA

MicroRNAs (miRNAs) are a further key category of epigenetic mediators of stem cell dysfunction. They are a class of small noncoding RNAs composed of 18- to 25-bp nucleotides that act as posttranscriptional regulators of gene expression in all somatic and stem cells [84]. In stem cells, miRNAs play an important role in regulating self-renewal and differentiation by regulating the translation of specific mRNAs in both stem cells and differentiating daughter cells [85]. For instance, miR-17 regulates osteoblast differentiation of MSCs [86], whereas miR-290-295 cluster seems to promote embryonic stem cell self-renewal, maintenance of pluripotency, and differentiation [87]. These results suggest the use of these miRNAs as clinical biomarkers of stem cell senescence and promise to be of potential use as therapeutic agents in stem cell aging.

5.2.3 Cellular reprogramming and induced pluripotent stem cells

The total reprogramming of the epigenetic setting contributes to cell fate conversion and lineage determination. Cellular reprogramming of aged somatic cells toward induced pluripotent stem cells (iPSCs) can be induced by a variety of reprogramming strategies, the most common of which involve overexpression of the four Yamanaka factors OCT4, SOX2, KLF4, and MYC [88]. Those pluripotency factors hold the potential to reprogram cells to a younger state at the cellular, tissue, and organismal levels, suggesting that aging may be a reversible process [89]. Indeed, the nuclear reprogramming allows the editing and resetting of the cellular clock by removing the molecular and cellular characteristics of senescent cell aging, including telomere size, gene expression profile, oxidative stress, and mitochondrial metabolism [90, 91].

In a recent study, the authors demonstrated that the in vivo cyclic expression of reprogramming factors extended life span in a premature aging mouse model and was beneficial to the health of physiologically aged mice [92].

Moreover, the reprogramming of aged somatic cells to stem cells can be used as an alternative source of cells for transplantation and for genetic editing. Indeed, the possibility to derive iPSCs from aging-related pathological cells has permitted investigators to develop recombination-based therapeutic strategies to edit genetic defects accountable for early and accelerated aging [93]. For instance, human iPSC-based models for aging-related degenerative diseases have been tested to understand the disease dynamics in Parkinson's disease, Alzheimer's disease, and progeroid laminopathies [94].

Age-related epigenetic modifications offer a model in which the memory of aging could be potentially altered or canceled to restore stem cell function. It is still unclear whether rejuvenation strategies more clinically viable such as modulating circulating factors or senolysis, which do not involve reprogramming to a fully pluripotent state, really reset the whole aged epigenome. It is possible that the stem cells exposed to these interventions of rejuvenation show some functional characteristics of the younger cells but still retain a memory of aging. Further analysis of the changes in epigenetic features of stem cells exposed to such interventions will help answer this important question.

6 Conclusions

Stem cell aging is influenced by many intrinsic and extrinsic cellular pathways, which often show crosstalk in determining stem cell function. Some signals seem to be involved more widely than others and are therefore possible targets for rejuvenation therapeutic approaches. In fact, as we have seen in this chapter, in some cases, the aging phenotypes can be reversed to restore stem cell regenerative function. However, whether these strategies truly restore stem cell functionality to a young state or instead induce a "pseudo-youth" state in which treated cells maintain an epigenetic memory of their true age remains an open and important question. The rejuvenation interventions that target the aging mechanisms in age-related diseases are particularly promising for the treatment of many impactful diseases, including neurodegeneration, sarcopenia, and heart failure.

Although, undoubtedly, there is still much to be discovered about stem cell function, aging, and treatment pathways, the data available so far clearly encourage further studies that will hopefully permit the development of effective therapeutic approaches that may, someday, ameliorate our aging.

References

[1] N.E. Sharpless, R.A. DePinho, How stem cells age and why this makes us grow old, Nat. Rev. Mol. Cell Biol. 8 (9) (2007) 703–713.
[2] P. Burra, D. Arcidiacono, D. Bizzaro, T. Chioato, R. Di Liddo, A. Banerjee, et al., Systemic administration of a novel human umbilical cord mesenchymal stem cells population accelerates the resolution of acute liver injury, BMC Gastroenterol. 12 (2012) 88.
[3] A. Banerjee, D. Bizzaro, P. Burra, R. Di Liddo, S. Pathak, D. Arcidiacono, et al., Umbilical cord mesenchymal stem cells modulate dextran sulfate sodium induced acute colitis in immunodeficient mice, Stem Cell Res. Ther. 6 (2015) 79.
[4] J. Oh, Y.D. Lee, A.J. Wagers, Stem cell aging: mechanisms, regulators and therapeutic opportunities, Nat. Med. 20 (8) (2014) 870–880.
[5] M.B. Schultz, D.A. Sinclair, When stem cells grow old: phenotypes and mechanisms of stem cell aging, Development 143 (1) (2016) 3–14.
[6] D.L. Jones, T.A. Rando, Emerging models and paradigms for stem cell ageing, Nat. Cell Biol. 13 (5) (2011) 506–512.
[7] A. Behrens, J.M. van Deursen, K.L. Rudolph, B. Schumacher, Impact of genomic damage and ageing on stem cell function, Nat. Cell Biol. 16 (3) (2014) 201–207.
[8] C.M. McCay, F. Pope, W. Lunsford, G. Sperling, P. Sambhavaphol, Parabiosis between old and young rats, Gerontologia 1 (1) (1957) 7–17.
[9] F.C. Ludwig, R.M. Elashoff, Mortality in syngeneic rat parabionts of different chronological age, Trans. N. Y. Acad. Sci. 34 (7) (1972) 582–587.
[10] J.M. Ruckh, J.W. Zhao, J.L. Shadrach, P. van Wijngaarden, T.N. Rao, A.J. Wagers, et al., Rejuvenation of regeneration in the aging central nervous system, Cell Stem Cell 10 (1) (2012) 96–103.
[11] M. Sinha, Y.C. Jang, J. Oh, D. Khong, E.Y. Wu, R. Manohar, et al., Restoring systemic GDF11 levels reverses age-related dysfunction in mouse skeletal muscle, Science 344 (6184) (2014) 649–652.
[12] I.M. Conboy, M.J. Conboy, A.J. Wagers, E.R. Girma, I.L. Weissman, T.A. Rando, Rejuvenation of aged progenitor cells by exposure to a young systemic environment, Nature 433 (7027) (2005) 760–764.
[13] S.A. Villeda, J. Luo, K.I. Mosher, B.D. Zou, M. Britschgi, G. Bieri, et al., The ageing systemic milieu negatively regulates neurogenesis and cognitive function, Nature 477 (7362) (2011) 90–U157.
[14] B.M. Carlson, J.A. Faulkner, Muscle transplantation between young and old rats—age of host determines recovery, Am. J. Physiol. 256 (6) (1989) C1262–C1266.
[15] Z.F. Song, J.W. Wang, L.M. Guachalla, G. Terszowski, H.R. Rodewald, Z.Y. Ju, et al., Alterations of the systemic environment are the primary cause of impaired B and T lymphopoiesis in telomere-dysfunctional mice, Blood 115 (8) (2010) 1481–1489.
[16] A. Eggel, T. Wyss-Coray, A revival of parabiosis in biomedical research, Swiss Med. Wkly. 144 (2014) w13914.
[17] D.T. Scadden, Nice neighborhood: emerging concepts of the stem cell niche, Cell 157 (1) (2014) 41–50.
[18] M. Boyle, C. Wong, M. Rocha, D.L. Jones, Decline in self-renewal factors contributes to aging of the stem cell niche in the Drosophila testis, Cell Stem Cell 1 (4) (2007) 470–478.
[19] L. Pan, S. Chen, S. Weng, G. Call, D. Zhu, H. Tang, et al., Stem cell aging is controlled both intrinsically and extrinsically in the Drosophila ovary, Cell Stem Cell 1 (4) (2007) 458–469.
[20] T. Sato, J.H. van Es, H.J. Snippert, D.E. Stange, R.G. Vries, M. van den Born, et al., Paneth cells constitute the niche for Lgr5 stem cells in intestinal crypts, Nature 469 (7330) (2011) 415–418.
[21] J.V. Chakkalakal, K.M. Jones, M.A. Basson, A.S. Brack, The aged niche disrupts muscle stem cell quiescence, Nature 490 (7420) (2012) 355–360.
[22] S.J. Morrison, A.C. Spradling, Stem cells and niches: mechanisms that promote stem cell maintenance throughout life, Cell 132 (4) (2008) 598–611.
[23] S.J. Forbes, N. Rosenthal, Preparing the ground for tissue regeneration: from mechanism to therapy, Nat. Med. 20 (8) (2014) 857–869.
[24] S.W. Lane, D.A. Williams, F.M. Watt, Modulating the stem cell niche for tissue regeneration, Nat. Biotechnol. 32 (8) (2014) 795–803.
[25] A.J. Wagers, The stem cell niche in regenerative medicine, Cell Stem Cell 10 (4) (2012) 362–369.

[26] C. Franceschi, J. Campisi, Chronic inflammation (inflammaging) and its potential contribution to age-associated diseases, J. Gerontol. A Biol. Sci. Med. Sci. 69 (2014) S4–S9.

[27] D. Jurk, C. Wilson, J.F. Passos, F. Oakley, C. Correia-Melo, L. Greaves, et al., Chronic inflammation induces telomere dysfunction and accelerates ageing in mice, Nat. Commun. 5 (2014) 4172.

[28] H.Y. Chung, B. Sung, K.J. Jung, Y. Zou, B.P. Yu, The molecular inflammatory process in aging, Antioxid. Redox Signal. 8 (3–4) (2006) 572–581.

[29] J.P. de Magalhães, J. Curado, G.M. Church, Meta-analysis of age-related gene expression profiles identifies common signatures of aging, Bioinformatics 25 (7) (2009) 875–881.

[30] D.J. Baker, T. Wijshake, T. Tchkonia, N.K. LeBrasseur, B.G. Childs, B. van de Sluis, et al., Clearance of p16Ink4a-positive senescent cells delays ageing-associated disorders, Nature 479 (7372) (2011) 232–236.

[31] M.E. Carlson, C. Suetta, M.J. Conboy, P. Aagaard, A. Mackey, M. Kjaer, et al., Molecular aging and rejuvenation of human muscle stem cells, EMBO Mol. Med. 1 (8–9) (2009) 381–391.

[32] F.S. Loffredo, M.L. Steinhauser, S.M. Jay, J. Gannon, J.R. Pancoast, P. Yalamanchi, et al., Growth differentiation factor 11 is a circulating factor that reverses age-related cardiac hypertrophy, Cell 153 (4) (2013) 828–839.

[33] C.T. Lutz, L.S. Quinn, Sarcopenia, obesity, and natural killer cell immune senescence in aging: altered cytokine levels as a common mechanism, Aging (Albany NY) 4 (8) (2012) 535–546.

[34] C. Elabd, W. Cousin, P. Upadhyayula, R.Y. Chen, M.S. Chooljian, J. Li, et al., Oxytocin is an age-specific circulating hormone that is necessary for muscle maintenance and regeneration, Nat. Commun. 5 (2014) 4082.

[35] C. Elabd, A. Basillais, H. Beaupied, V. Breuil, N. Wagner, M. Scheideler, et al., Oxytocin controls differentiation of human mesenchymal stem cells and reverses osteoporosis, Stem Cells 26 (9) (2008) 2399–2407.

[36] L. Katsimpardi, N.K. Litterman, P.A. Schein, C.M. Miller, F.S. Loffredo, G.R. Wojtkiewicz, et al., Vascular and neurogenic rejuvenation of the aging mouse brain by young systemic factors, Science 344 (6184) (2014) 630–634.

[37] L.M. Pacheco, L.A. Gomez, J. Dias, N.M. Ziebarth, G.A. Howard, P.C. Schiller, Progerin expression disrupts critical adult stem cell functions involved in tissue repair, Aging (Albany NY) 6 (12) (2014) 1049–1063.

[38] A. Infante, A. Gago, G.R. de Eguino, T. Calvo-Fernández, V. Gómez-Vallejo, J. Llop, et al., Prelamin A accumulation and stress conditions induce impaired Oct-1 activity and autophagy in prematurely aged human mesenchymal stem cell, Aging (Albany NY) 6 (4) (2014) 264–280.

[39] W.E. Balch, R.I. Morimoto, A. Dillin, J.W. Kelly, Adapting proteostasis for disease intervention, Science 319 (5865) (2008) 916–919.

[40] M. Bucciantini, E. Giannoni, F. Chiti, F. Baroni, L. Formigli, J. Zurdo, et al., Inherent toxicity of aggregates implies a common mechanism for protein misfolding diseases, Nature 416 (6880) (2002) 507–511.

[41] R.I. Morimoto, A.M. Cuervo, Protein homeostasis and aging: taking care of proteins from the cradle to the grave, J. Gerontol. A Biol. Sci. Med. Sci. 64 (2) (2009) 167–170.

[42] R.C. Taylor, A. Dillin, Aging as an event of proteostasis collapse, Cold Spring Harb. Perspect. Biol. 3 (5) (2011) a004440.

[43] I. Moreno-Gonzalez, C. Soto, Misfolded protein aggregates: mechanisms, structures and potential for disease transmission, Semin. Cell Dev. Biol. 22 (5) (2011) 482–487.

[44] M.R. Warr, M. Binnewies, J. Flach, D. Reynaud, T. Garg, R. Malhotra, et al., FOXO3A directs a protective autophagy program in haematopoietic stem cells, Nature 494 (7437) (2013) 323–327.

[45] M. Mortensen, E.J. Soilleux, G. Djordjevic, R. Tripp, M. Lutteropp, E. Sadighi-Akha, et al., The autophagy protein Atg7 is essential for hematopoietic stem cell maintenance, J. Exp. Med. 208 (3) (2011) 455–467.

[46] Ö.H. Yilmaz, P. Katajisto, D.W. Lamming, Y. Gültekin, K.E. Bauer-Rowe, S. Sengupta, et al., mTORC1 in the Paneth cell niche couples intestinal stem-cell function to calorie intake, Nature 486 (7404) (2012) 490–495.

[47] M. Laplante, D.M. Sabatini, mTOR signaling in growth control and disease, Cell 149 (2) (2012) 274–293.

[48] C. Chen, Y. Liu, P. Zheng, mTOR regulation and therapeutic rejuvenation of aging hematopoietic stem cells, Sci. Signal. 2 (98) (2009) ra75.

[49] D. Harman, Free radical theory of aging: dietary implications, Am. J. Clin. Nutr. 25 (8) (1972) 839–843.

[50] S. Pervaiz, R. Taneja, S. Ghaffari, Oxidative stress regulation of stem and progenitor cells, Antioxid. Redox Signal. 11 (11) (2009) 2777–2789.

[51] A. Stolzing, E. Jones, D. McGonagle, A. Scutt, Age-related changes in human bone marrow-derived mesenchymal stem cells: consequences for cell therapies, Mech. Ageing Dev. 129 (3) (2008) 163–173.

[52] K. Ito, A. Hirao, F. Arai, S. Matsuoka, K. Takubo, I. Hamaguchi, et al., Regulation of oxidative stress by ATM is required for self-renewal of haematopoietic stem cells, Nature 431 (7011) (2004) 997–1002.

[53] J.H. Paik, Z. Ding, R. Narurkar, S. Ramkissoon, F. Muller, W.S. Kamoun, et al., FoxOs cooperatively regulate diverse pathways governing neural stem cell homeostasis, Cell Stem Cell 5 (5) (2009) 540–553.

[54] H.F. Yuan, C. Zhai, X.L. Yan, D.D. Zhao, J.X. Wang, Q. Zeng, et al., SIRT1 is required for long-term growth of human mesenchymal stem cells, J. Mol. Med. (Berl) 90 (4) (2012) 389–400.

[55] H.S. Kim, K. Patel, K. Muldoon-Jacobs, K.S. Bisht, N. Aykin-Burns, J.D. Pennington, et al., SIRT3 is a mitochondria-localized tumor suppressor required for maintenance of mitochondrial integrity and metabolism during stress, Cancer Cell 17 (1) (2010) 41–52.

[56] K. Brown, S. Xie, X. Qiu, M. Mohrin, J. Shin, Y. Liu, et al., SIRT3 reverses aging-associated degeneration, Cell Rep. 3 (2) (2013) 319–327.

[57] K. Ito, A. Hirao, F. Arai, K. Takubo, S. Matsuoka, K. Miyamoto, et al., Reactive oxygen species act through p38 MAPK to limit the lifespan of hematopoietic stem cells, Nat. Med. 12 (4) (2006) 446–451.

[58] R.V. Kondratov, O. Vykhovanets, A.A. Kondratova, M.P. Antoch, Antioxidant N-acetyl-L-cysteine ameliorates symptoms of premature aging associated with the deficiency of the circadian protein BMAL1, Aging (Albany NY) 1 (12) (2009) 979–987.

[59] L. Drowley, M. Okada, S. Beckman, J. Vella, B. Keller, K. Tobita, et al., Cellular antioxidant levels influence muscle stem cell therapy, Mol. Ther. 18 (10) (2010) 1865–1873.

[60] G. Failla, The aging process and cancerogenesis, Ann. N. Y. Acad. Sci. 71 (6) (1958) 1124–1140.

[61] J. Vijg, Somatic mutations and aging: a re-evaluation, Mutat. Res. 447 (1) (2000) 117–135.

[62] H.J. Curtis, The Somatic Mutation Theory of Aging, 1965.

[63] D.J. Rossi, D. Bryder, J. Seita, A. Nussenzweig, J. Hoeijmakers, I.L. Weissman, Deficiencies in DNA damage repair limit the function of haematopoietic stem cells with age, Nature 447 (7145) (2007) 725–729.

[64] C.E. Rube, A. Fricke, T.A. Widmann, T. Furst, H. Madry, M. Pfreundschuh, et al., Accumulation of DNA damage in hematopoietic stem and progenitor cells during human aging, PLoS One 6 (3) (2011) 9.

[65] B.M. Moehrle, H. Geiger, Aging of hematopoietic stem cells: DNA damage and mutations? Exp. Hematol. 44 (10) (2016) 895–901.

[66] M.A. Shammas, Telomeres, lifestyle, cancer, and aging, Curr. Opin. Clin. Nutr. Metab. Care 14 (1) (2011) 28–34.

[67] B. Bernardes de Jesus, E. Vera, K. Schneeberger, A.M. Tejera, E. Ayuso, F. Bosch, et al., Telomerase gene therapy in adult and old mice delays aging and increases longevity without increasing cancer, EMBO Mol. Med. 4 (8) (2012) 691–704.

[68] M. Jaskelioff, F.L. Muller, J.H. Paik, E. Thomas, S. Jiang, A.C. Adams, et al., Telomerase reactivation reverses tissue degeneration in aged telomerase-deficient mice, Nature 469 (7328) (2011) 102–106.

[69] S.E. Artandi, S. Alson, M.K. Tietze, N.E. Sharpless, S. Ye, R.A. Greenberg, et al., Constitutive telomerase expression promotes mammary carcinomas in aging mice, Proc. Natl. Acad. Sci. U. S. A. 99 (12) (2002) 8191–8196.

[70] E. González-Suárez, E. Samper, A. Ramírez, J.M. Flores, J. Martín-Caballero, J.L. Jorcano, et al., Increased epidermal tumors and increased skin wound healing in transgenic mice overexpressing the catalytic subunit of telomerase, mTERT, in basal keratinocytes, EMBO J. 20 (11) (2001) 2619–2630.

[71] R. Jaenisch, A. Bird, Epigenetic regulation of gene expression: how the genome integrates intrinsic and environmental signals, Nat. Genet. 33 (Suppl) (2003) 245–254.

[72] I. Beerman, D.J. Rossi, Epigenetic control of stem cell potential during homeostasis, aging, and disease, Cell Stem Cell 16 (6) (2015) 613–625.

[73] M. Buscarlet, A. Tessier, S. Provost, L. Mollica, L. Busque, Human blood cell levels of 5-hydroxymethylcytosine (5hmC) decline with age, partly related to acquired mutations in TET2, Exp. Hematol. 44 (11) (2016) 1072–1084.

[74] M.F. Fraga, M. Esteller, Epigenetics and aging: the targets and the marks, Trends Genet. 23 (8) (2007) 413–418.

[75] S. Peleg, F. Sananbenesi, A. Zovoilis, S. Burkhardt, S. Bahari-Javan, R.C. Agis-Balboa, et al., Altered histone acetylation is associated with age-dependent memory impairment in mice, Science 328 (5979) (2010) 753–756.

[76] V. Krishnan, M.Z. Chow, Z. Wang, L. Zhang, B. Liu, X. Liu, et al., Histone H4 lysine 16 hypoacetylation is associated with defective DNA repair and premature senescence in Zmpste24-deficient mice, Proc. Natl. Acad. Sci. U. S. A. 108 (30) (2011) 12325–12330.

[77] L. Fontana, Interventions to promote cardiometabolic health and slow cardiovascular ageing, Nat. Rev. Cardiol. 15 (9) (2018) 566–577.

[78] M.G. Novelle, A. Ali, C. Diéguez, M. Bernier, R. de Cabo, Metformin: a hopeful promise in aging research, Cold Spring Harb. Perspect. Med. 6 (3) (2016) a025932.

[79] S.C. Bridgeman, G.C. Ellison, P.E. Melton, P. Newsholme, C.D.S. Mamotte, Epigenetic effects of metformin: from molecular mechanisms to clinical implications, Diabetes Obes. Metab. 20 (7) (2018) 1553–1562.

[80] T. Wang, B. Tsui, J.F. Kreisberg, N.A. Robertson, A.M. Gross, M.K. Yu, et al., Epigenetic aging signatures in mice livers are slowed by dwarfism, calorie restriction and rapamycin treatment, Genome Biol. 18 (1) (2017) 57.

[81] F. Pietrocola, F. Castoldi, M. Markaki, S. Lachkar, G. Chen, D.P. Enot, et al., Aspirin recapitulates features of caloric restriction, Cell Rep. 22 (9) (2018) 2395–2407.

[82] Y. Li, W. Zhang, L. Chang, Y. Han, L. Sun, X. Gong, et al., Vitamin C alleviates aging defects in a stem cell model for Werner syndrome, Protein Cell 7 (7) (2016) 478–488.

[83] M. Agathocleous, C.E. Meacham, R.J. Burgess, E. Piskounova, Z. Zhao, G.M. Crane, et al., Ascorbate regulates haematopoietic stem cell function and leukaemogenesis, Nature 549 (7673) (2017) 476–481.

[84] D.P. Bartel, MicroRNAs: target recognition and regulatory functions, Cell 136 (2) (2009) 215–233.

[85] E.M. Heinrich, S. Dimmeler, MicroRNAs and stem cells: control of pluripotency, reprogramming, and lineage commitment, Circ. Res. 110 (7) (2012) 1014–1022.

[86] Y. Liu, W. Liu, C. Hu, Z. Xue, G. Wang, B. Ding, et al., MiR-17 modulates osteogenic differentiation through a coherent feed-forward loop in mesenchymal stem cells isolated from periodontal ligaments of patients with periodontitis, Stem Cells 29 (11) (2011) 1804–1816.

[87] A.J. Gruber, W.A. Grandy, P.J. Balwierz, Y.A. Dimitrova, M. Pachkov, C. Ciaudo, et al., Embryonic stem cell-specific microRNAs contribute to pluripotency by inhibiting regulators of multiple differentiation pathways, Nucleic Acids Res. 42 (14) (2014) 9313–9326.

[88] K. Takahashi, S. Yamanaka, Induction of pluripotent stem cells from mouse embryonic and adult fibroblast cultures by defined factors, Cell 126 (4) (2006) 663–676.

[89] C. Soria-Valles, C. López-Otín, iPSCs: on the road to reprogramming aging, Trends Mol. Med. 22 (8) (2016) 713–724.

[90] T.A. Rando, H.Y. Chang, Aging, rejuvenation, and epigenetic reprogramming: resetting the aging clock, Cell 148 (1–2) (2012) 46–57.

[91] L. Lapasset, O. Milhavet, A. Prieur, E. Besnard, A. Babled, N. Aït-Hamou, et al., Rejuvenating senescent and centenarian human cells by reprogramming through the pluripotent state, Genes Dev. 25 (21) (2011) 2248–2253.

[92] A. Ocampo, P. Reddy, P. Martinez-Redondo, A. Platero-Luengo, F. Hatanaka, T. Hishida, et al., In vivo amelioration of age-associated hallmarks by partial reprogramming, Cell 167 (7) (2016) 1719–1733.e12.

[93] W. Zhang, J. Qu, G.H. Liu, J.C.I. Belmonte, The ageing epigenome and its rejuvenation, Nat. Rev. Mol. Cell Biol. 21 (3) (2020) 137–150.

[94] G.H. Liu, B.Z. Barkho, S. Ruiz, D. Diep, J. Qu, S.L. Yang, et al., Recapitulation of premature ageing with iPSCs from Hutchinson-Gilford progeria syndrome, Nature 472 (7342) (2011) 221–225.

Chapter 17

Role of biological markers in stem cell aging and its implications in therapeutic processes

Sivanandane Sittadjody[a], Aamina Ali[a], Thilakavathy Thangasamy[b], M. Akila[c], R. Ileng Kumaran[d], and Emmanuel C. Opara[a]

[a]Wake Forest Institute for Regenerative Medicine, Wake Forest University School of Medicine, Winston-Salem, NC, United States, [b]Department of Human Biology, Forsyth Tech Community College, Winston-Salem, NC, United States, [c]College of Nursing, JIPMER, Puducherry, India, [d]Biology Department, Farmingdale State College, Farmingdale, NY, United States

1 Introduction: Stem cells and their therapeutic values

Over the past several decades, with increasing average life expectancy, chronic diseases with no lasting cures have become prevalent. The major risk factor tied to this significant increase in the incidence of chronic diseases appears to be aging [1]. Often looked at as the biological fountain of youth, stem cell research has shown great promise and potential as a novel treatment approach. The ability of stem cells to perpetuate both specialized and naïve cells has given rise to its utility in the field of regenerative medicine [2]. However, as stem cells continue to be studied and experimented with, stem cell aging and senescence have posed new challenges. Knowledge about the mechanisms of stem cell aging would not only helps us to have a better understanding on the age-dependent risk for a disease, but also the role of stem cells in the aging process itself [3]. In this chapter, we will discuss different types of stem cells, intrinsic and extrinsic mechanisms involved in their aging process, the biomarkers of stem cell aging and its clinical implications.

Different types of stem cells exist in our body, varying widely based on their functions and abilities. Depending on the environment, stem cells have two critical functions: self-renewal and differentiation. Self-renewal can be defined as the ability of stem cells to divide symmetrically and indefinitely in a controlled manner that allows it to retain its developmental potential [4]. This means that although a stem cell may proliferate into more cells, it does not give rise to a mature cell type. On the other hand, differentiation may be defined as stem cells' capability to divide into either a single or diverse cell lineages. Through differentiation, stem cells may specialize and mature to become a part of various tissues and organs.

The extent to which a stem cell can differentiate and the environment from which it is derived can organize stem cells into different types. During the early stages of embryonic development, when the zygote begins to divide and create the blastocyst, each cell possesses stem cell property and is capable of giving rise to the whole organism and thus termed as "totipotent stem cells." Totipotent stem cells divide into identical stem cells that form the embryonic cells (i.e., germline layers) and the extraembryonic cells (i.e., the placenta). As gestation continues, "pluripotent stem cells" allow for organ development and tissue specialization. Unlike totipotent stem cells, pluripotent stem cells may only contribute to embryo development, yielding what is known as "embryonic stem cells" (ESCs). Due to the expression of telomerase, ESCs have been found to have unlimited self-renewal capabilities, although the mechanisms underlying this phenomenon are unknown [4]. However, even with its striking similarities with cancer, embryonic stem cells (ESCs) exhibit shorter cell cycles that contribute to its efficient maintenance of genetic integrity and resistance of any malignant transformations [4]. Near the end of gestation, pluripotent stem cells begin to yield "multipotent stem cells." Multipotent stem cells are much more specialized than pluripotent stem cells in that they differentiate terminally. "Adult stem cells" or "somatic stem cells" are multipotent stem cells that can be identified as tissue-specific stem cells such as neural stem cells, mesenchymal stem cells, and epidermal stem cells. This differentiation is possible due to asymmetric cell divisions of adult stem cells yielding a daughter stem cell and progenitor cell. The daughter stem cell maintains the pool of adult stem cells, while the progenitor propagates to mature tissue depending on its niche. Compared to embryonic stem cells, adult stem cells have longer cell cycles and contribute to mature tissue development and healing persisting throughout an organism's lifetime [4].

Multipotent stem cells can be further classified based on the niches from which they are derived. Unlike ESCs, adult stem cells can have multiple origins depending on their location within a tissue or the niche from which they are derived. Niches with altered microenvironment and organization in response to biological functions of tissues contribute to the impaired adult stem cell differentiation [5]. Accordingly, the multipotent stem cells include hematopoietic stem cells (HSCs), hair follicle stem cells (HFSCs), intestinal stem cells (ISCs), mesenchymal stem cells (MSCs), muscle stem cells (MuSCs), neural stem cells (NSCs), and germline stem cells (GSCs) [including oogonial stem cells (OSCs) and spermatogonia stem cells (SSCs)]. Both HSCs and MSCs can be found in various tissues all over the body but differ in their utility both biologically and clinically. HSCs differentiate into all cellular components comprised in the blood, such as red blood cells, white blood cells, and platelets. HSCs have been the longest studied stem cells dating back to the 1990s and have gained a lot of attention by pharmaceutical and biotechnology companies [6]. As the name suggests, NSCs are responsible for differentiating into neurons and all supporting cells such as astrocytes. NSCs are responsible for the creation and maintenance of neurons and connected nervous system. On the other hand, MSCs propagate into bone, adipose tissue, and muscles. MSCs are immune evasive cells that have been found to suppress host immune response [7]. Moreover, there is significant interest in MSCs for the regeneration due to their trophic capabilities [6].

Based on their respective niches, HSCs and MSCs can also be derived from cord blood and tissue. A distinctive hallmark of umbilical cord stem cells is their high proliferative potential resembling pluripotent stem cells as well as naïve or inexperienced antigen character. For this reason, umbilical cord stem cells have profound therapeutic effects allowing for allogeneic transplantations [8]. The HSCs derived from umbilical cord blood can be used to treat several blood and metabolic-related conditions, blood cancers, and other disorders without the interference of the adaptive immune system. Likewise, contained in both umbilical cord tissue and placental tissue is a surplus of naïve MSCs that can be used for tissue replacement, disease modeling, and tissue engineering [9].

2 Hallmarks of stem cell aging

From an evolutionary standpoint, stem cells can be seen as a mechanism to maintain biological homeostasis, fostering optimal fitness of organisms at reproductive age [10]. However, as we age, several extrinsic and intrinsic factors have been understood to influence the aging and senescence of stem cells, as illustrated in Fig. 1. The hallmarks of aging process of an organism postulated by Lopez-Otin et al. [11] and revisited recently by Guerville et al. [12] include theory of "dysfunctional stem cells" as a causative factor for an organism's aging, where stem cell senescence and consequent stem cell exhaustion are two critical steps. Stem cells share some of the aging mechanisms with somatic cells and have some mechanisms unique to themselves. Age-related changes in stem cells are both functional and numerical. Understanding and generalizing stem cell aging have posed challenges due to the diversity of stem cells and their various abilities. Stem cells naturally possess special mechanisms to counteract the aging process including various defense mechanisms to preserve their long telomere, increased protein homeostasis (proteostasis), and diminished ROS generation caused by toxic substances. For stem cell therapies and regenerative medicine approaches, we need a comprehensive understanding on the cell aging and various biological molecules involved in it.

3 Biological markers (biomarkers) of stem cell aging

As per the great Irish physicist William Thomson's statement, "If you cannot measure it, you cannot improve it," measuring physiological phenomena including the aging process is a pivotal step in assessing the quality of stem cells and their therapeutic potentials. In order to measure the aging process in stem cells, it requires indicators or biomarkers, the candidate molecules of pathways that lead to aging process. Identification of suitable biomarkers of stem cell aging is crucial to predict the success of their therapeutic approach. The major challenge in selecting biomarkers of stem cell aging is that there exists several candidate molecules, which participate in multiple pathways of aging. Moreover, existing variations in the aging mechanism between different adult stem cells make the process even more difficult. Another factor to be considered for the success of the stem cell therapy is that the aging phenomena need to be assessed in both (a) the donor stem cells and (b) recipient's systemic factors and the recipient site microenvironment status. In this chapter, biomarkers of both intrinsic mechanism of stem cells and the extrinsic markers that would influence the stem cell functions are compiled. The biomarker selection has been made based on some of the key criteria that (a) the biomarkers have to be reported and applicable in clinical cases; (b) whenever biomarkers are not available from human studies for a particular aging mechanism, biomarkers from animal studies have been listed; (c) specific attention has been given to those noninvasive biomarkers for their favorable applications; and (d) any suggestive biomarkers needs to be validated before their application in human cases. Since the stem cell therapy involves either autologous stem cells or donor stem cells, it is feasible to measure the

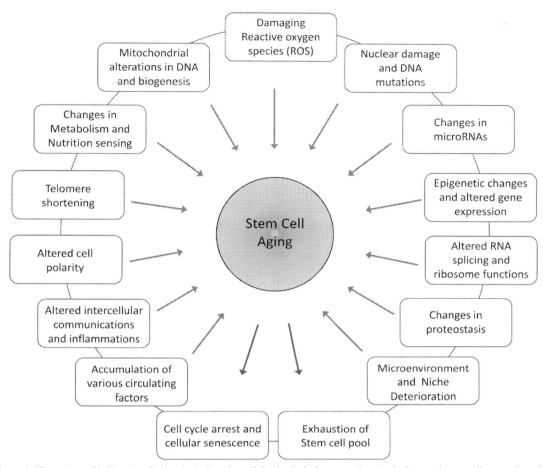

FIG. 1 Schematic illustration of hallmarks of aging including the cellular intrinsic factors and extrinsic factors that contribute to the aging process in the adult stem cells.

biomarkers of aging at the cellular level. At the same time, quantifying the circulatory factors that negatively influence the transplanted stem cell function in the recipient would aid in devising any approach to correct and counteract aging process for a successful transplant. However, the greatest challenge would be predicting the tissue niche at the transplant site, where the stem cells are going to end up and differentiate into functional cells.

3.1 Exhaustion of stem cell pools

Among various hallmarks of aging in an organism, "stem cell theory of aging" is a critical one, because stem cells are the stock cells that regenerate and repair any damages incurred in an organ system. Some of the stem cells include, but not restricted to, HSCs in bone marrow, HFSCs in skin, ISCs of intestinal crypts, MSC residing in several different tissues including bone marrow and adipose tissue, MuSCs in muscles, and NSC among the neural tissues. Adult stem cells that are present in every tissue and organ have been reported to decline with age. Although considered as immortal, even the stem cells undergo damages upon exposed to various external and internal insults. These damages accumulate over time, eventually lead to impaired stem cells, and a regress in the number in various tissues with age. As a result of the aging process, stem cells diminish not only in number, but also in function. Maintaining robust stem cell pools is very critical for both extended life span and healthy span of an organism including humans. Therefore, the self-repairing capacity of each tissue is determined by the pool of stem cell population that resides in that tissue. When these pool of stem cells undergo any functional attrition due to aging, it would reflect on the regenerative ability of the tissue to repair itself.

HSCs are the commonly studied stem cells to test the theory of stem cell pool exhaustion and aging. In an animal study, HSCs increase in cell number with advancement in age probably due to diminished capacity of asymmetric division [13]. In another mice study, there was no difference observed in the cycling activity of the HSCs with age [14]. However, the aged animal in these studies is shown to experience a functional depletion of HSC pool. Unlike these animal studies, a hematopoiesis analysis in a 115-year-old woman has shown a decline in HSC clones with age from about 10, 000 clones to fewer thousands [15]. Stem cell exhaustion is also reported in other stem cells such as MuSCs, NSCs, and GSCs [16–19]. The decline in the number of stem cells in particular tissue and the deterioration in their functions are linked to various mechanisms in the hallmarks of aging which are discussed in the following sections.

3.2 Oxidative stress

Free radical (oxidative stress) theory of aging has been the long accepted one ever since Harman [20] proposed in the 1950s as the most plausible explanation for aging process whether it is at the organism level or at the molecular level. The theory illustrates the existence of a fine balance between the free radicals generated in the form of Reactive Oxygen Species (ROS) and the antioxidant system in order to prevent any deterioration caused within the cells. When the free radical theory of aging was revised later, mitochondrial function was linked with the generation of ROS, thereby contributing to oxidative stress [21]. This ROS generation and associated oxidative stress increase with age [22, 23]. Some of the causative agents include (a) radiations (such as UV and X-rays), (b) damage-causing endogenous metabolites, and (c) uncoupled status between ROS generation and free radical-scavenging antioxidants. Therefore, the stem cells are unable to cope up with the generation of free radicals and result in oxidative stress [9, 24–26]. This age-related decline in the defense mechanism against the oxidative stress leads to dysregulation of cellular detoxifying pathways, which eventually results in activation of cell death signaling pathways such as apoptosis, necrosis, and autophagy [27]. Controlled by antioxidant molecules, excess ROS or depletion of ROS can have adverse effects on stem cell ability to differentiate, self-renewal capabilities, and senescence [28]. However, recent studies have found that ROS has a more complicated role than reported. It plays a central role in various pathways and responses maintaining homeostasis.

Evaluating the status of oxidative stress and free radical scavengers such as antioxidants in both the donor stem cell samples and the recipient's system would provide a promising development in successful stem cell therapies. Even though quantifying oxidative stress by measuring ROS in the clinical samples is challenging due to the transient and unstable nature of the free radicals, altered intracellular molecules such as (a) the downstream biomolecules (DNA, lipids, and proteins) that undergo damage due to the ROS, and (b) antioxidants (enzymatic and non-enzymatic antioxidants) could serve as biomarkers to assess the overall status of oxidative stress. Table 1 provides a brief list of ROS, antioxidants, and potential biomarkers that could be used in clinical setup for stem cell therapies.

Activation of FoxO family transcription factors such as FoxO-1, -3, -4 through PTEN-AKT-mTOR pathway is known to protect the stem cells from the damages induced by ROS through upregulation of antioxidant enzyme superoxide dismutase 2 (SOD2) [125–128]. Similarly, DNA damage-induced signaling through ataxia-telangiectasia mutated (ATM) pathway is also known to regulate FoxO-mediated defense in stem cells [129, 130]. FoxO transcription factors induce expression of PSDM11, which in turn regulates the expression of various stem cell markers such as *DPPA4, DPPA2, NANOG, OCT4, POU5F1, SOX2, TERT, UTF1, and ZFP42* [131].

3.3 Genomic instability

DNA damage caused by both extrinsic factors (harmful radiations and chemicals) and intrinsic factors (oxidative stress) has to be balanced by their counteracting intracellular repair mechanisms [132, 133]. Some of the damages occur at the level of nuclear DNA include formation of DNA adducts, modifications of nitrogenous base or sugar residues, appearance of single-strand or double-strand breaks, and cross-linking of DNA strands. Among these alterations, DNA double-strand breaks cleave the backbone of the DNA molecule and hence are considered as a lethal lesion when they are not corrected by the repair mechanisms [134, 135]. Depending on the type of DNA damage, at least six different DNA repair pathways exist to safeguard the integrity of DNA molecule [136, 137]. Activation of ATM and/or ATR pathway leads to the phosphorylation of histone H2AX (p-H2AX), and accumulation of p-H2AX could serve as biological indication of DNA damage. The DNA damage could also be quantified using comet tail assay to assess DNA strand breaks. Phospho-H2AX and comet tail assay would serve as simple and feasible biomarkers to check the genomic integrity in stem cells. Decline in DNA helicase would be another biomarker that would reflect the replication stress in stem cells [138]. Increase in *SIRT6* expression which codes for sirtuin 6 has been reported to counteract the aging process by enhancing DNA repair mechanisms [139]. In addition to the DNA damage, change in nuclear lamina, and alterations in the scaffold molecules such as lamins, would result in the alteration of nuclear architecture, which eventually leads to genomic instability [140]. Alterations in A-type lamins and decline in the levels of lamins B have been associated with cell senescence, suggesting their utility as biomarkers to predict stem cell aging [141].

During stem cell division, chromosomal aberrations result in the formation of micronuclei, which could be measured to assess any chromosomal damage in the stem cells. The percentage of micronucleated stem cells from a small portion of the

TABLE 1 Biomarkers of oxidative stress damage.

I. Measurement of ROS directly using fluorogenic probes [29–37]

1. 5-(and-6)-carboxy-2′,7′-dichlorodihydrofluorescein (DCFDA) to measure H_2O_2, OH^-, and ROO^- [36]
2. Dihydroethidium (DHE) to measure O_2^- [30, 35]
3. d-ROMs test to measure hydroxy peroxides (R-OOH) in serum [38]

II. Measurement of molecules undergoing ROS-induced damage

1. Protein damages
    (a) Protein carbonyl content [39–46]
    (b) Advanced oxidation protein products (AOPPs) [47, 48]
2. Lipid damages: Lipid peroxidation and resultant damage to cell membranes
    (a) Malondialdehyde (MDA) [49–57]
    (b) 8-iso prostaglandin F2alpha (8-isoPGF2α) [58, 59]
    (c) 4-hydroxy-2-nonenal (4-HNE) [60, 61]
    (d) Conjugated dienes (CDs) [62–64]
    (e) Lipid hydroperoxides (LOOHs) [65]
    (f) Oxidized low-density lipoproteins (oxLDLs) [66–68]
3. DNA damages
    (a) 8-hydroxy-2′-deoxyguanosine (8-OHdG) could be detected in circulation [69–75]
    (b) Thymidine glycol (TG) as intracellular markers and 8-hydroxy thymidine glycol (8-OHTG) in circulation [76]
    (c) DNA breaks (single- and/or double-strand breaks) as intracellular marker [32, 37]
    (d) DNA base modifications and intracellular marker [77]
    (e) DNA repair enzymes (human 8-oxoguanine-DNA-glycosylase (hOGG) and apurinic/apyrimidinic endonuclease (APE)) [78–81]

III. Measurement of antioxidant

1. Enzymatic antioxidants
    (a) Superoxide dismutase (SOD) [82–88]
    (b) Catalase [82, 89–97]
    (c) Glutathione peroxidase (GPx) [98–104]
    (d) Glutathione-S-transferase (GST) [94, 97, 98, 105–107]
2. Nonenzymatic antioxidants
    (a) Glutathione (GSH) [102, 108–110]
    (b) Vitamin A [59, 111–115]
    (c) Vitamin C [116–118]
    (d) Vitamin E [97, 119, 120]
3. Total antioxidant status (TAS) [121–124]

stem cell preparation (minimum 2000 stem cells) could be detected with the help of an automatic microscopic scoring or by an experienced pathology scorer [142, 143]. Any increase in the percentage of micronuclei due to donor age or prolonged propagation in vitro could serve as a suitable biomarker to predict any chromosomal damage in the stem cell [144, 145].

3.4 Telomere attrition

Telomere shortening theory of stem cell aging is one of the foremost mechanisms proposed for the decline in stem cell population and function. One of the unique hallmarks of stem cells is their self-renewal capabilities, which are closely tied to telomerase and telomere length [146]. The telomere region of chromosome is protected by a repetitive DNA sequence of TTAGGG referred to as chromosome capping or telomere. During the cell division process, the above-mentioned telomere shortening is called telomere attrition and it is the most evaluated hallmark of stem cell aging. As the longest living proliferating cell type in the body, adult stem cells are sensitive to telomere dysfunction. According to the Hayflick limit, there is a finite limit of cell divisions (around 50 to 70) human cells adhere to due to telomere shortening limitations [147]. For this reason, telomere dysfunction is observed to be strongly correlated with stem cell aging. In order to maintain DNA integrity and prevent telomere shortening, cells exhibit two major checkpoints: replicative senescence and crisis. The first checkpoint, replicative senescence, occurs when cells exhibit low levels of telomeric dysfunction and cause permanent cell cycle arrest. On the other hand, crisis checkpoint is characterized by high levels of telomeric dysfunction and massive chromosomal instability leading to cell apoptosis [146]. Voluminous work (till date around 8000 studies have been published as indexed in PubMed) in the field of telomere research provides reliable methods to quantify telomere attrition. Telomere attrition could be measured by one of the two methods: (a) through the conventional method of measuring telomere length using choice of Southern blot technique [148] or quantitative polymerase chain reaction (qPCR) technique [149], and (b) by quantifying the level of telomerase enzyme, the reverse transcriptase enzyme that proofreads and restores the telomere length. Telomere attrition has been reported to be linked with the production of progerin, an aberrant prelamin A isoform that affects nuclear integrity and thereby leading to stem cell aging [150].

Involvement of many intrinsic factors operates in the process of stem cell aging and senescence. In order to better understand these factors, in vivo experimental models using telomerase knockout mice have been used. Telomerase enzyme itself is composed of a telomerase RNA component (TERC) and telomerase reverse transcriptase (TERT). Active in most cancer cells and stem cells, telomerase is a reverse transcriptase enzyme used to extend telomeric sequences and thus, prolong cellular proliferative capabilities [151, 152]. Compared to mice, humans have longer life spans and exhibit greater telomere shortening, suggesting that insufficient telomerase activity with age can influence stem cell aging [153–155]. Through comparative studies between the telomerase knockout mice and wild-type mice, scientists have observed the protective and detrimental effects that various intrinsic factors can have on stem cell functions [146]. Therefore, TERT activity in the extracts of cultured stem cells could serve as biomarker. Genes involved in the mismatch repair pathway play a pivotal role in recognizing and correcting the telomere dysfunction. For this reason, the deletion of genes such as exonuclease 1 has been studied and reported to contribute for the improved life span of stem cells in knockout mice compared to wild-type mice [146]. Impairment in DNA damage sensing mechanism reduced prolonged functionality of adult stem cells. p53, the tumor suppressor, is another intrinsic factor that is involved in stem cell aging process. The deletion of p53 protein results in a shortened life span due to cancer formation in both wild-type and telomerase knockout mice. On the contrary, expression of p53 gene also resulted in shortened life span due to higher checkpoint controls, eventually resulting in long-term loss of functionality in adult stem cells [146]. However, telomere attrition is not the sole phenomenon that drives stem cell aging. In many studies including human subjects and mice, aging has been reported that is independent of telomere damage [156, 157]. Therefore, additional parameters need to be assessed using additional biomarkers to predict the stem cell aging process.

3.5 Epigenetic modifications

In addition to alterations in the DNA sequence including mutations, there are other changes in DNA that lead to altered gene expression profile in stem cells and thereby cause aging [11]. Epigenetics changes and their effects on the regulation of gene expression have also been found to play a pivotal role in stem cell aging. According to Waddington's epigenetic landscape theory, stem cells undergo cellular decision-making processes influencing its differentiation and final fate [158]. However, during this differentiation process and general aging process, stem cells can be affected by the numerous epigenetic changes. These changes include (a) methylation of certain nucleic acid base in DNA, (b) modification of histone such as deacetylation, and (c) remodeling of chromatin, together referred to as epigenetic modifications. Common epigenetic modifications such as histone acetylation and DNA methylation have been found not only to affect the rate of aging but also influence the lineage by skewing its differentiation path and overall function [159]. The expression of histone itself has been reported to decrease with age in stem cells [160]. In addition to the levels of histones, some of the activation modifications of histones such as H3K4me3 and H4K16Ac decrease, while the repressive modification of H3K27me3 increases in stem cells with age [160, 161]. These histone levels and histone modifications could also be used as biomarkers to assess the epigenetic changes in stem cells.

The level of sirtuin 1 (SIRT1), a histone deacetylase (HDAC), decreases with stem cell aging and this could serve as a biomarker [162]. In addition to SIRTs, the levels of other HDACs and components of chromatin modifiers such as switch/sucrose non-fermentable (SWI-SNF) complex, polycomb repressive complex (PRC), and DNA methyltransferase also gets altered with stem cell aging [163, 164]. DNA methylation at the CpG cluster plays a major role in altered gene expression, and several groups reported a predictor of donor stem cell aging status called "DNA methylation clocks" (DMCs) [165, 166]. The DMC was developed by analyzing the DNA methylation in the whole blood [167] and has been used in various clinical studies [168–170]. Even though the DMC was employed in clinical studies to predict the difference between true chronological aging and epigenetic aging, thereby correlating with survivability and mortality, the same method could be used to detect the epigenetic aging of stem cells and their quality.

3.6 Altered microRNA profile

MicroRNAs (miRNAs) are small noncoding RNAs that play a role in the regulation of gene expression mainly at posttranscriptional level. By binding to the 3′ untranslated region (3′UTR) of target RNA, the miRNAs cause degradation of the nascent transcript by RNA-induced silencing complex (RISC). In humans, 60% of the genes are regulated by about 1000 miRNAs coded by the human genome. The diverse role of miRNAs in all biological and metabolic processes including development, cell proliferation, apoptosis, cell cycle regulation, lipid metabolism, signal transduction, aging, and diseases [171]. Certain miRNAs are known to repress the potency of various stem cell types [172–177]. miR-145 is known to suppress the pluripotency of human ESCs by targeting *Oct4*, *Sox2*, and *Klf4* [178]. Similarly, the differentiation of mouse ESC has been shown to be regulated by miRNAs such as miR-134, miR-296, and miR-470 through targeting the genes of *Nanog*, *Oct4*, and *Sox2* [179]. miRNAs play a role not only in differentiation of stem cells into desired tissue cells but also in defending the stem cells against aging process. Defense against the aging process is one among various functions of miRNAs, through key mechanisms such as regulation of ROS, DNA repair, and apoptosis.

3.7 RNA splicing and defective ribosomal machineries

Studies have shown that aging could also be driven by defect in the RNA splicing machinery and defective splicing, which could be utilized as biomarkers to predict stem cell aging. In humans, one of the different components of splicing apparatus, splicing factor 1 (SFA-1), has been reported to be affected and leads to the accumulation of defective splicing complexes with age. In turn, these defective RNA splicing mechanisms have been reported to be reduced by caloric restriction (CR) and mTOR complex 1 pathway components as a part of their antiaging effects [180]. Together, this suggests the significance of homeostasis in the RNA splicing mechanism in stem cell aging. Downstream to RNA splicing, ribosomes are the actual protein synthetic machinery, which are comprised of ribosomal RNA (rRNA) and ribosomal proteins. Ribosomes, like any other cell organelle, are targeted by aging pathways. Within the stem cells, an upregulation of ribosomes (through increase in transcription of ribosomal protein genes and rRNA genes by hypomethylating the genes) has been linked to accelerated aging process [161].

3.8 Loss of proteostasis

The protein homeostasis (or) homeostasis of proteomes in the intracellular compartment in order to maintain various cellular functions is referred to as "proteostasis," which has a huge impact on stem cell aging. Proteostasis is maintained by several quality control mechanisms such as (a) chaperone-guided refolding of proteins, (b) ubiquitin-mediated protein degradation, and (c) lysosomal pathways (called as autophagy). As mentioned earlier, stem cells can often go through asymmetric division during maturation. Altered proteins during this defective process due to dysregulation of proteases can affect not only the functionality of stem cells but also the aging of an organism [181, 182]. This has been proven in many animal models, that chaperone and heat shock proteins can correlate with longevity of stem cell life span [183, 184].

Due to continuous division, there is an increase in intracellular stress within stem cells, which leads to an accumulation of misfolded proteins. Normally, these misfolded proteins are either corrected or eliminated by various in-built proteostasis mechanisms; when these dysregulated proteostasis mechanisms fail, they become a causative factor for stem cell aging. The levels of proteins that prevent protein aggregation or misfolding could be used as biomarkers of stem cell aging. One such protein that prevents protein aggregation and precipitation is clusterin (CLU), which also exists in the soluble form as well as called sCLU (also referred to as apolipoprotein J) [185]. A few studies used the levels of CLU components in serum or in intracellular compartment to correlate the loss of proteostasis mechanism with the aging process [186–190]. Because of the existence of soluble form in secretions and circulation, sCLU appears to be a suitable candidate to assess the status of proteostasis mechanisms, thereby predicting stem cell aging.

3.9 Change in cell polarity

Asymmetric cell division observed in stem cell is an adaptive mechanism to prevent any accumulation of damaged components within the cells. Asymmetric segregation of damaged proteins in addition to enhanced proteostasis precedes the asymmetric cell division. As a result of asymmetric division, the damaged components such as damaged DNA, carbonylated proteins, replicating circular DNA, and damaged organelles are distributed to the daughter cells that are destined to differentiate, whereas the daughter cells that stay as stem cell retain its integrity [191–194]. Therefore, asymmetrically dividing stem cells need to maintain polarity, which has been reported to be lost in several stem cell types [195]. In addition, age-related changes in the Wnt signaling have been reported to affect the polarity of HSCs and satellite cells [196, 197]. Altered polarity in the distribution of cell organelles could be utilized as a biomarker to predict the disrupted cell polarity and associated aging phenotype.

3.10 Mitochondrial dysfunction

Mitochondrial theory of aging is another causative mechanism for stem cell aging, and mitochondrial function has been reported to diminish with cellular age [198]. In addition to the increased ROS production in mitochondria, defects in mitochondrial integrity such as mitochondrial DNA (mtDNA) mutations and decline in mitochondria biogenesis are the known mechanisms responsible for the dysfunction of mitochondria. Apart from the mutations, deletion of whole mtDNA, unlike in the case of nuclear DNA, is the main alteration that affects the integrity of mitochondria, thereby causing stem cell aging [199]. Most of the mtDNA mutations are caused by replication errors and oxidative stress by compromising the mtDNA repair mechanisms and further aggravate the aging microenvironment of mitochondria [200]. To ensure healthy mitochondrial genome, stem cells must be screened for mutations or alternatively the stem cells could be collected from younger donor for a better clinical outcome.

Decline in mitochondrial biogenesis is another causative factor for mitochondrial dysfunction. Peroxisome proliferator-activated receptor gamma (PPARγ) coactivator-1 alpha and 1 beta (PGC-1α and 1β) are the known master regulators of mitochondrial biogenesis, and their repression by p53 is the resultant of telomere attrition [201]. SIRT1 is known to eliminate any damaged mitochondria employing the autophagy process through transcriptional coactivator PGC-1α [202]. Similarly, SIRT3, a deacetylase enzyme, targets many enzymes involved in mitochondrial functions and one of the key enzymes regulated by SIRT3 is manganese SOD (a major mitochondrial antioxidant enzyme), thereby controlling the ROS generation [203, 204]. These reports suggest sirtuins (SIRT1 and SIRT3) as potential biomarkers of mitochondrial biogenesis to predict stem cell aging.

A stress-induced cytokine, growth differentiation factor 15 (GDF-15), is a member of transforming growth factor (TGF)-β superfamily, which has been reported to be produced by organ systems including lung, kidney, and liver during their aging process [205]. In a Swedish clinical study, higher levels of circulating GDF-15 have been reported to be associated with age-related diseases such as cardiovascular issues and cancer, independent of the telomere length and other cytokines such as interleukin-6 (IL-6) [206]. GDF-15 could be considered as a potential biomarker to detect the mitochondrial dysfunction since it has been used as a diagnostic marker in predicting inherited mitochondrial diseases [207]. GDF-15 could be measured in stem cell culture media and also in the serum samples of the stem cell donors to evaluate the mitochondrial function. Although the increase in GDF-15 expression and secretion is in response to dysregulated energy metabolism and linked to the mitochondrial disease, their direct link with mitochondrial dysfunction-related aging is still unclear.

Apelin, a myokine induced by exercise, is also suggested to be another biomarker connected with mitochondrial dysfunction [208]. Apelin is known to increase the muscle function by regulating mitochondriogenesis and also by inducing muscle stem cells. In addition, their regulation on muscle function is also through other pathways of aging such as autophagy and inflammation [209].

3.11 Deregulated nutrient sensing and cellular metabolism

Nutrient sensing mechanism in mammals involves the growth hormone (GH) axis, which includes GH produced from anterior pituitary and the downstream mediators of GH action, insulin-like growth factor (IGF) system (earlier referred to as somatomedins) and its components. IGF system components include (a) ligands: IGF-I, IGF-II; (b) receptors: IGF receptors (type I and type II); (c) IGF-binding proteins (IGFBPs 1 through 6); and (d) IGFBP protease system [210, 211]. Interestingly, the IGF receptor type I through which IGFs mediate their action shares a lot of structural and functional similarities with insulin receptor (IR). In addition, a crosstalk between the ligands and receptors of IGF and insulin signaling mechanisms leads to terming these two pathways together as insulin-IGF signaling (IIS) pathway [212]. Along with the intracellular signaling molecules of IIS pathway, the external factors such as insulin and IGF system components are crucial predictors of nutrient sensing that could be measured in the stem cell culture medium and in circulation. IIS pathway has been reported to increase the aging process in stem cells. A plenty of studies are available that reported the measurement of IIS components in both culture medium and circulation to predict the normal function of the system. Alterations in the components of IIS have been reported as metabolic aging and found correlated with various phenomena related to aging [210, 211, 213].

Like most cells in the body, stem cells rely on oxidative phosphorylation and glycolysis as a means to generate energy. Some of the stem cells that reside in a hypoxic environment rely on glycolysis (anaerobic process) with reduced exposure to ROS [181, 214, 215]. In stem cell populations derived from mice and flies, caloric restriction (CR) has been found to be a longevity-extending intervention for cell life [16, 216, 217]. Nutrient sensing pathways that mediate cellular life span and CR include not only IIS pathways, but also mammalian target of rapamycin (mTOR) signaling, sirtuins, and adenosine monophosphate (AMP)-activated kinase (AMPK). Dysregulation of this nutrient sensing has been determined as one of the hallmarks of stem cell aging although their exact mechanisms in stem cell metabolism are not currently clear. As mentioned previously, organisms with decrease in IIS pathway live longer due to slower rates of cell growth and metabolism. mTOR is responsible for sensing amino acid levels and regulating anabolic metabolism via mTOR complexes 1 and 2 (mTORC1 and mTORC2). Through genetically modified mouse models, it has been observed that mTORC1 downregulation is correlated with increased life span, making it as a critical mediator of longevity ([218, 219]. However, there are also several detrimental effects of downregulating the IIS and mTOR pathways such as insulin resistance, poor wound healing, and cataracts, there by prompting studies to focus on the mechanism behind these pathways [220].

Sirtuins and AMP sensors collaborate to sense energy state via nicotinamide adenine dinucleotide (NAD^+) levels and AMP levels, respectively. Unlike IIS and mTOR pathways which need to be downregulated, the upregulation of sirtuins and AMP sensors is correlated with cellular longevity through positive feedback loops [221]. Sirtuins function as nutrient sensor that signals the nutrient scarcity and catabolism. In other words, the activation of sirtuin signaling mimics CR, which is known to contribute to the longevity of organism and stem cells [222]. SIRT1 is the product of *sirt1* gene, and the expression of this biomolecule could be measured as a predictor of stem aging [223]. To validate the link between SIRTs and stem cell functions, *Sirt1* gene has been reported to crucially play in the regeneration of skeletal muscle from their stem cells [224]. SIRT1 originally described as nuclear protein has been recently reported to be detected in circulation as well using commonly employed techniques including ELISA, Western blot, and surface plasmon resonance [225]. SIRT1 has been used as a biomarker in a few clinical cases to study the mechanism of aging process and health living in the elderly population [226, 227].

3.12 Niche deterioration

The microenvironment in which the stem cells reside in a tissue is referred to as "niche." Niche provides a suitable platform for the interaction between the stem cells and various extrinsic signals. The signal could be mediated directly through a cell-to-cell communication or cell-to-matrix interaction. Additionally, the signaling by diffusible ligands also regulates the functions of stem cells. The interactions between (a) one stem cell and another, (b) stem cells and matrix, and (c) ligand and the stem cells in a given niche determine whether the stem should stay quiescent or divide to self-renew or differentiate [228]. Alterations such as changes in niches and microenvironment have also been suspected to play an important role in stem cell aging. Accumulation of excess ROS in a tissue is an example of niche deterioration and cellular injury and contribute to stem cell aging. The cell culture conditions and preexisting microenvironment at the transplant site of the stem cell recipient are referred to and discussed as niche (present and future, respectively) under this topic.

An aged niche at the transplant site of a recipient will fail to send proper signals of morphogens and growth factors needed for the proper grafting and function of the transplanted stem cells. Thus, such a preexisting condition not only impairs the function of the stem cell but also induces aging in the same. Another major extrinsic factor that influences the rate of stem cell aging is the onset of disease. The type of illness, the age of onset, and its persistence in one's life all influence adult stem cell self-renewal capabilities. For example, treatments such as chemotherapy have been found to deplete the population of HSCs leaving patients in need of a bone marrow transplant or other sources of stem cells to remove cancer entirely [146]. Likewise, diet has been found to have many direct and indirect influences on niches through nutrient signaling pathways and metabolic pathways. Although the long-term and short-term effects of diet are still unclear, induced systemic factors such as changes in the microbiome, immune responses via inflammation, and DNA methylation levels from carcinogenic exposures have been found to be connected with dietary intake [10]. Markers of inflammation (which is discussed in Section 3.14 of this chapter) also alter the composition of niche and favor stem cell aging and therefore, could result in an aging niche [229].

3.13 Circulating factors

Apart from the stem cell niche (both present and future), attentions need to be paid on the circulating factors of the recipient. By understanding these biomarkers of aging, we can gain insight into both the rate of aging and its underlying process [230]. Certain circulating biomolecules in serum (referred as serum markers) could serve as biomarkers of stem cell aging. Some of the key serum markers include decreased hemoglobin and antioxidant contents, as well as an increased albumin levels. Although helpful in making an early diagnosis, serum markers are not specific to stem cell aging alone but rather indicative of many age-related co-morbidities [231]. Other circulatory factors have also been found to influence stem cell aging such as secondary mediators of the somatotropic axis in humans known as the IIS pathway. As a downstream mediator of anterior pituitary-released growth hormones (GH), the IIS pathway (which has already been discussed in Section 3.11 of this chapter) plays a critical role in stem cell aging and alterations have been reported in age-related studies [210, 211]. The longevity-extending effects of CR in mice are believed to be mediated through reduced signaling from these molecules. Increase in levels of TGF-β with age is also known to impair the functions of stem cells of muscle and NSCs [232]. On the contrary, GDF-11 has been suggested to improve the function of stem cells and their levels are found to decline with age [233]. The therapeutic benefits of CR on counteracting stem cell aging are believed to be mediated through the secretion of various systemic factors only [234]. Hence, alterations in circulating systemic factors also could be used as biomarkers of stem cell aging.

3.14 Altered intercellular communications and accumulation of inflammasomes

Another contributing factor for stem cell aging is the alteration in the intercellular communications between cells. Altered intercellular communication is indicative of stem cell aging, and some of the communications between the cells are mediated through gap junctions, and it has been found to be an important contributor to what is known as contagious aging [235]. Contagious aging is characterized by the induction of senescence through chronic inflammation. For this reason, lifespan manipulating techniques carried on one tissue can be found to influence the life span of other surrounding tissues [236].

Inflammatory markers such as an increase in cytokines like interleukin-6 (IL-6) and tumor necrosis factor alpha (TNFα) as well as C-reactive protein (CRP) have been linked to stem cell senescence. IL6 and TNFα, although frequently involved with acute-phase inflammatory responses, are indicative of chronic illnesses and monitoring of stem cell aging [231]. Aging of the immune system also known as immunosenescence is characterized by the aging of various tissues and niche deterioration. Other biomarkers related to hormonal changes and musculoskeletal changes have also been debated as potential markers of stem cell senescence but are generally known as predictors of the aging process in general. A marker of HSCs and $CD34^+$ progenitor cells has been found to decline with age. A higher number of $CD34^+$ cells for elderly ages of 80 years have been found to be better indicators of lifespan longevity than other markers (i.e., cardiovascular risk factors and inflammatory markers). Interestingly, higher levels of adult stem cells have been observed among elderly population with a decline in their pluripotent capacity. This means that during aging process, progenitor cells give rise to more progeny as a way to compensate for the declining stem cell pool. A major difference in these progenitors from younger progenitors is the size of the clones, contributing to a finite capacity for division.

The systemic accumulation of inflammation factor in a chronic fashion with age is called inflammaging. This inflammaging is the driving force involved in the alteration of intercellular communications [11, 237]. Several hallmarks of stem cell aging, including loss of proteostasis and cell cycle arrest, trigger this immune response mainly due to the intracellular accumulation of misfolded proteins and senescence-associated secretory phenotypes (SASPs) [238]. The link between inflammatory factors and age-related changes has been reported in a large body of literature [239]. Inflammasome pathways, mainly by the proinflammatory cytokines such as IL-1b and IL-18, are reported to mediate this inflammatory signals to trigger the aging process [240]. The circulating levels of the proinflammatory cytokines in both the donor and recipient could serve as biomarker to predict the inflammaging-associated impact on stem cell function. In addition to inflammaging phenomenon, immunosenescence that includes quantitative and functional changes in various components of both the innate and adaptive immune systems contributes to the stem cell aging process [241]. Based on the expression of 57 immune response genes, an "IMM-AGE" score was derived by screening 135 healthy elderly individuals in a clinical study conducted by Alpert et al., [242], and this IMM-AGE scoring helped in drawing a trajectory of immunosenescence and their prognostic value.

3.15 Cell cycle arrest and cellular senescence

Cell cycle arrest in stem cells is a survival mechanism to ensure the cellular integrity of actively dividing stem cells. Stem cells, like any other dividing cell, undergo arrest, where the stem cell tries to rectify any errors occurred during the previous division. Even though cell cycle arrest is a preventative mechanism to avoid propagating any mutated stem cell, a prolonged cell cycle arrest of stem cells on the other hand would lead to stem cell senescence. A balance between proliferation and cell cycle is maintained by the tight regulation in the spatiotemporal expression of mitogens and tumor suppressor proteins. In addition to p53 discussed earlier, p16^{INK4a} is another tumor suppressor, and when increased in stem cells due to epigenetic de-repression of *ink4/a*rk locus, it results in a prolonged cell cycle arrest, thereby causing stem cell senescence [243]. Increased expression of *p16INK4a* has been reported to be positively associated with the aging process [244, 245]. These studies suggest that *p16INK4a* expression quantified by RT-PCR could be used as a potential biomarker to detect cell cycle arrest and stem cell senescence.

A state of constant cell cycle arrest in stem cells in a tissue is coupled with several phenotypic changes in the tissue niche such as production proinflammatory factors and matrix metalloproteases (MMPs), altogether referred to as senescence-associated secretory phenotype (SASP) [246]. The SASP affects the function of the recipient tissue (where the stem cells will be transplanted) by spreading the senescence to the neighboring cells as well. It is believed that SASP induced by cell cycle-arrested stem cell is an adaptive mechanism to prevent the proliferation of aged and damaged stem cells. SASP in the recipient site of stem cell transplant will affect the function of transplanted stem cells and hence, would be useful if measured in the circulation of the recipient. Therefore, systemic measurement of SASP would be an interesting biomarker to predict the aging microenvironment of the recipient site during a stem cell therapy.

4 Potential therapeutic interventions

Fig. 2 provides some of the potential interventions to counteract the aging mechanism in stem cells. Through reducing the levels of ROS, the stem cell aging caused by the oxidative stress could be reversed by modulating the metabolic and redox status. The use of antioxidants such as *N*-acetyl-L-cysteine (NAC), a precursor of GSH-mediated free radical scavenging system, has been shown to ameliorate the ROS-induced damages [247–249]. Similarly, water-soluble vitamin C could be

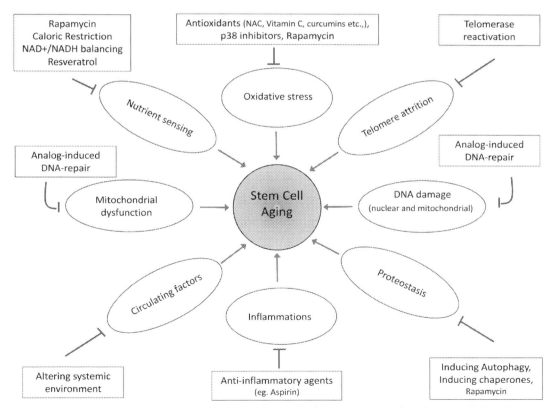

FIG. 2 Suggested strategies to target the pathways of stem cell aging either to ameliorate or to reverse them for stem cell therapies.

included in cell culture media as antioxidant to quench free radicals and counteract oxidative stress. Another approach tried in animal model is the use of polyphenolic antioxidants such as curcumin as a part of rejuvenation therapy prior to transplanting HSCs yielded a better outcome [250]. However, a better understanding of the endogenous sources of ROS and their target cell organelles/molecules is necessary to develop intervention therapies using ROS-modulating agents in order to counteract the damages inflicted by oxidative stress.

Telomerase reactivation is one approach that has been tested in animal models and shown to reverse the degenerative phenotypes in stem cells including reduced foci of DNA damage and decreased apoptosis. Although the telomerase reverting approach in stem cells is attractive due to its benefits. However, there are possibilities of inducing malignancies by this approach, since the same catalytic subunit of TERT is a known to promote cancer cell generation by suppressing their replicative senescence [251–253]. Nonetheless, there are studies in which pharmacological activation or viral transduction of telomerase systemically delays the aging process without any incidence of cancer in mice [254]. Further investigations in DNA damage repair/response pathways including the analogous inducible settings would be helpful in developing rejuvenating therapies using these DNA repair pathways. Artificial reinforcement of DNA repair mechanism as shown in mouse models could help to delay stem cell aging [11, 255]. Keyes and Fuchs recently have provided a list of transcriptional fingerprints that are associated with stem cell aging. In their reports, they have pointed that genes regulating cell adhesion, glycoproteins, and ribosome function are known to increase with stem cell aging, whereas genes governing DNA repair, DNA replication, and cell cycle are downregulated with age in stem cells [256].

Nutrient sensing is another area where intervention could be applied. Genetic manipulation decreasing the IIS pathway has been found to extend life span of several organisms, suggestive of its role in stem cell aging [257]. Rapamycin treatment was found to extend their longevity and therefore suggested as one of the robust chemical interventions against stem cell aging [219]. Even though CR has been found to be a successful strategy in animal models to reduce the impact of stem cell aging and extend the longevity, it is challenging to apply in clinical cases. Since an imbalance between the levels of intracellular NAD+ and NADH is one of the causative factors for stem cell aging through mitochondrial dysfunction [258], strategies to manipulate their intracellular levels in humans could be of therapeutic benefits. Similarly, a naturally occurring small molecule, resveratrol, is known to induce metabolic benefits similar to calorie restriction regulating mitochondria biogenesis mediated through SIRT1 and PGC-1α [259]. Induction of heat shock proteins in animal model through pharmacological agents has been reported to ameliorate the disrupted proteostasis and delay the progression of dystrophic pathologies associated with stem cell aging [260]. Studies using rejuvenation strategies to correct defects in intercellular communication, including anti-inflammatory agents (i.e., aspirin) and manipulation of intestinal microbiome, are underway in hopes of being translated to clinical medicine. Rejuvenating the systemic environment of an older stem cell recipient with circulating factors to mimic the microenvironment of a younger individual would be another potential approach. Several of these suggested intervention approaches have been tested and studied in animal models, which require further investigations before translating them into clinical cases.

5 Conclusion and perspectives

As the limitless potentials of stem cells continued to be explored, understandings on stem cell aging has proven to be valuable. In this current chapter, the role of aging biomarkers in the success of stem cell transplantation has been reviewed in relation to the donor age and the recipient age. By understanding the biomarkers related to stem cell aging, not only we can gain better comprehension into the aging process but also the fundamentals about the rates at which stem cells age. Quantitative analysis of stem cell aging can also give insight into predictive tools in terms of life span and understanding of human health [230]. By identifying targets of stem cell senescence, we may also be able to improve regenerative therapies and stratified medicine [146].

Reports are very limited on biomarkers for stem cell aging to predict and intervene in the stem cell aging process. Based on the biomolecules involved in various mechanisms of stem cell aging, the authors have provided a list of potential biomarkers representing various phenomena of stem cell aging. Owing to the complexity of the aging process and existence of cell-to-cell variations among different adult stem cells, it is not possible to compile a complete list of biomarkers. However, the authors have made an attempt to provide as many biomarkers as possible, which could be utilized to predict the aging status of the donor stem cells and also the niche in the recipient. However, one should tailor the provided information to apply in their stem cell therapy based on the stem cell type and status of niche in the recipient (both systemic and transplant sites).

References

[1] T. Niccoli, L. Partridge, Ageing as a risk factor for disease, Curr. Biol. 22 (17) (2012) R741–R752.
[2] E. Vlashi, F. Pajonk, Cancer stem cells, cancer cell plasticity and radiation therapy, Semin. Cancer Biol. 31 (2015) 28–35.
[3] N.E. Sharpless, G. Schatten, Stem cell aging, J. Gerontol. A Biol. Sci. Med. Sci. 64 (2) (2009) 202–204.

[4] A. Krtolica, Stem cell: balancing aging and cancer, Int. J. Biochem. Cell Biol. 37 (5) (2005) 935–941.
[5] E. Fuchs, T. Tumbar, G. Guasch, Socializing with the neighbors: stem cells and their niche, Cell 116 (6) (2004) 769–778.
[6] A.I. Caplan, Are all adult stem cells the same? Regen. Eng. Transl. Med. 1 (1) (2015) 4–10.
[7] J.A. Ankrum, J.F. Ong, J.M. Karp, Mesenchymal stem cells: immune evasive, not immune privileged, Nat. Biotechnol. 32 (3) (2014) 252–260.
[8] M.N. Fernandez, Improving the outcome of cord blood transplantation: use of mobilized HSC and other cells from third party donors, Br. J. Haematol. 147 (2) (2009) 161–176.
[9] L. Weinberger, et al., Dynamic stem cell states: naive to primed pluripotency in rodents and humans, Nat. Rev. Mol. Cell Biol. 17 (3) (2016) 155–169.
[10] M. Ermolaeva, et al., Cellular and epigenetic drivers of stem cell ageing, Nat. Rev. Mol. Cell Biol. 19 (9) (2018) 594–610.
[11] C. Lopez-Otin, et al., The hallmarks of aging, Cell 153 (6) (2013) 1194–1217.
[12] F. Guerville, et al., Revisiting the hallmarks of aging to identify markers of biological age, J. Prev Alzheimers Dis. 7 (1) (2020) 56–64.
[13] M.C. Florian, et al., Cdc42 activity regulates hematopoietic stem cell aging and rejuvenation, Cell Stem Cell 10 (5) (2012) 520–530.
[14] H. Geiger, G. de Haan, M.C. Florian, The ageing haematopoietic stem cell compartment, Nat. Rev. Immunol. 13 (5) (2013) 376–389.
[15] H. Holstege, et al., Somatic mutations found in the healthy blood compartment of a 115-yr-old woman demonstrate oligoclonal hematopoiesis, Genome Res. 24 (5) (2014) 733–742.
[16] M. Cerletti, et al., Short-term calorie restriction enhances skeletal muscle stem cell function, Cell Stem Cell 10 (5) (2012) 515–519.
[17] M. Boyle, et al., Decline in self-renewal factors contributes to aging of the stem cell niche in the Drosophila testis, Cell Stem Cell 1 (4) (2007) 470–478.
[18] A.V. Molofsky, et al., Increasing p16INK4a expression decreases forebrain progenitors and neurogenesis during ageing, Nature 443 (7110) (2006) 448–452.
[19] P. Sousa-Victor, et al., Geriatric muscle stem cells switch reversible quiescence into senescence, Nature 506 (7488) (2014) 316–321.
[20] D. Harman, Aging: a theory based on free radical and radiation chemistry, J. Gerontol. 11 (3) (1956) 298–300.
[21] D. Harman, The biologic clock: the mitochondria? J. Am. Geriatr. Soc. 20 (4) (1972) 145–147.
[22] J. Sastre, et al., Mitochondria, oxidative stress and aging, Free Radic. Res. 32 (3) (2000) 189–198.
[23] J. Sastre, F.V. Pallardo, J. Vina, Mitochondrial oxidative stress plays a key role in aging and apoptosis, IUBMB Life 49 (5) (2000) 427–435.
[24] S. De Barros, et al., Aging-related decrease of human ASC angiogenic potential is reversed by hypoxia preconditioning through ROS production, Mol. Ther. 21 (2) (2013) 399–408.
[25] G. Kasper, et al., Insights into mesenchymal stem cell aging: involvement of antioxidant defense and actin cytoskeleton, Stem Cells 27 (6) (2009) 1288–1297.
[26] A. Stolzing, et al., Age-related changes in human bone marrow-derived mesenchymal stem cells: consequences for cell therapies, Mech. Ageing Dev. 129 (3) (2008) 163–173.
[27] D.D. Haines, B. Juhasz, A. Tosaki, Management of multicellular senescence and oxidative stress, J. Cell. Mol. Med. 17 (8) (2013) 936–957.
[28] E. Cianflone, et al., Adult cardiac stem cell aging: a reversible stochastic phenomenon? Oxidative Med. Cell. Longev. 2019 (2019) 5813147.
[29] D. Andres, et al., Changes in antioxidant defence systems induced by cyclosporine A in cultures of hepatocytes from 2- and 12-month-old rats, Biochem. Pharmacol. 59 (9) (2000) 1091–1100.
[30] L. Benov, L. Sztejnberg, I. Fridovich, Critical evaluation of the use of hydroethidine as a measure of superoxide anion radical, Free Radic. Biol. Med. 25 (7) (1998) 826–831.
[31] N.S. Chandel, et al., Role of oxidants in NF-kappa B activation and TNF-alpha gene transcription induced by hypoxia and endotoxin, J. Immunol. 165 (2) (2000) 1013–1021.
[32] Y. Chen Wongworawat, et al., Chronic oxidative stress increases the integration frequency of foreign DNA and human papillomavirus 16 in human keratinocytes, Am. J. Cancer Res. 6 (4) (2016) 764–780.
[33] S. Cohen, et al., Reactive oxygen species and serous epithelial ovarian adenocarcinoma, Cancer Res. J. 4 (6) (2016) 106–114.
[34] F.L. Meyskens Jr., et al., Aberrant redox regulation in human metastatic melanoma cells compared to normal melanocytes, Free Radic. Biol. Med. 31 (6) (2001) 799–808.
[35] H.M. Peshavariya, G.J. Dusting, S. Selemidis, Analysis of dihydroethidium fluorescence for the detection of intracellular and extracellular superoxide produced by NADPH oxidase, Free Radic. Res. 41 (6) (2007) 699–712.
[36] P. Ubezio, F. Civoli, Flow cytometric detection of hydrogen peroxide production induced by doxorubicin in cancer cells, Free Radic. Biol. Med. 16 (4) (1994) 509–516.
[37] V.M. Williams, et al., Human papillomavirus type 16 E6* induces oxidative stress and DNA damage, J. Virol. 88 (12) (2014) 6751–6761.
[38] R. Trotti, M. Carratelli, M. Barbieri, Performance and clinical application of a new, fast method for the detection of hydroperoxides in serum, Panminerva Med. 44 (1) (2002) 37–40.
[39] D.A. Butterfield, et al., In vivo oxidative stress in brain of Alzheimer disease transgenic mice: requirement for methionine 35 in amyloid beta-peptide of APP, Free Radic. Biol. Med. 48 (1) (2010) 136–144.
[40] S. Mehrabi, et al., Oxidatively modified proteins in the serous subtype of ovarian carcinoma, Biomed. Res. Int. 2014 (2014) 585083.
[41] F. De Marco, et al., Oxidative stress in HPV-driven viral carcinogenesis: redox proteomics analysis of HPV-16 dysplastic and neoplastic tissues, PLoS One 7 (3) (2012) e34366.
[42] M. Rajesh, et al., Determination of carbonyl group content in plasma proteins as a useful marker to assess impairment in antioxidant defense in patients with Eales' disease, Indian J. Ophthalmol. 52 (2) (2004) 139–144.
[43] V. Kolgiri, V.W. Patil, Protein carbonyl content: a novel biomarker for aging in HIV/AIDS patients, Braz. J. Infect. Dis. 21 (1) (2017) 35–41.

[44] C.S. Mesquita, et al., Simplified 2,4-dinitrophenylhydrazine spectrophotometric assay for quantification of carbonyls in oxidized proteins, Anal. Biochem. 458 (2014) 69–71.
[45] R.L. Levine, et al., Determination of carbonyl content in oxidatively modified proteins, Methods Enzymol. 186 (1990) 464–478.
[46] B.S. Berlett, E.R. Stadtman, Protein oxidation in aging, disease, and oxidative stress, J. Biol. Chem. 272 (33) (1997) 20313–20316.
[47] C.J. Alderman, et al., The role of advanced oxidation protein products in regulation of dendritic cell function, Free Radic. Biol. Med. 32 (5) (2002) 377–385.
[48] V. Witko-Sarsat, et al., Advanced oxidation protein products as a novel marker of oxidative stress in uremia, Kidney Int. 49 (5) (1996) 1304–1313.
[49] D. Giustarini, et al., Oxidative stress and human diseases: origin, link, measurement, mechanisms, and biomarkers, Crit. Rev. Clin. Lab. Sci. 46 (5–6) (2009) 241–281.
[50] T. Yoshioka, et al., Lipid peroxidation in maternal and cord blood and protective mechanism against activated-oxygen toxicity in the blood, Am. J. Obstet. Gynecol. 135 (3) (1979) 372–376.
[51] H. Ohkawa, N. Ohishi, K. Yagi, Assay for lipid peroxides in animal tissues by thiobarbituric acid reaction, Anal. Biochem. 95 (2) (1979) 351–358.
[52] J.A. Buege, S.D. Aust, Microsomal lipid peroxidation, Methods Enzymol. 52 (1978) 302–310.
[53] M. Mihara, M. Uchiyama, Determination of malonaldehyde precursor in tissues by thiobarbituric acid test, Anal. Biochem. 86 (1) (1978) 271–278.
[54] K. Yagi, Lipid peroxides and human diseases, Chem. Phys. Lipids 45 (2–4) (1987) 337–351.
[55] P.J. Marshall, M.A. Warso, W.E. Lands, Selective microdetermination of lipid hydroperoxides, Anal. Biochem. 145 (1) (1985) 192–199.
[56] D. Lapenna, et al., Reaction conditions affecting the relationship between thiobarbituric acid reactivity and lipid peroxides in human plasma, Free Radic. Biol. Med. 31 (3) (2001) 331–335.
[57] H. Esterbauer, K.H. Cheeseman, Determination of aldehydic lipid peroxidation products: malonaldehyde and 4-hydroxynonenal, Methods Enzymol. 186 (1990) 407–421.
[58] R. Yu, et al., Method development and validation for ultra-high pressure liquid chromatography/tandem mass spectrometry determination of multiple prostanoids in biological samples, J. AOAC Int. 96 (1) (2013) 67–76.
[59] Y. Kataria, et al., Retinoid and carotenoid status in serum and liver among patients at high-risk for liver cancer, BMC Gastroenterol. 16 (2016) 30.
[60] E. Skrzydlewska, et al., Lipid peroxidation and antioxidant status in colorectal cancer, World J. Gastroenterol. 11 (3) (2005) 403–406.
[61] M. Jo, et al., Oxidative stress is closely associated with tumor angiogenesis of hepatocellular carcinoma, J. Gastroenterol. 46 (6) (2011) 809–821.
[62] K.S. Rao, R.O. Recknagel, Early onset of lipoperoxidation in rat liver after carbon tetrachloride administration, Exp. Mol. Pathol. 9 (2) (1968) 271–278.
[63] R.D. Situnayake, et al., Measurement of conjugated diene lipids by derivative spectroscopy in heptane extracts of plasma, Ann. Clin. Biochem. 27 (Pt. 3) (1990) 258–266.
[64] H.T. Facundo, et al., Elevated levels of erythrocyte-conjugated dienes indicate increased lipid peroxidation in schistosomiasis mansoni patients, Braz. J. Med. Biol. Res. 37 (7) (2004) 957–962.
[65] J. Nourooz-Zadeh, J. Tajaddini-Sarmadi, S.P. Wolff, Measurement of plasma hydroperoxide concentrations by the ferrous oxidation-xylenol orange assay in conjunction with triphenylphosphine, Anal. Biochem. 220 (2) (1994) 403–409.
[66] J. Frijhoff, et al., Clinical relevance of biomarkers of oxidative stress, Antioxid. Redox Signal. 23 (14) (2015) 1144–1170.
[67] S.B. Cheng, et al., Changes of oxidative stress, glutathione, and its dependent antioxidant enzyme activities in patients with hepatocellular carcinoma before and after tumor resection, PLoS One 12 (1) (2017), e0170016.
[68] H. Itabe, Oxidized low-density lipoprotein as a biomarker of in vivo oxidative stress: from atherosclerosis to periodontitis, J. Clin. Biochem. Nutr. 51 (1) (2012) 1–8.
[69] H.J. Helbock, et al., DNA oxidation matters: the HPLC-electrochemical detection assay of 8-oxo-deoxyguanosine and 8-oxo-guanine, Proc. Natl. Acad. Sci. U. S. A. 95 (1) (1998) 288–293.
[70] S. Toyokuni, et al., Quantitative immunohistochemical determination of 8-hydroxy-2'-deoxyguanosine by a monoclonal antibody N45.1: its application to ferric nitrilotriacetate-induced renal carcinogenesis model, Lab. Investig. 76 (3) (1997) 365–374.
[71] M.K. Shigenaga, et al., In vivo oxidative DNA damage: measurement of 8-hydroxy-2'-deoxyguanosine in DNA and urine by high-performance liquid chromatography with electrochemical detection, Methods Enzymol. 186 (1990) 521–530.
[72] A. Plachetka, et al., 8-Hydroxy-2'-deoxyguanosine in colorectal adenocarcinoma—is it a result of oxidative stress? Med. Sci. Monit. 19 (2013) 690–695.
[73] S. Kondo, et al., Overexpression of the hOGG1 gene and high 8-hydroxy-2'-deoxyguanosine (8-OHdG) lyase activity in human colorectal carcinoma: regulation mechanism of the 8-OHdG level in DNA, Clin. Cancer Res. 6 (4) (2000) 1394–1400.
[74] S. Kondo, et al., Persistent oxidative stress in human colorectal carcinoma, but not in adenoma, Free Radic. Biol. Med. 27 (3–4) (1999) 401–410.
[75] T. Akcay, et al., Increased formation of 8-hydroxy-2'-deoxyguanosine in peripheral blood leukocytes in bladder cancer, Urol. Int. 71 (3) (2003) 271–274.
[76] K. Ito, et al., Serum antioxidant capacity and oxidative injury to pulmonary DNA in never-smokers with primary lung cancer, Anticancer Res. 32 (3) (2012) 1063–1067.
[77] M. Dizdaroglu, Quantitative determination of oxidative base damage in DNA by stable isotope-dilution mass spectrometry, FEBS Lett. 315 (1) (1993) 1–6.
[78] F. Yamamoto, et al., Ubiquitous presence in mammalian cells of enzymatic activity specifically cleaving 8-hydroxyguanine-containing DNA, Jpn. J. Cancer Res. 83 (4) (1992) 351–357.
[79] Y.J. Park, et al., Genetic changes of hOGG1 and the activity of oh8Gua glycosylase in colon cancer, Eur. J. Cancer 37 (3) (2001) 340–346.

[80] D. Li, et al., Oxidative DNA damage and 8-hydroxy-2-deoxyguanosine DNA glycosylase/apurinic lyase in human breast cancer, Mol. Carcinog. 31 (4) (2001) 214–223.

[81] L. Chaisiriwong, et al., A case-control study of involvement of oxidative DNA damage and alteration of antioxidant defense system in patients with basal cell carcinoma: modulation by tumor removal, Oxidative Med. Cell. Longev. 2016 (2016) 5934024.

[82] Y. Oyanagui, Reevaluation of assay methods and establishment of kit for superoxide dismutase activity, Anal. Biochem. 142 (2) (1984) 290–296.

[83] P. Kakkar, B. Das, P.N. Viswanathan, A modified spectrophotometric assay of superoxide dismutase, Indian J. Biochem. Biophys. 21 (2) (1984) 130–132.

[84] M. Nishikimi, N. Appaji, K. Yagi, The occurrence of superoxide anion in the reaction of reduced phenazine methosulfate and molecular oxygen, Biochem. Biophys. Res. Commun. 46 (2) (1972) 849–854.

[85] E.F. Roth Jr., H.S. Gilbert, The pyrogallol assay for superoxide dismutase: absence of a glutathione artifact, Anal. Biochem. 137 (1) (1984) 50–53.

[86] S. Marklund, G. Marklund, Involvement of the superoxide anion radical in the autoxidation of pyrogallol and a convenient assay for superoxide dismutase, Eur. J. Biochem. 47 (3) (1974) 469–474.

[87] H.P. Misra, I. Fridovich, The role of superoxide anion in the autoxidation of epinephrine and a simple assay for superoxide dismutase, J. Biol. Chem. 247 (10) (1972) 3170–3175.

[88] J.M. McCord, I. Fridovich, Superoxide dismutase. An enzymic function for erythrocuprein (hemocuprein), J. Biol. Chem. 244 (22) (1969) 6049–6055.

[89] L. Goth, A simple method for determination of serum catalase activity and revision of reference range, Clin. Chim. Acta 196 (2–3) (1991) 143–151.

[90] A.K. Sinha, Colorimetric assay of catalase, Anal. Biochem. 47 (2) (1972) 389–394.

[91] H. Aebi, Catalase in vitro, Methods Enzymol. 105 (1984) 121–126.

[92] D.P. Nelson, L.A. Kiesow, Enthalpy of decomposition of hydrogen peroxide by catalase at 25 degrees C (with molar extinction coefficients of H_2O_2 solutions in the UV), Anal. Biochem. 49 (2) (1972) 474–478.

[93] R.F. Beers Jr., I.W. Sizer, A spectrophotometric method for measuring the breakdown of hydrogen peroxide by catalase, J. Biol. Chem. 195 (1) (1952) 133–140.

[94] J.K. Strzelczyk, et al., The activity of antioxidant enzymes in colorectal adenocarcinoma and corresponding normal mucosa, Acta Biochim. Pol. 59 (4) (2012) 549–556.

[95] G. Guner, et al., Evaluation of some antioxidant enzymes in lung carcinoma tissue, Cancer Lett. 103 (2) (1996) 233–239.

[96] A.L. Margaret, E. Syahruddin, S.I. Wanandi, Low activity of manganese superoxide dismutase (MnSOD) in blood of lung cancer patients with smoking history: relationship to oxidative stress, Asian Pac. J. Cancer Prev. 12 (11) (2011) 3049–3053.

[97] J. Grace Nirmala, R.T. Narendhirakannan, Detection and genotyping of high-risk HPV and evaluation of anti-oxidant status in cervical carcinoma patients in Tamil Nadu State, India—a case control study, Asian Pac. J. Cancer Prev. 12 (10) (2011) 2689–2695.

[98] H. Czeczot, et al., Glutathione and GSH-dependent enzymes in patients with liver cirrhosis and hepatocellular carcinoma, Acta Biochim. Pol. 53 (1) (2006) 237–242.

[99] P.A. Pleban, A. Munyani, J. Beachum, Determination of selenium concentration and glutathione peroxidase activity in plasma and erythrocytes, Clin. Chem. 28 (2) (1982) 311–316.

[100] D.E. Paglia, W.N. Valentine, Studies on the quantitative and qualitative characterization of erythrocyte glutathione peroxidase, J. Lab. Clin. Med. 70 (1) (1967) 158–169.

[101] L.R. Bennedsen, E.G. Sogaard, J. Muff, Development of a spectrophotometric method for on-site analysis of peroxygens during in-situ chemical oxidation applications, Water Sci. Technol. 70 (10) (2014) 1656–1662.

[102] G.L. Ellman, Tissue sulfhydryl groups, Arch. Biochem. Biophys. 82 (1) (1959) 70–77.

[103] D.G. Hafeman, R.A. Sunde, W.G. Hoekstra, Effect of dietary selenium on erythrocyte and liver glutathione peroxidase in the rat, J. Nutr. 104 (5) (1974) 580–587.

[104] J.T. Rotruck, et al., Selenium: biochemical role as a component of glutathione peroxidase, Science 179 (4073) (1973) 588–590.

[105] W.H. Habig, M.J. Pabst, W.B. Jakoby, Glutathione S-transferases. The first enzymatic step in mercapturic acid formation, J. Biol. Chem. 249 (22) (1974) 7130–7139.

[106] I. Hubatsch, M. Ridderstrom, B. Mannervik, Human glutathione transferase A4-4: an alpha class enzyme with high catalytic efficiency in the conjugation of 4-hydroxynonenal and other genotoxic products of lipid peroxidation, Biochem. J. 330 (Pt. 1) (1998) 175–179.

[107] M.R. Nogues, et al., Parameters related to oxygen free radicals in human skin: a study comparing healthy epidermis and skin cancer tissue, J. Invest. Dermatol. 119 (3) (2002) 645–652.

[108] M.L. Hu, Measurement of protein thiol groups and glutathione in plasma, Methods Enzymol. 233 (1994) 380–385.

[109] J. Sedlak, R.H. Lindsay, Estimation of total, protein-bound, and nonprotein sulfhydryl groups in tissue with Ellman's reagent, Anal. Biochem. 25 (1) (1968) 192–205.

[110] E. Beutler, O. Duron, B.M. Kelly, Improved method for the determination of blood glutathione, J. Lab. Clin. Med. 61 (1963) 882–888.

[111] G. Taibi, C.M. Nicotra, Development and validation of a fast and sensitive chromatographic assay for all-trans-retinol and tocopherols in human serum and plasma using liquid-liquid extraction, J. Chromatogr. B Anal. Technol. Biomed. Life Sci. 780 (2) (2002) 261–267.

[112] K.W. Miller, N.A. Lorr, C.S. Yang, Simultaneous determination of plasma retinol, alpha-tocopherol, lycopene, alpha-carotene, and beta-carotene by high-performance liquid chromatography, Anal. Biochem. 138 (2) (1984) 340–345.

[113] L.A. Kaplan, et al., Simultaneous, high-performance liquid chromatographic analysis of retinol, tocopherols, lycopene, and alpha- and beta-carotene in serum and plasma, Methods Enzymol. 189 (1990) 155–167.

[114] D. Gackowski, et al., Further evidence that oxidative stress may be a risk factor responsible for the development of atherosclerosis, Free Radic. Biol. Med. 31 (4) (2001) 542–547.
[115] D. Zhu, et al., Quantitative analyses of beta-carotene and retinol in serum and feces in support of clinical bioavailability studies, Rapid Commun. Mass Spectrom. 20 (16) (2006) 2427–2432.
[116] D. Ivanovic, et al., Reversed-phase ion-pair HPLC determination of some water-soluble vitamins in pharmaceuticals, J. Pharm. Biomed. Anal. 18 (6) (1999) 999–1004.
[117] M. Okamura, An improved method for determination of L-ascorbic acid and L-dehydroascorbic acid in blood plasma, Clin. Chim. Acta 103 (3) (1980) 259–268.
[118] V. Zannoni, et al., A rapid micromethod for the determination of ascorbic acid in plasma and tissues, Biochem. Med. 11 (1) (1974) 41–48.
[119] I.D. Desai, Vitamin E analysis methods for animal tissues, Methods Enzymol. 105 (1984) 138–147.
[120] S.A. Hashim, G.R. Schuttringer, Rapid determination of tocopherol in marco- and microquantities of plasma. Results obtained in various nutrition and metabolic studies, Am. J. Clin. Nutr. 19 (2) (1966) 137–145.
[121] O. Erel, A novel automated direct measurement method for total antioxidant capacity using a new generation, more stable ABTS radical cation, Clin. Biochem. 37 (4) (2004) 277–285.
[122] O. Erel, A new automated colorimetric method for measuring total oxidant status, Clin. Biochem. 38 (12) (2005) 1103–1111.
[123] G. Bartosz, Total antioxidant capacity, Adv. Clin. Chem. 37 (2003) 219–292.
[124] G. Cao, R.L. Prior, Comparison of different analytical methods for assessing total antioxidant capacity of human serum, Clin. Chem. 44 (6 Pt. 1) (1998) 1309–1315.
[125] S. Melov, et al., Mitochondrial disease in superoxide dismutase 2 mutant mice, Proc. Natl. Acad. Sci. U. S. A. 96 (3) (1999) 846–851.
[126] M.M. Juntilla, et al., AKT1 and AKT2 maintain hematopoietic stem cell function by regulating reactive oxygen species, Blood 115 (20) (2010) 4030–4038.
[127] C. Chen, et al., TSC-mTOR maintains quiescence and function of hematopoietic stem cells by repressing mitochondrial biogenesis and reactive oxygen species, J. Exp. Med. 205 (10) (2008) 2397–2408.
[128] J. Zhang, et al., PTEN maintains haematopoietic stem cells and acts in lineage choice and leukaemia prevention, Nature 441 (7092) (2006) 518–522.
[129] K. Miyamoto, et al., Foxo3a is essential for maintenance of the hematopoietic stem cell pool, Cell Stem Cell 1 (1) (2007) 101–112.
[130] K. Ito, et al., Reactive oxygen species act through p38 MAPK to limit the lifespan of hematopoietic stem cells, Nat. Med. 12 (4) (2006) 446–451.
[131] D. Vilchez, et al., Increased proteasome activity in human embryonic stem cells is regulated by PSMD11, Nature 489 (7415) (2012) 304–308.
[132] A.A. Moskalev, et al., The role of DNA damage and repair in aging through the prism of Koch-like criteria, Ageing Res. Rev. 12 (2) (2013) 661–684.
[133] J.H. Hoeijmakers, DNA damage, aging, and cancer, N. Engl. J. Med. 361 (15) (2009) 1475–1485.
[134] A. Ciccia, S.J. Elledge, The DNA damage response: making it safe to play with knives, Mol. Cell 40 (2) (2010) 179–204.
[135] S.P. Jackson, J. Bartek, The DNA-damage response in human biology and disease, Nature 461 (7267) (2009) 1071–1078.
[136] S.P. Jackson, Sensing and repairing DNA double-strand breaks, Carcinogenesis 23 (5) (2002) 687–696.
[137] R.D. Kennedy, A.D. D'Andrea, DNA repair pathways in clinical practice: lessons from pediatric cancer susceptibility syndromes, J. Clin. Oncol. 24 (23) (2006) 3799–3808.
[138] J. Flach, et al., Replication stress is a potent driver of functional decline in ageing haematopoietic stem cells, Nature 512 (7513) (2014) 198–202.
[139] Y. Kanfi, et al., The sirtuin SIRT6 regulates lifespan in male mice, Nature 483 (7388) (2012) 218–221.
[140] T. Dechat, et al., Nuclear lamins: major factors in the structural organization and function of the nucleus and chromatin, Genes Dev. 22 (7) (2008) 832–853.
[141] A. Freund, et al., Lamin B1 loss is a senescence-associated biomarker, Mol. Biol. Cell 23 (11) (2012) 2066–2075.
[142] M. Fenech, Cytokinesis-block micronucleus cytome assay, Nat. Protoc. 2 (5) (2007) 1084–1104.
[143] M. Fenech, A lifetime passion for micronucleus cytome assays—reflections from Down Under, Mutat. Res. 681 (2–3) (2009) 111–117.
[144] L. Migliore, et al., Association of micronucleus frequency with neurodegenerative diseases, Mutagenesis 26 (1) (2011) 85–92.
[145] S. Bonassi, et al., The HUman MicroNucleus project on eXfoLiated buccal cells (HUMN(XL)): the role of life-style, host factors, occupational exposures, health status, and assay protocol, Mutat. Res. 728 (3) (2011) 88–97.
[146] Z. Song, Z. Ju, K.L. Rudolph, Cell intrinsic and extrinsic mechanisms of stem cell aging depend on telomere status, Exp. Gerontol. 44 (1–2) (2009) 75–82.
[147] L. Hayflick, The limited in vitro lifetime of human diploid cell strains, Exp. Cell Res. 37 (1965) 614–636.
[148] M. Kimura, et al., Measurement of telomere length by the Southern blot analysis of terminal restriction fragment lengths, Nat. Protoc. 5 (9) (2010) 1596–1607.
[149] R.M. Cawthon, Telomere measurement by quantitative PCR, Nucleic Acids Res. 30 (10) (2002), e47.
[150] K. Cao, et al., Progerin and telomere dysfunction collaborate to trigger cellular senescence in normal human fibroblasts, J. Clin. Invest. 121 (7) (2011) 2833–2844.
[151] C.W. Greider, E.H. Blackburn, Identification of a specific telomere terminal transferase activity in Tetrahymena extracts, Cell 43 (2 Pt. 1) (1985) 405–413.
[152] E.H. Blackburn, Structure and function of telomeres, Nature 350 (6319) (1991) 569–573.
[153] S.R. Ferron, et al., Telomere shortening in neural stem cells disrupts neuronal differentiation and neuritogenesis, J. Neurosci. 29 (46) (2009) 14394–14407.

[154] I. Flores, et al., The longest telomeres: a general signature of adult stem cell compartments, Genes Dev. 22 (5) (2008) 654–667.
[155] T. de Lange, Shelterin-mediated telomere protection, Annu. Rev. Genet. 52 (2018) 223–247.
[156] R. Anderson, et al., Length-independent telomere damage drives post-mitotic cardiomyocyte senescence, EMBO J. 38 (5) (2019), e100492.
[157] J. Birch, et al., DNA damage response at telomeres contributes to lung aging and chronic obstructive pulmonary disease, Am. J. Phys. Lung Cell. Mol. Phys. 309 (10) (2015) L1124–L1137.
[158] M.H. Kim, M. Kino-Oka, Bioprocessing strategies for pluripotent stem cells based on waddington's epigenetic landscape, Trends Biotechnol. 36 (1) (2018) 89–104.
[159] I. Beerman, et al., Functionally distinct hematopoietic stem cells modulate hematopoietic lineage potential during aging by a mechanism of clonal expansion, Proc. Natl. Acad. Sci. U. S. A. 107 (12) (2010) 5465–5470.
[160] L. Liu, et al., Prognostic value of EZH2 expression and activity in renal cell carcinoma: a prospective study, PLoS One 8 (11) (2013), e81484.
[161] D. Sun, et al., Epigenomic profiling of young and aged HSCs reveals concerted changes during aging that reinforce self-renewal, Cell Stem Cell 14 (5) (2014) 673–688.
[162] J.A. Smith, R. Daniel, Stem cells and aging: a chicken-or-the-egg issue? Aging Dis. 3 (3) (2012) 260–268.
[163] A.E. Kofman, J.M. Huszar, C.J. Payne, Transcriptional analysis of histone deacetylase family members reveal similarities between differentiating and aging spermatogonial stem cells, Stem Cell Rev. Rep. 9 (1) (2013) 59–64.
[164] S.M. Chambers, et al., Aging hematopoietic stem cells decline in function and exhibit epigenetic dysregulation, PLoS Biol. 5 (8) (2007), e201.
[165] S. Horvath, DNA methylation age of human tissues and cell types, Genome Biol. 14 (10) (2013) R115.
[166] G. Hannum, et al., Genome-wide methylation profiles reveal quantitative views of human aging rates, Mol. Cell 49 (2) (2013) 359–367.
[167] A.E. Field, et al., DNA methylation clocks in aging: categories, causes, and consequences, Mol. Cell 71 (6) (2018) 882–895.
[168] R.E. Marioni, et al., The epigenetic clock is correlated with physical and cognitive fitness in the Lothian Birth Cohort 1936, Int. J. Epidemiol. 44 (4) (2015) 1388–1396.
[169] Y. Zhang, et al., DNA methylation signatures in peripheral blood strongly predict all-cause mortality, Nat. Commun. 8 (2017) 14617.
[170] R.E. Marioni, et al., DNA methylation age of blood predicts all-cause mortality in later life, Genome Biol. 16 (2015) 25.
[171] N. Li, et al., microRNAs: important regulators of stem cells, Stem Cell Res. Ther. 8 (1) (2017) 110.
[172] B.J. Reinhart, et al., The 21-nucleotide let-7 RNA regulates developmental timing in Caenorhabditis elegans, Nature 403 (6772) (2000) 901–906.
[173] L. Sinkkonen, et al., MicroRNAs control de novo DNA methylation through regulation of transcriptional repressors in mouse embryonic stem cells, Nat. Struct. Mol. Biol. 15 (3) (2008) 259–267.
[174] Y. Wang, et al., Embryonic stem cell-specific microRNAs regulate the G1-S transition and promote rapid proliferation, Nat. Genet. 40 (12) (2008) 1478–1483.
[175] C.H. Lin, et al., Myc-regulated microRNAs attenuate embryonic stem cell differentiation, EMBO J. 28 (20) (2009) 3157–3170.
[176] P. Neveu, et al., MicroRNA profiling reveals two distinct p53-related human pluripotent stem cell states, Cell Stem Cell 7 (6) (2010) 671–681.
[177] P.R. Tata, et al., Identification of a novel epigenetic regulatory region within the pluripotency associated microRNA cluster, EEmiRC, Nucleic Acids Res. 39 (9) (2011) 3574–3581.
[178] N. Xu, et al., MicroRNA-145 regulates OCT4, SOX2, and KLF4 and represses pluripotency in human embryonic stem cells, Cell 137 (4) (2009) 647–658.
[179] Y. Tay, et al., MicroRNAs to Nanog, Oct4 and Sox2 coding regions modulate embryonic stem cell differentiation, Nature 455 (7216) (2008) 1124–1128.
[180] C. Heintz, et al., Splicing factor 1 modulates dietary restriction and TORC1 pathway longevity in C. elegans, Nature 541 (7635) (2017) 102–106.
[181] D. Vilchez, M.S. Simic, A. Dillin, Proteostasis and aging of stem cells, Trends Cell Biol. 24 (3) (2014) 161–170.
[182] H. Koga, S. Kaushik, A.M. Cuervo, Protein homeostasis and aging: the importance of exquisite quality control, Ageing Res. Rev. 10 (2) (2011) 205–215.
[183] S. Alavez, et al., Amyloid-binding compounds maintain protein homeostasis during ageing and extend lifespan, Nature 472 (7342) (2011) 226–229.
[184] W.C. Chiang, et al., HSF-1 regulators DDL-1/2 link insulin-like signaling to heat-shock responses and modulation of longevity, Cell 148 (1–2) (2012) 322–334.
[185] V. Vanhooren, et al., Protein modification and maintenance systems as biomarkers of ageing, Mech. Ageing Dev. 151 (2015) 71–84.
[186] J.C. Lambert, et al., Genome-wide association study identifies variants at CLU and CR1 associated with Alzheimer's disease, Nat. Genet. 41 (10) (2009) 1094–1099.
[187] D. Harold, et al., Genome-wide association study identifies variants at CLU and PICALM associated with Alzheimer's disease, Nat. Genet. 41 (10) (2009) 1088–1093.
[188] F. Song, et al., Plasma apolipoprotein levels are associated with cognitive status and decline in a community cohort of older individuals, PLoS One 7 (6) (2012), e34078.
[189] M. Thambisetty, et al., Association of plasma clusterin concentration with severity, pathology, and progression in Alzheimer disease, Arch. Gen. Psychiatry 67 (7) (2010) 739–748.
[190] M. Riwanto, et al., Altered activation of endothelial anti- and proapoptotic pathways by high-density lipoprotein from patients with coronary artery disease: role of high-density lipoprotein-proteome remodeling, Circulation 127 (8) (2013) 891–904.
[191] P. Katajisto, et al., Stem cells. Asymmetric apportioning of aged mitochondria between daughter cells is required for stemness, Science 348 (6232) (2015) 340–343.
[192] M.R. Bufalino, B. DeVeale, D. van der Kooy, The asymmetric segregation of damaged proteins is stem cell-type dependent, J. Cell Biol. 201 (4) (2013) 523–530.

[193] R. Higuchi, et al., Actin dynamics affect mitochondrial quality control and aging in budding yeast, Curr. Biol. 23 (23) (2013) 2417–2422.
[194] H. Aguilaniu, et al., Asymmetric inheritance of oxidatively damaged proteins during cytokinesis, Science 299 (5613) (2003) 1751–1753.
[195] J. Cheng, et al., Centrosome misorientation reduces stem cell division during ageing, Nature 456 (7222) (2008) 599–604.
[196] M.J. Conboy, A.O. Karasov, T.A. Rando, High incidence of non-random template strand segregation and asymmetric fate determination in dividing stem cells and their progeny, PLoS Biol. 5 (5) (2007), e102.
[197] M.C. Florian, et al., A canonical to non-canonical Wnt signalling switch in haematopoietic stem-cell ageing, Nature 503 (7476) (2013) 392–396.
[198] Y. Wang, S. Hekimi, Mitochondrial dysfunction and longevity in animals: untangling the knot, Science 350 (6265) (2015) 1204–1207.
[199] C.B. Park, N.G. Larsson, Mitochondrial DNA mutations in disease and aging, J. Cell Biol. 193 (5) (2011) 809–818.
[200] A.W. Linnane, et al., Mitochondrial DNA mutations as an important contributor to ageing and degenerative diseases, Lancet 1 (8639) (1989) 642–645.
[201] E. Sahin, R.A. DePinho, Axis of ageing: telomeres, p53 and mitochondria, Nat. Rev. Mol. Cell Biol. 13 (6) (2012) 397–404.
[202] J.T. Rodgers, et al., Nutrient control of glucose homeostasis through a complex of PGC-1alpha and SIRT1, Nature 434 (7029) (2005) 113–118.
[203] X. Qiu, et al., Calorie restriction reduces oxidative stress by SIRT3-mediated SOD2 activation, Cell Metab. 12 (6) (2010) 662–667.
[204] A. Giralt, F. Villarroya, SIRT3, a pivotal actor in mitochondrial functions: metabolism, cell death and aging, Biochem. J. 444 (1) (2012) 1–10.
[205] S. Patel, et al., GDF15 provides an endocrine signal of nutritional stress in mice and humans, Cell Metab. 29 (3) (2019) 707–718.e8.
[206] F.E. Wiklund, et al., Macrophage inhibitory cytokine-1 (MIC-1/GDF15): a new marker of all-cause mortality, Aging Cell 9 (6) (2010) 1057–1064.
[207] Y. Fujita, et al., Secreted growth differentiation factor 15 as a potential biomarker for mitochondrial dysfunctions in aging and age-related disorders, Geriatr Gerontol Int 16 (Suppl. 1) (2016) 17–29.
[208] C. Vinel, et al., The exerkine apelin reverses age-associated sarcopenia, Nat. Med. 24 (9) (2018) 1360–1371.
[209] J. Masoumi, et al., Apelin, a promising target for Alzheimer disease prevention and treatment, Neuropeptides 70 (2018) 76–86.
[210] R. Ilangovan, et al., Dihydrotestosterone is a determinant of calcaneal bone mineral density in men, J. Steroid Biochem. Mol. Biol. 117 (4–5) (2009) 132–138.
[211] S. Sittadjody, et al., Age-related changes in serum levels of insulin-like growth factor-II and its binding proteins correlate with calcaneal bone mineral density among post-menopausal South-Indian women, Clin. Chim. Acta 414 (2012) 281–288.
[212] K. Jazbec, M. Jez, M. Justin, P. Rozman, Molecular mechanisms of stem cell aging, Slov. Vet. Res. 56 (1) (2019) 5–12.
[213] N. Barzilai, et al., The critical role of metabolic pathways in aging, Diabetes 61 (6) (2012) 1315–1322.
[214] Y.Y. Jang, S.J. Sharkis, A low level of reactive oxygen species selects for primitive hematopoietic stem cells that may reside in the low-oxygenic niche, Blood 110 (8) (2007) 3056–3063.
[215] M.B. Schultz, D.A. Sinclair, When stem cells grow old: phenotypes and mechanisms of stem cell aging, Development 143 (1) (2016) 3–14.
[216] W. Mair, et al., Dietary restriction enhances germline stem cell maintenance, Aging Cell 9 (5) (2010) 916–918.
[217] J. Chen, C.M. Astle, D.E. Harrison, Hematopoietic senescence is postponed and hematopoietic stem cell function is enhanced by dietary restriction, Exp. Hematol. 31 (11) (2003) 1097–1103.
[218] S.C. Johnson, P.S. Rabinovitch, M. Kaeberlein, mTOR is a key modulator of ageing and age-related disease, Nature 493 (7432) (2013) 338–345.
[219] D.E. Harrison, et al., Rapamycin fed late in life extends lifespan in genetically heterogeneous mice, Nature 460 (7253) (2009) 392–395.
[220] O. Renner, A. Carnero, Mouse models to decipher the PI3K signaling network in human cancer, Curr. Mol. Med. 9 (5) (2009) 612–625.
[221] N.L. Price, et al., SIRT1 is required for AMPK activation and the beneficial effects of resveratrol on mitochondrial function, Cell Metab. 15 (5) (2012) 675–690.
[222] S.J. Mitchell, et al., The SIRT1 activator SRT1720 extends lifespan and improves health of mice fed a standard diet, Cell Rep. 6 (5) (2014) 836–843.
[223] M. El Assar, et al., Better nutritional status is positively associated with mRNA expression of SIRT1 in community-dwelling older adults in the Toledo study for healthy aging, J. Nutr. 148 (9) (2018) 1408–1414.
[224] A.P. Sharples, et al., Longevity and skeletal muscle mass: the role of IGF signalling, the sirtuins, dietary restriction and protein intake, Aging Cell 14 (4) (2015) 511–523.
[225] R. Kumar, et al., Sirtuin1: a promising serum protein marker for early detection of Alzheimer's disease, PLoS One 8 (4) (2013), e61560.
[226] S. Yanagisawa, et al., Decreased serum sirtuin-1 in COPD, Chest 152 (2) (2017) 343–352.
[227] R. Kumar, et al., Identification of serum sirtuins as novel noninvasive protein markers for frailty, Aging Cell 13 (6) (2014) 975–980.
[228] S.J. Morrison, A.C. Spradling, Stem cells and niches: mechanisms that promote stem cell maintenance throughout life, Cell 132 (4) (2008) 598–611.
[229] J. Doles, et al., Age-associated inflammation inhibits epidermal stem cell function, Genes Dev. 26 (19) (2012) 2144–2153.
[230] X. Xia, et al., Molecular and phenotypic biomarkers of aging, F1000Res 6 (2017) 860.
[231] A.A. Saedi, et al., Current and emerging biomarkers of frailty in the elderly, Clin. Interv. Aging 14 (2019) 389–398.
[232] J.R. Pineda, et al., Vascular-derived TGF-beta increases in the stem cell niche and perturbs neurogenesis during aging and following irradiation in the adult mouse brain, EMBO Mol. Med. 5 (4) (2013) 548–562.
[233] L. Katsimpardi, et al., Vascular and neurogenic rejuvenation of the aging mouse brain by young systemic factors, Science 344 (6184) (2014) 630–634.
[234] M. Lavasani, et al., Muscle-derived stem/progenitor cell dysfunction limits healthspan and lifespan in a murine progeria model, Nat. Commun. 3 (2012) 608.
[235] G. Nelson, et al., A senescent cell bystander effect: senescence-induced senescence, Aging Cell 11 (2) (2012) 345–349.
[236] J. Durieux, S. Wolff, A. Dillin, The cell-non-autonomous nature of electron transport chain-mediated longevity, Cell 144 (1) (2011) 79–91.
[237] C. Franceschi, M. Bonafe, Centenarians as a model for healthy aging, Biochem. Soc. Trans. 31 (2) (2003) 457–461.

[238] M.T. Heneka, et al., Neuroinflammation in Alzheimer's disease, Lancet Neurol. 14 (4) (2015) 388–405.
[239] L. Ferrucci, E. Fabbri, Inflammageing: chronic inflammation in ageing, cardiovascular disease, and frailty, Nat. Rev. Cardiol. 15 (9) (2018) 505–522.
[240] F. Martinon, K. Burns, J. Tschopp, The inflammasome: a molecular platform triggering activation of inflammatory caspases and processing of proIL-beta, Mol. Cell 10 (2) (2002) 417–426.
[241] J. Nikolich-Zugich, The twilight of immunity: emerging concepts in aging of the immune system, Nat. Immunol. 19 (1) (2018) 10–19.
[242] A. Alpert, et al., A clinically meaningful metric of immune age derived from high-dimensional longitudinal monitoring, Nat. Med. 25 (3) (2019) 487–495.
[243] J. Krishnamurthy, et al., Ink4a/Arf expression is a biomarker of aging, J. Clin. Invest. 114 (9) (2004) 1299–1307.
[244] W.R. Jeck, A.P. Siebold, N.E. Sharpless, Review: a meta-analysis of GWAS and age-associated diseases, Aging Cell 11 (5) (2012) 727–731.
[245] Y. Liu, et al., Expression of p16(INK4a) in peripheral blood T-cells is a biomarker of human aging, Aging Cell 8 (4) (2009) 439–448.
[246] J.P. Coppe, et al., The senescence-associated secretory phenotype: the dark side of tumor suppression, Annu. Rev. Pathol. 5 (2010) 99–118.
[247] R.V. Kondratov, et al., Antioxidant N-acetyl-L-cysteine ameliorates symptoms of premature aging associated with the deficiency of the circadian protein BMAL1, Aging (Albany NY) 1 (12) (2009) 979–987.
[248] M. Abe, et al., Comparison of the protective effect of N-acetylcysteine by different treatments on rat myocardial ischemia-reperfusion injury, J. Pharmacol. Sci. 106 (4) (2008) 571–577.
[249] A.M. Sadowska, Y.K.B. Manuel, W.A. De Backer, Antioxidant and anti-inflammatory efficacy of NAC in the treatment of COPD: discordant in vitro and in vivo dose-effects: a review, Pulm. Pharmacol. Ther. 20 (1) (2007) 9–22.
[250] R. Khatri, et al., Reactive oxygen species limit the ability of bone marrow stromal cells to support hematopoietic reconstitution in aging mice, Stem Cells Dev. 25 (12) (2016) 948–958.
[251] E. Gonzalez-Suarez, et al., Increased epidermal tumors and increased skin wound healing in transgenic mice overexpressing the catalytic subunit of telomerase, mTERT, in basal keratinocytes, EMBO J. 20 (11) (2001) 2619–2630.
[252] S.E. Artandi, et al., Constitutive telomerase expression promotes mammary carcinomas in aging mice, Proc. Natl. Acad. Sci. U. S. A. 99 (12) (2002) 8191–8196.
[253] A. Canela, et al., Constitutive expression of tert in thymocytes leads to increased incidence and dissemination of T-cell lymphoma in Lck-Tert mice, Mol. Cell. Biol. 24 (10) (2004) 4275–4293.
[254] B. Bernardes de Jesus, M.A. Blasco, Potential of telomerase activation in extending health span and longevity, Curr. Opin. Cell Biol. 24 (6) (2012) 739–743.
[255] D.J. Baker, et al., Increased expression of BubR1 protects against aneuploidy and cancer and extends healthy lifespan, Nat. Cell Biol. 15 (1) (2013) 96–102.
[256] B.E. Keyes, E. Fuchs, Stem cells: aging and transcriptional fingerprints, J. Cell Biol. 217 (1) (2018) 79–92.
[257] A. Ortega-Molina, et al., Pten positively regulates brown adipose function, energy expenditure, and longevity, Cell Metab. 15 (3) (2012) 382–394.
[258] A.P. Gomes, et al., Declining NAD(+) induces a pseudohypoxic state disrupting nuclear-mitochondrial communication during aging, Cell 155 (7) (2013) 1624–1638.
[259] R. Gredilla, et al., Caloric restriction decreases mitochondrial free radical generation at complex I and lowers oxidative damage to mitochondrial DNA in the rat heart, FASEB J. 15 (9) (2001) 1589–1591.
[260] S.M. Gehrig, et al., Hsp72 preserves muscle function and slows progression of severe muscular dystrophy, Nature 484 (7394) (2012) 394–398.

Chapter 18

Alternative stromal cell-based therapies for aging and regeneration

Dikshita Deka*, Alakesh Das*, Meenu Bhatiya, Surajit Pathak, and Antara Banerjee
Department of Medical Biotechnology, Faculty of Allied Health Sciences, Chettinad Academy of Research and Education (CARE), Chettinad Hospital and Research Institute (CHRI), Chennai, India

1 Introduction

Aging is a biological process that is depicted by the gradual degeneration of the physiological function and declining health. According to Ernst Mayrl, an eminent evolutionary biologist, there are mainly two differences between living and nonliving beings: first, the property of self-reproduction of living organisms and second, the evolution of organisms over time [1].

The physiological integrity of the cell is lost progressively during the aging of the cell that leads to the deterioration of various functions like a sudden loss of weight and contraction power of the muscles including reduced bone density, alteration of the cardiovascular system, decline in the function of cognitive and state of proinflammation in the organism that also increases the probability of death and leads to various risk factors related to aging disorders such as cancer, diabetes, cardiovascular disorders, and neurodegenerative diseases [2].

The aggregation of cellular impairment in a time-dependent manner is the general cause of aging along with the association of genetic and environmental component contributing in cellular senescence as depicted in Fig. 1. At the molecular level, the cellular aging can be elucidated primarily with nine hallmarks that contribute to the process of aging and determine its phenotypic changes. They are epigenetic alterations, genomic instability, degenerated sensing of nutrients, decrease in the proteostasis, mitochondrial dysfunction, cellular senescence, telomere attrition, stem cell exhaustion, and altered intercellular communication. Each hallmark ideally emphasizes on the criteria that—it should manifest aging normally and not accelerated by experimental aggravation [3].

2 Aging in unicellular organisms

The senescence in microbial species is analyzed in two main ways: conditional senescence and replicative aging. In conditional aging analyses, the feasibility of the gross populace is traced and longevity is evaluated using chronological time. On the contrary, replicative senescence provides the information of the single-cell level and longevity is measured as several divisions before the death phase [4].

The bacteria that divide symmetrically lack morphological distinction, and they age even in the optimal condition. If a cell gets divided into identical daughters, each daughter gets clonal aging that should eventually lead to the death of the populace, but this usually never happens, even the symmetrically dividing species are not completely identical. In *E. coli*, during each division, one cell pole is synthesized by de novo, which is one daughter inherits the old pole while the other gets the newly synthesized pole [5]. The evidence involves poles of older cells as a site for the accumulation of damaged components. In bacteria, poles are considered an important area with pertinence to cytoskeleton organization, the formation of organelles, and surface structures along with partitioning of the chromosome. Accumulative destruction of lipid or protein complexes at the site of unicellular cell poles gradually decreases their capacity for the segregation of chromosomes [6].

3 Aging in multicellular organisms

The biological aging can be specified by the integral decline in the functioning of the multicellular organism over time like homeostasis, fertility, and resistance to the diseases as well as wear and tear, organ and cellular senescence that may precede

---

* Equal contribution.

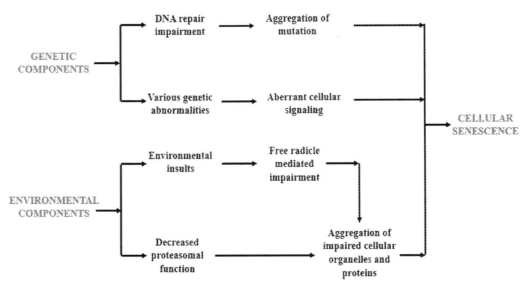

FIG. 1 Association of genetic and environmental component in the mechanism of cellular senescence.

the death of an organism and developmental maturation. Predominantly multicellular aging involves cellular and extracellular organelles. The aging procedure reckons that both cellular and extracellular components endure age-liable changes. The cellular attribute is required for viable phenotype production of the individual cells.

It has been reported that oxidative metabolism produces ROS (reactive oxygen species) as the secondary metabolic product that often destructs the cellular macromolecules. Proteins are apprehended as vital macromolecules to undergo oxidative moderation, and the assembling of the modified proteins is the specific property of cells undergoing aging and be a reason for the decrease in the function of the proteasome. Thus, age-associated alterations in the protein oxidation process emerge to be the combination of complicated biochemical alterations at the level of primary antioxidant defense, repair of oxidant-induced damage, and oxidant production [7].

4 Age-related genes and their role

Genes are believed to be an extremely important tool in evaluating the safety and efficacy of interference in preclinical and clinical settings. The understanding of genes and pathways that are abnormally regulated in the process of senescence and age-associated diseases has sternly expanded in recent years. Some of the genes involved in aging and age-related diseases are listed in Table 1.

The overall deviation in the immune system, influencing both adaptive and innate immune responses, has evolved as one of the pertinent hallmarks of aging activity and immunological factors were some of the markers evoked for aging. Aging also causes prominent changes in immune cell function and phenotypes provoking alterations in PRR (pattern recognition receptor) and reduction in phagocytic activity of neutrophils, dendritic cells, monocytes, and macrophages that are essential for recognizing the antigens [25]. Aggregation of senescent cells is a supplementary cause of age-related phenotypes in many organs and tissues [26] as the cells are active metabolically and obtained secretory phenotype recognized to be SASP (senescence-associated secretory phenotype) [25]. Some prominent genes involved in the aging process are detailed in the following sections:

IL-6 (Interleukin-6) is known to be the most eminent unit of SASP engaged in inflammation and is an important biomarker for the chronic inflammatory phenotype observed in the process of aging and age-associated diseases. It is manifested by different tissues like bone marrow, appendix, and prostate. *IL-6* is predominantly processed at inflammation sites and encourages a transcriptional inflammatory response through *IL-6 receptor-α* [25].

CX3CL1 functions as a chemo-attractant for NK cells, T-cells, and monocytes [27] that undergo alterations in signaling with increasing of age. *CX3CL1* receptor expression in monocytes corresponds positively with dementia and is negatively associated with diabetes and anemia [28]. The expression of *CX3CL1* is amplified by *TNF-α* (tumor necrosis factor-α) and *IFN-γ* (interferon-γ) that are found in elevated levels in rheumatoid arthritis patients that stimulates the migration of T-cells, monocytes, and osteoclast precursors in the joints [29].

TABLE 1 Genes involved in aging and age-related diseases.

Literature	Gene	Genes and the diseases associated with	Role of the genes in aging-related diseases/metabolic dysfunctions
Bartali et al. [8]	α-klotho	This protein is a circulating factor detectable in serum, declines with age. Hyperexpression of this gene increases the life span.	Cellular homeostasis Cell signaling Controls insulin sensitivity and involved in aging
Antonelli et al. [9]	CXCL10	Elevated serum levels in various aging cohorts. Elevated in rheumatoid arthritis, stimulates neurodegenerative diseases. Increased in cancer, promoting tumor growth.	SASP component reduces mitochondrial function and induces apoptosis. Reduces cellular proliferation.
Bauer et al. [10]	MMP7	Elevated expression in plasma and urine in renal fibrosis. Elevated MMP7 expression is found in tumors and metastasis.	Regulates the cleaves of ECM and basement membrane proteins Enhances MMP7 correlated with extensive tissue remodeling and organ dysfunction
Feger et al. [11]	FGF23	Increased in different renal disorders and CVDs. Associated with aging, liver diseases.	Pleiotropic action phosphate, mineral, and iron homeostasis. FGF23 acts with α-klotho as coreceptor.
Cappellari et al. [12]	CD14	Enhances CD14+/CD16+ monocytes in frail individual. Elevated expression are correlated with coronary phenotype in diabetic individuals with coronary artery disease; Reduced levels of CD14 and CD16 in mild AD patients; Decreased terminal differentiation of CD14+/CD16+ monocytes in RA shift toward CD14+/CD16+.	Surface antigen preferentially expresses phagocytes. Mediates innate immune responses to bacterial lipopeptides.
Constans and Conri [13]	sVCAM1/ sICAM1	Both are related to the increased odds of injurious falls, frailty sVCAM is associated with cognitive damage, elevated cerebrovascular resistance. sVCAM1 is associated with hypertension, vascular inflammation, and systemic endothelial dysfunction. Both are used as risk predictors of cardiovascular events.	sVCAM1 and sICAM1 are markers for endothelial inflammation. sICAM1 is released from aging cells through microvesicles.
Anuurad et al. [14]	Pentraxin	Pentraxin is elevated in blood with senescence process. Pentraxin is a pivotal biomarker for inflammatory processes.	Stimulates fibroblast differentiation;d Dysregulates inflammation and complement activation;p Plays a function in angiogenesis and tissue remodeling;p Pentraxin levels are related to leukocyte telomere length.
Bornheim et al. [15]	Vimentin	Vimentin contributes to chondrocyte stiffness, α- and β-cell dysfunction in type 2 diabetes. Changed expression in various cancers and contributes toward the enhanced survival of various cancers cells.	Stimulated by TGFβ1 and TNFα, cleaved and activated by calpain and osteopontin enhances vimentin stability Regulates actin dynamics Vimentin filaments play a function in active force development and contraction
Chow et al. [16]	FGF21	*Potential biomarker in:* Mitochondrial diseases, metabolic disorders, diabetes, cancer, osteoarthritis, rheumatoid arthritis. Linked to aging, premature aging, and life span.	*Pleiotropic action* Adipokine, mitokine, myokine, and neuroendocrine Regulated by Inflammation, fibrosis, alcohol, vitamin-D, glucose, ER, starvation, or fasting

Continued

TABLE 1 Genes involved in aging and age-related diseases—cont'd

Literature	Gene	Genes and the diseases associated with	Role of the genes in aging-related diseases/metabolic dysfunctions
Chai et al. [17]	Leptin	Controls body weight, inflammation Related to diabetes induction Promote cancer cell proliferation and different tumor development Leptin level elevates with aging Leptin instigates bone formation	Control appetite, modulate energy expend Involved in apoptosis, angiogenesis, cell proliferation, and cellular senescence
Huo et al. [18]	CX3CL1	Elevated concentrations detected in synovial fluid of patients with rheumatoid arthritis and osteoarthritis Stimulates accumulation of the receptor and attracts cytotoxic effector T-cells or NK cells Reduces cancer invasiveness	Soluble form responsible for chemo attracting T-cells, NK cells, and monocytes CX3CR1 defines peripheral blood cytotoxic effector lymphocytes and is a direct target of p53 Increases proliferation of endothelial cells and enhances the migration of endothelial progenitor cells in ischemic penumbra CX3CL1/CX3CR1 expression is decreased in the aged brain
Geiser et al. [19]	TGF B	Various correlations of TCFβs with several disorders have been newly discovered or elucidated in much more detail than before. Many of these are related to aging.	TGFβ1 is a secreted protein that functions in many cellular activities, involving the regulation of cell growth, cell proliferation, cell differentiation and apoptosis.
Chaker et al. [20]	IGF-I	Reduced IGF-I-level regulating hematopoietic stem cell protection, self-renewal, and regeneration Lacking encoded protein in mice shows generalized organ hypoplasia that includes under the development of CNS and developmental defects in muscle, bone, and reproductive systems	Cell proliferation, cell differentiation, cell death Lipid metabolic process, Protein metabolic process
Miskin et al. [21]	PLAU	Associated with the pathogenesis of late onset Alzheimer's disease (AD) Elevated levels have additional complication in diabetes patients	Expressed and secreted from senescent cells and controls cell proliferation Overproduction of PLAU in brain reduced food consumption and elevated longevity
Peine et al. [22]	ST2	Elevated in several senescent conditions Elevated in type 2 diabetes Associated with advanced and metastatic gastric cancer	Induces inflammation Potentiates macrophage response to LPS Modulates T-cell function and differentiation
Chien et al. [23]	AHCY	Clinical marker to evaluate rheumatoid arthritis patients with high cardiovascular risk. AHCY (adenosylhomocysteinase) levels relate to Alzheimer's disease, Parkinson's disease, neurologic impairment after stroke, and impaired cognitive function	AHCY activates inflammatory pathways including NFκB; AHCY induces oxidative, ER stress, mitochondrial dysfunction, and apoptosis AHCY accelerate senescence
Demirci et al. [24]	Resistin	Elevated in coronary artery disease, coronary syndrome, and peripheral arterial disease. Resistin levels are correlated with enhanced threat of acute cerebral infarction. Serum Resistin level is increased in subjects with metabolic syndrome and may be associated with the severity of it.	Inflammation; Cell proliferation; apoptosis; reduced mitochondrial content; reduced brown adipose tissue activity

Pentraxin is another important gene associated with frailty and is a *TNF-α* instigated protein engrossed in complement activation. The amount of *pentraxin* amplifies with aging and related to the length of leukocyte telomere [30]. It also acts as a negotiator of neurogenesis and is included in cerebral ischemia.

Among the variant aging factors, mitochondrial dysfunction turns up to be one of the essential hallmarks of the aging process, with a decrease in the rate of electron transfer and hindrance in ATP synthesis. The programmed cell death is also tightly related to dysfunction of the mitochondria that increases with the process of aging [31].

Myomitokine also known as *GDF-15* is a cytokine that is known to initiate the GFRAL (GDNF family receptor alpha-like) receptor associated with apoptosis and stress [32]. The level of *GDF-15* is assumed to be elevated with the progression of senescence and disease condition. It is involved in estimating decrease in physical health, resistance to insulin, diabetes, and other age-associated diseases.

Fibronectin type III domain containing 5 (*FNDC5*) is a transmembrane protein that experiences proteolytic transformation and generates myokine irisin. The reduced level of irisin in aging could forecast atherosclerosis, sarcopenia, and related to osteoporotic fractures.

Calcium is known to play an eminent role in intracellular and extracellular signaling pathways. Small alterations in calcium level may cause immense dysfunction of signaling cascades. S100B is a calcium-binding protein B that regulates the calcium signaling cascade. Overexuberance of S100B unveils premature aging. S100B-instigated repression of *p53* confers the progression of cancer, also its increased level is perceived in Alzheimer's disease (AD), Down's syndrome, and disorders of the central nervous system [33].

SMP30 (senescence marker protein 30) also known as regucalcin is acclaimed to decline with aging in different tissues like brain, prostate, and heart and due to these reduced levels, it influences cellular aging, fibrotic, and decrepitude of the cells. Regucalcin also regulates calcium signaling proteins that are instigated by oxidative stress. Autoantibodies antagonistic toward Regucalcin are utilized as a marker for the diagnosis of age-related diseases.

5 Age-related diseases

Age-associated diseases are predominantly observed to increase in frequency with an advancement in the process of senescence. Some of these diseases are Alzheimer's diseases, arthritis, cancer, hypertension, [33] skin abnormalities, etc. Age-related disorders are listed in Fig. 2.

5.1 Alzheimer's disease (AD)

It is a neurodegenerative disease that usually commences steadily and gradually intensifies with time. AD is distinguished as a protein misfolding disorder. At the molecular extent, both aging and AD involve the gene *Apolipoprotein E*. Alterations in DNA methylation are linked with AD, and more indicatively, *ANK 1* was perceived to be related to the neuropathology.

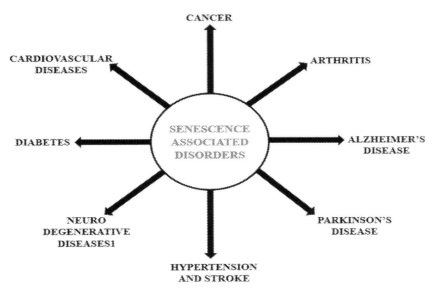

FIG. 2 Age-related disorders.

Moreover, histone acetylation plays a significant role in the field of AD as it was observed to be significantly decreased in both mouse and human models of AD [34].

5.2 Parkinson's disease

Parkinson's disease (PD) is instigated by discrete factors that degrade the populace of dopaminergic neurons. It is a moderately progressive neurodegenerative disease with an elevated prevalence in elderly people. During the proceeding of aging, the dopamine (DA) neurons may encounter physiological alterations, which is analogous to the modifications that these cells undergo before their degeneration in PD. Moreover, the inconsistent activity of the glial cells is inculpated in the advancement of PD. The PD and the aged brain undergo intricate alterations in the activity of microglia and astrocytes generating a low-level neuroinflammation. Astrocytes synthesize trophic factors that shield the dopaminergic neurons from degeneration, which gets deteriorate in both PD and aging [35].

5.3 Cancer

Aging is a biological progress that occurs effectively in all living organisms that can be designated by escalating deprivation in the organ functionality and the renewal ability of the tissue along with the decline in the activity of the immune system. Immune observation is a paramount factor in hindering cancer advancement; along with it, immune senescence is also a supreme factor linking aging and tumorigenesis. The inflammatory domain is assumed to initiate a protumorigenic state making aged individuals prone to oncogenic affront. One of the crucial phenomena leading to the instigation of cancer and its advancement is DNA destruction that is due to its unrelieved attack with the genotoxic agents. Decoding the mechanisms of the molecular activities can help to develop a future treatment for cancer [36, 37].

5.4 Colon cancer

Colon cancer is the most common cancer with complex genetic-environmental interactions. According to data from GLOBOCAN 2018, colon cancer is the 4th most incident, 3rd most commonly diagnosed, and 3rd leading cause of cancer-related death worldwide. There are around 1,096,000 new cases of colon cancer reported in 2018. In the United States, colon cancer accounts for the death of over 608,700 people each year. However, colon cancer can be curable if diagnosed at an early stage. The death rate by colon cancer has been falling due to novel and effective screening techniques and treatment improvements. Colon cancer develops as a noncancerous growth called a polyp that grows gradually, over 10 to 20 years, from the inner wall of the colonic part of the large intestine composed of glandular, epithelial cells. [38].

An adenomatous polyp is the commonest colon cancer. It can be developing due to certain acquired series of genetic or epigenetic mutations. Fig. 3 depicts differentially expressed genes unique to late onset of colorectal tumor. Approximately one-third of all people grow one or more adenomas. Cancer from the colon's inner lining is called adenocarcinoma that accounts for around 96% among all colon cancer [39].

5.4.1 Colon Cancer risk factors

The risk factors associated with colon cancer are stated in Fig. 4 and explained in the following:

(a) *Age*: Age is one of the risk factors of colon cancer. Approximately 90% of colon cancer are diagnosed after age 50. The median age for colon cancer diagnosis is 68 years in men and 72 years in women. The reported prevalence of colon cancer occurrence rates decreased by 4.6% per year in 65 years, by 1.4% in 50–64 years aged, but increased by 1.6% per year in people younger than age 50 [40].
(b) *Eating habits and lifestyle*: People who consume a diet with low nutrition, heavy animal protein (red meat), saturated fat calories, heavy alcohol consumption, and poor physical activity are prone to the development of the disease. Physical exercise is closely linked to reduced colon cancer incidence. Findings strongly suggest that people who are less active are having 25% of greater risk of developing colon tumors than the active people. Dietary behaviors are likely to have a significant effect on an injury, through particular dietary ingredients, and indirectly, through oversupply and obesity. Dietary fiber decreases the incidence of colon cancer. Red and processed meat consumption raises the incidence of all colon cancers. In 2015, the International Agency for Research on Cancer classified processed meat and red meat as "carcinogenic" and "probably carcinogenic" to human beings based on the evidence related to colon cancer risk [41]. In November 2009, the International Agency for Research provided evidence that tobacco and smoking causes colon cancer. The tobacco, smoking, and alcohol consumption shows a stronger link to rectal cancer than colon cancer [42–44].

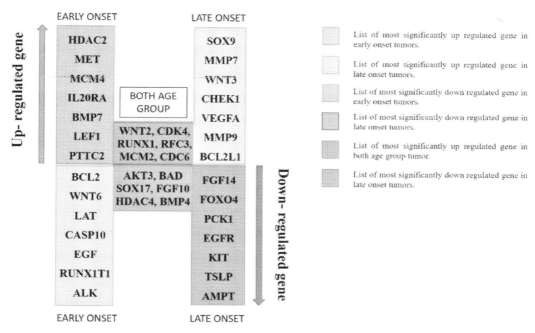

FIG. 3 Differentially expressed genes unique to late onset of colorectal cancer.

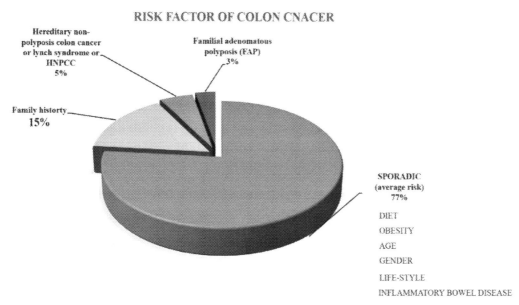

FIG. 4 Risk factors associated with colon cancer.

(c) *Inflammatory bowel disease*: Chronic inflammatory bowel disease, a condition of colon inflammation over a long period highly increases the risk of developing colon cancer compared to other normal people in the population [45]. Chronic ulcerative colitis causes inflammation in the inner lining of the colon and the risk of developing colon cancer begins to increase after 8 to 10 years of untreated colitis. An individual with the first-degree relative with colon cancer approximately doubles the risk of colon cancer growth if diagnosed with colon cancer before the age of 55. Approximately 3.1 million Americans have been diagnosed with inflammatory bowel disease that is particularly common in people with the poorest education and the highest deprivation.

(d) *Familial adenomatous polyposis (FAP)*: Familial adenomatous polyposis (FAP), second most common genetic syndrome represents less than 1% of all colon cancer. FAP identifies the countless number of colon polyps beginning at the age of 10–12 years. FAP can be spontaneous, it might not affect people not having a family history of FAP. Attenuated FAP is a milder version of FAP developing fewer than 100 polyps of the colon [46].

5.4.2 Diagnosis and treatment of colon cancer

Treatment of colon cancer depends on the stage, size, location of the polyp, or tumor. The diagnosis involves surgery, medicine, radiation therapy, and chemotherapy.

(a) *Colonography*: Colonography is the most prevalent colon cancer screening technique used in the USA. This technique is used for direct visual inspection of the entire large intestine.
(b) *Colonoscopy*: This is the most common surgical treatment performed to remove the polyps or affected part of the colon. Side effects of colonoscopy include fatigue, diarrhea, constipation, frequent bowel movements, gas, and bloating.
(c) *Chemotherapy*: Treatment regime includes systemic chemotherapy or regional chemotherapy. Drugs widely used for colon cancer treatment are 5-fluorouracil (5-FU), Capecitabine (Xeloda), irinotecan (Camptosar), oxaliplatin (Enoxaparin), trifluridine, and tipiracil (Lonsurf). The chemotherapeutic drugs 5-fluorouracil (5-FU) capecitabine, and oxaliplatin, are most often used in colon cancer treatment. The side effects of chemotherapy depend on the type, drug dosage, and length of treatment. Some most common side effects of chemotherapy are sore mouth, hair loss, vomiting, fatigue, diarrhea, nausea, loss of appetite, low blood cell counts [47].
(d) *Radiation therapy*: Beam with high-energy radiation used radiation therapy treatment to destroy cancer cells and prevent cancer cell to multiplying. Irritation of skin, nausea, vomiting, rectal discomfort, and sore swelling, bladder inflammation, and sexual complications are some common side effects of radiation therapy. Some of these adverse effects go away when therapies have been finished, but others may be permanent, such as genital issues and a degree of rectal and/or bladder irritation. Furthermore, the radiation therapy to the pelvic region can affect the ovaries and may cause infertility.

5.4.3 Applications of stem cell therapy in cancer

Mesenchymal stem cells (MSCs) for colon cancer

The microenvironment of tumor plays a crucial role in the promotion and development of tumor and research suggests that Mesenchymal stem cells (MSCs) can control cellular targets contributing to tumor development, which is an important aspect of tumor microenvironment maintenance [48]. MSCs are the spindle-shaped plastic adherent cells derived from bone marrow, adipose tissue, with multipotent in vitro differentiation ability. The MSCs differentiate into tumor-associated fibroblast (TAF)-like cells after migration to the tumor microenvironment, which are predominating tumor promoter stromal cells that affect the development and survival of colon cancer, while it is still controversial. Potential research in this area will concentrate on identifying MSC's therapeutic efficacy by clinical studies and better understanding of MSC's physiology to elucidate the dynamics of such therapeutic observations [49].

MSCs as carrier in colon cancer therapy

MSCs can be used as vehicles for the delivery of antitumor agent due to their homing property, and it could be beneficial in MSC-based tumor therapy. Since modified MSCs are attracted toward tumor stroma, they can deliver specific anticancer drug or agent to specific tumor sites [50]. In a recent study, in vitro adipose tissue-derived MSCs (AT-MSCs) administered in mice treated with 5-fuorocytosine caused tumor growth inhibition and, as a result, AT-MSCs are able to switch the CD genes to the tumor site and induce antitumor activity. Yet in another study, in HCT-15 and DLD-1 colon cancer cell lines, apoptosis-inducing ligand (TRAIL)-associated TNF transgenic human MSC suppressed colon cancer development through apoptosis induction [51]. MSCs as a carrier are an excellent MSC-based tumor therapy approach for treating colon cancer.

5.5 Skin cancer

The skin is the largest and complex structure of the body. In 2012, the American cancer society has reported about 1.5 million skin cancer cases and 12,190 skin cancer-related death cases. In 2012, the World Health Organization (WHO) reported more than 200,000 cases of malignant melanoma and 6500 cases of mortality related to malignant melanoma. The types of skin cancer are basal cell carcinoma (BCC), squamous cell carcinoma (SCC), and melanoma. The cutaneous BCC is more prevalent and disfiguring but curable. Cutaneous SCC is less common than BCC but may lead to death. Melanoma, the most common skin malignancy is less prevalent than BCC and SCC but accounting for the maximum number of deaths [52].

5.5.1 Risk factors of skin cancer

Multiple risk factors exist for all types of skin cancer, which includes exogenous and endogenous risk factor. The exogenous risk factors are UV radiation, which is one of the main risk factors of skin cancer, accounting for at least 65% of melanomas. Skin cancers are also attributed to chronically injured or nonhealing wounds that arise at the prior injury site. The incidence of skin malignancy is 0.1–2.5% in scar tissues.

5.5.2 Applications of stem cell therapy in skin disorders

(a) *For Vitiligo*: Vitiligo is a disorder caused due to loss of melanocytes, the cells that produce melanin. These melanocytes may decease or stop functioning in vitiligo disorder. Vitiligo repigmentation depends upon melanocyte's availability. Repigmentation occurs due to melanocyte stem cell (MelSC) migration toward infundibulum [53]. There are three major melanocyte sources: hair follicle unit, vitiligo lesion border, and unaffected melanocytes in depigmented skin areas. The hair follicles are the main providers of pigment cells. Adipose-derived stem cell (ADSC) therapy is reported to have therapeutic potential for vitiligo. The ADSCs have the potential to enhance melanocyte migration and proliferation whereas reduce differentiation process of the melanocytes. Melanocytes are rarely undergoing mitosis therefore mitogenic factors are required in transplantation treatments [54]. ADSCs can be a good source of growth factors if coadministration with melanocytes compared to the administration with melanocytes alone.

(b) *In Cutaneous GVHD (cutaneous graft-versus-host disease)*: Cutaneous GVHD is an immune-mediated reaction and a major complication of allogeneic hematopoietic stem cell transplantation (HSCT). It can affect 40% to 60% of patients and accounts for 15% of mortality after HSCT [55]. The skin and gastrointestinal tract are major organs affected by acute GVHD. The GVHD was classified into two categories based on onset time and severity of disease: first, acute GVHD, and second, chronic GVHD [56]. The chronic cutaneous GVHD further classified into sclerotic and lichenoid GVHD on the basis of skin lesion types and onset stages. Erythematous maculopapular rash is the first sign of acute cutaneous GVHD and maybe a strong indicator for diagnosing GVHD. During the starting stage of persistent GVHD emerges lichenoid GVHD, which form lichen planus-like skin inflammation, which then forms sclerotic GVHD with strong similarity with scleroderma. Conventionally, initial research has shown that the proliferation and differentiation of T lymphocyte can be effectively inhibited by MSCs by various direct and indirect mechanisms, such as direct cell-to-cell contact, activation of cell cycle arrest [57], secretion of hepatocyte growth factor (HGF), transforming growth factor $\beta 1$ (TGF-$\beta 1$) [58], and prostaglandin E2 (PGE2), as soluble mediators and indirectly by regulating immune cells [59].

(c) *In Psoriasis*: Psoriasis is a chronic autoimmune skin disease that affects about 2%–4% of the world's population. Psoriasis is considered as commonest affected inflammatory skin disorder and psoriasis vulgaris, also known as plaque psoriasis, is the most common type of psoriasis. Recent data suggesting that mesenchymal allogeneic stem cell therapy would be effective in the treatment of psoriasis because MSCs of psoriasis patients exhibits impaired anti-inflammatory activity against T_H cells [60]. Still, few trials have shown the efficacy of MSCs as a preventive agent against psoriasis.

(d) *In Atopic Dermatitis/Eczema*: Atopic dermatitis is also known as atopic eczema. The abnormal allergic immune responses result eczematous skin lesions, a representative inflammatory keratopathy. The major pathophysiological characteristics of atopic dermatitis include impairment of skin barrier function and acute skin inflammation. Atopic dermatitis immunological pathways are frequently defined T_{H2}-mediated pathological inflammatory responses and elevated serum IgE and eosinophils [61]. It is accepted that inflammatory responses powered by T_{H2} and T_{H22} are responsible for acute atopic dermatitis, whereas T_{H1} regulates chronic atopic dermatitis [62]. The clinical trials with atopic dermatitis patients with moderate-to-severe atopic dermatitis have examined the therapeutic efficacy of hUCB-MSCs (human umbilical cord blood-mesenchymal stem cells). The single administration of hUCB-MSCs showed dose-dependent therapeutic effectiveness in the manifestation of atopic dermatitis. About 55% of the patients exhibited a 50% reduction in the eczema area in high-dose hUCB-MSCs treatment group without any noteworthy side effects.

6 Factors affecting stromal cell quality due to aging

(a) *Cell damage and telomere shortening*: The aggregation of cellular impairment is similar to all cells with senescence. This impairment is correlated with the exposure to chemical mutagens, ionizing radiation, ROS and, in the case of a strongly multiplying populace such as progenitor and stromal cells (SCs), repeated replications of the DNA. As the SCs are capable of producing new cells all through the life span of the organism, the repair mechanisms of the DNA that is observed in all cells are especially crucial in this cell portion. Hereof, the gradual telomere shortening is an intrinsic alteration that eventually confines the counts of divisions that a cell can go through over its life span. A subdivision of cells, containing SCs, express telomerase, an enzyme that allows the expansion of the sequences of telomere, through averting the shortening of telomere and increasing their dividing potentiality. Analysis comprising the SC transplantation from aged mice into the bone marrow of young mice, and vice versa, observed that the age-associated alterations in self-renewal capability were highly cell-intrinsic and, recently, the role of DNA impairment and shortening of telomere in this course was resolved using transgenic mice deficient in Ku80, mTR, or XPD [63]. Combination of all these data recommended that even though DNA impairment and shortening of telomere may not play a crucial part in SC homeostasis while aging these procedures might have a considerable effect on the capacity of the aged SCs to contribute to tissue regeneration after disorder or injury.

(b) *Cell cycle regulators*: Various pieces of evidence have reported that the significance of cell-intrinsic cell cycle regulators in regulating SC kinetics while senescence has become apparent. With all the proliferating cells, compact tight organization of the advancement over the cell cycle is crucial for managing the cell division rate. As the G1-S cell cycle transformation, and its control by the family of Rb proteins, is among the most intensely examined of these regulatory mechanisms. Upstream of the family of Rb proteins in this regulatory cascade is the cyclin-dependent kinase inhibitor *p16Ink4a*, which confines SC's self-renewal in aging systems containing HSCs (hematopoietic stem cells) [64] and NSCs (neural stem cells) [65]. This shows that even though *p16Ink4a* has a small role in the young, it has a suppressive action in the aged cells. Usually, these pieces of evidence show that the damaged self-renewal capability with senescence is the outcome of the effective upregulation of suppressive cascade instead of the loss of a permissive cascade. Combining together, these shreds of evidence showed that SC quiescence with senescence is a perfectly regulated mechanism and important to assure that SCs do not become tumorigenic. It remains to be persistent which factors are accountable for altering these intrinsic regulators of SC multiplication and these factors are affected by either damage or disorder. Nevertheless, it is evident that the intentional reduction in the capacity of aged SC to self-renew has the capability to greatly alter SC-mediated regeneration of tissues.

(c) *Signaling proteins*: The extent to which SC role is controlled by intrinsic mechanisms in comparison with extrinsic cues, and to what degree these factors are detachable, is still argued. These types of alterations could be considered as two sides of the same coin; thereby, intrinsic SC alterations show the reaction of the SCs to the changed niche, or the intrinsic SC alterations modify the SC reaction to an unaltered niche. For example, the later alteration is the downregulation of the EGFR (epidermal growth factor receptor) noticed in the neural progenitor populace in vivo with senescence. Signaling via this cascade is crucial to regulate the multiplication of the progenitor cells in the NSC niche. Actually, the decrease in neurogenesis during senescence is complementary with a reduction in the count of neuronal progenitor cells expressing specific signaling molecules of the cascade. Appealingly, the administration of exogenous EGF (epidermal growth factor) to the CNS (central nervous system) leads to only a moderate enhancement in the count of progenitor cells within the aged brain. The interaction between extrinsic and intrinsic cues is further represented by the evidence that the NSC microenvironment in the aged brain has reduced TNF-α, the endogenous ligand for EGFR [66].

(d) *Epigenetics regulations*:
  (a) *DNA methylation*: DNA methylation has been recommended as the prognosticator for human age in several genomewide methylation analyses, but the interplay is complex. Advancement in age is correlated with escalating global hypomethylation, but various loci are also hypermethylated, like various tumor suppressor genes. Cells from mice and patients with progeroid syndromes (PS) also illustrated the pattern of methylation that reflects those observed in the normal process of senescence. A meta-analysis of around 13,000 participants exhibited that DNA methylation-associated biomarkers, also known as "epigenetic age," assumed all-cause fatality independent of chronological age, race, and additional risk factors [67]. "Epigenetic age" is a potent predictor of the chronological age illustrated by Horvath [68] and Hannum [69] and is associated with the degree of methylation on a definite count of CpG dinucleotide markers [67].
  (b) *Histone modification*: Histones can be modified by reversible and covalent, methylations, sumoylation, acetylations, ADP ribosylation, phosphorylations, and ubiquitylation. This will change the electric charge of amino acid residues on histones and cause the negatively charged DNA to bind looser or tighter, leading to the tightly transcribed silenced heterochromatin or actively transcribed loose euchromatin. These epigenetic alterations also cause changes in the functioning of the transcription regulation machinery. The alterations can arise at several distinct sites on the histone tails and are regulated by diverse enzymes, comprising the demethylases, acetyltransferases, and methyl transferase deacetylases. This regulation is very compound and includes several diverse players, but various analyses correlate histone alteration to both cancer and aging, and studies are still progressing regarding drugs that can block the enzymes associated. Aging is also correlated with particular complex modifications of histones and chromatin distribution, with a deterioration of repression over constitutive heterochromatin loci and accessory gain of facultative heterochromatin in different genomic regions.

7 Intrinsic limitations of aged stem cells

In different tissues, the effective SCs remain in old age, but are often imperil in their capacity to achieve their functions [70]. Overcoming the intrinsic changes to SC activities is important for interventions aimed to engage endogenous SCs for regenerative function. In the muscle, current analysis has drop light on the transformation of normally inactive SCs that can be stimulated during regeneration of muscle toward a completely senescent condition that hinders their stimulation in geriatric animals. SC inactivity is a strongly controlled procedure that assures that the progenitor populace maintains a long-lasting capability to be promptly stimulated in reaction to injury. At geriatric period, inactive muscle SCs are observed in a pre-senescent state that is correlated with the deterioration of epigenetic gene silencing (GS) of the *INK4a* locus and the subsequent increase of the cell cycle inhibitor *p16INK4a*. These cells are incapable to achieve their regenerative activities and will endure complete aging upon regenerative pressure [71]. Interestingly, the homeostasis between age and inactivity relies on mechanisms that avoid the aggregation of intracellular impairment. Inactive muscle SCs exhibit an uninterrupted high basal autophagy level, which is essential for the removal of impaired proteins and organelles. A deterioration to manage the autophagy flux in old satellite cells causes elevation in the ROS levels, lesser expression of *INK4a* and thus mucle SCs eventually enter into aging [72]. Age-associated alterations in the epigenome also influence the regulation of transcription and the structure of chromatin, therefore disturbing the activity of SC. MSCs derived from the aged individuals exhibit a generalized decrease in the heterochromatin-related *H3K9me3* mark and a suppression of proteins associated with the maintenance of heterochromatin [73]. Profiling the epigenome of senescent HSCs reported that the stimulating *H3K4me3* mark is higher in genes associated with regulating the identity of HSC, while differentiation-stimulating genes are progressively suppressed with age. However, if endogenous SCs are to be utilized for tissue repair in aged patients, approaches to reverse some of the illustrated age-associated intrinsic changes that damage the regenerative potential have to be formulated. Intervening at the gene regulation level, via silencing the expression of *p16INK4a*, was reported to be efficient in improving quiescence and muscle SC activity of geriatric mice [71].

8 Role of stromal cells in aging

Stromal cells are cells of the connective tissue that can be from any organ, for example, bone marrow, ovary, uterine mucosa, and prostate, that supports the purpose of parenchymal cells of the specific organ. Pericytes and fibroblast are very common among the stromal cells.

Stromal cells belong to a class of heterogeneous cells that can fulfill a broad area of extremely important roles in both health as well as disease. They play a major character during regeneration, development, immune responses, tissue injury, cancer, and other pathologies by offering various biological processes in distinct tissues. Phenotype and function of stromal cells depend on the microenvironment of the specific tissues, but they also can shape the integrity, organization, and dynamics of their microenvironment. Their multidisciplinary functions in both organs and tissues are gaining attention because of its elegant studies and cutting-edge methods [74–77].

Among the stromal cells, the most studied are the MSCs. In accordance with the International Society for Cellular Therapy (ISCT), "MSC" should stand for mesenchymal stromal cells. Bone marrow and adipose tissues are the main sources of MSCs and they can also be found in placenta, umbilical cord blood, dental pulp and other extra embryonic tissues. They are also present in non-mesenchymal vascularized sources and heart tissues [78, 79].

The process of hematopoiesis relies upon the maintenance of mesenchymal stromal cells inside the bone marrow (BM). Hence, alterations in the compartment of the hematopoietic cells that appears during aging and development apparently correlate with the changes in the formation of the stromal cell microenvironment. The microenvironment may maintain the inductive and constitutive formation of blood cells reacting to a physiological condition, but it is unclear that the favorable hematopoietic microenvironment is regulated either by alteration in the population of the stromal cell or by alterations in the cell's activity. The activity of the SCs can be due to cell-to-cell proximity with hematopoietic cells by expressing ECMs [80, 81] and by certain factors that control differentiation and proliferation of the hematopoietic cells. MSCs are defined as a population of a heterogeneous type of cells, and different subtypes of them may have distinct activity. Besides the heterogeneity in the morphology, the stromal cell lines are reported to show functional heterogeneity in maintaining myeloid progenitor cells in vitro. Masuo Obinata et al. reported that the established BM-derived stromal cell lines demonstrated distinct stimulatory function on the accelerated inflation of lineage-restricted myeloid progenitors, and therefore, it is implied that each of the stromal cells in the BM may support a desirable hematopoietic microenvironment for the progenitor [82]. Iguchi et al. indicated that several BM-derived stromal cell lines particularly activate large colony formation of macrophages, granulocytes, and erythrocytes. Excluding the erythrocytes, which needed erythropoietin in inclusion to the stromal cell layers, no externally added cytokines were needed for the activation of the large colony formation of macrophages and granulocytes. Thus, the cytokine-independent pathways of granulopoiesis are believed to occur in bone marrow, and the stromal cells may intervene information required for activation for the expansion and differentiation of granulocytes via cell-to-cell association.

In higher organisms, the normal somatic cells have only a restricted division ability when cultured in vitro [83]. Cells in culture decrease their capacity of division in a regulated, structured manner, like the characteristic of a differentiation marker in a digressing system of differentiation divisions and proliferation [84] although it remains uncertain why progenitor cells are decreased during subcultivation. According to the idea of the differentiation theory, it implies that both errors in the genetic program and DNA replication are at least moderately affected by exogenous factors. Age-related alterations of the lymphohematopoietic system, particularly of its lymphoid part, are well reported. However, the mechanisms of these alterations remain uncertain. Generally, the difficulty is to identify the external and internal factors, regulating the lymphoid and hematopoietic cell activity during aging. The stromal tissue appears to be one of the eminent sources of external signals, which can generate the age-dependent alterations in lymphohemopoiesis [85]. According to a study done by Stephan et al., the capacity of stromal cells to regulate the development of pro-B cells also gets decreased with age. Since the progression from pro-B cell to the pre-B stage is particularly dependent on the components generated by stromal cells, it is believed that stromal cells are one of the eminent targets of downregulation in senescence and their function in the deterioration of BM lymphopoiesis should be carefully reviewed. The report has also been there, of an age-associated deterioration in the quantity of available IL-7 in the conditioned media (CM) from LTBMC-B (long-term bone marrow culture system from B lymphocytes) [86]. Therefore, since stromal cells are the sole source of IL-7 in the bone marrow, it is predicted that decreased production of IL-7 by stromal cells is the principal underlying mechanism for the decrease in stromal cell activity in aged animals and plays a primary role in their reduced numbers of pre-B cells. In histological studies of various tissues, it was reported that there are changes in the composition of the cell according to age. The cellularity of the human dermis is known to decrease during aging. In the oral epithelia of the rat and the duodenal crypts of the mouse, it has been described that there is a reduction in the progenitor population. In the distal gut of the mouse, the reduction of crypts was observed, which is associated with the deterioration of stem cells.

Generally, it can be stated that aging in the replicating system is attended by alterations in the composition of tissues, resulting in a decrease of progenitor cells that is accountable for the replacement of nondividing, but functionally active cells. However, the reasons behind the loss of progenitor cells remain unclear.

9 The pros and cons of stromal cell therapy

Several analyses have been supervised to investigate the safety of MSC-based therapies. Clinical trials reported that in vitro culture of human MSCs is less prone to unfavorable alterations. The safety and effect of MSC therapy were also inspected by Karussis et al. in individuals with multiple sclerosis (MS) and amyotrophic lateral sclerosis (ALS). However, more comprehensive analyses and observations concerning the safety of utilizing MSC therapies are essential. However, an analysis documented that the utilization of autologous adipose tissue-procured MSCs (AT-MSCs) in an individual with chronic kidney disorder ensued not only in the advancement of renal activity but also in the atrophy of the tubules and fibrotic scarring of the interstitial tissue, which could recommend nephrotoxicity of the applied MSCs [87]. Another group analyzed the efficacy of the allogeneic therapy of MSCs provided to the aortas of patients with acute kidney damage postcardiac surgery. No distinctions were detected among the control and treated group in association with the improvement of the activity of the renal or in the development of unfavorable events. Despite many cons for utilizing MSCs in clinical areas, there are, however, a few queries that are required to be solved for the successful utilization of MSCs. One of them includes achieving sufficient count of the cells. Inappropriately, during the in vitro culture, cells at greater passages age because of the reduced activity of the telomerase. Additionally, during the long-term in vitro culture, MSCs lose their capacity to differentiate and initiate to display morphological alterations. Even more particularly, long-term in vitro culture might cause enhanced possibility of malignant conversion [88].

However, when utilizing SC-based therapies, all probable undesirable impacts should be counted. The threats related with tumorigenesis post-SC transplantation are broadly reviewed in the literature. Several components may influence the possible tumorigenesis after transplantation of MSCs, comprising the age of the donor, host tissue, mechanisms that regulate the nature of the MSCs at the target site, and growth regulators expressed by the tissue of the recipient. Also, manipulations and long-period in vitro cultures of MSCs can induce chromosomal impairment and genetic instability. Various aggregated components can yield a reaction in the form of an instinctive tumor transformation. Individuals who are transplanted with SCs often go through radiotherapy or chemotherapy for a long-term, due to which their immune system does not function efficiently, which may also be correlated with the threat of tumorigenesis.

10 Pitfalls in stem cell therapy for aging and regeneration

The inadequacy of a principle insight of the extrinsic and intrinsic alterations that arise while aging is among the most important barrier to the advancement of regenerative medicine approaches that will prove efficient in the aging populace. There is a count of important barriers that must be deliberated as this knowledge is pursued. One example of the pitfall is illuminated in the investigation of neural regeneration after stroke. However, stroke is usually common in aged patients; most investigations of stroke are done in young animals. Moreover, various investigations have shown a deterioration in regeneration after brain injury in the aged brain. Recovery of young rats is faster in comparison with aged rats and also can go through complete cure followed by ischemic cortical injury, but aged animals can recover only about 70% of their prestroke motor function.

Even though some of the mechanisms that govern SC aging are common between various systems, the large variations between the tissue-specific SC will certainly convert into differences in the manner they age. Much of our awareness regarding SC aging arises from the hematopoietic system, chiefly because HSCs are the best featured of the adult systems. Both Neural stem cells (NSCs) and Hematopoietic stem cells (HSCs) divide at a slow rate during the whole life span of the animal, frequently producing progeny that restore cells that are lost. As this role is uncommon to all tissue-specific SC, the large literature pool associated with these types of cells may be partial against SC that multiplies similarly and thus baffles our insight of SC aging. An array of stem cell-associated approaches are presently being analyzed for cell replacement. The assigning of the endogenous tissue particularly SCs that are already available in the impaired tissue provides enormous capacity; however, the most frequently utilized strategy includes the SC transplantation. For any SC transplantation-associated approach, the crucial concern is the age of both the recipient milieu and the donor tissue. A current analysis utilizing heterochronic parabiosis (linked circulatory systems of old and young adult mice) documented that circulating systemic agents from the young mouse enhanced the engraftment capability of the endogenous HSCs in the senescent mice and, further, that this impact was regulated by cells in the senescent HSC niche [89].

Strategies that use pluripotent SCs and aimed differentiation approaches for transplantation have long been a target of SC research. However, to the well-reported ethical considerations surrounding the utilization of pluripotent embryonic stem cells (ESCs), the work is surrounded with concerns varying from the complications of regulating differentiation, gaining convenient integration into particular tissues, the threat of tumorigenesis, and the probability of rejection due to histoincompatibility. Efforts to address the capacity for rejection have caused the advancement of patient-histocompatible SCs using approaches. Most currently, the finding that completely differentiated cells can be instigated to convert cells with pluripotent SC-like characteristics, defined as induced pluripotent stem cells (iPS cells), has extremely enhanced the work in the field. Although this area is in its inception, the very real probability exists that modifying senescent somatic cells to iPS cells may present difficulties beyond those that will be confronted when utilizing cells from younger individuals. Thus, a more comprehensive insight into the alterations that arise in cells during senescence is crucial when concerning the success of these approaches [90].

11 Stem cell-derived conditioned medium (CM) as therapy

Senescence is a process in which intrinsic and extraneous components govern a continuous loss of physiological structures and functions. One of the eminent indications of aging can be observed in the skin of an individual due to the deterioration in skin integrity. The major symptoms of aging are the emergence of wrinkles and loss of elasticity; however, in photoaging, the indications of aging are attended with various other symptoms such as discoloration, wrinkles, hyperkeratosis, irregular pigmentation, loss of elasticity, and other several neoplasms. The fibroblasts of the skin generate MMPs (matrix metalloproteinases) and collagen of which the MMPs can particularly degenerate almost the entire component of the Extracellular matrix (ECM) of the skin and perform an eminent role in photoaging of the skin. Moreover, elastin is a pivotal protein constituent of tissues that assists the skin to recover to its original stage when it is activated. It helps in supporting strength and elasticity that play a vital role in the reparation of tissue in the body. Hence, rejuvenation of the skin has evolved to be the major focus, in which several treatments in medical and cosmeceutical products are utilized for antisenescence.

In current years, the CM derived from stem cells is gaining greater attention in the area of regenerative medicine that is accredited to several pharmacologically active factors present in the CM. Treatments focusing on enhancing the activity of wound healing are immediately required. Fig. 5 portrays some of the therapeutic potentials of stem cell-derived conditioned medium. Recent studies have described that adipose tissue can form an optimum microenvironment to help wound healing [91], which is in conferment with the outcome that the quality and speed of wound healing were notably enhanced when administered with extract of the adipose tissue. Moreover, the study done by Campbell et al. [92] utilized a CM derived from adipose tissue to co-culture with keratinocytes and fibroblast feeder cells and achieved an increased emergence of cultured epidermal autografts. Another in vitro study reported that the CM can notably increase ADSCs cell development and proliferation of EC (endothelial cells). More accurately, in the hemostasis phase, *PDGF, TGF-A1*, and *TGF-2*, epidermal growth factor and insulin-like growth factors in the CM can instigate platelets and endothelial cells to form blood clot and *TNF-α, TGF-β, FGF IL-1*, and *FGF* can activate leukocytes, keratinocytes, macrophages, and neutrophils to eradicate foreign particles, dead cell debris in inflammation phase [93]. Therefore, the CM can be characterized as the

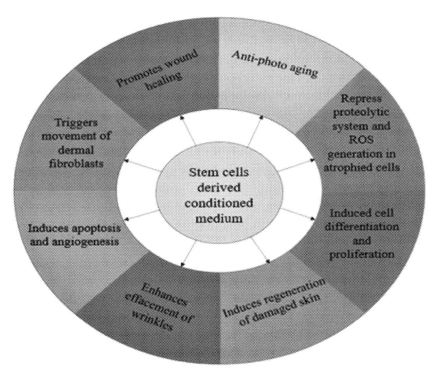

FIG. 5 Therapeutic potential of stem cell-derived conditioned medium.

medium containing the released soluble factors of the stem cells through the course of their culture, which is having both therapeutic and differentiation capacity. Hence, it can be capable to make the cells differentiate into different lineages and helps in the restoration of the damaged cells. Several studies have explored the SC origin CM as a vehicle for anti-aging and skin moisturizing. The function of moisturizers is not only to enhance the content of water of the skin but also to prevent the skin. Moisturizers furnish the skin with constituents that maintain the barrier integrity and water content and help the cells to perform its function normally. Notably, lipids are the vital constituents for both water retention activity and permeability barrier activity in the external surface of the skin. The outer layer consists of a condensed web of keratin that helps to hydrate the skin by retaining water evaporation. Interestingly, it is reported that the mice group treated with MSC CM displayed necessarily decreased contents of keratin in comparison with the UVB group. Thus, it is proved that the topical application of MSC CM developed notable prevention of skin barrier activity by reducing the transepidermal water loss (TEWL) and enhancing hydration of the skin, thereby decreasing the skin wrinkling. Hence, it may perform as an effective skin-preventing agent [94]. Moreover, the histological examination reported that UVB exposure can lead to the instigation of pathological characteristics in the skin correlated with photoaging. Animals that are irradiated with UVB exhibited a considerable enhancement in dermal thickness, epidermal thickness, hyperkeratosis, and infiltration of inflammatory cells in the dermis. However, a decrease amount of epidermal hyperplasia was observed which could be indicated by a lower value of dermal thickness, epidermal thickness, and hyperkeratosis, and also, the amount of inflammatory cell infiltration in the skin was observed to be moderate in the animals treated with CM formulation.

The study done by Amirthalingam et al., 2019, has described that the CM obtained from MSCs of human bone marrow enhances effacement of wrinkles and skin moisturization. Hence, due to the presence of the growth factors in the medium, CM-BMMSCs (conditioned medium from bone marrow MSCs) are playing a vital role in skin rejuvenation and antiaging. Moreover, the conditioned medium is further able to prevent fibroblast of the skin from tert-butyl hydroperoxide (tbOH)-stimulated oxidative stress and radiation of UV-B [95]. Sriramulu et al., 2020, reported that CM from umbilical cord-derived MSCs prevents skin against photoaging instigated by UVA and UVB radiation and is an encouraging agent for skin anti-photoaging treatments [94].

Furthermore, CM-hMSCs (condition medium from human MSCs) can trigger the movement of dermal fibroblasts and enhance the level of expression of the genes for the process of wound healing. Hence, CM-MSCs can be utilized as a treatment for the improvement of cutaneous wound healing [96]. On the contrary, studies reported that the secretory factors of ATMSC can instigate collagen formation and movement of fibroblasts during the wound healing process through stimulating the growth factors or dermal fibroblasts [97]. During the cell regeneration, CM-ATMSCs (conditioned-medium adipose tissue derived from MSCs) further showed essentially to escalate the number of cells in the Gap1 phase while decreasing the number of cells in the synthesis and Gap2/mitosis phases in cell cycle analysis. CM-ATMSCs are also known to restrict melanoma (skin cancer) growth by modifying cell cycle distribution and encouraging them to undergo apoptosis in vitro. The WJ-MSCs (Wharton's Jelly MSC) dispense the discrete cytokines of wound healing, like *CTGF, TGF-βs, VEGF, FGF-2,* and *HIF-1α,* that can stimulate the synthesis of collagen and also sirtuin-1 (*SIRT-1*) as antiaging genes, but passaging of the cells may influence the growth factors (*VEGF*) and cytokine (*IL1-α, IL-6,* and *IL-8*)-level secretion in human WJ-MSC [95].

Studies were done providing evidence that the SC-derived CM includes several important cytokines and growth factors that may be beneficial for skin health care. These proteins derived from stem cells were suitable to develop a stable skin-care product. The formulation prepared from the CM was harmless to the human primary cells, and it prevents the primary cell DNA from the formation of CPD (cyclobutene pyrimidine dimers). The CM obtained from several SCs can be utilized in the antiaging treatment and hair growth-enhancing cosmetic formulations. Moreover, therapies and products that can diminish (or) reverse (or) block this process of aging are very essential materials for doing research and product development. There is enormous confirmation that CM, which is a cell-free bioactive substrate, plays an eminent role in discrete pathophysiological processes via the paracrine/autocrine mechanism. Various studies reported the immense potential application of CM from discrete tissues. In association with ample sources and time-saving procedures, CM holds an immense capability to be extensively practiced in biomedicine. In addition, as the complex formation of CM, a pivotal step in this direction would involve the study of each constituent and the fundamental mechanism in definitive conditions, therapeutic and potential diagnostic application.

12 Cell-free therapy with extracellular vesicles

Extracellular vesicles (EVs) function as natural vectors and relocate their content to other tissues and cells. Therefore, EVs are a promising means for the transfer of biologically active compounds for therapeutic effect. A cell-free strategy averts the threats of indefinite growth of the cell and formation of tumor, as the EVs are not multiplying. EV may be derived from unaltered or genetically altered human SC and utilize them rather than the parental SC. The major obstruction for the therapeutic utilization of EV is the tedious isolation process using expensive equipment and tough enrichment procedures. The generation of purified EVs is definite but not adequate for wide clinical utilization. The approved procedure for segregation of EV suggested by the International Society of Extracellular Vesicles (ISEV) comprises sequential sedimentation followed by ultracentrifugation. Another technique for deriving EVs is FFFF (flow field-flow fractionation) [98]. However, this technique needs accession and is not extensive. Currently, commercially available kits are utilized for the segregation of EVs, based on immunomagnetic separation (IMS) [99] or precipitation of EV with farther centrifugation at lesser speeds. IMS of EV on the basis of definite surface marker(s) determines a high-purity fraction of EV. Additionally, despite the high distinction, these processes are costly. The process of precipitation of EV resolves the limitation of the higher cost of the separation method. However, neither of these methods resolved the major limitation of minimal yields. EV discharge is enhanced by oxidative impairment, hypoxic environment, irradiation, and the activity of the complement system. In order to enhance the generation of EV, it has been recommended to expose the cells to these conditions. However, these exposures alter the molecular structure of EV, which can influence their biological characteristics.

13 Tissue microenvironmental cues and influences in cell-free therapies

Regardless of either Induced pluripotent stem cells (iPSCs) or endogenous somatic SCs are utilized, the capacity to regenerate the somatic tissues is influenced by the tissue microenvironment or systemic environment, or it has become evident in current years that regulating the tissue microenvironment and systemic parameters in senescent or diseased conditions can specifically influence the process of regeneration. Extracellular signals can affect all the features of SC activities: inactivation, multiplication, multipotency, and differentiation. These signals are procured from either the systemic milieu, reaching the SC via the vasculature, or the local milieu—the niche. The niche can be considered as tissue, ECM, or any cell that exists in close proximity to the SC populace and that impacts SC biology. Examples of niche components comprise other immune cells, stromal cells, and somatic cells, innervating neuronal fibers, the vasculature, and ECM. Nevertheless, the structure niche differs among the various somatic SC types, indicating the varied structure of the host tissues and physiological requirements of their resident SCs and somatic cells [100]. Signals procured from the blood that impacts SC activities comprise the soluble molecules produced by any tissue in the body, which can be either growth factors, hormones, or any other signaling molecules or immune-procured signals produced by infiltrating immune cells. These signals may affect the SC activity directly or via regulation of the local milieu. Indirect action involves alterations to the makeup of the resident immune cell populace, regulation of somatic or stromal cell activity, and distress of the secretory phenotype of these cells [101].

Both the microenvironment and the systemic milieu, but also SCs themselves, alter with age, and these alterations restrict the progress of regenerative processes. Approaches aimed at developing the quality of this interplay in coordination with definite SC-directed interventions are likely to increase the success of regenerative therapies. This may involve recognizing and utilizing immune cell-procured factors that activate a definite feature of the regenerative process or aiming the immune cells themselves with instructive cue to regulate regeneration. Such strategies are specifically pertinent in conditions in which the immune milieu is compromised, particularly in degenerative disorders that are correlated with inflammation, and in aged patients, which often display constant low-level inflammation. Furthermore, favorable regeneration can help from the eradication of damaging local or systemic cues that aggregate with age.

14 Next generation of cell-based therapies for aging and regeneration

Previous achievements in the advancement of therapeutically significant manipulations of young adult SCs affirm the enormous capability that an insight of the mechanisms that regulate the behavior of SC may hold. Examples of such capable approaches involve transplantation analyses of hematopoietic stem cells (HSCs) that are genetically altered to overexpress the homeobox gene *HOXB4*, in which there were widely increased counts of HSCs in the transplant individual, with no increase in the tumor risk [102]. Similarly, the relevant neural regeneration and increased functional recovery followed by infusion of growth factor into the NSC niche of the adult rat brain following stroke additionally affirms to the promise for the advancement of SC-based therapies [103]. These achievements will essentially determine the backbone upon which the next-generation regenerative therapies are built as we recognize translating our understanding into the treatment of the elderly. Hence, although our current understanding about SC aging contributes more possibility than pause, it also indicates that an initial step in the advancement of regenerative medicine approaches for the elderly needs a better insight into the process of senescence.

15 Conclusion and perspectives

Advancement in regenerative medicine (RM) and the approach of SC therapies, now proceeding into the clinical trials, will probably illuminate the present obstruction to the utilization of SCs to treat disorders. Since the aged individuals have more possibility to experience constant degenerative conditions, which would be treated from the regenerative therapies, it is necessary to focus on the barriers enforced by the senescence to SC-based therapeutic interventions. Our developing insight of the primary mechanisms of organism senescence has unveiled the critical obstructions in the repair potential of aged tissues and features the essentiality to coordinate SC-definite interventions with systemic inflection and niche to enhance regenerative success.

In the work explained here, we highlighted the interventions with a definite success in preclinical analyses. These comprise suppression of inflammatory signaling, inflection of immune cell phenotypes, excision of senescent cells, and suppression of various intracellular signaling cascades that are continually stimulated during senescence. While so far, the translational achievements of these previous analyses will need handling with two major queries, first, the cascades aimed by these interventions are often pivotal regulators of the physiological regeneration and their acute stimulation is crucial for the process of regeneration. Thus, therapeutic approaches should target to regulate the cascade activity for avoiding chronic activation rather than complete inhibition. Secondly, local and systemic transfer of effector molecules may have effects in several cell types concurrently, sometimes with opposing results. Thus, recognizing the aims of each intervention is a pivotal step to assure that the clinical translation of these cell-free-based therapies considers the probable ill-effects. Therefore, further in-depth analyses of specific model systems will help in addressing these queries.

Acknowledgments

The authors are thankful to Chettinad Academy of Research and Education (CARE) for providing the infrastructural support and to SERB, DST, Govt. of India, for providing the financial support to complete this piece of work.

Funding

This work was supported by the grants sanctioned to Dr. Antara Banerjee (PI) from the SERB, DST, Govt. of India, with Sanction File No. ECR/2017/001066.

References

[1] J.C. Macedo, S. Vaz, E. Logarinho, Mitotic dysfunction associated with aging hallmarks, in: Cell Division Machinery and Disease, Springer, Cham, 2017, pp. 153–188, https://doi.org/10.1007/978-3-319-57127-0_7.

[2] F. Rodier, D. Zhou, G. Ferbeyre, Cellular senescence, geroscience, cancer and beyond, Aging (Albany NY) 10 (9) (2018) 2233, https://doi.org/10.18632/aging.101546.

[3] D. Gems, L. Partridge, Genetics of longevity in model organisms: debates and paradigm shifts, Annu. Rev. Physiol. 75 (2013) 621–644, https://doi.org/10.1146/annurev-physiol-030212-183712.

[4] G.E. Holmes, C. Bernstein, H. Bernstein, Oxidative and other DNA damages as the basis of aging: a review, Mutat. Res. 275 (3–6) (1992) 305–315, https://doi.org/10.1016/0921-8734(92)90034-M.

[5] M. Florea, Aging and immortality in unicellular species, Mech. Ageing Dev. 167 (2017) 5–15, https://doi.org/10.1016/j.mad.2017.08.006.

[6] C. Stephens, Senescence: even bacteria get old, Curr. Biol. 15 (8) (2005) 308–310, https://doi.org/10.1016/j.cub.2005.04.006.

[7] T. Grune, R. Shringarpure, N. Sitte, K. Davies, Age-related changes in protein oxidation and proteolysis in mammalian cells, J. Gerontol. Ser. A Biol. Med. Sci. 56 (11) (2001) 459–467, https://doi.org/10.1093/gerona/56.11.B459.

[8] B. Bartali, R.D. Semba, A.B. Araujo, Klotho, FGF21 and FGF23: novel pathways to musculoskeletal health? J. Frailty Aging 2 (4) (2013) 179, https://doi.org/10.14283%2Fjfa.2013.26.

[9] A. Antonelli, M. Rotondi, P. Fallahi, S.M. Ferrari, A. Paolicchi, P. Romagnani, et al., Increase of CXC chemokine CXCL10 and CC chemokine CCL2 serum levels in normal ageing, Cytokine 34 (1–2) (2006) 32–38, https://doi.org/10.1016/j.cyto.2006.03.012.

[10] Y. Bauer, E.S. White, S. de Bernard, P. Cornelisse, I. Leconte, A. Morganti, et al., MMP-7 is a predictive biomarker of disease progression in patients with idiopathic pulmonary fibrosis, ERJ Open Res. 3 (1) (2017), https://doi.org/10.1183/23120541.00074-2016.

[11] M. Feger, P. Hase, B. Zhang, F. Hirche, P. Glosse, F. Lang, et al., The production of fibroblast growth factor 23 is controlled by TGF-β2, Sci. Rep. 7 (1) (2017) 1–7, https://doi.org/10.1038/s41598-017-05226-y.

[12] R. Cappellari, M. D'Anna, B.M. Bonora, M. Rigato, A. Cignarella, A. Avogaro, et al., Shift of monocyte subsets along their continuum predicts cardiovascular outcomes, Atherosclerosis 266 (2017) 95–102, https://doi.org/10.1016/j.atherosclerosis.2017.09.032.

[13] J. Constans, C. Conri, Circulating markers of endothelial function in cardiovascular disease, Clin. Chim. Acta 368 (1–2) (2006) 33–47, https://doi.org/10.1016/j.cca.2005.12.030.

[14] E. Anuurad, B. Enkhmaa, Z. Gungor, W. Zhang, R.P. Tracy, T.A. Pearson, et al., Age as a modulator of inflammatory cardiovascular risk factors, Arterioscler. Thromb. Vasc. Biol. 31 (9) (2011) 2151–2156, https://doi.org/10.1161/ATVBAHA.111.232348.

[15] R. Bornheim, M. Müller, U. Reuter, H. Herrmann, H. Büssow, T.M. Magin, A dominant vimentin mutant upregulates Hsp70 and the activity of the ubiquitin-proteasome system and causes posterior cataracts in transgenic mice, J. Cell Sci. 121 (22) (2008) 3737–3746, https://doi.org/10.1242/jcs.030312.

[16] W.S. Chow, A. Xu, Y.C. Woo, A.W. Tso, S.C. Cheung, C.H. Fong, et al., Serum fibroblast growth factor-21 levels are associated with carotid atherosclerosis independent of established cardiovascular risk factors, Arterioscler. Thromb. Vasc. Biol. 33 (10) (2013) 2454–2459, https://doi.org/10.1161/ATVBAHA.113.301599.

[17] S. Chai, W. Wang, J. Liu, H. Guo, Z. Zhang, C. Wang, et al., Leptin knockout attenuates hypoxia-induced pulmonary arterial hypertension by inhibiting proliferation of pulmonary arterial smooth muscle cells, Transl. Res. 166 (6) (2015) 772–782, https://doi.org/10.1016/j.trsl.2015.09.007.

[18] L.W. Huo, Y.L. Ye, G.W. Wang, Y.G. Ye, Fractalkine (CX3CL1): a biomarker reflecting symptomatic severity in patients with knee osteoarthritis, J. Investig. Med. 63 (4) (2015) 626–631, https://doi.org/10.1097/JIM.0000000000000158.

[19] A.G. Geiser, J.J. Letterio, A.B. Kulkarni, S. Karlsson, A.B. Roberts, M.B. Sporn, Transforming growth factor beta 1 (TGF-beta 1) controls expression of major histocompatibility genes in the postnatal mouse: aberrant histocompatibility antigen expression in the pathogenesis of the TGF-beta 1 null mouse phenotype, Proc. Natl. Acad. Sci. 90 (21) (1993) 9944–9948, https://doi.org/10.1073/pnas.90.21.9944.

[20] Z. Chaker, S. Aïd, H. Berry, M. Holzenberger, Suppression of IGF-I signals in neural stem cells enhances neurogenesis and olfactory function during aging, Aging Cell 14 (5) (2015) 847–856, https://doi.org/10.1111/acel.12365.

[21] R. Miskin, O. Tirosh, M. Pardo, I. Zusman, B. Schwartz, S. Yahav, et al., αMUPA mice: a transgenic model for longevity induced by caloric restriction, Mech. Ageing Dev. 126 (2) (2005) 255–261, https://doi.org/10.1016/j.mad.2004.08.018.

[22] M. Peine, R.M. Marek, M. Löhning, IL-33 in T cell differentiation, function, and immune homeostasis, Trends Immunol. 37 (5) (2016) 321–333, https://doi.org/10.1016/j.it.2016.03.007.

[23] S.J. Chien, T.C. Chen, H.C. Kuo, C.N. Chen, S.F. Chang, Fulvic acid attenuates homocysteine-induced cyclooxygenase-2 expression in human monocytes, BMC Complement. Altern. Med. 15 (1) (2015) 61, https://doi.org/10.1186/s12906-015-0583-x.

[24] S. Demirci, A. Aynalı, K. Demirci, S. Demirci, B.C. Arıdoğan, The serum levels of resistin and its relationship with other proinflammatory cytokines in patients with Alzheimer's disease, Clin. Psychopharmacol. Neurosci. 15 (1) (2017) 59, https://doi.org/10.9758/cpn.2017.15.1.59.

[25] A.L. Cardoso, A. Fernandes, J.A. Aguilar-Pimentel, M.H. de Angelis, J.R. Guedes, M.A. Brito, et al., Towards frailty biomarkers: candidates from genes and pathways regulated in aging and age-related diseases, Ageing Res. Rev. 47 (2018) 214–277, https://doi.org/10.1016/j.arr.2018.07.004.

[26] D.J. Baker, T. Wijshake, T. Tchkonia, N.K. LeBrasseur, B.G. Childs, B. Van De Sluis, J.L. Kirkland, J.M. van Deursen, Clearance of p16 Ink4a-positive senescent cells delays ageing-associated disorders, Nature 479 (7372) (2011) 232–236, https://doi.org/10.1038/nature10600.

[27] B.A. Jones, M. Beamer, S. Ahmed, Fractalkine/CX3CL1: a potential new target for inflammatory diseases, Mol. Interv. 10 (5) (2010) 263, https://doi.org/10.1124%2Fmi.10.5.3.

[28] C.P. Verschoor, J. Johnstone, J. Millar, R. Parsons, A. Lelic, M. Loeb, et al., Alterations to the frequency and function of peripheral blood monocytes and associations with chronic disease in the advanced-age, frail elderly, PLoS One 9 (8) (2014), https://doi.org/10.1371/journal.pone.0104522, 104522.

[29] J.H. Ruth, M.V. Volin, G.K. Haines III, D.C. Woodruff, K.J. Katschke Jr., J.M. Woods, et al., Fractalkine, a novel chemokine in rheumatoid arthritis and in rat adjuvant-induced arthritis, Arthritis Rheum. 44 (7) (2001) 1568–1581, https://doi.org/10.1002/1529-0131(200107)44:7<1568::AID-ART280>3.0.CO;2-1.

[30] S. Pavanello, M. Stendardo, G. Mastrangelo, M. Bonci, B. Bottazzi, M. Campisi, et al., Inflammatory long pentraxin 3 is associated with leukocyte telomere length in night-shift workers, Front. Immunol. 8 (2017) 516, https://doi.org/10.3389/fimmu.2017.00516.

[31] M. Pollack, C. Leeuwenburgh, Apoptosis and aging: role of the mitochondria, J. Gerontol. Ser. A Biol. Med. Sci. 56 (11) (2001) 475–482, https://doi.org/10.1093/gerona/56.11.B475.

[32] J. Corre, B. Hébraud, P. Bourin, Concise review: growth differentiation factor 15 in pathology: a clinical role? Stem Cells Transl. Med. 2 (12) (2013) 946–952, https://doi.org/10.5966/sctm.2013-0055.

[33] A.V. Belikov, Age-related diseases as vicious cycles, Ageing Res. Rev. 49 (2019) 11–26, https://doi.org/10.1016/j.arr.2018.11.002.

[34] R.H. Swerdlow, J.M. Burns, S.M. Khan, The Alzheimer's disease mitochondrial cascade hypothesis: progress and perspectives, Biochim. Biophys. Acta (BBA) - Mol. Basis Dis. 1842 (8) (2014) 1219–1231, https://doi.org/10.1016/j.bbadis.2013.09.010.

[35] J.A. Driver, G. Logroscino, J.M. Gaziano, T. Kurth, Incidence and remaining lifetime risk of Parkinson disease in advanced age, Neurology 72 (5) (2009) 432–438, https://doi.org/10.1212/01.wnl.0000341769.50075.bb.

[36] C. López-Otín, M.A. Blasco, L. Partridge, M. Serrano, G. Kroemer, The hallmarks of aging, Cell 153 (6) (2013) 1194–1217, https://doi.org/10.1016/j.cell.2013.05.039.

[37] M. Collado, M.A. Blasco, M. Serrano, Cellular senescence in cancer and aging, Cell 130 (2) (2007) 223–233, https://doi.org/10.1016/j.cell.2007.07.003.

[38] S.J. Winawer, A.G. Zauber, The advanced adenoma as the primary target of screening, Gastrointest. Endosc. Clin. N. Am. 12 (1) (2002) 1–9, https://doi.org/10.1016/S1052-5157(03)00053-9.

[39] S.L. Stewart, J.M. Wike, I. Kato, D.R. Lewis, F. Michaud, A population-based study of colorectal cancer histology in the United States, 1998–2001, Cancer 107 (S5) (2006) 1128–1141, https://doi.org/10.1002/cncr.22010.

[40] A.M. Noone, K.A. Cronin, S.F. Altekruse, N. Howlader, D.R. Lewis, V.I. Petkov, et al., Cancer incidence and survival trends by subtype using data from the surveillance epidemiology and end results program, 1992–2013, Cancer Epidemiol. Biomarkers Prev. 26 (4) (2017) 632–641, https://doi.org/10.1158/1055-9965.

[41] V. Bouvard, D. Loomis, K.Z. Guyton, Y. Grosse, F.E. Ghissassi, L. Benbrahim-Tallaa, et al., Carcinogenicity of consumption of red and processed meat, Lancet Oncol. 16 (16) (2015) 1599–1600, https://doi.org/10.1016/S1470-2045(15)00444-1.

[42] J.W. Welsh, D. Mahadevan, R. Ellsworth, L. Cooke, D. Bearss, B. Stea, The c-Met receptor tyrosine kinase inhibitor MP470 radiosensitizes glioblastoma cells, Radiat. Oncol. 4 (1) (2009) 1–10, https://doi.org/10.1186/1748-717X-4-69.

[43] V. Walter, L. Jansen, M. Hoffmeister, A. Ulrich, J. Chang-Claude, H. Brenner, Smoking and survival of colorectal cancer patients: population-based study from Germany, Int. J. Cancer 137 (6) (2015) 1433–1445, https://doi.org/10.1002/ijc.29511.

[44] Y.J. Choi, S.K. Myung, J.H. Lee, Light alcohol drinking and risk of cancer: a meta-analysis of cohort studies, Cancer Res. Treat. 50 (2) (2018) 474, https://doi.org/10.4143%2Fcrt.2017.094.

[45] M.W. Lutgens, M.G. van Oijen, G.J. van der Heijden, F.P. Vleggaar, P.D. Siersema, B. Oldenburg, Declining risk of colorectal cancer in inflammatory bowel disease: an updated meta-analysis of population-based cohort studies, Inflamm. Bowel Dis. 19 (4) (2013) 789–799, https://doi.org/10.1097/MIB.0b013e31828029c0.

[46] L. Ricciardiello, D.J. Ahnen, P.M. Lynch, Chemoprevention of hereditary colon cancers: time for new strategies, Nat. Rev. Gastroenterol. Hepatol. 13 (6) (2016) 352, https://doi.org/10.1038/nrgastro.2016.56.

[47] T. Watanabe, T.T. Wu, P.J. Catalano, T. Ueki, R. Satriano, D.G. Haller, et al., Molecular predictors of survival after adjuvant chemotherapy for colon cancer, N. Engl. J. Med. 344 (16) (2001) 1196–1206, https://doi.org/10.1056/NEJM200104193441603.

[48] G. Jothimani, S. Sriramulu, Y. Chabria, X.F. Sun, A. Banerjee, S. Pathak, A review on theragnostic applications of microRNAs and long non-coding RNAs in colorectal cancer, Curr. Top. Med. Chem. 18 (30) (2018) 2614–2629, https://doi.org/10.2174/1568026619666181221165344.

[49] S. Kidd, E. Spaeth, J.L. Dembinski, M. Dietrich, K. Watson, A. Klopp, et al., Direct evidence of mesenchymal stem cell tropism for tumor and wounding microenvironments using in vivo bioluminescent imaging, Stem Cells 27 (10) (2009) 2614–2623, https://doi.org/10.1002%2Fstem.187.

[50] A. Banerjee, Y. Chabria, R.K. NR, J. Gopi, P. Rowlo, X.F. Sun, et al., Role of tumor specific niche in colon cancer progression and emerging therapies by targeting tumor microenvironment, Adv. Exp. Med. Biol. (2019) 1–16, https://doi.org/10.1007/5584_2019_355.

[51] J. Luetzkendorf, L.P. Mueller, T. Mueller, H. Caysa, K. Nerger, H.J. Schmoll, Growth inhibition of colorectal carcinoma by lentiviral TRAIL-transgenic human mesenchymal stem cells requires their substantial intratumoral presence, J. Cell. Mol. Med. 14 (9) (2010) 2292–2304, https://doi.org/10.1111/j.1582-4934.2009.00794.x.

[52] M. Watson, D.M. Holman, K.A. Fox, G.P. Guy, A.B. Seidenberg, B.P. Sampson, et al., Preventing skin cancer through reduction of indoor tanning: current evidence, Am. J. Prev. Med. 44 (6) (2013) 682–689, https://doi.org/10.1016/j.amepre.2013.02.015.

[53] J.H. Lee, D.E. Fisher, Melanocyte stem cells as potential therapeutics in skin disorders, Expert. Opin. Biol. Ther. 14 (11) (2014) 1569–1579, https://doi.org/10.1517/14712598.2014.935331.

[54] K. Vinay, S. Dogra, Stem cells in vitiligo: current position and prospects, Pigment Int. 1 (1) (2014) 8, https://doi.org/10.4103/2349-5847.135430.

[55] M. Jagasia, M. Arora, M.E. Flowers, N.J. Chao, P.L. McCarthy, C.S. Cutler, et al., Risk factors for acute GVHD and survival after hematopoietic cell transplantation, Blood 119 (1) (2012) 296–307, https://doi.org/10.1182/blood-2011-06-364265.

[56] J.L. Ferrara, J.E. Levine, P. Reddy, E. Holler, Graft-versus-host disease, Lancet 373 (9674) (2009) 1550–1561, https://doi.org/10.1016/S0140-6736(09)60237-3.

[57] S. Glennie, I. Soeiro, P.J. Dyson, E.W. Lam, F. Dazzi, Bone marrow mesenchymal stem cells induce division arrest anergy of activated T cells, Blood 105 (7) (2005) 2821–2827, https://doi.org/10.1182/blood-2004-09-3696.

[58] K. English, J.M. Ryan, L. Tobin, M.J. Murphy, F.P. Barry, B.P. Mahon, Cell contact, prostaglandin E2 and transforming growth factor beta 1 play non-redundant roles in human mesenchymal stem cell induction of CD4+ CD25Highforkhead box P3+ regulatory T cells, Clin. Exp. Immunol. 156 (1) (2009) 149–160, https://doi.org/10.1111/j.1365-2249.2009.03874.x.

[59] A.E. Aksu, E. Horibe, J. Sacks, R. Ikeguchi, J. Breitinger, M. Scozio, et al., Co-infusion of donor bone marrow with host mesenchymal stem cells treats GVHD and promotes vascularized skin allograft survival in rats, Clin. Immunol. 127 (3) (2008) 348–358, https://doi.org/10.1016/j.clim.2008.02.003.

[60] A. Campanati, M. Orciani, V. Consales, R. Lazzarini, G. Ganzetti, G. Di Benedetto, et al., Characterization and profiling of immunomodulatory genes in resident mesenchymal stem cells reflect the Th1-Th17/Th2 imbalance of psoriasis, Arch. Dermatol. Res. 306 (10) (2014) 915–920, https://doi.org/10.1007/s00403-014-1493-3.

[61] D. Simon, L.R. Braathen, H.U. Simon, Eosinophils and atopic dermatitis, Allergy 59 (6) (2004) 561–570, https://doi.org/10.1111/j.1398-9995.2004.00476.x.

[62] M. Suárez-Fariñas, N. Dhingra, J. Gittler, A. Shemer, I. Cardinale, C. de Guzman Strong, et al., Intrinsic atopic dermatitis shows similar TH2 and higher TH17 immune activation compared with extrinsic atopic dermatitis, J. Allergy Clin. Immunol. 132 (2) (2013) 361–370, https://doi.org/10.1016/j.jaci.2013.04.046.

[63] D.J. Rossi, D. Bryder, J.M. Zahn, H. Ahlenius, R. Sonu, A.J. Wagers, et al., Cell intrinsic alterations underlie hematopoietic stem cell aging, Proc. Natl. Acad. Sci. 102 (26) (2005) 9194–9199, https://doi.org/10.1073/pnas.0503280102.

[64] V. Janzen, R. Forkert, H.E. Fleming, Y. Saito, M.T. Waring, D.M. Dombkowski, et al., Stem-cell ageing modified by the cyclin-dependent kinase inhibitor p16 INK4a, Nature 443 (7110) (2006) 421–426, https://doi.org/10.1038/nature05159.

[65] A.V. Molofsky, S.G. Slutsky, N.M. Joseph, S. He, R. Pardal, J. Krishnamurthy, et al., Increasing p16 INK4a expression decreases forebrain progenitors and neurogenesis during ageing, Nature 443 (7110) (2006) 448–452, https://doi.org/10.1038/nature05091.

[66] D. Piccin, C.M. Morshead, Wnt signaling regulates symmetry of division of neural stem cells in the adult brain and in response to injury, Stem Cells 29 (3) (2011) 528–538, https://doi.org/10.1002/stem.589.

[67] B.H. Chen, R.E. Marioni, E. Colicino, M.J. Peters, C.K. Ward-Caviness, P.C. Tsai, et al., DNA methylation-based measures of biological age: meta-analysis predicting time to death, Aging (Albany NY) 8 (9) (2016) 1844, https://doi.org/10.18632%2Faging.101020.

[68] S. Horvath, DNA methylation age of human tissues and cell types, Genome Biol. 14 (10) (2013) 3156, https://doi.org/10.1186/gb-2013-14-10-r115.

[69] G. Hannum, J. Guinney, L. Zhao, L. Zhang, G. Hughes, S. Sadda, et al., Genome-wide methylation profiles reveal quantitative views of human aging rates, Mol. Cell 49 (2) (2013) 359–367, https://doi.org/10.1016/j.molcel.2012.10.016.

[70] M.B. Schultz, D.A. Sinclair, When stem cells grow old: phenotypes and mechanisms of stem cell aging, Development 143 (1) (2016) 3–14, https://doi.org/10.1242/dev.130633.

[71] P. Sousa-Victor, S. Gutarra, L. García-Prat, J. Rodriguez-Ubreva, L. Ortet, V. Ruiz-Bonilla, et al., Geriatric muscle stem cells switch reversible quiescence into senescence, Nature 506 (7488) (2014) 316–321, https://doi.org/10.1038/nature13013.

[72] L. García-Prat, M. Martínez-Vicente, E. Perdiguero, L. Ortet, J. Rodríguez-Ubreva, E. Rebollo, et al., Autophagy maintains stemness by preventing senescence, Nature 529 (7584) (2016) 37–42, https://doi.org/10.1038/nature16187.

[73] W. Zhang, J. Li, K. Suzuki, J. Qu, P. Wang, J. Zhou, et al., A Werner syndrome stem cell model unveils heterochromatin alterations as a driver of human aging, Science 348 (6239) (2015) 1160–1163, https://doi.org/10.1126/science.aaa1356.

[74] C. Niehage, J. Karbanová, C. Steenblock, D. Corbeil, B. Hoflack, Cell surface proteome of dental pulp stem cells identified by label-free mass spectrometry, PLoS One 11 (8) (2016), https://doi.org/10.1371/journal.pone.0159824, 0159824.

[75] J. Galipeau, M. Krampera, J. Barrett, F. Dazzi, R.J. Deans, J. DeBruijn, et al., International Society for Cellular Therapy perspective on immune functional assays for mesenchymal stromal cells as potency release criterion for advanced phase clinical trials, Cytotherapy 18 (2) (2016) 151–159, https://doi.org/10.1016/j.jcyt.2015.11.008.

[76] D. Iacobazzi, M.M. Swim, A. Albertario, M. Caputo, M.T. Ghorbel, Thymus-derived mesenchymal stem cells for tissue engineering clinical-grade cardiovascular grafts, Tissue Eng. A 24 (9–10) (2018) 794–808, https://doi.org/10.1089/ten.tea.2017.0290.

[77] C. Sassoli, L. Vallone, A. Tani, F. Chellini, D. Nosi, S. Zecchi-Orlandini, Combined use of bone marrow-derived mesenchymal stromal cells (BM-MSCs) and platelet rich plasma (PRP) stimulates proliferation and differentiation of myoblasts in vitro: new therapeutic perspectives for skeletal muscle repair/regeneration, Cell Tissue Res. 372 (3) (2018) 549–570, https://doi.org/10.1007/s00441-018-2792-3.

[78] E.M. Horwitz, K. Le Blanc, M. Dominici, I. Mueller, I. Slaper-Cortenbach, F.C. Marini, et al., Clarification of the nomenclature for MSC: the International Society for Cellular Therapy position statement, Cytotherapy 7 (5) (2005) 393–395, https://doi.org/10.1080/14653240500319234.

[79] L. Zimmerlin, J.P. Rubin, M.E. Pfeifer, L.R. Moore, V.S. Donnenberg, A.D. Donnenberg, Human adipose stromal vascular cell delivery in a fibrin spray, Cytotherapy 15 (1) (2013) 102–108, https://doi.org/10.1016/j.jcyt.2012.10.009.

[80] A.D. Campbell, M.W. Long, M.S. Wicha, Haemonectin, a bone marrow adhesion protein specific for cells of granulocyte lineage, Nature 329 (6141) (1987) 744–746, https://doi.org/10.1038/329744a0.

[81] M. Obinata, R. Okuyama, K.I. Matsuda, M. Koguma, N. Yanai, Regulation of myeloid and lymphoid development of hematopoietic stem cells by bone marrow stromal cells, Leuk. Lymphoma 29 (1–2) (1998) 61–69, https://doi.org/10.3109/10428199809058382.

[82] J.I. Kameoka, N. Yanai, M. Obinata, Bone marrow stromal cells selectively stimulate the rapid expansion of lineage-restricted myeloid progenitors, J. Cell. Physiol. 164 (1) (1995) 55–64, https://doi.org/10.1002/jcp.1041640108.

[83] H.E. Swim, R.F. Parker, Culture characteristics of human fibroblasts propagated serially, Am. J. Epidemiol. 66 (2) (1957) 235–243, https://doi.org/10.1093/oxfordjournals.aje.a119897.

[84] H.R. Hirsch, The dynamics of repetitive asymmetric cell division, Mech. Ageing Dev. 6 (1977) 319–332, https://doi.org/10.1016/0047-6374(77)90033-1.

[85] R.M. Gorczynski, M.P. Chang, M. Kennedy, S. MacRae, K. Benzing, G.B. Price, Alterations in lymphocyte recognition repertoire during ageing. I. Analysis of changes in immune response potential of B lymphocytes from non-immunized aged mice, and the role of accessory cells in the expression of that potential, Immunopharmacology 7 (3–4) (1984) 179–194, https://doi.org/10.1016/0162-3109(84)90035-3.

[86] R.P. Stephan, V.M. Sanders, P.L. Witte, Stage-specific alterations in murine B lymphopoiesis with age, Int. Immunol. 8 (4) (1996) 509–518, https://doi.org/10.1093/intimm/8.4.509.

[87] J.S. Kim, J.H. Lee, O. Kwon, J.H. Cho, J.Y. Choi, S.H. Park, et al., Rapid deterioration of preexisting renal insufficiency after autologous mesenchymal stem cell therapy, Kidney Res. Clin. Pract. 36 (2) (2017) 200, https://doi.org/10.23876%2Fj.krcp.2017.36.2.200.

[88] G.V. Røsland, A. Svendsen, A. Torsvik, E. Sobala, E. McCormack, H. Immervoll, et al., Long-term cultures of bone marrow–derived human mesenchymal stem cells frequently undergo spontaneous malignant transformation, Cancer Res. 69 (13) (2009) 5331–5339, https://doi.org/10.1158/0008-5472.CAN-08-4630.

[89] D. Piccin, C.M. Morshead, Potential and pitfalls of stem cell therapy in old age, Dis. Model. Mech. 3 (7–8) (2010) 421–425, https://doi.org/10.1242/dmm.003137.

[90] B.A. Gilchrest, A review of skin ageing and its medical therapy, Br. J. Dermatol. 135 (6) (1996) 867–875, https://doi.org/10.1046/j.1365-2133.1996.d01-1088.x.

[91] M. Keck, D. Haluza, D.B. Lumenta, S. Burjak, B. Eisenbock, L.P. Kamolz, et al., Construction of a multi-layer skin substitute: simultaneous cultivation of keratinocytes and preadipocytes on a dermal template, Burns 37 (4) (2011) 626–630, https://doi.org/10.1016/j.burns.2010.07.016.

[92] C.A. Campbell, B.A. Cairns, A.A. Meyer, C.S. Hultman, Adipocytes constitutively release factors that accelerate keratinocyte proliferation in vitro, Ann. Plast. Surg. 64 (3) (2010) 327–332, https://doi.org/10.1097/SAP.0b013e318199f82c.

[93] S.A. Guo, L.A. DiPietro, Factors affecting wound healing, J. Dent. Res. 89 (3) (2010) 219–229, https://doi.org/10.1177%2F0022034509359125.

[94] S. Sriramulu, A. Banerjee, G. Jothimani, S. Pathak, Conditioned medium from the human umbilical cord-mesenchymal stem cells stimulate the proliferation of human keratinocytes, J. Basic Clin. Physiol. Pharmacol. 1 (2020), https://doi.org/10.1515/jbcpp-2019-0283.

[95] R. Noverina, W. Widowati, W. Ayuningtyas, D. Kurniawan, E. Afifah, D.R. Laksmitawati, et al., Growth factors profile in conditioned medium human adipose tissue-derived mesenchymal stem cells (CM-hATMSCs), Clin. Nutr. Exp. 24 (2019) 34–44, https://doi.org/10.1016/j.yclnex.2019.01.002.

[96] L. Bussche, R.M. Harman, B.A. Syracuse, E.L. Plante, Y.C. Lu, T.M. Curtis, et al., Microencapsulated equine mesenchymal stromal cells promote cutaneous wound healing in vitro, Stem Cell Res Ther 6 (1) (2015) 66, https://doi.org/10.1186/s13287-015-0037-x.

[97] B.S. PARK, K.A. Jang, J.H. SUNG, J.S. PARK, Y.H. Kwon, K.J. Kim, et al., Adipose-derived stem cells and their secretory factors as a promising therapy for skin aging, Dermatol. Surg. 34 (10) (2008) 1323–1326, https://doi.org/10.1111/j.1524-4725.2008.34283.x.

[98] D. Kang, S. Oh, S.M. Ahn, B.H. Lee, M.H. Moon, Proteomic analysis of exosomes from human neural stem cells by flow field-flow fractionation and nanoflow liquid chromatography−tandem mass spectrometry, J. Proteome Res. 7 (8) (2008) 3475–3480, https://doi.org/10.1021/pr800225z.

[99] M.P. Oksvold, A. Neurauter, K.W. Pedersen, Magnetic bead-based isolation of exosomes, in: RNA Interference, Humana Press, New York, NY, 2015, pp. 465–481, https://doi.org/10.1007/978-1-4939-1538-5_27.

[100] C.S. Bjornsson, M. Apostolopoulou, Y. Tian, S. Temple, It takes a village: constructing the neurogenic niche, Dev. Cell 32 (4) (2015) 435–446, https://doi.org/10.1016/j.devcel.2015.01.010.

[101] D.L. Jones, A.J. Wagers, No place like home: anatomy and function of the stem cell niche, Nat. Rev. Mol. Cell Biol. 9 (1) (2008) 11–21, https://doi.org/10.1038/nrm2319.

[102] G. Sauvageau, U.N. Thorsteinsdottir, C.J. Eaves, H.J. Lawrence, C. Largman, P.M. Lansdorp, et al., Overexpression of HOXB4 in hematopoietic cells causes the selective expansion of more primitive populations *in vitro* and *in vivo*, Genes Dev. 9 (14) (1995) 1753–1765, https://doi.org/10.1101/gad.9.14.1753.

[103] B. Kolb, C. Morshead, C. Gonzalez, M. Kim, C. Gregg, T. Shingo, et al., Growth factor-stimulated generation of new cortical tissue and functional recovery after stroke damage to the motor cortex of rats, J. Cereb. Blood Flow Metab. 27 (5) (2007) 983–997, https://doi.org/10.1038%2Fsj.jcbfm.9600402.

Chapter 19

Stem cell-based therapeutic strategy in delaying prion disease

Sanjay Kisan Metkar, Koyeli Girigoswami, and Agnishwar Girigoswami

Medical Bionanotechnology, Faculty of Allied Health Sciences, Chettinad Hospital and Research Institute (CHRI), Chettinad Academy of Research and Education (CARE), Kelambakkam, Tamil Nadu, India

1 Introduction

Prion diseases are classified as fatal neurodegenerative diseases that result from the deposition of the prion protein in its misfolded state. This insoluble misfolded state of the prion protein is known as the infectious prion protein (PrP^{Sc}) [1]. This appears as spongiform lesions in the brain and central nervous system (CNS) of the affected individual. The diseases have different manifestations such as transmissible spongiform encephalopathies in cattle (bovine spongiform encephalopathy, BSE). In humans, it has different forms such as fatal familial insomnia, Gerstmann-Straussler-Scheinker disease, kuru, and Creutzfeldt Jakob disease (CJD) [2]. The name PrP^{Sc} was derived from the disease scrapie in sheep, where these animals developed similar symptoms to those of BSE in cattle. PrP^{C} is a normal host-encoded glycoprotein that is responsible for many cellular functions. However, it can get modified by PrP^{Sc} and when exposed to cells, it can develop the disease in the host organism [3, 4].

2 The innovation of the prion protein

The cause of CJD was unidentified for several years. The disease was identified to progress fast after onset, causing death within a few months without raising an immune response. It was noted that the scrapie and CJD agent was highly resistant to ionizing as well as UV radiations [5]. Such findings suggested that these agents have unusual properties. Proteins were identified as discrete biological molecules, and their implication in cellular procedures was accepted as early as the 18th century [6]. A century later, a "fundamental substance" was biochemically characterized by the Dutch chemist Gerhardus Johannes Mulder, and Jons Jakob Berzelius named it protein [5]. Merino sheep, a sheep of Spanish origin, first exhibited some unusual behaviors such as excessive licking, altered gaits, and excessive itching that compelled it to scrape against fences. Later, these clinical symptoms were termed scrapie [5]. In the mid-20th century, pathologists Creutzfeldt and Jakob portrayed a neurodegenerative disease that had symptoms similar to scrapie and called it transmissible spongiform encephalopathy (TSE) [7]. The Hershey and Chase experiment in 1952 proved that virus replication needs heritable material, that is, nucleic acid [8]. One year later, Watson and Crick explained the genetic code and defined the exact structure of DNA [9]. Further, in 1953 Crick explained protein synthesis and defined the term "central dogma of molecular biology," that is, information encoded in DNA can be synthesized, stored, and used by the organism to replicate itself [10]. TSE was an exception from the central dogma in molecular biology, as the prion protein replicated itself without the involvement of the genetic code, which was proved by the mechanisms based on the theory proposed by J.S. Griffith [11]. Stanley Prusiner anticipated this proteinaceous contagious pathogen as a prion in 1982 [1]. Fig. 1 focuses on finding the prions in the timeline, which began in the early 18th century.

In 1920, Hans Gerhard Creutzfeldt and Alfons Maria Jakob depicted a human neurological disorder of unclear etiology that would annoy established researchers for the coming 60 years [12]. First, scrapie was described as a "slow virus" by the scientists Cuille and Chelle in 1938 [13]. Scientists described another human neurological disorder that was similar to CJD and scrapie and was found in the Fore tribe of Papua New Guinea; they called it kuru [14–17].

In 1944, Veterinarian W.S. Gordon utilized formalin to inactivate the louping-sick virus infection found in the cerebrum and spleen of the infected animals. At this point, he utilized these treated tissues to immunize healthy animals. Formalin could inactivate the virus but failed to inactivate the scrapie agent. Animals that were vaccinated with the scrapie agent died within 2 years due to scrapie [15, 18]. Hadlow suggested that kuru was transmissible similarly to scrapie, which was

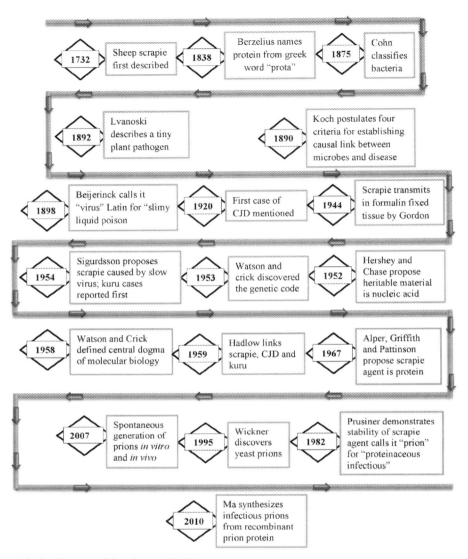

FIG. 1 Significant events in the discovery of the prion protein [5].

established in Sigurdsson's (1954) study; kuru was a moderate infection [19]. A.G. Dickinson suggested that nucleic acid was responsible for scrapie infection and identified a gene that could control the incubation period of scrapie in some mouse strains [20]. After Gordon failed to inactivate this scrapie agent using formalin, many scientists from different fields of research have tried to deactivate the scrapie agent by exposure to ionizing and UV irradiation, excessive heat, and treating under high pressures [18, 21, 22]. A few perceptive scientists, including Tikvah Alper, I.H. Pattison, and J.S Griffith, hypothesized that the agent that causes scrapie could originate from protein [5, 23]. Such a hypothesis was opposite to the central dogma of molecular science. In 1966, Tikvah Alper tried to determine the scrapie genomic size and its inactivation by using ionizing radiation. In this study, she concluded that the causative agent was not simply inactivated after exposure to high amounts of UV irradiation. Thus, she concluded that the agent must be replicating independent of any nucleic acid [24]. In 1967, Pattison introduced an additional proof that the scrapie was protein in nature, based on his experiments [25]. However, J.S. Griffith was the leading researcher who gave a bold statement that scrapie is completely made up of protein and proposed three mechanisms for protein to be extremely infectious [26].

A few scientists competed with Griffith's example and gathered information that kept on recommending that the scrapie agent is based on a protein [5]. Prusiner (1982) coined the term "prion," an irresistible proteinaceous molecule, for which he won the Nobel Prize in 1998 [1]. Prusiner et al. proposed a "protein-only hypothesis" by successfully isolating prion amyloid from the diseased animals and developed a method to inactivate the infectious agent. After a few years, the scientific community accepted Prusiner's protein-only hypothesis phenomena [27].

The agent responsible for scrapie was confirmed as a prion, but the source of this proteinaceous agent was not confirmed then. Further, Prusiner isolated the prion particle, which was recalcitrant from the mRNA transcript of the prion gene that coded the sequence PrP 27–30 [5]. This sequence was present in both infected and healthy individuals, which made supportive evidence of a protein-only hypothesis. Further, Jiyan Ma produced the recombinant infectious prion protein using a bacterial expression system that supported the prion hypothesis strongly. The scientific community had accepted the fact that ordinary, host-encoded PrP^C can be converted into infectious PrP^{Sc}, the main causative agent of the prion diseases [27].

3 Therapeutic approaches for managing prion disease

Prion infections are uncommon and sometimes-fatal neurodegenerative disorder portrayed by a unique, protein-only pathogenesis. To treat prion infections, so far the strategies developed are:

(i) The inhibitors of PrP^{Sc}, like small molecules.
(ii) Antibodies against the prion protein.
(iii) Regulating the prion gene.
(iv) The response of unfolded protein regulation.
(v) Targeting the heterologous prion proteins.
(vi) Stem cell-based therapy.

In vitro and in vivo studies have shown how to dissociate prion amyloids produced by the prion peptide PrP 106–126 using enzymes of the serine protease family, lumbrokinase (LK), and serratiopeptidase (SP) [28]. The role of extremolytes isolated from hot springs was also demonstrated in vitro for the dissociation of PrP 106–126 amyloids [29]. The enzymes LK and SP were also functional in dissociating insulin amyloids both in vivo and in vitro [30–32]. Zinc oxide nanoflowers have shown the potential to degrade insulin amyloids [33]. TSEs are considered incurable due to the late diagnosis, which leads to already-prominent severe brain lesions and thereby decreases the therapeutic window. Recent therapeutic strategies target the inhibition of PrP^{Sc} aggregation and cease the conversion of PrP^C to PrP^{Sc}, but are unable to reverse the brain lesions already caused. Thus, it is necessary to develop treatment strategies that can induce functional recovery of the tissues damaged due to the disease through cell replacement. Stem cell therapy and regenerative medicine are very promising for the possible treatment of many neurodegenerative diseases through the replacement of cells [34, 35]. The different types of stem cells used for transplantation are neural stem cells (NSCs), which are derived from pluripotent stem cells; fetal neural stem cells (fNSCs); embryonic stem cells (ESCs) or induced pluripotent stem cells (iPSCs); neuronal precursor cells (NPCs); mesenchymal stem cells (MSCs); and microglial cells, as depicted in Fig. 2. Stem cells are also categorized based on their capacity of differentiation into various types of cells in vitro as well as in vivo.

In this chapter, we will discuss the different stem cell-based therapies in managing prion disease.

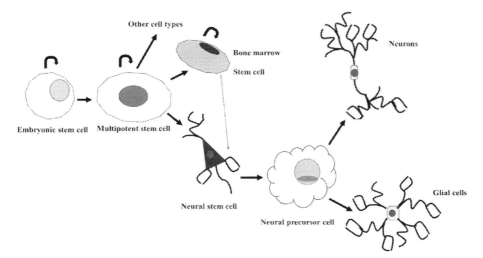

FIG. 2 Differentiation of neural cells: Embryonic stem cells can differentiate and give rise to almost all cells of the body; especially they can offer differentiate to neural progenitor cells. The capacity of differentiation of neural undifferentiated cells is then limited to neuronal and glial heredity. Bone marrow stem cells (BMSCs) can also transdifferentiate into neural stem cells.

4 Adult neural stem cells in therapeutic interventions of prion disease

Adult neurogenesis has been poorly understood in the case of prion ailments in comparison with other neurodegenerative diseases. In mice, PrP^C categorically controls the endogenous neurogenesis in the subventricular zone (SVZ) and dentate gyrus (DG). In vivo and in vitro studies have shown that there was expression of PrP^C in multipotent neural precursors as well as in matured neurons, although there was no measurable PrP^C found in the glia [36]. Adult neurogenesis also promotes endurance and axon focusing of olfactory sensible neurons [37]. The self-renewal capacity of NSCs to differentiate into astrocytes, neurons, and oligodendrocytes suggests its potential as a therapeutic agent in neurodegenerative diseases. Therapeutic strategies involving grafting of stem cells have shown promising results in the treatment of neurodegenerative diseases such as lateral sclerosis and Huntington's disease [38, 39]. Endogenous NSCs in the brains of adults can produce neurons and glia, which gives hope to researchers for the possibility of replacing the damaged cells. To attain this, stimulation of the endogenous progenitors in situ is necessary. Also, they can be modified genetically in vivo so that they can deliver the antiprion molecules [40]. In TSEs, the circumstances may be more complicated than in physiological conditions. There can be two possible reasons for this. First, during neuronal separation, the expression of PrP^C increases, which could be transitioned over into PrP^{Sc} resulting in further propagation of prions in the brain. Second, PrP^C advances into neural precursor multiplication during the neurogenesis of the embryos and adults, and is by all accounts associated with neuronal survival [41]. Therefore, the transition of PrP^C into PrP^{Sc} might halt neurogenesis and contribute to neurodegeneration [42]. Additionally, the autoremedial capacity of the brain is not sufficient to improve the network of the neurons because only a small part of the new cells will participate and survive. The prominent output showed that neural stem cells (NSC) aid in replicating prions and its alteration represents the pathogenic event that progresses into prion disorders [43]. The study also demonstrated that NSC helps in replicating prions and regulates the neuronal density. This kind of progress was observed in adult neural stem cells infected by prions, using in vitro conditions. Halting such prion propagation in such cells can improve the endogenous neurogenesis to repair the injured neuronal system [44]. The mechanism behind the enhancement of adult neurogenesis associated with prion disease is yet to be elucidated. Interestingly, the inhibition of PrP^{Sc} propagation in NSC possibly slows the progression of the disease by allowing PrP^C to perform neuroprotective or neurogenic functions. This is achieved through gene engineering utilizing a recombinant viral vector that can express antiprion molecules such as antibodies, dominant negative PrP mutants, and anti-PrP oligonucleotides for RNA interference [45–47]. The effectiveness of the neurons that are produced from endogenous NSCs is present only during the initial stages of prion disease and diminishes as the disease progresses [48]. Moreover, at the beginning, there was a stimulation of adult neurogenesis that produced new neurons to replace the lost or damaged cells, but these neurons were in their infant stage and could not achieve maturation. Eventually, most of the neurons die and cannot give effective damage replacement [43, 44, 49]. An enzyme named NEIL3, which is a glycosidase used in DNA repair, was knocked out in mice to study its role in adult neurogenesis and the propagation of prion disease. The results showed that mice that lacked the NEIL3 enzyme showed a reduced clinical phase of prion propagation and altered adult neurogenesis, but the time of survival was not affected. This study suggested that adult neurogenesis cannot influence the fatal consequence of prion disease, but can only delay the appearance of the clinical signs [49]. The exact mechanism through which NEIL3 can contribute to prion disease-related neuroprotection is not clear yet.

A major benefit of RNAi as a remedial methodology in prion disease is that it can suit every known strain of prion ailment. Inside any species, the essential grouping of PrP^C and PrP^{Sc} is equivalent for all strains; in this way, RNAi can be an active treatment for all prion variations. Using lentiviral-mediated RNA interference (RNAi) designed against the native prion protein (PrP), White et al. reported the first therapeutic application that led to neuronal rescue, the prevention of symptoms, and increased survival in mice that had established prion disease. They designed a viral vector that expressed shRNA sequences that were processed by the Dicer. Also, the antisense strand or guide strand was laden into the RNA-induced silencing complex (RISC), which can target a specific mRNA to then be degraded, as shown in Fig. 3 [50]. It is thought to be critical to evaluate whether this kind of approach may restrain prion accumulation/aggregation, yet additionally preserve and uphold neurogenesis to enable extra cell-based systems.

5 Fetal neural stem cell-based therapeutic approaches in prion diseases

Though stem cell graft-based technology has just been utilized to break down the pathophysiology related to prion issues, just a few examinations detailed the utilization of stem cells for remedial purposes [51]. A study reported the transplantation of fNSCs extracted from embryos of prion-resistant knockout (koPrP) mice to extend the survival of the hippocampal pyramidal cell in the mouse model, which was the scrapie-infected (C57Bl6/VM) wild type (wt) but had no symptoms of the disease. The fNSCs were injected 150 days postinoculation with the prion protein. It was found

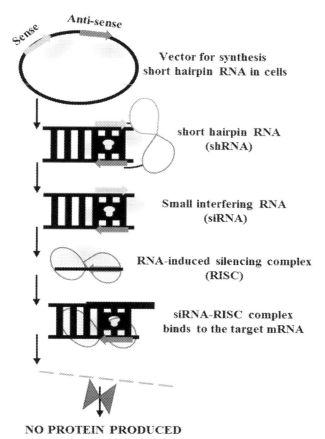

FIG. 3 RNAi as a remedial methodology in prion disease.

that functional neurons increased nearly 1.5 times in the pyramidal cell layer after day 21 postgraft as well as at the end of the study (250 dpi) in comparison to control mice. However, there was a 50% death of CA1 pyramidal cells after 180 days of the 250-day period of incubation. The number of neurons and the depth of CA1 were quantitated using an image analysis system. The grafted group acquired more CA1 neurons compared to the controls. Also, a greater CA1 depth was observed in the graft-receiving group compared to the control group that was administered only with medium. This study provided insight into the new late-stage therapy for TSEs, also known as prion disease. However, this study did not provide any proof that the transplantation of fetal NSCs can become a promising strategy in prion disease treatment [52]. In another study, fNSC was derived from three types of mice: (i) wild-type mice that express wtPrP, (ii) prion-resistant transgenic mice that can express porcine PrP (poPrP), and (iii) koPrP mice. It was found that NSC grafts were able to prolong the incubation time significantly by 20% and enhance survival time by 13% in RML prion-infected mice [53–55]. The beneficial effect was observed when fNSCs were grafted well in advance of the clinical symptoms, which indicates that successful treatment can happen only in a particular time window. The improvement in incubation time and rate of survival was evident for all types of fNSCs, irrespective of their origin (wtPrP, koPrP, and poPrP). The findings were unanticipated because ex vivo prions can infect and further replicate only in wt NSC, but not in koPrP NSC and poPrP NSC [53, 54].

Researchers have described an autologous and heterologous fNPC-centered therapy applied on transgenic mice (TgCJD mice) that mimics the mutation of the human E200K PrP gene that is responsible for a type of genetic CJD. These transgenic mice exhibit neurological deterioration along with aging, so fNPC grafting was done in newborn TgCJD mice, where the fNPC was derived from the brains of embryos of TgCJD or wt mice. After a follow-up of 10 months, a significant delay in disease progression was observed for both types of fNPC compared to nontransplanted mice. Both types of fNPC could prolong the incubation time of the disease by 35%. Also, there was an increase in the number of endogenous NSCs in the brains of grafted animals. This suggested that there was a proneurogenic effect exhibited by the grafted cells and there was no PrP^{Sc} accumulation in the endogenous NSCs of TgCJD mice. Moreover, there was no transmission of prions when the NSCs from TgCJD mice were used as a graft for wt mice [56].

6 Embryonic stem cells (ESCs) and induced pluripotent stem cells

ESCs are derived from the inner cell mass of the blastocyst stage, and they have tremendous potential for cell substitution treatments. In fact, they can multiply inconclusively in vitro along with holding the pluripotency to produce different cell types. ESCs from mice can proficiently differentiate in vitro to produce neural precursors that can further differentiate to give rise to functional astrocytes, neurons, and oligodendrocytes [57, 58]. Investigations into human ESCs have likewise now started in a few nations where they have incorporated human ESCs (hESCs) in the brains of rodents after differentiation. Recently, some progress happened on three crucial formative procedures: (i) control of the cell cycle; (ii) control of the fate of the stem cells; and (iii) control of the early stages of the differentiation of neurons [59].

PrP^C is the normal cellular isoform and its expression plays an important role in susceptibility to prion infection. The expression of this cellular isoform is not restricted within the nervous system, but rather it is found in other varieties of cells and tissues where it has different functions. PrP^C also aids with the capacity of long-term repopulation in hematopoietic stem cells [60, 61], and is known to be associated with the regulation of renewal versus differentiation of hESCs grown in culture [62]. In a study using seven types of hESC lines, the cells were cultured and were screened for the gene expression of prion protein gene codon 129. The normal cellular prion protein (PrP^C) mRNA was expressed in the hESC, but the accumulation of the protein was very low in the self-renewing population. When these hESCs were allowed undirected differentiation, the PrP^C protein expression was upregulated in the major embryonic lineages. Further, the self-renewing HESC populations were exposed to infectious prion proteins for both humans and animals. Following exposure to the infected prions, the cells rapidly consumed the material. When the source of infection was removed from the medium, the concentration and extent of the infectious prion protein inside the cells dropped rapidly. The outcome of this study suggested that hESCs can identify and rapidly take up the abnormal prion protein and immediately process it for clearance. This was a promising finding for the removal of infectious prion proteins from the cellular system [63].

The previous studies have been set apart by a significant technical breakthrough in cell replacement-based therapy. The Shinya Yamanaka group obtained pluripotent cells from the somatic cells of mice [64] by inducing the four translation factors—Sox2 (sex determining region Y), Oct4 (octamer-binding transcription factor 4), Klf4 (Krüppel-like factor 4), and cMyc (Myelocytomatosis)—into the fibroblasts. These transformed cells were termed IPS (induced pluripotent stem), and they displayed self-restoration and pluripotent properties for the ESCs [65, 66]. Considering this research, the generation of histocompatible cells becomes an important tool in the field of regenerative medicine [67]. Human cerebral organoids derived from human IPS cells act as a model system for infectious prion propagation. Two sporadic CJD prion subtypes were infected in the organoids to mimic the prion diseased condition. It was found that the organoids initially took up these infectious proteins, followed by clearance. However, there was a reemergence of the self-seeding activity of the prion protein to form insoluble protein aggregates, which indicated de novo propagation. Thus, for the purpose of research on prion disease and testing putative therapeutics, human cerebral organoids can act as a model system [68]. Other in vitro systems that have been identified for prion disease study are astrocytes, which are derived from human-induced pluripotent stem cells (iPSCs), [69] and the immortalized mouse neuronal astrocyte cell line (C8D1A) [70].

Despite the fact that ES and IPS comprise commanding tools in cell therapy, there are a few hindrances for the success of graft approaches. These include potential karyotypic variations from the norm because of their enhanced limit of proliferation and the instability of the genome; decreased cell survival; rejection due to immune response, although the brain is an immune-privileged organ where foreign grafts can survive for a longer time [71]; and the occurrence of teratomas. Researchers have demonstrated that the immunogenicity of the grafted cells can be reduced by (i) removing the MHC (major histocompatibility complex), (ii) removing the components related to immune response, or (iii) treatment with cyclosporine. Finally, it is important to predifferentiate the cells to be grafted in vitro before injecting into the brain so that there is enhanced cell survival and the absence of undifferentiated ES cells or any precursors that are still in a dividing state, which may give rise to tumors when they start multiplying in vivo [72].

7 Mesenchymal stem cells

Mesenchymal stem cells that are derived from various sources such as the umbilical cord, bone marrow, adipose tissue, etc., are used for cell therapy in animal models designed for neurodegenerative diseases and disorders associated with CNS. There are different methods of cell delivery, such as intravenous, intracerebroventricular, intracarotid, etc. [73]. The application of MSC in managing prion disease has been reviewed [74, 75].

Although many researchers have exploited the use of MSCs in the management of different neurodegenerative diseases, only a few have investigated the outcome of human MSC transplantation in prion-infected mice. The immortalized human MSCs (hMSCs) derived from the bone marrow, which can express the Lac Z gene, were transplanted to wt mice that were infected to create prion disease with the Chandler or Obihiro scrapie strains. The region of transplantation was one side of the thalamus or hippocampus. The presence of these hMSCs was found even 3 weeks postinjection. The cells were not only present at the site of the lesions, but they also migrated to the contralateral side. The study revealed that the prion-infected mice brain extracts could effectively promote the chemotaxis of the transplanted hMSCs in vitro and could identify the hMSCS in vivo at the site of the brain lesions of the infected mice. A small amount of these transplanted hMSCs could differentiate into cells that expressed neuronal, oligodendrocyte, and astrocyte markers. On the other hand, the accumulation of PrP^{Sc}, spongiosis, and astrocytosis due to the destruction of neurons was more severe in mice infected with the Obihiro strain. There was a positive correlation with the migration of hMSC and the severity of lesions in these mice. In contrast to the infected mice, the noninfected mice retained very few hMSCs, suggesting that the grafted hMSCs were well maintained in the infected mice and they were attracted toward the areas of the brain with lesions, whereas in noninfected mice, hMSCs were not necessary and were mostly eliminated. In the microenvironment of prion-infected mice, the expressions of different trophic factors (such as VEGF, BDNF, NGF, NT3, and CNTF) produced by the migrating hMSCs were stimulated due to the brain lesions. These tropic factors were higher in mice with prion infection compared to mock-infected mice. Although it is known that hMSCs can differentiate in certain conditions to neural lineages, the study was extended after 3 weeks to monitor whether the grafted hMSCs could give rise to astrocytes as well as neurons. Few LacZ-positive cells were found to express GFAP (an astrocyte marker) or MAP2 (a neuronal marker), which indicated that the grafted hMSCs could integrate and further get differentiated into neural lineages. There was a prolonged survival of Chandler-infected mice at 90 dpi; they were grafted with hMSCs but the increase was very small (5.3%) compared to the nongrafted mice. Thus, the summation of all these mechanisms could improve the lifespan but was unable to arrest the progression of the disease [76, 77]. The same researchers have identified some chemokine receptors that were responsible for hMSC migration toward the brain lesions [78].

In another study, researchers executed the autologous transplantation of compact bone marrow-derived MSCs (CB-MSC) in prion mouse models and evaluated the therapeutic efficacy. The CB-MSCs were isolated from the femur and tibia of the mouse infected with the Chandler prion strain. The transplantation was done in the hippocampus region, and at 120 dpi, there was a slight but significant increase in survival rate (>5%). The migration capacity of the CB-MSCs to the brain extracts isolated from prion-infected mice was established through an in vitro cell migration assay. This transplantation promoted the activation of microglial cells, as shown by the enhanced Aif1 gene expression that encodes the microglial IBA-1 protein. However, there was no influence on the rate of accumulation of PrP^{Sc} in the brain of the infected mouse [79].

8 Microglial cells

The microglial cells get proliferated and activated during the pathological conditions of the brain because they are considered the brain's innate immune cells. Microglia maintain tissue homeostasis and the regulation of synaptic plasticity to maintain the neuroinflammatory response. They have a neuroprotective role and play an important role in attenuating inflammation, repairing neuronal damage, and aiding the regenerative process. On the contrary, microglia can initiate neurotoxic pathways leading to neurodegeneration. Due to this dual role played by the microglia, many researchers have tried to study their role in the onset of the diseased condition [80–83]. To investigate the generation of microglia in prion disease, mice were injected with the RML prion strain. They were then transplanted with BM cells, which were previously transduced by a murine stem cell virus (MSCV)-based retroviral vector that encodes a boosted green fluorescent protein (GFP). There was a 50% replacement of microglia cells that were resident in the brain with the GFP-positive microglial cells derived from BM well before the clinical manifestation of the prion disease. However, the pathogenesis of the scrapie prion was not enhanced, thereby indicating that the prions could not replicate among the GFP-positive microglial cells. This showed that these cells can act as a cargo to deliver antiprion molecules [84]. Researchers have previously used murine Ra2 microglial cell line-derived microglial cells that expressed antiprion scFv antibodies as a grafting cell. These cells were grafted at week 7 (asymptomatic phase) in the brain of the wt mice, which were inoculated with the 22 L scrapie strain. There was a significant improvement in the survival rate of the mice in comparison to those animals that were grafted with GFP-expressing cells [45, 85]. It was observed that the direct infusion of anti-PrP antibodies could not improve the survival time of the mice. Instead, there was increased survival for Ra2 microglia graft-transplanted mice infected with the 22 L prion. This suggested that the microglia migrated to the regions that were further away from the injection site; this could be attributed to the survival. This improved survival was only observed when the grafting was done before disease manifestation, that is, at 7 weeks. When the grafting in mice took place through the clinical phase, that is, 13 weeks postinfection, of the 22 L scrapie prion or if the mice were inoculated by the Chandler scrapie strain, there was no change in survival time [86].

9 Discussion

Stem cell biology research has always been interesting for deciphering the treatment strategies for diseases related to the brain and spinal cord. There have been intensive investigations by several researchers revealing the therapeutic potential of stem cells and their progenitors in repairing brain damage caused by various neurodegenerative diseases. Regenerative medicine has a promising role in the management of prion disease. However, just a handful of studies have been done to reveal the efficacy of stem cell transplantation in delaying prion disease or improving the survival rate. The different strains of prions that cause infection for the onset of the disease also responded in different fashion to stem cell graft therapy. Moreover, the time of injection of the transplanted stem cells also decided the effectiveness of the therapy. The scarcity of preclinical trials in regenerative medicine strategies related to prion diseases indicates the complexity of this disease. Moreover, there are many issues that need to be addressed before approaching regenerative medicinal therapy, such as the source of the graft cell, the ethical concerns for using fetal or ESCs, the cargo used to deliver the grafts, confirming the safety over the long term, and the response of the donor cells. It may be possible that some kinds of stem cell lines already have an endogenous virus or there can be a contamination from exogenous viruses. These stem cells may themselves secrete virus particles or can express the viral antigens on their cell surface. On the other hand, some biotechnological products such as murine feeder cells, bovine fetal serum, etc., may also contain the prions that can pose a high risk [87]. This may also transmit serious diseases to the stem cell recipient, so the stem cell bank should assure a safe product to be used for therapy.

The available research on stem cell therapy for delaying prion disease or repairing damage caused by this disease indicates that just cell-based therapies are insufficient to combat the disease. Early diagnosis is required to initiate the treatment, prevent brain lesions, and repair them. This shows that multimodal approaches are necessary to manage prion disease. These approaches can include pharmacological drugs, modulation of the immune response to reduce brain inflammation, gene therapy, stem cell or multiple cell therapy, and agents that can stimulate endogenous neurogenesis.

Acknowledgment

The authors are grateful to the Chettinad Academy of Research and Education for providing infrastructural support.

References

[1] S.B. Prusiner, Novel proteinaceous infectious particles cause scrapie, Science 216 (4542) (1982) 136–144.
[2] M. Imran, S. Mahmood, An overview of human prion diseases, Virol. J. 8 (1) (2011) 1–9.
[3] S.B. Prusiner, Prions, Proc. Natl. Acad. Sci. U. S. A. 95 (23) (1998) 13363–13383.
[4] S.B. Prusiner, Molecular biology and pathogenesis of prion diseases, Trends Biochem. Sci. 21 (12) (1996) 482–487.
[5] M.D. Zabel, C. Reid, A brief history of prions, Pathog. Dis. 73 (9) (2015) ftv087.
[6] D.S. Schwarz, M.D. Blower, The endoplasmic reticulum: structure, function and response to cellular signaling, Cell. Mol. Life Sci. 73 (1) (2016) 79–94.
[7] J.D. Beckham, K.L. Tyler, Infectious disease: developments in the field of Creutzfeldt-Jakob disease, Rev. Neurol. Dis. 4 (3) (2007) 168.
[8] A.D. Hershey, M. Chase, Independent functions of viral protein and nucleic acid in growth of bacteriophage, J. Gen. Physiol. 36 (1) (1952) 39–56.
[9] J.D. Watson, F.H. Crick, Molecular structure of nucleic acids; a structure for deoxyribose nucleic acid, Nature 171 (4356) (1953) 737–738.
[10] F. Crick, Central dogma of molecular biology, Nature 227 (5258) (1970) 561–563.
[11] C. Soto, Prion hypothesis: the end of the controversy? Trends Biochem. Sci. 36 (3) (2011) 151–158.
[12] R. Henry, F.A. Murphy, Etymologia: Creutzfeldt-Jakob disease, Emerg. Infect. Dis. 23 (6) (2017) 956.
[13] D.B. Adams, Evolutionary biology and the risk of scrapie disease in sheep, Open Vet. J. 8 (3) (2018) 282–294.
[14] D.C. Gajdusek, V. Zigas, Kuru: clinical, pathological and epidemiological study of an acute progressive degenerative disease of the central nervous system among natives of the Eastern Highlands of New Guinea, Am. J. Med. 26 (3) (1959) 442–469.
[15] C.L. Jeffries, K.L. Mansfield, L.P. Phipps, P.R. Wakeley, R. Mearns, A. Schock, et al., Louping ill virus: an endemic tick-borne disease of Great Britain, J. Gen. Virol. 95 (Pt 5) (2014) 1005–1014.
[16] J. Collinge, M.S. Palmer, A.J. Dryden, Genetic predisposition to iatrogenic Creutzfeldt-Jakob disease, Lancet 337 (1991) 1441–1442.
[17] J. Collinge, J. Whitfield, E. McKintosh, A. Frosh, S. Mead, A.F. Hill, S. Brandner, D. Thomas, M.P. Alpers, A clinical study of kuru patients with long incubation periods at the end of the epidemic in Papua New Guinea, Philos. Trans. R. Soc. Lond. B Biol. Sci. 363 (1510) (2008) 3725–3739.
[18] W.S. Gordon, Advances in veterinary research, Vet. Rec. 58 (47) (1946) 516–525.
[19] W.J. Hadlow, Kuru likened to scrapie: the story remembered, Philos. Trans. R. Soc. Lond. B Biol. Sci. 363 (1510) (2008) 3644.
[20] C.M. Eklund, R.C. Kennedy, W.J. Hadlow, Pathogenesis of scrapie virus infection in the mouse, J. Infect. Dis. 117 (1) (1967) 15–22.
[21] P. Hedlin, R. Taschuk, A. Potter, P. Griebel, S. Napper, Detection and control of prion diseases in food animals, ISRN Vet. Sci. 2012 (2012) 254739.
[22] G.D. Hunter, G.C. Millson, Studies on the heat stability and chromatographic behaviour of the scrapie agent, J. Gen. Microbiol. 37 (1964) 251–258.

[23] T. Alper, W.A. Cramp, D.A. Haig, M.C. Clarke, Does the agent of scrapie replicate without nucleic acid? Nature 214 (5090) (1967) 764–766.
[24] I.H. Pattison, Resistance of the scrapie agent to formalin, J. Comp. Pathol. 75 (1965) 159–164.
[25] I.H. Pattison, K.M. Jones, The possible nature of the transmissible agent of scrapie, Vet. Rec. 80 (1) (1967) 2–9.
[26] J.S. Griffith, Nature of the scrapie agent: self-replication and scrapie, Nature 215 (5105) (1967) 1043–1044.
[27] S. Perrett, J. Ma, F. Wang, Prion disease and the 'protein-only hypothesis', Essays Biochem. 56 (2014) 181–191.
[28] S.K. Metkar, S. Ghosh, A. Girigoswami, K. Girigoswami, The potential of Serratiopeptidase and Lumbrokinase for the degradation of prion peptide 106-126-an in vitro and in silico perspective, CNS Neurol. Disord. Drug Targets 18 (9) (2019) 723–731.
[29] M. Kanapathipillai, S.H. Ku, K. Girigoswami, C.B. Park, Small stress molecules inhibit aggregation and neurotoxicity of prion peptide 106–126, Biochem. Biophys. Res. Commun. 365 (4) (2008) 808–813.
[30] S.K. Metkar, A. Girigoswami, R. Murugesan, K. Girigoswami, Lumbrokinase for degradation and reduction of amyloid fibrils associated with amyloidosis, J. Appl. Biomed. 15 (2) (2017) 96–104.
[31] S.K. Metkar, A. Girigoswami, R. Murugesan, K. Girigoswami, In vitro and in vivo insulin amyloid degradation mediated by Serratiopeptidase, Mater. Sci. Eng. C 70 (2017) 728–735.
[32] S.K. Metkar, A. Girigoswami, R. Vijayashree, K. Girigoswami, Attenuation of subcutaneous insulin induced amyloid mass in vivo using Lumbrokinase and Serratiopeptidase, Int. J. Biol. Macromol. 163 (2020) 128–134.
[33] A. Girigoswami, M. Ramalakshmi, N. Akhtar, S.K. Metkar, K. Girigoswami, ZnO nanoflower petals mediated amyloid degradation—an in vitro electrokinetic potential approach, Mater. Sci. Eng. C 101 (2019) 169–178.
[34] C.G. Song, Y.Z. Zhang, H.N. Wu, X.L. Cao, C.J. Guo, Y.Q. Li, M.H. Zheng, H. Han, Stem cells: a promising candidate to treat neurological disorders, Neural Regen. Res. 13 (7) (2018) 1294.
[35] J.M. Slack, What is a stem cell? Wiley Interdiscip. Rev. Dev. Biol. 7 (5) (2018) e323.
[36] A.D. Steele, J.G. Emsley, P.H. Özdinler, S. Lindquist, J.D. Macklis, Prion protein (PrPc) positively regulates neural precursor proliferation during developmental and adult mammalian neurogenesis, Proc. Natl. Acad. Sci. U. S. A. 103 (9) (2006) 3416–3421.
[37] L.E. Parrie, J.A. Crowell, G.C. Telling, R.A. Bessen, The cellular prion protein promotes olfactory sensory neuron survival and axon targeting during adult neurogenesis, Dev. Biol. 438 (1) (2018) 23–32.
[38] V. Silani, L. Cova, M. Corbo, A. Ciammola, E. Polli, Stem-cell therapy for amyotrophic lateral sclerosis, Lancet 364 (9429) (2004) 200–202.
[39] S.B. Dunnett, A.E. Rosser, Cell therapy in Huntington's disease, Neuro Rx 1 (4) (2004) 394–405.
[40] A. Consiglio, A. Gritti, D. Dolcetta, A. Follenzi, C. Bordignon, F.H. Gage, A.L. Vescovi, L. Naldini, Robust in vivo gene transfer into adult mammalian neural stem cells by lentiviral vectors, Proc. Natl. Acad. Sci. U. S. A. 101 (41) (2004) 14835–14840.
[41] X. Roucou, M. Gains, A.C. LeBlanc, Neuroprotective functions of prion protein, J. Neurosci. 75 (2) (2004) 153–161.
[42] L. Westergard, H.M. Christensen, D.A. Harris, The cellular prion protein (PrPC): its physiological function and role in disease, Biochim. Biophys. Acta 1772 (6) (2007) 629–644.
[43] A. Relaño-Ginès, A. Gabelle, C. Hamela, M. Belondrade, D. Casanova, C. Mourton-Gilles, S. Lehmann, C. Crozet, Prion replication occurs in endogenous adult neural stem cells and alters their neuronal fate: involvement of endogenous neural stem cells in prion diseases, PLoS Pathog. 9 (8) (2013), e1003485.
[44] A. Relaño-Ginés, S. Lehmann, C. Crozet, Prion diseases and adult neurogenesis: how do prions counteract the brain's endogenous repair machinery? Prion 8 (3) (2014) 240–246.
[45] K. Fujita, Y. Yamaguchi, T. Mori, N. Muramatsu, T. Miyamoto, M. Yano, H. Miyata, A. Ootsuyama, M. Sawada, H. Matsuda, R. Kaji, Effects of a brain-engraftable microglial cell line expressing anti-prion scFv antibodies on survival times of mice infected with scrapie prions, Cell. Mol. Neurobiol. 31 (7) (2011) 999.
[46] K. Toupet, V. Compan, C. Crozet, C. Mourton-Gilles, N. Mestre-Francés, F. Ibos, P. Corbeau, J.M. Verdier, V. Perrier, Effective gene therapy in a mouse model of prion diseases, PLoS One 3 (7) (2008) e2773.
[47] M.D. White, G.R. Mallucci, Therapy for prion diseases: insights from the use of RNA interference, Prion 3 (3) (2009) 121–128.
[48] D. Gomez-Nicola, S. Suzzi, M. Vargas-Caballero, N.L. Fransen, H. Al-Malki, A. Cebrian-Silla, J.M. Garcia-Verdugo, K. Riecken, B. Fehse, V.H. Perry, Temporal dynamics of hippocampal neurogenesis in chronic neurodegeneration, Brain 137 (8) (2014) 2312–2328.
[49] C.M. Jalland, K. Scheffler, S.L. Benestad, T. Moldal, C. Ersdal, G. Gunnes, R. Suganthan, M. Bjørås, M.A. Tranulis, Neil3 induced neurogenesis protects against prion disease during the clinical phase, Sci. Rep. 6 (2016) 37844.
[50] R. Ghosh, S.J. Tabrizi, Gene suppression approaches to neurodegeneration, Alzheimers Res. Ther. 9 (1) (2017) 82.
[51] A. Relano-Gines, A. Gabelle, S. Lehmann, O. Milhavet, C. Crozet, Gene and cell therapy for prion diseases, Infect. Disord. Drug Targets 9 (1) (2009) 58–68.
[52] K.L. Brown, J. Brown, D.L. Ritchie, J. Sales, J.R. Fraser, Fetal cell grafts provide long-term protection against scrapie induced neuronal loss, Neuroreport 12 (1) (2001) 77–82.
[53] H. Büeler, A. Aguzzi, A. Sailer, R.A. Greiner, P. Autenried, M. Aguet, C. Weissmann, Mice devoid of PrP are resistant to scrapie, Cell 73 (7) (1993) 1339–1347.
[54] G.A. Wells, S.A. Hawkins, A.R. Austin, S.J. Ryder, S.H. Done, R.B. Green, I. Dexter, M. Dawson, R.H. Kimberlin, Studies of the transmissibility of the agent of bovine spongiform encephalopathy to pigs, J. Gen. Virol. 84 (4) (2003) 1021–1031.
[55] A. Relaño-Ginés, S. Lehmann, A. Bencsik, M.E. Herva, J.M. Torres, C.A. Crozet, Stem cell therapy extends incubation and survival time in prion-infected mice in a time window–dependant manner, J. Infect. Dis. 204 (7) (2011) 1038–1045.
[56] K. Frid, O. Binyamin, N. Fainstein, G. Keller, T. Ben-Hur, R. Gabizon, Autologous neural progenitor cell transplantation into newborn mice modeling for E200K genetic prion disease delays disease progression, Neurobiol. Aging 65 (2018) 192–200.

[57] S.H. Lee, N. Lumelsky, L. Studer, J.M. Auerbach, R.D. McKay, Efficient generation of midbrain and hindbrain neurons from mouse embryonic stem cells, Nat. Biotechnol. 18 (6) (2000) 675–679.
[58] M. Meyer, A.D. Ducray, H.R. Widmer, Restorative neuroscience: concepts and perspectives, Swiss Med. Wkly. 138 (11–12) (2008).
[59] J.H. Kim, J.M. Auerbach, J.A. Rodríguez-Gómez, I. Velasco, D. Gavin, N. Lumelsky, S.H. Lee, J. Nguyen, R. Sánchez-Pernaute, K. Bankiewicz, R. McKay, Dopamine neurons derived from embryonic stem cells function in an animal model of Parkinson's disease, Nature 418 (6893) (2002) 50–56.
[60] C.C. Zhang, A.D. Steele, S. Lindquist, H.F. Lodish, Prion protein is expressed on long-term repopulating hematopoietic stem cells and is important for their self-renewal, Proc. Natl. Acad. Sci. U. S. A. 103 (7) (2006) 2184–2189.
[61] D.G. Kent, M.R. Copley, C. Benz, S. Wöhrer, B.J. Dykstra, E. Ma, J. Cheyne, Y. Zhao, M.B. Bowie, Y. Zhao, M. Gasparetto, Prospective isolation and molecular characterization of hematopoietic stem cells with durable self-renewal potential, Blood 113 (25) (2009) 6342–6350.
[62] Y.J. Lee, I.V. Baskakov, Treatment with normal prion protein delays differentiation and helps to maintain high proliferation activity in human embryonic stem cells, J. Neurochem. 114 (2) (2010) 362–373.
[63] Z. Krejciova, S. Pells, E. Cancellotti, P. Freile, M. Bishop, K. Samuel, G. Robin Barclay, J.W. Ironside, J.C. Manson, M.L. Turner, P. De Sousa, Human embryonic stem cells rapidly take up and then clear exogenous human and animal prions in vitro, J. Pathol. 223 (5) (2011) 635–645.
[64] K. Takahashi, K. Okita, M. Nakagawa, S. Yamanaka, Induction of pluripotent stem cells from fibroblast cultures, Nat. Protoc. 2 (12) (2007) 3081.
[65] K. Takahashi, S. Yamanaka, Induction of pluripotent stem cells from mouse embryonic and adult fibroblast cultures by defined factors, Cell 126 (4) (2006) 663–676.
[66] K. Takahashi, S. Yamanaka, A decade of transcription factor-mediated reprogramming to pluripotency, Nat. Rev. Mol. Cell Biol. 17 (3) (2016) 183.
[67] J. Yu, M.A. Vodyanik, K. Smuga-Otto, J. Antosiewicz-Bourget, J.L. Frane, S. Tian, J. Nie, G.A. Jonsdottir, V. Ruotti, R. Stewart, I.I. Slukvin, Induced pluripotent stem cell lines derived from human somatic cells, Science 318 (5858) (2007) 1917–1920.
[68] B.R. Groveman, S.T. Foliaki, C.D. Orru, G. Zanusso, J.A. Carroll, B. Race, C.L. Haigh, Sporadic Creutzfeldt-Jakob disease prion infection of human cerebral organoids, Acta Neuropathol. Commun. 7 (1) (2019) 1–2.
[69] Z. Krejciova, J. Alibhai, C. Zhao, R. Krencik, N.M. Rzechorzek, E.M. Ullian, J. Manson, J.W. Ironside, M.W. Head, S. Chandran, Human stem cell–derived astrocytes replicate human prions in a PRNP genotype–dependent manner, J. Exp. Med. 214 (12) (2017) 3481–3495.
[70] W. Tahir, B. Abdulrahman, D.H. Abdelaziz, S. Thapa, R. Walia, H.M. Schätzl, An astrocyte cell line that differentially propagates murine prions, J. Biol. Chem. (2020), jbc-RA120.
[71] A. Louveau, T.H. Harris, J. Kipnis, Revisiting the mechanisms of CNS immune privilege, Trends Immunol. 36 (10) (2015) 569–577.
[72] V.L. Joers, M.E. Emborg, Preclinical assessment of stem cell therapies for neurological diseases, ILAR J. 51 (1) (2010) 24–41.
[73] G. Paul, S.V. Anisimov, The secretome of mesenchymal stem cells: potential implications for neuroregeneration, Biochimie 95 (12) (2013) 2246–2256.
[74] D.R. Mediano, D. Sanz-Rubio, B. Ranera, R. Bolea, I. Martín-Burriel, The potential of mesenchymal stem cell in prion research, Zoonoses Public Health 62 (3) (2015) 165–178.
[75] A. Relaño-Ginés, S. Lehmann, C. Crozet, Cell-based therapy against prion diseases, Curr. Opin. Pharmacol. 44 (2019) 8–14.
[76] C.H. Song, O. Honmou, N. Ohsawa, K. Nakamura, H. Hamada, H. Furuoka, R. Hasebe, M. Horiuchi, Effect of transplantation of bone marrow-derived mesenchymal stem cells on mice infected with prions, J. Virol. 83 (11) (2009) 5918–5927.
[77] Y. Li, J. Chen, X.G. Chen, L. Wang, S.C. Gautam, Y.X. Xu, M. Katakowski, L.J. Zhang, M. Lu, N. Janakiraman, M. Chopp, Human marrow stromal cell therapy for stroke in rat: neurotrophins and functional recovery, Neurology 59 (4) (2002) 514–523.
[78] C.H. Song, O. Honmou, H. Furuoka, M. Horiuchi, Identification of chemoattractive factors involved in the migration of bone marrow-derived mesenchymal stem cells to brain lesions caused by prions, J. Virol. 85 (21) (2011) 11069–11078.
[79] Z. Shan, Y. Hirai, M. Nakayama, R. Hayashi, T. Yamasaki, R. Hasebe, C.H. Song, M. Horiuchi, Therapeutic effect of autologous compact bone-derived mesenchymal stem cell transplantation on prion disease, J. Gen. Virol. 98 (10) (2017) 2615–2627.
[80] G.W. Kreutzberg, Microglia: a sensor for pathological events in the CNS, Trends Neurosci. 19 (8) (1996) 312–318.
[81] A. Aguzzi, C. Zhu, Microglia in prion diseases, J. Clin. Invest. 127 (9) (2017) 3230–3239.
[82] C. Zhu, U.S. Herrmann, J. Falsig, I. Abakumova, M. Nuvolone, P. Schwarz, K. Frauenknecht, E.J. Rushing, A. Aguzzi, A neuroprotective role for microglia in prion diseases, J. Exp. Med. 213 (6) (2016) 1047–1059.
[83] A. Williams, P.J. Lucassen, D. Ritchie, M. Bruce, PrP deposition, microglial activation, and neuronal apoptosis in murine scrapie, Exp. Neurol. 144 (2) (1997) 433–438.
[84] J. Priller, M. Prinz, M. Heikenwalder, N. Zeller, P. Schwarz, F.L. Heppner, A. Aguzzi, Early and rapid engraftment of bone marrow-derived microglia in scrapie, J. Neurosci. 26 (45) (2006) 11753–11762.
[85] G. Donofrio, F.L. Heppner, M. Polymenidou, C. Musahl, A. Aguzzi, Paracrine inhibition of prion propagation by anti-PrP single-chain Fv miniantibodies, J. Virol. 79 (13) (2005) 8330–8338.
[86] C.H. Song, H. Furuoka, C.L. Kim, M. Ogino, A. Suzuki, R. Hasebe, M. Horiuchi, Effect of intraventricular infusion of anti-prion protein monoclonal antibodies on disease progression in prion-infected mice, J. Gen. Virol. 89 (6) (2008) 1533–1544.
[87] F. Cobo, P. Talavera, A. Concha, Diagnostic approaches for viruses and prions in stem cell banks, Virology 347 (1) (2006) 1–10.

Chapter 20

Traditional medicine for aging-related disorders: Implications for drug discovery

Antara Banerjee, M.S. Pavane*, L. Husaina Banu*, A. Sai Rishika Gopikar, K. Roshini Elizabeth, and Surajit Pathak

Department of Medical Biotechnology, Faculty of Allied Health Sciences, Chettinad Academy of Research and Education (CARE), Chettinad Hospital and Research Institute (CHRI), Chennai, India

1 Introduction

Aging is the multistep process, and it is a time-dependent phenomenon categorized by the reduced ability of the system to cope with stress caused by exogenous and endogenous factors, and the factors may be physical, chemical, or biological.

Traditional medicine consists of medical aspects of traditional knowledge that was passed from generation to generation before the emergence of modern medicine. In the contemporary context, it has become an alternative medicine. Traditional medicine not only includes edible and nonedible materials, but also compiles various therapeutic methods. Traditional medicine is the indigenous or folk medicine. It makes use of the traditional knowledge. Traditional knowledge is the knowledge system embedded in cultural traditions of regional, local, and social units with common norms, religion, and values and shares a common identity. The term *aesthetics*, a general term concerned with beauty, has now crossed the boundaries of cosmetics to become an integral part of the medical field and gained the title of aesthetic medicine (AM). It has become a standardized part in therapeutics in improvising the physical appearance. Patients can be treated with invasive or noninvasive methods accordingly. Advanced aesthetic medicine includes surgical interventions such as facelifts, radiofrequency ablation, liposuction, and breast implants. Even nonsurgical practices exist like chemical peels, nonsurgical liposuction, and radiofrequency skin tightening.

A detailed literature survey from scientific databases such as Scopus, PubMed/MEDLINE, Web of Science, and Google was undertaken, and this critical review work comprises of the prevalence of the use of traditional medicines and traditional practitioners in the world, particularly in India. In this chapter, we briefly commented on the potential benefits of traditional medicine in aging-related disorders and stem cell regeneration.

Indian traditional medicine has been classified into the following types based on the usage:

1.1 Siddha medicine

Siddha medicine is an Indian traditional medicinal approach. Agasthya is considered as the father of Siddha medicine. A unique feature is that the medicinal formulations have been maintained very secretly till date. Siddha medicine is the cumulation of ancient medicinal practices, spiritual disciplines, alchemy, and mysticism. The practitioners are termed as Siddhars. Numerous ancient philosophical tenets of Siddha medicine are relevant to the modern practice. Siddha medicine flourished in Indus Valley Civilization. It emerged in South India when the Dravidian people, who were inhabitants of Indus Valley, migrated to South India [1]. Siddha medicine is mainly dedicated to spiritual treatment. The main objective of this aspect is to preserve and prolong biolife. It is believed that there is a strong interlink between the macrocosm of external world and microcosm of corporeal being. Mainly, three elements are signified here, which are believed to be the fundamentals of the human constitution. They are Vata (air), Pita (fire), and Kapha (water), which are collectively known as humors. Vata refers to contraction, expansion, and motion in humans; Pita refers to hunger, thirst, sleep, and beauty; and Kapha refers to blood, bile, semen, sweat, and other glandular secretions. It is said that a pathology arises when there is an imbalance among them. Preparations made of sulfur and mercury were popular; however, they are considered to be toxic as stated in some reports.

---

* Equal contributing authors.

1.1.1 Mineral- and metal-based drugs in Siddha medicine

Herbal agents are also used, which can be grouped as Thadhu (inorganic materials), Thavaram (herbal products), and Jangamam (animal products) [2]. Inorganic materials are classified as shown in Table 1.

1.1.2 Clinical applications of Siddha medicine

Skin sometimes acts as a reflection of the health status of living being. A vast array of skin conditions relate to internal pathology of vital organs like liver, immune system, gut, kidney, and neural organs. It was traced that this connection among skin, immune system, emotional health, and gut is established by microbiome. For example, psoriasis or atopic dermatitis

TABLE 1 Thadhu agents and reported benefits.

Sl. no	Thadhu agents	Nature of the agent	Reported benefits
1	Uppu	Water-soluble inorganic materials that give out vapor when put in fire	**Sirucinni Uppu** (a herbal salt extract from *Acalypha fruticosa*)—A statistical informatics shows the usage of Sirucinni Uppu in reducing the symptoms of peptic ulcer like sore tongue and flatulence. **Moongil uppu**— a formulation with cardamom, *Piper longum*, jaggery, and clove to treat inflammation, bronchial ailments, menstrual irregularities, fungal infections like ringworms, painful joints, fever, and diarrhea. **Gunma Uppu Chooranam** is a classical medicine in powder form, effective in treating menstrual disorders, porumal (flatulence), gastritis, sootagavali (dysmenorrhea), sobai (anemia), agattuvayu (gastralgia), and painful micturition (Mooraththira krichcharam).
2	Pashanam	Agents that do not dissolve in water but emit vapor when fired	Navapashanam bead is available and known for its enormous cosmic energy, which is believed to be a curative of health ailments. It is actually made of nine poisonous herbs that have turned into a powerful drug [3].
3	Uparasam	Similar to pashanam but differ in action	Annabethi is a common hydrochemical drug, which is popularly known as green vitriol (chemical name is ferric sulfate). Naturally Annabethi is collected in hills and also synthesized, which is green in color with crystal form, nauseous astringent taste and has solubility in water. It is commercially prepared by mixing iron wire with sulfuric acid and evaporating the solution to crystallization. Annabethi chenthuram is commonly used in Siddha system of medicine for treating anaemia, fever, and dysentery [4].
4	Loham	Not dissolved in water but melt when fired	Loham refers to the metals used in traditional medicines, which include copper, zinc, gold, mercury, lead, and silver (Rajata). Rajata bhasma is less costly and abundantly available and has similar properties of gold like medhya and sathirvayaskarnam and hence can be a replacement of gold in traditional therapeutics. Ingestion of Ashudh Rajatha causes increase in body temperature, infertility, constipation, and body pains [5]. Hence, it must be processed either by purification of metal (shodhana) or by incineration (marana) for invasive use [6]. It is used in the management of disorders like neurological conditions, psychological conditions, memory loss, dry cough, digestive disorders, skin diseases, and other infectious conditions like urinary tract infections [7].
5	Rasam	Substances that are soft	**Pudina Rasam** treats vomiting, anorexia, mandaagni (poor digestion), cold, cough, liver malfunctions, and gall bladder disorders. **Sata Pushpa Rasam** can be used to address body pain, swelling, loss of appetite, constipation, fever, and stomach ailments. No toxic effects on rats under study were observed for the oral recommendations of Rasa chendooram up to 100 mg/kg of body weight [8].
6	Ghandhagam	Substances that are insoluble in water like sulfur	Sulfur which is traditionally called as Gandhagam is available in the forms of pills, rasayana, chenduram, etc. Gandhagam is an ingredient in Siddha medicine formulations and preparations like Gowri chinthamani chendooram for osteoarthritis management. It plays a prominent role in the treatment of ailments like leucoderma, respiratory problems, flatulence, diarrhea, rheumatic fever, poisonous attacks like bites, eye disorders, hepatomegaly, gastric ulcers, and ascites [9, 10].

has a link with gut and emotional status of humans. Siddha medicines are fast reacting though they are natural approaches. Siddha medicine is noted for treating psoriasis. Vetapalai Thailam is a lubricant made of soaking sunlight (leaf of *Wrightia tinctoria* is indigenously called as Vetpaliilai [11]). Adults having eczema are given laxatives for 3 days at the initial stages of the therapy. From the 4th day, the administration of specific herbo-mineral drugs was reported to be beneficial. Rasaganthi mezhuku is abundantly used as herbo-mineral drug [12]. Milakai thailam is externally applied on the skin for scabies-possessing patients. Pinda thailam is applied on the skin to reduce the lesions of ichthyosis (ichthys: fish as there are dry, fish-scale-like appearance on skin). Oral administration of Kunkiliya vennai which has a combination of sesame oil and gum of *Shorea robusta* is followed to the patients suffering from acne problems. Puzhu vettu is applied to the affected spots of alopecia areata patients. Siddha medicine even opts for the invasive procedures for treating alopecia though it is not practically done in the contemporary text [13].

1.2 Unani medicine

Unani medicine is another form of traditional medicine, which is incorporated in the alternative forms of treatments in India. This Arabic style of medicinal system is originated in Greece and gained popularity in India in the 10th century. Unani is an Arabic term depicting the meaning "Greek." Unani is mainly based on the principles set by Hippocrates and Galen and other Persian and Arab scholars. Unani medicine practitioners firmly state that the biological status of an individual is the cumulative balance of the main four humors in the body, i.e., blood, phlegm, yellow bile, and black bile [14]. The traditional knowledge also believes that certain environmental conditions like poor water quality and inappropriate composition of gases in the air are some of the factors in health disturbances. The same concept can be seen even in the Unani system. Unique herbs are often used in Unani therapies. One of the most famous formulae is "Khamira Abresham Hakim Arshad Wala." The main constituents of this herbal combination are cardamom, citron, saffron, and Indian bay leaf. This formula plays a key role in cardiovascular issues by controlling blood pressure and angina and also in keeping up the brain health status by securing brain from free radicals. Other herbal formulations include "Majoon Suranjan" in the treatment of rheumatoid arthritis and "Kohl-Chikni Dawa" in the treatment of cataracts that are frequently diagnosed in diabetic patients.

1.2.1 Orofacial disease management by Siddha medicine

Usool-e-Ilaj refers to "Unani principles for treatment." Unani physicians follow the herbal formulations in Usool-e-Ilaj for the prevention of tooth ailments and management of orofacial conditions. Antibacterial, analgesic effects and anti-inflammatory properties in orofacial treatments are due to the effects of some famous herbal formulations; they are "Sunune Zard, Habbe Gule Aak, Majoon Azaraqi, Majoon Suranjan, Buzidan, Sunune Mujalli, and Sunune Mulook [15]." Few herbs that play a vital role in orofacial therapies are described in Table 2.

1.3 Ayurveda

It has the origin from Vedas. Charaka Samhita is believed as the father of Ayurveda. The term Ayurveda means knowledge of life. Ayu and Veda are Sanskrit terminologies referring to life and knowledge or science. It not only deals with treatments but also compiles pathogenicity, diagnosis, lifestyle, and disease prevention [23]. The priority is given to herbal medications. However, yoga and meditation are included in this sector. Ayurveda views the dominance of Kapam (water) in childhood, Vatham (air) in old age, and Pitham (fire) in adulthood. It mainly focuses on 20 gunas or qualities of the human life. It comprises Panchkarma methods used to exfoliate and detoxify the body and also skin for aesthetic purposes [24]. Ayurvedic components fetch a great help in addressing the ailments after medical intervention. Some of such Ayurvedic components are listed in Table 3.

1.3.1 Analysis of Ayurvedic approaches in contemporary context

Though public, scientific, and medical communities have widely accepted Ayurveda as effective and logical approach for human health dealings, the need for advanced modern medications and therapies has elevated for safety, fast, efficacy, and quality therapies and ways of healing. However, some researchers found that Ayurvedic approaches may not be successful to meet the contemporary medical aspects and there are less chances of chief benefits and the need for separate methodologies for Ayurveda researches is eminent. Technologies enhancing data analytics can aid Ayurveda and traditional-based medical approaches in making it "evidence-based" by disclosing opaque information related to its chemical interactions, side effects, longevity, and efficacy. Ayurveda provides a shade for preventive, predictive, participatory, and personalized medicine. Survey and analytical views say that 75% of the Indian population avail Ayurveda either as sole treatment or as an

TABLE 2 Herbal products for orofacial disease management.

S. no	Herbal products (common name)	Scientific name	Compounds present	Medicinal properties
1	Miswak	Salvadora persica	Benzyl isothiocyanate, chlorides, fluorides, tannins, vitamins, alkaloids, essential oils, and silica	A bactericide against oral pathogens causing periodontal disease. Protects from cariogenic and genotoxic compounds [16] Aids in remineralization of enamel [17]
2	Haldi	Curcuma longa	Curcumin, demethoxycurcumin, and bisdemethoxycurcumin are collectively termed as curcuminoids (phenolic compound). An important component curcumin (gives yellow color to turmeric)	Cutaneous diseases such as acne, atopic dermatitis, facial photoaging radiodermatitis, oral lichen planus, pruritus, psoriasis, and vitiligo [18] Turmeric eliminates dental-related pain and swelling [19] and also periodontitis gingivitis [20]
3	Pomegranate	Punica granatum	**Juice:** Anthocyanins; pomegranate pericarp (peel, rind): phenolic punicalagins, gallic acid and other fatty acids, catechin; pomegranate **Leaves:** Tannins (punicalin and punicafolin), and flavones glycosides **Flower:** Gallic acid, ursolic acid, triterpenoids **Roots and bark:** Ellagitannins including punicalin and punicalagin and numerous piperidine alkaloids	*Anti-inflammatory properties* Can be used as mouth rinse as it has antimicrobial property Pomegranate methanolic extracts used to control oral pathogens [21]
4	Clove	Syzygium aromaticum	The active ingredient in clove is eugenol oil	Antiseptic against infectious diseases like periodontal disease due to the antimicrobial activities against oral bacteria (Clove is also commercially used in food processing units as a natural additive or antiseptic to increase the shelf life due to its effective antimicrobial nature against some food-borne pathogens [22])

additional therapy along with conventional medications. Big data analysis from a single group shows that in 86% of the patients having chronic ailments, 76% of the population gained complete or partial relief. 0.9% of the population represented aggravation in symptoms. In this study, the data were taken digitally from 353,000 patients by 300 Ayurvedic practitioners both in telecommunications and in-person consultations [14].

Long-term researches in Ayurveda include typical traditional approaches in medicine and cannot be easily incorporated in efficient treatments, especially for treating advanced stages in any disease. Although they are effective in the treatment of nerve, muscle disorders, etc., side effects were observed in some cases. As it is a natural therapy, it takes more time for curing an ailment. The research mainly includes the principles of anti-aging outlined in Ayurveda: Vayasthapana (age defying), Sandhaniya (cell regeneration), Tvachya (nurturing), Varnya (brighten skin glow), Vranaropana (healing), Shothahara (antiinflammatory), Tvagrysayana (retarding aging), and Tvachagnivardhani (strengthening skin metabolism).

2 Aging

The process of becoming older is known as aging. It includes psychological and social changes. One of the greatest risk factors for human diseases is aging. The cause of aging may be due to the occurrence of DNA damage. DNA oxidation or DNA methylation results in DNA damage. Cellular senescence is the process that causes permanent arrest in the proliferation of cells. It plays an important role in aging and age-related diseases. Early cellular senescence results in premature aging. Premature aging/aging can occur when intracellular and extracellular mechanisms change due to error in the cellular machinery resulting from mutations or environmental factors. Aging is of three types. It includes hormonal aging, genetic

TABLE 3 Ayurvedic components for various ailments alongside of mainstream treatments.

S. no	Herbs and their origin	Mainstream medical treatment options	Reported use against diseases	Effects or indications
1	Curcumin	Cisplatin (cis-diamminedichloroplatinum(II), CDDP) in chemotherapy	Hematological toxicity, dermatological, hepatobiliary, gastrointestinal disorders, hepatic toxicity, neurotoxicity, and upper respiratory disorders [25]	In some cases, curcumin can replace cisplatin as a chemotherapeutic drug [26]
2	Rasayana Avaleha: An Ayurvedic preparation has the composition of *Emblica officinalis*, *Withania somnifera*, *Tinospora cordifolia*, *Glycirrhiza glabra*, *Leptadenia reticulata*, *Ocimum sanctum*, and *Piper longum* [27]	Radiotherapy	Acute adverse reactions like nausea, vomiting, skin reaction, mucositis, and fatigue occur. Among the chronic long-term effects, xerostomia, tastelessness, edema, and damage to other organs may occur	Antiaging, restoring power, improving vitality, and other properties, to nourish all tissue elements properly and restore the basic homeostatic balance
3	*Punica granatum* seed oil and *Croton lechleri* resin extract [28]	With pregnancy, weight change or lack of skin elasticity	Striae distensae	Anti-inflammatory activity
4	Triterpenoids of *Calendula officinalis* or marigold	In cesarean delivery	Cesarean wound	Anti-inflammatory and anti-inflammation properties, antivirus, antimicrobial and antifungal activity, anticancer, antioxidant, and healing function [29]. Epithelial reconstruction of surgical wounds accelerates the wound healing process
5	Cesarean wound	Antibiotic ointments	*Aloe vera* gel	High stability, nontoxicity, hydrophilicity, biodegradability, gel-forming properties, and ease of chemical modification [30]
6	Cesarean wound	Steroids	Oily extracts of *Hypericum perforatum* and *Calendula arvensis* [31]	Antibacterial, anti-inflammatory and antioxidant activities
7	Scars	Due to surgery, or wounds due to diabetes	Rhubarb and tamarind	Antiscar activity, antihypertrophic scarring activity [32]

aging, and environmental aging. Genetic aging is associated with genetic alterations. Environmental aging is the acceleration of the aging process by environmental factors. Hormonal aging refers to a decrease in the hormonal level and causes an imbalance with the increase in age. Some genes and transcription factors are associated with aging. For example, *SIRT-6* (Sirtuin 6), the longevity gene, plays multiple roles like processes related to aging, maintenance of telomere, and DNA repair [33]. *DAF-16* is the transcription factor that controls the expression of genes, which are in association with aging. By controlling the gene expression, this transcription factor either reduces the life span or extends the life span. Gradual fall in the functioning of the endocrine gland is associated with aging. Secretion of human growth hormone (HGH) in combination with insulin growth factor 1 (*IGF-1*) decreases, and as a result, the aging process accelerates. The tendency of the genome to undergo changes in its position is known as genome instability. DNA damage is caused by both exogenous and endogenous factors. This leads to genome instability. As a result, cell cycle stresses and gene regulation changes, which may initiate premature aging. Chromosomes have telomeres, which serve as protective coating to prevent the DNA from damage, which is tightly packed in the chromosomes. Telomeres are the chronological markers for aging. As the cells divide repeatedly, the length of the telomere decreases. Stress, smoking, obesity, and lack of physical activity affect the length of the telomere. As a consequence of epigenetic alterations, changes in gene expression, disturbance in broad genome sequence, and disturbance in the epigenomic landscape result, the aging process surges. Epigenetic alterations affect the structure of DNA, and as a result, genome expression changes, which is associated with aging [34]. It has been recognized that mitochondria also play a major role in the aging process. Mitochondria is not only involved in energy production but also plays an essential role in calcium balance, cellular metabolism, interconversion of dietary substrates, etc. Mitochondrial dysfunction affects all the above mentioned processes. Mutation in mitochondrial DNA causes mitochondrial dysfunction. Mitochondrial DNA shows 10 times greater mutation rate than nuclear DNA and less repairing capacity [35].

3 Aging-related disorders

Aging causes alterations in the physiological and biochemical mechanisms of human that paves the way to most of the diseases and increases susceptibility for the phenotypical expression of disorders. There are many diseases that are directly or indirectly related to aging, and few of them are as follows:

3.1 Colon and rectal cancer

Colorectal cancer is a cancer that develops from the colon or rectum. Colorectal cancer is the third most prevalent form of cancer, and occurrence is recorded mostly in United States, Australia, Europe, and even Asian countries like China and India. The fact that many cancer patients face the mortality shows that awareness of risk factors and therapies for colorectal cancer must be given high priority in medical research. Many analysis states that colorectal cancer is mostly found in the people above 40 years and the risk is still higher in people above 50 years. 90% of colorectal cancer cases occur in people aged from 50 to 70 years, which clearly mentions that age is a nonmodifiable risk factor for colorectal cancer [36]. Alterations in biological mechanisms along with the age show a clear relation with the colorectal cancer development. Some of them are cellular senescence, prolonged exposure to oxidative stress, and increased epigenetic alterations. These alterations disturb the biochemical activities of normal colorectal cells. One such example is that *Wnt* signaling is improper because of DNA methylation in cells due to age factor that plays an important role in tumorigenesis [37]. Hence, it is highly recommended to integrate geriatric assessment in treating senior citizens with cancer. Carcinogenesis also disturbs the epigenetic maintenance system changing the pattern of epigenetic clock. An analysis found that the Haworth clock is the most accurate one among all epigenetic clocks for predicting the chronological age in the normal colon. In therapeutics for colorectal cancer, one must keep in mind the unique nature of this tissue, which has an intimate interaction with the microbiome of the gut and digestion byproducts and also recommends that tissue-specific epigenetic clock must be developed for greater assessment of colorectal cancer. Chemotherapeutic drugs like pembrolizumab, nivolumab, and other allopathic therapies are generally used to treat colon cancer. In advanced stages, surgical resection is the only option left. Along with the conventional therapies, adjuvant therapies are recommended for the management of the disease.

Numerous benefits against colon cancer progression are rendered by the medicinal plants. Many beneficial actions have been reported against the reduction of DNA oxidation, induction of superoxide dismutase, arresting of the cell cycle in S phase, which lead to apoptosis. Additionally decrease in the expression of *PI3K*, P-Akt protein, and MMP; reduction of antiapoptotic *Bcl-2* and *Bcl-xL* proteins; and decrease in proliferating cell nuclear antigen (*PCNA*), cyclin A, cyclin D1, cyclin B1, and cyclin E with the administration of active components of plant or its products were also investigated upon. Plant compounds also increase the expression of both the cell cycle inhibitors *p53*, *p21*, and *p27*, and the *Bax, caspase 3, caspase 7, caspase 8,* and *caspase 9* protein levels [38].

3.2 Neurodegenerative diseases

Neurodegenerative diseases are mostly characterized by the accretion of neuronal DNA damage. It is suggested that both neurodegeneration and premature aging are associated with impaired DNA repair mechanism. Hutchinson-Gilford progeria syndrome and Werner's syndrome are the two progeroid syndromes that are categorized by their clinical features impersonating the clinical features of early physiological aging. Molecular studies have revealed that transformed DNA damage responses and reduced cell proliferation are responsible for the pathogenesis of both the syndromes. Many studies suggest that age-related neurodegenerative diseases like Parkinson's disease, Huntington's disease, Alzheimer's disease, and amyotrophic lateral sclerosis are developed due to impaired DNA repair mechanism and are predominantly arising from impaired base excision repair pathway. Age-related neurodegenerative diseases are most probably caused due to the generation of endogenous reactive oxygen species from aerobic metabolism. The noxious O_2 and $OH-$ radicals have been occupied in age-related disorders, diabetes mellitus, and other diseases like atherosclerosis and cancer, causing peroxidation of membrane lipid, nonspecific glycation of protein, and cross-linking of proteins, finally affecting the functions of organelles and leading to cell death. Our biological system has the capability of bringing about the repair mechanism for the removal of reactive oxygen species (ROS). The human brain is vulnerable to oxidative stress, which is in association with age-related brain dysfunction, due to the build-up of key compounds required for oxidative damage and decreased functioning of the antioxidant defense system. Functional disruption results from protein oxidation, which is identified to be associated with augmented oxidation of specific proteins and has been concerned in the development of age-related neurodegenerative disorders such as Parkinson's disease and Alzheimer's disease.

Aging not only makes a person susceptible to age-related neurodegenerative diseases like Alzheimer's and Parkinson's diseases but also contributes to mitochondrial dysfunction and ATP production mechanisms. These mechanisms should be balanced, and the balance is maintained by mitophagy as well as proteolysis. Mitophagy is a unique form of autophagy process that controls the turnover of malfunctioning mitochondria and organelles that assist in synthesizing ATP, the energy currency of the cell. Aging gradually reverberates the depletion of mitophagy and other autophagy types, which imbalance the homeostasis. Hence, treatments in neurodegenerative diseases should also focus on the mitophagic pathways. The diagnosis of the extent of alteration in reducing mitophagy and planning for mitophagy functionality regulation can be as an adjuvant treatment.

Food and nourishment also contribute in promoting the aging and occurrence of progeroid syndromes. For example, advanced glycation end products (AGEs) consumed in diet accelerate natural aging, inflammation, diabetes, neurodegeneration, and other related consequences [39].

3.3 Facial wrinkling

Cutaneous aging is a result of both intrinsic aging and extrinsic aging on skin cells. In intrinsic aging, molecular changes occur in the system as time passes, and in extrinsic aging, molecular changes occur in the system due to environmental factors. In intrinsic aging, atrophy of dermis and epidermis occurs, reduction of fibroblast, elastin and collagen I, III, and IV and the number of mast cells also decreases. In extrinsic aging, extensor surface of arms, chest, and face is more sensitive to longtime UV exposure. The network of elastin fiber disrupts, and a large number of amorphous elastin materials accumulate. In dermal-epidermal junction (DEJ), the level of fibrin-rich microfibrils gets reduced. *Fibrillin-1* is one of the fibrin-rich microfibril which provides physical support to nonelastic and elastic connective tissues in the system [40]. Both intrinsically aged skin and extrinsically aged skin become stiffened and wrinkled. In intrinsic aging, not only the levels of elastin, fibrillin, and collagen I, III, and IV decrease, but also the level of oligosaccharide fraction decreases. As a result, the skin loses its capacity to recollect bound water [41]. Intrinsic aging results from various events as follows: (a) reduced matrix level in the dermis, (b) increased level of enzyme expression, which degrades the matrix of collagen, and (c) decreased proliferation capacity of skin cells. There are three components involved in extrinsic aging. They are glycosaminoglycans, collagen fibers, and elastic fibers. Reduced percentage of glycosaminoglycans leads to the dehydration of the skin, which may be one of the factors in the formation of wrinkles. Collagen VII also plays a major role in protecting dermal-epidermal junction. If it is getting damaged, it brings about adverse effects on the skin. The induction of matrix metalloproteinases (MMP) and the formation of reactive oxygen species are the fundamental aspects of cutaneous aging. The main source of reactive oxygen species is mitochondria. As a result of mutation in mitochondrial DNA, enormous amount of reactive oxygen species is secreted, which leads to the aging of the skin. Matrix metalloproteinases (MMP) is the chief structural protein in the extracellular matrix (ECM). Among the different types of MMPs, *MMP-1* initiates the breakdown of type 1 collagen, which in turn damages the extracellular matrix.

3.4 Graying of hair

Graying of hair is a natural and common phenomenon of aging. There are various mechanisms and reasons for graying of hair. Loss, failure, malfunction, and differentiation defects of melanocytes lead to graying of hair. Oxidative stress from the UV rays damages the melanocytes of the hair follicles. The UV rays result in the release of reactive oxygen species (ROS) that damages the molecules, resulting in mutations in the nucleic acid. These ROS are said to be the basic cause of aging from the past ages. It is proven that the B cell lymphoma 2 gene (*BCL-2*), first identified in a chromosomal translocation of follicular lymphoma deficiency, resulted in the graying of hair. Aging is also influenced by mitochondria. As age progresses, the mutations in the mitochondria increase, resulting in the loss of respiratory functions. These mitochondrial mutations have a tendency to rise the production of ROS, which leads to a variety of aging phenotypes like wrinkled skin, graying of hair, and decreased fertility.

However, the recent research is focused on the premature graying of hair that is caused due to the genetic predisposition factor and various other factors including hypothyroidism, anemia, and vitamin B12 deficiency. Premature hair graying is related to oxidative stress, emotional stress, smoking, alcohol consumption, etc. Premature graying of hair can be controlled by change in lifestyle like diet, exercises, weight control, and low alcohol consumption.

4 Clinical applications of traditional medicine

The use of traditional medicine is aimed not only toward developing attractive physical appearance but also toward achieving long life and good health. The effects of traditional medicine depend upon the product's composition and its form of application. Traditional medicines are safe, and they are well known for their holistic actions. Various Ayurvedic medications are available for smoothing skin's imperfections, ageless skin, increasing its hydration level, tonifying the skin, and restoring of radiant and healthy look. Traditional medicine is reported in stabilizing hormones and metabolism. Traditional medicine is more affordable than modern medicine. The main cause of aging is DNA damage. This damaged DNA could be partially reversed by traditional medicine. Natural products from traditional medicine like essential oils, glucans, carotenoids, glutathione, and alkaloids have the ability to donate electron and scavenge the free radicals. For example, extracts from *Geranium sanguine* are rich in polyphenolic compounds. These polyphenolic compounds are known to exhibit antimutagenic capacity and free radical scavenging capacity. Another example is essential oils in ginger that act as antioxidants. Most of the antioxidants involve in DNA repair mechanism. Traditional medicines are known to exert less side effect [42, 43]. The relation between antiaging and traditional medicine like Ayurveda, Siddha, and Unani is gaining momentum in the aesthetic and health/medicine category. Siddha medicine is also known to have its own importance in anti-aging. Sukkilam (one of the seven humors) has the ability to rejuvenate and act as an anti-aging agent. Unani medicine contemplates aging as a natural and assured process. Unani medicine deals with diet, lifestyle, drugs, and regimens for healthy life. Thus, the use of traditional medicine has a lot of potential in properties like anti-aging. Some of the Ayurvedic herbs that have been reported to slow down aging include turmeric, Ginseng, and Ashwagandha.

The universal, biological process of aging has always been considered in research. The goal of anti-aging has now become the recent trend in all research studies. Various techniques, principles, and strategies are being initiated for prolonged life span with the use of various traditional methods and techniques.

5 Applications of traditional medicine in aesthetics

Aesthetic medicine is generally described as a branch of medicine where the aim is to construct and reconstruct the physical appearance. Some of the applications of aesthetic medicine include cosmetic dermatology that covers anti-aging. Anti-aging medicine offers nutrition, vitamin supplementation, antioxidants, etc., which can be used in therapies of other age-related diseases. Some of the applications of herbal products are detailed below:

5.1 *Ocimum tenuiflorum*

Ocimum tenuiflorum or Tulasi (common name) is a perennial plant belonging to the family of Lamiaceae, native to the Indian subcontinent and widespread as a cultivated plant all over the Southeast Asian tropics. Tulasi is an antioxidant that helps in preventing the signs of aging. Basil or Tulasi is protective against oxidative damage and reduces free radicals and also balances antioxidant enzymes. Components of Tulasi leaf extract include oleanolic acid, ursolic acid, rosmarinic acid, eugenol, linalool, etc. Application of the basil involves in retaining skin moisture levels, reduces skin roughness and scaliness, prevents wrinkles, and makes the skin smooth. *Ocimum tenuiflorum* plays a major role in Ayurveda and Siddha practices for its supposed treatment of inflammatory diseases [44].

5.2 Tinospora cordifolia

Tinospora cordifolia or giloy (common name), which is known as anti-aging herb, is an herbaceous plant that belongs to the family of Menispermaceae found throughout the tropical areas of India. It is a climbing shrub. The leaves are 10–20 cm in diameter. Giloy contains anti-aging properties that reduce pigmentation, pimples, and signs of aging such as wrinkles and fine lines. Components of giloy belong to different classes such as alkaloids, diterpenoid lactones, glycosides, steroids, aliphatic compounds, and polysaccharides. Active components of giloy include 11-hydroxymustakone, *N*-methyl-2-pyrrolidone, *N*-formylannonain, cordifolioside A, magnoflorine, tinocordiside, and syringin. Giloy acts as a detoxifying agent and treats skin issues such as acne, eczema, and leprosy.

5.3 Centella asiatica

Centella asiatica or gotu kola (common name) is a tropical medicinal plant belonging to the family of Apiaceae, found in the Southeast Asian countries such as India, Sri Lanka, China, Indonesia, and Madagascar. These plants grow during August to September and are purple in color, with the aroma of tobacco leaves and heights ranging from 5 to 15 cm. It is used in skin treatments for a wide spectrum of skin conditions. Leaf extract is taken to cure dysentery and to improve memory, whereas the surface region of active components of gotu kola includes asiatic acid, madecassic acid, asiaticoside acid, asiaticoside A, and asiaticoside B. Gotu kola has been reported to enhance the process of repairing of the veins and broken capillaries and to even is involved in the breakdown of cellulite by shrinking the connective tissues that bind fat cells under the skin. Gotu kola has ability to decrease inflammation and also to boost collagen. A study in 2012 was conducted on rats to determine the effects of gotu kola on wound healing; later results state that the rats that received the treatment from the extracts of *Centella asiatica* have fully developed keratinization and epithelialization [45].

5.4 Panax ginseng

Ginseng (generic name) is a perennial umbel plant. Ginseng is cultured in Japan, Korea, China, Russia, and Canada. 10°C and 40–50 in. of annual precipitation are required for growing ginseng. It plays a major role in anti-aging. Ginseng leaf stem has properties of anti-inflammation, antioxidant, antitumor, antifatigue, skin protection, and immune modulation. Ginseng root is most widely used part of the plant for anti-aging; these are usually harvested at 4–6 years of age during autumn. It consists of ginsenosides as the major bioactive component. Ginseng helps in maintaining the skin's elasticity, lengthens the life span of skin cells, activates the skin, and prevents the skin's cells from aging. Seasonal fluctuations, geographical differences, and age variations may affect the ginsenoside content in ginseng leaf [46].

5.5 Wolfberry

Wolfberry is a fruit of *L. barbarum* belonging to the species of *boxthorn* in the nightshade family, *Solanacae*. It is the most common herb used in many Asian countries like China. It is also named as *Gouqizi* in China, where this fruit is beneficial for eye and liver. Wolfberry is regarded as an herb that has little side effects and good therapeutic effects. A number of studies suggest that wolfberry possesses anti-aging properties. Wolfberry has various components like zeaxanthin, carotene, polysaccharides, and small molecules like betaine, cerebroside, p-coumaric, and various vitamins. Wolfberry is appreciated for its properties like increase in insulin-like growth factor and inhibition of oxidative stress. Polysaccharides from *L. barbarum* are the active components of this fruit. Wolfberry helps in the reduction of plasma triglyceride levels and increases plasma cAMP and SOD levels, which help in maintaining healthy body [47].

5.6 Atriculum lappa

Articum lappa is commonly called as greater burdock. These are the Eurasian species belonging to the family *Asteraceae*. The roots of this plant are used in various aspects of traditional medicine as a diuretic, diaphoretic, and also a blood purifying agent. The seeds isolated from *Atriculum lappa* resulted in the identification of various compounds like arctiin, arctigenin, matairesinol, and diarctigenin ligands. This fruit extract is important in reducing wrinkles in vivo. Arctiin had the highest activity in terms of procollagen and MMP-1 inhibition, which help in the treatment of wrinkles. It helps in the regeneration of dermal matrix, offering treatment for aging. This fruit contains hyaluronan synthase 2 gene expression, which determines procollagen and hyaluronan synthesis, thereby reducing wrinkle volume. These produce ligands, which act as secondary metabolites that provide a wide range of bioactivity [48].

5.7 Chamomile

Chamomile or camomile is a plant belonging to the family *Asteracae*. There are two species that include German chamomile (*Chamomila recutita*) and Roman chamomile (*Chamaemelum nobile*). The dried flowers of the plant are used for various purposes in cosmetics. The principle component that is extracted from the flowers of chamomile is terpenoids. Chamomile is used as a herbal treatment for various skin conditions. It is used as antiallergic, antioxidant, and analgesic. Active component of chamomile contains terpenoids (bisoprolol, matricin, and chamazulene), flavonoids (luteolin, rutin, and apigenin), hydroxycoumarins, and mucilages. It has anti-inflammatory action and also assist in wound healing. This has been widely used in cosmetic products like soothing moisturizers, cleansers, and color-enhancing hair products. This has been used as a popular ingredient in various aroma therapies and hair care. This is known to have a soothing and softening effect on the skin [49].

5.8 Soy

Soy (*Glycine max*) belongs to the species of legume. Soy is commonly found in East Asia and has numerous benefits. Soy has important application as antioxidant, anticarcinogenic, and antiproliferative activities. It helps to reduce hyperpigmentation and improve the elasticity of the skin. It contains additional benefits like oil control, skin moisturizer, and hair growth control. The isoflavones in soy control various physiological processes like antitumor, antimenopausal, osteoporosis, and antiaging. It also contains other components like saponins, essential amino acids, phytosterols, calcium, potassium, and iron. Genistein, a soy isoflavone, upregulates the expression of antioxidant genes that regulated the better appearance of skin and less wrinkling. The proteases help in the lightening of skin and reduce unwanted body and facial hair. Therefore, this has become popular to a wide variety of skincare products [50].

5.9 Plant extracts in treating diseases

Plant extracts and their bioactive components regulate the malfunctioning genetic and biochemical mechanisms induced by aging, which are nonmodifiable risk factors for many diseases like cancer, diabetes, and neural degeneration.

Studies on the saponin-enriched ethanol extract from *Momordica charantia* plant have reported the alleviation of fat accumulation in the nematode *C. elegans* (*Caenorhabditis elegans*). Parallelly, these studies also state that *Momordica charantia* extracts (MCEs) are effective in maintaining longevity and neuroprotection in the organism and also alleviate age-related pigmentation and enhance physiological mechanism, which ultimately increase the life span of the nematode. Physiological analysis revealed that the levels of lipid oxidation and ROS were dropped down. Genetic studies manifest that these extracts upregulate the expression of the *sod-3*, *sod-5*, *clt-1*, *clt-2*, *hsp-16.1*, and *hsp-16.2* genes, which possess longevity extension properties and *skn-1* activity required for lifespan extension [51]. All biomarkers indicated that the antiaging and antistress mechanisms of MCEs were due to its firm antioxidant properties. The treatment with these extracts showed great stress-resistant effects and might serve as a lead nutraceutical in geriatric research. In a study on the alloxan-induced hyperglycemic mice which were administered with saponin fraction (SF) extracted from *M. charantia* (500mg/kg), the blood glucose decreased ($P<0.05$), and the level of insulin secretion and glycogen synthesis was elevated ($P<0.05$, $P<0.01$) [52]. Hence, these extracts can be used as antidiabetic and hypoglycemic agent. *Ananas sativa and Moringa oleifera* are great antioxidants and health-promoting botanicals with anti-inflammatory, anticancer, and anti-aging features. The bioactive compound in pineapple, bromelain, promotes wound healing, and it is a component of postsurgical applications due to its anti-inflammatory property. Seventy-two male Wistar albino rats were orally administered with *Ferula elaeochytris Korovin* (FE) root extract over 8 weeks and gave positive results in the studies to analyze their anti-inflammatory, antidiabetic, and antioxidant properties [53].

Vitamin C has therapeutic impact on atherosclerosis, cancer, diabetes, metal toxicity, and neurodegenerative diseases. Although most vitamin C is totally absorbed in the small intestine, the amount of absorbed vitamin C decreases as intraluminal concentrations increase [54]. The effect of flavonoids on vitamin C uptake in vivo is not certain because of the low plasma bioavailability of flavonoids. Thus, the interaction of flavonoids with vitamin C would be expected to occur in the intestinal region for active uptake. Kiwi fruit is an abundant source of vitamin E, and one animal model study has exhibited that vitamin E is able to preserve vitamin C in vivo [55]. Onions and tea are significant dietary sources of flavones and flavonols, the subgroups of flavonoids which are reported to have antioxidant effects and other biochemical impacts in the treatment of numerous health ailments such as Alzheimer's disease, cancer, and atherosclerosis and also as an adjuvant therapy for diabetes, inflammation, endothelial dysfunction, and dyslipidemia.

EGb 761 is a standardized extract of *Ginkgo biloba* (commonly known as maidenhair tree) leaves that has demonstrated shielding properties against vascular and neuronal damage. In randomized, placebo-controlled clinical trials and meta-analyses in adults with dementia, the extract showcased positive effects regarding cognition, global change, and behavior, which are better than those shown with placebo [56].

The main extracts from Epilobium genus plants are flavonoids, which comprise flavonol 3-O-glycosides, tannins (macrocyclic ellagitannins—oenothein A and B), tiny amounts of sterols along with some fatty acids, and seed oil comprising campesterol, β-sitosterol, brassicasterol, and stigmasterol, which show significant role in addressing benign prostate hyperplasia (BPH) and lower urinary tract symptoms (LUTS) in males [57].

Treatment of the complex of terpenoids from *Abies sibirica* terpenes plant on colorectal adenocarcinoma cell line Caco-2 and human pancreas adenocarcinoma cell line AsPC-1 indicated twofold elevation of expression levels of GADD45B/G genes and 1.5-fold elevation of GADD45A (growth arrest and DNA damage-inducible genes), and tumor-suppressive regulators of MAPK-signaling cascades *DUSP5*, *DUSP1*, and *DUSP6*, nerve growth factor receptor (*NGFR*), connective tissue growth factor (*CTGF*), and growth differentiation factor 15 (*GDF15*) gene expressions that induce carcinogenesis are not increased after *Abies sibirica* terpenes supplementation [58].

Plant seeds, fruits, leaves, and roots of soybean, grape, green tea, olive, pomegranate, and garlic tested in vivo and in vitro models confirmed the fact that they are the most effective plants against colorectal cancer.

Panax notoginseng flower extracts are reported to induce apoptotic effect on HCT-116 human colorectal cancer cell line, thus proving that the extracts have antiproliferative and anticancer properties [59].

6 Insights from traditional medicine in translational stem cell research

Herbal preparations with medicinal values cannot be characterized in clinical trials solely, and subsequent identification of chemical compounds and their molecular effects can be served as surrogate biomarkers in the assessment of their therapeutic efficacy. Numerous clinical trials are carried out by merging together techniques along with the scientific knowledge of herbal and plant extracts that resulted in affordable and less toxic alternative therapeutics. Stem cells provide the core beneficiary factors in terms of clinical and research applications. They aim to provide a cure for various medical conditions. It has been investigated that the yield and quality of stem cells are improved by Ayurvedic formulations. Many kinds of herbs and plants like *Panax ginseng and Ginkgo biloba* have been tested on mesenchymal stem cells with either individual or mixtures of crude herbal extracts to investigate the effects and mechanisms of these on stem cell growth and differentiation [60]. But to apply these herbal extracts as drugs in stem cell therapy in regenerative medicine will require high-end research with cutting-edge technologies to provide a better understanding into their mechanisms of effective pathways. Thus, this area needs the concept of "standardization" compiling a set of scientific experiments for the stem cells to grow, proliferate, and differentiate. The effect of Chinese herbs combined with stem cell therapy has reported therapeutic effects on health ailments like Parkinson's disease, myocardial infarction, stenosis, and osteoporosis [61]. Nowadays, it is a major concern on evaluations that assure reproducibility, quality, safety, and efficacy that will confirm the safety of herbal products in the global market [62]. Herbal extracts in traditional medicine are nontoxic, highly available, and cost-effective alternatives for therapeutic application globally. Many scientists do not allow the indigenous and traditional medicines in clinical applications as there is a lack of molecular evidence. Hence, strong evidence and their mechanisms standardize the application of herbal extracts and medicines in clinical therapies and biopharmaceutical production.

6.1 Mesenchymal stem cells in aging and role of traditional medicines

Mesenchymal stem cells (MSCs) perform numerous roles in the cellular environment and in the body. MSCs have self-renewal and differentiation capacity and immunomodulatory functions and support other cell maintenance. The important property of MSCs is epithelial-mesenchymal transition. Mesenchymal stem cells have multilineage differentiation potential, the capability of high expansion, and most importantly immunomodulatory functions. These properties make them good candidates for cell therapy applications [63]. MSCs play a major role in dermal repair, especially chronic wounds and diabetic wounds, and also, they serve as the guardians against inflammatory reactions [64, 65]. MSCs can repair the chronic wounds or wounds due to injury or accident and diabetic wounds by the secretion of growth factors and matrix proteins. Recent studies discovered that the human placental membrane is the rich source of mesenchymal stem cells, which can be used for regeneration and repair of tissues [66]. However, mesenchymal stem cells may lose the properties and capabilities of repairing wounds with the increase in age or the MSC population may decrease in the bone marrow niche. As the age increases, MSCs lose their differentiation potential, regenerative capacity, and immunomodulatory effects. The cause of the senescence of MSCs is induced by oxidative stress, telomere shortening, etc. [67]. The senescence of mesenchymal stem cells can be detected by various markers. The most widely used marker is senescence-associated β-galactosidase (SA-β-gal). In senescent MSCs, the percentage of SA-β-gal upsurged during the aging process. During the aging process, cytosolic pH is altered and lysosomal activities also increased, and this leads to an increase in percentage of SA-β-gal [67, 68]. Recent studies revealed that the senescence-associated lysosomal α-L-fucosidase (SA-α-Fuc) is the novel marker of senescent mesenchymal stem cells. This is more robust and accurate than the SA-β-gal. The other characteristic features of senescent MSCs are flattened or enlarged morphology, and in G1 phase, growth is arrested [67, 69]. As the impact of aging, senescent MSCs undergo mitochondrial fusion. Due to these limitations, MSCs cannot be used for cellular therapies, as these MSCs undergo senescence. The aging-associated senescence of mesenchymal stem cells can be reduced or restored by the administration of some traditional medicines/formulations. One such Ayurvedic medicine is Dhanwantharam kashayam. This Ayurvedic medicine is made by using extracts of plant's roots, leaves, fruits, etc. More than 40–45 components have been used to make this Ayurvedic medicine. *Sida cordifolia*—Bala (roots), *Hordeum vulgare*—Yava (barley grain), *Aegle marmelos*—Bilva (Indian bael), etc., are some of its main herbal ingredients. This kashayam was tested in mesenchymal stem cells (in vitro) and could improve the growth of MSCs, increase proliferation rate and delay senescence, and does not create toxicity in the mesenchymal stem cells [70]. *Tinospora cordifolia* and *Withania somnifera* are the two herbs that are extensively used in Ayurveda. These herbs have anti-aging and rejuvenating potential. Leaf extract of *Tinospora cordifolia* and root extract of *Withania somnifera* were taken and tested for senescent MSCs in vitro to detect whether these plant extracts can delay senescence and increase proliferation potential in MSCs. These plant extracts efficiently increased proliferation potential and delayed senescence, and it was tested and proved using β-galactosidase senescent assay and MTT assay, and the levels of various senescent markers are also tested [71]. Oxidative stress is one of the factors that induce senescence in MSCs as the age increases. This oxidative stress can be reduced by using the extract of *Cirsium setidens*. *Caesalpinia sappan* L. plant extract due to the presence of flavonoids is known to have high antioxidant activity that can decline the senescence of human mesenchymal stem cells.

6.2 Potential role of herbal remedies in stem cell therapy: Proliferation and differentiation

Herbal remedies are those in which the extracts are obtained from various parts such as fruits, leaves, and roots of the potential herbs as efficient remedy for stem cell therapy, especially proliferation and differentiation. Herbal remedies are known to cause least side effects and minimum toxicity than other synthetic drugs. The potential herbal extracts that reportedly induce proliferation and differentiation are present in the various parts of the following plants like *Salvia miltiorrhiza*, *Curcumin longa L*, and naringin in citrus fruits and grapes.

6.2.1 Salvia miltiorrhiza

Salvia miltiorrhiza is the red sage, which is also called as Danshen. The roots of Salvia miltiorrhiza are used to isolate herbal extract for the treatment of cardiovascular disease. This plant was most widely used to treat cardiovascular diseases in Asian countries. The active components present in the roots of *Salvia miltiorrhiza* extract like flavonoids, terpenoids, and salvianolic acid increase the viability of MSCs. *Salvia miltiorrhiza* extract at the concentration of 0.0001–100 μg/mL increases the viability of MSCs in patients with ischemic stroke of the brain, and thereby, it reduces the severity of the disease [72]. *Salvia miltiorrhiza* extract has the property to promote differentiation in the stem cells. Likewise, human mesenchymal stem cells (hMSCs) were differentiated into neural-like cells with substantial changes in morphology, when *Salvia miltiorrhiza* extract had been tested on hMSCs. The positive effects of *Salvia miltiorrhiza* were proved by testing the neuronal markers of neural cells, which had been differentiated from hMSCs. *β-Tubulin*, *Nestin*, glial fibrillary acidic protein, and neurofilament are the strong positive markers of neural cells, which were observed in the differentiated cells. When hMSCs are induced with *Salvia miltiorrhiza* extract, the neural cell marker called neurite outgrowth-promoting protein expression was noticeably increased [73]. 5 μ/mL of *Salvia miltiorrhiza* extract had been used in induced pluripotent stem cells (iPSCs) to promote differentiation into neural cells. This idea was experimented upon the use of the *Salvia miltiorrhiza* iPSCs in the ischemic stroke patients. *Salvia miltiorrhiza* extract markedly increased the level of neural markers like *Nestin* and microtubule-associated protein 2 (*MAP-2*) in vitro. The level of MAP-2 was increased in in vivo even when transplanted in *Salvia miltiorrhiza*-treated rats.

6.2.2 Curcumin longa L.

Curcumin longa L. is the flowering plant, in which curcumin is the natural polyphenolic compound, which shows antiinflammatory and antioxidant effects. Curcumin exerts its effects by altering several targets [74]. Stem cells, for its survival and cell repairing, require antioxidant mechanism widely. Curcumin, due to its antioxidant property, is the potential herbal extract for stem cell therapy research [73]. The differentiation of human mesenchymal stem cells into osteoblasts is used to be blocked by hydrogen peroxide (H_2O_2), and it is also inhibited by ROS formation and Wnt/β-catenin pathway. These are weakened and diminished by the treatment with curcumin. The heme oxygenase-1 (HO-1) is an enzyme, which has positive effect in the differentiation of mesenchymal stem cells into osteoblasts. The alkaline phosphatase (ALP) activity and *RUNX* gene expressions are also associated with HO-1 in stimulating the differentiation of MSCs into osteoblasts. The activities of HO-1, ALP, and Runx gene expressions are enhanced by curcumin [75]. From dried curcumin rhizome, the ethanol extraction was introduced in the hMSCs to check the efficacy of curcumin in enhancing proliferation and differentiation. As a result, hMSCs were proliferated and differentiated into endothelial progenitor cells. This was proved using the level of cell surface markers like vascular endothelial growth factor receptor 2 (*VEGFR-2*), CD33, and CD134. The level of these markers was increased when hMSCs were differentiated and proliferated into endothelial progenitor cells with curcumin administration [73].

6.2.3 Naringin

Naringin is the active component present in the extracts of citrus fruits and grapes. The chemical formula of naringin is naringenin 7-*O*-neohesperidose, a flavonoid. Reports of its anticancerous and antioxidant properties have been documented. It is widely used to treat diseases like osteoarthritis and osteoporosis. The use of naringin is the potent herbal remedy for enhancing the proliferation of stem cells by inducing pro-osteogenic effects. The effect of naringin had been tested in vitro in human bone marrow stem cells (hBMSCs) for proliferation. The dose of 1–100 μg/mL of naringin increased the proliferation of hBMSCs. Meanwhile, the dose of 200 μg/mL of naringin was reported to have toxic effects in hBMSCs, so the number of cells decreased. The activity of *ALP* was also increased at the dose of 100 μg/mL. The levels of osteogenic differentiation markers like *osteocalcin, osteopontin, collagen-1,* and *ALP* were increased when hBMSCs were treated with naringin at the dose of 100 μg/mL [76]. The differentiation induction potential of naringin had been tested in human amniotic fluid-derived stem cells (hAFSCs) as well. hAFSCs were differentiated into osteoblasts when 100 μg/mL of naringin was used. The osteogenic differentiation of hAFSCs was proved by testing the increased level of osteogenic differentiation markers like *ALP, Cyclin D1,* and proliferation protein *β-catenin.*

7 Conclusion

This chapter is mainly focused on aging and stem cell therapies and the role of traditional medicines as an alternative treatment for age-related diseases. The importance of aging is not only in aesthetic perspective but also a risk factor for the initiation of several ailments. Due to aging, there are alterations in biochemical mechanisms in the body, which affect the normal physiology, thereby inducing diseases. So the therapies in the management of age-related diseases should mainly focus on the control of altering mechanisms and mutations caused by these alterations. Traditional medicines like Siddha, Unani, and Ayurveda can be used as an alternative or adjuvant therapy for the regulation of biochemical characteristics that change with aging. Conventional medicines as well as treatments can target one problem at a time. Hence, for the multiple targets, the patient consumes various types of drugs in his lifetime, which in turn exerts toxicity and is harmful. Plentiful polyphenols and phytochemicals found in the plants can act on a wide range of targets, thereby giving multiple target activity with negligible or minimum side effects. A proper analysis of herbal components in all aspects like dose-related adverse drug reactions, market studies, and biocompatibility plays a vital role in standardizing the impact of Ayurvedic products in the medical field. Even the analysis of herbal formulations along with the pharmacological intervention is necessary. The herbal compounds should be thoroughly analyzed in such a way that they should not cause any short- and long-term adverse effects to the person who is undergoing allopathic treatment and should not minimize the pharmacological mechanisms and properties in the treatment of a particular disease. Research in standard and automated methods to analyze the herbal components and their derivatives in clinical aspects and their potentiality in therapies and all-round effects on an individual stimulates the establishment of standardized Ayurvedic intervention in adjuvant therapies.

Mesenchymal stem cells are widely used for wound healing. As the age increases, the wound healing capability of mesenchymal stem cells gets reduced due to senescence. The level of various most notable senescence markers like SA-beta gal and SA-alpha-Fuc, CD90, and CD 105 is increased in mesenchymal stem cells. This makes the reduced usage of MSCs for therapies. But, the extracts of traditional medicines delay the senescence of mesenchymal stem cells by various mechanisms and enhance the proliferation and differentiation of mesenchymal stem cells, which have been proved by testing the level of various cell surface markers and gene expressions. This increases the chances of using MSCs for wound healing process and other stem cell therapies in the future. Even though the herbal remedies have positive effects, it has some drawbacks. The crude extracts of herbs had shown the best results. But the efficacy of the herbal extracts may vary when it is formulated as a drug by mixing with other components.

8 Future perspectives

The combination of principles of modern molecular medicine with certain ideas of empirical traditional formulations may be beneficial in translational medicine in general. Proper attempts should be made for narrowing down to the specific part of the plant, which contains the bioactive compound. This ultimately gives a clear idea regarding the dose of plant material required, and also, the part alone can be used in the separation techniques to have maximum extraction. Focus of recent research is now on various herbs and their bioactive components to control the reduction of telomere length, and antioxidative stress formulations in delaying the senescence in stem cells. Studies are required, which need to document more number of potent herbs that have effects to delay senescence of MSCs and induce proliferation and differentiation of stem cells. Before moving for drug formulation, these herbal extracts can be tested in animal models like rodents, *Drosophila melanogaster*, and *Caenorhabditis elegans* to check in vivo effects. Efforts to expedite stem cell research can be translated from bench to bedside with interdisciplinary collaboration of academics, clinical environment science, biotechnology-related fields, marketing studies, and patient experience. Crowdsourcing applications aid in fueling the stable survival of stem cell therapies and other aging-related treatments and drugs in clinical and massive applications. Hurdles to translational medicine exist due to the lack of an interdisciplinary workforce in an organization.

The active and potential compounds and components with anti-inflammatory, antioxidant, and anti-aging properties can be extracted from the Ayurvedic plants, and their derivatives can be used to design and synthesize antiaging creams, drugs, ointments, oral pills, etc. These compounds may thereby pave the way for the development and designing of commercial products enhancing the health status of an individual.

To achieve more progress in the development of Ayurvedic and other traditional products for treating age-related diseases and also other diseases along with products for aesthetic-related therapies, a deeper and greater understanding of the active components which have anti-aging and other medicinal properties is needed. A focus on finding out new and purified active components from different plant resources and comparing these components for better selection for each disease therapy and management is necessary. Analysis of the atoms present in the compound is not enough, and their structural arrangements and bonding analysis are required. It can be done by analytical techniques like chemo-informatics (computer modeling) and spectroscopy studies. Cell culture studies after the basic analysis by these techniques are appreciable to get a deeper understanding of the compounds. The drug synthesis from plant extracts must contain crisp and simple steps to lessen the market cost of these drugs. Focusing on simplifying techniques in bioactive compound identification, compound extraction, and drug designing and production enhances the feasibility in pharmaceutical marketing and business. Molecular docking is a very good molecular simulation for basic studies in exploring the mechanisms of activity, toxicity, and side effects of drugs by using software tools and can be used before wet-lab procedures. It is used to automatically identify potential target proteins. The NAPRALERT database contains information on 43,879 species of higher plants encompassing the details of their chemical, ethnomedical, and pharmacologic uses. Through these databases, it is possible to correlate. Thus, it is possible to correlate the pharmacological uses and identify plants having beneficial anticancer, antiaging, and other suitable effects. A pivotal step in translating research into clinical therapy is thorough analysis of the therapeutic method, which comprises side effects, and variation studies among different races, genetic populations, genders, habitants, age-groups, etc. Using simple, feasible, fidelity techniques and minimizing the steps of compound extraction, drug designing, and production lead to market-friendly and massive drug production. The current advances in improvements in mainstream clinical/surgical interventions and the development of pharmaceutical drugs have combined to sweep mainstream medicine toward a more technological approach to health care. The use of ethnomedical information has contributed to health care worldwide, even though efforts to use it have been sporadic.

9 Approaches toward drug discovery

Nonetheless, indigenous practices have been critically imperative in the discovery of many new drugs. Unfortunately, traditional practitioners face various difficulties and challenges. The intimidation is both from the loss of biodiversity that depletes the natural sources that make up their pharmacopeia and from encroachment by the outside world that may wipe out their cultures. Though an impressive number of elements have been either isolated from herbs or synthesized on the basis of natural lead compounds. Generally a single herb or formula may contain many phytochemical constituents such as terpenoids, flavonoids, alkaloids, etc. But several reports stress that these compounds may function alone or in synergy with one another to produce the desired pharmacological effect, which should be taken into account before they are advised. It is noteworthy that every plant and its parts are decorated with novel compoundsthat are of therapeutic importance due to the novel bioactive components present. Knowledge accumulated in traditional medicine, therefore, plays an important role in enhancing the success rate of drug discovery from herbal medicine. While many of the age-old traditions of usage of traditional medicine for various diseases may reflect very careful trials and observations, they are prone to errors as a result of unreliable transmission and anecdotal reports.

Hence, with the advancement of technologies, the research in drug discovery from natural products needs to develop robust and viable lead molecules and they may be further characterized by techniques like HPLC, NMR spectroscopy, and GC-MS to elucidate their structural constituents, which are of utmost importance. It is suggestive that multidisciplinary collaborative research, with synergistic inputs from network pharmacology and big data, will be possible to explain the beneficial effect of traditional medicine in halting many ailments. However, the principle of ecological ethics should be upheld by preserving biodiversity while exploiting natural resources for drug discovery.

Acknowledgment

The authors are thankful to the Chettinad Academy of Research and Education (CARE) for providing the infrastructural support and to SERB, DST, Govt. of India, and CARE for providing the financial support to complete this piece of work.

Funding

This work was supported by the grants sanctioned to Dr. Antara Banerjee (PI) from the SERB-DST Govt. of India with the sanction file no. ECR/2017/001066.

Consent for publication

All the authors hereby give consent to publish this piece of work.

Conflict of interest

All the authors certify that none of the authors have any commercial association that might pose a conflict of interest in connection with the publication of the submitted chapter.

References

[1] B. Subbarayappa, Siddha medicine: an overview, Lancet 350 (9094) (1997) 1841–1844.
[2] J. Michael, R. Aja, C. Padmalatha, Antibacterial potential of some herbo-mineral siddha preparation: an alternative medicine for enteric pathogens, J. Chem. Pharm. Res. 3 (3) (2011) 572–578.
[3] B. Ravishankar, V. Shukla, Indian systems of medicine: a brief profile, Afr. J. Tradit. Complement. Altern. Med. 4 (3) (2008) 319–337.
[4] M. Ganguli, R. Singh, S. Jamadagni, A. Mitra, S. Upadhyay, J. Hazra, Toxicological evaluation of Kasisa Bhasma, an ayurvedic organo metallic preparation, Int. J. Res. Ayurveda Pharm. 3 (3) (2012) 381–386.
[5] P. Devi, K. Rana, A review article on the use of Rajata Bhasma in the management of various diseases, World J. Pharm. Res. 8 (6) (2019) 462–470.
[6] K. Pardeshi, V. Kadibagil, B. Ganti, Therapeutic potential of Rajata (silver) Bhasma: a review, Int. J. Res. Ayurveda Pharm. 8 (5) (2017) 126–129.
[7] P. Aiello, M. Sharghi, S. Mansourkhani, A. Ardekan, L. Jouybari, N. Daraei, et al., Medicinal plants in the prevention and treatment of colon cancer, Oxidative Med. Cell. Longev. 2019 (2019) 1–51.
[8] M. Jayabharathi, M. Mohamed Mustafa, P. Sathiyarajeswaran, Acute and sub-acute toxicity study on Siddha drug Rasa Chendooram, Int. J. Adv. Res. Biol. Sci. 4 (11) (2017) 2348–8069.
[9] J. Arunachalam, Researches on mercurial preparations: the prime requirement for their acceptance in medical world, Ayu 36 (2) (2015) 118–124.
[10] B. Devi, G. Reddy, G. Kumar, P. Sathiyarajeswaran, P. Elankani, Herbomineral formulation's safety and efficacy employed in siddha system of medicine: a review, Int. Res. J. Pharm. 10 (1) (2019) 16–24.
[11] P. Rajalakshmi, M. Abeetha, R. Devanathan, Psychochemical analysis of Gandhagam before and after purification, Int. J. Curr. Pharm. Res. 2 (4) (2010) 975–1491.
[12] S. Shukla, S. Saraf, S. Saraf, Fundamental aspect and basic concept of siddha medicines, Syst. Rev. Pharm. 2 (1) (2011) 48–54.
[13] J. Thas, Siddha medicine—background and principles and the application for skin diseases, Clin. Dermatol. 26 (1) (2008) 62–78.
[14] H. Singh, S. Bhargava, S. Ganeshan, R. Kaur, T. Sethi, M. Sharma, et al., Big data analysis of traditional knowledge-based ayurveda medicine, Prog. Prev. Med. 3 (5) (2018), e0020.
[15] H. Sudhir, Role of unani system of medicine in management of orofacial diseases: a review, J. Clin. Diagn. Res. 8 (10) (2014) ZE12–ZE15.
[16] M. Abhary, A. Al-Hazmi, Antibacterial activity of Miswak (*Salvadora persica* L.) extracts on oral hygiene, J. Taibah Univ. Sci. 10 (4) (2016) 513–520.
[17] W.O. Mariam, S.I. Dalia, Ion release and enamel remineralizing potential of miswak, propolis and chitosan nano-particles based dental varnishes, Pediatr. Dent. J. 29 (1) (2019) 1–10.

[18] A. Vaughn, A. Branum, R. Sivamani, Effects of turmeric (*Curcuma longa*) on skin health: a systematic review of the clinical evidence, Phytother. Res. 30 (8) (2016) 1243–1264.

[19] S. Gunduz, E. Mozioglu, H. Yilmaz, Biological activity of curcuminoids isolated from Curcuma longa, Rec. Nat. Prod. 2 (1) (2008) 19–24.

[20] S. Sood, M. Nagpal, Role of curcumin in systemic and oral health: an overview, J. Nat. Sci. Biol. Med. 4 (1) (2013) 3–7.

[21] H. Deswal, Y. Singh, H. Grover, A. Bhardwaj, Pomegranate as a curative therapy in medical and dental sciences: a review, Innov. J. Med. Sci. 4 (2) (2016) 15–18.

[22] Q. Liu, X. Meng, Y. Li, C. Zhao, G. Tang, H. Li, Antibacterial and antifungal activities of spices, Int. J. Mol. Sci. 18 (6) (2017) 1283.

[23] B. Patwardhan, Bridging Ayurveda with evidence-based scientific approaches in medicine, EPMA J. 5 (1) (2014) 19.

[24] G. Meulenbeld, A history of Indian medical literature, Indo-Iran. J. 45 (4) (2002) 358–361.

[25] L. Astolfi, S. Ghiselli, V. Guaran, M. Chicca, E. Simoni, E. Olivetto, et al., Correlation of adverse effects of cisplatin administration in patients affected by solid tumours: a retrospective evaluation, Oncol. Rep. 29 (4) (2013) 1285–1292.

[26] S. Chen, W. Gao, M. Zhang, J. Chan, T. Wong, Curcumin enhances cisplatin sensitivity by suppressing NADPH oxidase 5 expression in human epithelial cancer, Oncol. Lett. 18 (2) (2019) 2132–2139.

[27] P. Vyas, A. Thakar, M. Baghel, A. Sisodia, Y. Deole, Efficacy of Rasayana Avaleha as adjuvant to radiotherapy and chemotherapy in reducing adverse effects, Ayu 31 (4) (2010) 417–423.

[28] C. Bogdan, S. Iurian, I. Tomuta, M. Moldovan, Improvement of skin condition in striae distensae: development, characterization and clinical efficacy of a cosmetic product containing *Punica granatum* seed oil and *Croton lechleri* resin extract, Drug Des. Devel. Ther. 11 (2017) 521–531.

[29] F. Jahdi, A. Khabbaz, M. Kashian, M. Taghizadeh, H. Haghani, The impact of calendula ointment on cesarean wound healing: a randomized controlled clinical trial, J. Family Med. Prim. Care 7 (5) (2018) 893–897.

[30] J. Hamman, Composition and applications of aloe vera leaf gel, Molecules 13 (8) (2008) 1599–1616.

[31] S. Lavagna, D. Secci, P. Chimenti, L. Bonsignore, A. Ottaviani, B. Bizzarri, Efficacy of Hypericum and Calendula oils in the epithelial reconstruction of surgical wounds in childbirth with caesarean section, Farmaco 56 (5–7) (2001) 451–453.

[32] Q. Ye, S. Wang, J. Chen, K. Rahman, H. Xin, H. Zhang, Medicinal plants for the treatment of hypertrophic scars, Evid. Based Complement. Alternat. Med. 2015 (2015) 101340.

[33] Y. Kanfi, S. Naiman, G. Amir, V. Peshti, G. Zinman, L. Nahum, et al., The sirtuin SIRT6 regulates lifespan in male mice, Nature 483 (7388) (2012) 218–221.

[34] S. Gonzalo, Epigenetic alterations in aging, J. Appl. Physiol. 109 (2) (2010) 586–597.

[35] R. Haas, Mitochondrial dysfunction in aging and diseases of aging, Biology 8 (2) (2019) 48.

[36] E. Kuipers, W. Grady, D. Lieberman, T. Seufferlein, J. Sung, P. Boelens, et al., Colorectal cancer, Nat. Rev. Dis. Primers 1 (1) (2015) 15065.

[37] G. Jothimani, R. Di Liddo, S. Pathak, M. Piccione, S. Sriramulu, A. Banerjee, Wnt signaling regulates the proliferation potential and lineage commitment of human umbilical cord derived mesenchymal stem cells, Mol. Biol. Rep. 47 (2) (2020) 1293–1308.

[38] H. Wang, T. Khor, L. Shu, Z. Su, F. Fuentes, J. Lee, et al., Plants vs. cancer: a review on natural phytochemicals in preventing and treating cancers and their druggability, Anti Cancer Agents Med. Chem. 12 (10) (2012) 1281–1305.

[39] R. Ramasamy, S. Vannucci, S. Yan, K. Herold, S. Yan, A. Schmidt, Advanced glycation end products and RAGE: a common thread in aging, diabetes, neurodegeneration, and inflammation, Glycobiology 15 (7) (2005) 16R–28R.

[40] A. Langton, M. Sherratt, C. Griffiths, R. Watson, A new wrinkle on old skin: the role of elastic fibres in skin ageing, Int. J. Cosmet. Sci. 32 (5) (2010) 330–339.

[41] E. Naylor, R. Watson, M. Sherratt, Molecular aspects of skin ageing, Maturitas 69 (3) (2011) 249–256.

[42] C. Shen, J. Jiang, L. Yang, D. Wang, W. Zhu, Anti-ageing active ingredients from herbs and nutraceuticals used in traditional Chinese medicine: pharmacological mechanisms and implications for drug discovery, Br. J. Pharmacol. 174 (11) (2016) 1395–1425.

[43] S. Sen, R. Chakraborty, Revival, modernization and integration of Indian traditional herbal medicine in clinical practice: importance, challenges and future, J. Tradit. Complement. Med. 7 (2) (2017) 234–244.

[44] K. Das Mahapatra, B. Kumar, A review on therapeutic uses of *Ocimum sanctum linn* (tulsi) with its pharmacological actions, Int. J. Res. Ayurveda Pharm. 3 (5) (2012) 645–647.

[45] K. Gohil, J. Patel, A. Gajjar, Pharmacological review on Centella asiatica: a potential herbal cure-all, Indian J. Pharm. Sci. 72 (5) (2010) 546–556.

[46] Y. Yang, C. Ren, Y. Zhang, X. Wu, Ginseng: an nonnegligible natural remedy for healthy aging, Aging Dis. 8 (6) (2017) 708–720.

[47] R. Chang, K. So, Use of anti-aging herbal medicine, *Lycium barbarum*, against aging-associated diseases. What do we know so far? Cell. Mol. Neurobiol. 28 (5) (2007) 643–652.

[48] S. Su, M. Wink, Natural lignans from *Arctium lappa* as antiaging agents in *Caenorhabditis elegans*, Phytochemistry 117 (2015) 340–350.

[49] K. Janmejai, S. Eswar, S. Gupta, Chamomile: a herbal medicine of the past with a bright future (review), Mol. Med. Rep. 3 (6) (2010) 895–901.

[50] B. Borrás, J. Gambini, M. Gómez-Cabrera, J. Sastre, F. Pallardó, G.E. Mann, V. Jose, Genistein, a soy isoflavone, up-regulates expression of antioxidant genes: involvement of estrogen receptors, ERK1/2, and NFκB, FASEB J. 20 (12) (2006) 2136–2138.

[51] Y. Zhao, D. Wang, Formation and regulation of adaptive response in nematode *Caenorhabditis elegans*, Oxidative Med. Cell. Longev. 2012 (2012) 1–6.

[52] C. Han, Q. Hui, Y. Wang, Hypoglycaemic activity of saponin fraction extracted from *Momordica charantia*in PEG/salt aqueous two-phase systems, Nat. Prod. Res. 22 (13) (2008) 1112–1119.

[53] N. Eser, A. Yoldaş, A. Yigin, N. Yumusak, A. Bozkurt, U. Kokbas, A. Mustafa, The protective effect of Ferula elaeochytris on age-related erectile dysfunction, J. Ethnopharmacol. 258 (2020) 112921.

[54] M. Abdullah, R.T. Jamil, F.N. Attia, Vitamin C (ascorbic acid), in: StatPearls, StatPearls Publishing, Treasure Island, FL, 2019.

[55] C. Anitra, C. Margreet, Synthetic or food-derived vitamin C—are they equally bioavailable? Nutrients 5 (11) (2013) 4284–4304.

[56] K. McKeage, K. Lyseng-Williamson, Ginkgo biloba extract EGb 761 in the symptomatic treatment of mild-to-moderate dementia: a profile of its use, Drugs Ther. Perspect. 34 (8) (2018) 358–366.

[57] R. Kujawski, J. Bartkowiak-Wieczorek, M. Ożarowski, A. Bogacz, J. Cichocka, M. Karasiewicz, B. Czerny, P.M. Mrozikiewicz, Current knowledge on phytochemical profile of Epilobium sp. raw materials and extracts. Potential benefits in nutrition and phytotherapy of age-related diseases, Herba Polonica 57 (4) (2011) 33–44.

[58] A. Kudryavtseva, G. Krasnov, A. Lipatova, B. Alekseev, F. Maganova, M. Shaposhnikov, M. Fedorova, A. Snezhkina, A. Moskalev, Effects of *Abies sibirica terpenes* on cancer- and aging-associated pathways in human cells, Oncotarget 7 (50) (2016) 83744–83754.

[59] C.Z. Wang, X. Luo, B. Zhang, W. Song, M. Ni, S. Mehendale, J.T. Xie, H.H. Aung, T.C. He, C.S. Yuan, Notoginseng enhances anti-cancer effect of 5-fluorouracil on human colorectal cancer cells, Cancer Chemother. Pharmacol. 60 (1) (2006) 69–79.

[60] V. Udalamaththa, C. Jayasinghe, P. Udagama, Potential role of herbal remedies in stem cell therapy: proliferation and differentiation of human mesenchymal stromal cells, Stem Cell Res Ther 7 (1) (2016) 110.

[61] P. Lin, L. Chang, P. Liu, S. Lin, W. Wu, W.W. Chen, W.S. Chen, C.H. Tsai, T.W. Chiou, H.J. Harn, Botanical drugs and stem cells, Cell Transplant. 20 (1) (2011) 71–83.

[62] F.O. Kunle, E. Omoregie, A. Ochogu, Standardization of herbal medicines—a review, Int. J. Biodivers. Conserv. 4 (3) (2012) 101–112.

[63] A. Dehghanifard, M. Shahjahani, M. Soleimani, N. Saki, The emerging role of mesenchymal stem cells in tissue engineering, Int. J. Hematol. Oncol. Stem Cell Res. 7 (1) (2013) 46–47.

[64] E. Maranda, L. Rodriguez-Menocal, E. Badiavas, Role of mesenchymal stem cells in dermal repair in burns and diabetic wounds, Curr. Stem Cell Res. Ther. 12 (1) (2016) 61–70.

[65] A. Banerjee, S. Pathak, G. Jothimani, S. Roy, Anti-proliferative effects of combinational therapy of *Lycopodium clavatum* and Quercetin in colon cancer cells, J. Basic Clin. Phsiol. Pharmacol. 31 (4) (2020) 1–12.

[66] S. Maxson, E. Lopez, D. Yoo, A. Danilkovitch-Miagkova, M. LeRoux, Concise review: role of mesenchymal stem cells in wound repair, Stem Cells Transl. Med. 1 (2) (2012) 142–149.

[67] Y. Li, Q. Wu, Y. Wang, L. Li, H. Bu, J. Bao, Senescence of mesenchymal stem cells, Int. J. Mol. Med. 39 (4) (2017) 775–782.

[68] A. Bertolo, M. Baur, J. Guerrero, T. Pötzel, J. Stoyanov, Autofluorescence is a reliable in vitro marker of cellular senescence in human mesenchymal stromal cells, Sci. Rep. 9 (1) (2019) 2074.

[69] A. Brandl, M. Meyer, V. Bechmann, M. Nerlich, P. Angele, Oxidative stress induces senescence in human mesenchymal stem cells, Exp. Cell Res. 317 (11) (2011) 1541–1547.

[70] S. Warrier, N. Haridas, S. Balasubramanian, A. Jalisatgi, R. Bhonde, A. Dharmarajan, A synthetic formulation, Dhanwantharam kashaya, delays senescence in stem cells, Cell Prolif. 46 (3) (2013) 283–290.

[71] A. Sanap, B. Chandravanshi, P. Shah, G. Tillu, A. Dhanushkodi, R. Bhonde, K. Joshi, Herbal pre-conditioning induces proliferation and delays senescence in Wharton's Jelly Mesenchymal Stem Cells, Biomed. Pharmacother. 93 (2017) 772–778.

[72] R. Kim, S. Lee, C. Lee, H. Yun, H. Lee, M. Lee, J. Jongmin Kim, J.Y. Jeong, K. Baek, W. Chang, Salvia miltiorrhiza enhances the survival of mesenchymal stem cells under ischemic conditions, J. Pharm. Pharmacol. 70 (9) (2018) 1228–1241.

[73] W. Widowati, C. Sardjono, L. Wijaya, D. Lakshmitawati, F. Sandra, Extract of Curcuma longa L. and (-)-Epigallo Catechin-3-Gallate enhanced proliferation of adipose tissue-derived mesenchymal stem cells (AD-MSCs) and differentiation of AD-MSCs into endothelial progenitor cells, J. US-China Med. Sci. 9 (1) (2012) 22–29.

[74] B. Ferguson, H. Nam, R. Morrison, Curcumin inhibits 3T3-L1 preadipocyte proliferation by mechanisms involving post-transcriptional p27 regulation, Biochem. Biophys. Rep. 5 (2016) 16–21.

[75] S. Sharifi, S. Zununi Vahed, E. Ahmadian, S. Maleki Dizaj, A. Abedi, S. Hosseiniyan Khatibi, M. Samiei, Stem cell therapy: curcumin does the trick, Phytother. Res. 33 (11) (2019) 2927–2937.

[76] P. Zhang, K. Dai, S. Yan, W. Yan, C. Zhang, D. Chen, B. Xu, Z.W. Xu, Corrigendum to: effects of naringin on the proliferation and osteogenic differentiation of human bone mesenchymal stem cell, Eur. J. Pharmacol. 607 (1–3) (2009) 1–5.

Index

Note: Page numbers followed by *f* indicate figures and *t* indicate tables.

A

Acarbose, 97
Activins, 72
Adaptive immunity, 92–93, 93*t*
Adipose-derived MSC (AT-MSC), 205
Adipose-derived stem cells (ADSC), 1, 259
Adulthood, 61
Adult neural stem cells, 274
Adult neurogenesis, 274
Adult stem cells, 1–2, 158–159, 231
Advanced glycation end products (AGEs), 189, 287
Aerobic exercise, 195
Aesthetic medicine (AM), 281
Aesthetics, 281
Age-associated osteoporosis, 194–195
Aging, 1–3, 90–91, 187–190, 188*f*
  and cardiomyocytes, 147–149
    in vivo reprogramming, 148–149
    mechanism in, 148–149
    reprogramming-induced senescence, 148
    stability and potential regeneration, 148
    tissue remodeling mechanism, 148
  definition, 61
  drugs for, 97–98
  free radical theory, 3–4
  and human embryonic stem cells (hESCs), 18–32, 19*f*
  immunity in, 92–93
  prevention of, 97–98
  risk factor for, 10
  sex differences, 90
  stem cells in, 91
Aging gut, 180
Aging-induced stem cell dysfunction, 203
  factors responsible for, 206–211
    changes in stem cell metabolic intake, 211
    DNA damage, 208, 209*f*
    epigenetics, 210–211
    microenvironment, 206–207, 207*f*
    mitochondrial dysfunction, 208–209
    proteostasis dysfunction, 209–210
    stem cell aging and gender, 211
    stem cell exhaustion, 206
  germline stem cells (GSCs), 204
  hematopoietic stem cells (HSCs), 203–204, 208
  intestinal stem cells (ISCs), 204
  mesenchymal stem cells (MSCs), 205
  models for, 206
  neural stem cells (NSCs), 205
  skeletal muscle stem cells, 204–205
  therapeutic approaches for, 211–214, 223–230
    caloric restriction (CR), 214
    cellular reprogramming of iPSCs, 213
    intrinsic aging factors and, 225–226
    mitochondrial function improvement, 226
    nuclear aging, as potential therapeutic targets, 226–228
    oxidative metabolism, regulation of, 226
    parabiosis, 211–212, 212*f*
    protein homeostasis improvement, 225–226
    regulation of autophagy, 225–226
    rejuvenation of stem cell niche, 224–225
    rejuvenation research, 223–224
    retrotransposons, 213
    targeting extracellular aging factors, 224–225
    telomere lengthening, 214
Airway basal stem/progenitor cells, 56*t*
Akt mediated cell survival, 149, 150*f*
Alzheimer's disease (AD), 255–256
Amacrine cells, 118
Anabolism, 129
Androgens, 193
Angiogenesis, in cardiac aging, 169–170
  cardiac stem cells for therapy, 170–172
    cardiac stem cell population, 171–172
    cardiomyogenic/angiogenic stimulants, 173
    pro-angiogenic potential of, 172–173
    targeting angiogenesis, 172–173, 172*f*
  cellular senescence in, 170
  factors influencing, 169–170
  vascular changes in, 170
Antiaging, 97, 97*f*
Antioxidants, 226
Apelin, 238
Apolipoprotein E, 255–256
Apolipoprotein J, 237–238
Apoptosis, 150, 190
Arf/p53 pathway, 94
Articum lappa, 289
Asherman syndrome (AS), 27–31
Aspirin, 97
AstroRx, 27
Ataxia-telangiectasia mutated (ATM) pathway, 234
Atg, 130
Atopic dermatitis/eczema, 259
Atrial natriuretic peptide (ANP), 147
Atrophic age-related macular degeneration, 26
Autologous CSC transplantation therapy, 172–173
Autophagy, 21, 93–94, 129, 149, 237
  in bone cell senescence, 193
  regulation of, 225–226
  of satellite cells, 130–132, 131*f*
Ayurveda, 283–284, 285*t*
  analysis of, 283–284

B

Basic fibroblast growth factor (bFGF), 24
β-catenin, 76–77, 196–197
B cells, 92
BCL2-interacting protein 3 (BNIP-3), 149
Beclin-I, 149
β-III tubulin/Tuj-1 (tubulin-beta-3 chain), 11–14*t*
Biological markers (biomarkers), 232–241
  accumulation of inflammasomes, 240
  altered intercellular communications, 240
  altered microRNA profile, 237
  cell cycle arrest and cellular senescence, 241
  change in cell polarity, 238
  circulating factors, 240
  deregulated nutrient sensing and cellular metabolism, 238–239
  epigenetic modifications, 236–237
  exhaustion of stem cell pools, 233–234
  genomic instability, 234–236
  loss of proteostasis, 237–238
  mitochondrial dysfunction, 238
  niche deterioration, 239–240
  oxidative stress, 234, 235*t*
  potential therapeutic interventions, 241–242, 241*f*
  RNA splicing and defective ribosomal machineries, 237
  telomere attrition, 236
Biorhythm, and gut stem cells, 179–180
Bioscaffold, 58
Blastocyst, 9, 69
Blastomere, 32
Blastomere-derived hESCs, 18
Blood stem cells, 103
Bone aging, 190
Bone cell senescence
  bone frailty improvement, therapeutic target to, 198–199
  extrinsic factors, 193–195
    excess glucocorticoids, 194
    physical activity, 194–195
    sex hormone deficiency, 193
  factors inducing, 188*f*

Bone cell senescence *(Continued)*
  intrinsic factors, 190–193
    autophagy, 193
    epigenetic modifications, 191–192
    oxidative stress, 190–191
    telomere shortening, 192–193
  markers, 235t
  skeletal aging, 190, 191f
  pathways regulating, 196–198
    p53/p21 signaling pathway, 197–198
    p16/Rb signaling pathways, 197–198
    secretory phenotype and senescent markers on, 195–196
    Wnt signaling pathway, 196–197, 197f
Bone marrow, 57
Bone marrow-derived mesenchymal stem cells (BMMSC), 1, 56t
Bone-marrow stem cells, 211
Bone morphogenetic proteins (BMPs), 72, 160, 178, 189
Bone remodeling, 189f
Brachyury/T (T-box transcription factor T), 11–14t
Brain natriuretic peptide (BNP), 147
Buthionine sulfoximine (BSO), 191
Butyrate, 181–182

C

Calcium, 255
Caloric restriction (CR), 183, 237, 239
  for aging-induced stem cell dysfunction, 214
Cancer, 256
Cancer Cell Line Encyclopedia (CCLE), 65t
Canonical Wnt/β-catenin signaling, 151–152
Capecitabine (Xeloda), 258
Carcinogenesis, 286
Cardiac aging, angiogenesis in, 169–170
  cardiac stem cells for therapy, 170–172
    cardiac stem cell population, 171–172
    cardiomyogenic/angiogenic stimulants, 173
    pro-angiogenic potential of, 172–173
    targeting angiogenesis, 172–173, 172f
  cellular senescence in, 170
  factors influencing, 169–170
  vascular changes in, 170
Cardiac atrial appendage stem cells (CASCs), 173
Cardiac healing, 150
Cardiac MSCs, 163
Cardiac progenitor cells (CPCs), 169–171
Cardiac-resident stem cell-mediated regeneration, 152–153
Cardiac senescence, definition of, 149
Cardiac stem cell lineage, 171, 171f
Cardiac stem cell population, 171–172
Cardiac stem cell therapy, 170–172
  cardiac stem cell population, 171–172
  cardiomyogenic/angiogenic stimulants, 173
  pro-angiogenic potential of, 172–173
  targeting angiogenesis, 172–173, 172f
Cardiogenesis, 157
Cardiomyocyte differentiation markers, 11–14t
Cardiomyocytes, 147
  aging and, 147–149
    in vivo reprogramming, 148–149
    mechanism, 148–149
    reprogramming-induced senescence, 148
    stability and potential regeneration, 148
    tissue remodeling mechanism, 148
  in vitro, generation of, 159f
  regeneration of, 149–153
    angiogenesis and vascularization, 150
    cardiac-resident stem cell-mediated regeneration, 152–153
    cardiomyogenesis, 151
    cell-to-cell communication, 150
    inflammation reduction, 150
    miRNA-mediated regeneration, 152, 152f
    molecular mechanisms, proliferation and cell cycle, 151–152, 151f
    survival and protection, 149
  signaling pathways, in self-renewal and differentiation of, 160
Cardiomyogenesis, 151, 171f
Cardiomyogenic/angiogenic stimulants, 173
Cardiosphere-derived cell (CDC) differentiation, 163
Cardiovascular disease (CVD), 169
  age-related risk factors for, 169
Cardiovascular diseases, 157
Catabolism, 129
cBioPortal, 66t
Cell-based therapies, 169
Cell cycle arrest, 241
Cell cycle regulators, 260
Cell damage, 259–260
Cell polarity, change in, 238
Cell replacement-based therapy, 276
Cell senescence, 148
Cell-to-cell communication, cardiomyocytes regeneration in, 150
Cellular-based therapies, 157
Cellular damage, 190–191
Cellular metabolism, 238–239
Cellular Myc (c-Myc), 11–14t, 149
Cellular reprogramming, 57
  induced pluripotent stem cells, 213, 227–228
Cellular senescence, 96, 157–158, 206, 241, 284–286
  in cardiac aging, 170
  genetic and environmental component, 252f
Cellular senescence and aging in bone, 187
  bone frailty improvement, therapeutic target to, 198–199
  extrinsic factors, 193–195
    excess glucocorticoids, 194
    physical activity, 194–195
    sex hormone deficiency, 193
  factors inducing, 188f
  intrinsic factors, 190–193
    autophagy, 193
    epigenetic modifications, 191–192
    oxidative stress, 190–191
    telomere shortening, 192–193
  markers, 235t
  skeletal aging, 190, 191f
  pathways regulating, 196–198
    p53/p21 signaling pathway, 197–198
    p16/Rb signaling pathways, 197–198
    secretory phenotype and senescent markers on, 195–196
    Wnt signaling pathway, 196–197, 197f
Centella asiatica, 289
Cerebral organoids, 32
Chamomile, 290
Chaperone-mediated autophagy, 130
Chaperones, 22
Chemotherapy, 258
Cholinesterase (ChE), 25
Chromatin immunoprecipitation (ChIP) sequencing, 70
Chromatin regulation, 96
Chronic degenerative disease, 28–30t
Chronic endocrine diseases, 189
Chronic psychological stress, 170
Circulatory factors, 240
  satellite cells, 132–133
c-Jun N-terminal kinase (JNK), 160–161, 204
Clinical Proteomic Tumor Analysis Consortium (CPTAC), 65t
Clonal hematopoiesis (CH), 107
Collagen VII, 287
Colon and rectal cancer, traditional medicine for, 286
Colon cancer, 256–258, 257f
  diagnosis and treatment of, 258
  risk factors, 256–257, 257f
  stem cell therapy in, 258
Colon epithelium, 179
Colonography, 258
Colonoscopy, 258
Compact bone marrow-derived MSCs (CBMSC), 277
Conditioned-medium adipose tissue derived from MSCs (CM-ATMSCs), 264–265
Conditioned medium from bone marrow MSCs, 264
Condition medium from human MSCs (CM-hMSCs), 264–265
Cone-rod homeobox (CRX), 11–14t
Conjunctiva, 115–116
Conjunctival stem cells, 116
  aging, 116
Core-binding factor subunit alpha-1 (Cbfa1), 192
Cornea, 114
Corneal stem cells, 114–115
  aging effect, 115
  endothelial stem cells, 115
  stromal stem cells, 115
C-reactive protein (CRP), 240
Creutzfeldt Jakob disease (CJD), 271
Cross-age transplantation, 223–224
Curcumin longa L., 292–293
Cutaneous aging, 287
Cutaneous basal cell carcinoma (BCC), 258
Cutaneous graft-*versus*-host disease (GVHD), 259
Cutaneous squamous cell carcinoma, 258
C-X-C chemokine receptor type 4 (CXCR4), 11–14t

CX3CL1 functions, 252
Cyclin-dependent kinase inhibitor 1A (Cdkn1a), 192–193
Cytokines, satellite cells and, 133

D

DAPT, 163
Defective ribosomal machineries, 237
Degenerative cardiovascular diseases, 27
Degenerative meniscus, 28–30t
Dendritic epidermal T cells (DETCs), 89
Dermal-epidermal junction (DEJ), 287
Dermomyotome, 126
Developmental pluripotency associated 5 (DPPA5), 11–14t
Dexamethasone, 194
Diabetes mellitus (DM), 31
Diapause, 69
DICER1 (dsRNA endoribonuclease), 11–14t
Differentiation
 definition, 231
 of stem cells, 161
Differentiation-resistant-ESCs (DR-ESCs), 73
Dishevelled (DVL), 77
DNA damage, 204–205, 284–286
 in aging-induced stem cell dysfunction, 208, 209f
 HSC aging, 105–106
 repair to restore stem cell function, 226
DNA damage response (DDR), 148
DNA damage theory, 5
DNA damaging, 182
DNA methylation, 181–182, 227, 236–237, 260, 286
DNA methylation clocks (DMCs), 237
Downstream regulatory element antagonist modulator (DREAM), 23
DPPA3, 11–14t
Drosophila model, for stem cell dysfunction, 206
Dysfunctional stem cells, 232

E

EGb 761, 290–291
Embryoid bodies (EBs), 161, 163
Embryonic stem cells (ESCs), 1–2, 9, 69, 91, 158–159, 213, 231
 in prion diseases, 276
Endocytosis, 164
Endogenous mechanisms, 169–170
Endogenous NSCs, 274
Endometrial degeneration, 27–31
Endothelial dysfunction, 169
Endothelial progenitor cells (EPCs), 3, 150
Endothelial stem cells, 115
Engelbreth-Holm-Swarm (EHS) cells, 10
Eotaxin, 225
Epaxial muscles, 126
Ephrin-B signaling, 178
Epiblast stem cells (epiSCs), 71
Epigenetic age, 260
Epigenetic alterations, 5
 chemicals against, 227

Epigenetic clock, 191–192
Epigenetic drift, 191–192
Epigenetic erosion, 210–211
Epigenetic gene silencing (GS), 260–261
Epigenetic modifications, 236–237
 in bone cell senescence, 191–192
Epigenetic reprogramming, 227–228
 cellular reprogramming and induced pluripotent stem cells, 227–228
 chemicals against epigenetic alteration, 227
 microRNAs, 227
Epigenetics, 90, 135
 in aging-induced stem cell dysfunction, 210–211
 HSC aging and, 107
 p38 and, 138
 $p16^{Ink4A}$ and, 136–137
 regulations, 260
 satellite cell aging, 135–138
Epigenomic reprogramming, 151
Epigenomics, 62–63
ERK1/2 signaling pathway, 75–76
ESC-derived cardiomyocytes, 161, 163
Estrogen, 189
Estrogen deficiency, 189, 193
Estrogenrelated receptor-beta (ESRRB), 11–14t
Estrogen supplementation, 211
Exosomes, 173
Extracellular matrix (ECM), 55, 148, 177–178
 satellite cells, 132
Extracellular vesicles (EV), 182
 cell-free therapy with, 265

F

Facial wrinkling, traditional medicine for, 287
Familial adenomatous polyposis (FAP), 257
Familial hypertrophic cardiomyopathies (FHC), 147
Fecal microbiota transplantation, 183
Fetal glucocorticoids, 25
Fetal growth factors, 24–25
Fetal insulin, 24
Fetal neural stem cell-based therapeutic approaches, 274–275
FGF receptors (FGFRs), 25
Fibrillin-1, 287
Fibroadipogenic progenitors (FAPs), 128
Fibroblast growth factor (FGF), 16, 160
 in proliferation, 25
 satellite cells, 133
Fibroblast-induced cell-extracellular matrix signaling, 150
Fibroblasts, 2
Fibronectin type III domain containing 5 (*FNDC5*), 255
Fibrosis, 55–56, 132
Firebrowse, 66t
5-Fluorouracil (5-FU), 258
Forkhead box P3 (FOXP3) mechanism, 181–182
Forkhead box protein A2/hepatocyte nuclear factor 3-beta (FoxA2/HNF-3B), 11–14t
FoxO, 21–22
Free radical theory, 3–4, 234

G

GATA-binding factor 6 (GATA6), 11–14t
GATA binding protein 4 transcription factor (GATA-4), 11–14t
Gene cocktail therapy, 213
Genetic aging, 284–286
Genetically unmodified hESCs, 18
Genomic instability, 234–236
Genomic medicine, 61
Genomics, 62, 62f
Germline stem cells (GSCs), 90, 204, 232
Giloy, 289
Ginseng, 289
Glial cells, 118
Glucocorticoid-induced osteoporosis, 194
Glucocorticoids, 25, 194
 in bone cell senescence, 194
Glucose-regulated proteins (GRPs), 22
Glycolysis-OXPHOS balance, 94
Gotu kola, 289
Grape exosome-like nanoparticles (GELN), 183
Graying of hair, traditional medicine for, 288
Green fluorescent protein (GFP), 277
Growth differentiation factor 3 (GDF3), 11–14t
Growth differentiation factor 15 (GDF-15), 238
Growth factors, 24
 satellite cells, 133
Growth hormone (GH), 183
Gut dysbiosis, 182
Gut microbiome, 180–182
 on epigenetics of stem cell aging, 181–182
 on immunity of stem cell aging, 181
 on intestinal stem cell regulation, 182f
Gut stem cells
 aging gut, 180
 and biorhythm, 179–180
  mitochondrial function in, 179–180
 gut microbiome, 180–182
  on epigenetics of stem cell aging, 181–182
  on immunity of stem cell aging, 181
  on intestinal stem cell regulation, 182f
 immune system, 180
 intestinal stem cell, 177–178
 microbiota, 180
 regulatory mechanisms, 178–179
 stem cells, 180
 tentative interventional perspectives, 183, 184f

H

Hair follicle stem cells (HFSCs), 89
Hair graying, traditional medicine for, 288
Hayflick limit, 4, 113
Hayflick limit replication theory, 190
Heart diseases, 158–159
Heart dysfunction, 169–170
Heart failure (HF), 147
Heat-shock proteins (HSPs), 22
Hematopoiesis, 261
Hematopoietic progenitor cells (HPCs), 103

Hematopoietic stem cells (HSCs), 5, 10, 91, 103, 158–159, 203–204, 208, 232–234
  aging, 103, 104f
    cell-extrinsic mechanism, 108
    cell-intrinsic mechanisms, 105, 106f
    consequence of, 108–109
    DNA damage, 105–106
    epigenetics, 107
    functional changes, 105, 106f
    gene expression, 104, 105f
    hypercholesterolemia, 108
    impaired autophagy, 107
    mechanism, 104–105
    miRNA, 107
    mitochondrial activity, 107
    rejuvenation, 109–110, 109f
    senescence and polarity, 106–107
    signaling pathways, 108
    single-cell analysis, 108
  changes in, 103
  clonal hematopoiesis, 107
  in wound healing, 56t
Hematopoietic stem cell transplantation (HSCT), 259
Hepatocyte growth factor (HGF), 133
Herbal products, orofacial disease management, 284t
hESCs. See Human embryonic stem cells (hESCs)
Heterochronic parabiosis, systemic rejuvenation by, 212f
Hexosamine biosynthetic pathway (HBP), 177
High-mobility group box proteins, 20–21
High molecular weight HSP, 22
Hippo-YAP signaling pathway, 151–152, 151f
Histone deacetylase 1(HDAC1), 74
Histone methylation, 192
Histone modifications, 260
  HSC aging, 107
Hormonal aging, 284–286
HSPD1, 22
Human amniotic fluid-derived stem cells (hAFSCs), 293
Human embryonic stem cells (hESCs), 9, 158–159, 161, 276
  in age-related diseases, 26–31
  aging and, 18–19, 19f
    autophagy, 21
    factors, 20–23
    heat-shock proteins and chaperones, 22
    metabolic and epigenetic alterations, 22–23
    miRNAs in, 23
    oxidative stress, 21
    proteostasis, 21–22
    signaling pathways, 20
    telomere loss, 20–21
  clinical application, 28–30t
  ethical constraints, 32–33
  FGF signaling, 16
  leukemia inhibitory factor (LIF) signaling pathway, 15
  metabolic and epigenetic regulation, 17
  organoids from, 31–32, 32f
  properties, 10
  regulatory framework/guidelines, 33t
  signaling pathways, 19f
  sources and regulations for generation, 17–18
  in tissue regeneration, 25–26
  transcriptional regulatory circuits, 11–14t
  Wnt signaling pathway, 15–16
Human Genome Project, 62
Human mesenchymal stem cells (hMSCs), 276–277, 292
Human telomerase reverse transcriptase (hTERT), 11–14t
Humors, 281
Hutchinson-Gilford progeria syndrome, 208, 287
Hypaxial muscles, 126
Hypercholesterolemia, 108
Hypoxia, 171–173
Hysteroscopy, 27–31

I

ICM-derived hESCs, 10
"2i" inhibitors, 160–161, 162f
IIS pathway, 240
IL-10, 181
Immune cells, 92, 92f
Immune senescence, 92
Immune system, 180
Immunity, 92–93, 93t
  during aging
    cellular senescence, 96
    Notch signaling, 96
    WNT signaling, 96
  aging-related changes, 92f, 94, 95f
Impaired angiogenesis, 169–170
Induced pluripotent stem cells (iPSCs), 32, 148, 158–160, 292
  cellular reprogramming, 213, 227–228
  epigenetic modifiers and gene editing in, 170
  in prion diseases, 276
Infectious prion protein (PrP^{Sc}), 271
Inflammaging, 53–54, 93
Inflammasomes, accumulation of, 240
Inflammation, 53–54
Inflammation reduction, cardiomyocytes regeneration in, 150
Inflammatory bowel disease, 257
Innate immunity, 92–93, 93t
Inner cell mass, 69
Insert zone, 119
Insulin, fetal, 24
Insulin-like growth factors (IGFs), 24
  satellite cells, 133
Insulin-producing cells (IPCs) differentiation markers, 11–14t
Integrins, 164
Interfibrillar mitochondria (IFM), 208–209
Interleukin-6 (IL-6), 252
International Cancer Genomics Consortium (ICGC), 65t
International Society for Stem Cell Research (ISSCR), 18
Intestinal stem cell (ISC), 177–180
Intestinal stem cells (ISCs), 89, 204, 232
Intrauterine adhesions (IUAs), 27–31

Intrinsic aging, 287
Intrinsic cellular processes, 169–170
In vitro fertilization-embryo transfer (IVF-ET), 32
In vivo lineage tracing technique, 178
In vivo reprogramming, 148–149
Irinotecan (Camptosar), 258
Ischemic CVDs, 169
Islet cell transplantation, 31

J

Janus kinases (JAKs), 69–70

K

Kapha, definition of, 281
Keratocytes, 114
Klf4 gene, 149
Kruppel-like factors (KLF), 11–14t

L

Lamellar cells, 119
Late adulthood, 61
Left-right determination factor 2 (LEFTY2), 11–14t
Lefty proteins, 72
Lens, 116–117
Lens epithelial stem/progenitor cells, 117
  aging, 117–118
Lentiviral-mediated RNA interference (RNAi), 274
Leukemia inhibitory factor (LIF), 15, 69, 161, 162f
LIF/JAK/STAT3 signaling pathway, 69–72
Limbal epithelial stem cells, 114
Limbal region, 114
LIM/homeobox protein 2 (LHX2), 11–14t
Lin28, 23
LinkedOmics, 66t
Long interspersed nuclear elements (LINE-1) gene products, 213
Long-term hematopoietic stem cells (LT-HSCs), 108–109
Low molecular weight HSP, 22
Lumbrokinase (LK), 273
Lymphocytes, 92
Lymphohematopoietic system, 261–262

M

Macroautophagy, 130, 131f
Macroglia, 118–119
Macular degeneration, 26, 28–30t
Mast cells, 150
Matrigel, 10
Matrix metalloproteinases (MMP), 55, 55f, 287
M-cadherin, 126–127
Mechanotransduction pathways, 164
Meis-1 (Meis homeobox 1), 151–152
Melanocytes, 89
Melanocyte stem cell (MelSC), 259
Mesenchymal stem cells (MSCs), 1–2, 10, 115, 192, 195, 205, 232, 261–262
  in aging and role of traditional medicines, 291–292

as carrier in colon cancer therapy, 258
for colon cancer, 258
in prion diseases, 276–277
senescence, 65
Mesoderm-derived satellite cells, 128
Metabolism, 129
Metabolomics, 62, 64
Metformin, 97
Microautophagy, 130
Microbiota, 180
Microenvironment, 57
satellite cell aging and, 132–133
Microglia, 118–119
Microglial cells, 277
Micropetrosis, 190
Micro-RNA (miRNA), 227, 237
in cardiac regeneration and cardiac conditions, 151f
in hESC aging, 23
HSC aging and, 107
Middle adulthood, 61
Minor HSPs, 22
miRNA-mediated regeneration, 152, 152f
Mitochondrial biogenesis, 238
Mitochondrial damage, 91, 94
Mitochondrial DNA (mtDNA), 238
Mitochondrial dysfunction, 208–209, 238, 284–286
Mitochondrial function, in gut and stem cell interplay, 179–180
Mitochondria, of satellite cells, 129, 130f
Mitophagy, 287
MIXL1, 11–14t
Molecular switches, 152
Molecular Taxonomy of Breast Cancer International Consortium (METABRIC), 65t
Momordica charantia, 290
Mononuclear cells, 150
Muller glial cells, 118–119
Multiomics, 61–62
data integration and application, 64–65, 65t
stem cell-based, 65
Multipotent stem cells, 231–232
Murine ESCs (mESCs), 9
Musashi1/MSI1, 11–14t
Muscle stem cells (MuSCs), 54, 128–129, 232
WNT signaling, 96
in wound healing, 56t
Mutational theory, 226
Myofibers, 132
Myogenic determination, 128
Myogenic differentiation (MyoD), 127, 127t
Myogenic factor 5 (Myf5), 127, 127t
Myogenic regulatory factors (MRFs), 127, 127t
Myogenin, 127, 127t
Myokines, 194–195
Myomitokine, 255
Myosin heavy chain 6 (MYH6), 11–14t
Myosin heavy chain (MHC) proteins, 147

N

N-acetyl-L-cysteine (NAC), 226, 241–242
"Naive" hESCs, 10
promoters, 11–14t
Naive human stem cell medium (NHSM), 71
Nanog, 10, 11–14t
Naringin, 293
Natural killer cells, 92–93
Neoangiogenesis, 169
Nestin (NES), 11–14t
Nestor-Guillermo progeria syndrome, 208
NetGestalt, 66t
Neural cells, differentiation of, 273, 273f
Neural differentiation markers, 11–14t
Neural stem cells (NSCs), 90, 205, 232
Neurodegenerative diseases
hESCs for, 26–27, 28–30t
traditional medicine for, 287
Next-generation sequencing, 62
Niche, 128
components, 224
deterioration, 239–240
microenvironment of, 224
rejuvenation of
aging and rejuvenating circulating factors, 225
inflammaging and senolysis, modulation of, 224–225
Niche cells (Paneth cells), 177–178
Nicotinamide N-methyltransferase (NTMT), 17
Nkx2.5, 11–14t
NKX6.1, 11–14t
Nodal, 72
Nodal pathway, 72
Noggin, 74
Noncellular ECM, 55
Nonhomologous end-joining (NHEJ), 105–106
Nordihydroguaiaretic acid, 97
Notch signaling, 20, 96, 161–164
satellite cell aging, 134–135
Novel lineage tracing techniques, 179
Nuclear aging, as potential therapeutic targets, 226–228
Nuclear factor erythroid 2-related factor 2 (NRF2), 3–4, 21
Nucleotide excision repair (NER), 105–106
Numb family proteins (NFPs), 163–164
Nutrient sensing, 96, 238–239, 242

O

OASIS, 66t
Ocimum tenuiflorum, 288
Oct4 gene, 149
Oct4/POU5F1, 11–14t
Ocular stem cells
conjunctival stem cells, 115–116
corneal stem cells, 114–115
lens stem cells, 116–118
retinal stem cells, 118–119
trabecular meshwork stem cells, 119–120
Old HSCs, 103, 104f
Omics, 61–64
3Omics, 66t
Omics Discovery Index, 65t

Oncogene-induced senescence (OIS), 148
Oncogenic stress, 187
Oogonial stem cells (OSCs), 204
Organoids, 31–32, 32f
Orthodenticle homeobox 2 (Otx2), 11–14t
Osteoblastogenesis, 192
Osteoblasts, 194
Osteoclastogenesis, 192
Osteocytes, 194, 196
Osteoporosis
age-associated, 194–195
glucocorticoid-induced, 194
Osterix (Osx), 192
Oxaliplatin (Enoxaparin), 258
Oxidative metabolism, regulation of, 226
Oxidative phosphorylation (OXPHOS), 17, 239
Oxidative stress, 3–4, 21, 190, 196–197, 206–207, 234, 235t, 291–292
in bone cell senescence, 190–191
Oxygen, 17
Oxytocin, 225

P

p38, 138
Paintomics 3, 66t
Panax ginseng, 289
Panax notoginseng flower extracts, 291
Pan-Cancer Atlas, 65
Pancreas/duodenum homeobox protein 1 (PDX1), 11–14t
Parabiosis, 132, 223
aging-induced stem cell dysfunction, 211–212, 212f
Paracrine mechanisms, 173
Parkinson's disease (PD), 256
Parthenotes, 158
Pax3, 126–127
Pax6, 11–14t
Pax7, 126–127
Pentraxin, 255
Peripheral blood mononuclear cells (PBMCs), 90
Phenomics, 64
p16^{Ink4A}, 136–137
Pita, definition of, 281
Plant extracts, 290–291
Platelet-derived growth factor (PDGF), 157–158, 173
Pluripotent stem cells (PSCs), 158, 160, 231, 263
p53 phosphorylation, 197–198
p53/p21 signaling pathway, 197–198
p16/Rb signaling pathways, 197–198
PR-domain zinc-finger protein 14 (PRDM14), 11–14t
"Primed" hESCs, 10
Prion diseases
adult neural stem cells in, 274
classification of, 271
embryonic stem cells (ESCs), 276
fetal neural stem cell-based therapeutic approaches in, 274–275
induced pluripotent stem cells, 276
mesenchymal stem cells in, 276–277

Prion diseases *(Continued)*
  microglial cells in, 277
  prion protein, 271–273, 272f
  stem cell therapy for, 278
  therapeutic approaches for, 273
Prion protein, 271–273, 272f
Progerin, 225
Progeroid syndromes (PS), 260
Protein homeostasis, 225–226
Protein-only hypothesis, 272
Proteomics, 62–64, 64f
Proteostasis, 21–22, 93–94
  dysfunction, 209–210
  loss of, 237–238
Psoriasis, 259

Q

Quantitative polymerase chain reaction (qPCR) technique, 236
Quiescent satellite cells, 125

R

Radiation therapy, 258
Rapamycin, 97, 227, 242
Reactive nitrogen species (RNS), 208
Reactive oxygen species (ROS), 190–191, 206–208, 226, 234, 252
Recombinant Nrg-1, 151–152
Red fluorescent protein (RFP), 163
Reduced expression 1/zinc-finger protein 42 homolog (Rex1/ZFP42), 11–14t
Regenerative medicine, 278
Regenerative stem cell therapy, 57–58
Regucalcin, 255
Regulatory mechanisms, in gut stem cells, 178–179
Replicative senescence, 187
Reprogrammed oocytes, 158
Reprogramming-induced senescence (RIS), 148
Retina, 118
Retinal differentiation markers, 11–14t
Retinal stem cells, 118
Retinitis pigmentosa, 28–30t
Retrodots, 117–118
Retrotransposons, 213
RNA induced silencing complex (RISC), 274
RNA interference (RNAi), 274, 275f
RNA splicing, 237
Runt-related transcription factor 2 (Runx2), 192

S

Salvia miltiorrhiza, 292
Sanger sequencing, 62
Sarcomeric proteins, 161
Sarcopenia, 125, 132
Satellite cells, 125, 126f, 212
  age-dependent changes, 128–129
  aging
    and metabolism, 129–132
    and microenvironment, 132–133
    Notch signaling pathway, 134–135
    Wnt/β-catenin signaling, 135
  autophagy of, 130–132, 131f
  circulatory factors, 132–133
  cytokines, 133
  epigenetic events
    in differentiation state, 137–138
    in proliferation state, 137
    in quiescence state, 135–137
  extracellular matrix, 132
  growth factors, 133
  mitochondria of, 129, 130f
  origin of, 126
  quiescence state, 128
  telomeres, 133–135
  transcription factor-mediated regulation, 126–127
Self-renewal
  definition of, 231
  promoters, 11–14t
Senescence, 61, 90–91
  cellular, 96
  HSC aging, 106–107
  immune, 92
  "omics" in, 61–64
Senescence-associated heterochromatin foci (SAHF), 148
Senescence-associated secretory phenotype (SASP), 54, 148–149, 169–170, 180, 187–189, 195–196, 206–207, 241
Senescence-messaging secretome (SMS), 187–189
Senolysis, 224–225
Serratiopeptidase (SP), 273
Serum markers, 240
Sex differences, aging, 90
Sex hormone deficiency, 193
Shelterin, 133, 134t
Short chain fatty acids (SCFAs), 181–182
Siddha medicine, 281–283
  clinical applications of, 282–283
  mineral- and metal-based drugs in, 282
  orofacial disease management by, 283
Signaling pathways, 78–79, 79f
  ERK1/2, 75–76
  HSC aging, 108
  LIF/JAK/STAT3, 69–72
  mechanotransduction pathways, 164
  Notch signaling system, 161–164
  in self-renewal and differentiation of cardiomyocytes, 160
  stem cells, 158
    culture and therapy, 158–160
    differentiation of, 161
  TGF-β/Smad, 72–74
  Wnt/β-catenin, 76–78, 160–161, 162f
Signaling proteins, 260
Single-blastomere biopsy, 18
Single-cell analysis, HSC aging, 108
Sirtuin, 20, 22, 149, 226
Sirtuin 1 (SIRT1), 205, 237–238
Skeletal aging, 190, 191f
  pathways regulating, 196–198
  p53/p21 signaling pathway, 197–198
  p16/Rb signaling pathways, 197–198
  secretory phenotype and senescent markers on, 195–196
  Wnt signaling pathway, 196–197, 197f
Skeletal muscles, 125
Skeletal muscle stem cells, 125–128, 204–205
Skeletal stem/progenitor cell (SSPC), 199
Skin cancer, 258–259
  risk factors of, 258–259
  stem cell therapy in, 259
SMADs, 72
SMP30 (senescence marker protein 30), 255
SNF2-related CBP activator protein (SRCAP), 138
Soluble form of clusterin (sCLU), 237–238
Somatic cell nuclear transfer (SCNT) method, 158
Somatic stem cells, 158, 231
Somatomedins, 238–239
Somite, 126
Southern blot technique, 236
Sox2 (SRY (sexdetermining region Y)-box 2), 11–14t
SOX17, 11–14t, 103
Sox2 gene, 149
Soy *(Glycine max)*, 290
Spermatogonial stem cells (SSCs), 204
Splicing factor 1 (SFA-1), 237
Sp7 transcription factor, 192
SRY-box transcription factor 1 (SOX1), 11–14t
SRY-box transcription factor 9 (SOX9), 11–14t
Stage-specific embryonic antigen3/4 (SSEA3/4), 11–14t
Stargardt's macular dystrophy, 26, 28–30t
STAT3, 15–16, 70
STAT3C, 70
Stem cell aging, 91
  epigenetics and, 90
  hallmarks of, 232, 233f
  and inflammation, 53–54
Stem cell-derived conditioned medium (CM), 263–265, 264f
Stem cell dysfunction, aging-induced, 203
  factors responsible for, 206–211
    changes in stem cell metabolic intake, 211
    DNA damage, 208, 209f
    epigenetics, 210–211
    microenvironment, 206–207, 207f
    mitochondrial dysfunction, 208–209
    proteostasis dysfunction, 209–210
    stem cell aging and gender, 211
    stem cell exhaustion, 206
  germline stem cells (GSCs), 204
  hematopoietic stem cells (HSCs), 203–204, 208
  intestinal stem cells (ISCs), 204
  mesenchymal stem cells (MSCs), 205
  models for, 206
  neural stem cells (NSCs), 205
  skeletal muscle stem cells, 204–205
  therapeutic approaches for, 211–214, 223–230
    caloric restriction (CR), 214
    cellular reprogramming of iPSCs, 213
    intrinsic aging factors and, 225–226
    mitochondrial function improvement, 226

nuclear aging, as potential therapeutic targets, 226–228
oxidative metabolism, regulation of, 226
parabiosis, 211–212, 212f
protein homeostasis improvement, 225–226
regulation of autophagy, 225–226
rejuvenation of stem cell niche, 224–225
rejuvenation research, 223–224
retrotransposons, 213
targeting extracellular aging factors, 224–225
telomere lengthening, 214
Stem cell exhaustion, 206
Stem cell pool exhaustion, 233–234
Stem cells, 1, 158
in aging, 2–3
culture and therapy, 158–160
differentiation of, 161
types, 1–2
Stem cell therapy, 158
Stroma, 114
Stromal cell-based therapies, 251
age-related diseases, 255–259, 255f
Alzheimer's disease (AD), 255–256
cancer, 256
colon cancer, 256–258, 257f
Parkinson's disease (PD), 256
skin cancer, 258–259
age-related genes and their role, 252–255, 253–254t
aging
in multicellular organisms, 251–252
in unicellular organisms, 251
extracellular vesicles, cell-free therapy with, 265
factors affecting stromal cell quality due to aging, 259–260
intrinsic limitations of aged stem cells, 260–261
next generation of, 266
pitfalls in, 262–263
pros and cons of, 262
stem cell-derived conditioned medium (CM), 263–265, 264f
stromal cells in aging, 261–262
tissue microenvironmental cues and influences, 265–266
Stromal-cell-derived factor (SDF)-1, 173
Stromal stem cells, 115

T
T-box 3 (TBX3), 11–14t
T cells, 92
Telomerase, 4, 20–21, 158–159
Telomerase reactivation, 242
Telomerase reverse transcriptase (TERT), 4, 236
Telomerase RNA component (TERC), 236
Telomere, 4, 20–21, 190, 226

attrition, 96, 236
dysfunction, 236
lengthening, 214
satellite cells and, 133–135
Telomere dysfunction-induced foci (TIFs), 192
Telomere loss, 20–21
Telomere shortening, 4, 259–260
in bone cell senescence, 192–193
Telomere shortening theory, 236
Telomere theory, 4
Tenascin-C, 173
TGF-β/Smad pathway, 72–74
Thadhu agents, 282t
The Cancer Genome Atlas (TCGA), 65t
Therapeutically Applicable Research to Generate Effective Treatments (TARGET), 65t
Thymosin β4 (Tβ4), 163
Tinospora cordifolia, 289
Tipiracil (Lonsurf), 258
Tissue degeneration, 54–55
Tissue remodeling mechanism, 148
Totipotent stem cells, 231
TRA1–60/80 (T-cell receptor alpha locus/podocalyxin), 11–14t
Trabecular meshwork (TM), 119
regeneration of, 119–120
Trabecular meshwork stem cells, 119
Traditional medicine, 281
in aesthetics, 288–291
Articum lappa, 289
Centella asiatica, 289
Chamomile, 290
Ocimum tenuiflorum, 288
Panax ginseng, 289
plant extracts, 290–291
soy *(Glycine max)*, 290
Tinospora cordifolia, 289
wolfberry, 289
aging, 284–286
aging-related disorders, 286–288
colon and rectal cancer, 286
facial wrinkling, 287
graying of hair, 288
neurodegenerative diseases, 287
Ayurveda, 283–284, 285t
analysis of, 283–284
clinical applications of, 288
Siddha medicine, 281–283
clinical applications of, 282–283
mineral- and metal-based drugs in, 282
orofacial disease management by, 283
in translational stem cell research, 291–293
herbal remedies in, 292–293
mesenchymal stem cells in, 291–292
Unani medicine, 283
Transcription factor AP-2-gamma (TFAP2C), 11–14t
Transcription factor CP2-like protein 1 (TFCP2L1), 11–14t

Transcriptomic profiling, 57
Transcriptomics, 63, 63f
Transforming growth factor (TGF), 24
Transforming growth factor-β (TGF-β), 24, 189
satellite cells and, 133
Transient amplifying cells (TACs), 114, 177
Transmissible spongiform encephalopathy (TSE), 271, 274
Trifluridine, 258
Trophectoderm, 69
Troponin T type2 (TNNT2), 11–14t
Tubulin-beta-3 chain, 11–14t

U
Ubiquitin proteasome system (UPS), 93–94
UCSC Xena, 66t
Unani medicine, 283
Uncoupling protein 2 (UCP2), 17
UPR^{mito}, 129
Usool-e-Ilaj, 283
Uterus degeneration, 28–30t

V
Vascular cells, 118
Vascular cell senescence, 169–170
Vascular endothelial growth factor (VEGF), 173
Vata, definition of, 281
Vitagenes, 183
Vitamin C, 290
Vitiligo, 259

W
Werner's syndrome, 287
Wharton's Jelly MSC (WJ-MSCs), 264–265
Wnt3, 11–14t
Wnt/β-catenin signaling pathways, 160–161, 162f, 191
Wnt canonical signaling pathway, 189
Wnt signaling, 15–16, 76–78
immunity and, 96
satellite cell aging, 135
skeletal aging, 196–197, 197f
Wolfberry, 289
Wound healing, 53
models, 56t
stem cell aging and, 53–58

X
X-inactive specific transcript (XIST), 11–14t

Y
Yamanaka factors, 148–149
Young HSCs, 103, 104f

Z
Zygote, 69

Printed in the United States
by Baker & Taylor Publisher Services